Magmatism in the Ocean Basins

Geological Society Special Publications

Series Editor K. COE

GEOLOGICAL SOCIETY SPECIAL PUBLICATION NO 42

Magmatism in the Ocean Basins

EDITED BY
A. D. SAUNDERS & M. J. NORRY
Department of Geology
The University
Leicester LEI 7RH

1989

Published for
The Geological Society by
Blackwell Scientific Publications

OXFORD LONDON EDINBURGH
BOSTON MELBOURNE

Published for
The Geological Society by
Blackwell Scientific Publications
Osney Mead, Oxford OX2 0EL
(*Orders*: Tel. 0865-240201)
8 John Street, London WC1N 2ES
23 Ainslie Place, Edinburgh EH3 6AJ
3 Cambridge Center, Suite 208
Cambridge, Massachusetts 02142, USA
107 Barry Street, Carlton, Victoria 3053
Australia

First published 1989

Typeset by Setrite Ltd, Hong Kong
Printed and bound at
Cambridge University Press

DISTRIBUTORS

USA and Canada
Publishers' Business Services
PO Box 447
Brookline Village
Massachusetts 02147
(*Orders*: Tel. (617) 524-7678)

Canada
Oxford University Press
70 Wynford Drive
Don Mills
Ontario M3C 1J9
(*Orders*: Tel. (416) 441-2941)

Australia
Blackwell Scientific Publications
(Australia) Pty Ltd
107 Barry Street, Carlton
Victoria 3053
(*Orders*: Tel. (03) 347-0300)

British Library Cataloguing in Publication Data

Magmatism in the ocean basins.
1. Oceans. Beds. Magmatism.
I. Saunders, A.D. II. Norry, M.J.
III. Geological Society of London
IV. Series
551.46'08

ISBN 0-632-02384-8

Library of Congress Cataloging-in-Publication Data

Magmatism in the ocean basins.
(Geological Society special publication; no. 42)
Bibliography: p.
Includes index.
1. Magmatism. 2. Ocean bottom. I. Saunders, A. D.
II. Norry, M. J. III. Series.
QE461.M23784 1989 551.46'08 88–16828
ISBN 0-632-02384-8

Contents

Introduction

The ocean basins provide a unique opportunity to investigate magmatic processes and mantle composition. The absence of continental crust as a potential contaminant provides an untarnished sample of the deep Earth, although a full understanding of the messages carried in the sample is still far away. In part, this reflects the difficulty of sampling the ocean floor. Drilling, dredging and submersible operations are costly and remote; investigations of ocean islands provide information on only one, albeit important, province (*ie* that associated with hot-spots), and studies of exposed oceanic material, ophiolites, are always tempered by the realization that ophiolites may form in a back-arc environment and thus not provide a representative sample of the oceanic lithosphere.

However, lack of understanding about the physical and chemical processes that accompany the genesis of oceanic magmas (and indeed any magma) is generating extensive research. Recent advances in modelling of mantle components and dynamics ('chemical geodynamics'), modelling of magma formation and extraction, experimental studies of basalt formation and magmatic differentiation, and detailed investigations of ocean ridges and islands are all beginning to produce an integrated view of how the Earth works.

It was with these recent advances in mind that we convened a meeting at Leicester in January 1987. Over 40 papers were presented during the three days of the meeting; this volume of 19 papers offers a selection of the main topics covered. The emphasis of the meeting and this volume is on magmatic and mantle processes; we have purposely chosen to avoid studies of the structure of the oceanic lithosphere because these were covered by the Special Publication *Ophiolites and Oceanic Lithosphere*.

That ophiolites do indeed provide a model of oceanic lithosphere, notwithstanding uncertainty about the tectonic setting in which they have formed, is now generally accepted. In reviewing the magmatic processes responsible for ophiolite formation, I.G. Gass emphasizes the role of mantle melting, magma fractionation in subridge magma bodies and emplacement of the magmas as dykes and lavas. The processes presented, in particular the concept that magma production along the mid-ocean ridge (MOR) system is discontinuous in both space and time, are subsequently developed in the papers by H.J.B. Dick, J.H. Natland and S.H. Bloomer *et al.*

The underlying thermal reasons as to how and why the mantle beneath ridges and islands actually melts are summarized by R.S. White. Along normal MORs, the mantle passively migrates into the void left by the two separating plates and undergoes decompression-related melting as it ascends adiabatically. This, incidently, provides one of the least selective samples of the Earth's (suboceanic) mantle, in the form of adiabatically generated melts. Thickened passive margins, characterized by voluminous basaltic intrusions and effusions, appear to be related to plume or hot-spot activity, and the archipelogos and clusters of ocean islands are manifestations of hot jets of mantle material rising from the deeper mantle.

The nature of the mantle beneath the oceans may be evaluated directly either by studying on-land exposures (J.A. Spray) or by investigating regions where overlying crust is thin or displaced, such as in fracture zones (H.J.B. Dick). The study by Spray shows that the mantle is probably macroscopically heterogeneous and that some of these heterogeneities may well represent long-term events, *eg* oceanic crust recycled into the mantle at ancient subduction zones. Such macroscopic heterogeneity is also suggested by detailed chemical and isotopic studies of erupted MOR basalts, which imply that their sources may contain fertile and refractory components (J.H. Natland and G. Thompson *et al.*)

Extrapolation of basalt compositions to estimate the nature of the primary magma or even the mantle source involved in mid-ocean ridge basalt (MORB) genesis is fraught with uncertainty. There is still no consensus as to whether primitive MORBs are primary liquids derived from the mantle at relatively low pressures (≈10 kbar) (T. Fujii) or whether they represent liquids derived from high-magnesium picritic melts originating at much higher pressures in the mantle (≈20–25 kbar) (D. Elthon). In the dynamic environment of the ridge crest, a spectrum of primary liquids is likely to form; the critical issue is at what stage the melts finally leave the residue with which they were in equilibrium and the nature of any subsequent interactions and fractionation as they ascend to the surface.

On ocean islands, the effects of such fractionation may be extreme. Silicic magmas are common, with a resultant increase in explosivity and implications for the environment. The way in which silicic magmas evolve from, and apparently coexist with, their basaltic counterparts is highly controversial, apparently necessitating complex zoned magma bodies (M. Storey *et al.*). The felsic bodies may undergo high-temperature interaction between groundwater and magma (Z.A. Palacz & J.A. Wolff), resulting in significant increases in the strontium isotope chemistry of the magma. Even basaltic chambers on Iceland exhibit complex eruption histories and ridge jumps (J. Helgason), with implications for the evolution of MOR magmatic systems.

Ocean island and ocean plateau basalts are also important from the point of view of whole-rock chemistry. It is now generally accepted that the suboceanic mantle is compositionally heterogeneous, although the scale of the heterogeneities (metres, kilometres or thousands of kilometres?) is poorly constrained. The characteristics of the different mantle components, and suggestions for their origin, are presented by G.R. Davies *et al.*, D.J. Chaffey *et al.* and S.-s. Sun & W.F. McDonough. Four main mantle components may be identified on the basis of isotopes and trace elements: enriched mantle (EM1 and EM2, also termed the Dupal component); depleted MORB mantle (DMM); and HIMU mantle, with high time-integrated U-Pb ratios. The existence of a fifth component, namely primordial mantle, is debatable, even when rare gas data are considered (D.E. Fisher; S.-s. Sun & W.F. McDonough).

It appears that the bulk of the mantle has undergone differentiation and that there has been extensive recycling of oceanic lithosphere. The entrainment of subducted sediment or ancient subcontinental lithosphere appears to be a prerequisite for the formation of enriched Dupal-type ocean island basalts. The exact location of these components in the mantle is unknown, but the existence of mantle plumes beneath many ocean islands (*eg* R.S. White) indicates a deep source, perhaps on the 670 km discontinuity or in the lower mantle (S.-s. Sun & W.F. McDonough).

These papers, and that by J.-L. Joron & M. Treuil, indicate that mixing of the components, perhaps during recent convective stirring or even during the melting event that produces the magma, produces the spectrum of basalt types observed in the present-day oceans. The latter authors rule out the necessity for crustal recycling although this is not the conclusion reached by S.-s. Sun & W.F. McDonough.

Finally, B.J. Murton describes a suite of unusual rocks with boninitic affinities from the Troodos complex, Cyprus. That these rocks have not as yet been recovered from the major ocean basins suggests that their origin is due to the physicochemical processes occurring in the vicinity of subduction zones. Alternatively it may simply reflect that our sampling of the ocean floor, and all of its tectonic environments, is incomplete.

Acknowledgments

We should like to thank the reviewers of the papers for this volume for their valuable comments. Those staff of the Geology Department, University of Leicester, who gave much help and time to make the conference a success are also warmly thanked.

We particularly wish to thank the following reviewers for their efforts during the preparation of this volume: P. Baker, P. Browning, J.R. Cann, H.J.B. Dick, H. Downes, D. Elthon, M. Fisk, J.G. Fitton, I.L. Gibson, D.H. Green, S.R. Hart, C.J. Hawkesworth, I.A. Hill, M.J. Hole, C.H. Langmuir, R. MacDonald, D.P. Mattey, D.P. McKenzie, M.A. Menzies, J. Natland, R.W. Nesbitt, A. Nicolas, J.A. Pearce, M. Perfit, G. Rogers, D. Smythe, T. Staudacher, M. Storey, J. Tarney, M.F. Thirlwall, R.N. Thompson, G. Wörner.

A.D. Saunders
M.J. Norry
April 1988

Magmatic processes at and near constructive plate margins as deduced from the Troodos (Cyprus) and Semail Nappe (N Oman) ophiolites

I. G. Gass

SUMMARY:The Troodos and Semail Nappe ophiolites, both produced during Upper Cretaceous (Cenomanian) times at Neotethyan supra-subduction zone constructive margins, are sufficiently comparable to allow a generalized model of constructive-margin magmatic processes to be presented. Such processes can be traced from (i) the partial melting of lherzolitic mantle within rising asthenospheric diapirs and the formation and intramantle fractionation of picritic magmas through (ii) the feeding of these melt batches into high-level, open-system crustal magma chambers to produce phase-layered mafic and ultramafic cumulate assemblages to (iii) the upward injection of melts to form sheeted dyke complexes and eruptive sequences. Evidence suggests that there is a series of mantle diapirs, each limited in time and space, beneath a constructive margin, each feeding its own magma chamber, sheeted dyke complex and volcanic sequence. Thus the oceanic lithosphere is seemingly the product of numerous diapiric 'point' sources embellished by a variety of off-axis magmatic products. The spreading axes are interrupted by transform fault zones which, during extensional phases, are the sites for magmatic products that are geochemically different from those produced at the axes.

Evidence that ophiolites are fragments of oceanic lithosphere formed at fossil constructive plate margins and preserved by tectonic emplacement onto continental or island arc margins is increasingly convincing (Coleman 1977; Panayiotou 1980; Gass *et al*. 1984). In order to deduce the magmatic processes that operate at constructive margins, it is essential to study those ophiolites where the original oceanic lithosphere structures and rock types are well preserved. This paper is based on the study of two such ophiolites, the Troodos massif, Cyprus, and the Semail Nappe of northern Oman, both of which were formed in Neotethys at supra-subduction constructive margins during the uppermost Cretaceous (*c*. 85 Ma and 96–94 Ma respectively) and emplaced onto continental margins shortly (*c*. 10 Ma) thereafter.

Although there is no exposed base to the Troodos ophiolite (Wilson 1959) and the massif has been rotated anticlockwise through 90° since its formation (Moores & Vine 1971; Shelton & Gass 1980; Clube *et al*. 1985), geophysical evidence (Gass & Masson-Smith 1963; Makris *et al*. 1983) indicates that the ophiolite rests at a depth of between 12 and 20 km on attenuated continental crust. This suggests that the Troodos ophiolite was thrust onto a continental margin in a similar manner to the Semail Nappe. The annular geographical disposition of the various Troodos 'stratigraphic' units, in which the structurally highest extrusive unit occurs at the periphery of the massif and the deepest mantle sequence in the central highest part, has been produced by the domal uplift of the massif centred on its summit at Mount Olympus. Since 1982, studies of the Troodos massif have accelerated as a result of the investigations associated with the drilling of 4.5 km of cored boreholes by the International Crustal Drilling Project through the crustal sequence on the N flank of the massif.

That part of the Neotethyan oceanic lithosphere that was to become the Semail ophiolite was formed above a northeasterly inclined subduction zone which lay between 300 and 750 km from the Arabian passive continental margin. This new oceanic lithosphere was formed between 97.9 and 93.5 Ma ago (Tilton *et al*. 1981), detached from its *in situ* position 90.0±3.0 Ma ago (Searle 1980; Lanphere 1981) and emplaced as one of a series of nappes onto the continental margin during the Maastrichtian (85–73 Ma) (Alleman & Peters 1972). These far-travelled nappes were emplaced from NE to SW onto the Arabian continental margin (Lees 1928; Glennie *et al*. 1973, 1974; Lippard *et al*. 1986). As thrusting progressed southwestwards towards the Arabian foreland, thrust slices of Mesozoic ocean floor and continental rise sediments were incorporated into the nappe pile and, close to the base of the Semail Nappe, were locally metamorphosed to give the greenschist facies rocks. Thrusting developed progressively from an external (oceanic) area towards an internal continental region. As a result the farthest

From SAUNDERS, A. D. & NORRY M. J. (eds), 1989, *Magmatism in the Ocean Basins*, Geological Society Special Publication No. 42, pp. 1–15.

travelled nappe, the Semail ophiolite, occurs highest in the sequence, and the lowest thrust slices were the last to be detached and move. The involvement of successively lower thrust slices was produced by the characteristic 'piggy-back' style of thrusting described by Elliott (1976).

The Semail ophiolite forms a huge thrust sheet or nappe, the Semail Nappe, that is over 600 km long, up to 150 km wide and between 5 and 10 km thick. It crops out over an area of about 20 000 km^2 along the length of the Oman Mountains (Fig. 1). The sheet-like form of the nappe has been supported by a regional gravity survey (Shelton 1984) which indicates that the nappe extends beneath most of the coastal plain, probably with its tapering trailing edge close to the present Gulf of Oman coastline (Fig. 1*c*). Marked structural thinning and imbrication of the ophiolite along the western or southern edges of the outcrop suggest that the leading edge of the nappe during emplacement was not far beyond the present outcrop.

As a result of syn- and post-emplacement deformation the Semail Nappe has been broken up into about a dozen major tectonic 'blocks' of between 600 and 7000 km^2 in area (Fig. 1). Within each block the original internal 'Penrose' (Anon, 1972) stratigraphy of the ophiolite is more or less intact. The major blocks in the northern part of the mountains are separated either by cross-strike fault zones, some of which may have been seafloor structures, or by imbricate thrusts or reverse faults. The occurrence of late Cretaceous tectonic melanges along these fault zones shows that thinning and break-up of the nappe, largely by the upward protrusion of melange from beneath (Lippard *et al.* 1983; Robertson & Woodcock 1983), was a syn-emplacement event. The deeply incised relief and almost complete exposure of the ophiolite rocks in the rugged mountainous terrain, combined with the effects of tilting and folding of the nappe during and after emplacement, have provided complete stratigraphic sections through the ophiolite that are at least 16 km, and perhaps as much as 20 km, thick. Gravity data (Shelton 1984) suggest that the maximum structural thickness of the nappe in the centres of the largest blocks is between 5 and 10 km and that the basal thrust is more or less flat lying so that it cuts across usually eastward-dipping internal stratigraphy (see Fig. 1*c*).

These ophiolites are divided by a petrological Moho into (i) a lower 'mantle sequence', so called because it is believed to represent the upper part of the suboceanic mantle (Allen 1975; Smewing 1980), which is composed largely of variably serpentinized peridotites, mainly tectonized harzburgite with associated lherzolites and dunites, with a maximum exposed thickness that is estimated to be 10–12 km and (ii) a 'crustal sequence' consisting of a magmatic assemblage of cumulate peridotites and gabbros (the layered series) overlain by non-layered plutonic rocks (the high-level intrusives), a sheeted dyke complex and an extrusive sequence of lavas interbedded with and overlain by fine-grained pelagic sediments. These units, up to and including parts of the lava sequence, are cut by several suites of cross-cutting plutonic and hypabyssal rocks collectively termed the late intrusive complexes. Figure 2 is a schematic vertical section of the Semail ophiolite on which major rock types and units are marked. For detailed descriptions of all the Semail ophiolite units, the reader is referred to Lippard *et al.* (1986) and the *Journal of Geophysical Research* (1981), Volume 86, No. B4, where the findings of the University of California–USGS team studying the Semail Nappe along a strip 25 km wide S from Muscat are presented.

Hereafter, brief descriptions of the main rock types common to both ophiolites are given before magmatic processes are discussed in detail, first the partial melting of lherzolitic mantle and the passage and modification of these melts through the mantle, and then crustal magmatic processes.

Mantle processes

The *mantle sequence* consists of 85%–95% harzburgite, locally gradational into lherzolite, and 5%–15% dunite. The harzburgites are medium- to coarse-grained rocks composed of 75%–85% olivine ($Fo_{89.2-91.7}$) and 15%–25% orthopyroxene ($En_{91.2-83.4}$, $Fs_{10.2-7.8}$, $Wo_{7.6-0.5}$). Chrome spinel (chromite) is a ubiquitous accessory forming 0.5%–2% of the mode. Clinopyroxene, although generally low in abundance (<1%), can form up to 5% of the mode, with concomitant decrease in orthopyroxene content. Usually between 50% and 80% of the primary silicate minerals have been altered to mixtures of lizardite and chrysotile with accessory magnetite; many chromite grains display marginal oxidation.

A pervasive tectonite fabric (foliation) has been produced in the harzburgite by the planar alignment of the orthopyroxene and spinel grains. Although extremely homogeneous on a regional scale, the harzburgite displays a segregation banding subparallel to the foliation into olivine- and orthopyroxene-rich bands that are

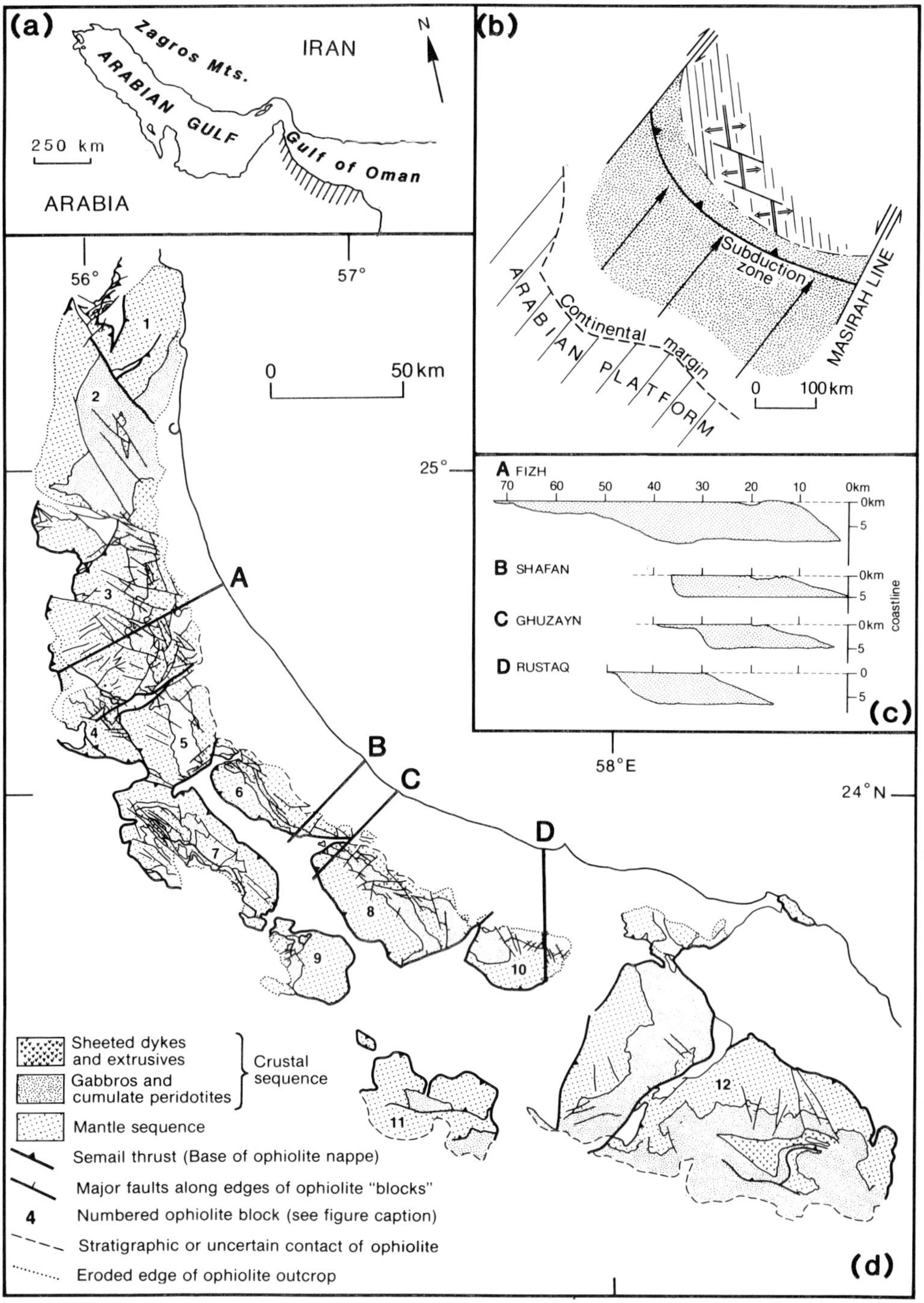

FIG. 1. The Semail ophiolite and its division into 12 major blocks: 1, Khor Fakkan; 2, Aswad; 3, Fizh; 4, West Jizi; 5, Salahi; 6, Sarami; 7, Wuqbah; 8, Haylayn; 9, Muqniyat; 10, Rustaq; 11, Bahlah; 12, Ibra (from Lippard *et al.* 1986). (*a*) Location diagram where the shading indicates the area shown in the main figure; (*b*) plate tectonic setting envisaged for NE Arabia in the mid-Cretaceous showing new oceanic crust (destined to become the Semail Nappe) forming NE (above) the subduction zone; (*c*) cross-sections of the Semail Nappe deduced from the gravity data showing the horizontal distance from the coastline (from Shelton 1984).

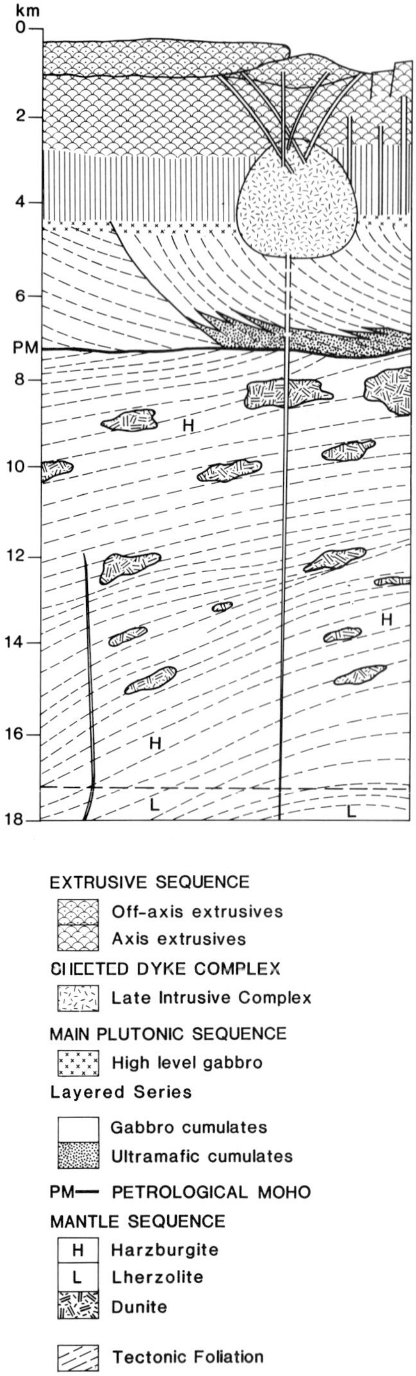

FIG. 2. Simplified ophiolite section based primarily on the Semail Nappe (Lippard *et al.* 1986). Note that (i) off-axis extrusives are related to the late intrusive complexes, (ii) some axis magma chambers form ultramafic cumulates whilst others do not, and (iii) dunitic pods are elongated parallel to the harzburgite foliation which cuts through them without divergence and the largest pods are close to (within 1.5 km of) the petrological Moho.

usually between 5 mm and 5 cm thick, and always less than 10 cm, and have diffuse margins. These bands or ratio layers can rarely be traced for more than a few metres across an outcrop. The banding is bilaterally symmetrical, and preferential mineral grading in any one direction does not occur; rather, the olivine- and orthopyroxene-rich bands grade into one another. As this compositional banding is usually parallel to the foliation, it is most probably the result of metamorphic segregation and can be thought of as gneissic banding. These structures were produced at high temperatures (1000–1200 °C) (Nicolas & Poirier 1976) by continuous and progressive deformation of *in situ* upper mantle immediately below a constructive plate margin.

Dunite bodies occur sporadically throughout the mantle sequence, and although they vary widely in shape and size they always have sharp contacts with the surrounding harzburgite. They are composed of more than 98% olivine (Fo_{92}) and less than 2% red-brown chrome spinel grains which are conspicuous on weathered surfaces. Although dunites with isolated chrome spinel grains forming only 1%–2% of the mode are most common, all petrographic varieties from dunite through chromitiferous dunite and olivine chromitite to chromitite occur. The tectonite fabric in the harzburgites generally continues into the dunites and is not deflected as it crosses the boundary. Indeed, although the length of the dunite bodies varies widely, from a few metres up to several hundreds of metres, their shape is directly related to the amount of deformation they suffered; when highly deformed they are markedly elongated in the foliation direction. In all cases the chromitite is enclosed in a dunitic host, although this may only be a few centimetres thick and intensely sheared. Usually, it is the largest dunite bodies that contain the largest chromitite masses, and in general these lie in the uppermost 1.5 km of the mantle sequence, just below the petrological Moho (Brown 1980, 1982).

Intrusive relations have been so widely observed that there can be little doubt that the dunites were precipitates from magmatic bodies within the harzburgite. For instance, Brown (1982) identified structures and texture suggesting that both the olivines and the chromite spinels were magma precipitates. In particular, the chromitite commonly occurred in layers that showed igneous sedimentation features such as cross-bedding, slumping and differentional compaction. Brown's identification of these textures as dominantly adcumulate stands up to subsequent investigations (Roberts 1985; Christiansen & Roberts 1986).

The following model is proposed to explain the rock types and structures within the mantle sequence. Relatively 'fertile' aluminous lherzolite of the asthenospheric upper mantle, convectively rising and diverging beneath a constructive margin, begins to melt at *c.* 50–70 km depth (O'Hara 1968) to produce a magnesium-rich picritic melt which rises adiabatically along with the enclosing high-temperature but sub-solidus residual mantle. The percentage of melt increases so that by the time the system has reached a depth of 20–30 km the melt fraction coalesces into discrete magma bodies or diapirs. As the magma–harzburgite system continues to rise convectively, only forsteritic olivine and chrome spinel are precipitated from the melt that remains during its adiabatic ascent within the expanded olivine stability volume of the hydrous basalt tetrahedron (O'Hara 1968). In this three-component system of (i) tectonized harzburgite, (ii) melt and (iii) crystal precipitates, sooner or later the melt will escape into an overlying magma chamber (see Fig. 5) leaving the precipitated olivine and chrome spinel to form the dunite–chromitite masses. Whilst the upper mantle was convecting at temperatures of *c.* 1000 °C, the system was subjected to continued and progressive deformation. Therefore not only the harzburgites but also the dunites show tectonite fabrics that were preserved as the rocks passed through the asthenosphere–lithosphere boundary during more or less horizontal flow away from the constructive margin. As dunitic bodies are largest in the uppermost 1.5 km of the mantle sequence, it can be inferred that these were once in the centre of the rising asthenospheric mantle where the degree of partial melting was highest.

Our present understanding is that the mantle sequence structures and textures were largely formed at high temperatures (1000–1200 °C) by plastic deformation, which is best considered as a flow process in which the transportation of matter is driven by stress. It is assumed (Nicolas & Violette 1982) that the mantle structures were produced by asthenospheric flow and once the mantle cooled below 1000 °C it was no longer ductile. Thus the limit to ductile deformation is the asthenosphere–lithosphere boundary and the mantle structures are those that were imposed during near-horizontal flow away from and in the immediate vicinity of the spreading axis.

Although earlier workers (*eg* Cann 1970, 1974; Greenbaum 1972) envisaged that convective overturn in the upper mantle beneath a constructive margin was normal to that margin along its length, evidence has been growing that this is probably not so. The earliest indication that such mantle rose in diapiric form was the evidence of Juteau *et al.* (1977) from the Antalya ophiolite of southern Turkey where mantle lineations radiate away from 'point' sources. Similar evidence from other ophiolites (Nicolas & Violette 1982) supports this view as do geophysical models on present-day constructive margins (Whitehead *et al.* 1984). Recent work on the mantle sequence structures in the Semail ophiolite (Bartholomew 1983) allows this model to be elaborated.

Bartholomew (1983) noted that 'for most areas specimens have been deformed in a plastic shear flow regime in which the shear (slip) direction had changed'. He identified an earlier deformation D_1 and a later deformation D_2 and noted that, in areas of intense D_2 deformation, foliation planes have a constant trend and no relicts of the earlier (D_1) shearing have been preserved. However, in areas where the D_2 foliation is weak, relict D_1 foliations are preserved. To explain the production of two shearing events (D_1 and D_2) Bartholomew proposed that, during horizontal asthenosphere flow away from a number of diapiric centres, the divergent flow systems would meet and interact with one another (Fig. 3). So, if the diapirs were active at different times, the younger ones would produce D_2 flow planes that would fold and shear those (D_1) of an earlier centre.

Laboratory studies by Elder (1977) suggest that the divergent flow from such centres should be horizontal and the mean orientation of the shear planes in the Semail ophiolite implies that they were. However, in the Semail ophiolite where diapirs have influence of the order of 10 km from the centre of divergence, these planes show no consistency in their shear sense. This suggested to Bartholomew (1983) that surges of asthenospheric mantle, moving at different rates and depths away from the diapiric centre, produced variations in both the shear sense and the intensity of deformation with depth and along the spreading axis.

The mantle sequences of these ophiolites consist predominantly of highly refractory clinopyroxene-poor harzburgite. In the Semail Nappe Brown (1982) and Browning (1984) record that in the thickest mantle section, the Fizh block, there is a change to less refractory clinopyroxene-bearing harzburgite and lherzolite at a depth of *c.* 10 km below the petrological Moho. Although these rocks are still depleted in basaltic components compared with 'fertile' mantle compositions, it has been calculated (Lippard *et al.* 1986) that fertile mantle would be encountered at *c.* 15 km depth (see Fig. 4). As it is thermally unlikely that mantle depletion occurred during horizontal flow away from the

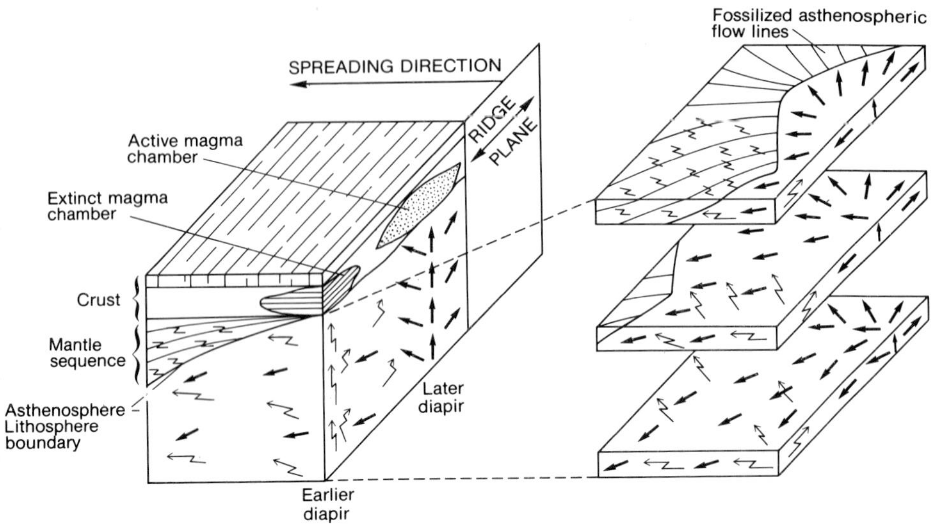

FIG. 3. Cut-away diagram illustrating the envisaged diapiric flow system in subspreading-axis asthenosphere showing how flow lines of an earlier diapir (thin arrows) would have been shear folded by those of a later diapir (bold arrows). After Bartholomew (1983).

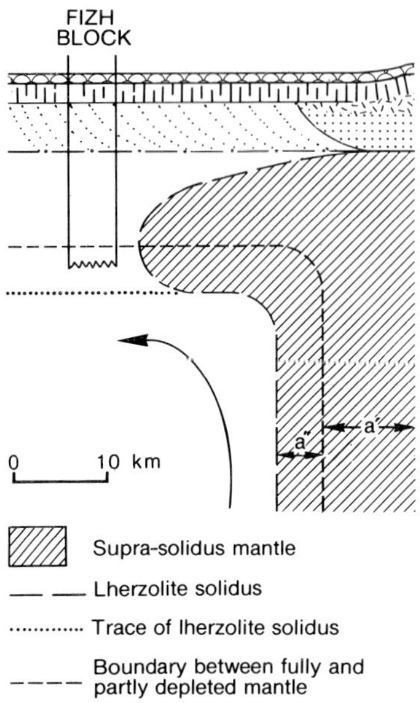

FIG. 4. Suggested spatial distribution of suprasolidus mantle in a half-section normal to a spreading axis. The highest temperature would be reached in the central part (a′) of the rising diapir and here extensive melting would leave fully depleted harzburgite. An outer zone (a″) would be progressively less depleted outward. In the Semail ophiolite the transfer from fully to partly depleted mantle is reached at about 10 km below the petrological Moho.

spreading axis, it is confidently inferred that this depletion took place during the vertical ascent of convecting mantle. A central zone (a′ in Fig. 4) of *c.* 20 km width was totally depleted of its basaltic melt fraction whilst a 5 km zone outside it (a″) was partially depleted.

Lippard *et al.* (1986) calculated that the Semail primary magmas had a composition of *c.* 15.5% MgO with Mg/(Mg+Fe)=0.78, 1200 ppm Cr and 350 ppm Ni which, they believe, was produced by 20%–30% melting of a slightly depleted lherzolite mantle. All calculated Semail and Troodos melts have high SiO_2 and low TiO_2 contents and high CaO/Al_2O_3 ratios compared with the parental melts from which mid-ocean ridge basalts (MORBs) were produced. Many authorities (Green 1973; Ford 1976; Green *et al.* 1979) suggest that these compositions are the result of high degrees of partial melting, at elevated P_{H_2O}, of previously depleted lherzolite. These proposals are compatible with a site of magma production in the mantle wedge above a subduction zone (Pearce *et al.* 1984).

Varying amounts of fractional crystallization during ascent will control whether the melt, when it escapes from the diapir, is basaltic or picritic in composition. If the magma, when it arrives in a subaxial high-level magma chamber, is close to the three-phase gabbro cotectic, then it will mix freely with melt already there and produce gabbro cumulates at or close to the base of the layered sequence. Conversely, an ultrabasic melt would be less likely to mix with the magma already in the chamber and would form a basal picritic layer from which basal ultramafic cumulates would precipitate. This model is compatible with field evidence that minor basic intrusives (pods and dykes) occur within the uppermost mantle sequence beneath layered sequences that have gabbroic basal layers, whereas, where these intrusives are ultrabasic, the basal layers are ultrabasic dunites and/or wehrlites (Browning 1984). A cartoon depicting the process whereby melt from intramantle magma diapirs is fed into a high-level chamber is given in Fig. 5.

Crustal processes

Directly overlying the mantle sequence, and separated from it by the petrological Moho, is the crustal sequence, a magmatic assemblage of cumulate peridotites and gabbros (the layered series) overlain by non-layered plutonic rocks (the high-level intrusives), a sheeted dyke complex and an extrusive sequence consisting of lavas interbedded with, and overlain by, fine-grained, manganoan, pelagic sediments. All models of ocean crust formation based on ophiolite studies (*eg* Greenbaum 1972; Cann 1974; Dewey & Kidd 1977; Gass & Smewing 1981;

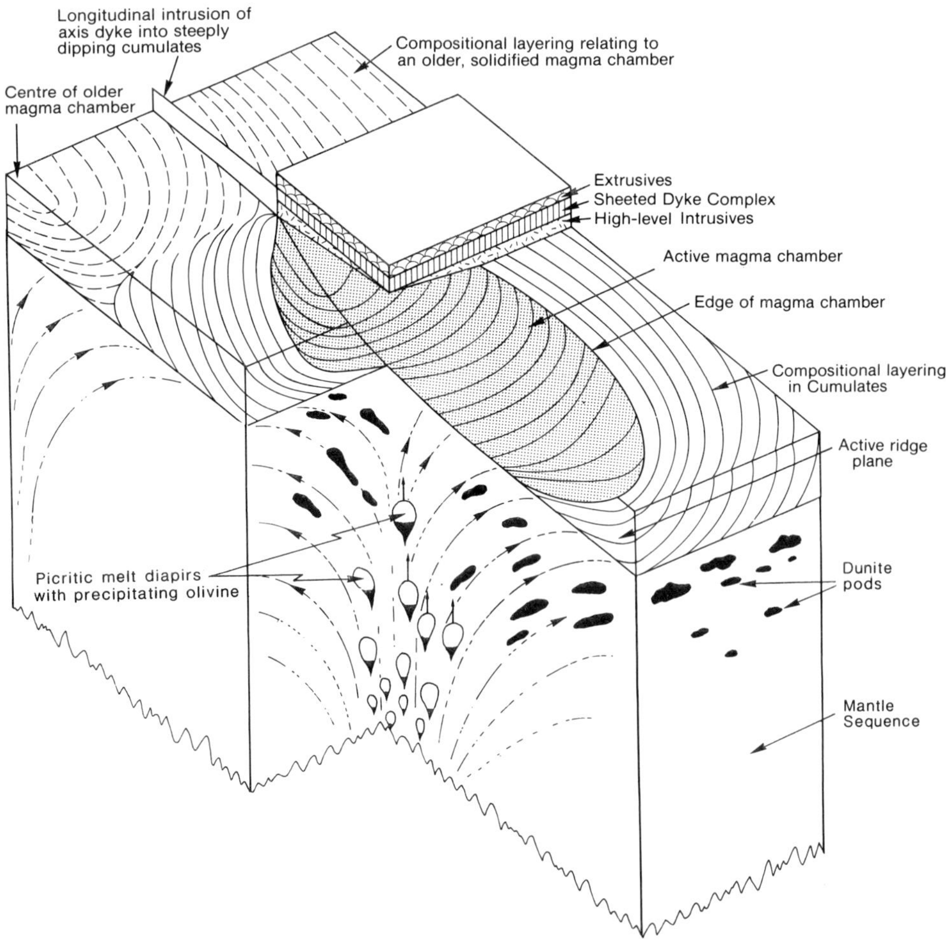

FIG. 5. Cartoon depicting the shape of an individual magma chamber and those of the layered plutonic rock derived therefrom in relation to overlying high-level intrusives, sheeted dykes and extrusives (from Browning 1982). Within the mantle sequence solid black patches represent dunitic pods, the half-black balloons represent rising picritic melt diapirs, the black balloons represent precipitating magnesium-rich olivine and the thin vertical arrows represent melt moving towards the overlying crustal chamber.

Browning 1984; Lippard *et al.* 1986) envisage the existence of crustal magma chambers beneath constructive plate margins. These chambers are considered to be dynamic long-lived open systems which are periodically replenished from mantle-derived melts and depleted by dyke injection and lava eruption. Within these chambers plutonic complexes consisting of phase-layered cyclic ultramafic–mafic units were most probably produced by repeated influxes into the chamber of relatively primitive (magnesium-rich) mantle-derived melts.

The *layered series* varies in thickness from 0.5 to 6.0 km and consists primarily of gabbros (74% in the Semail Nappe) with dunite and wehrlite being the other main rock types. The most persistent and characteristic feature of the series is rhythmic igneous layering. Individual layers vary between 0.5 cm and 2.0 m in thickness and can be traced for several hundreds of metres along strike. Only near the base of the sequence is the layering usually concordant with the plane of the underlying petrological Moho. Up-sequence, it becomes steeper and this, it has been suggested (Smewing 1980; Browning 1982; Rothery 1982, 1983), reflects the shape of the magma chamber walls (see Fig. 5). Also, the angle between the strike of the layering and that of the sheeted dykes varies from 0° to 90° (see Fig. 5). If, as widely accepted, the sheeted dyke strike represents the trace of the constructive plate margin, then the strike variation between the sheeted dykes and the layered series suggests the presence of numerous finite-length magma chambers along the spreading axis. Browning (1982) and Rothery (1982, 1983), on the basis of the criteria just given, calculated that the Semail Nappe crustal magma chambers were no more than 10 km long and had a half-width, normal to the spreading axis, of *c.* 2.5 km.

The primary minerals in the layered series are plagioclase, clinopyroxene, olivine, chrome spinel, orthopyroxene, hornblende and titanomagnetite. Generally, olivine and chrome spinel are early-formed cumulus phases, followed by clinopyroxenes and plagioclase, whereas hornblende, orthopyroxene and titanomagnetite occur as minor late-stage intercumulus phases. The most common crystallization sequence of cumulus-phase minerals in the Semail ophiolite is olivine (± chrome spinel) followed by clinopyroxene and then plagioclase (ol–cpx–plag) which gives rise to the rock sequence dunite, wehrlite, olivine-gabbro, gabbro from base to top of a cyclic unit. Two other, less common, orders of crystallization give rise to other rock sequences: ol–cpx–opx–plag gives rise to dunite, olivine-websterite, two-pyroxene olivine-gabbro, gabbronorite, and ol–plag–cpx gives rise to dunite, troctolite, olivine-gabbro, gabbro. Browning (1982) considered that contrasting magma compositions must be responsible for the different crystallization sequences, the magmas with higher CaO/Al_2O_3 ratios producing the most common ol–cpx–plag sequence. Geochemically, most specimens from the Semail layered series are extremely low in incompatible elements (potassium, phosphorus, zirconium, yttrium and rare earths) suggesting that they are adcumulates from which the intercumulus melt has been expelled.

Cryptic variation in the cumulus phases is limited (Fo_{92-67}; cpx Mg'_{94-74}; opx Mg'_{88-71}; plag An_{96-60}) (where $Mg' = 100\,Mg/(Mg + Fe)$). Within each ultramafic–mafic cyclic unit there is usually an upward trend to more fractionated mineral compositions, *ie* lower Fo, Mg and An contents (Fig. 6). At the base of each cycle is a marked and abrupt shift to less fractionated compositions, most marked in Cr′ (100 Cr (Cr+Fe)) in clinopyroxene (see Fig. 6). Browning (1982, 1984) noted that these 'zigzag' patterns of mineral composition occurred throughout a 900 m sequence (Wadi Bani Kharus) in the Semail Nappe, suggesting that they are characteristic of a magma chamber undergoing open-system fractionation.

The most probable explanation for the breaks in the fractionation trends, and the resetting of mineral phases to less-evolved compositions, is the repeated influx of less-differentiated melts into the magma chamber. Most of the ultramafic–mafic units have a sharp boundary between peridotite and gabbroic facies. This has been explained by Browning & Smewing (1981), quoting Huppert & Sparks' (1980*b*) magma density contrast model, as being due to the ponding of high-density picritic melts at the base of the chamber. Until these melts had sufficiently reduced their density, primarily by the precipitation of olivine to give dunitic layers, they could not mix with the melt already in the chamber. Once this mixing takes place, three-phase gabbros form. In other less common cases only gabbro cumulates form. In such cases it seems likely that the mantle melts injected into the chamber were already sufficiently fractionated to mix directly with the magma already there (Browning 1982, 1984). This explanation, relating the presence of ultramafic–mafic or just mafic units within the layered series to the composition of the incoming melts, is supported by the presence in the immediately subjacent mantle of dunitic bodies where ultramafic cumulates form the gabbroic bodies where ultramafic

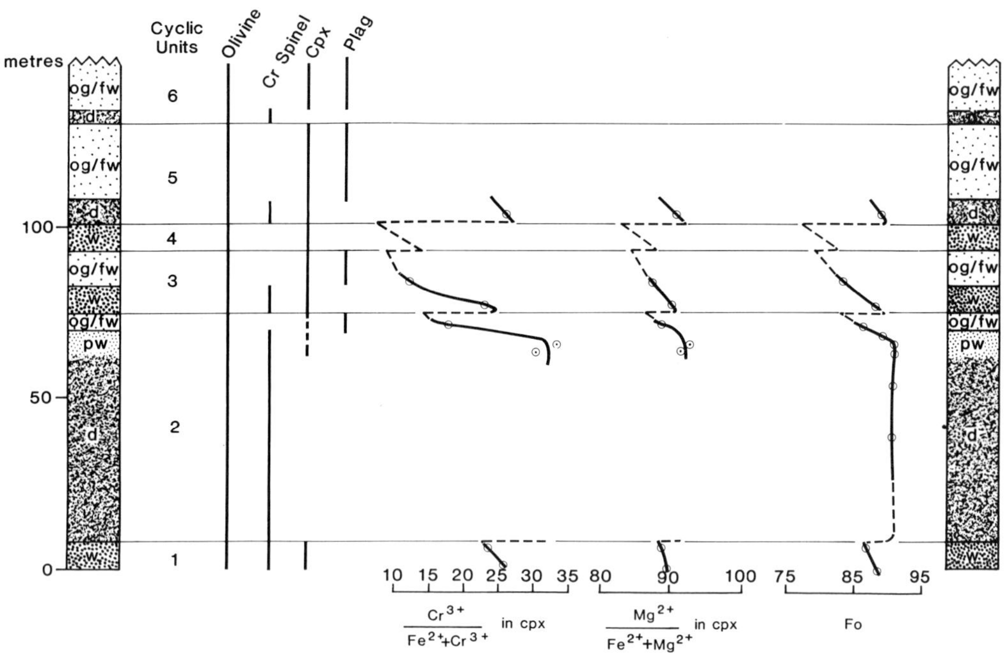

FIG. 6. Mineral chemistry of cyclic units in the Wadi Jizi section of the Semail ophiolite (from Smewing 1981, p. 2650, Fig. 3): og, olivine-gabbro; fw, feldspathic wehrlite; d, dunite; w, wehrlite; pw, poikilitic wehrlite. Note that the greatest compositional variation is for $Cr^{3+}/(Fe^{2+} + Cr^{3+})$ in clinopyroxenes (cpx).

cumulates are absent (Browning 1982).

If the open-system–magma-batch influx model is correct, it is unlikely that the depth of the melt from which these layered rocks formed was the same as the total thickness of the layered series, which varies from 0.5 to 6.0 km. Browning (1982, 1984) made calculations on magma chamber height based on the upward change in weight per cent NiO content in olivine (0.11–0.05) and Cr_2O_3 in clinopyroxene (0.28–0.12) over a single *c.* 55 m thick gabbro layer consisting of 12% ol, 38% cpx and 50% plag in the Wadi Bani Kharus section of the Semail ophiolite. Assuming perfect Rayleigh fractionation and that it was a closed system for the time of the cycle, Browning (1982, 1984) calculated that the layer would form in a tabular chamber from *c.* 100 m of melt; for chambers that were semicircular or triangular (walls dipping inwards at 30°) the calculated melt column heights were *c.* 160 m and 190 m respectively.

Browning (1982, 1984) suggests two possible mechanisms (see Browning 1984, p. 78, Fig. 5) to explain the order-of-magnitude thickness contrast between the melt column from which the gabbro layer formed and the total thickness of the layered series: either the depth of melt in the magma chamber at any one time was never more than 100–200 m and the thickness of the layered series was achieved by the repeated subsidence of 'cumulates' precipitated from such small melt volumes, or such a layer could form from one 100–200 m layer within a deeper magma body compositionally stratified by double diffusive convection (*eg* Huppert & Sparks 1980*a*). Browning (1984, pp. 80–81) favoured the compositionally stratified melt model because of Fo–An covariation contrasts between successive cycles.

Immediately above the layered series, and separating it from the sheeted dyke complex, there is a thin and irregular (0–500 m) unit of *high-level gabbros* that are distinguishable from those of the layered series by their variable texture and the absence of well-defined igneous layering. These rocks were first identified by Wilson (1959) from the Troodos massif and were interpreted by Bear (1960) as having been produced by a basic melt chilling against the magma chamber roof. Aldiss (1978), who studied such rocks in both Cyprus and the Oman, concluded that they were formed by the crystallization of hydrous melts in a magma chamber roof zone where magmatic volatiles concentrated and seawater convecting through the ocean crust entered the melt via the stoping of hydrothermally altered roof rocks. The role of seawater in the formation of these rocks has been confirmed by Gregory & Taylor (1981) and Stakes *et al.* (1984) for the Semail Nappe and by Sheppard (1977) and Heaton & Sheppard (1979) for Troodos.

The most widely accepted model for the formation of ophiolitic plagiogranites, a significant component of the high-level gabbros, is that they formed from residual melts (*c.* 10%) produced by the fractional crystallization, under hydrous conditions, of a sub-alkalic, low-potassium, tholeiitic magma (Coleman & Peterman 1975; Aldiss 1978, 1981). However, as both iron gabbros and quartz-rich plagiogranites occur within various parts of the unit, it seems probable that there were two fractionation trends operating in isolated magma pockets. One, which initially fractionated under relatively anhydrous (low fo_2) conditions, formed ferrogabbros whereas the other fractionated under more hydrous and oxidizing conditions (higher fo_2) to form quartz-rich plagiogranites.

As the name implies, the *sheeted dyke complex*, which forms a unit *c.* 1.5 km thick above the high-level gabbros, consists of 100% near-vertical dykes that have lava and gabbro screens in the upper and lower contact zones where they give way, respectively, to the extrusive sequence and the high-level gabbros or, in rare cases, to layered gabbros. Over 95% of the complex consists of parallel steeply dipping dykes 0.5–1.0 m thick whose relations to one another indicate emplacement by infilling of tension fractures rather than forceful injection.

Although the majority of dykes in the sheeted complex are relatively fractionated aphyric to sparsely phyric basalts with equigranular, fine-grained ophiolitic, sub-ophitic and intergranular textures, Bear (1960) was able to identify four petrographic types on Troodos that were confirmed geochemically by Desmet *et al.* (1978). For the Semail Nappe, three dyke types, directly related compositionally to lava group, have been identified by Pearce *et al.* (1981). Subsequently, Lippard *et al.* (1986, pp. 101–102) identified that dykes produced by axis magmatism belong to the sheeted complex whereas those of off-axis activity cross-cut the sheeted structure. Microprobe studies of clinopyroxene crystals from a drill core through the sheeted dyke complex on the N flank of Troodos (CY-4 of the Cyprus Deep Drilling Project) have identified two (high and low titanium) types (Alabaster, pers. comm.). This suggests that there are dyke types that correspond to the long-identified but poorly defined upper and lower divisions of the Troodos Pillow Lava Series.

The uppermost part of coherent ophiolites, which overlies the sheeted dyke complex, is an *extrusive sequence* consisting of pillowed and non-pillowed lavas occasionally intercalated with and overlain by deep-water sediments, all being the products of submarine volcanism and sedimentation at, or close to, a constructive plate margin. In the Semail ophiolite, field and geochemical evidence (see Lippard *et al.* 1986) clearly identifies that spreading axis volcanism produced the Geotimes unit. The unit is geochemically similar to the bulk of the sheeted dykes and like them formed from relatively evolved, slightly light-rare-earth-element-depleted, basaltic and andesitic melts produced in open-system magma chambers. It has not been possible to relate individual Geotimes sequences to specific magma chamber (layered gabbro) products. On Troodos the Lower Pillow Lavas have similar spreading axis geochemistry, and field work associated with the Troodos Deep Drilling Project confirms the axis nature of these volcanic rocks and that they seem to have the most evolved rocks at the base of the sequence, as originally identified by Gass & Masson-Smith (1963, p. 438, Fig. 5) on the basis of major oxide geochemistry. In addition, it has been shown that along the northern flank of Troodos volcanic sequences along river sections, some 5–10 km apart, commonly differ from one another, suggesting that they are sections through lava piles from different axis volcanoes. The geochemically inverted sequence in which dacitic lavas occur at the base could be due to compositional stratification in the underlying magma chamber which made silicic melts spatially more readily available for eruption than the denser more basic magmas.

Of the four other identifiable lava units in the Semail ophiolite, one (the Salahi unit) is mildly alkalic and is probably associated with plate collision and ophiolite obduction, whereas the three others, the Lasail, Alley and Cpx-phyric units, all have island arc geochemistry (Pearce *et al.* 1981) and were erupted respectively in off-axis oceanic seamount, intra-seamount and intra-seamount graben settings. In Oman, the recognition of off-axis eruptive activity was accompanied by the identification by Browning & Smewing (1981) of plutonic late intrusive complexes and hypabyssal equivalents (see Fig. 2). On Troodos, although it had been suggested (Gass & Smewing 1981) that the upper division of the Pillow Lava Series was the product of off-axis volcanism, this was put in doubt by the apparent geochemical continuity throughout the series. Recently, it has been recognized that off-axis plutons are far more common than previously thought (Malpas, pers. comm.). Furthermore, as the primitive geochemistry of these plutons is similar to that of the Upper Pillow Lavas, it has been proposed that the two are genetically related. This would explain the existence within the eruptive sequence of an upper low-titanium unit and a lower higher-titanium unit. This geochemical discrimination can also be made for dykes of the Troodos sheeted dyke complex where off-axis dykes are commonly parallel to the sheeted structure. As the disparate geochemistry of the two units cannot readily be explained by modification of a single magma, it seems that a distinct primitive melt was produced near the Troodos spreading axis. The primitive nature of this magma is explained in a similar manner to that proposed by Murton (this volume) to explain the boninitic character of the lavas in the southern Troodos transform fault zone, namely that these rocks are the products of partial melting of a harzburgitic upper mantle that had already undergone two phases of depletion, one at a major ocean constructive margin to produce MORBs and then one at a supra-subduction zone (SSZ) spreading axis.

Discussion

The wide acceptance of ophiolites as fragments of oceanic lithosphere formed primarily at constructive plate margins makes it a consensus model. However, several workers doubt if the detailed evidence on magmatic processes provided by the two best preserved ophiolites, the Troodos massif and the Semail Nappe, can be unconditionally accepted. The arguments are as follows:

(1) As the majority (99.98%) of oceanic lithosphere is subducted (Coleman 1977), obducted oceanic lithosphere must perforce by anomalous.

(2) these two ophiolites were formed at SSZ constructive margins (Pearce *et al.* 1984) and therefore cannot be accepted as indicating processes at major ocean-spreading axes such as the Mid-Atlantic Ridge and the East Pacific Rise. In support of this contention it has been pointed out that present-day oceanic crust produced at such SSZ sites does not have the same well-defined linear magnetic anomalies as those produced at major constructive margins.

(3) Because magma chambers beneath the slow-spreading Mid-Atlantic Ridge have never been detected geophysically, the role of magmatism in constructive-margin processes is suspect.

With regard to (1) McKenzie (1969) has calculated that, as oceanic lithosphere becomes colder and denser with age, only young low-density lithosphere can avoid subduction. Spray (1984) has shown that all ophiolites on which age data are available are young (<10 Ma old) when obducted. This is certainly the case with the Semail Nappe which was produced between 96 and 93 Ma and obduction started at 88 Ma (Lippard *et al.* 1986, p. 155, Fig. 4.11). Spray (1984) also demonstrated that undisrupted ophiolites have similar thicknesses (*c.* 15 km) and concluded that this was because their *in situ* position, prior to detachment, was immediately adjacent to a spreading axis. As only young oceanic lithosphere is obducted, it will only occur in major ocean basins when a spreading axis is close to a continental–oceanic destructive plate boundary, as is presently the case where the East Pacific Rise approaches the W coast of Central and North America. A more common obduction locale will be above subduction zones where new oceanic lithosphere formed at SSZ spreading axes will be close to an island arc or an oceanic–continental margin and can more readily be obducted thereon, especially if the area is one of regional compressional stress.

There is ample evidence that the volcanic products of SSZ spreading differ geochemically from those of MORBs (*eg* Saunders & Tarney 1979). The 'anomalous' geochemical character of SSZ ophiolites has been explained (*eg* Pearce 1980) as being due to the partial melting of already depleted peridotites of the SSZ mantle wedge under hydrous conditions provided by fluids emanating from the subducting slab. Under such conditions, the olivine stability volume in a basalt melt system will be enlarged and olivine will be precipitated throughout the adiabatic rise of melt batches. In the less hydrous (*c.* anhydrous) conditions beneath a major ocean-spreading axis the olivine stability volume would be smaller, olivine precipitation less conspicuous and dunitic masses in the mantle sequence smaller and less abundant. The effect of this difference on the composition of volcanic products is that those of SSZ ophiolites, unlike those of MORBs, are depleted in trace elements (especially nickel) compatible with olivine. Another variable resultant upon the different tectonic situations is that, whereas the major ocean-spreading processes can call upon mantle-wide convective processes, those above subduction zones can only 'sweep' the SSZ mantle wedge which will be some 200–400 km thick beneath an SSZ spreading axis. However, none of the geochemical variations between MORBs and SSZ ophiolite volcanics have so far been attributed to this difference. The hydrous nature of the parental mantle wedge would, with the same thermal energy available, make the percentage melting greater, and thereby make the melt abundance higher than that beneath major ocean-spreading axes. This should not preclude the diapiric upward movement of the uppermost mantle along such major axes. Indeed, the geophysical studies of Whitehead *et al.* (1984) suggest diapirism along sections of the East Pacific Rise.

If, as the ophiolite evidence suggests, the mantle diapirs are overlain by discrete crustal magma chambers each with its own suprajacent volcanic pile, then it might be expected that the strike of the dykes, transferring the melts from the high-level chambers to the surface, would be more irregular than hitherto identified from ophiolite studies. The 'sheeted' nature of the dyke complex and its commonly near-vertical 'pack of cards' structure has been taken to be the best indicator of the strike of the constructive margin and palaeovertical. If the mantle-diapir –individual high-level magma chamber model is correct, then the melt would be expected to radiate upwards from the chambers along planes (dykes) normal to the main tensional stress field. The linearity of ophiolite dyke strike, and the absence of any radial dyke pattern so common in terrestrial volcanoes, indicates that the principal tensional stress field, probably normal to the spreading axis, was dominant and that melts were passively injected from the magma chamber along tension planes. However, much of this melt must have moved sideways along the length of the constructive margin and could be expected to interfinger with, or have its configuration affected by, dykes emanating from adjacent centres. Although there are numerous thin late-stage cross-cutting dykes in both the Troodos and Semail sheeted dyke complexes and the only major dyke vergations so far identified are those seemingly related to movement along transform fault zones (Simonian & Gass 1978; Smewing 1980), field evidence for dyke interaction in ophiolites would be worth investigating in more detail.

Another argument used to query the relevance of SSZ ophiolite evidence on major ocean-spreading processes is the assertion that recent oceanic crust produced in marginal (SSZ) basins does not carry the same well-defined magnetic lineations as that of the major oceans; there is good evidence to the contrary. For instance, Weissel (1981, p. 223) states that 'Magnetic lineations in these [marginal basins of the Western Pacific] resemble lineations commonly associated with the world's mid-oceanic spreading system' and that 'almost all of these basins are underlain by crust which is similar to that of the major ocean basins'. Indeed, one of the first places where oceanic magnetic lineations were identified was the SSZ Scotia Sea where Barker & Hill (1981, p. 249) identify that 'well-developed oceanic magnetic lineaments confirm that the present spreading episode started about 8 Ma ago'. It seems therefore that the magnetic lineament evidence supports rather than contradicts that of the ophiolites.

Despite the overwhelming evidence from ophiolites and the geochemistry and petrology of oceanic drill core and dredge haul specimens that magmatic processes produce new oceanic crust at constructive plate margins, queries have been made on the grounds that no magma body has been geophysically detected beneath the slow-spreading Mid-Atlantic Ridge, although magma chambers have been identified for the faster-spreading East Pacific Rise (*eg* Orcutt *et al.* 1984). The first query in any evaluation of this objection must be, just how thick must a melt layer be before its presence can be detected by geophysical (seismic) surveys? Although searches through the literature and discussions with seismologists have failed to identify any precise opinion, there is a consensus view that, despite the many variables involved, a melt layer of over 500 m ought of be identifiable. It is therefore relevant that, in the recent magma chamber models (Browning 1984) discussed earlier, the layered series of peridotites and gabbros, which can be up to 6.0 km thick, could have been produced from melts that were never more than 100 m thick on the subsidence model or less than 500 m if double diffusive convection operated. Either way, the magmatic model is well within the seismic constraints.

The episodic continuum of magmatic processes at or near spreading axes has to be dependent and consequent upon the single-pass upwardly divergent convection in the upper mantle. The movement of melts from their generation within rising diapirs and their transfer into crustal magma chambers and subsequently to dyke systems and eruptive sequences is in response to space being produced by divergent convection. In view of the isolated magma chamber model and the calculation that 1.0 m dykes (the average dyke width of the sheeted dyke complex) would be injected once every 50–100 years, the picture is one of spatial and temporal magmatic periodicity with major spatial breaks occurring in spreading axis magmatism at transform fault zones where more primitive melts are produced (Murton, this volume). The temporal periodicity of axis magmatism is further complicated by off-axis activity which, like that in the transform zones (see Murton, this volume) produces primitive magma by partial melting of already twice-depleted upper mantle.

ACKNOWLEDGMENTS: This is a generalized rather over-simplified account based primarily on the studies of various research students who worked on these two ophiolites and whom I have had the privilege of supervising (Drs D. Greenbaum, J. D. Smewing, K. O. Simonian, D. T. Aldiss, D. A. Rothery, P. Browning, M. A. Brown, A. W. Shelton, S. Roberts, G. Stanger and B. J. Murton), many of whom subsequently became research colleagues. For the Oman work on which much of this paper is based, it represents a double distillate, first from PhD theses and related papers into an extended synopsis (Lippard *et al.* 1986) and then to this much shorter, more generalized contribution. I am most grateful to Dr Browning, Dr Roberts and Dr Rothery for their constructive comments on the first draft and to John Taylor and Pam Owen for cartographic and clerical services.

References

ALDISS, D. T. 1978. *Granitic Rocks of Ophiolites*. Unpublished PhD thesis, Open University, 198 pp.

—— 1981. Plagiogranites from the ocean crust and ophiolites. *Nature* **289**, 577–578.

ALLEMAN, F. & PETERS, T. 1972. The ophiolite–radiolarite belt of the Northern Oman mountains. *Eclogae Geological Helvetiae* **65**, 657–697.

ALLEN, C. R. 1975. *The Petrology of a Portion of the Troodos Plutonic Complex, Cyprus*. Unpublished PhD thesis, University of Cambridge.

ANON, 1972. Penrose field conference on ophiolites. *Geotimes* **17**, 24–25.

BARKER, P. F. & HILL, I. A. 1981. Back-arc extension in the Scotia Sea. *Philosophical Transactions of the Royal Society, London, Series A* **300**, 249–262.

BARTHOLOMEW, I. D. 1983. *The Primary Structures and Fabrics of the Upper Mantle and Lower Crust from Ophiolite Complexes*. Unpublished PhD thesis, Open University, 523 pp.

BEAR, L. M. 1960. The geology and mineral resources of the Akaki–Lythrodondha area. *Cyprus Geological Survey Department Memoir* **3**, 122 pp.

BROWN, M. A. 1980. Textural and geochemical evidence for the origin of some chromite deposits in the Oman ophiolite. *In*: PANAYIOTOU, A. (ed.) *Ophiolites: Proceedings of the International Ophiolite Symposium, Nicosia, Cyprus, 1979*, 714–721. Geological Survey Department, Cyprus.

—— 1982. *Chromite Deposits and their Ultramafic Host Rocks in the Oman Ophiolite*. Unpublished PhD thesis, Open University, 263 pp.

BROWNING, P. 1982. *The Petrology, Geochemistry and Structure of the Plutonic Rocks of the Oman Ophiolite*. Unpublished PhD thesis, Open University, 404 pp.

—— 1984. Cryptic variation within the Cumulate Sequence of the Oman ophiolite: magma chamber depth and petrological implications. *In*: GASS, I. G., LIPPARD, S. J. & SHELTON, A. W. (eds) *Ophiolites and Oceanic Lithosphere*. Geological Society of London Special Publication **13**, 71–82.

—— & SMEWING, J. D. 1981. Processes in magma chambers beneath spreading axes: evidence for magmatic associations in the Oman ophiolite. *Journal of the Geological Society, London* **138**, 279–280.

CANN, J. R. 1970. New model for the structure of the oceanic crust. *Nature* **226**, 928–930.

—— 1974. A model for oceanic crustal structure developed. *Geophysical Journal of the Royal Astronomical Society* **39**, 169–187.

CHRISTIANSEN, F. G. & ROBERTS, S. 1986. Formation of olivine pseudo-crescumulates by syntectonic axial planar growth during mantle deformation. *Geological Magazine* **123**, 73–79.

CLUBE, T. M. M., CREER, K. M. & ROBERTSON, A. H. F. 1985. Palaeorotation of the Troodos microplate, Cyprus. *Nature* **317**, 522–525.

COLEMAN, R. G. 1977. *Ophiolites: Ancient Oceanic Lithosphere*? Springer, Berlin, 229 pp.

—— & PETERMAN, Z. E. 1975. Oceanic plagiogranite. *Journal of Geophysical Research* **80**, 1099–1108.

—— *et al.* 1981. Oman Ophiolite. *Journal of Geophysical Research* **86**, 2495–2782 (special issue containing 18 papers on various aspects of the Oman (Samail) ophiolite).

DESMET, A., LAPIERRE, H., ROCCI, G., GAGNY, C. L., PARROT, J. -F. & DELALOYE, M. 1978. Constitution and significance of the Troodos sheeted complex. *Nature* **273**, 527–530.

DEWEY, J. F. & KIDD, W. S. F. 1977. Geometry of plate accretion. *Bulletin of the Geological Society of America* **88**, 960–968.

ELDER, J. W. 1977. Model of hydrothermal ore genesis. *In*: *Volcanic Processes in Ore Genesis*. Geological Society of London Special Publication **7**, 4–13.

ELLIOTT, D. 1976. The motion of thrust sheets. *Journal of Geophysical Research* **81**(5), 946–963.

FORD, C. E. 1976. Phase relations in the system $CaO-MgO-Al_2O_3-SiO_2-H_2O$. *Progress in Experimental Petrology* **3**, 266–272.

GASS, I. G. & MASSON-SMITH, D. 1963. The geology and gravity anomalies of the Troodos massif, Cyprus. *Philosophical Transactions of the Royal Society, London, Series A* **255**, 417–467.

—— & SMEWING, J. D. 1981. Ophiolites: obducted oceanic lithosphere. *In*: EMILIANI, C. (ed.) *The Sea*, **Vol. VII**, *Oceanic Lithosphere*. Wiley Interscience, New York, 339–362.

——, LIPPARD, S. J. & SHELTON, A. W. (eds) 1984. *Ophiolites and Oceanic Lithosphere*. Geological Society of London Special Publication **13**, 413 pp.

GLENNIE, K. W., BOEUF, M. G. A., HUGHES CLARK, M. W., MOODY-STUART, M., PILAAR, W. F. H. & REINHARDT, B. M. 1973. Late Cretaceous nappes in the Oman Mountains and their geological evolution. *Bulletin of the American Association of Petroleum Geology* **57**, 5–27.

——, ——, ——, ——, —— & —— 1974. The Geology of the Oman mountains. *Koninklijk Nederlands Geologisch Mijnbouwkundig Genoolschap Verdh.* **31**, Part 1 (text), 423 pp., Part 2 (illustrations), Part 3 (enclosures).

GREEN, D. H., HIBBERSON, W. O. & JAQUES, A. L. 1979. Petrogenesis of mid-ocean ridge basalts. *In*: MCELHINNEY, M. W. (ed.) *The Earth, its Origin, Structure and Evolution*. Academic Press, London, 265–297.

GREEN T. H. 1973. Experimental melting studies on a model upper-mantle composition at high pressure under water-saturated conditions. *Earth and Planetary Science Letters* **19**, 37–53.

GREENBAUM, D. 1972. Magmatic process at ocean ridges: evidence from the Troodos Massif, Cyprus. *Nature, Physical Sciences* **238**, 18–21.

GREGORY, R. T. & TAYLOR, H. P. 1981. An oxygen isotope profile in a section of Cretaceous oceanic crust, Samail ophiolite: evidence for $\delta^{18}O$ buffering of the oceans by deep (>5 km) seawater–hydrothermal circulation at mid-ocean ridges. *Journal of Geophysical Research* **86**, 2737–2755.

HEATON, T. H. E. & SHEPPARD, S. M. F. 1979. *Hydrogen and Oxygen Isotope Evidence for Seawater Hydrothermal Alteration and Ore Deposition, Troodos Complex, Cyprus*. Geological Society of London Special Publication **7**, 42–57.

HUPPERT, H. E. & SPARKS, R. S. J. 1980a. The fluid dynamics of a basaltic magma chamber replenished by influx of hot, dense, ultrabasic magma. *Contributions to Mineralogy and Petrology* **75**, 279–289.

—— & —— 1980*b*. Restrictions on the composition of mid-ocean ridge basalts: a fluid dynamical investigation. *Nature* **286**, 46–48.

JUTEAU, T., NICOLAS, A., DUBESSY, J., FRUCHARD, J. C. & BOUCHEZ, J. L. 1977. The Antalya ophiolite complex (western Taurides, Turkey): a structural model for an oceanic ridge. *Bulletin of the Geological Society of America* **88**, 1740–1748.

LANPHERE, M. A. 1981. K–Ar ages of metamorphic rocks at the base of the Samail ophiolite, Oman. *Journal of Geophysical Research* **86**, 2777–2782.

LEES, G. M. 1928. The geology and tectonics of Oman and parts of SE Arabia. *Quarterly Journal of the Geological Society, London* **110**, 121–152.

LIPPARD, S. J., GRAHAM, G. M., SMEWING, J. D. & SEARLE, M. P. 1983. Melanges associated with the Semail ophiolite in the Northern Oman Mountains allochthon, southwest Arabia. *In*: MCCALL, G. J. H. (ed.) *Ophiolitic Melanges*. Benchmark Papers in Geology, 300–308.

——, SHELTON, A. W. & GASS, I. G. 1986. *The Ophiolite of Northern Oman*. Geological Society of London Memoir **11** 178 pp.

MAKRIS, J., BEHLE, A., GIESE, P., GINZBURG, A., ABRAHAM, B., STEINMETZ, A., WHITMARSCH, B. & ELETHERIOU, S. 1983. Seismic refraction studies between Cyprus and Israel. *Geophysical Journal of the Royal Astronomical Society* **75**, 575–593.

MCKENZIE, D. P. 1969. Speculations on the consequences and causes of plate motion. *Geophysical Journal of the Royal Astronomical Society* **18**, 1–18.

MOORES, E. M. & VINE, F. J. 1971. Troodos Massif, Cyprus and other ophiolites as oceanic crust: evaluation and implications. *Philosophical Transactions of the Royal Society, London, Series A* **268**, 443–466.

NICOLAS, A. & POIRIER, J. D. 1976. *Crystalline Plasticity and Solid State Flow in Metamorphic Rocks*, Wiley, Chichester, 444 pp.

—— & VIOLETTE, J. F. 1982. Mantle flow at oceanic spreading centres: models derived from ophiolites. *Tectonophysics* **81**, 319–339.

O'HARA, M. J. 1968. Are ocean floor basalts primary magma? *Nature* **220**, 683–686.

ORCUTT, J. A., MCCLAIN, J. S. & BURNETT, M. 1984. Evolution of the ocean crust: results from recent seismic experiments. *In*: GASS, I. G., LIPPARD, S. J. & SHELTON, A. W. (eds) *Ophiolites and Oceanic Lithosphere*. Geological Society of London Special Publication **13**, 7–16.

PANAYIOTOU, A. (ed.) 1980. *Ophiolites: Proceedings of the International Ophiolite Symposium, Nicosia, Cyprus, 1979*, 781 pp. Geological Survey Department, Cyprus.

PEARCE, J. A. 1980. Geochemical evidence for the genesis and eruptive setting of lavas from Tethyan ophiolites. *In*: PANAYIOTOU, A. (ed.) *Ophiolites: Proceedings of the International Ophiolite Symposium, Nicosia, Cyprus, 1979*, 261–272. Geological Survey Department, Cyprus.

——, ALABASTER, T., SHELTON, A. W. & SEARLE, M. P. 1981. The Oman ophiolite as a Cretaceous arc–basin complex: evidence and implications *Philosophical Transactions of the Royal Society, London, Series A* **3**, 299–317.

——, LIPPARD, S. J. & ROBERTS, S. 1984. Characteristics and tectonic significance of suprasubduction zone ophiolites. *In*: KOKELAAR, B. P. & HOWELLS, M. F. (eds) *Marginal Basin Geology*, Geological Society of London Special Publication 16, 77–94.

ROBERTS, S. 1985. *The Role of Igneous Processes in the Formation of Ophiolitic Chromitite*. Unpublished PhD thesis, Open University.

ROBERTSON, A. H. F. & WOODCOCK, N. H. 1983. Zabyat Formation, Semail Nappe, Oman: sedimentation onto an emplacing ophiolite. *Sedimentology* **30**, 105–116.

ROTHERY, D. A. 1982. *The Evolution of the Wuqbah Block and the Applications of Remote Sensing in the Oman Ophiolite*. Unpublished PhD thesis, Open University, 414 pp.

—— 1983. The base of the sheeted dyke complex, Oman Ophiolite: implications for magma chambers of oceanic spreading axes. *Journal of the Geological Society, London* **140**, 287–296.

SAUNDERS, A. D. & TARNEY, J. 1979. The geochemistry of basalts from a back-arc spreading centre in the East Scotia Sea. *Geochimica et Cosmochimica Acta* **43**, 555–572.

SEARLE, M. P. 1980. *The Metamorphic Sheets and Underlying Volcanic Rocks beneath the Semail Ophiolite in Northern Oman Mountains of Arabia*. Unpublished PhD thesis, Open University, 213 pp.

SHELTON, A. W. 1984. *Geophysical Studies on the Northern Oman Ophiolite*. Unpublished PhD thesis, Open University, 353 pp.

—— & GASS, I. G. 1980. Rotation of the Cyprus microplate. *In*: PANAYIOTOU, A. (ed.) *Ophiolites: Proceedings of the International Ophiolite Symposium, Nicosia, Cyprus, 1979*, 61–65. Geological Survey Department, Cyprus.

SHEPPARD, S. M. F. 1977. Identification of the origin of ore-forming solutions by the use of stable isotopes. *In*: *Volcanic Processes in Ore Genesis*. Geological Society of London Special Publication **7**, 25–41.

SIMONIAN, K. O. & GASS, I. G. 1978. The Arakapas fault belt, Cyprus: a fossil transform fault. *Bulletin of the Geological Society of America* **89**, 1220–1230.

SMEWING, J. D. 1980. An Upper Cretaceous ridge–transform intersection in the Oman ophiolite. *In*: PANAYIOTOU, A. (ed.) *Ophiolites: Proceedings of the International Ophiolite Symposium, Cyprus, 1979*, 407–413. Geological Survey Department, Cyprus.

—— 1981. Mixing characteristics and compositional differences in mantle-derived melts beneath spreading axes: evidence from cyclically layered rocks in the ophiolite of North Oman. *Journal of Geophysical Research* **86**, 2645–2660.

SPRAY, J. G. 1984. Possible causes and consequences of upper mantle decoupling and ophiolite displacement. *In*: GASS, I. G., LIPPARD, S. J. & SHELTON, A. W. (eds) *Ophiolites and Oceanic Lithosphere*. Geological Society of London Special Publication **13**, 255–268.

STAKES, D. S., TAYLOR, H. P. JR & FISHER, R. L. 1984. Oxygen isotope and geochemical characterization of hydrothermal alteration in ophiolite complexes and modern oceanic crust. *In*: GASS, I. G., LIPPARD, S. J. & SHELTON, A. W. (eds) *Ophiolites and Oceanic Lithosphere*. Geological Society of London Special Publication **13**, 199–214.

TILTON, G. R., HOPSON, C. A. & WRIGHT, J. E. 1981. Uranium–lead isotopic ages of the Samail ophiolite, Oman, with applications to Tethyan Sea ridge tectonics. *Journal of Geophysical Research* **86**, 2763–2776.

WEISSEL, J. K. 1981. Magnetic lineations in marginal basins of the Western Pacific. *Philosophical Transactions of the Royal Society, London, Series A* **300**, 223–247.

WHITEHEAD, 1977 *Dynamics of Atmospheres and Oceans* **2**, 1.

WHITEHEAD, J. A., DICK, H. J. B. & SCHOUTEN, H. 1984. A mechanism for magmatic accretion under spreading centres. *Nature* **312**, 146–148.

WILSON, R. A. M. 1959. The geology of the Xeros–Troodos area. *Cyprus Geological Survey Department Memoir* **1**, 135 pp.

I. G. GASS, Department of Earth Sciences, the Open University, Milton Keynes, MK7 6AA, UK.

Asthenospheric control on magmatism in the ocean basins

Robert S. White

SUMMARY: Asthenospheric processes exert a dominant control over the volume and distribution of magmatism in oceanic basins. In particular, the temperature of the asthenosphere controls the amount of melt and hence the volume of igneous crust produced at rifted continental margins and oceanic spreading centres. Changes of as little as 100–150 °C in the asthenosphere temperature cause massive differences in the thickness of the igneous crust.

Where the asthenosphere wells up passively during rifting, such as at continental margins and oceanic spreading centres, partial melt is produced by adiabatic pressure release. Continental rifts near hot-spots are often emergent above sea level, with considerable extrusive and intrusive magmatism, whilst those away from hot-spots subside gently without volcanism. Oceanic spreading centres are fed by sublithosphere feeders that generate spreading cells typically 40–80 km long which operate independently of one another. At the intervening fracture zones, where the magma supply from the adjacent feeder is least, the crust thins, sometimes to less than 1 km.

Hot-spots are fed by mantle plumes. The partial melt and resultant surface volcanism comes from the rising plume which is only 100–200 km wide. The broad surrounding topographic swells, 1000–2000 km across, are dynamically supported by the mushrooming head of abnormally hot mantle beneath the lithosphere.

In this paper the effect of asthenospheric processes on the volume and the spatial and temporal distribution of magmatism in the ocean basins and their margins is discussed. The three main settings in which extensive magmatic activity occurs are examined in turn: the continent–ocean boundaries which lie in the region of the initial continental rift; the midplate swells or 'hot-spots'; and the spreading centres which lie within oceanic basins.

New ocean basins are formed by the rifting apart of pre-existing continental lithosphere. Sometimes this initial rift stage is accompanied by very considerable outpourings of basaltic lavas and by the addition of large quantities of intrusive rock at depth in the crust. However, in other cases there is very little igneous activity during the rift stage.

The problem at rifted continental margins is in understanding what factors control the presence or absence of magmatic activity. Details of the continental lithosphere which is being split to create a new ocean basin do not appear to be particularly important. The magmatism can be explained very simply by partial melting of asthenosphere as it rises to fill the space left by the thinning and separating lithosphere in the region of the rift. Relatively small variations in the temperature of the asthenosphere, of a magnitude and extent that we suspect exist from entirely separate evidence, can explain the production or absence of extensive volcanism.

A source of considerable intraplate volcanism, the islands which lie above hot-spots, are considered next. Why should there be such locally intense melt production in these intra-oceanic regions well away from plate boundaries? The geophysical evidence which suggests that the island swells are underlain by narrow but vigorous mantle plumes is discussed below. Considerable quantities of melt can be generated within the central rising mantle plume.

Lastly, the oceanic spreading centres are discussed. Globally, they are by far the most important source of basaltic volcanism. There are two sets of geophysical observations that need to be explained when this magmatism is considered. The first is that the average thickness of igneous crust produced at different spreading centres is the same everywhere in the world. Our theories of melt production need to explain this. The second set of geophysical observations concern the fine-scale structure of the spreading centres. The volcanism is known to occur within a relatively narrow central rift and in detail the spreading centres are segmented along strike like a string of pearls: the pearls represent the segments of normal crust, often with extremely thin crust (the string) in the intervening fracture zones. As shown below, both the broad-scale uniformity of igneous production and the local detailed variations in structure can be explained by asthenospheric processes.

The location and magnitude of igneous activity can be explained by melt production by adiabatic decompression of asthenosphere as it rises. In the case of hot-spots the upward move-

From SAUNDERS, A. D. & NORRY, M. J. (eds), 1989, *Magmatism in the Ocean Basins*, Geological Society Special Publication No. 42, pp. 17–27.

ment of the asthenosphere in a narrow plume is imposed by thermal convection. In the other two cases we consider, namely the rifted continental margins and oceanic spreading centres, the asthenosphere moves upwards passively to fill space left as the lithosphere above it, be it continental or oceanic, separates. The volume of melt generated depends primarily on the temperature of the asthenosphere.

Continental margins

Two end-members have been recognized in the development of rifted continental margins bordering ocean basins. In one end-member, typified by the Biscay margin of France, the continental stretching and thinning prior to formation of a new oceanic spreading centre occurs with little contemporaneous igneous activity. The gradually thinning and subsiding continental margin is marked by a series of normally faulted tilted blocks.

The other end-member is typified by the continental margins around the N Atlantic, such as those of Greenland, Norway and the NW UK (Fig. 1), where the continental rifting is accompanied by extensive intrusive and extrusive igneous activity. Another characteristic of these margins is that they often remain shallow during rifting, so that much of the extrusive igneous activity is in subaerial or shallow-marine conditions.

On the Biscay margin (Fig. 2), the continental crust thins gradually over a distance of several hundred kilometres, finally breaking to form new oceanic crust when it has thinned to about one-quarter or one-fifth of its original unstretched thickness. The Moho which marks the base of the crust is apparently the same original Moho everywhere. Apart from local uplift on the crests of the tilted fault blocks, the crust subsides downwards as it maintains local isostatic equilibrium in the simple way postulated by McKenzie (1978). The actual transition from stretched old continental crust to young entirely igneous oceanic crust occurs over a narrow region near the foot of the continental slope (Fig. 2, TZ).

On the volcanic margins such as those of Hatton Bank and the Vøring Plateau, the continental crust also thins across the margin (Fig. 2). Above the thinned continental crust the extrusive basalt flows form sequences of seaward-dipping reflectors (Hinz 1981). The dipping-reflector sequences are typically up to 4 km thick and may extend across the margin in one or more series up to 100 km wide. On the

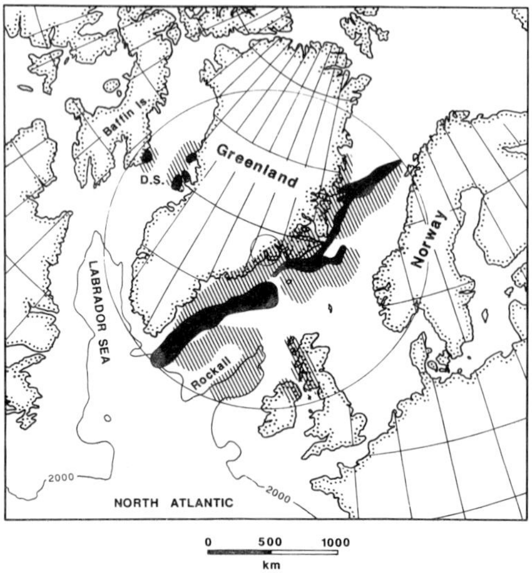

FIG. 1. Distribution of prominent seaward-dipping reflector sequences (solid shading) caused by extrusive basalt flows on the continental margins of the NE Atlantic shown on a reconstruction at magnetic anomaly 23 time, just after the onset of seafloor spreading. Hatching shows the extent of early Tertiary igneous activity in the region. Superimposed circles show the inferred position of the hot-spot plume beneath Greenland and the extent of the mushroom-shaped head of abnormally hot asthenospheric mantle. Data compiled from Nunns (1983); Smythe (1983); Larsen (1984); Mutter *et al.* (1984*a*); Roberts *et al.* (1984*a*,*b*); Ginzburg *et al.* (1985); Uruski & Parson (1985); Parson *et al.* (1986); Skogseid & Eldholm (1987); White *et al.* (1987*a*,*b*). Projection is equal area centred on the hot-spot plume.

Hatton margin, for example, the uppermost section of the dipping reflectors overlie thinned continental crust, whilst the most oceanward development is over oceanic crust (Fig. 2). Analyses of basalts from dipping reflectors on the Rockall continental margin are consistent with contamination by underlying continental lithosphere (Morton & Taylor 1987). Massive volumes of basaltic rock were extruded in a very short period to form these dipping-reflector sequences. The volcanic rocks of the N Atlantic margins shown in Fig. 1 were mostly extruded between magnetic anomalies 25 and 24B, a period of less than 5 Ma (White & McKenzie, in press), yet they extend along both sides of a continental split about 2000 km long. Their volume has been estimated to be about 2×10^6 km^3, comparable, for example, with the size of the Deccan volcanic province (Roberts *et al.* 1984*b*). In individual areas where more detailed

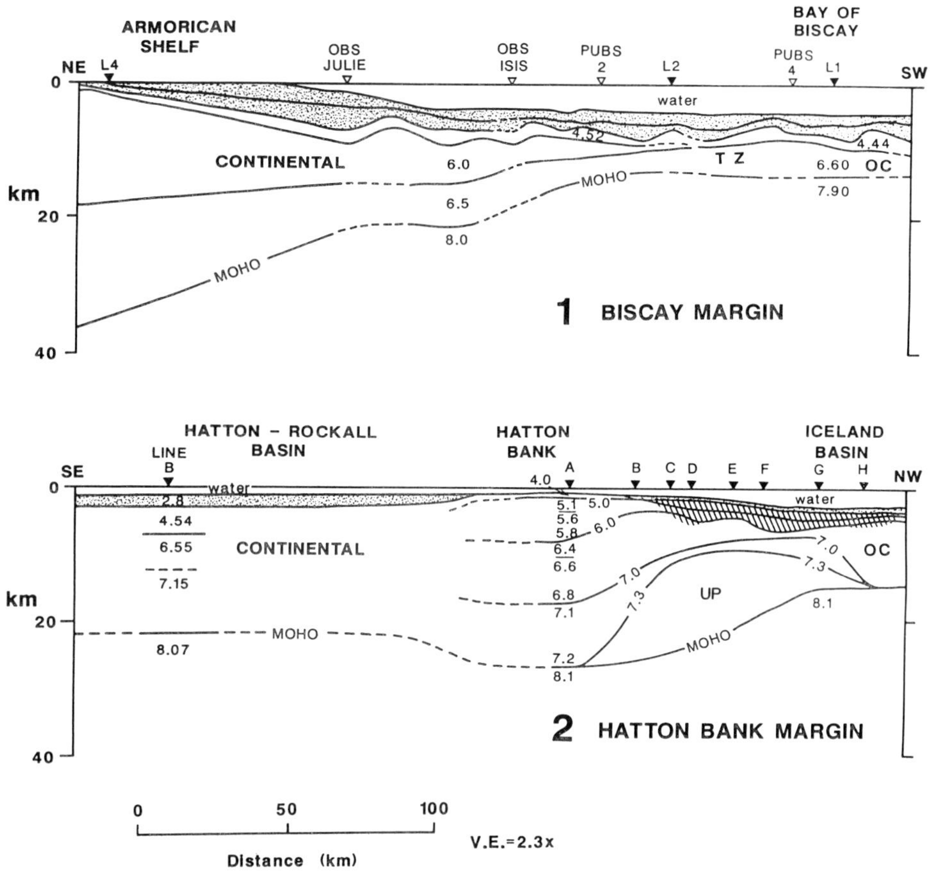

FIG. 2. Cross-sections across two typical end-members of continental margins: section 1, across the Biscay margin, is a classic tilted-block margin with no syn-rift volcanism (redrawn from Ginzburg *et al.* (1985); section 2, across the Hatton Bank margin, exhibits well-developed seaward-dipping reflectors caused by extrusive basalt flows (diagonal hatching), together with intrusive igneous underplating (UP) (Hatton Bank structure is from White *et al.* (1987*b*); Hatton–Rockall Trough from line B of Scrutton (1972)). Stipple shows post- and syn-rift sediments; OC, oceanic crust; TZ, transition zone between continental and oceanic crust. Both sections have the same vertical exaggeration of 2.3.

dating is possible, it appears that the entire dipping-reflector sequence may have been extruded in as little as 10^5 years (Roberts *et al.* 1984*b*).

The characteristic seaward-dipping upward-convex shape of the igneous reflectors suggests that they were formed by flows fed from vents to seaward of their present position. The flows are relatively flat along strike (White *et al.* 1987*a*), which suggests that they were fed from fissures aligned parallel to the strike of the initial split and the present margin. The seaward-dipping configuration can be explained by the loading of subsequently deposited lavas to seaward, as proposed by Palmason (1980) for similar geometric shapes of stacks of flows in Iceland and applied by Mutter *et al.* (1982) to the seaward-dipping reflector sequences of the Vøring Plateau on the Norwegian margin.

The extrusive basalts forming the dipping reflectors were deposited subaerially or in shallow-marine conditions (Murray 1984; Hyndman & Roberts 1987; Skogseid & Eldholm 1987). On the Hatton Bank margin individual flows can be traced several tens of kilometres laterally and a prominent unconformity within the extrusive igneous section may indicate emergence and erosion above sea level (White *et al.* 1987*a*).

Associated with the extrusive lavas on volcanic continental margins is large-scale addition of igneous rock to the lower crust (White *et al.*

1987*a*; LASE Study Group 1986; Mutter *et al.* 1988). On the Hatton Bank, for example, a pod of material over 10 km thick with unusually high seismic velocities of 7.3–7.4 km s^{-1} has been added beneath the rifted continental margin (Fig. 2). It is not possible to tell from the seismic data alone whether the igneous rock was underplated *en masse* or whether it was intruded into the lower crust as a series of sills and pods. Nevertheless, it is clear that an enormous volume of fresh igneous rock has been emplaced. The total volume of magma produced during splitting is therefore probably much larger than the estimates reported above from the dipping-reflector sequences alone. A subsidiary point worth noting is that the Moho under volcanic margins, although well developed and well defined, may not represent the original base of the continental crust. It may be a new feature marking the base of newly accreted igneous rock, which is much deeper and so cannot be used to estimate the amount of thinning and stretching as it can, for example, on the non-volcanic Biscay margin.

Why should apparently similar continental rifting processes produce such different results along the same continental margin? Under the Biscay margin the continental crust thins and subsides; under the N Atlantic margins the continental crust also thins, but enormous volumes of igneous rock are extruded and added to the crust with the margin initially remaining near or above sea level.

A simple explanation provided by White *et al.* (1987*a*) is that the only difference is that the N Atlantic is affected by a hot-spot, which is now centred under Iceland, whereas the more southerly region is in a region with normal asthenosphere temperatures unaffected by any hot-spots. The difference in asthenosphere temperature is only of the order of 150 °C, but this is sufficient to explain both the quantity of magma generated under the northern continental margins and the concomitant elevation.

When a continental region starts to rift, the underlying asthenosphere rises to fill the void left by lithosphere thinning. Partial melting occurs as the asthenosphere rises passively (McKenzie 1984; McKenzie & Bickle 1988). The amount of melt generated is controlled by the temperature of the asthenosphere and by the degree of lithosphere thinning. For ease of comparison all the temperatures quoted are potential temperatures, which are defined as the temperature the asthenosphere would have if brought to the surface adiabatically without melting. To convert potential to actual asthenosphere temperatures, the increase appropriate to the depth due to the adiabatic gradient of about 0.6 °C km^{-1} has to be added. The normal asthenosphere potential temperature is about 1280 °C. When continental lithosphere with a thickness of 100 km underlain by normal-temperature asthenosphere is stretched by a factor of 5, only about 2 km of partial melt is produced. Such is the case, for example, under the Biscay margin. However, when similar lithosphere underlain by asthenosphere with a potential temperature 100 °C higher is stretched by the same amount, nearly 10 km of melt is produced. If the potential temperature is increased to 150 °C above normal, nearly 15 km of melt can be produced simply by passive asthenosphere upwelling (White *et al.* 1987*a*).

As the melt is added to the crust it dramatically decreases the amount of initial subsidence caused by lithosphere thinning. Because the margin maintains isostatic balance with little flexural strength the addition of material to the crust acts immediately to counteract the subsidence that would otherwise occur due to lithosphere thinning. For a stretching factor of 5, the melt generated by asthenosphere temperatures of 1380 °C, some 100 °C above normal, causes the crust to sink only 1.5 km instead of subsiding the 2.3 km that it normally would for an initial lithosphere thickness of 100 km. For an asthenosphere potential temperature of 1430 °C, some 150 °C higher than normal, the surface of the crust will sink only 0.5 km for a stretching factor of 5 (White *et al.* 1987*a*). When the mantle density decrease caused by elevated temperatures and by extraction of melt is also considered, the initial subsidence is decreased by a further 1 km or more, for asthenosphere temperature anomalies of 100–150°C (White & McKenzie, in press). An asthenosphere potential temperature of 1430 °C causes elevation well above sea-level when the overlying lithosphere is stretched by a factor of 5.

So we have here a simple mechanism for explaining the major differences, both in amount of magma generated and in the subsidence or uplift, between the passive margins like the Biscay margin which is underlain by normal asthenosphere and those found in the N Atlantic which are affected by a hot-spot. Although the rising convective plume in the mantle under hot-spots is only 100 or 200 km wide, the material convected up in the central plume mushrooms out to form a hot region beneath the plate which is typically 1500–2000 km across, as shown in the next section. So the 2000 km length of the N Atlantic split which is dominated by volcanic activity is entirely consistent with the region over which asthenosphere temperatures are raised by heat advected in the hot-spot plume. The hot-spot lies directly beneath Iceland at present and formed the thick crust of the Greenland–Iceland–Faeroes ridge as the N Atlantic opened. The magmatism along the rifted margin generated by passive upwelling was enhanced by the elevated temperatures in the mushroom head of convected mantle, but only directly above the mantle plume in the Iceland–Faeroes ridge are plume-derived igneous rocks found.

Intraplate swells

There are numerous intraplate swells or 'hot-spots' within the oceanic basins. Perhaps the best known is the long-lived Hawaiian–Emperor chain which is well away from an oceanic spreading centre. Another is the Iceland hot-spot (Fig. 1), which sits astride the N Atlantic spreading centre. From the magmatic point of view, a notable feature of these intraplate swells is that the extrusive volcanic activity is restricted to a narrow region typically less than 200 km across, whereas the bathymetric and geoid anomaly of the swell extends over a region with a diameter of 1000–2000 km (Fig. 3).

Recent heat-flow measurements have shown that there are indeed heat-flow anomalies associated with the swells, although it was only with the development of multipenetration heat-flow probes that sufficient determinations could be made to measure such small heat-flow anomalies. The anomalies are typically of the order of 15%–20% with their maxima at the centre of the swells. The Hawaiian chain exhibits a heat-flow anomaly of 10–12 mW m^{-2} (von Herzen *et al.* 1982), the Bermuda swell has somewhat enhanced heat flow (Detrick *et al.* 1986) and the heat-flow anomaly over the centre of the Cape Verde Rise is about 16 mW m^{-2} (Courtney & White 1986) (Fig. 4). This paper concentrates on the Cape Verde Rise. This is a good hot-spot to model because the Atlantic plate is only moving relatively slowly at about 12 mm a^{-1} in the hot-spot frame of reference, whereas Bermuda is moving twice as fast, and Hawaii is racing at about 90 mm a^{-1} across its associated hot-spot. Since the Cape Verde Rise is almost stationary above the hot-spot, the motion of the underlying mantle can be modelled using an axisymmetric convection model.

There are four main observational constraints which can give information on the asthenosphere beneath the Cape Verde swell (Fig. 4): the bathymetric swell, after removal of the

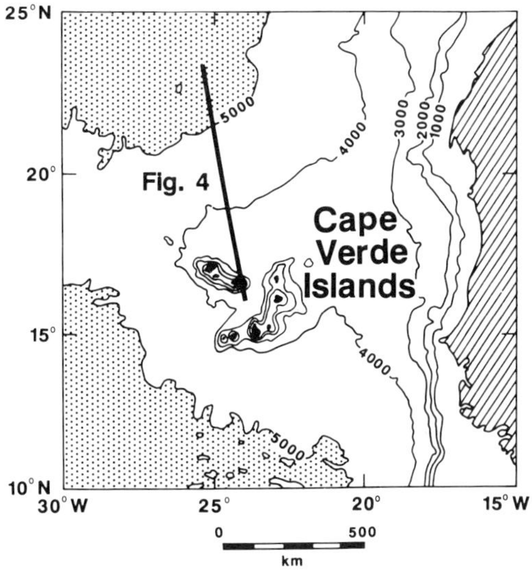

FIG. 3. Bathymetric map of the Cape Verde swell. Note that the volcanic islands are restricted to a 200 km wide region in the centre, while the swell extends across a region 1500 km in diameter. The solid line shows the location of the profiles shown in Fig. 4. Depths in metres.

effects of sediment loading, rises to a maximum of 1900±200 m above the surrounding normal seafloor of Mesozoic (*c.* 125 Ma) age; the geoid anomaly reaches almost 8 m across the swell; the heat flow increases from normal off the swell to 16 mW m^{-2} high at the centre; and the island volcanism has been active for at least 20 Ma and possibly for 40–50 Ma (Courtney & White 1986). The sustained volcanism allows us to apply a steady-state convective model to the Cape Verdes without having to worry about transient effects. The remaining three observed constraints, the bathymetry, geoid and heat-flow anomalies, can be predicted from the numerical axisymmetric convection computer models.

The best-fitting convection model, together with the predicted and observed anomalies, is shown in Fig. 4. No corrections have been made for crustal thickening of 2–4 km near the centre of the swell. The fit is remarkably good for such a simple model.

We can draw two main conclusions from this modelling. Firstly, the upwelling plume is only 100–200 km in diameter, and the island volcanism only occurs above this region. Secondly, the bathymetric swell is dynamically supported by the mantle convection. Because the hot asthenosphere mushrooms laterally when the rising plume is deflected by the overlying plate,

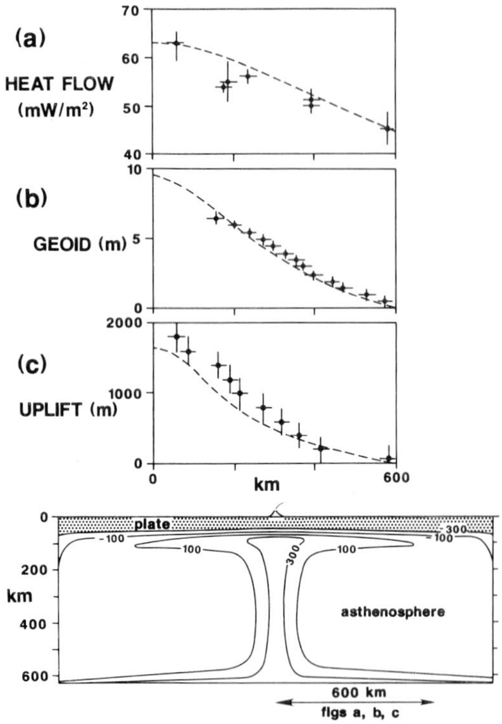

FIG. 4. Theoretical predictions of (*a*) heat flow, (*b*) geoid and (*c*) bathymetry anomalies (broken lines) calculated from an axisymmetric convecting model of a mantle plume beneath the Cape Verde swell, compared with observational constraints (crosses) along the radial profile marked in Fig. 3 (after Courtney & White 1986). Temperature anomalies in the lower panel are labelled in degrees Celsius with respect to the mean asthenosphere temperature.

it spreads sideways and causes uplift (and increased asthenosphere temperatures) over a region of 1500–2000 km in diameter. The magnitude of the temperature anomaly in the mantle plume is of the order of 100–200 °C, which is similar to the anomalous temperatures deduced from the volcanic rifted continental margins of the N Atlantic.

The convective plume model also explains well the observations from other hot-spots such as the Hawaiian chain. Volcanism, uplift and the onset of the geoid anomaly occur right at the nose of the Hawaiian chain, as the Pacific plate moves across the underlying asthenosphere plume. The melt generated in the rising plume thickens the crust from its normal value of 6.5–6.8 km to 18–19 km under the islands (Watts *et al.* 1985). Some of the thickening is by extrusive volcanism, and the rest by underplating which is probably similar to that found on the rifted continental margins. Like the rifting margins, the main phase of volcanic activity occurs very rapidly, probably in less than 1 Ma. The crust is thickened across a width of about 200 km above the plume (Watts *et al.* 1985), whereas the swell itself is some 1200 km wide. No heat-flow anomaly is seen near Hawaii itself, as the extra temperatures generated at the base of the conductive upper layer do not penetrate fully to the surface for 20 Ma.

The high temperatures in the upwelling plume are well able to generate the partial melt and resultant crustal thickening under the islands: 11–12 km beneath the Hawaiian chain, about 15 km beneath Iceland. However, it is harder to explain the fine details of the volume of melt generated under different hot-spots. The Cape Verde Rise, for example, exhibits much less crustal thickening than Hawaii. The most probable explanation is that relatively small potential temperature variations in the upwelling plume cause large differences in the amount of partial melt generated: a temperature anomaly of 100 °C generates approximately twice as much partial melt as an anomaly only 50 °C lower. Also, there are inevitably complex interactions between the partial melt and the matrix which have yet to be modelled. For example, the melting itself absorbs heat, and the rapidly upward-moving buoyant melt advects heat upwards. There is also the unknown effect of three-dimensional asthenosphere motions to consider. It is possible that it is easier for newly generated partial melt to penetrate through the fresh lithosphere brought across the plume by the fast-moving Pacific plate on which Hawaii sits than it is in the case of the slow-moving Atlantic plate above the Cape Verde plume. Nevertheless, it is clear that the convective asthenosphere model can explain all the observed geophysical features of hot-spots.

Oceanic spreading centres

Perhaps the most remarkable feature of the igneous crust generated at oceanic spreading centres is that everywhere, throughout the world's oceans, it has the same velocity structure and the same average thickness. This was realized in the early days of seismic studies of the oceanic crust using fairly crude interpretation techniques (Raitt 1963), and has since been confirmed by careful synthetic seismogram analysis of wide-angle seismic data (Spudich & Orcutt 1980; White 1984).

From the crustal thickness that is produced by partial melting of the asthenosphere as it wells up to fill the space left by the separating lithosphere plates we can infer the potential temperature of the underlying asthenosphere since the amount of melting is strongly dependent on the temperature (McKenzie 1984). The normal potential temperature is 1280 °C (McKenzie & Bickle 1988). The limited variations in average crustal thickness that are observed globally indicate that the normal asthenosphere potential temperature varies worldwide by only a few tens of degrees Celsius.

When a hot-spot sits astride an oceanic spreading centre, as in the case of Iceland, the increased asthenosphere temperatures lead to the generation of thicker than normal crust in the vicinity of the mantle plume (White *et al.* 1987*b*). Compositionally, also, the effects of the hot-spot can be detected, particularly in the magnesium content of the basalts which are especially sensitive to the asthenosphere temperature (Klein & Langmuir 1987; McKenzie & Bickle 1988; White, in press).

A second surprising feature of oceanic spreading centres is that the actual intrusion zone at the ridge crest is extremely narrow, typically less than 5–10 km wide, even on slow-spreading ridges such as the Atlantic. This is indeed fortunate, because it means that the seafloor spreading magnetic anomalies are preserved with a fidelity that makes their identification possible. Had the average intrusion zone been a lot wider, individual magnetic reversals would not have been preserved and it would have been a great deal harder to document seafloor spreading and the motions of the oceanic plates. However, it is still puzzling at first sight that the magmatism should be concentrated at such a narrow cleft in the lithosphere when the underlying asthenosphere upwelling must occur over a much broader region. Spiegelman & McKenzie (1987) have shown an elegant solution to this problem. As the lithosphere thickens away from the spreading centre it forms a wedge shape. The resultant corner flow in the underlying asthenosphere induces a pressure field which concentrates the partial melt generated in the upwelling asthenosphere towards the gap in the lithosphere at the spreading axis. So the average thickness and location of crust formed at oceanic spreading centres can be explained by the potential temperature and the flow in the asthenosphere. However, as shown below, the finer details of the magmatic variations along spreading centres and of the episodicity of magmatic cycles can also be explained by asthenospheric processes.

A striking feature of oceanic crust is that it is broken into ribbons typically 40–80 km wide by fracture zones which extend from margin to margin across entire basins. In recent years high-resolution seismic studies of the crustal structure have shown that the crust thins over distances of 20–30 km as fracture zones are approached (Fig. 5). Within the fracture zones themselves, very thin crust is found, sometimes as little as 1 km thick (White *et al.* 1984). The long-wavelength thinning is interpreted as being due to a decreasing supply of magma as it moves laterally along the ridge segment from a central feeder (White 1984). It is not surprising that the crust thins towards major fracture zones, such as the 130 km offset Oceanographer fracture zone (Fig. 5*a*) and the 320 km offset Vema fracture zone (Fig. 5*e*), since the spreading cell must terminate against the old thick lithosphere across the fracture zone. It is more surprising, however, that similar crustal thinning is also found where the fracture zone has only a very small offset, such as the 25 km offset Fracture Zone I (Fig. 5*a*), the 20 km offset Blake–Spur fracture zone (Mutter *et al.* 1984*a*) and an unnamed fracture zone at 45.5°N with 10 km offset reported by White & Matthews (1980). This suggests that the fracture zones occur near the ends of spreading segments where the magma supply from the central feeder is least.

The observation of thin crust regardless of the size of the fracture zone offset, together with the existence of differences in the fine-scale seafloor spreading magnetic anomalies between adjacent ridge segments, led Schouten & White (1980) to propose the existence of 'zero-offset' fracture zones marking the terminations of adjacent spreading centre cells. On a fine scale the adjacent cells spread independently of one another. Schouten & Klitgord (1982) showed from detailed magnetic anomaly patterns that the discontinuities at the ends of spreading cells remained on flow lines, or in other words maintained their relative locations on the spreading axis, for very long periods, even following changes in the sense of the offset across them. This 'memory' in the accreting plate boundary could not come from shallow crustal conditions, and again points to deeper asthenospheric control on the locations of the spreading cells.

Whitehead *et al.* (1984) showed by analogy with simple experiments using water and glycerine that regularly spaced diapirs of melt could be responsible for the characteristic spacing of magma intrusion zones on spreading axes. The diapirs are produced as a result of a modified Rayleigh–Taylor gravitational instability in the partial melt zone underlying the lithosphere. Schouten *et al.* (1985) extended this idea quanti-

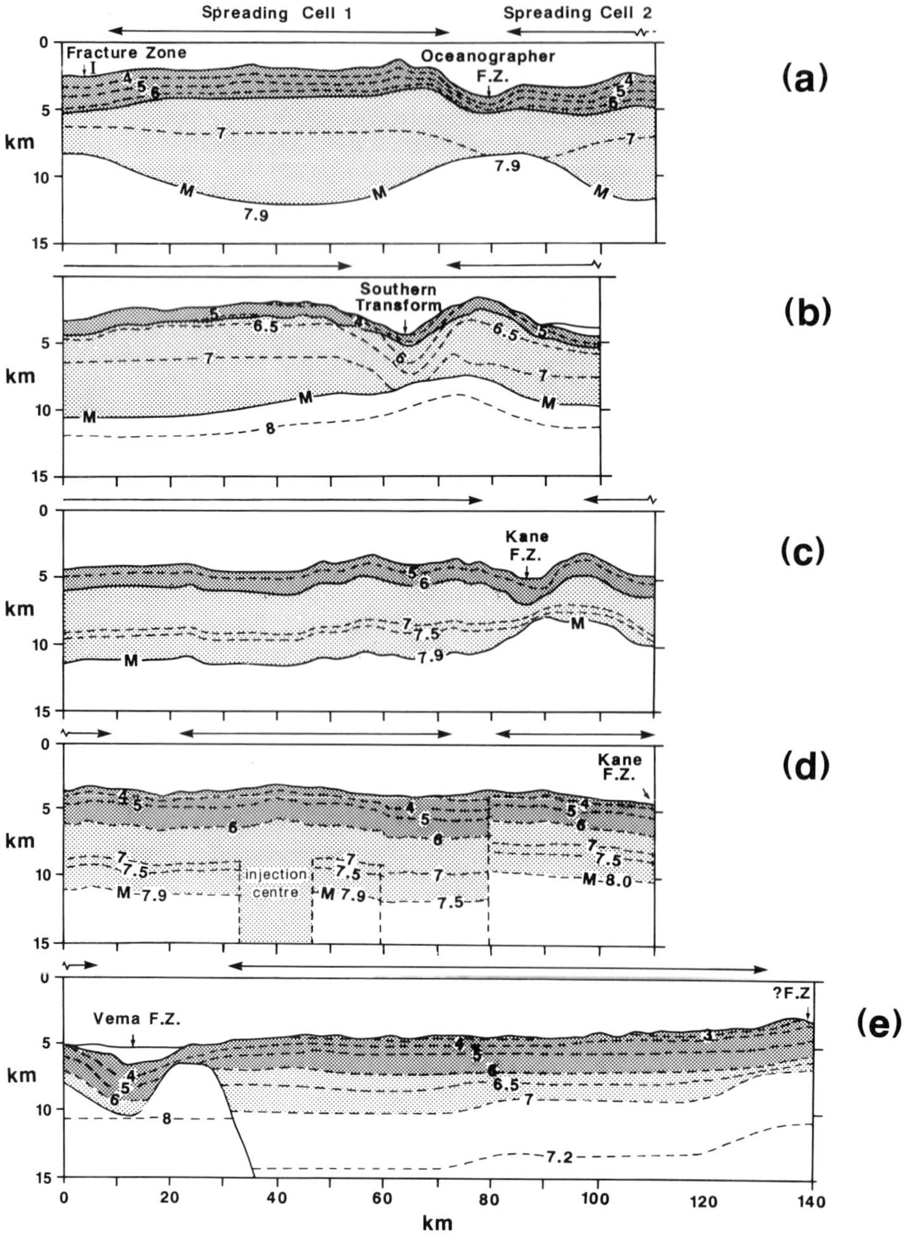

FIG. 5. Cross-sections redrawn at the same scale showing the velocity–depth structure parallel to the Mid-Atlantic Ridge spreading centre. Arrowed lines along the top of each section show the inferred extent of individual spreading cells. Stipple shows the extent of the crust recognized by the author, with M annotated where the Moho is developed. Dark stipple has crustal seismic velocities less than 6 km s^{-1}; light stipple has crustal velocities greater than 6 km s^{-1}. Note the long-wavelength crustal thinning over distances of 20–30 km from both small-offset (*eg* Fracture Zone I) and large-offset (*eg* Oceanographer, Charlie-Gibbs, Kane) fracture zones. (*a*) Oceanographer fracture zone from line B (Sinha & Louden 1983); (*b*) Charlie-Gibbs fracture zone from line 10617 (Whitmarsh & Calvert 1986); (*c*) Kane fracture zone from line EXP81 (Abrams *et al.*, in press); (*d*) Mid-Atlantic Ridge, 0 Ma crust from Purdy & Detrick (1986), with their interpretation of injection centre and spreading cells; (*e*) Vema fracture zone along 0 Ma crust from Louden *et al.* (1986).

tatively to show that the spacing of the intrusion centres caused by melt diapirs is proportional to the cube root of the spreading rate: this agrees with observations that the spreading cells on the fast-spreading (100–200 mm a^{-1}) East Pacific Rise are about twice as long as those on the slow-spreading (25 mm a^{-1}) Mid-Atlantic Ridge. Another quantitative prediction is that the frequency of magmatic events should be higher on the faster spreading ridges, as blobs of melt in the asthenosphere become unstable and rise at more frequent intervals. Again the frequency should be proportional to the cube root of the spreading rate, which is consistent with near-bottom studies which indicate that the frequency of magmatic activity on the East Pacific Rise is only a factor of 2 or 3 higher than on the slow-spreading Mid-Atlantic Ridge.

Apart from the spreading rate, the main control on the Rayleigh–Taylor instability is the contrast between the viscosities of the host mantle and the partially molten mantle region under the spreading centre. This is dependent on the amount of partial melt present, so again it comes back to being controlled by the temperature.

So we see that the asthenosphere temperature ultimately controls the total thickness of crust produced, its localization along the spreading axis in separately spreading cells and even the episodicity of magmatic events on oceanic spreading centres.

Conclusions

Asthenospheric processes, and especially the temperature, are dominant in controlling crustal magmatism in all the major settings in which it occurs in the ocean basins. Partial melting by decompression of passively rising asthenosphere produces the igneous crust beneath oceanic spreading centres and rifted continental margins. The quantity of melt and hence igneous crust generated depends critically on the potential temperature of the asthenosphere: differences of only a hundred degrees Celsius produce radically different volumes of igneous crust and, on the rifted continental margins, determine whether the margin subsides or remains near or above sea level during initial rifting. Under oceanic spreading centres not only the average thickness of crust but also its localization in individual spreading segments and the frequency of magmatic events are controlled by the instability of partial melt in the upper mantle. The consistency of thickness of the crust, its composition, spreading cell dimensions and frequency of magmatic events on spreading centres around the world point to remarkable consistency globally, to within a few tens of degrees Celsius, in the potential temperature of the asthenosphere.

Where extensive midplate volcanism occurs above 'hot-spots', the concordance between theoretical models and observed bathymetry, geoid and heat-flow anomalies across swells suggests that the volcanism is localized above a relatively narrow upwards-convecting mantle plume. Although volcanism is restricted to the region directly above the plume, the hot asthenosphere is deflected sideways when it reaches the overlying plate to give enhanced temperatures over a region 1000–2000 km in diameter. This not only causes dynamic uplift of the seafloor, producing the well-known midplate swells, but also, when the plume lies initially beneath a rifting continental region, causes extensive volcanism on the newly forming continental margins over distances of up to 2000 km.

Acknowledgments: Most of the work described in this paper has been done at sea and in Cambridge in collaboration with numerous research students and other colleagues. I am grateful to them all. I thank Barbara Dyson for typing the manuscript, and Phyl Fisher for her drafting skills. The Natural Environment Research Council funded all the work at sea. This is Department of Earth Sciences contribution number ES 1074.

References

Abrams, L. J., Detrick, R. S. & Fox, P. J. 1988. Morphology and crustal structure of the Kane Fracture Zone Transverse Ridge. *Journal of Geophysical Research* **93**, 3195–3210.

Courtney, R. C. & White, R. S. 1986. Anomalous heat flow and geoid across the Cape Verde Rise: evidence for dynamic support from a thermal plume in the mantle. *Geophysical Journal of the Royal Astronomical Society* **87**, 815–867; microfiche GJ87/l.

——, Von Herzen, R. P., Parsons, B., Sandwell, D. & Doughe, M. 1986. Heat flow on the Bermuda Rise and thermal models of mid-plate swells. *Journal of Geophysical Research* **85**, 3701–3723.

Ginzburg, A., Whitmarsh, R. B., Roberts, D. G., Montadert, L., Camus, A. & Avedik, F. 1985. The deep seismic structure of the northern continental margin of the Bay of Biscay. *Annales Geophysicae* **3**, 499–510.

Hinz, K. 1981. A hypothesis on terrestrial catastrophes: wedges of very thick oceanward dipping layers beneath passive continental margins. *Geologische Jahrbuch* **E22**, 3–28.

HYNDMAN, R. D. & ROBERTS, D. G. 1987. DSDP heat flow and models of the rifted west Rockall margin. *Journal of Geophysical Research* **92**, 9376–9384.

KLEIN, E. M. & LANGMUIR, C. H. 1987. Global correlations of ocean ridge basalt chemistry with axial depth and crustal thickness. *Journal of Geophysical Research* **92**, 8089–8115.

LARSEN, H. C. 1984. Geology of the East Greenland Shelf. *In*: SPENCER, A. M. *et al.* (eds) *Petroleum Geology of the North European Margin*. Graham & Trotman, London, 329–339.

LASE STUDY GROUP 1986. Deep structure of the US East Coast passive margins from large aperture seismic experiments (LASE). *Marine and Petroleum Geology* **3**, 234–242.

LOUDEN, K. E., WHITE, R. S., POTTS, C. G. & FORSYTH, D. W. 1986. Structure and seismotectonics of the Vema Fracture Zone. *Journal of the Geological Society, London* **143**, 795–806.

MCKENZIE, D. P. 1978. Some remarks on the development of sedimentary basins. *Earth and Planetary Science Letters* **40**, 25–32.

—— 1984. The generation and compaction of partially molten rock. *Journal of Petrology* **25**, 713–765.

—— & BICKLE, M. 1988. The volume and composition of melt generated by extension of the lithosphere. *Journal of Petrology* **29**, 625–679.

MORTON, A. C. & TAYLOR, P. N. 1987. Lead isotope evidence for the structure of the Rockall dipping reflector passive margin. *Nature* **326**, 381–383.

MURRAY, J. W. 1984. Paleogene and Neogene benthic foraminifers from Rockall Plateau. *In*: ROBERTS, D. G., SCHNITKER, D. *et al.* (eds) *Initial Reports of the Deep Sea Drilling Project* **81**, 503–534.

MUTTER, J. C., BUCK, W. R. & ZEHNDER, C. M. 1988. Convective partial melting I: a model for the formation of thick basaltic sequences during the initiation of spreading. *Journal of Geophysical Research* **93**, 1031–1048.

——, DETRICK, R. W. & NAT. STUDY GROUP 1984*a*. Multi-channel seismic evidence for anomalously thin crust at Blake–Spur Fracture Zones. *Geology* **12**, 534–537.

——, TALWANI, M. & STOFFA, P. L. 1982. Origin of seaward dipping reflectors in oceanic crust off the Norwegian margin by 'subaerial sea-floor spreading'. *Geology* **10**, 353–357.

NUNNS, A. G. 1983. Plate tectonic evolution of the Greenland–Scotland Rise. *In*: BOTT, M. H. P., SAXON, S., TALWANI, M. & THIEDE, J. (eds) *Structure and Development of the Greenland–Scotland Ridge*. Plenum Press, New York, 11–30.

PALMASON, G. 1980. A continuum model of crustal generation in Iceland: kinematic aspects. *Journal of Geophysics* **47**, 7–18.

PARSON, L. M., MASSON, D. G., MILES, P. R. & PELTON, C. D. 1986. *Structure and Evolution of the Rockall and East Greenland Continental Margins. Report of Work Undertaken by IOS in the period up to April 1985*. Institute of Oceanographic Sciences Report **223**, 71 pp.

PURDY, G. M. & DETRICK, R. S. 1986. Crustal structure of the Mid-Atlantic Ridge at 12°N from seismic refraction studies. *Journal of Geophysical Research* **91**, 3739–3762.

RAITT, R. W. 1963. The crustal rocks. *In*: HILL, M. N. (ed.) *The Sea*, Vol. 3. Wiley–Interscience, New York, 85–102.

ROBERTS, D. G., BACKMAN, J., MORTON, A. C., MURRAY, J. W. & KEENE, J. B. 1984*a*. Evolution of volcanic rifted margins: synthesis of Leg 81 results on the west margin of Rockall Plateau. *In*: ROBERTS, D. G., SCHNITKER, D. *et al.* (eds) *Initial Reports of the Deep Sea Drilling Project* **81**, 883–911.

——, MORTON, A. C. & BACKMAN J. 1984*b*. Late Paleocene–Eocene volcanic events in the northern North Atlantic Ocean. *In*: ROBERTS, D. G., SCHNITKER, D. *et al.* (eds) *Initial Reports of the Deep Sea Drilling Project* **81**, 913–923.

SCHOUTEN, H. & KLITGORD, K. D. 1982. The memory of the accreting plate boundary and the continuity of fracture zones. *Earth and Planetary Science Letters* **59**, 255–266.

—— & WHITE, R. S. 1980. Zero-offset fracture zones. *Geology* **8**, 175–179.

——, KLITGORD, K. D. & WHITEHEAD, J. A. 1985. Segmentation of mid-ocean ridges. *Nature* **317**, 225–229.

SCRUTTON, R. A. 1972. The crustal structure of Rockall Plateau microcontinent. *Geophysical Journal of the Royal Astronomical Society* **27**, 259–275.

SINHA, M. C. & LOUDEN, K. E. 1983. The Oceanographer fracture zone — I. Crustal structure from seismic refraction studies. *Geophysical Journal of the Royal Astronomical Society* **75**, 713–736.

SKOGSEID, J. & ELDHOLM, O. 1987. Early Cenozoic crust at the Norwegian continental margin and the conjugate Jan Mayen Ridge. *Journal of Geophysical Research* **92**, 11 471–11 491.

SMYTHE, D. K. 1983. Faeroe–Shetland Escarpment and continental margin north of the Faeroes. *In*: BOTT, M. H. P. SAXON, S., TALWANI, M. & THIEDE, J. (eds) *Structure and Development of the Greenland–Scotland Ridge*, Plenum Press, New York, 109–119.

SPIEGELMAN, M. & MCKENZIE, D. P. 1987. Simple 2-D models for melt extraction at mid-ocean ridges and island arcs. *Earth and Planetary Science Letters* **83**, 137–152.

SPUDICH, P. & ORCUTT, J. 1980. A new look at the seismic velocity structure of the oceanic crust. *Reviews of Geophysics and Space Physics* **18**, 627–645.

URUSKI, C. I. & PARSON, L. M. 1985. *A Compilation of Geophysical Data on the East Greenland Continental Margin and its Use in Gravity Modelling across the Continent–Ocean Transition*. Institute of Oceanographic Sciences Report **214**, 53 pp.

VON HERZEN, R. P., DETRICK, R. S., CROUGH, S. T., EPP, D. & FEHN, U. 1982. Thermal origin of the Hawaiian Swell: heat flow evidence and thermal models. *Journal of Geophysical Research* **87**, 6711–6723.

WATTS, A. B., TEN BRINK, U. S., BUHL, P. & BROCHER, T. M. 1985. A multichannel seismic study of lithospheric flexure across the Hawaiian–Emperor seamount chain. *Nature* **315**, 105–111.

WHITE, R. S. 1984. Atlantic oceanic crust: seismic structure of a slow spreading ridge. *In*: GASS, I. G., LIPPARD. S. J. & SHELTON, A. W. (eds) *Ophiolites and Oceanic Lithosphere*. Geological Society of London Special Publication **13**, 34–44.

—— Initiation of the Iceland plume and opening of the North Atlantic. In: TANKARD, A. J. & BALKWILL, H. (eds) *Extensional, tectonics and stratigraphy of the North Atlantic margins*, AAPG Memoir, in press.

—— & MATTHEWS, D. H. 1980. Variations in oceanic upper crustal structure in a small area of the northeastern Atlantic. *Geophysical Journal of the Royal Astronomical Society* **61**, 401–436.

—— & MCKENZIE, D. P. 1988. Magmatism at rift zones: the generation of volcanic continental margins and flood basalts. *Journal of Geophysical Research* **29**, 625–679.

——, DETRICK, R. S., SINHA, M. C. & CORMIER, M. H. 1984. Anomalous seismic crustal structure of oceanic fracture zones. *Geophysical Journal of the Royal Astronomical Society* **79**, 779–798.

——, SPENCE, G. D., FOWLER, S. R., MCKENZIE, D. P., WESTBROOK, G. K. & BOWEN, A. N. 1987*a*. Magmatism at rifted continental margins in the North Atlantic. *Nature* **303**, 439–444.

——, WESTBROOK, G. K., BOWEN, A. N., FOWLER, S. R., SPENCE, G. D., PRESCOTT, C., BARTON, P. J., JOPPEN, M., MORGAN, J. & BOTT, M. H. P. 1987*b*. Hatton Bank (northwest U.K.) continental margin structure. *Geophysical Journal of the Royal Astronomical Society* **89**, 265–272.

WHITEHEAD, J. A., JR, DICK, H. J. B. & SCHOUTEN, H. 1984. A mechanism for magmatic accretion under spreading centres. *Nature* **312**, 146–148.

WHITMARSH, R. B. & CALVERT, A. J. 1986. Crustal structure of Atlantic Fracture Zones, I: The Charlie–Gibbs F. Z. *Geophysical Journal of the Royal Astronomical Society* **85**, 107–138.

——, AVEDIK, F. & SAUNDERS, M. R. 1986. The seismic structure of thinned continental crust in the northern Bay of Biscay. *Geophysical Journal of the Royal Astronomical Society* **86**, 589–602.

R. S. WHITE, Bullard Laboratories, Department of Earth Sciences, Madingley Road, Cambridge CB3 0EZ, UK.

Upper mantle segregation processes: evidence from alpine-type peridotites

John G. Spray

SUMMARY: Studies of alpine-type peridotites indicate that the upper mantle is pervaded by various segregations whose origin cannot be exclusively attributed to the effects of crustal emplacement. The segregations typically occur as veins and layers (less than 1 m thick). They are common in the lherzolitic varieties of alpine-type peridotite but are of more restricted occurrence in true ophiolites. The segregations are represented by predominantly metamorphic mineral assemblages which span the garnet, spinel and plagioclase peridotite stability fields. Low-pressure segregations may remain igneous. Many segregations are varieties of aluminous clinopyroxenite with or without garnet, spinel or plagioclase. Segregations possess variable bulk chemistries and mineralogies so a number of different origins for them must be considered. The available evidence indicates that they are (i) metamorphosed trapped partial melts, melt residua or conduit platings, (ii) recycled and streaked-out ocean crust (via subduction zones), (iii) solid-state metamorphic differentiates or (iv) products of mantle metasomatism. The formation of segregations renders the mantle heterogeneous on a macroscopic scale, but this is combated in time by diffusive processes and segregation recycling at spreading centres, both of which contribute to its rehomogenization.

Investigations of alpine-type peridotites and xenoliths from basalts and kimberlites show that, to a limited extent, the Earth's uppermost mantle is lithologically varied. Compositions range from ultrabasic to basic with the former predominating as garnet, spinel or plagioclase lherzolite or as harzburgite, a condition dependent on the pressure and temperature of latest equilibration and the degree of partial melting realized. These four peridotite types occur in metamorphic form due to the effects of high-temperature solid-state flow in the mantle. This typically results in the generation of tectonite fabrics due to the preferential orientation of constituent minerals, especially olivine, so that foliations and lineations are created. In more extreme cases gneissose or mylonitic textures are generated owing to compositional layering or the focusing of shear deformation.

Aside from peridotites, a number of distinct and generally more basic 'segregations' pervade the upper mantle. Segregations can be distinguished from their peridotite hosts on the basis of mineralogical, chemical and isotopic criteria. It appears that segregations only constitute a minor component of the upper mantle, probably much less than 10% by volume. They typically form discordant veins less than 1 m thick or dykelets or concordant layers, bands, lenses or pods.

This work is primarily concerned with identifying generation mechanisms responsible for the formation of segregations using evidence from alpine-type peridotites rather than xenoliths. This is not to say that xenoliths are not of great importance, but rather to emphasize the necessary spatial control that alpine-type peridotites afford for assessing mantle structure.

Varieties of alpine-type peridotites

Alpine-type peridotites represent the largest mantle sample exposed at the Earth's surface, typically occurring in volumes of 10^2–10^4 km^3. Their size facilitates microscopic to megascopic evaluation of the relationships between petrological, chemical and structural processes occurring in the upper mantle, provided that the effects of emplacement overprinting and retrogression are not extensive. This is in contrast to xenoliths transported in kimberlites and alkali basalts which, although typically providing a deeper mantle sample, do not reveal gross structure because of their limited size, which is typically less than 30 cm diameter. Furthermore, xenoliths may undergo static annealing due to high-temperature ascent within magma at zero deviatoric stress. Conversely, alpine-type peridotites are emplaced under conditions of decreasing temperature and increasing deviatoric stress (Mercier 1985). The possibility of preserving original mantle textures is therefore more likely in the latter.

It has been the practice to divide alpine-type peridotites into two subtypes (Thayer 1960;

From Saunders, A. D. & Norry, M. J. (eds), 1989, *Magmatism in the Ocean Basins*, Geological Society Special Publication No. 42, pp. 29–40.

Green 1967; Den Tex 1969; Jackson & Thayer 1972): lherzolitic (also known as orogenic 'root zone' or 'high-temperature' peridotite) and harzburgitic (also known as ophiolitic or, somewhat misleadingly, 'low-temperature' peridotite). These have been interpreted as fragments of subcontinental and subocean mantle respectively (Nicolas & Jackson 1972). Whether they are *representative* of these areas of upper mantle is open to debate. Both subtypes are emplaced into or onto continental crust as bodies which are usually less than 15 km thick.

The lherzolite subtype is exemplified by the peridotite massifs of the western Mediterranean and western Alps (*eg* Lherz, southern France; Ronda, southern Spain; Beni Bouchera, northern Morocco; Lanzo, Italy). These mainly comprise garnet-, spinel- or plagioclase-bearing ol+opx+cpx assemblages (*ie* four-phase lherzolites). The harzburgitic subtype is typified by such ultramafic–mafic bodies as the Troodos ophiolite of Cyprus and the Semail ophiolite of Oman. These possess ol+opx mantle units and overlying crusts of basaltic overall composition. However, the two subtypes are merely end-members of a spectrum of peridotite types (Fig. 1). The majority of alpine-type peridotites are combinations of lherzolite and harzburgite overlain by variable thicknesses of basic crust. The pure lherzolite end-member possesses no crust, whilst the pure harzburgite end-member carries a crust *c*. 5–6 km thick (*ie* the ideal ophiolite). Thus there is an apparent inverse relationship between the degree of clinopyroxene depletion in the associated peridotite and the development of overlying crust, although this does not appear to be of a simple parent-daughter nature (*eg* Menzies 1984).

The variable composition and structure of alpine-type peridotites reveals the unrepresentative status of true ophiolites, a position distorted by the prominence given to ocean lithosphere–ophiolite comparisons within the plate tectonic paradigm. The term ophiolite, as defined by the Penrose conference (Anon. 1972), only applies to a very small fraction of alpine-type peridotites and hence the large mantle sample available at the Earth's surface. As a result, workers have appreciated the need to broaden the Penrose definition (*eg* Moores 1982).

The variety of alpine-type peridotite developed in a given geological situation depends on a number of factors, one of the most important being the depth of the lithosphere–asthenosphere boundary at the site of peridotite generation (Fig. 2). Formation of the harzburgitic subtype is favoured by the presence of high-level asthenosphere (*ie* at established spreading centres), whilst the development of the lherzolitic subtype is favoured by the presence of a less perturbed lithosphere–asthenosphere boundary (*ie* at failed continental or incipient rifts). Under steady-state rifting conditions the spreading rate will control the depth to the asthenosphere. Normal rates (more than 1 cm a^{-1}) will tend to yield harzburgitic and slow rates (less than 1 cm a^{-1}) lherzolitic varieties (Boudier & Nicolas 1985). Similarly, spreading rate can be related to crustal development (Klein & Langmuir 1987).

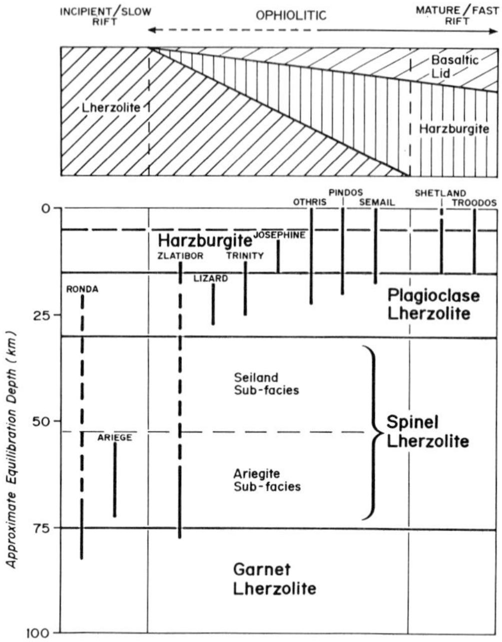

FIG. 1. Varieties of alpine-type peridotite distinguished by variation in lherzolite and/or harzburgite contents and absence or presence of 'oceanic' crust. The highest grade mantle facies recorded and partial re-equilibration ascent paths are indicated for selected examples.

Varieties of segregation

Mineralogy

Segregation classification poses a more difficult problem than that of the host peridotites. Both peridotites and segregations comprise the same minerals but in different proportions such that many segregations are basic rather than ultrabasic.

Figure 3 summarizes the main types of segregation based on their common occurrence in

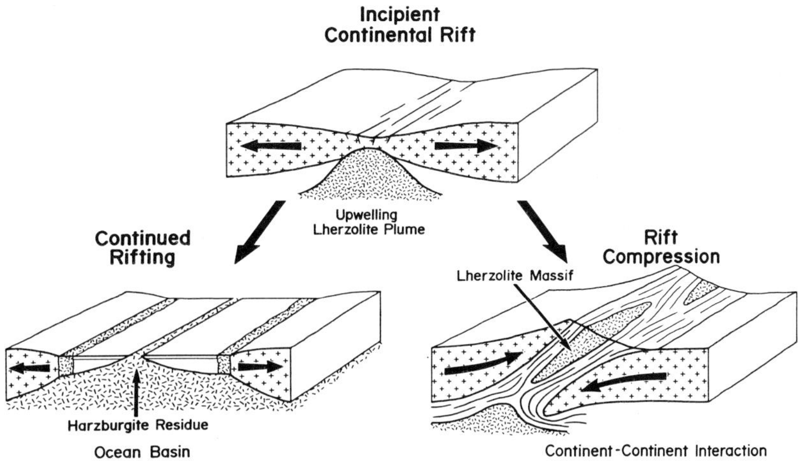

FIG. 2. Diagrammatic representation of the development of a continental rift into either an oceanic basin if extension continues or continental–continental interaction if initial extension is replaced by compression. Ophiolitic or lherzolitic alpine-type peridotites develop accordingly.

alpine-type peridotites. The classification scheme emphasizes the significance of clinopyroxene, which is an important constituent in many, though not all, segregations. Consequently, the pentagram of Fig. 3 has clinopyroxene at its centre, surrounded by olivine and orthopyroxene and the three aluminous phases plagioclase, spinel and garnet. Virtual monomineralic varieties occur as plagioclasite, dunite, chromitite and clinopyroxenite (± spinel). Orthopyroxenite segregations also occur, but these are usually subordinate to websterite (*eg* Lensch 1976). The eight groups represent an elaboration of the work of Lacroix (1901), although his original descriptions were restricted to those segregations occurring in the spinel lherzolite stability field (*ie* mainly the ariegites of group 6, Fig. 3).

The diagnostic mineralogy is accompanied by accessory amounts of some of the other six mineral phases. For example, minor quantities of orthopyroxene and/or plagioclase may accompany wehrlites. Accessory hydrous phases may also occur. These are typically high-temperature varieties of amphibole (*eg* pargasite, kaersutite) and mica (phlogopite). Hydrous phases predominate in group 6b, constituting the 'lherzite' subgroup of Lacroix (1901). A similar pentagram to that in Fig. 3 could be constructed with amphibole in place of clinopyroxene, with the possible exception of amphibole + olivine.

Plagioclase-rich segregations (group 1) typically occur as discordant veins within plagioclase lherzolite or harzburgite and thus appear to be

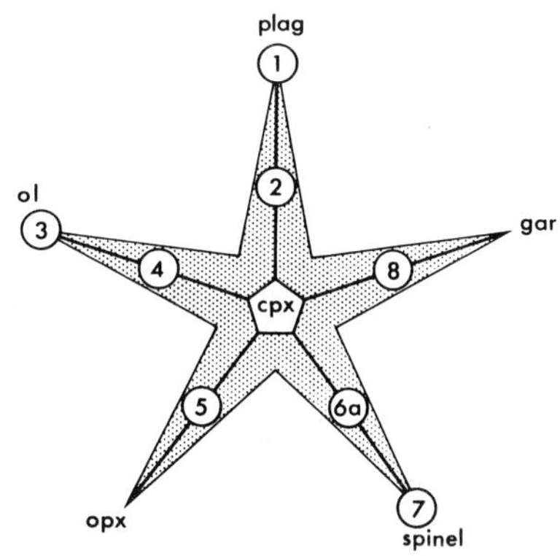

Segregation type	Diagnostic mineralogy
1. Plagioclasite	Plagioclase
2. Gabbro	Plag + clinopyroxene
3. Dunite	Olivine
4. Wehrlite	Olivine + clinopyroxene
5. Websterite	Clinopyroxene + orthopyroxene
6. Ariegite a. Anhydrous	Clinopyroxene + spinel
b. Hydrous	Amphibole and/or mica + spinel
7. Chromitite, etc.	Spinel group
8. Garnet clinopyroxenite	a. Garnet + diopside b. Garnet + fassaite c. Garnet + omphacite

FIG. 3. Classification scheme for segregations from alpine-type peridotites.

a relatively high-level phenomenon (*eg* Quick 1981*a*). The same applies to the gabbroic (group 2) segregations. Wehrlites (group 4) and websterites (group 5) tend to span the plagioclase and spinel lherzolite fields whereas the hydrous ariegites (group 6b) predominate in spinel lherzolite.

Dunite segregations (group 3) are of widespread occurrence and show considerable range in size. They may occur as centimetre thick veins and bands or kilometre-size pods (*eg* Quick 1981*b*) and do not appear restricted to a particular mantle facies, although they are more common in harzburgites than in lherzolites.

The higher-pressure segregations are represented by aluminous clinopyroxene-rich rocks of group 6a and the garnet clinopyroxenites of group 8. These are the main types of segregation developed in spinel and garnet lherzolites. In the strict sense, group 8c (eclogite) is limited to xenoliths and has not yet been recorded in alpine-type peridotites, although several workers have incorrectly referred to non-omphacitic garnet clinopyroxenite as eclogite (*eg* Bloxam & Allen 1960; Pamic 1977; Popevic 1977).

Texture

Segregations may also be classified on the basis of their texture. Whilst attempts have been made to classify fabrics in lherzolites (*eg* Spry 1969; Mercier & Nicolas 1975), little work has been carried out specifically on segregations, although attempts have been made (*eg* Nicolas & Jackson 1982). In this work they are merely classified as 'dynamic' or 'static', reflecting the predominance of metamorphic versus igneous textures respectively. Further work is needed in this field, but the feasibility of classifying tectonic fabrics in a mantle undergoing steady-state creep and cyclic solid-state recrystallization presents considerable difficulties.

Dynamic segregations generally occur concordant or subparallel to layering developed within host peridotites and may help to define mantle tectonite fabrics. Wherever segregations are abundant their interlayering with peridotite creates an ultrabasic–basic gneiss. The constituent minerals of dynamic segregations, especially the amphiboles and micas of group 6b (Fig. 3), may show preferred orientation, although relict phases (*eg* garnet porphyroclasts) may preserve metamorphic textures indicative of earlier subsolidus events. None retain primary igneous textures because solid-state dynamic recrystallization will have destroyed any that were present. This type of segregation is typically 1–20 cm thick, can be traced laterally for hundreds of metres and commonly occurs boudinaged and isoclinally folded (*eg* Beni Bouchera). The common scarcity of fold closures can be attributed to the transposition of earlier foliations (*eg* McKenzie 1979).

Static segregations tend to be discordant relative to host peridotite fabrics and their constituent minerals show little evidence of preferred orientation. Many are demonstrably high-level phenomena and are the products of partial melting occurring during the emplacement of the peridotite into the crust rather than *in situ* mantle anatexis. Consequently they tend to be more fractionated and may retain igneous mineralogies and textures. Many of the discordant gabbro intrusions in harzburgite fall into this category. Static segregations are typically less than 1 m thick. It has also been suggested that certain high-level discordant segregations (which would appear in groups 1, 2, 4 and 5) have been injected downwards from overlying crustal magma chambers and are not directly generated from the mantle (*eg* Reuber *et al.* 1982).

Chemistry

Chemical analyses of selected segregations from groups 2, 4 and 8 and their CIPW norms are shown in Table 1. All analyses have been recalculated to 100% free of volatiles and with iron converted to FeO. Samples 1–6 are from lherzolite massifs and samples 7–10 are from ophiolitic bodies. Their mean is shown in column 11, although there is considerable variation in certain oxides between the 10 samples (*eg* TiO_2). All possess high MgO (14%–23%) and variable FeO (3%–11%). Interestingly, the mean MgO content of 19% is close to that calculated by Elthon (1979) for average ophiolitic crust (which includes 'cumulate' peridotites located below the geophysical Moho) (column 12) and to the 'eclogite' residue of O'Hara *et al.* (1975) and the olivine basalt of Clarke (1970) from Baffin Bay (columns 13 and 14).

Figure 4 shows the 14 samples of Table 1 plotted with examples of a komatiite and a picrite on an MgO–Al_2O_3–CaO ternary diagram. This reveals a progressive decrease in MgO at a relatively constant CaO/Al_2O_3 ratio (0.9) from komatiite through to mid-ocean ridge basalt (MORB) with a marked decrease in CaO/Al_2O_3 ratio between sample 9 and MORB. The segregations, their mean and comparable 'primitive basalts' are shown enclosed within the area located between komatiite and picrite and more evolved MORB. Overall, the major-

TABLE 1. *Bulk chemical analyses of mafic segregations from selected alpine-type peridotites (columns 1–10) and their mean (11) compared with estimated compositions of primitive magma (12 and 13) and Baffin Bay olivine tholeiite (14).*

Wt%	1	2	3	4	5	6	7	8	9	10	11	12	13	14
SiO_2	47.30	48.79	49.73	47.56	41.94	47.11	44.33	44.07	44.95	44.51	46.15	47.85	46.17	45.70
TiO_2	0.19	0.31	0.69	0.26	2.45	0.07	0.52	0.07	0.09	0.85	0.55	0.59	0.81	1.22
Al_2O_3	13.08	11.94	10.88	12.37	13.30	14.82	13.97	14.52	19.73	13.15	13.81	12.08	13.85	10.50
FeO	6.40	4.91	5.79	4.64	10.68	6.22	4.50	11.06	2.73	10.84	6.80	8.98	9.35	11.24
MnO	0.17	0.12	0.10	0.10	0.14	0.14	0.08	0.20	0.07	0.21	0.13	0.12	0.16	0.18
MgO	21.09	22.60	20.57	19.23	18.20	16.05	23.07	15.26	14.14	20.62	19.06	17.82	16.30	19.14
CaO	10.47	9.95	10.88	13.81	11.30	14.70	12.96	13.74	16.49	8.61	12.32	11.22	11.93	9.98
Na_2O	0.97	0.78	1.01	1.50	1.57	0.87	0.46	1.03	1.77	1.14	1.11	1.31	1.28	1.41
K_2O	—	0.01	—	0.01	0.41	0.01	0.11	0.05	0.03	—	0.06	0.03	0.02	0.13
P_2O_5	—	—	—	—	—	0.01	—	—	—	0.06	0.01	—	0.03	0.14
Mg no.	85	89	87	88	79	82	91	71	90	78	84	79	77	77
CaO/Al_2O_3	0.80	0.83	1.00	1.12	0.84	0.99	0.93	0.95	0.84	0.65	0.90	0.93	0.86	0.95
CIPW norm														
Or	—	—	—	—	—	—	—	—	—	—	0.6	—	—	0.6
Ab	8.4	6.8	8.4	7.3	—	7.2	—	—	—	9.4	7.4	11.0	11.0	12.1
An	31.2	28.9	25.3	27.0	28.1	36.5	35.9	34.5	45.9	30.9	32.3	27.0	32.0	22.0
Lc	—	—	—	—	1.8	—	0.4	0.4	—	—	—	—	—	—
Ne	—	—	—	2.9	7.1	—	2.0	4.8	8.2	—	1.1	—	—	—
Di	16.6	16.1	22.7	32.8	22.3	29.1	22.4	27.3	28.3	9.6	23.1	23.0	21.9	21.4
Hy	10.2	21.4	21.7	—	—	—	—	—	—	4.2	—	8.7	0.4	1.3
Ol	33.0	25.6	20.2	29.0	36.3	27.0	38.6	32.9	19.1	44.2	34.6	29.1	32.9	39.7
Il	0.3	0.6	1.4	0.5	4.7	0.2	1.1	0.1	0.2	1.7	1.1	1.1	1.5	2.3

All iron has been converted to FeO and analyses recalculated to 100% free of volatiles, from which the CIPW norm has been calculated.

1 Olivine gabbro, Ronda lherzolite massif, Spain: mode, 33% plag, 27% ol, 23% cpx, 12% opx, 3% sp, 2% secondary phases; total includes 0.25% Cr_2O_3 and 0.08% NiO (Dickey 1970, sample R120).

2 Olivine gabbro, Ronda lherzolite massif, Spain: 35% cpx, 23% ol, 22% plag, 16% opx, 4% chr + sp, less than 0.5% brown amphibole; total includes 0.47% Cr_8O_3 and 0.11% NiO (Dickey 1970 sample R322).

3 Garnet pyroxenite, Ronda lherzolite massif, Spain: 50% cpx, 25% opx, 25% gar; total includes 0.25% Cr_2O_3 (Schubert 1977, sample 387b).

4 Spinel pyroxenite (websterite), Ronda lherzolite massif, Spain: 30% ol, 22% cpx, 20% plag, 5% sp, 3% hbl + serp; total includes 0.41% Cr_2O_3 and 0.11% NiO (Schubert 1977, sample 394).

5 Garnet–amphibole–spinel clinopyroxenite, Lherz lherzolite massif, France (Conquere 1971, sample ARG-1).

6 Garnet clinopyroxenite, Lherz lherzolite massif, France: 72% cpx, 24% gar, less than 0.5% sp, 4% secondary phases (Dickey 1970, sample L4).

7 Recrystallized 'basic' band, Lizard complex, SW England (Green 1964).

8 Garnet clinopyroxenite, Zlatibor massif, Yugoslavia (Popevic 1977, sample 1).

9 Low-pressure gabbroic segregation, Othris ophiolite, Greece (Menzies 1973, sample 3).

10 Garnet–amphibole clinopyroxenite, Bay of Islands ophiolite, Newfoundland (McCaig 1983, sample 3).

11 Mean of analyses 1–10 recalculated to 100%.

12 Calculated composition for bulk oceanic crust (Elthon 1979).

13 'Eclogite' extract from primary magma (O'Hara *et al.* 1975; Table 8, analysis 6); total includes 0.06% NiO and 0.03% Cr_2O_3.

14 Average of seven group II olivine basalts, Svartenhuk, Baffin Bay, Greenland (Clarke 1970, Table 1, column 4); total includes 0.23% Cr_2O_3 and 0.13% NiO.

element chemistry is in keeping with the segregations originating as melt residua or possibly as trapped partial melts (Fisk 1984).

The CIPW norms show that the segregations fall into two distinct compositional groups: saturated, olivine-rich, hypersthene-normative (Table 1, columns 1, 2, 3 and 10) and undersaturated, olivine-rich, nepheline-normative (Table 1, columns 4, 5, 7, 8 and 9), corresponding to tholeiitic and alkaline respectively. Columns 12, 13 and 14 are all olivine-rich and hypersthene-normative, but with low hypersthene in 13 and 14.

With regard to trace elements, it is well established that alpine-type peridotites are depleted in light rare earth elements (LREEs), a signa-

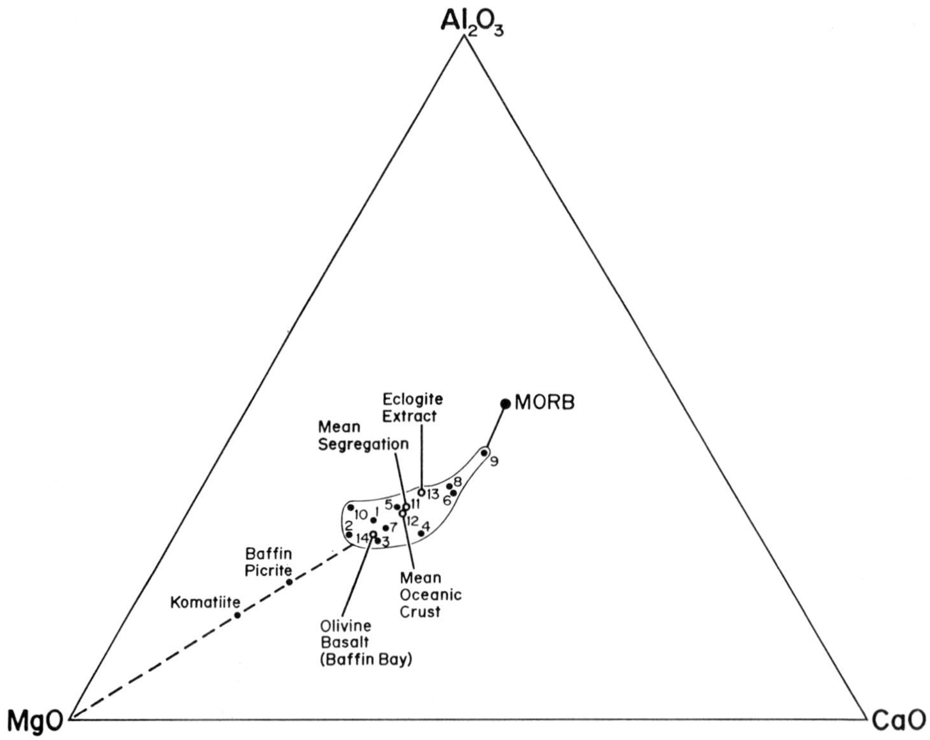

FIG. 4. $MgO-Al_2O_3-CaO$ ternary plot showing primitive segregations and primitive and evolved melts. In addition to the 14 analyses plotted from Table 1, the positions of a typical komatiite (Nesbitt & Sun 1976), picrite (Clarke 1970) and averaged MORB (Cann 1971) are shown for comparison.

ture confirmed by their potassium, rubidium and strontium concentrations and Rb/Sr ratios (*eg* Polvé & Allègre 1980; Prinzhofer & Allègre 1985). These characteristics are believed to result from 1%–5% melting of a primitive peridotitic mantle to leave a partly depleted residue. It might be expected, therefore, that segregations represent the complementary partial melt, but the available trace element and isotopic data do not support this suggestion.

Although REE concentrations are generally higher and more varied in segregations than in peridotites, pyroxenitic layers that have been analysed from the Lherz, Beni Bouchera and Lanzo also show LREE depletion like their peridotite hosts (Loubet & Allègre 1982). However, some of them do show HREE enrichment. On the basis of depleted trace element signatures, Frey (1980) also rejects a direct partial melt origin for pyroxenite and garnet pyroxenite segregations occurring as xenoliths in alkali basalt from Hawaii and, by comparison, certain mafic segregations in alpine-type peridotites. As a result, many workers now interpret segregations to be crystal cumulates derived by the interaction of partial melt with host peridotite rather than representing the actual partial melt itself. Alternatively, it has been suggested that certain segregations may form by the plating of liquidus crystals on conduit walls during melt ascent (Irving 1980).

The LREE depletion shown by both peridotites and segregations may be the manifestation of an earlier unrelated melting event (*ie* one that occurred during the early stages of mantle evolution). If this is the case it would be expected that subsequent melts generated from an already depleted source would themselves be depleted. Thus, some segregations could be orthomagmatic, as has been argued by Dickey *et al.* (1979) and Zindler *et al.* (1983).

The requirement for a potassium-, REE- and phosphorus-bearing source for the generation of, for example, MORB indicates that amphibole and/or mica are essential ingredients for the source magma (*eg* Roden *et al.* 1984). By the same reasoning, segregations which are bereft of hydrous phases are unlikely to be trapped

MORB parents. This probably excludes the majority of the segregations in groups 1–8 (Fig. 3).

Segregation processes

A number of processes can result in the formation of segregations in the upper mantle. Four mechanisms are considered here, summarized in Figs 5 and 6.

Melting

The generation of melt during adiabatic decompression of a rising mantle plume is potentially one of the most important mechanisms for generating segregations. Much of the veining preserved in lherzolite massifs has been attributed to this process as, for example, at Lherz (Lacroix 1901; Conquéré 1971) and Ronda (Dickey 1970; Schubert 1977; Obata 1980). High-pressure segregations attributed to partial melting have also been identified in the mantle sequences of certain ophiolitic bodies (Spray 1982), although these are less common than the lower-pressure pyroxenitic and gabbroic segregations which commonly pervade the mantle tectonites of ophiolites.

Melting may generate segregations in several ways: (i) as frozen trapped melt, (ii) as cumulates, by the gravity settling of crystals in mantle magma pockets, (iii) as reaction products between peridotite host and ascending partial melt and (iv) as liquidus phases plated on conduit walls during melt flow (Irving 1980). The geochemical evidence discussed above indicates that the majority of segregations originate by processes (iii) or (iv) rather than being cumulates or the actual melts themselves.

The generation of partial melts and their transport are subjects of considerable interest (*eg* Spera 1980; McKenzie 1984; Nicolas 1986). It is believed that partial melting begins in the garnet lherzolite stability field (*ie* at depths greater than 75 km) where the adiabatic ascent path crosses the lherzolite solidus. Because the generation and composition of melts produced during anatexis are both pressure and temperature dependent, it is to be expected that melt chemistries will change as a mantle plume ascends (*eg* Klein & Langmuir 1987). For example, komatiite may be generated at the deepest levels, whereas picrite, olivine tholeiite and tholeiite would be produced at progressively shallower depths (Fig. 5). Should any of these melts fail to enter the main melt stream they may remain trapped and become smeared out within the host peridotite to form streaky segregations.

The interaction of ascending melt with peridotite may result in the formation of a zoned interface as illustrated in Fig. 6*a*. In this example, a plagioclase lherzolite host has been depleted in plagioclase, clinopyroxene and then

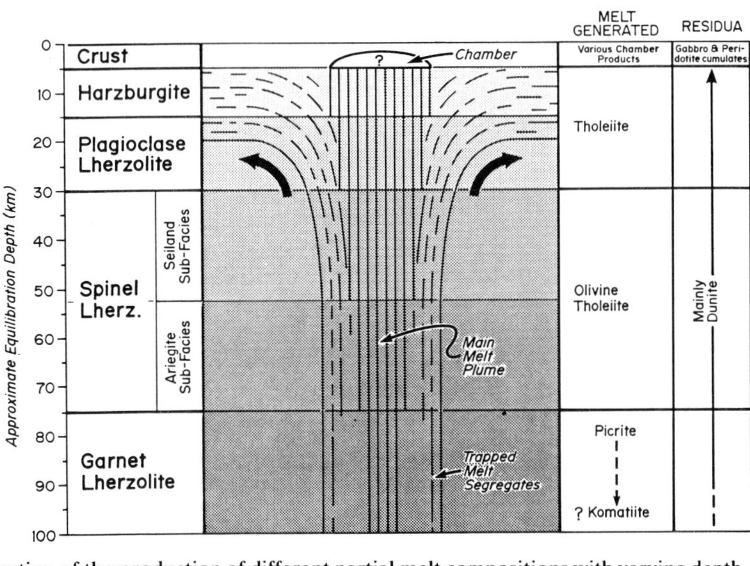

FIG. 5. Idealization of the production of different partial melt compositions with varying depth. The diagram suggests that certain melt fractions fail to enter the main melt plume and so remain trapped in the mantle as segregations. The deflection of mantle facies boundaries due to hot-body isotherm perturbation is not shown.

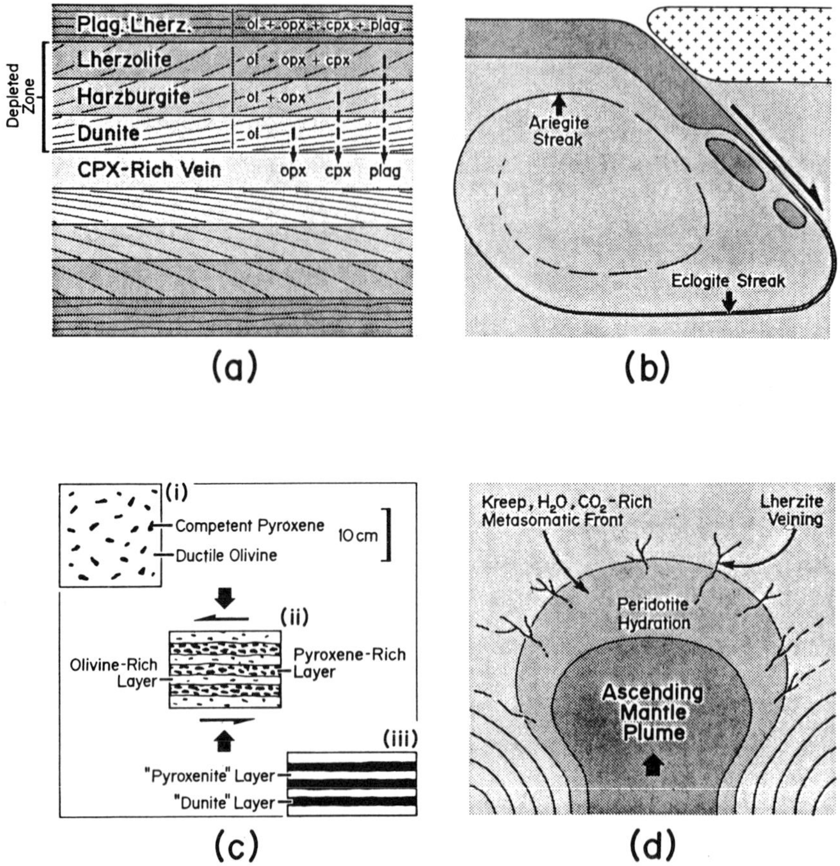

FIG. 6. (*a*) Segregation formation by partial melting of a plagioclase lherzolite. The depleted zone is typically about 1–5 cm thick on either side of the clinopyroxenite-rich vein. (*b*) Segregation formation by recycling and thinning ocean crust into eclogitic streaks which, in time, disappear by dissolution or by reprocessing at a spreading centre. (*c*) Segregation formation by solid-state differentiation due to a combination of flow differentiation and pressure solution creep. (*d*) Segregation formation by metasomatic impregnation of peridotite to form hydrous lherzolite or lherzite veins.

orthopyroxene in order of increasing melting point toward the conduit. If the melt were completely removed then a dunite vein would remain in its wake. Such 'wall–rock' reactions can be observed in many alpine-type peridotites (*eg* Quick 1981*a*, *b*).

Recycling of subducted ocean crust

When oceanic lithosphere is subducted the mafic crust undergoes a number of phase transformations accompanied by dehydration and partial melting to yield eclogite at depths greater than 50–70 km. This may then sink to the base of the upper mantle to form a perched eclogite layer owing to its greater density relative to peridotite (*eg* Anderson 1981). The layer is then supposedly reactivated by heating such that it ascends and re-enters the MORB source region. Alternatively the sinking eclogite sliver may become stretched and thinned into streaks by convective mixing (Fig. 6*b*). In time these streaks are either destroyed by diffusive processes or reprocessed at ocean ridges (Kellogg & Turcotte 1987).

An origin as recycled ocean crust has been suggested for a number of pyroxenitic segregations in lherzolite massifs (Allegre & Turcotte 1986). This provides an explanation for the formation of REE-depleted anhydrous segregations such as pyroxenites, group 6a ariegites and eclogites as an alternative or additional mechanism to their origin due to peridotite anatexis.

Metamorphic segregation

Compositional layering is a common feature of alpine-type peridotites. It is typically defined by layers rich in pyroxene and poor in olivine alternating with layers poor in pyroxene and rich in olivine. This mineralogical layering is usually less than 10 cm thick and more commonly 1–2 cm thick. The origin of the layering can be attributed to the effects of partial melting, solid-state metamorphic differentiation, or a combination of both these processes.

At subsolidus temperatures it is the mechanical properties of minerals that control their behaviour under stress, and in the case of ultrabasic rocks it has been shown that olivine deforms more readily than pyroxene. This is supported by the observation that olivine has commonly undergone recrystallization during plastic flow whilst coexisting pyroxenes show more brittle forms of deformation such as kinking and abrasion. This is corroborated by experimental work (*eg* Mercier 1985). Under shear these mineral competency differences can lead to segregation via strain concentration processes occurring in the more ductile phases. Such a segregation mechanism is referred to as flow differentiation (Fig. 6*c*).

It has also been suggested that dissolution and precipitation of pyroxene occur during the plastic flow of peridotite (Dick & Sinton 1979). This results in the mass transfer of pyroxene ingredients from high stress to low stress regions via a fluid phase such that pyroxene recrystallization and growth occur. This would be assisted by the presence of partial melt or some metasomatic fluid which could wet grain boundaries and facilitate the transport of ions in solution rather than in the solid state. This segregation mechanism is referred to as pressure solution creep (Fig. 6*c*).

Metasomatism

Metasomatism is considered to be an important process for generating 'enriched' mantle source areas necessary for the production of tholeiitic and especially alkaline magmas (*eg* Menzies & Murthy 1980). Such metasomatic fluids are rich in mainly iron, titanium, potassium, REEs, phosphorus, H_2O and CO_2 which are supposedly released during mantle degassing and diapirism. The result is the conversion of anhydrous peridotite to hydrous alkaline peridotite or its veining by metasomatic precipitates (Fig. 6*d*). Mineralogically, this manifests itself in the addition of amphibole and/or mica to the four-phase lherzolite assemblage (*eg* Boettcher & O'Neil 1980; Wilshire *et al.* 1980). The potassium, REE- and phosphorus-rich hydrous phases may occur as interstitial or poikiloblastic amphibole or mica grains in peridotite, or if they occur as distinct veins they constitute the lherzite segregations of group 6b (Fig. 3).

Summarizing remarks

The objective of this work has been to emphasize the segregated nature of the upper mantle as revealed primarily by alpine-type peridotites. Segregations occur on a number of scales and a wide variety of segregation types can be developed. A tentative classification scheme has been suggested based on segregation mineralogy and texture. A better scheme would be one which included isotopic and trace element criteria, but at present there are insufficient data to attempt this.

Four mechanisms have been suggested for generating segregations and two for destroying them. The formation of segregations renders the mantle heterogeneous on a macroscopic scale, but this is combated in time by diffusive processes and segregation recycling at spreading centres which both act to rehomogenize it.

It appears that few of the segregations developed in alpine-type peridotites possess the requisite chemistry for MORB production. However, some of the selected examples shown in Table 1 and Fig. 4 appear to be capable of yielding oceanic crust, though not necessarily MORB, if fractionated. The best contenders appear to be certain hydrous segregations of group 6. Otherwise it is clear that the majority of segregations are not sufficiently fertile to yield MORB. This is not surprising in view of the fact that alpine-type peridotites are most probably unrepresentative of typical suboceanic mantle. There is growing evidence that the ophiolitic subtype is generated above subduction zones. In this situation the presence of greater amounts of water from the dehydrating slab will tend to favour the production of more silica-saturated basalts and increased anatexis of the underlying peridotite to yield harzburgite (Pearce *et al.* 1984; Thompson 1987). The lherzolitic subtype is produced by incipient continental rifting and so is also unrepresentative compared with the steady-state spreading centres that produce MORB.

Notwithstanding these differences, alpine-type peridotites afford the largest and most accessible mantle sample available for studying the *processes* involved in melt generation, melt extraction and segregation formation, even though they may not directly relate to the production of typical MORB. Furthermore, whereas there is evidence indicating that certain segregations have been formed during the relatively high-level crustal emplacement of alpine-type peridotites and so may not be representative of true mantle processes, there is also convincing data showing that some segregations last equilibrated at considerable depth within the garnet, spinel or plagioclase lherzolite stability fields. This is particularly the case for the high-pressure group 6a (aluminous clinopyroxenite) and group 8 (garnet clinopyroxenite) segregations. It is these which are of most significance with regard to mantle segregation processes.

A further intriguing observation regarding the composition of alpine-type peridotites is that certain massifs comprise combinations of plagioclase, spinel and garnet lherzolites within very restricted thicknesses, *ie* much less than 15 km (*eg* Ronda, Spain; Beni Bouchera, Morocco). Such a range of mantle facies would normally span more than 75 km (Fig. 1). This suggests that certain alpine-type peridotites have been assembled from discrete mantle blocks or layers which became juxtaposed following their ascent.

ACKNOWLEDGMENTS: I am grateful to Adolphe Nicolas, Andy Saunders and a third anonymous referee for suggesting improvements to the manuscript. This work represents a development of research carried out in the Eastern Mediterranean, UK and Canada and was completed with support from NSERCC Grant A1455.

References

ALLÈGRE, C. J. & TURCOTTE, D. L. 1986. Implications of a two-component marble-cake mantle. *Nature* **323**, 123–127.

ANDERSON, D. L. 1981. A global geochemical model for the evolution of the Earth. *In*: O' CONNELL, R. J. & FYFE, W. S. (eds) *Evolution of the Earth*. Geodynamics Series Volume **15**, 6–18.

ANON. 1972. Penrose field conference on ophiolites. *Geotimes* **17**, 24–25.

BLOXAM, T. W. & ALLEN, J. B. 1960. Glaucophane schist, eclogite and associated rocks from Knockormal in the Girvan–Ballantrae Complex, south Ayshire. *Transactions of the Royal Society of Edinburgh* **64**, 1–27.

BOETTCHER, A. L. & O'NEIL, J. R. 1980. Stable isotope, chemical and petrographic studies of high pressure amphiboles and micas: evidence for metasomatism in the mantle source regions of alkali basalts and kimberlites. *American Journal of Science* **280A**, 594–621.

BOUDIER, F. & NICOLAS, A. 1985. Harzburgite and lherzolite subtypes in ophiolitic and oceanic environments. *Earth and Planetary Science Letters* **76**, 84–92.

CANN J. R. 1971. Major element variations in ocean floor basalts. *Philosophical Transactions of the Royal Society of London, Series A*, **268**, 495–505.

CLARKE, D. B. 1970. Tertiary basalts of Baffin Bay: possible primary magma from the mantle. *Contributions to Mineralogy and Petrology* **25**, 203–224.

CONQUÉRÉ, F. 1971. Les pyroxenolites à amphibole et amphibolites associées aux lherzolites du gisement de Lherz (Ariège, France): un exemple du rôle de l'eau au cours de la fusion partielle de lherzolites. *Contributions to Mineralogy and Petrology* **33**, 32–61.

DEN TEX, E. 1969. Origin of ultramafic rocks, their tectonic setting and history: a contribution to the discussion of the paper 'The origin of ultramafic and ultrabasic rocks' by P. J. Wyllie. *Tectonophysics* **7**, 457–488.

DICK, H. J. B. & SINTON, J. M. 1979. Compositional layering in alpine peridotites: evidence for pressure solution creep in the mantle. *Journal of Geology* **87**, 403–416.

DICKEY, J. S. 1970. Partial fusion products in alpine-type peridotites: Serrania de la Ronda and other examples. *Mineralogical Society of America Special Paper* **3**, 33–49.

—— OBATA, M. & SUEN, C. J. 1979. Chemical differentiation of the lower lithosphere as represented by the Ronda ultramafic massif, southern Spain. *In*: AHRENS, L. H. (ed.) *Origin and Distribution of the Elements*, Pergamon Press, Oxford, 587–595.

ELTHON, D. 1979. High magnesia liquids as the parental magma for ocean floor basalts. *Nature* **278**, 514–518.

FISK, M. R. 1984. Depths and temperatures of mid-ocean-ridge magma chambers and the composition of their source magmas. *In*: GASS, I. G., LIPPARD, S. J. & SHELTON, A. W. (eds) *Ophiolites and Oceanic Lithosphere*. Geological Society of London Special Publication **13**, 17–23.

FREY, F. A. 1980. The origin of pyroxenites and garnet pyroxenites from Salt Lake Crater, Oahu, Hawaii: trace element evidence. *American Journal of Science* **280A**, 427–449.

GREEN, D. H. 1964. The petrogenesis of the high-temperature peridotite at the Lizard, Cornwall. *Journal of Petrology* **5**, 134–188.

—— 1967. High-temperature peridotite intrusions. *In*: WYLLIE, P. J. (ed.) *Ultramafic and Related Rocks*. Wiley, New York, 212–222.

IRVING, A. J. 1980. Petrology and geochemistry of composite ultramafic xenoliths in alkalic basalts and implications for magmatic processes within the mantle. *American Journal of Science* **280A**, 389–426.

JACKSON, E. D. & THAYER, T. P. 1972. Some criteria for distinguishing between stratiform, concentric and alpine peridotite–gabbro complexes. *24th International Geological Congress*, Section 2, 289–296.

KELLOGG, L. H. & TURCOTTE, D. L. 1987. Homogenization of the mantle by convective mixing and diffusion. *Earth and Planetary Science Letters* **81**, 371–378.

KLEIN, E. M. & LANGMUIR, C. H. 1987. Global correlations of ocean ridge basalt chemistry with axial depth and crustal thickness. *Journal of Geophysical Research* **92**, 8089–8115.

LACROIX, A 1901. Les roches basiques accompagnant les lherzolites et les ophites des Pyrénées. *8th International Geological Congress, Paris, 1900*, 806–838.

LENSCH, G. 1976. Ariegite und websterite im lherzolitt van Balmuccia (Val Sesia/Zone von Ivrea). *Neues Jahrbuch für Mineralogie Abhandlungen* **128**, 189–208.

LOUBET, M. & ALLEGRE, C. J. 1982. Trace elements in orogenic lherzolites reveal the complex history of the upper mantle. *Nature* **298**, 809–814.

McCAIG, A. M. 1983. P–T conditions during emplacement of the Bay of Islands ophiolite complex. *Earth and Planetary Science Letters* **65**, 459–473.

McKENZIE, D. P. 1979. Finite deformation during fluid flow. *Geophysical Journal of the Royal Astronomical Society* **58**, 689–715.

—— 1984. The generation and compaction of partially molten rock. *Journal of Petrology* **25**, 713–765.

MENZIES, M. A. 1973. Mineralogy and partial melt textures within an ultramafic–mafic body, Greece. *Contributions to Mineralogy and Petrology* **42**, 273–285.

—— 1984. Chemical and isotopic heterogeneities in orogenic and ophiolitic peridotites. *In*: GASS, I. G., LIPPARD, S. J. & SHELTON, A. W. (eds) *Ophiolites and Oceanic Lithosphere*. Geological Society of London Special Publication **13**, 231–240.

—— & MURTHY, V. R. 1980. Mantle metasomatism as a precursor to the genesis of alkaline magmas — isotopic evidence. *American Journal of Science* **280A**, 622–638.

MERCIER, J-C. C. 1985. Olivine and pyroxenes. *In*: WENK, H-R. (ed.) *Preferred Orientation in Deformed Metals and Rocks: an Introduction to Modern Texture Analysis*. Academic Press, New York, 407–430.

—— & NICOLAS, A. 1975. Textures and fabrics of upper mantle peridotites as illustrated by xenoliths from basalts. *Journal of Petrology* **16**, 454–487.

MOORES, E. M. 1982. Origin and emplacement of ophiolites. *Reviews of Geophysics and Space Physics* **20**, 735–760.

NESBITT, R. W. & SUN, S. S. 1976. Geochemistry of Archaean spinifex-textured peridotites and magnesian and low-magnesian tholeiites. *Earth and Planetary Science Letters* **31**, 443–453.

NICOLAS, A. 1986. A melt extraction model based on structural studies in mantle peridotites. *Journal of Petrology* **27**, 999–1022.

—— & JACKSON, E. D. 1972. Répartition en deux provinces des peridotites des chaînes alpines longent La Méditerranée: implications géotectonique. *Bulletin Suisse Mineralogische und Petrographische* **52**, 479–495.

—— & JACKSON, M. 1982. High temperature dykes in peridotites: origin by hydraulic fracturing. *Journal of Petrology* **23**, 568–582.

OBATA, M. 1980. The Ronda peridotite: garnet-, spinel- and plagioclase lherzolite facies and the P–T trajectories of a high-temperature mantle intrusion. *Journal of Petrology*, **21**, 533–572.

O'HARA M., SAUNDERS, M. J. & MERCY, E. L. P. 1975. Garnet peridotite, primary ultrabasic magma and eclogite: interpretation of upper mantle processes in kimberlite. *Physics and Chemistry of the Earth* **9**, 545–558.

PAMIC, J. J. 1977. Variation in geothermometry and geobarometry of peridotite intrusions in the Dinaride Central Ophiolite Zone, Yugoslavia. *American Mineralogist* **62**, 874–886.

PEARCE, J. A., LIPPARD, S. J. & ROBERTS, S. 1984. Characteristics and tectonic significance of supra-subduction zone ophiolites. *In*: KOKELAAR, B. P. & HOWELLS, M. F. (eds) *Marginal Basin Geology*. Geological Society of London Special Publication **16**, 77–94.

POLVÉ, M. & ALLÈGRE, C. J. 1980. Orogenic lherzolite complexes studied by ^{87}Rb–^{87}Sr: a clue to understanding mantle convection processes? *Earth and Planetary Science Letters* **51**, 71–93.

POPEVIC, A. 1977. Zicni eklogit Bistrice. Juzni deo Zlatiborskog ultramafitskog masiva. *8th Yugoslavian Geological Congress*, **1**, 245–260.

PRINZHOFER, A. & ALLÈGRE, C. J. 1985. Residual peridotites and the mechanisms of partial melting. *Earth and Planetary Science Letters* **74**, 251–265.

QUICK, J. E. 1981*a*. Petrology and petrogenesis of the Trinity peridotite, an upper mantle diapir in the Eastern Klamath Mountains, Northern California. *Journal of Geophysical Research* **86**, 11 837–11 863.

—— 1981*b*. The origin and significance of large, tabular dunite bodies in the Trinity Peridotite, Northern California. *Contributions to Mineralogy and Petrology* **78**, 413–422.

REUBER, I., WHITECHURCH, H. & CARON, J. M. 1982. Setting of gabbroic dykelets in an ophilitic complex by hydraulic fracturing. *Nature* **296**, 141–143.

RODEN, M. K., HART, S. R., FREY, F. A. & MELSON, W. G. 1984. Sr, Nd and Pb isotopic and REE geochemistry of St Paul Rocks: the metamorphic and metasomatic development of an alkali basalt mantle source. *Contributions to Mineralogy and Petrology* **85**, 376–390.

SCHUBERT, W. 1977. Reaktionen im alpinotypen Peridotitmassiv von Ronda (Spanien) und seinen partiellen Schmelzprodukten. *Contributions to Mineralogy and Petrology* **62**, 105–220.

SPERA, F. J. 1980. Aspects of magma transport. *In*: HARGRAVES, R. B. (ed.) *Physics of Magmatic Processes*. Princeton University Press, Princeton, NJ, 265–323.

SPRAY, J. G. 1982. Mafic segregations in ophiolite mantle sequences. *Nature* **299**, 524–528.

SPRY, A. 1969. The interpretation of the textures of peridotites, eclogites and granulites. *In*: BROWN, D. A. (ed.) *Specialists' Meeting, Canberra 1968.* Special Publication of the Geological Society of Australia **2**, 307–321.

THAYER, T. P. 1960. Some critical differences between alpine-type and stratiform peridotite-gabbro complexes. *21st International Geological Congress, Copenhagen* **13**, 247–259.

THOMPSON, R. N. 1987. Phase-equilibria constraints on the genesis and magmatic evolution of oceanic basalts. *Earth-Science Reviews* **24**, 161–210.

WILSHIRE, H. G., PIKE, J. E. N., MEYER, C. E. & SCHWARZMAN, E. C. 1980. Amphibole-rich veins in lherzolite xenoliths, Dish Hill and Deadman Lake, California. *American Journal of Science* **280A**, 576–593.

ZINDLER, A., STAUDIGEL, H., HART S. R., ENDRES, R. & GOLDSTEIN, S. 1983. Nd and Sr isotopic study of a mafic layer from the Ronda ultramafic complex. *Nature* **304**, 226–230.

J. G. SPRAY, Department of Geology, University of New Brunswick, Fredericton, Canada E3B 5A3.

Partial melting of a lithologically heterogeneous mantle: inferences from crystallization histories of magnesian abyssal tholeiites from the Siqueiros Fracture Zone

James H. Natland

SUMMARY: Complex processes of magma mixing are revealed by the crystallization histories of several magnesian, porphyritic abyssal tholeiites dredged from the Siqueiros Fracture Zone at about 8°30′ N near its intersection with the East Pacific Rise. The compositions of chromian spinels, olivines, plagioclases and glass inclusions in these rocks indicate that polybaric crystallization and mixing occurred at high temperatures involving a range of refractory (low-TiO_2, low-Na_2O) melts. These later coalesced with cooler much less refractory components to produce the bulk rocks.

The crystallization histories and comparisons with experimental data indicate that mantle sources of these basalts were lithologically heterogeneous, ranging from fertile (lherzolitic) to refractory (in the extreme, harzburgitic) with respect to a basaltic melt fraction. The separate magma strains going into a typical magnesian abyssal tholeiite were produced by simultaneous partial melting of the different lithofacies, after which they coalesced and mixed. Such mixing is more likely than batch melting to explain the ranges in CaO/Al_2O_3, and abundances of Na_2O and TiO_2, within the fairly small range and low (non-picritic) magnesium numbers (0.75−0.64) exhibited by all liquids including those inferred from mineral compositions.

The early crystallization of minerals from refractory magma strains suggests that melting domains beneath spreading ridges consist of cores of refractory peridotite surrounded by cooler, more fertile peridotite. Sequential mixing of refractory and fertile magma strains is an unavoidable consequence of the ascent of magma batches from the interiors of complex melting regimes (diapirs) in the mantle. The diapirs may have been triggered into ascent by the buoyancy of low-density (low-iron) refractory peridotite.

Several of the Siqueiros magnesian abyssal tholeiites are among the most depleted basalts ever obtained from the eastern Pacific, and yet alkalic basalt was also dredged from the fracture zone a few kilometres away. The tholeiites are even more depleted than typical, more fractionated, summit eruptives of the East Pacific Rise. The contrast may arise because of the homogenizing effects of axial magma chambers, within which truly depleted tholeiites are mixed with enriched alkalic magmas instead of erupting separately, as they do in the fracture zone. Mantle in this region thus appears to be segregated into strongly contrasting depleted and enriched peridotites, with depleted peridotite also ranging from refractory to fertile lithofacies. Primary lithological variability in the mantle thus was probably imposed by one or more ancient melting events that left distinctively enriched and depleted reservoirs in their wake.

A general question in abyssal tholeiite petrogenesis is whether sources of potential primary magmas are fundamentally heterogeneous or homogeneous on the scale of domains of melting in the mantle. There is a consensus amongst isotope and trace-element geochemists that heterogeneity exists on a regional scale, but there is increasing evidence that the scale of heterogeneity can be local as well (*eg* Zindler *et al.* 1984). Some have argued that the observed variability of trace-element ratios in abyssal basalts need not necessarily reflect mantle heterogeneity but only post-melting differentiation processes (O'Hara 1977; O'Hara & Mathews 1981). Others, conceding regional variability, consider that local variability in trace-element ratios and bulk compositions of basalts can result from a two-stage melting process (Duncan & Green 1980, 1987) or, in principle, from a multistage process (Langmuir *et al.* 1977) acting on a locally homogeneous mantle.

A group of porphyritic abyssal tholeiites dredged from the Siqueiros Fracture Zone (Batiza *et al.* 1977) includes several of the least fractionated and most depleted basalts ever obtained from the eastern Pacific. These rocks are plausibly representative of parents to the more fractionated olivine tholeiites and ferrobasalts that have been sampled so abundantly on the nearby axial summit of the East Pacific Rise

From SAUNDERS, A. D. & NORRY, M. J. (eds), 1989, *Magmatism in the Ocean Basins*, Geological Society Special Publication No. 42, pp. 41−70.

(Natland & Melson 1980). The compositions of the Siqueiros basalts, of their phenocrysts, and of glass inclusions in those phenocrysts, all reported here, support complex histories of polybaric mixing and coalescence of diverse magma strains prior to injection into the crust. Such histories are probably typical of magmas supplied to axial magma chambers in this region, but because of mixing and fractionation within those chambers they are almost never evident in the evolved, usually nearly aphyric, basalts sampled from the ridge crest.

The diversity of parental abyssal tholeiites is most often ascribed to differences in depth and degree of partial melting. The two are probably related on a regional scale to contrasts in geothermal gradients from place to place (Klein & Langmuir 1987). However, the crystallization histories of the Siqueiros samples indicate that an additional very important factor was involved. This is source heterogeneity, not simply of the type involving trace elements and isotopes, but of the fundamental lithological constitution of the mantle, *ie* the mantle can be variably fertile or refractory with respect to a basaltic melt fraction, ranging from fairly sodic and aluminous lherzolite on the one hand perhaps to harzburgite on the other. This diversity exists on a scale small with respect to domains of partial melting beneath spreading ridges, so that quite different melt strains can be produced from the different mantle lithofacies under almost identical thermal conditions and at about the same depths. These melt strains coalesce to produce more or less typical parental abyssal tholeiites of the type described here. Such lithological heterogeneity is no more than that usually encountered in alpine peridotites (*eg* Dick 1977), or among serpentinized peridotites in many individual dredge hauls from fracture zones (*cf* Dick *et al.* 1984) or in suites of ultramafic xenoliths from ocean islands (*eg* Jackson 1968). The distinction here, however, is that the variability of the sources of the Siqueiros tholeiites must have existed *before* extraction of basalts. This obviously cannot be proved by direct examination of mantle samples, but might be approached from consideration of how parental basalts were assembled.

The objective of this paper is thus to outline the crystallization histories of several magnesian and porphyritic basalts from the Siqueiros Fracture Zone, showing how they may be used to understand the petrogenesis of parental abyssal tholeiites. This will lead to a consideration of the general process of extraction of magmas from the mantle and then to development of a geodynamic hypothesis relevant to ridge-axis segmentation and the contrasts between axial and fracture zone basalts along this portion of the East Pacific Rise.

Analytical procedures and some definitions

Unless otherwise noted, mineral compositions were obtained using the electron microprobe at Scripps Institution of Oceanography, with data normalized to US National Museum standards according to procedures outlined in Natland *et al.* (1983). Glass compositions were obtained either at the Smithsonian Institution (Natland & Melson 1980) or at Scripps using virtually identical procedures. High-resolution abundances of K_2O, MnO and P_2O_5 were obtained by electron microprobe at Scripps using enhanced beam current and a total of 120 s of counting. Values for these oxides were standardized to USNM glass standard VG-2 after calibration to least-squares regressions of determinations on glasses separately analysed by dc plasma techniques. The basalts discussed here were also analysed for major-oxide and trace-element abundances using dc plasma spectrometry at Lamont–Doherty Geological Observatory and are reported here with estimates of precision taken from Hekinian & Walker (1987). Isotopic data and abundances of rubidium, strontium, neodymium and samarium on several samples were obtained by J. D. Macdougall at Scripps Institution.

With respect to rock compositions, the term 'depleted' is used to refer to isotopic and geochemical attributes of the samples described (also 'enriched'), whereas the term 'refractory' refers primarily to melt fractions with unusually low abundances of Na_2O and TiO_2, as determined from mineral or glass analyses, whether or not the melts were depleted or enriched. The symbol Mg# is used for the ratio $Mg/(Mg + Fe^{2+})$, whether for glasses or spinels. Mg# depends on the ratio $Fe^{2+}/(Fe^{2+} + Fe^{3+})$, which is assumed to be 0.86 for glasses (following Presnall *et al.* 1979) and is computed by stoichiometry for the spinels.

Magnesian tholeiites and alkalic basalts from the Siqueiros Fracture Zone and the East Pacific Rise

The Siqueiros Fracture Zone is one of the few places where a suite of highly magnesian rocks has been recovered which almost fully spans the

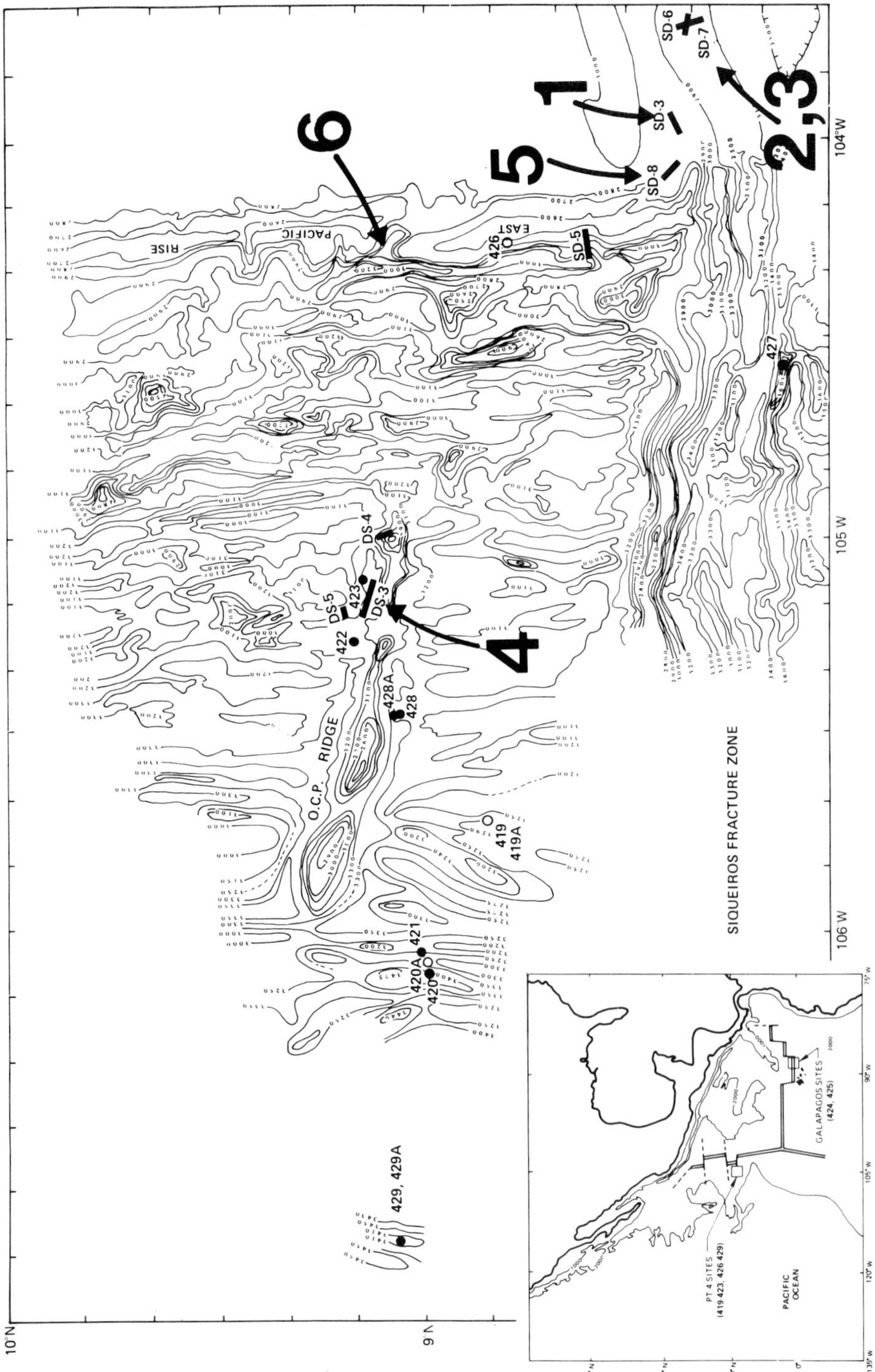

FIG. 1. Bathymetry of a portion of the Siqueiros Fracture Zone and the axis of the East Pacific Rise to the north from Rosendahl & Dorman (1980). Dredge stations and drill sites in the vicinity are indicated. The locations of the six samples discussed in the text are given by numbers and arrows. The contour interval is 100 m.

range of potential parental compositions likely to have supplied adjoining segments of the East Pacific Rise, where rocks are systematically more fractionated (Natland 1980*a*; Natland & Melson 1980). The rocks from the fracture zone include several mineralogically distinct olivine– or olivine–plagioclase–phyric tholeiites as well as alkalic basalt. These were sampled from within a few kilometres of each other on the floor and steep side slopes of the fracture zone (Fig. 1). Both major-oxide and trace-element abundances preclude any cogenetic relationships (Batiza *et al.* 1977; Natland & Melson 1980). Two additional samples are a 'transitional' basalt (Carlson *et al.* 1978; Johnson 1979; Batiza & Johnson 1980) sampled from a seamount ridge to the N and another more enriched basalt sampled from a non-transform offset of the Rise axis near 9° N (Natland & Macdougall 1986; Natland *et al.* 1986).

I adopt an earlier nomenclature (Natland 1980*b*) to refer to the samples. In a particular sequence, these are (i) plagioclase–olivine–phyric basalt; (ii) oceanite; (iii) olivine tholeiite; (iv) the 'transitional' basalt; and (v) and (vi) two more enriched basalts, termed 'alkalic' basalts, following Batiza *et al.* (1977). The latter are nepheline normative using whole-rock analyses and all iron calculated as FeO, whereas the glass compositions tend to be slightly hypersthene normative (Table 1). The alkalic glasses have elevated abundances of barium, zirconium, strontium, K_2O, TiO_2 and P_2O_5 (Table 2), at the extreme of compositions found in this region, and also the highest $^{87}Sr/^{86}Sr$ ratio (Carlson *et al.* 1978; Natland & Macdougall 1986).

The sequence cited is dictated by the locations, right to left, of the plotted points computed from Table 1 and numbered 1–6 on a ternary diagram projected from plagioclase onto the plane Ol–Di–SiO_2 (Fig. 2), based on the algorithm of Walker *et al.* (1979). Such a spread may be produced by variability in the depth of melting (Stolper 1980), degree of melting (Fujii & Scarfe 1985) or both of magmas parental to

TABLE 1. *Glass group compositions, East Pacific Rise, Siqueiros Fracture Zone*

Sample type	1	2	3	4	5	6
Dredge[a]	SD-3 (3)	SD-7C (1)	SD-7A (6)	DS4-6 (2)	SD-8 (3)	R13-1
Oxides (wt%)						
SiO_2	50.43	49.15	48.71	49.08	49.60	47.75
Al_2O_3	16.00	17.58	17.06	16.80	16.73	16.74
FeO*	9.23	8.01	10.07	9.05	9.88	9.42
MnO	0.106	0.149	0.160		0.167	0.167
MgO	8.50	9.58	8.07	07.82	5.78	6.36
CaO	12.38	12.14	11.33	10.65	9.54	9.79
Na_2O	2.16	2.38	2.76	3.21	3.63	3.33
K_2O	0.056	0.016	0.034	0.35	1.12	0.870
TiO_2	1.20	0.96	1.42	1.90	2.69	2.34
P_2O_5	0.145	0.072	0.105	0.25	0.530	0.429
	100.207	100.037	99.79	98.86	99.667	97.196
Mg#	0.656	0.712	0.624	0.642	0.548	0.584
CIPW norms[b]						
q	0.00	0.00	0.00	0.00	0.00	0.00
or	0.53	0.24	0.47	2.07	6.49	5.14
ab	18.26	20.12	23.34	27.14	30.69	28.15
an	33.68	37.15	33.91	30.38	26.09	28.15
ne	0.00	0.00	0.00	0.00	0.00	0.00
wo (di)	11.26	9.36	8.99	8.67	7.65	7.28
en (di)	6.82	6.12	5.25	5.33	4.28	4.27
fs (di)	3.80	2.58	3.30	2.83	3.05	2.65
en (hy)	12.23	6.06	4.13	2.45	0.92	0.56
fs (hy)	6.80	2.56	2.60	1.30	0.65	0.35
fo (ol)	1.42	8.11	7.45	8.13	6.40	7.66
fa (ol)	0.87	3.78	5.18	4.76	5.04	5.26
mt	1.90	1.81	2.07	1.87	2.04	2.13
il	2.28	1.83	2.70	3.62	5.12	4.45
ap	0.25	0.23	0.25	0.58	1.02	1.04

[a] The number of samples in each group is given in parentheses.
[b] Calculated assuming that $Fe^{2+}/Fe^{2-} + Fe^{3+} = 0.86$.

TABLE 2. *Plasma spectrometry and isotope dilution trace-element and isotopic data, East Pacific Rise, Siqueiros Fracture Zone*

Sample type	1	2	3	4	5	6	Std dev.
Dredge	SD-3	SD7-C	SD7-A	DS4-102	SD-8-1A	R13-1	
Oxides (wt%)							
SiO_2	49.3	47.14	47.6	49.13	48.63	48.40	±0.22
Al_2O_3	15.9	14.04	15.85	16.58	16.56	16.88	±0.14
Fe_2O_3	9.95	9.14	11.05	9.76	10.77	10.91	±0.05
MnO	0.162	0.145	0.174	0.161	0.167	0.169	±0.002
MgO	9.10	16.71	9.85	8.37	6.42	7.23	±0.06
CaO	12.1	9.91	10.8	10.70	9.43	9.68	±0.13
Na_2O	2.23	2.01	2.71	3.03	3.55	3.36	±0.03
K_2O	0.067	0.009	0.144	0.511	1.12	0.88	±0.007
TiO_2	1.16	0.80	1.30	1.67	2.54	2.28	±0.02
P_2O_5	0.103	0.077	0.108	0.271	0.530	0.435	±0.002
	100.072	99.981	99.586	100.183	99.717	100.224	
Trace elements (ppm)							
Ba^2	4.8	0.9	5.8	99.7	254	200	±1.0
Rb^1	0.33	0.09	0.39	(7.2)		15.24	
Sr^1	87	75.6	164	240	421	400	
Sr^2	95	64	160			391	±1.0
Sm^1	2.44	(2.45)	3.07	4.68	6.47	6.15	±2.3
Nd^1	6.49	(5.72)	9.16	17.20	28.10	25.51	±0.5
Zr^2	72	52	100	165	215	206	±4.0
Y^2	28	22	29	32	34	34	±0.4
V^2	268	194	234	249	250	251	±4.0
Sc^2	36.0	27.8	38.9	32.2	26.8	27.6	±1.0
Cr^2	386	982	445	269	142	187	
Ni^2	129	—	247	155	143	120	
$^{87}Sr/^{86}Sr_1$	0.702653 ±26	0.702614 ±24	0.702577 ±26	0.702810 ±32	0.702896 ±18	0.702902 ±26	
$^{143}Nd/^{144}Nd_1$	0.513113 ±10	0.513083 ±2	0.513005 ±14	0.513106 ±14	0.512993 ±34	0.512988 ±10	

Isotopic data and isotope dilution trace elements on hand-picked glasses were obtained in laboratory 1 (indicated as a superscript to the element and subscript to the isotope ratio) by J. D. Macdougall (sample types 1–3 and 6) or R. Carlson (sample types 4 and 5; Carlson *et al.* (1978)) at Scripps Institution of Oceanography. Major-oxide and remaining trace-element data were obtained by A. Biesada and J. Natland at laboratory 2, Lamont–Doherty Geological Observatory, by plasma-emission spectrometry, using hand-picked glasses for sample types 1, 4, 5 and 6, phenocryst-free whole-rock powder for sample type 3 and the phenocryst-bearing whole rock for sample type 2. All iron is reported as Fe_2O_3. Samarium and neodymium for SD7-C are isotope dilution data on phenocrystbearing whole-rock powder. Standard deviations give the repeatability of eight replicate analyses of basalt 10–17 of Hekinian & Walker (1987).

each of these lava types. The sequence 1–6 is also indicated for sample locations in Fig. 1.

The petrography of most of these basalts has been described elsewhere (Batiza *et al.* 1977; Natland 1980*b*). Here it is necessary chiefly to point out that all these rocks carry chromian spinel and olivine phenocrysts or microphenocrysts, none carry clinopyroxene phenocrysts and only one, sample type 1 of Fig. 2, has plagioclase phenocrysts. Olivine and spinel phenocrysts in the three tholeiites (sample types 1–3) are much larger (olivines to 0.5 cm) than in the transitional or alkalic basalts (sample types 4–6).

All the natural glass compositions are fairly magnesian (Table 1) with the oceanite (sample type 2) having Mg# in excess of 0.7, the highest so far found in the eastern Pacific. This sample was also described by Schrader *et al.* (1980) but their glass composition is incorrect (as summarized by Wright (1984)). The analysis of sample type 2 in Table 1 was obtained at the Smithsonian Institution and was confirmed by a replicate analysis using nearly identical procedures at Scripps Institution.

Spinels and olivines

Chromian spinels are useful petrogenetic indicators both of tectonic provenance and of conditions of crystallization (Dick & Bullen 1984).

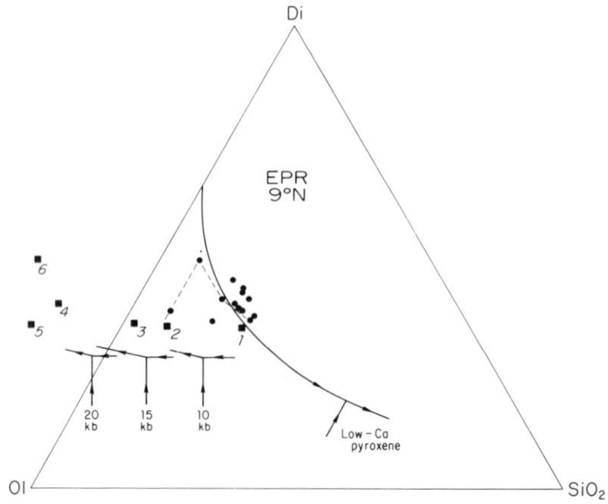

FIG. 2. Projection from plagioclase into the ternary *Ol–Di–SiO₂* calculated using the procedure of Walker *et al.* (1979) for the six samples discussed in the text (numbered squares), four of them from the Siqueiros Fracture Zone and additional olivine tholeiites and ferrobasalts from the adjacent segment of the East Pacific Rise to the N (Natland & Melson 1980). The high-pressure pseudo-invariant points of Stolper (1980) are also shown. The broken line indicates the likely fractionation path for most of the basalts, extending away from the olivine apex through the projected point for sample type 2 to the 1 atm cotectic and then down the cotectic to ferrobasalt compositions.

Here they provide evidence for polybaric crystallization and the attributes of parental precursors to their hosts.

Representative analyses of spinels are listed in Table 3 and of olivines in Table 4. In Fig. 3 the range of spinel compositions is plotted in terms of Cr# (=Cr/(Cr + Al)) and Mg#. The ratios in Fig. 3 are related to the compositions of coexisting olivines and glasses in the upper part of the figure by assuming that the most chromian spinel in each data array (with least or nearly least Mg#) crystallized from the host glass (as deduced from zoning relations); the glass was simultaneously crystallizing groundmass olivine, corresponding to the lower bound of forsterite content shown at the top.

The pertinent features of the diagram are (i) the range in Cr#, from aluminous chromian spinels to magnesiochromites, within several of the fields and (ii) the correspondence in Mg# in each of the fields with the Fo contents of olivines and the Mg# of host glasses. The Mg# of glasses and spinels and the Fo content of olivines all increase in the samples from right to left on the diagram, resulting in subparallel arrays in Cr#. If no other proof were available, this would certify that each of these basalts had a separate crystallization history and that none could be parental to any other.

Some of the individual spinels, and portions of other zoned spinels, are quite aluminous. The origin of spinels with Cr# < 0.3 in abyssal tholeiites has been attributed to unspecified but nevertheless elevated pressures (Sigurdsson & Schilling 1976; Fisk & Bence 1980). The latter authors' experimental data show that aluminous spinels do not form from magnesian abyssal tholeiite melts at low pressure, but such spinels have been synthesized at higher pressure in experiments on lunar basalts (Green *et al.* 1972). A contrary view has been expressed on the basis of reverse zoning (more aluminous rims) in certain spinels (Kempton *et al.* 1985), and kinetic factors may also produce aluminous rims on spinels at low pressure (J. Allan, pers. comm., 1986). Here, in both the oceanite and olivine tholeiite (sample types 2 and 3), the spinels have chromian cores surrounded by aluminous mantles and then by chromian skeletal overgrowths (Fig. 4).

Dick & Bullen (1984) interpreted the range in Cr# in spinels in both abyssal tholeiites and peridotites as resulting from a variation in the extent of partial melting. This probably occurs at a pressure of several kilobars for most abyssal peridotites and has been confirmed experimentally at 10 kbar by Fujii & Scarfe (1985). Assuming that pressures of several kilobars are required for abyssal tholeiites to crystallize aluminous spinels at all, then the chromian overgrowths on the spinels in Fig. 4 can be interpreted as a consequence of a reduction in pressure. The cores and mantles plausibly reflect mixing at sufficiently elevated pressure between magmas having different chromium contents, either because of differences in the extent of

TABLE 3. *Representative microprobe analyses of spinels in East Pacific Rise basalts*

	Sample type 1 Plagioclase–olivine–phyric basalt			Sample type 2 Oceanite				Sample type 3 Olivine-rich tholeiite		
	1	2	3	4	5	6	7	8	9	10
Oxides (wt%)										
SiO_2	0.08	0.07	0.07	0.11	0.10	0.10	0.07	0.10	0.05	0.02
TiO_2	0.25	0.18	0.35	0.30	0.29	0.30	0.30	0.36	0.40	0.42
Al_2O_3	42.88	39.10	31.10	43.28	40.88	37.87	33.04	48.29	45.65	41.02
Cr_2O_3	24.09	28.33	34.29	24.51	26.71	29.88	34.51	18.49	20.24	25.00
Fe_2O_3	3.59	4.02	4.79	3.44	3.08	3.49	3.46	3.47	4.36	4.35
FeO	10.16	9.97	11.83	9.36	9.50	9.69	9.81	11.61	11.43	12.51
MnO	0.12	0.19	0.22	0.13	0.13	0.13	0.15	0.11	0.17	0.20
MgO	18.70	18.41	16.10	19.40	18.89	18.52	17.76	18.60	18.32	17.12
CaO	0.08	0.06	0.15	0.10	0.06	0.08	0.01	0.04	0.04	0.03
NiO	0.16	0.20	0.14	0.25	0.21	0.24	0.24	0.31	0.24	0.24
	100.11	100.53	99.04	100.88	99.85	100.30	99.35	101.38	100.90	100.92
Structural formulae (32 oxygen)										
Si	0.02	0.02	0.02	0.02	0.02	0.02	0.02	0.02	0.01	0.00
Al	11.10	10.25	8.62	11.00	10.67	9.98	8.98	12.16	11.67	10.74
Cr	4.18 } 15.93	4.98 } 15.95	6.38 } 15.93	4.21 } 15.94	4.68 } 15.93	5.28 } 15.92	6.29 } 15.94	3.12 } 15.92	3.47 } 15.93	4.39 } 15.93
Fe^{3+}	0.59	0.67	0.85	0.56	0.51	0.59	0.60	0.56	0.71	0.73
Ti	0.04	0.03	0.06	0.05	0.05	0.05	0.05	0.06	0.07	0.07
Mg	6.12	6.11	5.64	6.28	06.23	6.17	6.10	5.92	5.92	5.66
Fe^{2+}	1.87	1.86	2.33	1.70	1.76	1.81	1.89	2.07	2.07	2.32
Mn	0.02 } 8.06	0.04 } 8.06	0.04 } 8.08	0.02 } 8.06	0.02 } 8.06	0.03 } 8.07	0.03 } 8.06	0.02 } 8.07	0.03 } 8.07	0.04 } 8.07
Ni	0.03	0.04	0.03	0.04	0.04	0.04	0.04	0.05	0.04	0.04
Ca	0.02	0.01	0.04	0.02	0.01	0.02	0.00	0.01	0.01	0.01
Type[a]	C	C	C	R	C	R	R	C	C	C
Cr#[b]	0.274	0.327	0.425	0.277	0.305	0.346	0.412	0.204	0.229	0.290
Mg#[b]	0.766	0.767	0.708	0.798	0.780	0.773	0.763	0.741	0.741	0.709
Olivine	—	—	—	—	—	—	—	—	—	—

[a] C, core; R, rim.
[b] Cr# = Cr/ (Cr + Al); Mg# = Mg/(Mg + Fe^{2+}) where Fe^{3+}/Fe^{2+} is calculated from stoichiometry.

TABLE 3. *Continued*

	Sample type 3 Olivine-rich tholeiite		Sample type 4 Transitional basalt			Sample 5 Alkalic basalt	Sample type 6 (R13−1) Alkalic basalt		
	11	12	13	14	15	16	17	18	19
Oxides (wt%)									
SiO_2	0.09	0.01	0.04	0.08	0.03	0.16	0.04	0.01	0.02
TiO_2	0.65	0.70	0.96	0.87	0.80	2.90	1.47	2.07	1.95
Al_2O_3	37.88	35.92	31.92	33.58	32.85	20.54	31.63	22.09	18.80
Cr_2O_3	26.30	27.51	30.40	28.35	28.96	28.42	25.18	32.62	37.05
Fe_2O_3	5.82	6.42	7.03	7.12	6.75	14.79	12.80	13.85	12.90
FeO	13.12	13.45	13.04	12.93	12.57	19.15	15.77	16.97	17.42
MnO	0.23	0.20	0.26	0.21	0.17	0.25	0.22	0.31	0.29
MgO	16.42	15.93	15.86	16.05	15.95	11.48	14.88	13.17	12.49
CaO	0.18	0.15	0.10	0.02	0.03	0.12	0.03	0.03	0.02
NiO	0.28	0.26	0.16	0.26	0.18	0.15	0.14	0.17	0.12
	100.97	100.56	99.77	99.47	98.29	97.96	102.16	101.29	101.06
Stuctural formulae (32 oxygens)									
Si	0.02	0.00	0.01	0.02	0.01	0.04	0.01	0.00	0.00
Al	10.07	9.68	8.79	9.20	9.12	6.20	8.63	6.37	5.53
Cr	4.69 } 15.88	4.97 } 15.87	5.62 } 15.83	5.21 } 15.83	5.39 } 15.86	5.75 } 15.40	4.61 } 15.74	6.31 } 15.61	7.31 } 15.63
Fe^3+	0.99	1.10	1.24	1.25	1.20	2.85	2.23	2.55	2.42
Ti	0.11	0.12	0.17	0.15	0.14	0.56	0.26	0.38	0.37
Mg	5.52	5.43	5.52	5.56	5.60	4.38	5.13	4.80	4.64
Fe^2+	2.47	2.57	2.55	2.51	2.47	4.10	3.05	3.47	3.63
Mn	0.04 } 8.12	0.04 } 8.13	0.05 } 8.17	0.04 } 8.16	0.03 } 8.14	0.05 } 8.59	0.04 } 8.26	0.06 } 8.39	0.06 } 8.37
Ni	0.05	0.05	0.03	0.05	0.03	0.03	0.03	0.03	0.02
Ca	0.04	0.04	0.03	0.00	0.01	0.03	0.01	0.01	0.01
Type[a]	S	S	S	S	S	In glass	In Ol	In Ol	In Ol
Cr#[b]	0.318	0.339	0.390	0.362	0.372	0.481	0.348	0.498	0.569
Mg#[c]	0.691	0.679	0.684	0.689	0.693	0.517	0.627	0.580	0.561
Olivine	—	—	—	—	$Fo_{87.4}$	—	$Fo_{83.5}$	$Fo_{83.5}$	$Fo_{83.5}$

[a] S, small crystal in glass; In Ol, in olivine.
[b] Cr# = Cr/(Cr+Al); Mg# = Mg/(Mg + Fe^{2+}) where Fe^{3+}/Fe^{+} is calculated from stoichiometry.

TABLE 4. *Representative electron microprobe analyses of olivines*

	Sample type 1 Plagioclase–olivine–phyric basalt			Sample type 2 Oceanite		
	1	2	3	4	5	6
Oxides (wt%)						
SiO_2	39.54	39.54	39.76	40.33	40.21	39.82
FeO	11.21	11.95	11.95	9.27	9.39	10.11
MnO	0.16	0.22	0.20	0.21	0.18	0.17
MgO	47.30	46.83	46.62	49.40	48.19	48.55
CaO	0.33	0.38	0.32	0.31	0.32	0.34
NiO	0.19	0.19	0.21	0.28	0.34	0.24
Cr_2O_3	0.03	0.08	0.07	0.03	0.04	0.05
	98.76	99.19	99.42	99.83	99.47	99.28
Olivine	$Fo_{88.3}$	$Fo_{87.5}$	$Fo_{87.4}$	$Fo_{90.5}$	$Fo_{90.3}$	$Fo_{89.5}$
type[a]	PC	PC (PL)	GM	PC	SPC	GM

	Sample type 3 Olivine-rich tholeiite			Sample type 4 Transitional/enriched basalts	Sample type 5 Transitional/enriched basalts	Sample type 6 Transitional/enriched basalts
	7	8	9	10	11	12
Oxides (wt%)						
SiO_2	39.56	39.01	38.74	40.03	39.03	39.21
FeO	12.78	13.15	13.50	12.16	16.40	15.41
MnO	0.22	0.23	0.19	0.19	0.27	0.26
MgO	46.18	46.61	45.68	47.59	43.92	44.39
CaO	0.29	0.29	0.33	0.26	0.26	0.26
NiO	0.32	0.34	0.19	0.18	0.18	0.19
Cr_2O_3	0.03	0.05	0.10	0.15	0.02	0.16
	99.28	99.76	98.90	100.56	100.08	99.89
Olivine	$Fo_{86.5}$	$Fo_{86.3}$	$Fo_{85.8}$	$Fo_{87.5}$	$Fo_{82.7}$	$Fo_{83.7}$
type[a]	PC	PR	GM	M	M	M

[a] P, phenocryst; C, core; R, rim; S, skeletal crystal; (PL), next to plagioclase in glomerocryst; GM, groundmass crystal; M, microphenocryst.

melting (Dick & Bullen, 1984) or, as argued here, because of different source compositions.

The arrays in Fig. 3 confirm for these abyssal tholeiites Irvine's (1965) treatment which showed that spinels with a wide variation in Cr# but a small variation in Mg# would be in equilibrium with olivines of specific composition at a given temperature (nominally 1200 °C). The arrays nearly parallel isopleths in Irvine's (1965) Mg–Fe^{2+} equipotential plot for spinels and olivines, as indicated for portions of those isopleths drawn for particular olivine compositions in Fig. 3. This implies that the spinels in any sample, whatever their Cr#, crystallized from liquids with a very narrow range in Mg#. If mixing produced the core-to-mantle zoning shown in Fig. 4, then the mixing end-members also had similar Mg# values. This interpretation is supported by the small variation in olivine Fo contents in each rock. The more aluminous mantles to the spinels probably crystallized from slightly cooler melts than the chromian cores. This is indicated by both the zoning sequences (Fig. 4) and the trends of spinel compositional arrays for sample types 1 and 2 shown in Fig. 3; these are slightly steeper than the isopleths indicated (the more chromium-rich portions of the crystals were in equilibrium with slightly more magnesian, *ie* higher-temperature, olivines).

On the basis of Fig. 3, the transitional and alkalic basalts clearly crystallized from less magnesian liquids than the tholeiites did, and spinels in only one of the samples exhibit a range in Cr#. None of the spinels in this sample is sufficiently aluminous to infer anything about crystallization at elevated pressure, although the existence of a range of compositions suggests a mixing history.

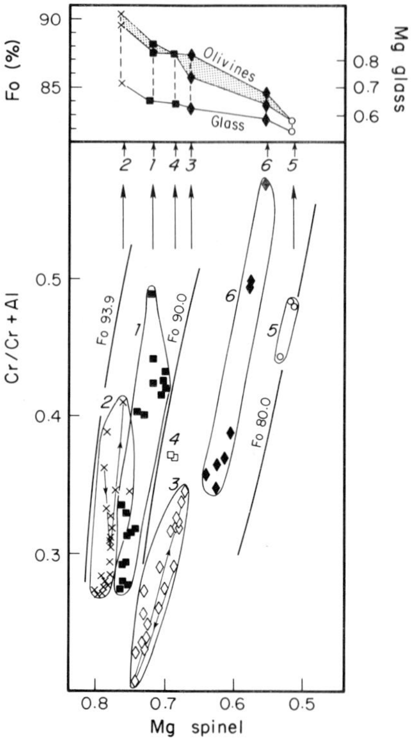

FIG. 3. Cr# versus Mg# for spinel compositions obtained in sample types 1–6. Lines with arrows within arrays give zoning sequences (cores to mantles to rims) of some individual spinels. The most chromium-rich spinels in each array are linked by vertical arrows to the upper panel, which gives ranges of olivine composition (left margin) and glass Mg# (right margin) in the same samples, as discussed in the text. The arrays are identified by the sample numbers used in Fig 1 and 2. Portions of three olivine–spinel isopleths computed for 1200 °C (Irvine 1965), with olivine compositions, are also plotted.

Plagioclases, melt inclusions and spinels in sample type 1

The plagioclase–olivine–phyric basalt (sample type 1) provides important information about the compositions of the liquids from which the spinels precipitated in all three Siqueiros tholeiites. Of the three tholeiites, this is the only one with plagioclase megacrysts (up to 1 cm across), and these comprise about 25% of the rock. The spinels in this sample occur exclusively in the plagioclases which, apart from their rims (An_{71-77}), are very calcic (An_{85-92}; Table 5). One glomerocryst with such plagioclases also contains olivine ($Fo_{88.5}$; Table 4). The spinels have variable Cr# (Fig. 3) and none is zoned. Spinels in plagioclase cores (which have rounded shapes, suggesting resorption, and which lack glass inclusions) are more aluminous than those in surrounding plagioclase rims (with glass inclusions). The sodic rims to the plagioclase megacrysts attest to lack of equilibrium between the host lavas and the feldspar mantles and cores at low pressure.

The glass inclusions, which were analysed by electron microprobe using a defocused (15–30 μm) beam, are all basaltic (Table 6), and all but one have high values of Mg# (0.69–0.74). Post-entrapment growth of the surrounding plagioclase-modified melt compositions is indicated (extracting calcium and aluminium but causing enrichments of MgO, FeO, TiO_2 and Na_2O in the inclusions). The values of Mg#, however, probably did not change significantly since both magnesium and iron should have partitioned almost exclusively into the inclusions. Plagioclases immediately adjacent to the inclusions are above An_{89} (Tables 5 and 6).

The most remarkable feature of the glass inclusions is that they contain extremely low contents of TiO_2 despite post-entrapment enhancement of this oxide by plagioclase crystal growth into the inclusions. If one assumes that TiO_2 and MgO enhancements were proportional to the amount of plagioclase subtracted from the inclusions, then an estimate of original little-modified TiO_2 contents can be made by assuming an arbitrary, but reasonable, initial MgO content for the inclusions and adjusting TiO_2 for the amount of plagioclase required to reduce MgO to that abundance. The MgO value chosen here, 10%, is close to that for several glass inclusions showing minimal effect of post-entrapment growth and is appropriate for liquids with Mg# about 0.74. The histogram provided at the top of Fig. 5 is based on 29 compositions of glass inclusions from several different plagioclases and is adjusted in this fashion. Clearly, the majority of the inclusions are extremely refractory. Figure 5 also shows that no reasonable proportion of olivine and plagioclase fractionation from any of them will produce the host glass. The most refractory estimated initial TiO_2 abundance is only 0.18%. This is extraordinarily low for a tholeiitic basaltic liquid, even granting the high Mg#. Low TiO_2 concentrations have been reported for glass inclusions in minerals from other abyssal tholeiites (Donaldson & Brown 1977; Dmitriev *et al.* 1979; Price *et al.* 1986), but this is the lowest.

The calcic plagioclases in the sample and their included chromian spinels, whether aluminous or not, must also have crystallized from the very low-TiO_2 melts represented by the

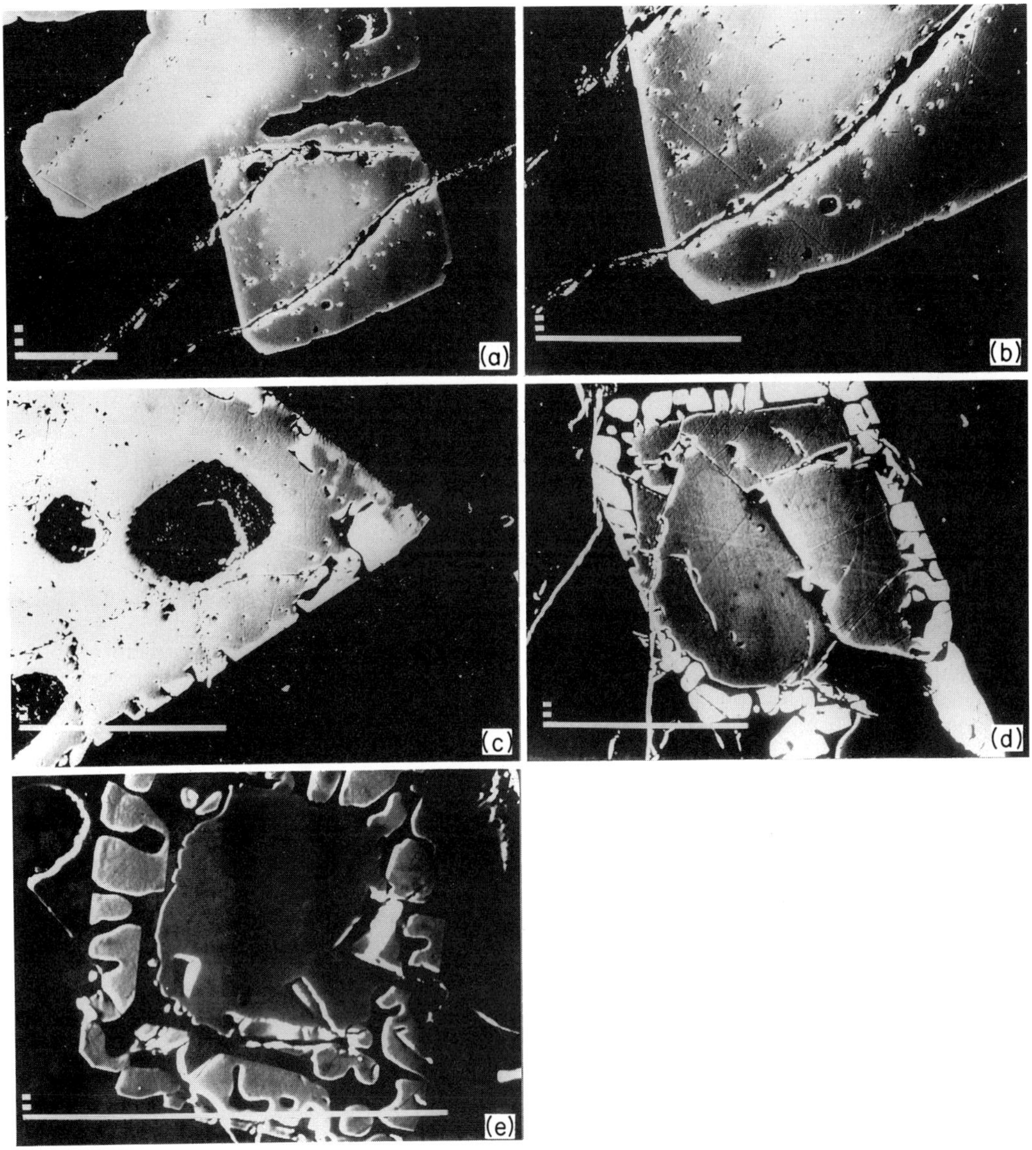

FIG. 4. Backscattered electron photomicrographs of spinels in the oceanite (*a–c*, sample type 2) and olivine tholeiite (*d* and *e*, sample type 3) showing zoning and overgrowth morphologies. Light shades define portions of the crystals with heavier mean atomic weights (more chromian and iron rich); dark shades define portions with lighter mean atomic weights (more aluminous and magnesian). Crystal cores grade to more aluminous outer portions (*a–d*) but have narrow more chromium-rich corners (*b*, enlarged from *a*) or skeletal rims (*c–e*). (*e*) A reduced-contrast image which sacrifices depiction of the zoning of the crystal's core but shows details of zoning in the skeletal rim. The core was partly rounded (resorbed) before the overgrowth, which is attached in this cross-section at only a few points (*ie* upper right), formed. The scale bar in all photomicrographs is 0.1 mm.

glass inclusions. The inclusions confirm for this sample the inference from the isopleths drawn in Fig. 3 that those melts had limited Mg# (0.71–0.74). About the same maximum Mg# can be inferred for melts in equilibrium with the most magnesian olivine ($Fo_{90.5}$) in the oceanite (sample type 2) using the equation of Roeder & Emslie (1970) and assuming that K_D = 0.30.

TABLE 5. *Representative electron microprobe analyses of plagioclases*

	Sample type 1				Sample type 2	
	1	2	3	4	5	6
Oxides (wt%)						
SiO_2	44.71	44.97	46.78	47.75	48.65	48.49
Al_2O_3	34.51	34.47	33.44	32.02	31.27	31.25
FeO	0.35	0.31	0.39	0.47	0.51	0.48
MgO	0.18	0.19	0.26	0.38	0.21	0.24
CaO	18.65	18.54	17.60	15.71	14.77	14.63
Na_2O	0.84	0.94	1.49	2.67	3.08	2.90
K_2O	0.02	0.01	0.01	0.05	0.02	0.01
	99.26	99.43	100.29	98.98	98.51	98.10
	$An_{92.5}$	$An_{91.6}$	$An_{86.7}$	$An_{76.5}$	$An_{72.6}$	$An_{73.6}$
Type[a]	PC(G)	PC(G)	PC	PR	GM	GM

[a] P, phenocrysts; C, core; R, rim; (G), next to glass inclusion; GM, groundmass crystal.

Similarly the maximum melt Mg# in the olivine tholeiite (using $Fo_{87.5}$) is only 0.664. This is unusually low for such an olivine-rich rock.

The titanium contents in the spinels of all three tholeiites are about the same, much less than in the transitional and alkalic basalts (Table 2). The simplest interpretation of this is that the titanium activity in melts was about the same when the spinels in all three samples crystallized, *ie* very low, comparable with that in the glass inclusions in the plagioclases of sample type 1.

The calcic plagioclases themselves are evidence for crystallization from refractory melts. Fisk (1985) argued that plagioclases as calcic as An_{90} or more could only have precipitated from melts having CaO/Na_2O (weight per cent, not molar ratio) >10, based on a plagioclase saturation surface calculated by Langmuir & Hanson (1981) from experiments of Drake (1972, 1976). The median composition reported in Table 6 for plagioclases adjacent to analysed glass inclusions is $An_{90.5}$. If Fisk's criterion is used, then initial Na_2O contents for melts from which the plagioclases precipitated had to be less than about 1.2%–1.4%.

Drake (1972, 1976), moreover, concluded that equilibrium constants using limited data at P(total)=10 kbar were statistically indistinguishable from the 1 atm data obtained in his study, although variable $P(H_2O)$ greatly modifies plagioclase-melt equilibria. Thus pressure was probably not a factor in producing the calcic plagioclases in sample type 1. Figure 5 shows that the glass inclusions were not antecedent to the more titanian host glass, in support of Fisk's argument. This is less evident from their Na_2O contents, which are high because of post-entrapment growth of plagioclase into the inclusions. Even so, least-modified Na_2O contents are less than 2% in most inclusions, and one inclusion listed in Table 6 has only 1.4% Na_2O. However, initial TiO_2 abundances in all the inclusions affirm that the plagioclases crystallized from uncommonly refractory melts, which we now can say were low in both TiO_2 *and* Na_2O content.

To summarize, based on the presence of variably aluminous zoned chromian spinels, the three porphyritic tholeiites evidently crystallized phenocrysts over a range of pressures. The zoning and diversity of the spinels and glass inclusions indicate that mixing at some elevated pressure was also involved. The compositions of olivines, spinels and glass inclusions indicate that the range of Mg# for liquids involved in such mixing in each sample was narrow, being no more magnesian than Mg#=0.71–0.74 for sample types 1 (plagioclase–olivine–phyric basalt) and 2 (oceanite). Liquids involved in the polybaric mixing and crystallization history of sample type 3 (olivine tholeiite) were even less magnesian (Mg#=0.66–0.62). The liquids involved in this polybaric mixing were also refractory, with very low TiO_2 contents (0.2%–0.8%) in all three cases, and low Na_2O contents (1.2%–1.4%) for at least the plagioclase-bearing sample, and by inference the others as well. Some hint of this is given by the composition of a glass inclusion in an olivine in the oceanite (Table 5, sample type 2).

Two other aspects of these crystallization histories deserve emphasis. The first is that *no hint* of genuinely magnesian (picritic) antecedent melts is given by the compositions of phenocrysts or glass inclusions in these samples, despite the diversity of compositions involved in mixing implied by the glass inclusions and

TABLE 6. *Least-modified glass inclusions in minerals, sample types 1–3*

Sample type	1	1	1	1	1	1	1	1	1	2	3	3
Oxides (wt%)												
SiO_2	49.24	51.01	51.67	50.01	50.11	48.82	48.80	48.82	50.29	48.85	47.72	48.53
Al_2O_3	12.53	15.64	14.64	15.00	15.33	10.66	15.44	14.86	16.23	16.87	18.12	17.48
FeO	10.63	8.30	7.81	9.25	9.05	11.34	8.80	9.35	9.63	8.48	8.92	9.19
MgO	12.24	10.28	10.73	10.77	11.06	13.17	10.47	11.09	8.96	8.99	7.71	7.83
CaO	12.17	12.85	13.46	12.60	13.24	12.09	12.38	12.44	11.92	13.38	11.91	11.88
Na_2O	1.62	1.99	1.84	1.88	1.80	1.40	1.61	1.86	2.24	1.80	2.70	2.51
K_2O	0.06	0.07	0.03	0.03	0.06	0.10	0.09	0.06	0.11	0.05	0.06	0.12
TiO_2	0.28	0.45	0.37	0.56	0.47	0.75	0.69	0.81	0.88	0.78	1.25	1.08
P_2O_5	0.00	0.10	0.01	0.10	0.11	0.11	0.05	0.11	0.15	0.09	0.06	0.15
	98.77	100.69	100.56	100.26	101.23	98.44	98.33	99.40	100.41	98.99	98.44	98.77
Mg#[a]	0.705	0.720	0.740	0.707	0.717	0.706	0.711	0.711	0.658	0.687	0.642	0.638
CaO/Al_2O_3	0.97	0.82	0.92	0.84	0.86	1.13	0.81	0.84	0.73	0.793	0.65	0.68
CaO/Na_2O	7.51	6.46	7.33	6.70	7.37	8.64	7.69	6.69	5.33	7.43	4.41	4.73
Adjacent plagioclase	$An_{91.3}$	$An_{89.5}$	$An_{91.1}$	$An_{90.4}$	$An_{91.9}$	$An_{91.6}$	—	$An_{91.5}$	$An_{87.2}$	—	—	—

[a] Calculated assuming that $Fe^{2+}/(Fe^{2+} + Fe^{3+}) = 0.86$.

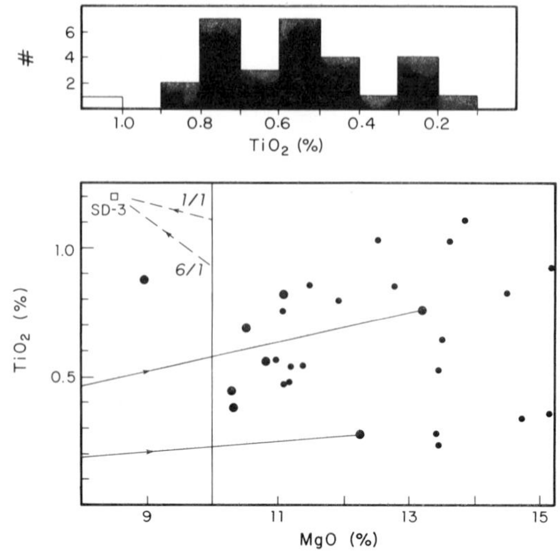

FIG. 5. TiO_2 content versus MgO content for glass inclusions in plagioclases from Siqueiros tholeiite SD-3, plagioclase–olivine–phyric basalt (sample type 1 of Figs 1–3). The inclusions have experienced post-entrapment plagioclase growth, enriching both TiO_2 and MgO. Large dots give the least-modified inclusions listed in Table 6. The original TiO_2 contents of these oxides in each inclusion can be estimated by assuming that both TiO_2 and MgO partition wholly into the inclusions and correcting along 'plagioclase subtraction' lines such as the one indicated to an arbitrary parental MgO value (here, 10%). The histogram at the top of the figure gives the resulting backtracked TiO_2 contents for all 29 inclusions plotted in the lower panel. The host glass (SD-3) is too titanian to be related even to the most titanian glass inclusions by anything less than a 6:1 ratio of plagioclase to olivine fractionation. A more reasonable 1:1 ratio shows that none of the inclusions are likely to be related to the host glass by simple fractionation.

spinels and despite the fact that two of the samples are olivine rich. It is as if, in each case, mixing occurred between almost identically fractionated magmas of about the same temperature yet having quite different compositions in terms of such parameters as CaO/Na_2O ratio and TiO_2 and chromium contents. On the face of it, this implies that the differences in melt compositions were *not* related to temperature differences (*ie* to degree of melting) in the mantle. For these rocks, then, the question of picritic primary magmas becomes entirely hypothetical. These abyssal tholeiites are fairly insistent that no liquids more magnesian than about Mg# = 0.75 were involved in their petrogenesis.

The second aspect of the crystallization histories concerns how, if refractory magmas were responsible for most of the phenocrysts in the rocks, they acquired considerably less refractory host glass compositions. This is considered next.

Aggregation of magmas to produce sample type 1

The pillow-rim composition of the plagioclase–olivine–phyric basalt (Table 1, sample type 1) is too titanian and sodic either to have formed by coalescence of magmas conforming exclusively to the compositions of glass inclusions in its plagioclase phenocrysts or to be derived from any one of them by reasonable proportions of plagioclase plus olivine fractionation (Fig. 5). These must have been counterbalanced by mixing with some magma batches even more titanian and sodic than the present host. Three candidates for these are (i) titanium-rich ferrobasalt, (ii) enriched compositions resembling the alkalic basalts from this region and (iii) equivalently sodic and titanian basalts with high Mg# but being otherwise depleted in composition. Figure 6 shows that typical ferrobasalts from the nearby crest of the East Pacific Rise do not lie on a mixing trend between the field of least-modified glass inclusions and sample type 1 host glass. Transitional to enriched compositions may lie on such a trend, but abundances of barium and rubidium for the rock (Table 2) are too low to allow much more than about 1%–2% of alkalic basalt sample type 5 into the mix; this would not significantly influence TiO_2 contents.

The alternative is for mixing with a quite sodic and titanian yet depleted basaltic magma

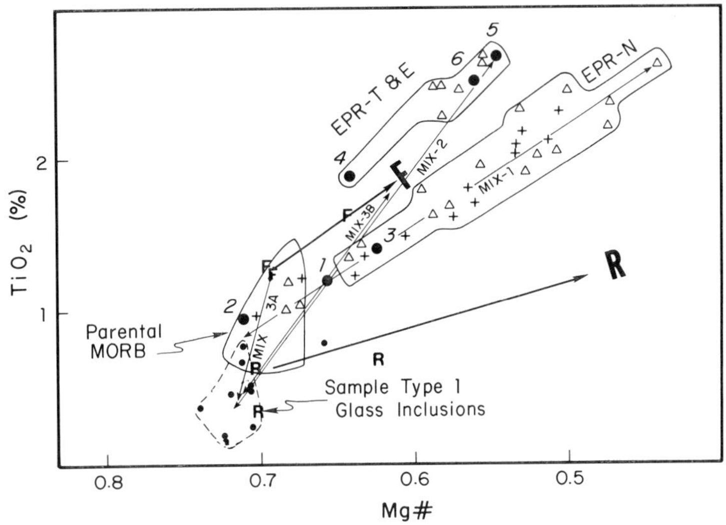

FIG. 6. TiO_2 content versus Mg# illustrating possible mixing relations between glass inclusions (small dots, dashed boundary) and various other basalts. The standard East Pacific Rise trend toward depleted ferrobasalts is labelled EPR-N (normal mid-ocean ridge basalt (MORB); +, data from Natland & Melson (1980); △, unpublished data of J. Natland). Sample types 1–6 of Figs 1–3 are numbered large dots. Transitional and alkalic basalts are in the field labelled EPR-T & E (transitional and enriched MORB). Data points based on Table 7 are labelled R (for refractory compositions) and F (for fertile compositions), as are general trends indicating fractionation pathways for the Indian Ocean suites containing the basalts. Mixing trends pertaining to sample type 1 suggest potential relationships between glass inclusions and East Pacific Rise ferrobasalt (MIX 1), the most enriched alkalic basalt (sample type 5) (MIX 2), fertile but depleted abyssal tholeiites from Table 7 (MIX 3a) and more fractionated basalts with the same (fertile but depleted) characteristics (MIX 3b). The field for parental abyssal tholeiites is based on 81 high Mg# (higher than 0.67) glass groups in the Smithsonian Institution's Data File and unpublished data.

of a type that thus far has not been sampled from this region. Such basalts are not unknown in the ocean basins, however. Two examples from the Central and SW Indian Ridges are indicated as fertile (type F) compositions in Table 7. These are similar to basalts from Cayman Trough and the Australian–Antarctic discordance, described as high-Na_2O small partial melts of the mantle by Klein & Langmuir (1987). One of these plots at the high-TiO_2 end of the field of parental (high-Mg#) abyssal tholeiite glasses shown in Fig. 6. Mixing between a composition like this and the cluster of glass inclusion compositions in the figure coupled with a modest amount of fractionation could have produced sample type 1 glass without significant enrichment in barium or rubidium.

Basalts with TiO_2 and Na_2O contents as low as those demonstrated or inferred for unmodified glass inclusions in sample type 1 plagioclases are also unknown in this region and almost never erupt on the seafloor. Two exceptions, one from the Indian Ocean (DSDP Site 212, Wharton Basin; Melson *et al.* (1976)) and the other from the Lau Basin (LAU 95–12; Hawkins & Melchior (1985)), are listed in Table 7. Both have high Mg# values and very low Na_2O and TiO_2 contents. Bryan (1979) reports another glass from the FAMOUS area, Mid-Atlantic Ridge, to have a TiO_2 content of only 0.50%. These compositions are clearly kindred to the glass inclusions, if not quite so refractory in the extreme.

In short, sample type 1 is *one rock* which coalesced from magnesian melt strains which matched or exceeded the known *global* range of high-Mg# parental abyssal tholeiite compositions in both Na_2O and TiO_2 contents. Subsequently, it only experienced moderate fractionation. Sample types 2 and 3 (oceanite and olivine tholeiite) coalesced in similar fashion from a range of magnesian melt strains, but experienced virtually no subsequent fractionation.

These are not isolated occurrences. Wilkinson (1982) summarized the wide distribution of abyssal tholeiites with similarly calcic plagioclase megacrysts. They occur in the FAMOUS area of the Mid-Atlantic Ridge (Donaldson & Brown 1977) and are strikingly abundant in basalt samples from DSDP Hole 504B near the

TABLE 7. *Representative glass compositions from refractory (R) and fertile (F) mantle*

Column	1	2	3	4	5	6
Samples[a]	DSDP 212 (1)	DSDP 236 (1)	LAU 95–12 (1)	AII93–5 6–1 (19)	CIRCE 110–3 (7)	CIRCE 87–1 (1)
Type	R	R	R	F	F	F
Oxides (wt%)						
SiO_2	51.33	53.38	50.42	50.80	50.88	50.69
Al_2O_3	15.60	14.43	15.75	17.27	16.66	17.32
FeO*	7.86	9.93	8.59	7.83	8.58	7.53
MgO	8.97	7.80	9.78	8.53	7.42	8.25
CaO	13.48	12.38	14.67	10.15	10.22	10.25
Na_2O	1.42	1.65	0.99	3.64	3.77	3.71
K_2O	0.05	0.06	0.009	0.21	0.20	0.19
TiO_2	0.61	0.61	0.34	1.23	1.62	1.28
P_2O_5	0.03	0.04	0.02	0.18	0.18	0.15
	99.35	100.28	100.57	99.81	99.47	99.37
Mg#[b]	0.703	0.619	0.702	0.693	0.642	0.695
Trace elements (ppm)						
Rb	—	1.3	0.15	2.0	1.7	—
Sr	65	56	12	181	214	—
Ba	—	—	1.2	28	20	—
Zr	22	—	—	—	154	—
Sm	1.08	1.19	—	—	4.2	—
Nd	3.4	3.2	—	—	14.0	—

[a] All samples except column 3 are from the Indian Ocean: Smithsonian Institution Glass Group Data File, T. O'Hearn, analyst. Column 3 is for a sample from the Lau Basin (Hawkins & Melchior 1985).
[b] Calculated assuming that $Fe^{2+}/(Fe^{2+} + Fe^{3+}) = 0.86$ (Presnall *et al.* 1979).
[c] Data sources are as follows: column 1, Frey *et al.* (1977); column 2, samarium and neodymium, Frey *et al.* (1980); column 3, Hawkins & Melchior (1985). Other data are isotope dilution determinations of rubidium, strontium, samarium and neodymium by John Mahoney, barium and zirconium by dc plasma by J. Natland. Barium in column 4 is from Price *et al.* (1986).
FeO*: total iron as FeO.

Costa Rica Rift (Natland *et al.* 1983; Kempton *et al.* 1985); one plagioclase there is as calcic as $An_{94.5}$. Yet, the Na_2O content of the least sodic glass group from this site is 1.81%, and the highest CaO/Na_2O ratio is only 7.38. All contain 'second-generation' (Natland 1979) plagioclase microphenocrysts and rims to megacrysts of An_{70-80}. A more extreme contrast is reported from the SE Indian Ridge (Price *et al.* 1986), where plagioclase megacrysts are $An_{89.6-91.0}$ in composition and two glass inclusions in the plagioclases have TiO_2 contents of 0.54% and 0.24% (the latter with Mg#=0.78). Yet the host glass has 2.97% Na_2O, a CaO/Na_2O ratio of merely 3.65, 1.58% TiO_2, Mg#=0.627, and rims to megacrysts of An_{67-74}.

Thus many abyssal tholeiites provide evidence for coalescence of refractory magma strains in their crystallization histories. In all these examples, the contributions of the low-Na_2O magmas from which the megacrysts crystallized must have been counterbalanced by considerably more sodic compositions to produce the host lavas. Surprisingly, neither in the Siqueiros samples nor in any of the other examples just cited is there mineralogical evidence for the sodic components. The indications of such components are entirely occult in the compositions of host glasses. All these phenomena reflect a general process in the petrogenesis of parental abyssal tholeiites.

Experimental studies and the problem of restricted Mg#

In seeking the explanation for these occurrences, it is useful to focus on the variable, but evidently extremely low, abundances of Na_2O and TiO_2 in the melt inclusions and some basalts, the typically much higher levels in most eruptive abyssal tholeiites, and the restricted range of Mg# in the inclusions and implied by the mixing-crystallization histories of the rocks.

TiO_2 and Na_2O contents correlate strongly in depleted high-Mg# abyssal tholeiite glasses (Fig. 7), and both titanium and sodium behave similarly with respect to early crystallizing phases (both are effectively excluded and hence concentrate into residual melts). This is especially so for titanium, which Bougault *et al.*

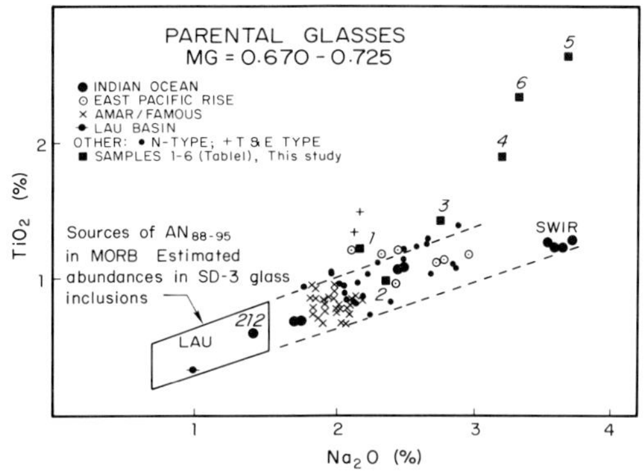

FIG. 7. TiO_2 content versus Na_2O content for abyssal tholeiites having Mg# > 0.67, taken from the Smithsonian Institution's Glass Group Data File. Symbols are given in the key. The broken lines bound data points for depleted (N-type) abyssal tholeiites. Sample types 1–6 of Table 1 are indicated by number. Only sample type 2 has Mg# = 0.67, but sample types 4–6 show that transitional alkalic compositions are enriched in TiO_2 relative to Na_2O compared with depleted abyssal tholeiites. The trapezoidal field at the lower left represents the estimated range of TiO_2–Na_2O covariation among unmodified glass inclusions in plagioclases of Siqueiros sample type 1 (plagioclase–olivine–phyric basalt). Full analyses for samples DSDP 212 and LAU 95–12, which fall within the field for glass inclusions, are given in Table 7.

(1983) consider to behave as a nearly hygromagmatophile (almost ideally incompatible) element during melting. In general, then, the comparative abundances of sodium and titanium in parental abyssal tholeiites probably reflect the same fundamental process.

Both experimental studies and evaluations of natural basalt glasses indicate that a primary control on Na_2O abundances can be extent of melting and that this is reflected in the compositions of magmatic lineages (Bryan & Dick 1982; Schouten *et al.* 1986; Klein & Langmuir 1987). But can simple degree of melting explain the variability described here? Experimental studies in which the extent of melting has been monitored (or at least allowed to vary) at assorted elevated pressures all show that Na_2O and TiO_2 contents diminish as melting progresses (Jaques & Green 1980; Takahashi & Kushiro 1983; Fujii & Scarfe 1985). Indeed, other aspects of parental abyssal tholeiite compositions, including their spread across diagrams such as Fig. 2 and the ratio CaO/Al_2O_3, are duplicated by extent-of-melting experiments at elevated pressure (Fujii & Scarfe 1985).

However, among these rocks and glass inclusions there are variations in TiO_2 contents by a factor of 6 and variations in Na_2O contents by a factor of nearly 3. To accomplish this by straightforward increases in degree of melting of any particular source material requires temperature contrasts of more than 200 °C and occurs at the expense of much variability in Mg#, to as high as 0.88, in the experiments of Jaques & Green (1980). The Siqueiros tholeiites, however, coalesced from melts having extreme contrasts in Na_2O and TiO_2 contents, yet very similar values of Mg#, and with almost no temperature contrasts evident from their crystallization histories.

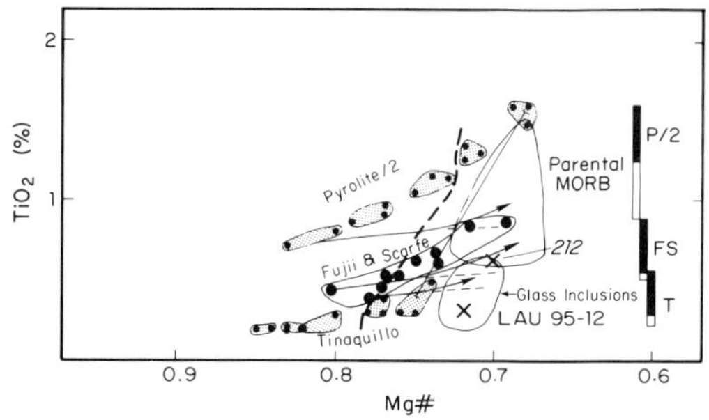

FIG. 8. TiO_2 content (%) versus Mg# for experimentally produced compositions (Jaques & Green 1980), the parental abyssal tholeiite field (bounding the same glass group compositions as in Fig. 7) and a field for the glass inclusions listed in Table 6. The lower series of shaded balloons enclose compositions produced in experiments using Tinaquillo lherzolite with one data point within each for 5, 10 and 15 kbar experiments, respectively right to left in the diagram. The upper series is a similar group based on experiments at the same pressures using model pyrolite, except that the Ti_2O contents have been halved to make them accord more closely with compositions relevant to abyssal tholeiites. Ranges in calculated extents of melting are indicated by the balloons. Compositions produced in the 'sandwich' experiments of Fujii & Scarfe (1985) lie above the Tinaquillo fields. Refractory basalt glasses from DSDP Site 212 and the Lau Basin (LAU 95–12) from Table 7 are also indicated. Curves with arrows give olivine-dominated fractionation pathways computed using the range of olivines shown by the solid bar along the abscissa. Dashed extensions give simple olivine-controlled pathways; curved lines above are estimates of the added influence of plagioclase and clinopyroxene during fractionation. The vertical bars on the right give estimated ranges of parental TiO_2 contents that can be produced by combining extent of melting and olivine-dominated fractionation for (in descending order) pyrolite halved, the Fujii & Scarfe glasses and Tinaquillo glasses. The solid portions of the bars give ranges that can be produced with the additional restriction that $CaO/Al_2O_3 < 1.0$ (primary melts lie to the right of the solid broken curve) based on Fig. 9.

The simultaneous dependence of Mg# and TiO_2 content on extent of melting is shown in Fig. 8, which includes fields for parental abyssal tholeiite glass compositions (based primarily on the Smithsonian Glass Group Data File compiled by W. G. Melson and T. O'Hearn) and for the glass inclusion data of Table 6. The glass inclusion field is in effect an extension of the parental field and hooks to slightly higher Mg# at low values of TiO_2.

Experimentally produced glass compositions derived from partial melting on two different starting materials used by Jaques & Green (1980) are also shown in Fig. 8. These are a refractory natural peridotite, Tinaquillo lherzolite, and the much more sodic and titanian peridotite, model pyrolite. The data points plotted are for 5, 10 and 15 kbar experiments with one point for each pressure enclosed within each small shaded field. Along each partial-melting trend towards higher Mg#, the small fields represent 50 °C increments of increasing temperature at a given pressure. To provide a closer match in TiO_2 content to natural parental abyssal tholeiites, the TiO_2 contents for the experiments on pyrolite compositions are halved (this is done simply for discussion purposes and assumes ideally incompatible behaviour for TiO_2 during partial melting). The two peridotites can be taken as potential extremes in the range of mantle compositions relevant to parental abyssal tholeiite petrogenesis, with pyrolite, if anything, being too sodic, and much too titanian, and Tinaquillo lherzolite being, for most compositions, too refractory (*eg* Thompson 1987).

Figure 8 shows that it is not possible to match the field of parental abyssal tholeiites simply by varying the extent of melting of any single mantle source. Even coupling partial melting of pyrolite halved with olivine-dominated fractionation (the curving arrow starting from the most magnesian of the shaded pyrolite fields) only encompasses the top half of the parental field. Doing the same for Tinaquillo lherzolite only encompasses the field for glass inclusions. Glass compositions produced in the somewhat different 'sandwich' experiments with peridotite by Fujii & Scarfe (1985) are as limited in this respect as those produced by partial melting of Tinaquillo lherzolite, and are only slightly more titanian.

A similar argument can be based on Fig. 9, CaO/Al_2O_3 ratio versus Na_2O content. The CaO/Al_2O_3 ratio is also sensitive to the degree of partial melting (Fujii & Scarfe 1985), so much so that it can be used to set an upper limit on the extent of melting relevant to the generation of parental abyssal tholeiites. Based again on the experiments of Jaques & Green (1980), who used modal estimates and mass balance calculations to specify both extents of melting and compositions, natural abyssal tholeiite parents (shown by the shaded field in Fig. 9) are encompassed by less than 25% partial melting of a pyrolitic type of source and less than 21% melting of a Tinaquillo type of source. Presumably, any intermediate composition would have about the same upper limit, as suggested schematically by the bold broken line. A corresponding line is shown in Fig. 8, indicating an upper limit of about Mg# = 0.78 for a Tinaquillo-type of source pertinent to the parental field and of Mg#=0.73 for a pyrolitic source. Partial-melting trends for pyrolite and Tinaquillo lherzolite determined for 5, 10 and 15 kbar bound the parental field in Fig. 9. These illustrate that the CaO/Al_2O_3 ratio is more sensitive to extent of melting at these percentages of melting than Na_2O content is, and that this effect is independent of pressure. Moreover, they demonstrate that only a very restricted portion of the total parental field for abyssal

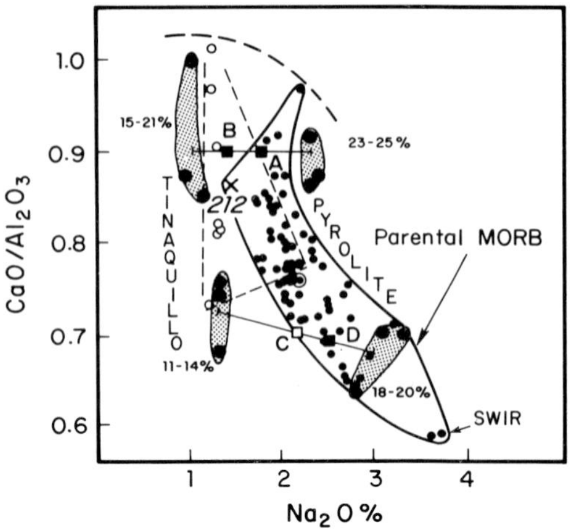

FIG. 9. CaO/Al_2O_3 content versus Na_2O content for parental abyssal tholeiites (●), the glass inclusion data from Table 3 (corrected to the condition that CaO/Na_2O = 10; see text for discussion), host sample SD-3 (sample type 1, Table 1; circled dot) and experimental data (shaded balloons with large dots, as in Fig. 8; Jaques & Green (1980)). Other conventions are as in Fig. 8. Possible mixing trends and calculated mixes A–D from Table 8 are also indicated. Refractory and fertile high-Mg# basalt compositions from Table 7 are plotted. The bold broken curve at the top, representing the upper limit in extent of melting for parental abyssal tholeiites, corresponds to the similar curve in Fig. 8.

tholeiites can be encompassed by partial melting of a single source composition. It is true that a source composition intermediate between Tinaquillo lherzolite and pyrolite is a more appropriate average source, but it will be no more successful than either of those extremes in encompassing the range of parental compositions.

The glass inclusion compositions of Table 5 are also plotted in Fig. 9, after correction of Na_2O contents to the minimum condition specified by Fisk (1985) that melts precipitating plagioclase of An_{90} composition had $CaO/Na_2O = 10$. These fall on the refractory (low-Na_2O) side of the host glass composition for sample type 1. The extremes of low- and high-Na_2O basalts in Table 7 are also shown, and these lie respectively toward the Tinaquillo and pyrolite sides of the parental field. Based on Fig. 6 and trace-element data, sample type 1 thus appears to be a hybrid between refractory compositions given by the glass inclusions and some more sodic basaltic magma such as those from the Indian Ocean, which plot in the lower right-hand corner of the parental field in Fig. 9. This and Fig. 8 together indicate that mixing took place between magmas derived by partial melting of variably sodic and titanian mantle sources, represented in the extreme by refractory Tinaquillo lherzolite on the one hand and by peridotite resembling model pyrolite, which is more fertile with respect to a basaltic melt fraction, on the other. The chemically contrasting melt strains were not produced by differences in the degree of partial melting but by similar degrees of partial melting acting on a lithologically heterogeneous source. The extent of melting of any particular source did not exceed 20%, as specified both by the CaO/Al_2O_3 ratio and fairly low Mg# of the inclusions, and the low Mg# implied by the olivines and spinel isopleths in Fig. 3.

Many fairly typical parental abyssal tholeiites could also result from such mixing. Mixing trends, simplified to straight lines, are indicated in Fig. 9, linking shaded data fields for Tinaquillo lherzolite and pyrolite. Four hybrid compositions (A–D in Fig. 9) are calculated (Table 8). With no more than 25% partial melting of pyrolite and 21% of Tinaquillo lherzolite, plausible parents to the full range of parental abyssal tholeiites can be produced by mixing. None of the calculated parental melts in Table 8 has Mg# >0.75. The mixing in each case is specified to occur at constant temperature at any given pressure. In short, both the range and the limitations in composition of parental abyssal tholeiites globally can be satisfied by such mixing; just such mixing is demonstrated by the Siqueiros tholeiites.

Other mixing relations might be construed, such as between a higher-temperature refractory magma (upper shaded Tinaquillo field in Fig. 9) and lower-temperature more fertile melts (lower shaded pyrolite field). The hotter melts would tend to resorb minerals in cooler intercepted melts, offering an explanation for the lack of mineralogical evidence for more sodic components in the Siqueiros tholeiites described here.

However, rather than supposing that the arrays of parental compositions shown in Fig. 8 and 9 result from mixing, could they not simply reflect mantle lithofacies heterogeneity directly? Within the parental field, certain compositions (such as the 'refractory' sample from DSDP Site 212 and the 'fertile' samples from the SW Indian Ridge) define extremes of the arrays. These examples suggest that there are important *regional* heterogeneities that might be sampled from place to place, as is already known from study of ultramafic rocks of fracture zones (*eg* Dick *et al.* 1984). If the parental field is representative, then its obliquity to the experimental data in Fig. 9 may have global significance.

A geodynamical hypothesis

Assuming now that the mantle is lithologically heterogeneous, the crystallization sequences described here also indicate how mantle lithologies may be distributed. The spinels in the basalts, their associated olivines and, in the one sample, calcic plagioclases all crystallized from melts derived from refractory mantle. Some of these melts were actually trapped as inclusions in the minerals. In each sample, the minerals represent the first, highest-temperature, stages of crystallization. All the mixing relations implied by the compositional variability and zoning of the spinels occurred exclusively among refractory magma strains. This is confirmed by the compositions of the glass inclusions. Yet in all three tholeiites the hybrid refractory compositions later mixed with cooler magmas which had to come from significantly more fertile mantle lithofacies.

Why should such sequences exist? Why not the opposite sequence of mixing between warmer magmas derived from fertile sources and cooler more refractory melts? If variously fertile mantle lithofacies are distributed uniformly, should there not be mineralogical evidence for this diversity at all stages of crystallization? Clearly, such evidence does not

TABLE 8. *Mixing calculations between idealized Tinaquillo (T) and pyrolite (P) partial-melt analogues to parental abyssal tholeiites*

	Mg#	Na_2O	CaO/Al_2O_3	T:P
Starting compositions				
Tinaquillo (T)	0.76	1.00	0.90	
Pyrolite (P)	0.72	2.20	0.90	
Hybrids[a]				
A[b]	0.75	1.36	0.90	70:30
B[b]	0.74	1.72	0.90	40:60
Starting compositions				
Tinaquillo (T)	0.74	1.20	0.73	
Pyrolite (P)	0.68	3.10	0.68	
Hybrids[a]				
C[b]	0.71	2.15	0.71	50:50
D[b]	0.70	2.53	0.70	30:70

[a] Mg# and CaO/Al_2O_3 assume mixing end-members have the same cation total of Mg + Fe and Ca + Al, which will not be strictly true among natural compositions.
[b] Plotted in Figs 8 and 9.

exist. The explanation for this must be that mantle sources are structured in such a way that the more refractory lithofacies coincide with high-temperature regions.

Imagine this in terms of the complex melting domains envisaged by O'Hara (1984). The mantle beneath spreading ridges is thought to well up and fill the void left by separating plates, displacing geotherms upward. This results in zones of partial melting concentrated directly beneath ridge axes but extending to cooler peripheral regions defined by the solidus of some assemblage of mantle lithofacies (Oxburgh 1980). The crystallization sequences in the Siqueiros tholeiites thus indicate that the cores or centres of these bodies are the sources of refractory melts and the peripheral regions the main sources of fertile melts. This is depicted in Fig. 10 in terms of mushroom-shaped lobes beneath individual spreading segments between the Siqueiros Fracture Zone and the 9°N non-transform offset of the East Pacific Rise. The longitudinal dimensions of the lobes are shown to correspond to distances between non-transform offsets with overlapping rifts, which are spaced perhaps 50–100 km apart between major transform faults (Macdonald *et al.*, 1984; Lonsdale 1985), or perhaps by the even shorter distances between even smaller offsets of the axial neovolcanic zone (Langmuir *et al.* 1986). The lobe shape is borrowed from the scaled modelling of lithothermal convection by Elder (1966, 1972) which agrees well with Oxburgh's (1980) deduced cross-sectional thermal structure beneath spreading ridges and which anticipated in many respects the work of Whitehead *et al.* (1984). In the conception of these last workers, ridges should be segmented in a manner defined by the ascent of individual convective cells spaced according to the development of Rayleigh–Taylor instabilities in the mantle. These are not 'plumes' in the manner often used with reference to linear island chains, but could be relatively persistent domains of mantle upwelling corresponding to individual volcanic centres defined geomorphologically (*eg* Francheteau & Ballard 1983; Lonsdale 1985) on the East Pacific Rise.

Melts produced within such domains must pool into local magma bodies or 'network(s) of connected melt veins and gashes' (Nicolas 1986, p. 999) before they achieve sufficient mass and buoyancy (pressure) to escape. The larger of such networks will develop in the warmer interiors of the melt zones and thus will carry the signature of somewhat greater degrees of partial melting and the particular stamp of their source composition (refractory peridotite). These rising melts on ascent must necessarily intersect smaller bodies of melt at cooler temperature (wavy arrows in Fig. 10); the latter will contain proportionally more of smaller melt fractions and bear the fertile compositional stamp of their particular sources. Vertical ascent from melt pockets at the periphery of melt domains will not involve melts produced in the

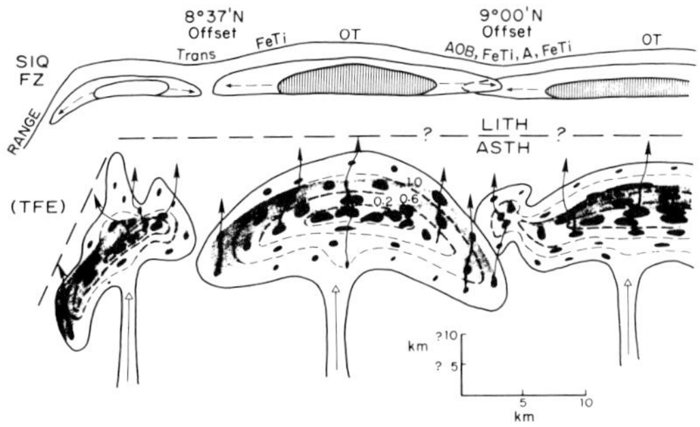

FIG. 10. Hypothetical melt domains (mushroom-shaped tops of diapirs, after Elder (1966)) beneath a segmented portion of the East Pacific Rise, bounded on the left by the Siqueiros Fracture Zone. The black blotches are coalesced bodies of melt, larger and more extensive toward the interior of each diapir, which travel upward (wiggly arrows), encountering and entraining melt batches produced in lower-temperature portions of the region of melting. The contours are designed to suggest where melts of different TiO_2 contents are likely to be most abundant. TiO_2 contents of the melts are indicated, based on the likely distributions of refractory mantle (toward the interior of the melt region) and fertile mantle (toward the periphery). Surficial rock distributions and locations of offsets are based on Langmuir *et al.* (1986), Macdonald *et al.* (1984) and unpublished analytical data.

interiors; hence there should be a tendency for more sodic and titanian compositions to erupt above the extremities, indicated in Fig. 10 as corresponding to structural discontinuities on the seafloor such as fracture zones or non-transform offsets.

Why should refractory mantle be concentrated at the centre of such hypothetical bodies? This deduction is perhaps more reasonable when it is considered that refractory peridotite, being comparatively depleted in a basaltic melt fraction, has lower iron and therefore at appropriate depths reduced proportions of dense phases such as garnet. Under appropriate thermal conditions such mantle will be more buoyant than fertile mantle (*eg* O'Hara, 1975; Boyd & McAllister, 1976; Jordan, 1979) and hence will tend to rise through it and even coalesce as more massive and rapidly ascending bodies of it encounter smaller more slowly rising pockets. In this way, primary lithological variability of the mantle may contribute important density contrasts which will accelerate convective processes beneath spreading ridges under appropriate rheological conditions. The distribution of such materials at depth will also influence the spacing of instabilities as they develop into diapirs beneath spreading ridges (*cf* Whitehead *et al.* 1984). This may exert some influence on the structural segmentation of spreading ridges.

Another consequence of this general hypothesis is that, although some contrast in temperature across an individual melting domain must exist, it need not be nearly so great as the 200°C or more implied by the full range of extent-of-melting experiments on homogeneous mantle analogues (Jaques & Green 1980). Refractory cores to individual diapirs need be no more than about 50°C warmer than peripheral more fertile regions at the same depth to give the range of melts necessary to produce the rocks described here. Similarly, if, as Fig. 9 suggests, regional variability in parental basalt compositions reflects diversity of mantle sources more than extent of partial melting (contrary to the opinion of Klein & Langmuir (1987)), then lateral temperature gradients at any depth in an ocean basin (such as the North Atlantic) also need be no greater than this. Extrapolation of the local geodynamic hypothesis just presented thus suggests that hot-spot regions may be characterized by refractory mantle (in effect being very large domains influenced by the distribution of comparatively buoyant low-density peridotite). Such regions need be only somewhat warmer than those dominated by more sodic denser iron-rich peridotite. A regional evaluation based on this hypothesis is in progress.

Trace elements and isotopes

The singular geochemical feature of the Siqueiros tholeiites (sample types 1–3) is their unusually low abundances of potassium, rubidium and barium (Table 6). Sample type 2, the oceanite, is perhaps *the* most depleted abyssal tholeiite ever obtained from the vicinity of a spreading ridge, in terms of its low barium and especially rubidium abundances.

Yet all these Siqueiros tholeiites erupted within a few kilometres of one of the most enriched basalts so far obtained from the East Pacific Rise or its fracture zones (sample type 5). The rocks are isotopically different as well (Table 2). Similar enriched basalts were obtained at the 9°N offset (sample type 6) together with more typically depleted tholeiites. The mantle in this region therefore must be extraordinarily partitioned into depleted and enriched sectors, in the manner described for mantle beneath young eastern Pacific seamounts by Zindler *et al.* (1984), but perhaps to an even greater degree in terms of extremes of compositional diversity.

Obviously, for all the Siqueiros tholeiites, the primary mantle lithofacies variability that gave rise to their bulk compositions has almost nothing to do with the distribution of enriched components in the mantle. That is a separate aspect of mantle heterogeneity in this region. Both fertile and refractory mantle sources here are *depleted* in composition. To the extent that the isotopic differences between enriched mantle and both types of depleted mantle (fertile and refractory) required great lapses of time (hundreds of millions of years) to develop (Zindler *et al.* 1984), we may surmise that all aspects of the mantle heterogeneity existed a long time ago. That is, neither the refractory nor the fertile mantle in this region was segregated from enriched mantle in the recent past.

There are less extreme but still significant contrasts in abundances and ratios of trace elements between the extremely depleted tholeiites of the Siqueiros Fracture Zone and more typically evolved tholeiites of the adjacent northerly segment of the East Pacific Rise. Figure 11 shows this in terms of a histogram of chondrite-normalized $(K/Ti)_n$, computed using high-resolution plasma and microprobe data for potassium abundances in basalt glasses. The ratio, which for glasses is free of the consequences of alteration, is comparable in many respects with Th/Hf, La/Sm or Nb/Zr ratios (*eg* R.N. Thompson *et al.* 1984) in that it is very little modified by fractionation or partial melting and thus provides information about mantle sources. Together with the low abundances of rubidium and barium, it implies that a number of Siqueiros tholeiites (including several in addition to the three considered in detail here) originated from mantle sources considerably more depleted, on the face of it, than sources to parents of fractionated axial basalts.

However, this may be an artefact of magma chamber mixing between truly depleted magmas, similar in composition to the Siqueiros tholeiites, and small amounts of alkalic basalt. Table 9 shows the results of simple binary mixing calculations between hypothetical truly depleted and enriched components, based on the analyses given in Table 6. Table 9 is designed primarily to show qualitative relationships and does not rigorously model relationships between specific basalts. Thus addition of only 1%–2% of alkalic basalt to the hypothetical truly depleted tholeiite significantly modifies Ba/Rb, K/Rb, $(Ba/K)_n$, $(Ba/Ti)_n$, K/P and $(K/Ti)_n$ to ratios representative of estimated average axial abyssal tholeiite (see the explanation to Table 9). There will even be a detectable increase in $^{87}Sr/^{86}Sr$, from about 0.7024 to 0.7026. Certain

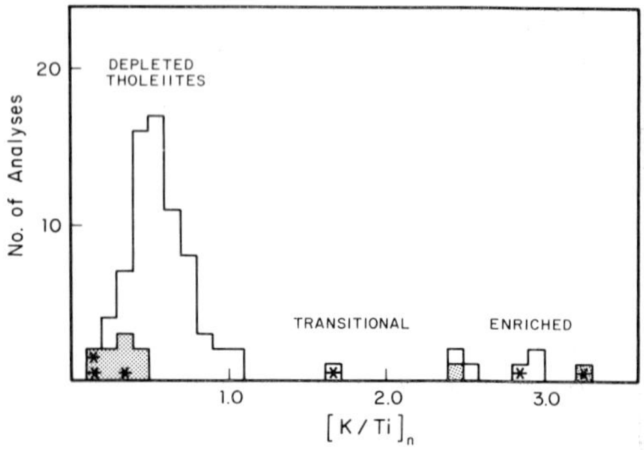

FIG. 11. A histogram chondrite normalized $(K/Ti)_n$ for basalts from the East Pacific Rise (not shaded) and Siqueiros Fracture Zone. The diagram is based on high-resolution determinations of potassium and titanium by dc plasma electron microprobe procedures (see text discussion of analytical methods). Most samples are from summit regions and overlapping rifts of the East Pacific Rise at 12.9°N, 9°N, 5.5°N and 3.9°N (J. Natland, unpublished data).

elemental ratios (Ba/Rb, K/Rb and Ba/K) are greatly modified by even trivial amounts of enriched material (less than 2%) but change little thereafter. Others such as K/P and $(K/Ti)_n$ require more of the alkalic component to match average axial basalt (Table 9, asterisks). The K/P and $(K/Ti)_n$ ratios probably specify the proportion of enriched peridotite (5%–7%) in the mantle more accurately than the ratios involving potassium, rubidium and barium.

The mechanism for this mixing is simply to assume that mantle lying more directly beneath the spreading ridge is as diverse as that which supplied depleted and enriched basalts to the Siqueiros Fracture Zone. Yet there, as opposed to the fracture zone, such magmas were obliged to pass into and through major crustal reservoirs where mixing and extensive fractionation necessarily occur. This suggestion is shown by the placement of crustal magma chambers in Fig. 10.

On this basis, the mantle may consist mainly of very depleted peridotites of various lithologies but constituting in aggregate perhaps 93%–95% of the total volume. Together, these produce truly depleted magmas having both fertile and refractory major-element characteristics. The remaining 5%–7% of the mantle is enriched peridotite which gives rise to alkalic melts. At axial discontinuities such as fracture zones and non-transform offsets, the diverse magmas can erupt separately or with very little mixing between them. Elsewhere, the existence

TABLE 9. *Trace-element abundances and ratios in abyssal tholeiites estimated from mixing between truly depleted and enriched basalts of the Siqueiros Fracture Zone*

	100% truly depleted	Abundances calculated from proportions					100% most enriched	Average East Pacific Rise basalt
		98:2	96:4	94:6	93:7	90:10		
Ba (ppm)	(0.5)	5.5	10.5	15.5	18.0	25.5	250	
Rb (ppm)	(0.01)	0.41	0.81	1.21	1.41	2.01	(20)	
K_2O (wt%)	0.008	0.030	0.052	0.074	0.084	0.117	1.10	
Sr (ppm)	70	77	83	90	93	103	400	
P_2O_5 (wt%)	0.080	0.088	0.097	0.105	0.109	0.122	0.500	
Zr (ppm)	50	53	57	60	62	67	215	
TiO_2 (wt%)	0.80	0.83	0.87	0.90	0.92	0.97	2.50	
Ba/Rb	50 *	13.4	13.0	12.8	12.8	12.7	12.5	* ~14
K/Rb	6642 *	607	533	508	495	483	457	* >1000
$(Ba/K)_n$	0.131 *	0.384	0.423	0.439	0.449	0.457	0.476	* 0.197±0.041
$(Ba/Ti)_n$	0.009	0.099 *	0.181	0.258	0.293	0.394	1.50	* 0.133±0.053
K/P	0.190	0.649	1.02	1.34 *	1.64	1.82	4.19	* 1.47 ±0.38
$(K/Ti)_n$	0.072	0.259	0.428	0.588	0.615 *	0.863	3.15	* 0.653±0.175
$^{87}Sr/^{86}Sr$	0.70245	0.70250	0.70254	0.70258	0.70259	0.70264	0.70293	

Truly depleted and most enriched abundances are based on Table 2 except for values in parentheses which are hypothetical for truly depleted, and interpolated for most enriched. The asterisks indicate interpolated positions of average East Rise basalt in the sequences of calculated mixing ratios. Ratios are given as calculated from elemental abundances or normalized to chondrites (subscript n) using chondritic abundances given in Thompson *et al.* (1984). Ratios of Ba/Rb and K/Rb for average East Pacific Rise basalt are estimated from unpublished data of J.D. Macdougall (Rb) and J. Natland (K and Ba).

of large crustal magma chambers compels mixing, and no tholeiite emerges without a small alkalic increment in its makeup.

The alternative to this is to suppose that average mantle beneath spreading segments of the East Pacific Rise is fundamentally more enriched than that supplying the Siqueiros Fracture Zone. But even if this is the case, there are still significant variations in trace-element ratios even among ferrobasalts, so that some base level of heterogeneity must still exist. It is simpler to suppose that mixing is not totally efficient and that even possibly steady-state axial magma-chambers do not impose monotonous trace-element and isotopic ratios on fractionated tholeiitic basalts.

Support for this is provided by Fig. 12, in which strontium and neodymium isotopes are plotted against a calculated 'parental' Sm/Nd ratio, based on isotope dilution data. Since Sm/Nd drops by about 10% as fractionation proceeds (with Mg# dropping in glasses from 0.7 to 0.4), a simple linear interpolation (using Mg# as an index) can be used to calculate the parental ratio. The effect is small, but making the correction clarifies relationships in Fig. 12.

The trends for the data in Table 2 (and some unpublished data) are nearly identical with those of Zindler *et al.* (1984) for Seamount 6. Moreover, even within a suite of samples from a 4 km length of a normal spreading segment at 21°N (Macdougall & Lugmair 1986) there are subtle but clear trends in the same direction. As $^{87}Sr/^{86}Sr$ changes from 0.7024 to 0.7026 in this group of samples, there is a small but detectable enrichment in light rare earth elements, given by Sm/Nd. An isotopically more enriched component is clearly present in some samples of what can be considered a typical depleted abyssal tholeiite suite. This is one advantage to working exclusively with hand-picked glasses and determining isotopic ratios to six significant figures! These results corroborate the inferences from Table 8.

Other portions of the Pacific, and other oceans, are different, as indicated by the data fields in Fig. 12. However, no abyssal tholeiite, especially not a fractionated one that passed through a major magma storage complex in the crust, is likely to have escaped a mixing history among antecedent parental magma stems. Part of this history can be detected using high-precision isotopic and trace-element determinations. At various places on the East Pacific Rise, including the Siqueiros–9°N region, isotopically somewhat enriched mantle provided melt components which mixed with depleted magmas to produce the typical range of abyssal tholeiites. Aspects of this have been suggested on a number of grounds before (Carlson *et al.* 1978; Johnson 1979; Batiza & Johnson 1980), but the close spatial association of distinctly enriched and depleted eruptives along the East Pacific Rise (*eg* Langmuir *et al.* 1986; G. Thompson *et al.* 1986) plus these isotopic relationships point to mixing between diverse enriched and depleted parental magma stems as a pervasive and fundamental process in all abyssal tholeiite petrogenesis. At the East Pacific Rise, axial magma chambers almost certainly play a fundamental role in this mixing.

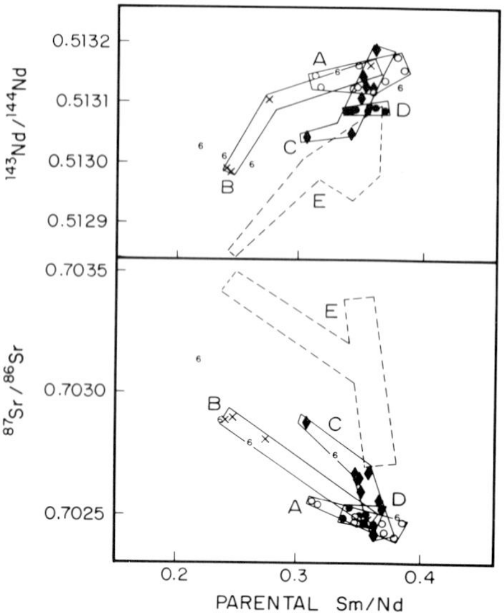

FIG. 12. $^{143}Nd/^{144}Nd$ and $^{87}Sr/^{86}Sr$ versus parental Sm/Nd (corrected for fractionation, as discussed in the text) for East Pacific Rise basalts: X, present data (Table 2; (field B); 6, samples from Seamount 6 of Zindler *et al.* (1984). Other Rise axis samples are as follows: field A, +21°N (○ Macdougall & Lugmair (1986); field C, near Easter microplate (○ Macdougall & Lugmair 1986); field D, + 29°–34°S (●, Macdougall & Lugmair 1985). Field E (no data points) bound determinations for samples from the Indian Ocean (Mahoney & Natland 1985 and unpublished). Isotopic enrichment coupled with enrichment in light rare earth elements (reduction in Sm/Nd) characterizes field B and the Seamount 6 samples, but is also subtly evident in field A and perhaps field D. Samples in field C (with one exception) exhibit isotopic variations without much change in Sm/Nd, a pattern also observed for most samples in the Indian Ocean.

One consequence to such mixing is as follows. If, as Table 8 suggests, very small increments of alkalic magmas mixed with tholeiites have such remarkable influences on trace-element ratios, then we almost certainly cannot characterize the geochemistry of the genuinely depleted portions of the mantle by using typical compositions of ridge basalts, or an average of them. Abundances and ratios of large-ion lithophile elements (K, Rb, Ba, Cs etc.) in particular will be dominated by the alkalic component, even in nominally depleted basalts. Only very rare basalts, such as the Siqueiros tholeiites discussed here, provide an indication of how extremely depleted the great bulk of the mantle beneath the East Pacific Rise may truly be. Ratios of Ba/Rb and K/Rb (and those involving other large-ion lithophile elements such as caesium) may be far *greater* for truly depleted mantle (Table 9) than is indicated by most abyssal tholeiites, *eg* Hofmann & White 1983). Extremely precise analytical procedures will be needed to determine such trace-element abundances and ratios in truly depleted abyssal tholeiites. Within such basalts, however, isotopic data already indicate some heterogeneity. Values of $^{87}Sr/^{86}Sr$ for the three Siqueiros tholeiites (Table 2) are slightly higher than those for several otherwise less depleted tholeiites from the Rise axis further N (Natland & Macdougall 1986). This supports the interpretation of Klein *et al.* (1986) that there is a 'significant geochemical discontinuity' at this fracture zone.

Discussion

That the mantle beneath spreading ridges is geochemically and isotopically heterogeneous has been recognized for a long time. However, to say that the mantle is lithologically heterogeneous on either a local or a regional scale raises some complicated issues. Lithological heterogeneity among peridotites from fracture zones or ophiolites is no proof that the oceanic mantle was similarly heterogeneous *before* the basalts that make up the ocean crust were extracted. Even if the general isotopic depletion of abyssal tholeiites was induced by some ancient melting events — perhaps extraction of continental crust — would those events have produced significant lithological variability and could that variability have persisted through several aeons of mantle convection? Langmuir (1980), for example, argued that extraction of average continental crust (roughly a granodiorite in composition) would not greatly have changed the bulk composition of mantle sources to abyssal tholeiites.

However, it may not be appropriate to consider that ancient continental crust containing komatiites, major layered intrusions, andesites and granites should be treated as an average composition in terms of its potential influence on mantle sources to ridge basalts. Moreover, fluid dynamics studies of the effects of turbulent versus laminar flow indicate that even 2 billion years of mantle convection will not completely dissipate zones of heterogeneity as small originally as 10 km in diameter (Gurnis 1986). Subcontinental mantle itself is depicted as being less dense (more refractory) than that beneath the ocean basins (Jordan 1979), apart from exhibiting lithological heterogeneity in xenolith suites. To the extent that micro-continental fragments have been left in the middle of some ocean basins (*eg* Seychelles, NW Indian Ocean), we may even surmise that associated subcontinental mantle can be caught up in magmatism at some spreading ridges. On the grossest scale, then, convection has barely rehomogenized the mantle at all, and fairly recent tectonic events may actually have added lithological heterogeneity of ancient vintage to mantle sources of abyssal tholeiites.

That melts with very low abundances of TiO_2 and Na_2O play a role in the petrogenesis of abyssal tholeiites is also not a new discovery. The nature of that role is still uncertain. One opinion, for example, is that such melts arise as a result of two stages of melting in the mantle (Duncan & Green 1980, 1987), *ie* an initial stage of melting creates refractory mantle, and then a second stage pulls refractory melts from it. Dynamic partial melting as described by Langmuir *et al.* (1977) is similar to this except that it is a more elaborate multistage process and requires that variable proportions of melt be left behind in residual mantle to become incorporated in later melting events. Moreover, zone refining is required to cause variable enhancements of rare earths and other trace elements.

A third conception is that of Elthon & Casey (1985) who propose that refractory melts, in appropriate circumstances, can be parental to more typical abyssal tholeiites. They suggest that TiO_2 and Na_2O enrichments in the latter are caused by open-system processes in repeatedly replenished axial magma chambers (*eg* O'Hara & Mathews 1981) and that the appropriate refractory parents are picritic in composition.

Partial melting of a lithologically heterogeneous mantle followed by mixing will produce virtually the same geochemical effects as does dynamic partial melting. In most respects, in fact, the concepts are identical, except that the isotopic data for the eastern Pacific suite considered here suggest that the melting and remelting events are separated by long periods of time. This does not preclude dynamic partial melting from occurring today either in the eastern Pacific or elsewhere. However, dynamic melting involves divergence of magma compositions from a common source whereas the process described here is that of convergence of compositions from a diverse source. Dynamic partial melting is also at present only a model for explaining geochemical differences. Crystallization histories provide additional conditions for any such model to satisfy, namely the sequences of assembly of primary magma strains and attributes of their compositions (such as values of Mg#) at the time of coalescence. The process, even though it is complex, is orderly and consistent from sample to sample, even considering lavas from the FAMOUS area in the N Atlantic (*cf* Sigurdsson & Schilling 1976; Donaldson & Brown 1977) for which the hypothesis of dynamic melting was proposed. How dynamic partial melting by itself could achieve such order and consistency is not obvious.

The case for incremental melting and against extended batch melting, however, has lately been given considerable credence from a surprising direction, namely indications of $^{230}Th-^{238}U$ isotopic disequilibrium in abyssal tholeiites (*eg* Condomines *et al.* 1981; Newman *et al.* 1983). This, McKenzie (1985) has argued, could not exist if the mantle involved in melting beneath spreading ridges ever retained more than a very few per cent of the melt it was producing at any given time. Melting will continue, so that in the end perhaps 20% of the original bulk of the partially melted mantle is extracted to produce the basaltic and gabbroic carapace of the ocean crust (McKenzie 1984). But extraction of melts in small portions rather than pure batch melting must now be considered a firm deduction supporting quantitative estimates of the rheological properties of the mantle during melting.

To some extent, however, arrival of magmas at the crust in bulk, to cause inflation events which induce eruptions, mitigates this conclusion. Retention of no more than 2%−3% of melt throughout the melting process would only be possible if incremental melts escaped with perfect efficiency to the seafloor as they were produced. Such small amounts are contradicted by estimates of extents of retained melts in both alpine peridotites (*eg* Dick 1977) and abyssal ultramafic rocks (Dick *et al.* 1984). Also, however, studies of alpine peridotites show that melts clearly coalesce into substantial networks of magma before they fracture overlying rock and abruptly evacuate the mantle (Nicolas 1986). More properly, then, continued melting amounts to reaction of mantle rock in sustained contact with coalescing streams or pools of magma. Although equilibrium will not exist in such a situation, the phases in the mantle rock undergoing melting at such times will be in contact with a substantially greater proportion of melt than the 2%−3% estimated by McKenzie (1984, 1985) from his equations. Those equations apply most specifically to the earliest stages of melting, and to peripheral portions of melt domains which are near the solidus of the mantle assemblage present (Fig. 10). That a signal of Th−U isotopic disequilibrium survives in abyssal tholeiites is because mixing between melts produced in different portions of melting domains is unavoidable. It exists because incremental melts, produced in peripheral portions of melting domains, are entrained by streams of magma ascending from the interiors of such domains (Fig. 10; *cf* Spiegelman & McKenzie (1987)) in a particular and consistent sequence, as revealed by crystallization histories. Whether magma chamber mixing with alkalic constituents influences Th−U isotopic disequilibrium among abyssal tholeiites is not yet established.

One important aspect of the consistency of crystallization histories is the fairly low (non-picritic) Mg# of coalesced magma strains that can be deduced from any aspect of the rocks. Also, the refractory glass inclusions in at least one sample were coprecipitating plagioclase, olivine and spinel. Thus low TiO_2 and Na_2O contents do not require picritic (or 'komatiitic') parentage, contrary to the suggestions of both Elthon & Casey (1985) and Fisk (1985). Indeed, magmas that precipitate extremely calcic plagioclase are, by definition, non-picritic, as the glass inclusions reported in Table 5 prove. Such minerals and their glass inclusions constitute most of the evidence for refractory magmas in the ocean crust.

Do average, typically sodic and titanian, abyssal tholeiites have refractory parents, as suggested by Elthon & Casey (1985)? The crystallization histories cited here say yes, in part, but only as components. Evidently quite a range of more and less sodic magma strains can coalesce at high temperatures and over a range of pressures to produce an individual 'parental' composition; crustal magma chambers were not involved in polybaric coalescence. More probably, magmas injected into crustal magma chambers at the East Pacific Rise were already as titanian and sodic as the Siqueiros tholeiites. The situation at the Bay of Islands ophiolite discussed by Elthon & Casey (1985) appears to have been different.

Conclusions

Phase relations in a suite of magnesian tholeiites from the Siqueiros Fracture Zone show that parental abyssal tholeiites can be produced by coalescence of melt batches derived from a lithologically heterogeneous mantle. Source peridotites varied from refractory (perhaps harzburgitic) to fertile (lherzolitic) compositions and provided melt increments under very nearly identical physical conditions. Refractory increments coalesced first, and variously precipitated chromian spinels, olivines and plagioclases over a range of pressures on their ascent from melt regions. These magma batches were later counterbalanced by more sodic and titanian magmas which mixed with the refractory magma stems to produce the host lavas.

Experimental data show that both less fertile and more fertile peridotites can partially melt simultaneously through much the same ranges in $P-T$ conditions. Thus quite diverse melt strains can be produced in the same region of the mantle, provided that it is lithologically heterogeneous to begin with. Mixing of such melt strains can explain the full range in CaO/Al_2O_3, Na_2O and TiO_2 contents in parental (high-Mg#) abyssal tholeiites and their glass inclusions, without requiring extraordinary degrees of melting, excessive lateral or vertical temperature gradients or extremely magnesian precursors. Fundamental lithological heterogeneity of the mantle may also exist on a regional basis, producing important contrasts in basalt compositions from place to place.

The crystallization sequences indicate that the highest-temperature magma stems came from refractory sources, which thus are more abundant in the interiors of melting domains than at their peripheries. Ascent of such melts through overlying more fertile regions of the mantle produces the mineralogical relations and bulk lava compositions described. Mantle diapirs beneath spreading ridges thus may be triggered, or their rate of ascent at least accelerated, by the buoyancy of low-density (low-iron) peridotite.

The magnesian Siqueiros tholeiites are among the most depleted basalts ever obtained from the eastern Pacific, yet they erupted in close proximity to alkalic basalt. More fractionated tholeiites from the East Pacific Rise axis are somewhat less depleted than the Siqueiros tholeiites. This may be a consequence of persistent mixing of depleted and enriched magmas in axial magma chambers, whereas the different lava types are free to erupt separately at ridge-axis discontinuities such as fracture zones.

Sources to magnesian basalts in this region are thus strongly partitioned into a range of refractory to fertile yet extremely depleted peridotites, which constitute the bulk of the mantle, and a smaller increment (perhaps 5%–7%) of enriched peridotite. Isotopic data indicate that all these heterogeneities were established during one or more prior melting events which initiated distinctive isotopic reservoirs a long time ago.

ACKNOWLEDGMENTS: Readers who skip ahead to this point will appreciate my indebtedness to the writings and thought of J. M. Rhodes and M. Dungan concerning mixing, of Henry Dick concerning abyssal peridotites and spinels, of Martin Fisk on plagioclases and spinels, and of Mike O'Hara concerning melting domains in the mantle. Roy Fujita at Scripps Institution and Sandra Bromble at Lamont–Doherty Geological Observatory assisted greatly with the microprobe studies. Critical portions of the data were obtained by John Melchior and J. D. Macdougall. The impetus for the paper was provided by discussions with Charlie Langmuir and Emily Klein and most of it was written while I was enjoying the hospitality of the Petrology Laboratory during a summer at Lamont, where the plasma analyses were also accomplished. I appreciate more than I can express the critical comments of Bill Bryan, Charlie Langmuir, David Walker, Godrey Fitton, Andy Saunders and an anonymous reviewer on various long and complex drafts of the manuscript. I especially thank Andy Saunders for making possible my participation in this volume. The work was supported by US National Science Foundation Grants OCE83–08696 and OCE85–10526.

References

BATIZA, R. & JOHNSON, J. 1980. Trace element and isotopic evidence for magma mixing in transitional and alkalic basalts from the East Pacific Rise at 8°N. *In*: ROSENDAHL, B., HEKINIAN, R. *et al.* (eds) *Initial Reports of the Deep Sea Drilling Project* **54**, 63–69.

——, ROSENDAHL, B. & FISHER, R. L. 1977. Evolution of oceanic crust, 3, Petrology and chemistry of basalts from the East Pacific Rise and Siqueiros transform fault. *Journal of Geophysical Research* **82**, 265–276.

BOUGAULT, H., VILLEMAN, B., JORON, J.-L. & TREUIL, M. 1983. Titanium: hygromagmaphile and/or non-hygromagmaphile element? *Eos (Transactions of the American Geophysical Union)* **64**, 33.

BOYD, F. R. & MCALLISTER, R. H. 1976. Densities of fertile and sterile garnet peridotites. *Geophysical Research Letters* **3**, 509–512.

BRYAN, W. B. 1979. Regional variation and petrogenesis of basalt glasses from the FAMOUS area, Mid-Atlantic Ridge. *Journal of Petrology* **20**, 293–325.

—— & DICK, H. J. B. 1982. Contrasted abyssal basalt liquidus trends: evidence for mantle major element heterogeneity. *Earth and Planetary Science Letters* **58**, 15–26.

CARLSON, R. W., MACDOUGALL, J. D. & LUGMUIR, G. W. 1978. Differential Sm/Nd evolution in oceanic basalts. *Geophysical Research Letters* **5**, 229–232.

CONDOMINES, M., MORAND, P. & ALLEGRE, C. J. 1981. $^{230}Th-^{238}U$ radioactive disequilibrium in tholeiites from the FAMOUS zone (Mid-Atlantic Ridge, 36°50′N): Th and Sr isotopic geochemistry. *Earth and Planetary Science Letters* **55**, 247–256.

DICK, H. J. B. 1977. Partial melting in the Josephine Peridotite, I. The effect on mineral composition and its consequence for geobarometry and geothermometry. *American Journal of Science* **277**, 801–832.

—— & BULLEN, T. 1984. Chromian spinel as a petrogenetic indicator in abyssal and alpine-type peridotites and spatially associated lavas. *Contributions to Mineralogy and Petrology* **86**, 54–76.

——, —— & BRYAN, W. B. 1984. Mineralogic variability of the uppermost mantle along mid-ocean ridges. *Earth and Planetary Science Letters* **69**, 88–106.

DMITRIEV, L., SOBOLEV, A. V. & SUSCHEVSKAYA, N. M. 1979. The primary melt of the oceanic tholeiite and the upper mantle composition. *In*: TALWANI, M., HARRISON, C. G. A. & HAYES, D. E. (eds) *Deep Drilling Results in the Atlantic Ocean: Ocean Crust*. Maurice Ewing Series **2**. American Geophysical Union, Washington, 302–313.

DONALDSON, C. H. & BROWN, R. W. 1977. Refractory megacrysts and magnesium-rich melt inclusions within spinel in oceanic tholeiites: indicators of magma mixing and parental magma compositions. *Earth and Planetary Science Letters* **37**, 81–89.

DRAKE, M. J. 1972. *The distribution of Major and Trace Elements between Plagioclase Feldspar and Magmatic Silicate Liquid: an Experimental Study*. PhD thesis, University of Oregon, 190 pp.

—— 1976. Plagioclase-melt equilibria. *Geochimica et Cosmochimica Acta* **40**, 457–465.

DUNCAN, R. A. & GREEN, D. H. 1980. The role of multi-stage melting in the formation of oceanic crust. *Geology* **8**, 22–26.

—— & —— 1987. The genesis of refractory melts in the formation of oceanic crust. *Contributions to Mineralogy and Petrology* **96**, 326–342.

ELDER, J. 1966. Penetrative convection: its role in volcanism. *Bulletin Vocanologique* **29**, 327–343.

—— 1972. *The Bowels of the Earth*. Oxford University Press, Oxford.

ELTHON, D. & CASEY, J. 1985. The very depleted nature of certain primary mid-ocean ridge basalts. *Geochimica et Cosmochimica Acta* **49**, 289–298.

FISK, M. 1985. Depths and temperatures of mid-ocean-ridge magma chambers and the composition of their source. *In*: GASS, I. G., LIPPARD, S. J. & SHELTON, A. W. (eds) *Ophiolites and Oceanic Lithosphere*. Geological Society of London Special Publication **13**, 17–23.

—— & BENCE, A. E. 1980. Experimental crystallization of chrome spinel in FAMOUS basalt 527–1–1. *Earth and Planetary Science Letters* **48**, 111–123.

FRANCHETEAU, J. & BALLARD, R. D. 1983. The East Pacific Rise near 21°N and 20°S: inferences for along-strike variability of axial processes of the mid-ocean ridge. *Earth and Planetary Science Letters* **64**, 93–116.

FREY, F. A., DICKEY, J. S., THOMPSON, G. & BRYAN, W. B. 1977. Eastern Indian Ocean DSDP sites: correlations between petrography, geochemistry, and tectonic setting. *In*: HEIRTZLER, J. R. *et al.* (eds) *Indian Ocean Geology and Biostratigraphy*. American Geophysical Union Monograph. American Geophysical Union, Washington, 189–257.

——, ——, ——, —— & DAVIES, H. L. 1980. Evidence for heterogeneous primary MORB and mantle sources, NW Indian Ocean, *Contributions to Mineralogy and Petrology* **74**, 387–402.

FUJII, T. & SCARFE, C. M. 1985. Compositions of liquids coexisting with spinel lherzolite at 10 kbar and the genesis of MORBs. *Contributions to Mineralogy and Petrology* **90**, 18–28.

GREEN, D. H., RINGWOOD, A. E., WARE, N. G. & HIBBERSON, W. O. 1972. Experimental petrology and petrogenesis of Apollo 14 basalts. *Proceedings of the 3rd Lunar Science Conference, Geochimica et Cosmochim Acta*, Suppl. 2, MIT Press. **1**, 197–206.

GURNIS, M. 1986. How long can a passive chemical heterogeneity survive in a convecting, unlayered mantle? *Eos (Transactions of the American Geophysical Union)* **67**, 1257.

HAWKINS, J. W. & MELCHIOR, J. 1985. Petrology of Mariana Trough and Lau Basin basalts. *Journal of Geophysical Research* **90**, 11431–11468.

HEKINIAN, R. & WALKER, D. 1987. Diversity and spatial zonation of volcanic rocks from the East Pacific Rise near 21°N. *Contributions to Mineralogy and Petrology* **96**, 265–280.

HOFMANN, A. W. & WHITE, W. M. 1983. Ba, Rb, and Cs in the earth's mantle, *Zeitschrift für Naturforschung* **38a**, 256–266.

IRVINE, T. N. 1965: Chromian spinel as a petrogenetic indicator, Part I, Theory. *Canadian Journal of Earth Sciences* **2**, 648–671.

JACKSON, E. D. 1968. The character of the lower crust and upper mantle beneath the Hawaiian Islands. *Report of the 23rd International Geological Congress, Prague, 1968* **1**, Academia, Prague, 135–150.

JAQUES, A. L. & GREEN, D. H. 1980. Anhydrous melting of peridotite at 0–15 kb pressure and the genesis of tholeiitic basalts. *Contributions to*

Mineralogy and Petrology **73**, 287–310.

JOHNSON, J. 1979. Transitional basalts and tholeiites from the East Pacific Rise, 9°N. *Journal of Geophysical Research* **84**, 1635–1652.

JORDAN, T. H. 1979. Mineralogies, densities, and seismic velocities of garnet lherzolites and their geophysical implications. *In*: BOYD, F. R. & MEYER, H. O. A. (eds) *The Mantle Sample: Inclusions in Kimberlites and other Volcanics, Proceedings of the 2nd International Kimberlite Conference*, Vol. 2. American Geophysical Union, Washington, 1–14.

KEMPTON, P. D. AUTIO, L. K., RHODES, J. M., HOLDAWAY, M. J., DUNGAN, M. A. & JOHNSON, P. 1985. Petrology of basalts from Hole 504B, Deep Sea Drilling Project Leg 83. *In*: ANDERSON, R. N., HONNOREZ, J., BECKER, K. *et al.* (eds) *Initial Reports of the Deep Sea Drilling Project* **83**, 129–164.

KLEIN, E. M. & LANGMUIR, C. H. 1987. Ocean ridge basalt chemistry, axial depth, crustal thickness, and temperature variations in the mantle. *Journal of Geophysical Research* **92**, 8089–8115.

——, —— BENDER, J., BATIZA, R. & ZINDLER, A. 1986. Nd and Sr isotopic composition of volcanic glasses from the East Pacific Rise (8°N–12°N). *Eos (Transactions of the American Geophysical Union)* **67**, 1255.

LANGMUIR, C. H. 1980. The undepleted major-element character of the MORB source: evidence and implications. *In*: PRESNALL, D. C., HALES, A. L. & FREY, F. A. (Convenors) *The Generation of the Oceanic Lithosphere*. Chapman Conference. American Geophysical Union, Washington, Abstracts with Program.

—— & HANSON, G. N. 1981. Calculating mineral-melt equilibria with stoichiometry, mass balance, and single-component distribution coefficients. *In*: NEWTON, R. C., NAVROTSKY, A. & WOOD, B. J. (eds) *Thermodynamics of Minerals and Melts*. Advances in Physical Chemistry 1. Springer, New York, 247–271.

——, BENDER, J. F. & BATIZA, R. 1986. Petrologic and tectonic segmentation of the East Pacific Rise, 5°30′–14°30′N. *Nature* **322**, 422–429.

——, ——, BENCE, A. E., HANSON, G. N. & TAYLOR, S. R. 1977. Petrogenesis of basalts from the FAMOUS area: Mid-Atlantic Ridge. *Earth and Planetary Science Letters* **36**, 133–156.

LONSDALE, P. 1985. Linear volcanoes along the Pacific–Cocos plate boundary, 9°N to the Galapagos triple junction. *Tectonophysics* **116**, 255–279.

MACDONALD, K., SEMPERE, J.-C. & FOX, P. J. 1984. East Pacific Rise from Siqueiros to Orozco Fracture Zones: along-strike continuity of axial neovolcanic zone and structure and evolution of overlapping spreading centers. *Journal of Geophysical Research* **89**, 6049–6069.

MACDOUGALL, J. D. & LUGMAIR, G. W. 1985. Extreme isotopic homogeneity among basalts from the southern East Pacific Rise: mantle or mixing effect? *Nature* **313**, 209–211.

—— & —— 1986. Sr and Nd isotopes in basalts from the East Pacific Rise: significance for mantle heterogeneity. *Earth and Planetary Science Letters* **77**, 273–284.

MAHONEY, J. & NATLAND, J. H. 1985. Domains of isotopic and basalt glass compositions at Indian Ocean Ridges. *Eos (Transactions of the American Geophysical Union)* **66**, 1108.

MCKENZIE, D. 1984. The generation and compaction of partially molten rock. *Journal of Petrology* **25**, 713–765.

——, 1985. ^{230}Th–^{238}U disequilibrium and the melting processes beneath ridge axes. *Earth and Planetary Science Letters* **72**, 149–159.

MELSON, W. G., VALLIER, T. L., WRIGHT, T. L., BYERLY, G. R. & NELEN, J. A. 1976. Chemical diversity of abyssal volcanic glass erupted along Pacific, Atlantic, and Indian Ocean sea-floor spreading centers. *Geophysics Monograph* **19**. American Geophysical Union, Washington, 351–368.

NATLAND, J. H. 1979. Crystal morphologies in basalts from DSDP Site 395, 23°N, 46°W, Mid-Atlantic Ridge. *In*: MELSON, W. G., RABINOWITZ, P. D. *et al.* (eds) *Initial Reports of the Deep Sea Drilling Project* **45**, 423–445.

——, 1980*a*. Effect of axial magma chambers beneath spreading centers on the compositions of basaltic rocks. *In*: ROSENDAHL, B. R., HEKINIAN, R. *et al.* (eds) *Initial Reports of the Deep Sea Drilling Project* **54**, 833–850.

—— 1980*b*. Crystal morphologies in basalts dredged and drilled from the East Pacific Rise near 9°N and the Siqueiros Fracture Zone. *Initial Reports of the Deep Sea Drilling Project* **54**, 605–633.

——, ADAMSON, A. C., LAVERNE, C., MELSON, W. G. & O'HEARN, T. 1983. A compositionally nearly steady-state magma chamber at the Costa Rica Rift: evidence from basalt glass and mineral data, Deep Sea Drilling Project Sites 501, 504, and 505. *In*: CANN, J. R., LANGSETH, M., HONNOREZ, J., VON HERZEN, R. P., WHITE, S. M. *et al.* (eds) *Initial Reports of the Deep Sea Drilling Project* **69**, 811–859.

——, LANGMUIR, C. H., BENDER, J. F., BATIZA, R. & HOPSON, C. 1986. Petrologic systematics in the vicinity of the 9° N non-transform offset, East Pacific Rise. *Eos (Transactions of the American Geophysical Union)* **67**, 1254.

—— & MACDOUGALL, J. D. 1986. Parental abyssal tholeiites and alkali basalts at the East Pacific Rise near 9° N and the Siqueiros Fracture Zone. *Eos (Transactions of the American Geophysical Union)* **67**, 410–411.

—— & MELSON, W. G. 1980. Compositions of basaltic glasses from the East Pacific Rise and Siqueiros Fracture Zone, near 9°N. *In*: ROSENDAHL, B. R., HEKINIAN, R. *et al.* (eds) *Initial Reports of the Deep Sea Drilling Project* **54**, 705–723.

NEWMAN, S., FINKEL, R. C. & MACDOUGALL, J. D. 1983. ^{230}Th–^{230}U disequilibrium systematics in oceanic tholeiites from 21°N on the East Pacific Rise. *Earth and Planetary Science Letters* **65**, 17–33.

NICOLAS, A. 1986. A melt extraction model based on

structural studies in mantle peridotites. *Journal of Petrology* **27**, 999–1022.
O'HARA, M. J. 1975. Is there an Icelandic mantle plume? *Nature* **253**, 708–710.
—— 1977. The geochemistry of lavas erupted from a magma chamber undergoing fractional crystallization with periodic addition and intermingling of more primitive magma. *Nature* **266**, 503–507.
—— 1984. Importance of the 'shape' of the melting regime during partial melting of the mantle. *Nature* **314**, 58–62.
—— & MATHEWS, R. E. 1981. Geochemical evolution in an advancing, periodically replenished, periodically tapped, continuously fractionating magma chamber. *Journal of the Geological Society, London* **138**, 237–277.
OXBURGH, E. R. 1980. Heat flow and magma genesis. *In*: HARGRAVES, R. B. (ed.) *Physics of Magmatic Processes*. Princeton University Press, Princeton, 161–199.
PRESNALL, D. C., DIXON, J. R. O'DONNELL, T. H. & DIXON, S. A. 1979. Generation of mid-ocean ridge basalts. *Journal of Petrology* **20**, 3–35.
PRICE, R. C., KENNEDY, A. K., RIGGS-SNEERINGER, M. & FREY, F. A. 1986. Geochemistry of basalts from the Indian Ocean triple junction: implications for the generation and evolution of Indian Ocean ridge basalts. *Earth and Planetary Science Letters* **78**, 379–396.
ROEDER, P. L. & EMSLIE, R. F. 1970. Olivine–liquid equilibrium. *Contributions to Mineralogy and Petrology* **29**, 275–289.
ROSENDAHL, B. R. & DORMAN, L. M. 1980. Summary of the geology and geophysics of the East Pacific Rise in the vicinity of the Siqueiros Fracture Zone. *In*: ROSENDAHL, B. R., HEKINIAN, R. *et al.* (eds) *Initial Reports of the Deep Sea Drilling Project* **54**, 23–36.
SCHOUTEN, H., DICK, H. J. B. & KLITGORD, K. D. 1987. Migration of mid-ocean ridge volcanism. *Nature* **326**, 835–839.
SCHRADER, E., ROSENDAHL, B. & FURBISH, W. 1980. Picritic basalts from the Siqueiros Transform Fault. *In*: ROSENDAHL, B. R. HEKINIAN, R. *et al.* (eds) *Initial Reports of the Deep Sea Drilling Project* **54**, 71–78.
SIGURDSSON, H. & SCHILLING, J.-G. 1976. Spinels in Mid-Atlantic Ridge basalts: chemistry and occurrence. *Earth and Planetary Science Letters* **29**, 7–20.
SPIEGELMAN, M. & MCKENZIE, D. 1987. Simple 2-D models for melt extraction at mid-ocean ridges and island arcs. *Earth and Planetary Science Letters* **83**, 137–152.
STOLPER, E. 1980. A phase diagram for mid-ocean ridge basalts: preliminary results and implications for petrogenesis. *Contributions to Mineralogy and Petrology* **74**, 13–27.
TAKAHASHI, E. & KUSHIRO, I. 1983. Melting of a dry peridotite at high pressures and basalt magma genesis. *American Mineralogist* **68**, 859–879.
THOMPSON, G., BRYAN, W. B., BALLARD, R., HAMURO, K. & MELSON, W. G. 1986. Axial processes along a segment of the East Pacific Rise, 10°–12°N. *Nature* **318**, 429–433.
THOMPSON, R. N. 1987. Phase-equilibria constraints on the genesis and magmatic evolution of oceanic basalts. *Earth-Science Reviews* **24**, 161–210.
——, MORRISON, M. A., HENDRY, G. L. & PARRY, S. J. 1984. An assessment of the relative roles of crust and mantle in magma genesis: an elemental approach. *Philosophical Transactions of the Royal Society of London, Series A* **310**, 549–590.
WALKER, D., SHIBATA, T. & DELONG, S. E. 1979. Abyssal tholeiites from the Oceanographer Fracture Zone II. Phase equilibria and mixing. *Contributions to Mineralogy and Petrology* **70**, 111–125.
WHITEHEAD, J. A. JR, DICK, H. J. B. & SCHOUTEN, H. 1984. A mechanism for magmatic accretion under spreading centers. *Nature* **312**, 146–148.
WILKINSON, J. F. G. 1982. The genesis of mid-ocean ridge basalts. *Earth-Science Reviews* **18**, 1–57.
WRIGHT, T. L. 1984. Origin of Hawaiian tholeiite. *Journal of Geophysical Research* **89**, 3233–3252.
ZINDLER, A., STAUDIGEL, H. & BATIZA, R. 1984. Isotope and trace element geochemistry of young Pacific seamounts: implications for the scale of upper mantle heterogeneity. *Earth and Planetary Science Letters* **70**, 175–195.

J. H. NATLAND, Geological Research Division, Scripps Institution of Oceanography A-015, La Jolla, CA 92093, USA.

Abyssal peridotites, very slow spreading ridges and ocean ridge magmatism

H. J. B. Dick

SUMMARY: The SW Indian and American–Antarctic Ridges are two of the world's slowest spreading ocean ridges (less than 1 cm a^{-1}), making them the low end-members for rate of ocean ridge magma supply. Two-thirds of the rocks dredged at the numerous large offset transforms along the ridges are residual mantle peridotites. Gabbroic rocks, however, representing layer 3 and possible palaeo-magma chambers are rare. This suggests a highly segmented crustal structure, with anomalously thin crust near fracture zones that may consist of only a thin veneer of pillow basalt erupted over mantle peridotite.

The dredged peridotites underwent high degrees of melting, spanning the range believed to produce abyssal basalt. Their depleted compositions show that the melt was almost entirely removed. At the same time, the spatially associated basalts have a large range of compositions, similar to those from the rift valleys, requiring extensive shallow-level fractional crystallization. Since there is little evidence for magma chambers at these fracture zones, it is concluded that melts formed in the underlying mantle flowed laterally through the mantle beneath the crust towards a magmatic centre at the mid-point of an adjacent ridge segment. Magma was then subsequently intruded down the rift valley fissure system from the magmatic centre to erupt onto the fracture zone floor. Alternatively, the melt was drained from a mantle diapir beneath the midpoint of a ridge segment, prior to lateral flow of the residual peridotite beneath the ridge axis to the fracture zone. These processes suggest behaviour of the partially molten layer beneath ocean ridges analogous to Rayleigh–Taylor fluid instability, where a light less viscous fluid layer floating upwards in a denser medium goes unstable and drains at regularly spaced points into protrusions which rise rapidly to the surface. Evidence for such dynamically driven non-uniform melt flow in the mantle is seen in locally-abundant plagioclase peridotites, where the plagioclase crystallized from impregnated trapped melt. These rocks can contain up to 30% trapped melt, contrasting sharply with the typical abyssal peridotite which contains virtually none.

Basalts erupted along these ridges provide a classic case of trace- and major-element decoupling during magma genesis. Despite trace-element and isotopic diversity, basalts from individual ridge segments were derived from primary magmas with similar major-element compositions. These observations can be explained if melt flows locally through the depleted mantle at the end of melting towards the midpoint of a ridge segment. This would cause melts originating at different points in an initially heterogeneous mantle to migrate through and equilibrate with the same section of mantle immediately prior to segregation—which, for the most part, would homogenize the melt's major-element compositions. However, by virtue of the lever rule, this would have little effect on critical incompatible-trace-element or isotopic ratios of the migrating melts because of the very low incompatible-trace-element content of residual peridotite.

Ocean ridges, then, appear to be marked by strings of regularly spaced volcanic centres overlying instability points in the partially molten upwelling asthenosphere much as has been postulated for arc volcanism and early continental rifting. Unlike arcs, the asthenosphere upwells to the base of the crust and the magmatic centres undergo continuous extension. Thus, large volcanoes are not constructed, and instead, ribbons of basaltic crust form parallel to the spreading direction. This is most evident at the SW Indian and American–Antarctic Ridges because of their highly attenuated magma supply. Where the magma supply is more robust and the magma chambers are correspondingly larger, the chambers may merge and eliminate the surficial morphological and chemical expression of punctuated magmatism at ocean ridges.

The discovery of altered peridotites at ocean ridges has played a major role in models for ridge magmatism and crustal formation since it was first reported by Shand (1949). Hess (1962), for example, suggested in his seminal paper 'The history of the ocean basins' that mantle material came to the surface at the mid-ocean ridges and thus the oceanic crust was composed largely of hydrated mantle peridotite. Today, with more observation, serpentinized peridotite

From SAUNDERS, A. D. & NORRY, M. J. (eds), 1989, *Magmatism in the Ocean Basins*, Geological Society Special Publication No. 42, pp. 71–105.

is assigned a more modest role in the constitution of the ocean crust, but, as the inferred residue of mantle melting that generated the crust, they remain a key to understanding its evolution.

This paper presents field and laboratory data for abyssal peridotites, particularly for the American–Antarctic and SW Indian Ridges. These ridges represent the low end of the spreading-rate spectrum, and therefore the lowest rate of magma supply of any major ocean ridge system. The results presented here show that this has produced a very heterogeneous crustal structure. Specifically, the abundance of peridotite and scarcity of gabbro dredged at very slow slipping fracture zones show that the crust thins dramatically there, with the virtual disappearance of a gabbroic layer 3. This implies that ocean crust formation is not a uniform process creating a simple layer-cake of basalt, diabase and gabbro overlying the mantle. Rather, ocean ridge magmatism is punctuated as at an island arc or during early continental rifting (*eg* Mohr & Wood 1974; Marsh 1979). The volcanism is concentrated at magmatic centres, generally spaced between fracture zones at slower spreading ridges, as a consequence of non-uniform flow of melt out of the mantle.

Moreover, an examination of the compositions of abyssal peridotites has major implications for the formation of mid-ocean ridge magmas and suggests that the major- and trace-element cycles are fundamentally decoupled. The major-element composition of primary basalt responds largely to the overall composition of the melting column of mantle beneath the ridge and the depth to which it extends. The trace-element composition of primary basalt is more a function of short-wavelength variations in the local composition of the mantle entering the base of the melting column. Thus, primary abyssal basalt compositions show relatively short-wavelength variations in isotopic ratios and trace elements in time and space in comparison with major elements. Overall, trace-element signatures are more a function of parent mantle composition, while major-element signatures are more a function of mantle dynamics.

Abyssal peridotites

We are concerned here with variably altered and serpentinized peridotites dredged from the world ocean ridge system. In view of their ambiguous provenance, alpine-type peridotites and ultramafic rocks dredged from the landward walls of deep-sea trenches are not included in this class of ultramatic rocks. Abyssal peridotites are widely accepted as residues of pressure-release melting accompanying upward convection of the mantle beneath mid-ocean ridges. Emplaced near the surface or base of the crust, they represent the top of the mantle section drawn up between the spreading plates to form the lithosphere. As such they have ascended the furthest of any rocks in the mantle column, have undergone the most decompression, and must represent the most extreme residues and the end-point of the entire melting process.

Occurrence

Fracture zones

Abyssal peridotites have been most frequently sampled from oceanic fracture zones, where their emplacement is one of the great enigmas of marine geology. In theory, transform faults are simple strike-slip translational plate boundaries which offset ridge axes laterally in the direction of plate motion. In nature, however, fracture zones, which mark the present or palaeo-location of transform faults, are complex features often marked by deep valleys. Fracture zone physiography varies drastically with spreading rate, age offset, and the state of stress across the transform due to changes in spreading direction—which may put the fault into transient compression or extension along its length (Menard & Atwater 1968; van Andel *et al.* 1971).

Major fracture zones at slow and intermediate spreading ridges are marked by valleys 1–6 km deep and from 15 to 40 km wide measured from the tops of the valley walls. These valleys consist of two principal domains: the active transform fault, and its inactive extensions. In the extensions the opposing walls of the fracture zone valley have fundamentally different origins, one wall having and one wall not having passed through the transform and its related zone of tectonism. Figure 1 illustrates a hypothetical fracture zone with a large-offset slow slipping transform fault generalized from recent detailed surveys of the Kane (Pockalny *et al.*, in press) and Atlantis II (Dick *et al.* 1987) Fracture Zones. It provides a generalized model for many fracture zones at which peridotites occur.

Fracture zone valley walls are often dramatically steep (15°–40°), consisting of sloping benches broken by sets of imbricate high-angle normal faults often covered by large debris slides and talus ramps (*eg* Francheteau *et al.*

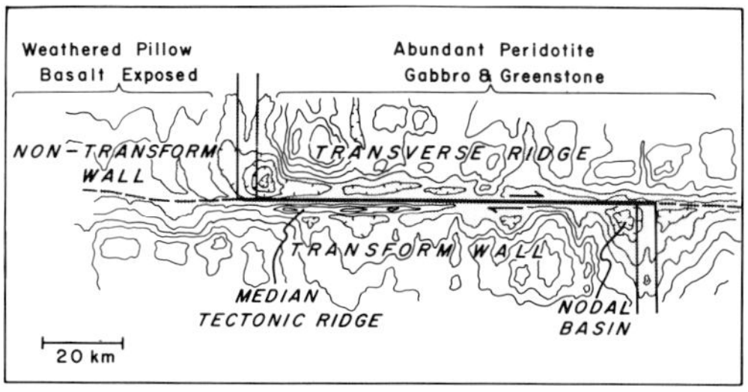

FIG. 1. Composite large offset slow slipping fracture zone and transform fault modelled after the Atlantis II and Kane Fracture Zones. Contour interval is 500 m. The inferred zone of present-day extension and volcanism along the ridge axis is shown by parallel lines, while the transform fault is shown by the heavy solid line and its inactive trace by the heavy broken line. Maximum depth is around 6500 m at the nodal deeps. These deeps are located on the transform side of the neovolcanic zones, immediately below the ridge–transform corners of the ridge–fracture zone intersections.

1976; Karson & Dick 1983; OTTER Scientific Team 1985). Individual fault scarps, however, are rarely seen exceeding 100–200 m. In many cases this faulting has produced a series of uplifted horsts, which are often connected by low saddles to produce steep transverse ridges parallel to the transform valley. The crests of these ridges may shoal to less than a thousand metres or even, as at St Paul's Rocks, to the sea surface. It is along the walls and crests of these ridges that abyssal peridotites are most frequently found. In other cases the walls of fracture zones may simply be formed by the ridge-parallel rift mountain valleys and ridges gently sloping downward to meet the opposing wall of the fracture zone. In either case, the fracture zone walls and ridges are generally cross-faulted by ridge-parallel structures such that they consist of alternating highs and lows with a blocky discontinuous aspect.

The floors of fracture zone valleys may be quite wide (5–15 km) and often have a median 'tectonic' ridge running parallel to the transform fault bisecting them. These ridges can constitute a small mountain range in themselves, standing as much as 2 km above the valley floor. Dredging along median tectonic ridges has recovered a variety of rock types, including serpentinite, and they are commonly thought to represent hydrated mantle diapirs uplifting faulted and extended crust along or near the transform fault (*eg* Thompson & Melson 1972; Macdonald *et al.* 1986).

Transforms are often bounded by transverse ridges or steep walls, while walls on the fracture zone extensions which do not pass through the transform frequently have gentler slopes, rarely shoaling more than a few thousand metres from the valley floor (Fig. 1). At the Atlantis II and Kane Fracture Zones newly formed crust spreading away from the transform at the ridge–transform intersection is only slightly uplifted to form a gentle sloping wall, while in the opposite direction the crust is uplifted many thousands of metres to form high transverse ridges. Dredging, photo-geological and submersible studies of these walls and intersections show that the steep transform walls expose largely metabasalt, diabase, gabbro and peridotite, representing the deeper levels of the ocean crust (Karson & Dick 1983; OTTER Scientific Team 1985; Dick *et al.* 1987). On the gentler non-transform walls of these two fracture zones only weathered pillow basalt has been found, representing the shallow layers of the ocean crust. Dick *et al.* (1981, 1987) have proposed a model to explain this asymmetric distribution of rocks as due to a crustal weld between new and old crust at the ridge–transform intersection. This weld dies out with distance from the transform and as the temperature increases and viscosity decreases with depth. Under the right conditions, the weld causes the shallow levels of the ocean crust near the fracture zone to spread preferentially away from the transform. The accompanying low-angle normal faulting, then, unroofs the ocean crust spreading towards the transform, which is uplifted to form the transform wall and transverse ridge, exposing plutonic basement along it (Dick *et al.* 1981).

Examination of Fig. 1 demonstrates the important point that rocks dredged from the walls of fracture zones may be situated well away from the actual transform fault plate boundary, as the horizontal distance from the crest of the wall may be 15–20 km from the actual transform fault. Thus, many of the rocks dredged from fracture zones originally formed beneath the median valley, well away from the transform plate boundary, and cannot be regarded as representing atypical 'transform' domain ocean crust.

In addition to peridotite, other rocks, including diabase, gabbro and their greenschist and amphibolite facies equivalents are commonly found at fracture zones. Weathered and metamorphosed pillow basalt, however, is by far the most common rock found. In some cases, dredges spaced systematically up fracture zone walls recover these rocks in apparent stratigraphic order, exposing 'cross-sections' of ocean crust (Engel & Fisher 1975; Bonatti &

Honnorez 1976). Francheteau *et al.* (1976), however, have correctly pointed out that the scale of individual faults on these walls is too small for such sections to represent simple cross-sections exposed on a single fault plane. Thus, these localities should be regarded as representing a tectonic process preferentially exposing deeper sections of the crust and mantle as the transform fault is approached (*eg* Dick *et al.* 1981). Elsewhere, dredges spaced up and down the transform walls, and for considerable distances along them, recover almost only serpentinized peridotite, suggesting that fracture zone walls and ridges can expose major mantle diapirs on the seafloor over regions of 1000 km^2 or more (Miyashiro *et al.* 1969; Thompson & Melson 1972; Bonatti 1976). Well-documented examples occur at the Romanche and Vema Fracture Zones on the Mid-Atlantic Ridge (Bonatti & Honnorez 1976), the Islas Orcadas Fracture Zone on the SW Indian Ridge (Sclater *et al.* 1978; Kimball *et al.* 1985) and the Owen Fracture Zone on the Carlsberg Ridge (Bonatti & Hamlyn 1978). Often, however, basalt, gabbro and peridotite are recovered in no apparent order, being jumbled together in the same dredge haul. This, the penetrative deformation and alteration found in many of the rocks, and their tectonic setting, suggests that large areas of fracture zones are tectonic melanges (*eg* Fox *et al.* 1976). Owing to the overall tectonic complexity of fracture zones, it is likely that in most cases pseudo-stratigraphic crustal cross-sections and mantle diapirs are localized discontinuous features of transform walls.

Rift mountains

Peridotite tectonites have been dredged or drilled at many localities on rift valley walls and in rift mountains well away from fracture zones (Aumento & Loubat 1971; CAYTROUGH 1979; Michael & Bonatti 1985; Karson *et al.* 1987). Michael & Bonatti (1985) made a systematic study of these peridotites, finding them to be very similar to fracture zone peridotites, possibly representing somewhat higher degrees of melting. Differences in composition compared with peridotites from adjacent fracture zones may be significant but are small in comparison with the overall variation found along individual ridge systems.

Where sampling was carefully coordinated with individual topographical features (*eg* Aumento & Loubat 1971) or outcrops followed by submersible (Karson *et al.* 1987), serpentinized peridotites are situated in ridge-parallel belts on faults zones. Hess (1955) discussed the similar occurrence of serpentinite bodies along faults in alpine orogenic belts and described them as 'tectonic watermelon seeds'. He suggested that deformed serpentinite because of its low density and cohesive strength, is highly mobile and tends to move up and along fault planes. This model is frequently used to explain the emplacement of serpentinite to shallow levels in the thin ocean crust, where water can percolate down major faults into the mantle (*eg* Aumento & Loubat 1971; Bonatti & Honnorez 1976; Casey 1986). A systematic traverse across the Mid-Atlantic Ridge at 45°N found that the relative proportions of the deeper crustal rocks, as well as serpentinized peridotite, increase with distance away from the rift valley floor out into the rift mountains (Aumento & Loubat 1971). This implies migration of serpentinite up faults with continued tectonism during rift mountain formation.

Statistical variations in abundance

There is a general awareness that the frequency with which peridotites (and other plutonic rocks) have been dredged from fracture zones along a given ridge is dependent on spreading rate. There are only a few reported occurrences of serpentinite on the fast spreading ridges at East Pacific Rise fracture zones (Eltanin Fracture Zone, Neprochnov, in Bonatti & Hamlyn (1980); Ecuador Fracture Zone, Anderson & Nishimori (1979); Garret Fracture Zone, Hebert *et al.* (1983)) and at one locality in its rift mountains at the Mathematician Seamounts (Vanko & Batiza 1983). This contrasts with the ubiquitous occurrence of peridotites in fracture zones and rift mountains at the slow spreading Mid-Atlantic and Central Indian ridges and their even greater abundance at very slow spreading ridges (Whitehead *et al.* 1984; Fisher *et al.* 1986; this paper).

To date, peridotites have been noted in the literature largely by their presence or absence in a dredge haul. Few data have been available on the proportions of different rock types, making precise determinations of relative abundances in different tectonic settings and the variation between ridges nearly impossible. Over the last 10 years, however, 11 635 kg of rock from 115 dredge hauls, including 60 from fracture zones, were described in detail during a systematic survey of the American–Antarctic and SW Indian ridges (Table 1). These statistics give a preliminary picture of the distribution of rocks there.

Overall, 42.3% (by weight) of the SW Indian and American–Antarctic ridge dredge hauls

TABLE 1. *SW Indian and American–Antarctic Ridge fracture zone dredge hauls*

Dredge No.	Start depth (m)	Location	Latitude	Longitude (E=+, W=−)	Peridotite (kg)	Perid Mylon (kg)	Sum peridotites	Dunite (kg)	Gabbro (kg)	Meta gabbro	Diabase (kg)	Meta diabase	Basalt	Meta basalt	Total (kg)
SW Indian Ridge fracture zone dredge hauls															
AII−107−35	584	W Bouvet FZ	54.723	0.803	12.20	0.00	12.10	9.00	0.00	4.00	8.00	0.00	14.90	0.00	48.00
AII−107−36	941	W Bouvet FZ	54.715	0.823	0.40	0.00	0.40	9.00	0.50	0.00	1.90	0.00	9.20	0.00	21.00
AII−107−37	1196	W Bouvet FZ	54.737	0.838	0.00	0.00	0.00	0.00	0.00	0.00	0.90	0.00	6.20	0.00	7.10
AII−107−39	5022	W Bouvet FZ	54.747	1.622	21.43	0.00	21.43	0.00	0.60	0.00	0.00	0.00	0.00	0.05	22.08
AII−107−40	3701	W Bouvet FZ	54.423	1.570	65.30	0.00	65.30	0.00	0.60	0.00	0.00	0.00	0.00	0.10	66.00
1011/76−56	4390	SE Islas Orcadas FZ	54.092	6.285	40.00	0.00	40.00	0.00	0.00	0.00	0.00	0.00	0.00	0.00	40.00
1011/76−58	3580	SE Islas Orcadas FZ	54.072	6.398	110.00	0.00	110.00	0.00	0.10	0.00	0.00	0.00	0.90	0.00	111.00
1011/76−59	2520	SE Islas Orcadas FZ	54.057	6.500	138.00	0.00	138.00	0.00	0.00	0.00	0.00	0.00	0.00	0.00	138.00
1011/76−60	2780	SE Islas Orcadas FZ	54.458	6.487	284.00	0.00	284.00	0.00	0.00	0.00	0.00	4.00	1.60	0.00	289.60
1011/76−62	2835	NW Islas Orcadas FZ	53.903	6.342	0.00	0.00	0.00	0.00	0.00	0.00	0.00	0.00	54.25	0.00	54.25
1011/76−63	3110	NW Islas Orcadas FZ	53.858	6.402	0.00	0.00	0.00	0.00	0.00	0.00	0.00	0.00	1.60	2.00	3.60
AII−107−60	5899	E Shaka FZ	53.428	9.152	4.22	48.68	52.90	0.15	0.00	0.00	4.50	0.00	4.10	0.00	61.65
AII−107−61	5073	E Shaka FZ	53.418	9.198	30.21	17.51	47.72	0.00	0.00	0.00	0.00	0.00	0.06	0.00	47.78
AII−107−63	4180	E Shaka FZ	53.388	9.338	7.12	1.97	9.09	0.00	0.00	0.00	0.00	0.00	0.30	3.89	13.28
AII−107−64	5024	W Shaka FZ	53.320	9.140	0.00	0.00	0.00	0.00	0.00	0.00	0.00	0.00	3.00	0.00	3.00
AII−107−65	4729	W Shaka FZ	53.238	9.193	0.00	0.00	0.00	0.00	0.00	0.00	0.00	0.00	11.70	0.30	12.00
AII−107−66	2619	W Shaka FZ	53.147	9.192	0.00	0.00	0.00	0.00	0.00	0.00	0.00	0.00	60.20	0.00	60.20
AII−107−67	5311	E Shaka FZ	53.240	9.460	0.30	0.00	0.30	2.50	2.00	0.00	0.04	0.76	0.95	22.05	28.60
PS86−4	2800	S FZ	52.153	14.189	132.70	0.00	132.70	0.00	0.00	9.60	0.00	0.00	1.50	2.50	146.30
PROT−5−10	1400	DuToit FZ	53.355	24.928	322.35	0.00	322.35	15.55	195.41	68.90	0.00	0.00	0.00	0.00	602.21
PROT−5−11	5250	DuToit FZ	53.160	25.488	0.00	0.00	0.00	0.00	0.00	0.00	2.80	1.70	4.10	4.60	13.20
PROT−5−12	4300	DuToit FZ	53.175	25.508	0.35	0.00	0.35	0.00	0.00	0.10	3.62	0.10	1.58	0.20	5.95
PROT−5−13	5300	DuToit FZ	53.095	25.537	178.70	0.00	178.70	0.00	0.40	10.00	1.05	0.00	0.00	0.00	190.15
PROT−5−14	4150	DuToit FZ	53.113	25.327	158.40	0.00	158.40	10.80	0.00	0.00	0.30	0.00	13.50	0.00	183.00
PROT−5−15	4400	Bain FZ	47.700	32.147	195.90	126.20	196.10	0.00	8.30	1.10	29.90	45.40	104.00	7.20	392.00
PROT−5−16	2500	Bain FZ	47.647	32.352	14.50	259.30	14.50	0.00	6.30	8.20	0.50	0.00	0.00	0.00	29.50
PROT−5−18	5100	P. Edward FZ	46.525	33.763	1.00	0.00	127.20	0.00	0.00	2.10	57.80	6.50	43.80	92.70	330.10
PROT−5−19	3900	P. Edward FZ	46.507	33.832	132.25	0.00	391.55	0.00	0.00	0.00	15.87	17.80	13.40	87.90	526.52
PROT−5−20	2950	P. Edward FZ	46.495	33.870	0.00	0.00	0.00	0.00	0.00	0.00	0.00	0.00	232.00	71.30	303.30
PROT−5−21	5040	P. Edward FZ	45.515	35.157	64.00	0.00	64.00	0.00	0.00	0.00	0.00	0.00	0.00	0.00	64.00
PROT−5−22	2350	P. Edward FZ	45.565	35.262	174.10	0.00	174.10	0.00	2.60	0.00	0.00	0.00	0.00	0.00	176.70
PROT−5−23	1650	P. Edward FZ	45.563	35.288	198.00	0.00	198.00	14.90	0.00	0.00	0.00	0.00	0.00	0.00	212.90
PROT−5−24	4400	P. Edward FZ	45.575	35.177	379.60	0.00	379.60	0.00	0.00	0.00	0.40	0.00	0.00	0.00	380.00
PROT−5−27	3800	Simpson FZ	43.612	39.308	0.00	0.00	0.00	0.00	0.00	0.00	0.00	0.00	136.60	44.10	180.70
PROT−5−29	4600	Discovery 11 FZ	42.685	41.890	2.96	0.00	2.96	0.35	1.53	10.90	22.82	1.15	64.64	0.25	104.60
PROT−5−30	3400	Discovery 11 FZ	42.687	41.985	127.00	0.00	127.00	0.00	0.00	7.35	0.00	0.00	218.25	12.10	364.70
PROT−5−31	2740	Discovery 11 FZ	42.657	42.128	11.35	0.00	11.35	30.60	72.30	96.30	0.00	0.40	0.00	11.00	221.95
PROT−5−32	4800	Discovery 11 FZ	41.683	42.680	0.00	0.00	0.00	0.00	0.10	0.00	0.00	0.00	0.00	0.00	0.10
PROT−5−34	3350	Discovery 11 FZ	41.510	42.522	53.80	0.00	53.80	0.00	0.00	0.00	0.00	0.00	0.00	139.35	193.15
PROT−5−35	4250	Discovery 11 FZ	41.623	42.708	0.00	56.50	0.00	0.00	4.70	0.00	0.00	0.00	0.00	0.00	4.70
PROT−5−38	2125	Indomed FZ	39.787	45.985	17.00	3.50	17.00	0.00	0.00	0.00	0.00	0.00	0.00	0.00	17.00
PROT−5−39	3400	Indomed FZ	39.850	46.208	563.40		619.90	10.50	27.60	16.60	0.00	0.00	0.00	7.40	682.00
PROT−5−40	2800	Indomed FZ	39.715	46.288	325.70		329.20	5.40	0.35	1.25	0.00	39.80	0.00	0.00	376.00
				Totals	3766.24	513.86	4280.00	108.75	323.39	236.40	150.40	117.61	1002.33	508.99	6727.87
					55.98	7.64	63.62	1.62	4.81	3.51	2.24	1.75	14.90	7.57	100.00

TABLE 1. (*cont.*)

Dredge No.	Start depth (m)	Location	Latitude	Longitude (E=+, W=−)	Peridotite (kg)	Perid Mylon (kg)	Sum peridotites	Dunite (kg)	Gabbro (kg)	Meta gabbro	Diabase (kg)	Meta diabase	Basalt	Meta basalt	Total (kg)
American–Antarctic Ridge fracture zone dredge hauls															
1011/76–45	3499	NE Conrad FZ	55.625	3.315	0.00	0.00	0.00	0.00	0.00	0.00	23.50	57.50	27.40	27.60	136.00
1011/76–47	2150	NE Conrad FZ	55.603	3.728	0.00	0.00	0.00	0.00	0.00	0.00	6.10	8.60	7.30	19.40	41.40
VULC–5–18	2488	S Conrad FZ	55.866	4.290	0.00	0.00	0.00	0.00	0.00	0.00	0.00	0.00	135.00	0.00	135.00
VULC–5–19	2835	S Conrad FZ	55.851	4.322	0.00	0.00	0.00	0.00	7.50	0.00	8.10	0.00	87.00	0.00	102.60
VULC–5–20	4782	S Conrad FZ	55.834	4.422	0.00	0.00	0.00	0.00	6.30	1.50	4.00	2.90	98.20	8.50	121.40
VULC–5–21	4456	S Conrad FZ	55.845	4.421	0.00	0.00	0.00	0.00	5.20	0.00	13.20	3.10	255.30	10.20	287.00
VULC–5–22	4560	N Conrad FZ	55.782	4.653	0.00	0.00	0.00	0.00	0.00	0.00	0.00	0.00	25.50	0.00	25.50
VULC–5–23	5378	S Conrad FZ	55.786	3.928	0.00	0.00	0.00	0.00	0.10	0.00	127.00	0.00	127.20	1.20	255.60
VULC–5–26	3984	S 56.6S FZ	56.613	4.448	0.00	0.00	0.00	0.00	0.40	0.00	2.00	0.00	3.60	7.40	13.40
VULC–5–32	3152	NE Bullard FZ	56.894	6.044	0.00	0.00	0.00	0.00	0.00	0.00	0.00	0.00	6.00	0.00	6.00
VULC–5–33	3944	NE Bullard FZ	56.959	6.126	0.00	0.00	0.00	0.00	0.00	0.00	4.00	0.00	66.20	0.00	70.20
VULC–5–34	3983	SE Bullard FZ	57.782	7.672	27.60	0.00	27.60	4.40	0.00	0.00	0.00	0.00	0.00	0.20	32.20
VULC–5–35	3479	SE Bullard FZ	57.950	7.812	514.80	0.00	514.80	0.20	2.20	0.20	0.00	0.10	4.90	3.00	525.40
VULC–5–36	3414	SE Bullard FZ	58.147	9.764	0.00	0.00	0.00	0.00	0.00	0.00	23.20	3.00	126.30	4.00	156.50
VULC–5–37	4104	SW Bullard FZ	58.430	15.662	2.10	0.00	2.10	2.50	0.00	0.00	13.10	0.00	71.10	9.20	98.00
VULC–5–38	3233	SW Bullard FZ	58.453	15.486	0.00	0.00	0.00	0.00	0.00	0.00	17.10	0.00	44.50	6.60	68.20
VULC–5–41	4645	N Vulcan FZ	59.087	16.808	60.00	0.00	60.00	1.30	0.00	0.00	0.00	0.00	0.00	0.20	61.50
				Totals	604.50	0.00	604.50	8.40	21.70	1.70	241.40	75.20	1085.50	97.50	2038.40
					29.66	0.00	29.66	0.41	1.06	0.08	11.84	3.69	53.25	100.00	100.00
SW Indian Ridge rift valley dredge hauls															
CH115–44	861	Speiss ridge crest	−54.670	0.020	0.00	0.00	0.00	0.00	0.00	0.00	0.00	91.00	0.00	0.00	91.00
PS86–2	3191	RV axial high	−52.201	−14.729	0.00	0.00	0.00	0.00	0.00	0.00	0.00	20.00	0.00	0.00	20.00
PS86–5B	3740	RV axial high	−52.276	−13.706	0.00	0.00	0.00	0.00	0.00	0.00	0.00	154.20	0.00	0.00	154.20
PS–5B	3740	RV axial high	−52.276	−13.706	0.00	0.00	0.00	0.00	0.00	0.00	0.00	154.20	0.00	0.00	154.20
PS86–7	4150	Ridge in RV	−52.758	−11.923	0.00	0.00	0.00	0.00	0.00	0.00	0.00	0.90	0.00	0.00	0.90
AII107–31	829	Speiss Ridge crest	−54.690	0.038	0.00	0.00	0.00	0.00	0.00	0.00	0.00	46.40	0.00	0.00	46.40
AII107–32	891	Speiss Ridge crest	−54.662	0.013	0.00	0.00	0.00	0.00	0.00	0.00	0.00	11.60	0.00	0.00	11.60
AII107–33	937	Speiss Ridge crest	−54.650	0.068	0.00	0.00	0.00	0.00	0.00	0.00	0.00	50.80	0.00	0.00	50.80
AII107–34	1060	Speiss Ridge crest	−54.635	0.127	0.00	0.00	0.00	0.00	0.00	0.00	0.00	16.90	0.00	0.00	16.90
AII107–46	2176	RV floor	−53.982	−3.593	0.00	0.00	0.00	0.00	0.00	0.00	0.00	0.15	0.00	0.00	0.15
AII107–47	2069	RV floor	−54.012	−3.552	0.00	0.00	0.00	0.00	0.00	0.00	0.00	43.35	0.00	0.00	43.35
AII107–48	1901	RV floor	−54.030	−3.522	0.00	0.00	0.00	0.00	0.00	0.00	0.00	16.25	0.00	0.00	16.25
AII107–56	3664	RV floor	−54.397	−5.137	0.00	0.00	0.00	0.00	0.00	0.00	0.00	33.40	0.00	0.00	33.40
1011/76–75	1760	Ridge axis	−52.905	−11.350	0.00	0.00	0.00	0.00	0.00	0.00	0.00	9.70	0.00	0.00	9.70
AG22–1	3700	RV floor	−52.293	−16.910	0.00	0.00	0.00	0.00	0.00	0.00	0.00	11.65	0.00	0.00	11.65
				Totals	0.00	0.00	0.00	0.00	0.00	0.00	0.00	0.00	660.50	0.00	660.50
					0.00	0.00	0.00	0.00	0.00	0.00	0.00	0.00	100.00	0.00	100.00

American–Antarctic Ridge rift valley dredge hauls															
CH115–38	3500	Median valley	−55.337	1.710	0.00	0.00	0.00	0.00	0.00	0.00	0.10	0.00	16.20	0.00	16.30
VULC5–24	3295	Median valley	−56.043	4.702	0.00	0.00	0.00	0.00	0.00	0.00	0.00	0.00	46.10	0.00	46.10
VULC5–25	3294	Median valley	−56.283	4.653	0.00	0.00	0.00	0.00	0.00	0.00	0.00	0.00	2.00	0.00	2.00
VULC5–27	2379	RV axial high	−56.894	6.044	0.00	0.00	0.00	0.00	0.00	0.00	0.00	0.00	47.90	0.00	47.90
VULC5–28	3098	RV floor	−56.959	6.126	0.00	0.00	0.00	0.00	0.00	0.00	0.00	0.00	17.00	0.00	17.00
VULC5–29	3589	RV floor	−57.055	6.067	0.00	0.00	0.00	0.00	0.00	0.00	0.00	0.00	17.00	0.00	17.00
VULC5–30	4074	RV floor	−57.468	7.011	0.00	0.00	0.00	0.00	0.00	0.00	0.00	0.00	121.30	0.00	121.30
VULC5–31	3808	RV axial high	−57.573	6.986	0.00	0.00	0.00	0.00	0.00	0.00	1.60	0.00	96.80	0.00	98.40
VULC5–39	3958	RV axial high	−58.651	16.211	0.00	0.00	0.00	0.00	0.00	0.00	0.00	0.00	196.40	0.00	196.40
VULC5–40	4476	RV floor	−58.989	16.317	0.00	0.00	0.00	0.00	0.00	0.00	0.00	0.00	77.70	0.00	77.70
VULC5–42	3984	RV axial high	−59.364	18.060	0.00	0.00	0.00	0.00	0.00	0.00	0.00	0.00	44.60	0.00	44.60
VULC5–43	2594	RV floor	−59.748	17.974	0.00	0.00	0.00	0.00	0.00	0.00	0.00	0.00	15.80	0.00	15.80
				Totals	0.00	0.00	0.00	0.00	0.00	0.00	1.70	0.00	698.80	0.00	700.50
					0.00	0.00	0.00	0.00	0.00	0.00	0.24	0.00	99.76	0.00	100.00
SW Indian Ridge rift mountain and rift valley wall dredge hauls															
CH115–50	1806	NE R.V. wall	−54.228	−4.052	0.00	0.00	0.00	0.00	0.00	0.00	0.00	0.00	10.00	0.00	10.00
AII107–31	973	Anna Maria Smt	−54.425	−4.732	0.00	0.00	0.00	0.00	0.00	0.00	0.00	0.00	68.40	0.00	68.40
AII107–53	2810	NE flank RV	−54.225	−5.175	0.00	0.00	0.00	0.00	0.00	0.00	14.00	0.00	4.05	0.00	18.05
AII107–55	2355	NE flank RV	−54.293	−5.308	0.00	0.00	0.00	0.00	0.00	0.00	0.80	0.00	13.40	7.52	21.72
AII107–57	3648	Rift valley wall	−54.045	−7.223	0.00	0.00	0.00	0.00	0.00	0.00	0.00	0.00	34.10	0.00	34.10
AII107–58	1507	Rift valley wall	−53.980	−7.340	30.37	0.00	30.37	0.00	0.00	0.12	4.85	0.15	9.30	25.98	70.77
AII107–59	3501	Rift valley wall	−54.018	−7.235	0.70	0.00	0.70	1.70	8.30	0.00	1.10	0.00	0.20	0.00	12.00
1011/76–76		Core in rift mtns	−52.500	−11.400	0.10	0.00	0.10	0.00	0.00	0.00	0.0	0.00	0.00	0.00	0.10
AG22–3	3800	Base RV S wall	−52.608	−18.040	0.00	0.00	0.00	0.00	0.00	0.00	13.80	0.00	23.68	0.00	37.48
AG22–5	3850	Base RV S wall	−52.760	−19.097	0.00	0.00	0.00	0.00	0.00	0.00	4.00	0.00	52.73	0.00	56.73
AG22–7	3330	RV S wall	−53.022	−21.833	0.00	0.00	0.00	0.00	0.00	0.00	0.00	0.00	1.00	0.00	1.00
AG22–8	3800	RV S wall	−53.105	−22.197	0.00	0.00	0.00	0.00	0.00	0.00	0.00	0.00	31.00	0.00	31.00
AG22–9	3860	RV S wall	−53.158	−22.830	0.00	0.00	0.00	0.00	0.00	0.00	0.00	0.00	5.95	0.00	5.95
AG22–12	3650	RV S wall	−53.393	−23.692	0.00	0.00	0.00	0.00	0.00	0.00	5.50	0.00	33.75	0.00	39.25
AG22–13	3870	RV S wall	−53.408	−24.758	0.00	0.00	0.00	0.00	0.00	0.00	0.00	0.00	94.30	0.00	94.30
PS86–3	3694	RV N wall	−52.151	−14.491	0.00	0.00	0.00	0.00	0.00	0.00	105.10	0.00	91.80	0.00	196.90
PS86–6	4460	Small RV high	−52.449	−13.076	171.90	0.00	171.90	0.00	0.00	0.00	0.00	0.00	28.81	0.00	200.71
PROT5–9	3100	Rift valley wall	−52.580	−26.335	0.00	0.00	0.00	0.00	0.00	0.00	0.15	0.00	1.00	16.10	1.25
PROT5–17	2380	Rift valley wall	−46.378	−33.508	1.30	0.00	1.30	0.00	79.00	0.00	65.10	13.60	37.30	0.00	212.40
PROT5–25	1920	Rift valley wall	−44.815	−35.847	0.00	0.00	0.00	0.00	0.00	0.00	16.00	0.00	436.30	0.00	452.30
PROT5–26	3050	Rift valley wall	−44.733	−37.070	2.05	0.00	2.05	0.00	0.00	0.00	0.00	0.00	0.15	3.14	2.20
PROT5–28	2750	Rift valley wall	−43.377	−39.852	0.00	0.00	0.00	0.00	0.00	0.00	171.90	0.00	710.66	0.00	885.70
PROT5–37	3150	Rift valley wall	−40.130	−45.760	0.00	0.00	0.00	0.00	0.00	0.00	0.00	0.00	6.10	43.20	6.10
PROT5–43	2640	Rift valley wall	−38.898	−46.892	0.00	0.00	0.00	0.00	0.00	0.00	0.00	0.00	0.20		43.40
				Totals	206.42	0.00	206.42	1.70	87.30	0.12	402.30	13.75	1694.18	96.04	2501.81
					8.25	0.00	8.25	0.07	3.49	.00	16.08	0.55	67.72	3.84	100.00

are variously serpentinized and weathered peridotite, 41.7% is basalt or metabasalt, while only 5.9% is gabbro and 9.1% is diabase, or their metamorphosed equivalents. In Fig. 2 histograms are plotted showing the relative abundances of rocks for different physiographic provinces. The fracture zone province includes the entire fracture zone valley and the flanking transverse ridges up to their crests. The rift valley province includes only the inner floor of the valley, where the dredges were located largely on small axial highs believed to correspond to the neovolcanic zone. The rift valley walls extend from the foot of the inner rift valley wall to the crest of the adjoining rift mountains.

Virtually the only rock recovered from the rift valley floors is fresh or weathered pillow basalt. In contrast, dredging of the rift valley walls recovered rocks representing a potential cross-section of the shallow lithosphere. There is an exponential decrease in abundance with inferred stratigraphic depth from basalt through diabase to gabbro and then an increase with peridotite. This is consistent with the proposal that deeper rocks are exposed by continued faulting and tectonism of the crust as it is uplifted into the rift mountains, with the greater abundance of peridotite possibly reflecting the relative ease with which it can migrate up faults (Hess 1955). Nevertheless, because of the relatively small number of dredges for rift valley walls and mountains, inferences based on their statistics should be viewed with great caution.

The proportions of rocks dredged from fracture zones, however, are strikingly different from those of other provinces. Peridotite, by far the most abundant rock, constitutes 55.7% followed by 30.7% basalt, while only 6.6% is gabbro and 6.7% is diabase. The relative abundance of peridotite is even higher if we exclude the Conrad Fracture Zone where eight widely dispersed dredge hauls recovered only basalt and diabase, suggesting, by comparison, that the crust may be anomalously thick at that fracture zone. For SW Indian Ridge fracture zones alone, dredging recovered 65.4% peridotite and dunite, 8.5% gabbro, 3.9% diabase and 22.6% basalt. The GEBCO bathymetric maps indicate that more than 40% of the ocean crust at the SW Indian Ridge lies within fracture zones (defined as the region between the summits of the flanking walls and transverse ridges). Thus, using the abundances in Table 1, we can estimate that peridotite constitutes at least 13% of the exposed crust along this ridge system (*ie* half the fracture zone abundance since we have yet systematically to sample the non-transform walls and the large majority of the dredge hauls are from transforms). The rift valley walls at the ridge–transform intersection show roughly the same relative abundances of rocks as the rest of the fracture zone domain. This is not surprising as uplift of crust from this region of the rift valley creates the transverse ridges and the walls of the transform valleys.

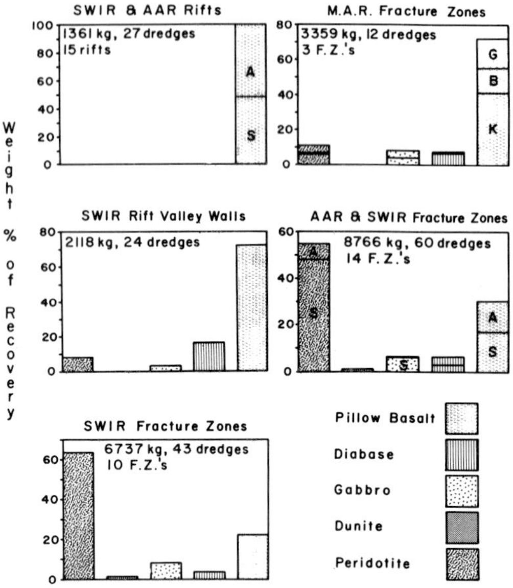

FIG. 2. Dredge statistics plotted from Table 1 for the SW Indian and American–Antarctic Ridges: A and S, American–Antarctic and SW Indian Ridges respectively; G, B and K, the Gibbs, B and Kane Fracture Zones respectively.

The American–Antarctic and SW Indian Ridges have the lowest spreading rates of any major ridge system (8 and 9 mm a^{-1} respectively) and so, for comparison, similar statistics are presented in Table 2 for three fracture zones on the faster spreading (around 14 mm a^{-1}) northern Mid-Atlantic Ridge. The northern Mid-Atlantic Ridge fracture zones have a far lower abundance of peridotitic rocks, roughly the same abundance of gabbroic rocks and a much greater proportion of basalts (Fig. 2). This strongly supports the concept that the abundance of peridotites exposed at ocean ridges increases with decreasing spreading rate. It also suggests that the crustal section is generally thicker beneath the northern Mid-Atlantic Ridge fracture zones than at slower spreading ridges. However, it is important to note that there are Mid-Atlantic Ridge fracture zones such as the Romanche, Vema and 43° N where peridotites are as abundant as at the SW Indian Ridge.

TABLE 2. *Northern Mid-Atlantic Ridge fracture zone dredge hauls*

Dredge No.	Start depth	Location	Latitude	Longitude	Peridotite (kg)	Pd myl	Sum peridotites (kg)	Dunite	Gabbro	Meta gabbro	Diabase	Meta diabase	Basalt	Meta basalt	Sum
Kane Fracture Zone															
AII96–3–15	2044	Kane FZ	23.544	45.090	0.00	0.00	0.00	0.00	0.00	0.00	2.30	0.00	25.30	0.00	27.60
AII96–3–10	2578	N wall Kane FZ	23.639	44.486	2.70	0.00	2.70	0.00	0.00	0.00	5.40	0.00	11.70	0.00	19.80
AII96–3–8	2860	N wall Kane FZ	24.011	46.186	0.00	0.00	0.00	0.00	0.00	0.00	8.60	0.00	163.00	0.00	171.60
AII96–3–3	3022	S wall Kane FZ	23.775	46.575	7.70	0.00	7.70	0.00	0.50	0.00	0.00	0.00	0.00	0.00	8.20
AII96–3–14	3611	S wall Kane FZ	23.604	45.026	0.00	0.00	0.00	0.00	9.00	0.00	85.00	0.00	32.90	0.00	126.90
Knorr79–2–26	3670	E side RV EI Kane	23.612	44.788	0.00	0.00	0.00	0.00	0.00	0.00	0.00	0.00	10.00	0.00	10.00
Knorr79–2–15	3682	S wall Kane FZ	23.612	45.098	0.00	0.00	0.00	0.00	0.00	0.00	0.40	0.00	3.80	0.00	4.20
AII96–3–18	3686	Kane FZ	23.524	44.819	0.00	0.00	0.00	0.00	0.00	0.00	0.00	0.00	5.40	0.00	5.40
AII96–3–17	3800	S wall Kane FZ	23.731	45.978	0.00	0.00	0.00	0.00	7.70	0.00	0.90	0.00	0.00	0.00	8.60
G104–27	3840	Rt edge EI Kane FZ	23.578	44.998	0.00	0.00	0.00	0.00	0.40	0.00	6.80	0.00	6.60	0.00	13.80
Knorr79–2–28	3840	E side RV EI Kane	23.617	44.820	0.00	0.00	0.00	0.00	0.00	0.00	0.00	0.00	73.00	0.00	73.00
G104–29	3880	N wall EI Kane FZ	23.390	44.802	0.00	0.00	0.00	0.00	0.00	0.00	0.00	0.00	2.60	0.00	2.60
Knorr79–2–25	3910	E side RV EI Kane	23.607	44.778	0.00	0.00	0.00	0.00	0.00	0.00	0.00	0.00	18.00	0.00	18.00
AII96–3–1	3918	S wall Kane FZ	23.805	46.575	60.80	0.00	60.80	0.00	28.30	0.00	0.00	0.00	0.00	0.00	89.10
Knorr79–2–27	3930	E side RV EI Kane	23.608	44.773	0.00	0.00	0.00	0.00	0.00	0.00	0.00	0.00	35.00	0.00	35.00
AII96–3–9	3982	N wall Kane FZ	23.608	44.478	0.00	0.00	0.00	0.00	1.10	0.00	0.00	0.00	0.00	0.00	1.10
AII96–3–12	4039	N wall Kane FZ	23.518	43.906	68.80	0.00	68.80	10.90	6.80	0.00	0.00	0.00	0.00	0.00	86.50
Knorr79–2–10	4225	S wall Kane FZ	23.533	44.957	0.00	0.00	0.00	0.00	0.00	0.00	0.00	0.00	217.00	0.00	217.00
G104–19	4240	Rt edge EI Kane FZ	23.515	44.982	0.00	0.00	0.00	0.00	0.00	0.00	7.30	0.00	187.40	0.00	194.70
AII96–3–7	4259	N wall Kane FZ	23.918	45.242	0.00	0.00	0.00	0.00	0.00	0.00	0.00	0.00	45.90	0.00	45.90
G104–20	4380	S wall Kane FZ	23.628	45.047	0.00	0.00	0.00	0.00	2.40	0.00	12.90	0.00	56.40	0.00	71.70
AII96–3–16	4450	N wall KZ	23.791	45.831	0.00	0.00	0.00	0.00	5.90	0.00	17.90	0.00	54.90	0.00	78.70
G104–37	4550	Rt edge WI Kane FZ	23.842	46.138	0.00	0.00	0.00	0.00	0.00	0.00	0.00	0.00	4.20	0.00	4.20
G104–25	4560	Rt edge EI Kane FZ	23.533	44.957	0.00	0.00	0.00	0.00	0.00	0.00	5.40	0.00	87.70	0.00	93.10
AII96–3–2	4569	S wall Kane FZ	23.833	46.572	0.70	0.00	0.70	0.00	0.50	0.00	0.00	0.00	0.00	0.00	1.20
G104–35	4590	Rt edge WI Kane FZ	23.853	46.223	0.00	0.00	0.00	0.00	0.00	0.00	0.00	0.00	20.50	0.00	20.50
G104–17	4820	Rt edge EI Kane FZ	23.620	45.035	68.66	0.00	68.66	0.00	10.00	0.00	46.30	0.00	43.05	0.00	168.01
G104–13	4870	N wall EI Kane FZ	23.653	44.988	0.00	0.00	0.00	0.00	0.66	0.00	4.05	0.00	138.00	0.00	142.71
G104–7		S wall WI Kane FZ	23.853	46.333	0.00	0.00	0.00	0.00	0.00	0.00	0.00	0.00	12.00	0.00	12.00
AII96–3–4	5095	S wall Kane FZ	23.852	46.388	0.00	0.00	0.00	0.00	0.00	0.00	0.00	0.00	48.40	0.00	48.40
AII96–3–6	5505	N wall Kane FZ	23.882	46.263	0.00	0.00	0.00	0.00	0.00	0.00	0.00	0.00	98.10	0.00	98.10
AII96–3–13	6180	S wall Kane FZ	23.601	44.965	0.00	0.00	0.00	0.00	65.80	0.00	19.90	0.00	0.00		85.70
					209.36	0.00	209.36	10.90	139.06	0.00	223.15	0.00	1400.85	0.00	1983.32
					10.56	0.00	10.56	0.55	7.01	0.00	11.25	0.00	70.63	0.00	100.00

TABLE 2. (*cont.*)

Dredge No.	Start depth	Latitude	Longitude	Peridotite (kg)	Pd myl	Sum peridotites (kg)	Dunite	Gabbro	Meta gabbro	Diabase	Meta diabase	Basalt	Meta basalt	Sum
Fracture Zone B, FAMOUS region														
AII73–47				7.47	0.00	7.47	0.00	7.54	0.00	0.00	0.00	0.00	0.00	15.01
AII77–52	2300	36.625	33.538	0.21	0.00	0.21	0.00	0.00	0.00	0.00	3.24	5.50	1.49	8.95
KNR42–121	2377	36.590	33.550	25.72	0.00	25.72	0.00	0.00	0.00	0.60	5.50	56.14	5.50	87.96
KNR42–129				0.00	0.00	0.00	0.00	0.06	0.00	2.16	0.00	57.80	0.00	60.02
KNR42–130				0.00	0.00	0.00	0.00	0.00	0.00	2.75	0.00	57.61	0.00	60.36
KNR42–131				0.00	0.00	0.00	0.00	0.00	0.00	0.00	0.00	55.91	0.00	55.91
KNR42–145				0.00	0.00	0.00	0.00	0.00	0.00	0.00	5.99	11.47	1.22	17.46
KNR42–146				0.00	0.00	0.00	0.00	0.00	0.00	5.01	0.00	30.51	0.00	35.52
				0.00	0.00	0.00	0.00	0.00	0.00	0.00	0.68	133.74	64.82	134.42
			33.40	0.00	33.40	0.00	7.60	0.00	10.52	15.41	408.68	73.03	548.64	
		6.09	0.00	6.09	0.00	1.39	0.00	1.92	2.81	74.49	13.31	100.00		
Gibbs Fracture Zone														
TR100–12	52.333	31.517	115.90	0.00	115.90	0.00	94.36	42.78	19.24	0.00	372.72	5.75	650.75	
TR100–13	52.483	31.567	19.10	0.00	19.10	0.00	0.00	0.00	0.00	0.00	193.70	1.25	214.05	
TR100–16	52.233	30.983	9.07	0.00	9.07	0.00	0.00	0.00	0.00	0.00	0.00	1.13	10.20	
				144.07	0.00	144.07	0.00	94.36	42.78	19.24	0.00	566.42	8.13	875.00
				16.47	0.00	16.47	0.00	10.78	4.89	2.20	0.00	64.73	0.93	100.00
Combined Mid-Atlantic Ridge fracture zone dredge hauls														
				386.83	0.00	386.83	10.90	241.02	42.78	252.91	5.41	2375.95	81.16	3406.96
				11.35	0.00	11.35	0.32	7.07	1.26	7.42	0.45	69.74	2.36	

Absence of layer 3 at slow slipping fracture zones

Gabbro, with the notable exception of the Atlantis II Fracture Zone (Dick *et al.* 1987), is only a minor lithofacies in the American–Antarctic and SW Indian Ridge dredge hauls—which is quite surprising considering that gabbro is thought to make up two-thirds of a normal oceanic crustal section. In Table 1 peridotite and basalt occur together in 11 dredge hauls while peridotite, basalt and gabbro occur together in 17 dredge hauls. In all but five of the latter, however, gabbro constitutes less than 10% of the dredge haul and largely represents coarse-grained slowly cooled basaltic dykes intruding the peridotites, not plutonic complexes.

Dunite associated with oceanic and alpine-type peridotites is commonly believed to be produced by wall-rock reaction and incongruent melting during melt migration through the shallow mantle (Dick 1977; Cassard *et al.* 1981; Quick 1981a,b; Dick & Bullen 1984) and by the early stages of basalt crystallization at or near the base of the ocean crust (Greenbaum 1972; Boudier & Coleman 1981; Hopson *et al.* 1981; Auge 1983). In many ophiolites the gabbroic section is also floored by a thick section of dunite hundreds of metres thick. This is thought to represent the base of a fossil magma chamber and the earliest stages of basalt crystallization in the crust. The scarcity of dunite associated with the abyssal peridotites in Table 1 is consistent with the overall scarcity of gabbro; this appears to demonstrate that magma chambers rarely formed beneath fracture zones or that the underlying mantle rarely served as a conduit for magma migrating out of the mantle to the crust at the SW Indian and American–Antarctic Ridges. This and the large abundance of basalt and peridotite at these fracture zones indicate that the crustal section must often have originally consisted of basalt erupted directly over peridotite.

Volcanism at very slow spreading ridge fracture zones

If magma chambers are generally absent beneath SW Indian and American–Antarctic Ridge fracture zones, basalts there should have more restricted and primitive compositions than in the rift valleys, since magma chambers should greatly enhance fractional crystallization. Le Roex *et al.* (1983, 1985), however, found no systematic difference in isotopic, trace- and major-element whole rock analyses of basalts in 45 dredge hauls from six fracture zones and eight rift valleys between 11.3°E and 18°W on these ridges, except that basalts from SW Indian Ridge fracture zones extend to more evolved compositions. New American–Antarctic Ridge basalt glass analyses for fracture zones and ridge axes, shown for example in Fig. 3, have similar large linear ranges of Mg/(Mg+Fe) and Ca/(Ca+Na), consistent with large variations in the degree of fractional crystallization. A systematic offset of individual suites to higher Ca/(Ca+Na) at fixed Mg(Mg+Fe) correlates roughly with latitude and with decreasing depth of the ridge axis and proximity to the Bouvet hot-spot to the E. This probably reflects variations in the primary melt composition due to increasing degree of mantle melting near the hot spot (Dick *et al.* 1984; Klein & Langmuir 1987). The scatter of the analyses for the Bullard Fracture Zone basalts, compared with the other glass suites, is probably due to the anomalous tectonic environment associated with formation of this exceptional 560 km long transform offset.

The large range of fractional crystallization prior to eruption of fracture zone basalts must have occurred either within the crust or during magma ascent through the mantle. If this happened beneath the fracture zones, then dredging should recover the residues of this fractional crystallization, either with the abundant mantle peridotites or as gabbroic remains of fossil magma chambers. This is the case since the dredged peridotites were originally part of the subcrustal mantle section, and if a gabbroic section had been present then tectonism accompanying emplacement of the peridotites to the seafloor would certainly have exposed the gabbros. Despite this, gabbro and dunite, representing possible residues of fractional crystallization, though present, are scarce. In addition, dunite found with or cross-cutting peridotite, which could represent *in situ* fractionation of melts rising through the mantle, is rarely seen. At the same time, the majority of intrusive veins in the peridotites have a ferrobasaltic or other highly evolved composition, while primitive veins appear to be scarce (Dick *et al.* 1982; Fisher *et al.* 1986; Bloomer *et al.*, this volume). This is striking since it requires 60%–70% fractional crystallization of a primitive abyssal basalt to form ferrobasalt. Thus primitive veins representing the earlier stages of basalt differentiation should be far more abundant if there was extensive *in situ* basalt fractionation in the underlying mantle. All this suggests, given the similarity of fracture zone and ridge basalts, that much of the fractionation occurred beneath magmatic centres in the ad-

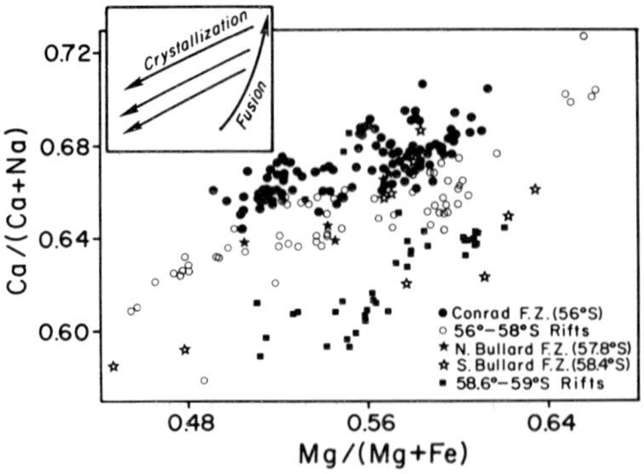

FIG. 3. Analyses of pillow basalt glasses from the American–Antarctic Ridge. Mg/(Mg+Fe) and Ca/(Ca+Na) are molecular ratios, and Fe is total iron. The inset shows the inferred basalt liquid trend for increasing degrees of partial melting of the mantle source and the contrasting trends for fractional crystallization for primary basalts produced by different degrees of partial melting. Thus, generally, the Conrad Fracture Zone basalts are products of the highest degree of mantle melting, and the 58.6–59°S rift basalts are products of the lowest degree of mantle melting. All the suites reflect similar large ranges in differentiation and fractional crystallization at shallow depths.

jacent rift valleys from which basalt was erupted down the rift valley fissure system to the fracture zone floor or intruded into the underlying peridotites. Locally, small ephemeral satellite magma chambers may have formed, accounting for the small amounts of dunite and gabbro dredged there.

This model is consistent with subaerial fissure eruptions where they have been studied in detail (Wright *et al.* 1968; Sigurdsson & Sparks 1978; Ryan 1987). In such systems, an eruptive cycle generally begins with inflation of a magma chamber underlying a major volcanic centre at 3–7 km beneath a linear fissure system such as the Askja caldera in Iceland or the Kilauea caldera in Hawaii. Subsequently, lateral injection of magma down the fissure system is accompanied by earthquake swarms migrating down the rift to the point at which magma vents to the surface. Magma transport over 70 km has occurred down the east Kilauea rift zone and up to 70 km in Iceland. There appear to be two styles of activity in Iceland, one where major eruptions are associated with major rifting episodes and a second style where small to medium eruptions occur near the central volcano during tectonically quiescent periods. In both the Iceland and Ethiopian rifts there appears to be a regular (*eg* 40 km) spacing of active central volcanoes, suggesting similarly spaced melt diapirs rising from a gravitationally unstable layer of low density melt at the base of the lithosphere (Mohr & Wood 1974; Sigurdsson & Sparks 1978).

Studies of basalts erupted along subaerial fissure systems show that they are highly variable in composition in space and time. Wolfe (1988) found that, in the early stage of a recent major eruption of Kilauea Volcano, highly evolved magmas were initially erupted from the rift, but were later replaced by more primitive magma from the summit caldera as eruption proceeded. During the 1965 eruption of Kilauea Volcano, however, lavas erupted successively down the rift were shown to be differentiated from lavas erupted further up the rift (Wright *et al.* 1968). Elsewhere, evolved magmas have erupted at the volcanic centre simultaneously with more primitive lavas down the fissure system (*eg* Gronvold 1988). In some cases the more evolved magmas appear to have resulted from *in situ* fractionation of magma stored in satellite chambers along the rift, and in other cases to represent older magmas forced from an inflating central chamber.

What is important is that the fissure systems are capable of randomly delivering relatively unfractionated or highly evolved magmas anywhere along their length during a robust eruption. If this is also the case for ocean ridge systems, as seems reasonable, then the spatial resolution for mantle heterogeneities that can be detected using the variability of ridge basalts is limited to the length scale of magmatic segmentation and fissuring, which is believed to be of the order of 30–70 km (Schouten *et al.* 1985).

Mineralogy and petrology of abyssal peridotites

Abyssal peridotites are generally coarse-grained tectonites with equigranular, protoclastic or porphyroclastic textures interpreted to be the result of plastic deformation during and after melting (*eg* Bonatti & Hamlyn 1980; Dick *et al.* 1984). They are characteristically highly altered, with 20%–100% serpentine replacing olivine and pyroxene, often with any remaining olivine heavily altered to clay. In addition, small amounts of higher temperature alteration products, including hornblende, tremolite, cummingtonite, chlorite, talc and metamorphic diopside and olivine, are common, often forming hydrothermal veins (Aumento & Loubat 1971; Dick 1979; Kimball *et al.* 1985). The alteration sequence indicates rapid cooling of peridotite, intruded at 600–800 °C to shallow depths near the ridge axis, by seawater (Bonatti & Hamlyn 1978; Dick 1979; Kimball *et al.* 1985). About 5% of all dredged peridotites are extremely fine-grained mylonites (Table 1) similar to those described by Melson *et al.* (1972) from St Paul's Rocks on the transverse ridge of St Paul's Fracture Zone. Foliations in the latter are parallel to the Mid-Atlantic Ridge rather than the fracture zone, suggesting mylonitization during faulting, exposure and uplift to the transverse ridge from beneath the rift valley.

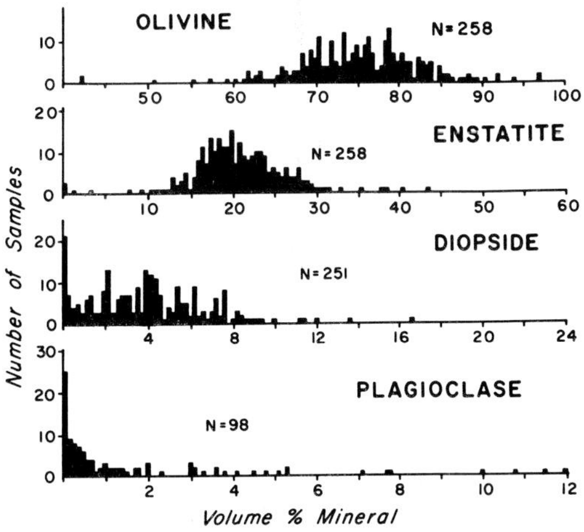

FIG. 4. Histograms showing the abundance of minerals in abyssal peridotites from the Atlantic, Indian and Caribbean Oceans. Localities and averages for the data are given by Dick *et al.* (1984). *N* is the subset of the sample population containing a particular phase, with the total population equaling 258 samples from 42 dredge hauls collected at 21 localities along six different ocean ridge systems.

Modal mineralogy

Shown in Fig. 4 are histograms of 258 modal analyses of peridotites from 41 dredge hauls from 22 fracture zones in the Atlantic, Caribbean and Indian Oceans. Modal data for representative localities along the American–Antarctic and SW Indian Ridges are given in Appendix 1. Each analysis represents about 2000 point counts at 1 mm spacing. No histogram is shown for spinel, which is ubiquitous, typically comprising 0.5% of the rock. All the data represent 'primary' modes, where characteristic pseudomorphs are counted with the relict mineral phase to estimate the original mineral proportions in the rock. This can be done with fair precision in roughly half the thin sections examined, while the remainder are rejected as the original replaced phases cannot be determined with confidence. No correction is made for differential volume expansion during serpentinization. The nature of the alteration is so variable that this simply cannot be done with confidence. In some peridotites, pyroxene is preferentially replaced relative to olivine, in others olivine relative to pyroxene and in some the serpentinization is uniform. A large proportion of the olivine pseudomorphs are clay, which is unlikely to represent an isochemical replacement and therefore may even represent an isovolumetric replacement.

Although the average abyssal peridotite contains 0.5% plagioclase, it is bimodally distributed (Fig. 4): nearly absent in the peridotites from 80% of the dredge hauls, and ubiquitous in the remainder. The average plagioclase-free peridotite contains 75% olivine, 21% enstatite, 3.5% diopside and 0.5% spinel. The typical plagioclase peridotite contains roughly 2.5% plagioclase and slightly more pyroxene relative to olivine than plagioclase-free peridotites. The most plagioclase-rich peridotites examined are from the Romanche Fracture Zone, with an average of 8.4%, and up to 17%, plagioclase. The majority of the remaining plagioclase peridotites come from the Argo Fracture Zone (averaging 3.2% plagioclase), the Marie Celeste Fracture Zone (2.1%), a single dredge haul from the Atlantis II Fracture Zone (2.5%) and a single dredge haul from the walls of the rift valley at the mid-Cayman spreading centre (2.3%). At most localities, such as the Atlantis II Fracture Zone, plagioclase peridotites are found in only one or two dredge hauls, even where numerous other dredges contain peridotites. The exception to this rule seems to be the Romanche Fracture Zone, where the typical dredge haul seems to be plagioclase peridotite.

Although there are also large variations locally, the average peridotite modal composition varies systematically along the Atlantic, Central Indian and SW Indian Ocean ridges, defining three separate regional melting trends (Dick *et al.* 1984; Michael & Bonatti 1985). Those from the vicinity of mantle 'hot-spots' corresponding to regional bathymetric highs, like the Azores, are the most depleted in basaltic components (low modal diopside and enstatite), while peridotites dredged away from hot-spots, where the ridges deepen, become systematically less depleted. This shift in peridotite composition correlates with variations in the major-element composition of spatially associated basalt, with basalts away from hot-spots reflecting lower degrees of mantle melting (Dick *et al.* 1984).

Primary mineral composition

Mineral data for four representative SW Indian and American–Antarctic Ridge abyssal peridotite localities are given in Appendix 2. All the analyses were made using natural standards at the electron microprobe facilities at the Massachusetts Institute of Technology and the University of Rhode Island. Pyroxene analyses were made on fused glasses of hand-picked optically clear diopside and enstatite to homogenize exsolution lamellae to determine a 'primary' composition. Short (10–15 s) fusion times on an iridium strip were used to minimize iron loss. The most serious correction was for chromium, which decreased rapidly with fusion time, and the chromium data should be taken as representing minimum values.

Due to possible errors in the value of internal standards used for Mg and Fe during pyroxene analysis, these values must be regarded as preliminary. This should only lead to a small systematic error in bulk rock compositions computed in this paper (see Appendix 2).

The overall compositions of individual minerals in abyssal peridotites are relatively restricted within dredge hauls and at individual localities (Appendix 2). However, as can be seen by comparing Appendices 1 and 2, mineral compositions vary along the ridges, becoming systematically more refractory and depleted in magmaphile elements as the average peridotite becomes depleted in modal pyroxene (Dick *et al.* 1984). The relatively uniform mineral compositions at a given locality appear to reflect equilibrium at high temperatures with melts of similar major-element composition, while the local variability of modal compositions is produced on an outcrop scale owing to the mechanical phenomena of melt formation, segregation and migration. Much of the variation in mineral composition that does exist at a locality is probably due to local pockets of trapped melt which crystallized in the peridotite (*eg* sample Vulc 5:35–3—see the section on plagioclase peridotites) or subsolidus re-equilibration of rocks whose modal, and therefore bulk, composition varies (*eg* Komor *et al.* 1985).

Variations in bulk rock composition

Although whole-rock X-ray fluorescence data are available for abyssal peridotites, they are of limited use because of the generally high degree of alteration. It is therefore useful to compute bulk-rock compositions free of alteration effects from mineral and modal data. Although this technique omits interstitial trace phases not seen in a particular thin section, this is not a significant problem for major elements in residual peridotites. The calculation is made using appropriate mineral densities extrapolated from the published end-member densities and the mineral data given in Appendix 2. It is worth noting that, despite the ambiguities of the technique, the values computed can be quite precise. This is often true for low-concentration elements ordinarily difficult to analyse in whole rocks, which are concentrated in individual minerals where they are easily determined by microprobe. An example is sodium, which resides almost exclusively in diopside in spinel peridotites. In many of our computations we have had to use plagioclase compositions from other abyssal peridotites of similar modal composition. This is done since plagioclase is rarely preserved in the peridotites and its original presence is usually based on its characteristic chlorite and prehnite pseudomorphs. This is not likely to be a problem for the average compositions in Table 3. With less than 0.2% modal plagioclase in the average modes, even a large error in the assumed plagioclase composition produces only a very small difference in the concentration of any element, including sodium, in the calculated bulk-rock composition.

Although the computed chromium values should be treated as minimum values, the relative differences computed between localities are reasonably accurate as the fusion times used were nearly constant. The poorest computed values are for oxides such as NiO and MnO, which occur only in minor concentrations in any phase, because the analytical technique and the quality of the standards used for concentrations less than 0.5 wt% have a large inherent error. Thus the absolute concentrations of MnO and NiO are probably determined to no better than $1\sigma = \pm 0.05$ wt%. Once again the precision is much better than the accuracy; thus the numbers are useful for examining relative differences in concentration.

Spinel peridotites

Shown in Table 3 are computed average bulk compositions for the Vulcan and Bullard Fracture Zones (American–Antarctic Ridge) and the Bouvet and Islas Orcadas Fracture Zones (SW Indian Ridge). These peridotites, though a quarter contain small amounts of plagioclase, have very refractory compositions, particularly with respect to those incompatible elements which are key constituents of basaltic lavas. They therefore contain little 'potential' basalt component. In Fig. 5 the concentrations of

TABLE 3. *Computed average spinel peridotite compositions*

	American–Antarctic Ridge						SW Indian Ridge					
	Oliv.	En.	Di.	Sp.	Plag.[a]	Rock	Oliv.	En.	Di.	Sp.	Plag.[a]	Rock
Vulcan Fracture Zone							*Bouvet Fracture Zone*					
Mode	71.13	20.56	7.33	0.96	0.03	100.0	81.55	16.05	1.58	0.48	0.18	99.8
SiO_2	40.53	53.63	50.11		(44.88)	43.65	40.49	55.47	51.52			42.97
TiO_2		0.05	0.26	0.15		0.031		0.02	0.00	0.09		0.004
Al_2O_3		5.16	6.37	49.1	(34.36)	1.96		3.17	4.70	32.55		0.790
FeO	9.66	5.87	3.41	13.42	(0.09)	8.43	8.87	5.71	3.01	17.57		8.27
MnO	0.16	0.12	0.07	0.12		0.145	0.12	0.13	0.11	0.17		0.121
MgO	49.18	31.27	18.21	19.21	(0.09)	42.87	49.09	33.56	18.66	15.11		45.83
CaO	0.04	2.43	19.40		(18.39)	1.99	0.03	2.14	21.11			0.755
Na_2O		0.05	0.56		(1.00)	0.053		0.01	0.05			0.005
K_2O		0.01	0.00		(0.03)	0.006		0.01	0.02			0.002
Cr_2O_3		0.67	1.11	16.68		0.357		0.64	1.20	33.99		0.252
NiO	0.31			0.41		0.223	0.30			0.13		0.245
Sum	99.9	99.3	99.5	99.1	(98.8)	99.50	98.9	100.9	100.4	99.6		99.244
Bullard Fracture Zone							*Islas Orcadas Fracture Zone*					
Mode	74.69	19.98	4.67	0.61	0.07	100.0	72.83	21.77	4.60	0.70	0.10	100.0
SiO_2	40.66	54.48	50.82			43.74	40.37	54.19	50.8			43.67
TiO_2		0.01	0.09	0.11		0.007		0.06	0.17	00.06		0.022
Al_2O_3		4.13	5.19	43.81		1.33		4.98	6.13	48.8		1.71
FeO	9.23	5.67	3.32	14.86		8.25	9.40	5.78	2.96	12.4		8.31
MnO	0.15	0.15	0.09	0.14		0.147	0.14	0.12	0.10	0.09		0.133
MgO	49.4	32.1	18.9	17.9		44.29	49.23	31.55	17.65	19.08		43.61
CaO	0.04	2.41	19.18			1.45	0.03	1.98	19.7			1.41
Na_2O		0.03	0.23			0.018		0.07	0.61			0.045
K_2O		0.01	0.05			0.004		0.01	0.02			0.003
Cr_2O_3		0.80	1.20	22.53		0.332		0.64	1.40	18.17		0.314
NiO	0.33			0.26		0.247	0.28			0.28		0.205
Sum	99.8	99.8	99.1	99.6		99.82	99.5	99.4	99.5	98.9		99.43

[a] Plagioclase composition is from a Marie Celeste Fracture Zone harzburgite (Antp 89-HD9: 78.3% olivine, 19.5% enstatite, 1.25% diopside, 0.23% spinel and 0.68% plagioclase).

the various elements in the peridotites are plotted with respect to distance from the Speiss Ridge segment of the SW Indian Ridge, which is believed to mark the present-day location of the Bouvet hot-spot. There is a clear correlation of increasing compatible-element and decreasing incompatible-element concentrations in the peridotites with proximity to the hot-spot as expected from the mineral and modal variations described above and the shift in the major-element composition of the ridge basalt (*eg* Fig. 4). A notable feature is that chromium decreases rather than increases with proximity to the Bouvet hot-spot, correlating inversely with the abundances of the compatible elements magnesium and nickel. This confirms that chromium behaves incompatibly during mantle melting as previously suggested by some workers (*eg* Kurat *et al.* 1980; Dick & Fisher 1984).

The potential basaltic component of a peridotite is best inferred from the concentration of incompatible elements not normally retained in simple residues of melting. For major elements, the most useful is sodium, because it is most likely to behave like an incompatible trace element. For example, a typical primitive basalt contains roughly 2 wt% Na_2O, while the peridotites contain only 0.053%–0.005% Na_2O. There is only sufficient sodium, then, to form 2.5%–0.25% primitive basalt in these peridotites. This 'potential' basalt represents either basaltic component retained in the residual minerals at the end of melting or melt trapped interstitially in the peridotite. Sodium, in particular, does not behave exactly as a totally incompatible element, and some is probably retained in pyroxene at low degrees of melting. The amount of 'potential' basalt in the peridotites, implied from their sodium contents in

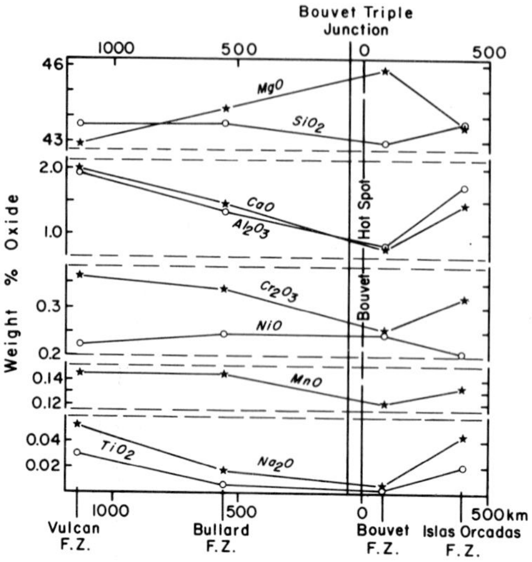

FIG. 5. Average residual mantle peridotite compositions plotted with distance from the present-day location of the Bouvet hot-spot, believed to be at the Speiss Ridge segment of the SW Indian Ridge, just E of the Bouvet Triple Junction and to the W of Bouvet Island. Compositions are computed as discussed in the text using the average mineral and modal compositions for each locality.

Table 3, correlates inversely with the concentration of compatible elements and with distance from the Bouvet hot-spot (Fig. 5). This correlation is compelling evidence, then, that the higher proportions of 'potential' basalt in the average peridotites in Table 3 reflect lower degrees of melting of the residual peridotite rather than less efficient melt removal or late-stage trapping of melt. By contrast, Maaloe & Aoki's (1977) average ocean island spinel peridotite xenolith contains 0.27 wt% sodium—sufficient to form 13.5% basalt by simple melting.

The efficiency of melt removal is a key parameter for modelling basalt generation and is defined as the proportion of the melt extracted from a peridotite divided by the total amount formed. The latter quantity determines the composition of the melt, which is a direct function of the degree of melting. The efficiency of melt extraction, however, is a function of the amount of melt dynamically stable in a peridotite, which determines the minimum trapped at the end of melting, and the velocity with which melt migrates by permeable flow relative to that of the mantle from which it forms. Overall, with only 2.5% – 0.25% potential basalt left in the typical abyssal peridotite, melt removal is clearly very efficient beneath ocean ridges. The best estimate of the efficiency, however, is the amount of potential basalt left in the most depleted abyssal peridotites, since the amount of basalt components retained in the residual minerals is minimized and it can be assumed that most of the potential basalt represents trapped melt. The Bouvet Fracture Zone peridotites are some of the most depleted ever sampled at an ocean ridge and are residues of 25% – 30% melting (Dick *et al.* 1984). The average Bouvet peridotite contains only 0.25% potential basalt, as measured by its sodium content, demonstrating that melt removal was locally 99% efficient — to the degree that the last melt formed resembles abyssal basalt.

Plagioclase peridotites

The origin of plagioclase in abyssal peridotites has been the subject of debate. Hamlyn & Bonatti (1980) proposed that it is the product of reaction of enstatite, diopside and spinel to form olivine and plagioclase with decreasing pressure in ascending mantle peridotite as it moves from the spinel to the plagioclase peridotite facies. Such an origin, however, requires plagioclase peridotites to be simple residues of melting. Others have pointed out that plagioclase peridotites are compositionally distinct from plagioclase-free peridotites and have suggested that they formed by impregnation of depleted residual peridotite by either *in situ* or transient melt (Church & Stevens 1971; Menzies 1973; Menzies & Allen 1974; Quick 1981a,b; Dick & Bullen 1984; Dick & Fisher 1984; Nicolas & Dupuy 1984; Boudier & Nicolas 1986).

Table 4 gives representative computed compositions for two equatorial Mid-Atlantic Ridge plagioclase peridotites from the Romanche Fracture Zone. This is the only region for which there are sufficient data to characterize these rocks adequately, although they are petrographically similar to plagioclase peridotites dredged along the SW Indian and American–Antarctic Ridges. The composition of the plagioclase peridotites is strikingly less refractory than that of the spinel peridotites in Table 3, with approximately an order of magnitude more sodium and titanium. As shown in Fig. 6, the concentration of titanium (and other incompatible elements) appears to increase linearly with increasing modal plagioclase from the concentrations in plagioclase-free spinel peridotites. The systematic shift in bulk composition with modal plagioclase, particularly for el-

TABLE 4. *Computed plagioclase peridotite compositions*

	Oliv.	En.	Di.	Sp.	Plag.	Rock
AII20–17–63						
Mode	32.8	29.9	19.5	00.6	17.1	99.9
SiO_2	40.93	55.20	51.26		49.68	48.62
TiO_2		0.13	0.40	0.71		0.117
Al_2O_3		3.26	4.64	23.73	32.08	8.36
FeO	9.22	5.52	2.75	19.36	0.15	5.12
MnO	0.17		0.06	0.17		0.066
MgO	50.30	32.56	17.46	14.52	0.08	28.65
CaO	0.05	1.82	21.07		14.61	7.50
Na_2O		0.05	0.41		3.40	0.773
K_2O			0.01		0.02	0.007
Cr_2O_3		0.36	0.88	40.12		0.449
NiO	0.26			0.11		0.082
Sum	100.93	98.90	98.94	99.32	100.02	100.49
Magnesium number	0.907	0.913	0.919	0.654		0.909
				$^{a}Cr_{53.1}$	$An_{70.3}$	
AII20–17–96						
Mode	68.8	17.1	2.8	0.3	11.0	100.0
SiO_2	40.44	55.73	51.29		48.26	44.26
TiO_2		0.37	0.63	1.48		0.084
Al_2O_3		2.91	5.51	23.54	32.60	4.97
FeO	9.83	6.04	3.44	26.27	0.15	7.78
MnO	0.17	0.14	0.13	0.09		0.14
MgO	49.44	32.48	17.20	12.25	0.11	39.13
CaO	0.05	1.63	20.61		15.87	2.96
Na_2O			0.19		2.39	0.318
K_2O		0.0	0.0		0.03	0.004
Cr_2O_3		0.23	0.36	34.90		0.126
NiO	0.25			0.19		0.168
Sum	100.18	99.53	99.36	98.72	99.41	99.94
Magnesium number	0.900	0.905	0.899	0.550		0.900
				$Cr_{49.8}$	$An_{78.4}$	

[a] Cr=Cr/(Cr+Al)

ements such as titanium which are only trace constituents of plagioclase, strongly argues that these are hybrid rocks with varying amounts of trapped impregnated basaltic melt.

Since the normative plagioclase content of abyssal basalts is actually quite uniform between 55% and 60% (Bryan & Dick 1982) and is unlikely to change much with pressure for basaltic melts in equilibrium with olivine and plagioclase (Presnall *et al.* 1978), we can use the plagioclase content of a peridotite to estimate the minimum amount of trapped melt. Thus plagioclase peridotite AII20–17–96 (Table 4) should contain roughly 20% trapped melt. As can be seen in Table 5, subtraction of 80% residual peridotite from its bulk composition does yield an approximate basaltic composition, consistent with a simple impregnation model. Despite a careful search, the maximum amount of plagioclase identified in any plagioclase–bearing peridotite is 17%, corresponding to roughly 30% trapped melt for a basalt with 55% normative or 60 vol.% plagioclase. This is close to the theoretical limit for melt in a grain-supported matrix. If more melt than this were trapped, gravitational settling of minerals would result in segregation of melt from the peridotite. Figure 6 shows a mixing line for a primitive basalt with 0.43 wt% TiO_2 and the average abyssal peridotite of Dick & Fisher (1984). As can be seen, the plagioclase peridotites scatter around this line, and within the limits of analytical uncertainty fit the impregnation model.

In contrast, a simple equilibrium melting model clearly shows that these rocks do not lie even close to an expected residual path, ruling out an origin as the simple residues of variable degrees of partial melting (Fig. 6). In the melting

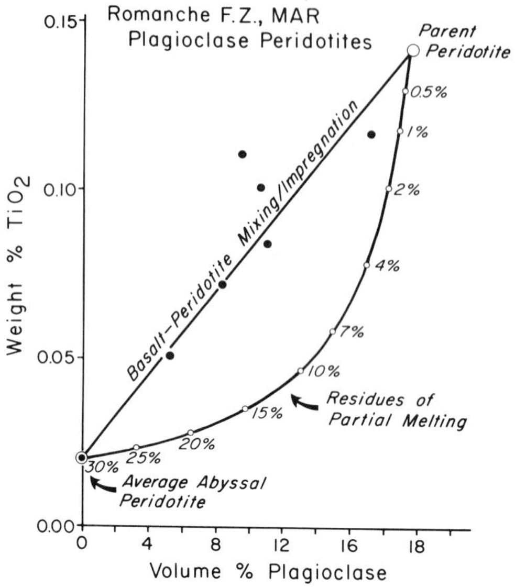

FIG. 6. Computed compositions of Romanche Fracture Zone plagioclase peridotites and the average abyssal residual spinel peridotite of Dick & Fisher (1984). A mixing impregnation line for a hypothetical basalt, with 60% normative plagioclase and 0.47 wt% TiO_2, and the residual peridotite which bests fit the data is shown. For comparison, a simple equilibrium melting model for a parent peridotite containing 30 wt% of the same basalt is given. The concentrations of TiO_2 in the inferred basalt and the residual peridotite were used to estimate the partition coefficient for titanium. It is assumed that the fraction of plagioclase entering the melt is 60%. Assuming a smaller plagioclase fraction increases the misfit of the melting model to the plagioclase peridotite data, as would a fractional, rather than an equilibrium, melting model.

model, we assume constant proportions of plagioclase (60 wt%) going into the melt and that titanium, which is not an essential structural constituent in any phase, shows behaviour approximately following Henry's law. This, and assuming peridotites from the same locality have similar melting histories, justifies application of the simple batch equilibrium equation for computing the change in titanium with partial melting. Fractional melting models would produce an even more extreme exponential decrease in titanium. The actual melting curve might vary somewhat for more sophisticated equilibrium or fractional melting models, but would not change the general conclusion that these peridotites are hybrid rocks produce by late impregnation of residual peridotite by melt migrating through the shallow mantle.

Melt distribution in the uppermost oceanic mantle

The distribution of melt in residual abyssal peridotites is bimodal. Despite conclusive evidence that they have undergone high degrees of melting, 80% of abyssal peridotites are plagioclase-free spinel harzburgites or lherzolites containing little trapped melt: evidence that at most a few per cent, and probably much less, can be held in stable mechanical equilibrium in the mantle beneath a ridge. This implies very effective forces driving melt segregation in the mantle, making simple equilibrium melting unlikely, and favours buoyancy-driven permeable flow models where melt migrates upwards faster than the mantle in which it forms (two-phase flow) (Sleep 1975; Ahern & Turcotte 1979; Maaloe & Scheie 1984; McKenzie 1984).

Plagioclase peridotites with significant trapped melt are either abundant or virtually absent in peridotite dredge hauls. This mimics the distribution of plagioclase in alpine-type peridotites, where plagioclase is largely absent except for the rare massif where it is locally abundant (*eg* the Alpine Lanzo Massif (Boudier & Nicolas 1977) or the Cordilleran Trinity Ophiolite (Quick 1981a,b). On average, abyssal plagioclase peridotites contain 2.5% plagioclase, representing about 4% trapped melt. At some localities like the Romanche Fracture Zone the average amount of trapped melt is much higher. This suggests that the abundance of crystallized trapped melt at a few localities represents a dynamic phenomenon resulting from the non-uniform flow of melt through the mantle and its channelization. Field observations in the Lanzo peridotite by Boudier & Nicolas (1977) show that where plagioclase is particularly abundant in a peridotite there is considerable evidence for melt segregation into discrete bodies and flow out of the mantle. The numerous podiform dunites found in alpine-type peridotites and their associated chromitite deposits are now widely accepted as having formed as crystal residues from such ascending bodies of melt (Dick 1977; Quick 1981a,b).

There are a number of possible sources for the melt which impregnated abyssal plagioclase peridotites. It could be essentially *in situ* melt generated from the impregnated peridotite itself (*eg* Menzies 1973; Menzies & Allen 1974). Boudier & Nicolas (1986), for example, have suggested that plagioclase peridotites are formed under conditions where the mantle ascending beneath a ridge spreading at a rate of less than a centimetre a year or near a transform fault meets conditions where heat loss by con-

TABLE 5. *Direct computation of trapped melt composition*

	Plag. Pd.−0.8 Sp. Pd. (AII20−17−96)	Recalculated to 100%	Primitive abyssal basalt
SiO_2	9.52	47.4	48.6
TiO_2	0.064	0.32	0.61
Al_2O_3	3.98	19.78	16.3
FeO	1.20	5.99	8.69
MnO	0.028	0.14	0.15
MgO	2.94	14.63	10.2
CaO	1.97	9.78	12.3
Na_2O	0.391	1.95	1.90
K_2O	0.012	0.06	0.07
Sum	20.101	100.00	98.82
Mg/(Mg+Fe)		0.813	0.676

duction to the surface becomes more efficient than the heat supply by advection, leading to melt crystallizing *in situ* rather than being expelled to the overlying crust. As 80% of abyssal peridotites dredged from fracture zones and the rift mountains at slow spreading ridges are plagioclase free, however, it is clear that this mechanism alone is insufficient to explain what are essentially local concentrations of trapped melt.

Whitehead *et al.* (1984) have suggested that plagioclase peridotites result when gravitational instability in the partially molten zone at shallow depth produces dynamically driven flow of partially molten mantle into a protrusion which forms a mantle diapir. This diapir then ascends comparatively rapidly towards the crust. In the normal case, melt flowing more rapidly by porous flow than the crystal mush as a whole concentrates at the top of the protrusion, from whence it segregates and rises out of the mantle. Occasionally, a protrusion forming at relatively shallow depths could produce a diapir which is emplaced into relatively cool lithosphere. Such a situation may occur near a transform fault or periodically beneath the crust at a very slow spreading ridge after a prolonged period of amagmatic extension has produced a significant thickness of lithosphere beneath the ridge owing to conductive cooling. When emplaced into relatively cool lithosphere much of the melt may crystallize at the margins of the diapir as it is cooled against the lithosphere before it can escape.

The data reported here, however, make it clear that typically melt is nearly entirely removed prior to emplacement of mantle peridotite to the base of the crust, rather than being trapped *in situ*. This suggests that the melt trapped in plagioclase peridotites is transient melt, migrating through an upwardly convecting melting mantle. Thus, plagioclase peridotites are a manifestation of the process by which melt is normally drained from the mantle beneath ridges. What is important is that plagioclase peridotites are direct evidence for dynamic gravitationally driven non-uniform melt flow and segregation beneath ocean ridges.

Major- and trace-element decoupling in primary basalts

The SW Indian and American−Antarctic Ridges appear to be classic examples of the decoupling of the major- and trace-element compositions of basalts. Le Roex *et al.* (1983, 1985) have studied the major-, isotopic and trace-element variability of SW Indian and American−Antarctic Ridge basalts in considerable detail in a transect along the ridge system that crosses over the Bouvet hot-spot. Unlike the Reykjanes Ridge near Iceland (Schilling 1973), however, they find no systematic gradient of the trace-element and isotopic basalt composition along the ridge with proximity to the hot-spot. Rather, isotopically and light-rare-earth enriched basalts, light-rare-earth- and isotopically depleted basalts and transitional varieties occur together at individual segments along the ridge. The proportion of depleted mid-ocean ridge basalts decreases, and that of enriched varieties increases, with proximity to the hot-spot. They attributed the eruption of such diverse magmas at individual spreading-centre segments to melting of a chemically heterogeneous mantle beneath the ridge and to the absence of steady-state magma chambers at very slow spreading ridges which might otherwise have homogenized the basalts.

Despite their often large trace-element and isotopic diversity, basalts from individual segments at the SW Indian and American–Antarctic Ridges appear to be the products of fractional crystallization of melts with very similar major-element compositions, reflecting a common major-element mantle source (Dick *et al.* 1984). An example is shown in Fig. 7 where the compositions of basalt glasses from a single segment of the SW Indian Ridge are plotted in a portion of the normative olivine–plagioclase–pyroxene ternary diagram. These glasses define a single tight trend in the plot which defines the liquid line of descent for the basalts erupted along this segment of the ridge. Bryan & Dick (1982) have shown that basalt suites from numerous other ridge segments around the world define similar liquid lines of descent which are systematically displaced from each other in composition space—each reflecting variable degrees of fractional crystallization of a different primary magma. Such shifts can best be explained by changes in the degree of melting and composition of the source region. Dick *et al.* (1984) have shown that the systematic shift in the liquidus trends along the ocean ridges in proximity to hot-spots correlates directly with increasing degrees of depletion of spatially associated mantle peridotites. This indicates that basalts from individual ridge segments, lying close to a single trend in olivine–plagioclase–pyroxene composition space, must have been derived from a *similar* major-element mantle source despite their differences in trace-element and isotopic composition. A number of glasses plotted in Fig. 7 are also annotated with the strontium isotopic composition found by Le Roex *et al.* (1983), showing a large range in composition. This range is even larger if whole-rock strontium isotopic analyses for the same dredge hauls are included. This isotopic variation reflects similar variability of other critical trace-element ratios such as Zr/Nb and varying rare-earth-element patterns and can only be explained if the basalts were derived from *different* isotopic and trace-element mantle sources.

Thus there is the apparent contradiction of lavas requiring mantle sources of similar major-element composition but very different trace-element and isotopic composition at individual ridge segments along the SW Indian and American–Antarctic Ridges. The same problem exists for basalts from other ridges as well. For example, glasses from the FAMOUS region of the Mid-Atlantic Ridge, representing some 40 dredge hauls, define an incredibly tight range in olivine–plagioclase–pyroxene space (Bryan 1979), reflecting widely varying degrees of fractional crystallization but a very uniform major-element mantle source region—despite considerable trace-element heterogeneity (*eg* Langmuir *et al.* 1977; Le Roex *et al.* 1983).

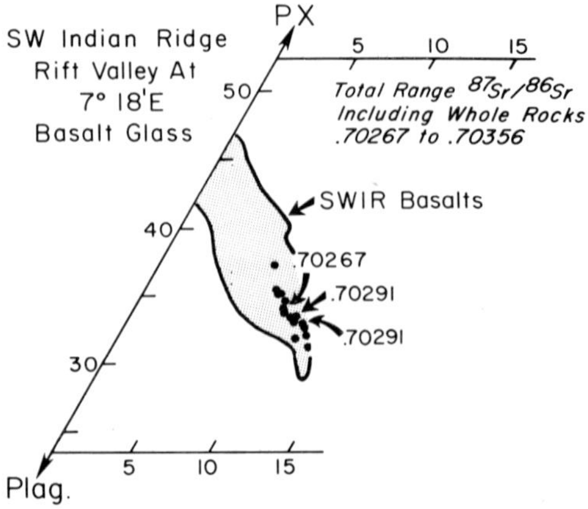

FIG. 7. Unpublished electron microprobe analyses of basalt glasses from a single rift valley segment of the SW Indian Ridge plotted in a portion of the normative olivine–plagioclase–pyroxene ternary. Individual glasses, shown by the arrows, are annotated with the measured strontium isotopic composition from Le Roex *et al.* (1983). Note that the total range of strontium isotopic compositions measured by Le Roex *et al.* (1983) for this one ridge segment is even larger when analyses of rocks without basalt glass are included. Shown for comparison is the field of composition for SW Indian Ridge basalt glasses from the Bouvet Triple Junction (1°W to 11°E).

The apparent contradiction is explicable in a continuously melting mantle where the melt migrates upwards at a substantially faster rate than the mantle matrix. At depth, as the ascending mantle begins to melt as it is drawn up between the diverging tectonic plates, the low partition coefficients of the incompatible trace elements dictate that they will be virtually removed from the source in the first few per cent melting. This produces an enriched melt leaving a residue still relatively rich in the less-incompatible basaltic components (*eg* Ca, Al, Fe, Si, Mg). Since little magma can be trapped in the melting peridotite, it ascends upwards, faster than the solid residue, through the overlying ascending mantle column. The overlying mantle, previously depleted in incompatible elements by earlier melting, can contribute only major elements, diluting the trace-element composition but not changing the relative proportions or isotopic composition significantly. The exact reverse is true for the major-element composition of the melt. As the melt ascends, decompression melting of the enclosing more slowly ascending mantle matrix must continue and contribute the major portion of the elemental makeup of the melt which finally evolves to that of tholeiitic basalt prior to eruption.

Thus, the major-element composition of the melt must reflect the composition of the entire mantle column through which it passes and the total degree of melting, which is controlled by the initial temperature and total depth of upwelling. Since the column is at least 70 km high (Ahern & Turcotte 1979), changes in the major-element characteristics of oceanic basalts will be small in time and space, explaining the remarkable gross correlation of the major-element composition of peridotites dredged from crust of varying age along fracture zones and of spatially associated zero-age basalts (Dick *et al.* 1984) along ocean ridges. The incompatible-trace-element heterogeneities are free to vary on the length scale for removal of the incompatible elements from a source peridotite entering the melting column—probably no more than about 10 km of upwelling once a peridotite passes its solidus. The smaller the percentage of melt that can be trapped in a partially molten zone prior to segregation, the shorter is the length scale of heterogeneity in the underlying mantle which will be preserved in the basalts erupted. This length scale must be fairly short, and the mantle quite heterogeneous, when it is considered that depleted (N-type, high Zr/Nb and Y/Nb ratios) and plume-enriched (T- and P-type, low Zr/Nb and Y/Nb ratios) mid-ocean ridge basalts have all been collected together on a zero-age central high at 7°E on the SW Indian Ridge (Le Roex *et al.* 1983).

The overall implication of all this is that the major-element composition of basalts is more a function of mantle dynamics than is the trace-element and isotopic composition, which is critically dependent on the composition of the source region at the base of the melting mantle column. This fundamental decoupling of the trace- and major-element cycles of oceanic basalts is clearly evident when their compositions in other environments and on different scales are examined. Thus, while there is no systematic trace-element or isotopic gradient for ridge basalts across the Bouvet hot-spot, there is a systematic major-element trend, with a shift to higher degrees of melting for residual peridotites and spatially associated basalts as the hot-spot is approached. This likely reflects deeper mantle up-welling and higher initial mantle temperatures near hot-spots (Dick *et al.* 1984; Michael & Bonatti 1985; Klein & Langmuir 1987). Unlike the situation at individual ridge segments, there is a general correlation between the major-element composition of ridge basalts and their isotopic composition along many ridges with proximity to mantle plumes; with more isotopically enriched basalts reflecting higher degrees of mantle melting in these regions. Although the Bouvet region appears to be an exception, the increasing proportion of enriched basalts near Bouvet does reflect a similar pattern of overall variation.

Exactly the opposite trend, however, has been seen by Batiza & Vanko (1984) for the composition of basalts erupted at small Pacific seamounts. Unlike many ocean ridge segments, chemically distinct basalt groups at individual seamounts define different major-element liquid lines of descent. The positions of these trends correlate directly with the isotopic and trace-element composition of the basalts. Yet it is the basalts whose major-element compositions reflect the lowest degrees of melting which appear to be derived from the geochemically enriched source, while those produced by the highest degrees of melting appear to have been derived from a geochemically depleted mantle source. Their explanation is that these basalts are the products of sequential melting of a locally heterogeneous mantle, where geochemically enriched veins melt first, producing enriched basalts, while higher degrees of melting produce more depleted basalts reflecting the overall mantle composition (Batiza & Vanko 1984; Zindler *et al.* 1984).

Thus, by comparing their major- and trace-element evolutions, the difference between the generation of small seamounts and major mantle plumes such as Iceland and Hawaii becomes evident. One represents a major mantle-melting anomaly, producing large volumes of isotopically enriched melt derived from a relatively deep mantle source, the other, local mantle heterogeneity and low rates of magma supply during seamount volcanism which inhibits magma mixing and homogenization.

It is worth noting that the correlation between major and trace elements changes even along ocean ridges. Klein & Langmuir (1987), for example, have shown that along the deepest section of the SE Indian Ridge, at the Antarctic discordance zone south of Australia, basalts are erupted with the geochemical trace-element and isotopic signatures normally associated with mantle hot-spots. The major element composition of these basalts, however, shows that they have been derived by very low degrees of mantle melting, consistent with the great depth of the ridge. Thus these basalts must represent upwelling and melting of unusually cold mantle (*ie* a cold spot (Klein & Langmuir 1987)) or the presence of anomalously undepleted mantle at relatively shallow depths.

The change in basalt major-element composition along ocean ridges appears to be discontinuous, with jumps between ridge segments observed along both slow and fast spreading ridges (Whitehead *et al.* 1984; Thompson *et al.* 1985). This argues that, unlike the small Pacific seamounts, there is a process which homogenizes ridge basalts with respect to major elements on the length scale of individual ridge segments. Lateral migration of melts through the mantle beneath the ridge segment towards a magmatic centre at its midpoint would be one mechanism which would homogenize the major-element composition of primary magmas originating at different points in an initially heterogeneous mantle by re-equilibration with a common residual mantle section. However, as the residual mantle has incompatible-element concentrations of the order of one-thousandth of that of the basalt, the critical trace-element concentrations and isotopic ratios of the basalt would be relatively unaffected by the re-equilibration. The residual mantle, on the other hand, would always have a trace-element and isotopic equilibration imposed by the last batch of melt to pass through it.

A geologic model for melt segregation and crustal formation at ocean ridges

There has been a rapid shift in thinking about ocean crust generation from essentially two-dimensional models of an ocean crust viewed as a simple uniform layer cake consisting of pillow basalt, sheeted dykes and gabbro to more complex three-dimensional models (Whitehead *et al.* 1984; Crane 1985; Schouten *et al.* 1985; Rabinowicz *et al.* 1987) which emphasize the lateral variability of the structure of the ocean crust along ocean ridges. This is the result of new seismic evidence for a laterally heterogeneous crust and high-resolution geomorphological and volcanological studies which show regular volcanic segmentation along the ocean ridges. The results presented here, and in preliminary form by Whitehead *et al.* (1984), augment these studies and greatly constrain models for magmatism and crustal formation in the oceans. The principal results include the following.

(1) The distribution of rocks dredged from fracture zones on very slow spreading ridges indicates even more extreme thinning of the ocean crust there than is shown by the seismic evidence at slow spreading ridges like the Mid-Atlantic Ridge. Close to two-thirds of the rock exposed in the transform valley is altered residual peridotite. Even if this exceptional exposure of peridotite could be accounted for by tectonic disruption of the ocean crust, the small abundance of gabbro in the dredge hauls indicates that no uniform gabbroic layer, corresponding to seismic layer 3 in 'normal' ocean crust, existed and that only ephemeral magma chambers formed rarely beneath the fracture zone. In addition, the scarcity of dunite associated with the dredged peridotites also argues that the underlying mantle was not a significant conduit for magmas from the mantle. The crustal section, then, is most likely to consist of a veneer of pillow basalts erupted directly over serpentinized mantle peridotite. Locally it is possible that the mantle was emplaced directly onto the seafloor where it forms crust only as a result of serpentinization.

(2) Overall, the peridotites dredged from the SW Indian and American–Antarctic Ridge fracture zones have undergone a high degree of melting and a severe depletion in basaltic components, reflecting the formation and removal of large volumes of melt (10%–30%). This is surprising in view of the apparent dramatic thinning of the ocean crust in the vicinity of the fracture zones. In view of the comparative scarcity of basalt and associated dunites, and the near absence of gabbros in many of these fracture zones, the melt formed in the mantle beneath the fracture zone must have flowed laterally in the mantle beneath the fracture zone to feed a magmatic centre located in the adjacent ridge segment. Alternatively, the peridotites dredged in the fracture zones may have been part of an upwelling mantle diapir centred beneath a ridge segment which were emplaced by shallow mantle flow parallel to the ridge axis to the fracture zone after melt segregation. Direct evidence for such a phenomenon has been seen in the disposition of magmatic ore deposits and diverging mantle flow lines in the residual mantle sections of some ophiolite complexes (*eg* Nicolas & Violette 1982).

(3) The basalts dredged from the fracture zones along very slow spreading ridges show a large range in composition, similar to that

for basalts dredged in the adjacent rift valleys, requiring extensive shallow-level fractional crystallization. In the absence of any evidence for a persistent magma chamber beneath the fracture zone, this requires that the basalts differentiated at and were erupted from a magmatic centre in the rift valley adjacent to the fracture zone floor. It is important to note in this context that on-land fissure systems can deliver relatively unfractionated or highly evolved magmas from a magmatic centre anywhere along their length during a robust eruption. This indicates that the length scale of magmatic segmentation is the limiting scale for sampling lateral mantle heterogeneities along ocean ridges by dredging rift valley basalts.

(4) The bimodal distribution of trapped melt in abyssal peridotites indicates that segregation and flow of melt out of the mantle was not uniform along ocean ridges. In the typical case the mantle is effectively (99%) drained of melt prior to its emplacement to the base of the crust. The local abundance of plagioclase peridotites, then, is evidence for the segregation and flow of melt due to some form of gravitational instability in the partially molten mantle beneath ocean ridges.

(5) The systematic covariation of the major-element composition of abyssal peridotites and spatially associated basalts with proximity to mantle plumes demonstrates the widely inferred direct cogenetic relationship between the two, and clearly demonstrates the first-order dependence of their composition on the local dynamics of mantle upwelling such as depth of convection and total heat available for melting.

(6) Despite trace-element and isotopic diversity, basalts from individual ridge segments were derived from primary magmas with similar major-element compositions. These observations can be explained if melt flows locally through the depleted mantle at the end of melting towards the midpoint of a ridge segment. This would cause melts originating at different points in an initially heterogeneous mantle beneath a ridge segment to equilibrate with a common section of mantle—which, for the most part, would homogenize only the major-element compositions of the melts with little effect on critical incompatible-trace-element or isotopic ratios.

A general model for magmatism at slow spreading ridges is shown in Plate 1 based on the results of this study extrapolated to slightly faster spreading rates (2 cm a^{-1}) and from seismic and volcanological data from slow and fast spreading ridges. In this model, ridge magmatism is viewed as the development of a series of regularly spaced tholeiitic shield volcanoes developed above long-lived instabilities in the upwelling partially molten asthenosphere. The model in some respects is strikingly like that developed by Marsh (1979) for island arcs. Major differences lie in the absence of a thick lithosphere overlying the melt production zone, the absence of a subduction component to the melt and continuous extension during volcanism. Owing to the extension, a steady-state situation arises where ribbons of crust parallel to the spreading direction are produced rather than a series of constructional volcanoes as at island arcs.

Following the geomorphological volcanological and seismic observations, the crustal segmentation shown extends beyond the obvious division by fracture zones. It includes regularly spaced bathymetric highs and intervening saddle points which may occur in conjunction with or between major transforms (Ballard & Francheteau 1982; Schouten & Klitgord 1982; Francheteau & Ballard 1983; Crane 1985; Schouten *et al.* 1985; MacDonald *et al.* 1987) and 'zero-offset' fracture zones which have the morphological expression of fracture zones but no offset of the ridge. The morphological highs mark the location of major magmatic centres along the ridge and are often associated with maxima of hydrothermal activity (Francheteau & Ballard 1983). The segmentation of volcanism appears to be on a length scale of 30–60 km and is similar to that found in island arcs. The spacing can be directly related to spreading rate, which can be predicted by gravitationally driven instability models for melt segregation beneath ocean ridges (Schouten *et al.* 1985).

Layer 3 is shown as the plutonic root zone of the shield volcanoes. In Plate 1 it is the crystallized remains of transient multiple magma chambers, the product of repeated injections of melt to form the mantle. An excellent example of such magma chambers are the multiple intrusions seen in the Norwegian Karmøy ophiolite (Pedersen 1986). As a consequence of the relatively rapid cooling of magmas in the crust at the low rates of magma supply associated with slow spreading ridges (*eg* Sleep 1975; Kuznir 1980) the individual magma chambers are small and ephemeral with their distal ends tapering out rapidly along strike. Secondary magma chambers, however, are likely to form in the crust some distance from the magmatic centre, fed by dyke injection down the fissure system of the rift valley and by occasional small batches of melt fed from the underlying mantle. This interpretation of layer 3 fits well with the observations along the SW Indian and American–Antarctic Ridges and is strongly supported by seismic studies in old Atlantic Ocean crust. Mutter *et al.* (1985), using multichannel seismics, have clearly identified pronounced thinning of the ocean crust and the apparent disappearance of layered cumulates at the small-offset (20 km) Blake Spur and two adjacent zero-offset fracture zones (Schouten & White 1980). They postulate that this 'accompanies a reduction in chamber size and persistence as the accretion center becomes more distant from its primary source of magma'. Their data make it clear that local thinning of the crust between magmatic centres represents more than the inhibiting effect of a large-offset fracture zone on magma generation and eruption, but rather reflects a fundamental facet of the way ocean crust is created. At greater rates of magma supply, such as the East Pacific Rise, where it is likely that there are large long-lived magma chambers (*eg* Sleep 1975), it is likely that magma chambers may extend for a considerable distance along strike and may even merge to form magmatic super-chambers, creating a more uniform gabbroic layer 3 and eliminating many of the surficial manifestations of segmented ridge volcanism. Detrick *et al.* (1987), for example, have identified a continuous reflector extending more than 90 km along the East Pacific Rise between 9°N and the Clipperton Fracture Zone which significantly exceeds the 68 km segment length predicted for this spreading rate (Schouten *et al.* 1985).

The basaltic crust constituting seismic layer 2 is produced by eruption and intrusion of melts from the magma chamber beneath the shield volcanoes down and along the rift valley fissure system to the fracture zone floor (Whitehead *et al.* 1984; Thompson *et al.* 1985). Magmas cooling at depth in the fissure system produce sheeted dykes. Although the slope of the median valley floors towards the fracture zones and between ridge segments appear gentle, of the order of 1°–4°, the elevation drop required to drive magma along a fissure system is small. The overall slope of Hawaii to the seafloor, for example, is only about 3.5°, while locally the Kilauea rift zone slopes only 1°–2°. In Iceland the Sveinagja fissure system, developed during rifting in 1874–1875, drained the Askja magma chamber, resulting in subsidence of the Oskjuvatn caldera and extrusion of lava 40–65 km to the N (Sigurdsson & Sparks 1978) down a 0.4° slope. By comparison, the overall slope of the Mid-Atlantic rift valley to its southern intersection with the Kane Fracture Zone is 1.7°. Just S of this intersection, the median valley contains a central linear volcanic high which slopes 3.7° down the valley over 15 km to the fracture zone floor.

In the model, the mechanism used to induce lateral melt migration is analogous to simple Rayleigh–Taylor instability. It supposes that a partially molten layer is created beneath the ridge axis where melt accumulates until the layer goes unstable and drains to feed overlying magma chambers in the crust. In the cartoon, this layer is shown as distinct from the lithosphere–athenosphere boundary, although there is no particular reason why the two boundaries could not generally coincide. It should also be recognized that there are a number of other potential forces which may focus magma segregation to a narrow zone beneath ridge axes; for example, pressure gradients arising due to corner flow of the matrix at spreading centres can cause melt to migrate toward the ridge axis, enabling the extraction of small melt fractions from a wide melting zone to produce a narrow zone of volcanism at the surface (Speigelman & McKenzie 1987).

Magmas rising from the instability points are shown in large kilometre-scale balloons. While computations show that melts might rise in such a manner through the asthenosphere beneath ocean ridges, they may also rise or be assisted through brittle fracture (Turcotte 1982). One possibility is that the stress concentration above a slowly rising diapir will cause sudden failure and brittle fracture of the roof, causing the magma to drain upward through dyke propagation until viscous head loss in the dyke halts fracture propagation. The subsequent rise and collection of the magma at the top of the dyke could then form a new magma chamber which would subsequently lead to repeated brittle fracture and draining till the melt reaches the crust. That segregated bodies of melt forming subcrustal magma chambers do exist beneath some types of oceanic spreading centres is shown by the presence of massive chromitite bodies in podiform dunites cross-cutting the residual mantle peridotites in many ophiolites. These have formed from melts rising through the mantle (*eg* Nicolas & Violette 1982). The distribution of such ore bodies is of considerable interest and has been found to be baffling to many geologists. One alpine peridotite will contain numerous economic deposits, while another

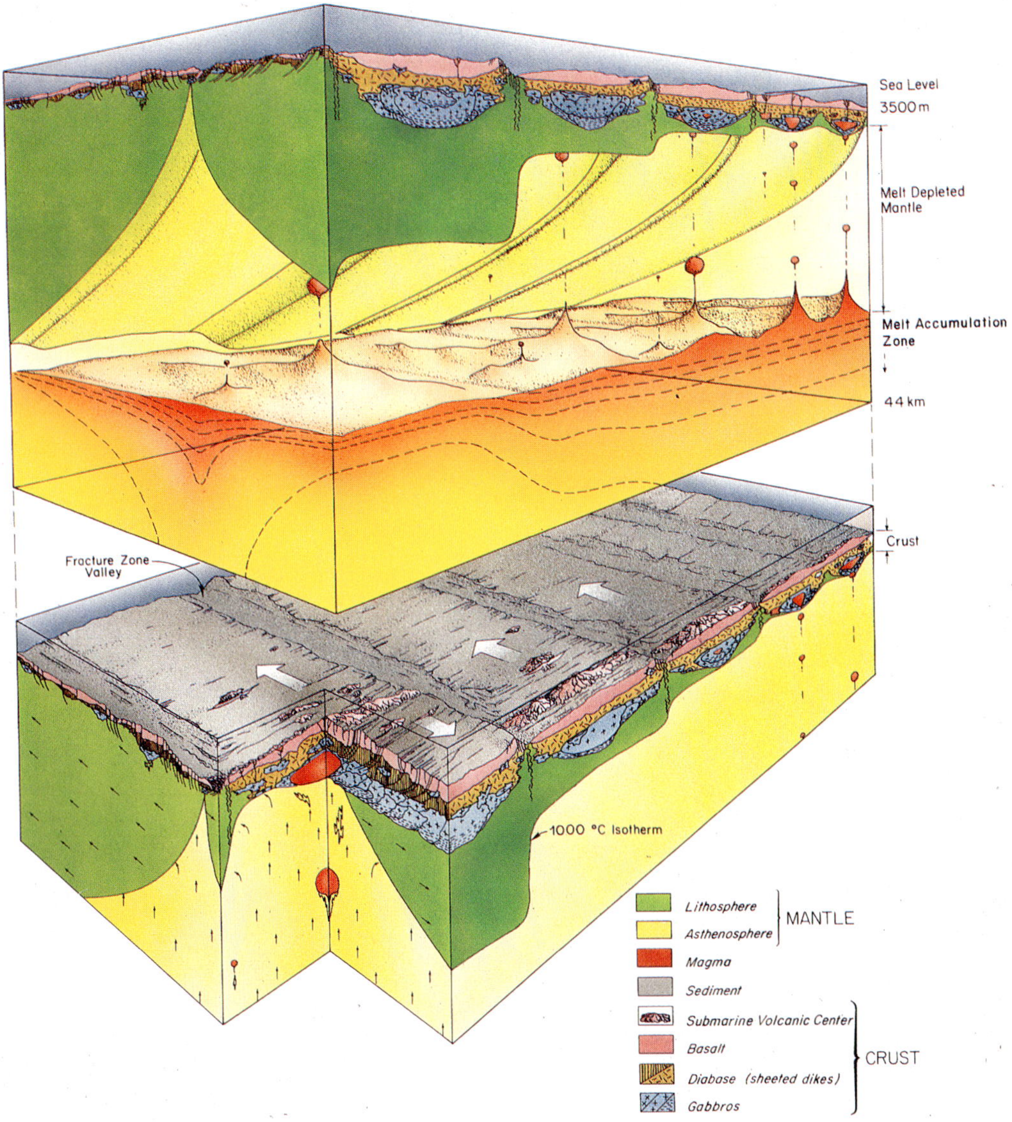

PLATE 1. Geological model for a slow spreading ocean ridge. Drawn for a spreading rate of about 20 mm yr^{-1}. Contours in the upper block are for inferred increasing flux of melt. Large white arrows in the lower block show the spreading direction. The lower surface in the upper diagram is an inferred partially molten zone.

similar peridotite will be devoid of them. In the model presented here, however, this distribution is explicable as such bodies would largely form only in a narrow zone concentrated above an instability focusing melt segregation in the mantle.

Another consideration, shown in Plate 1, is that as the partially molten layer drains this will affect the critical wavelength of the instability. At some point as the layer thins, it would be likely that secondary instabilities might form leading to intrusion of melt in regions away from the primary magmatic centres—beneath the floor of the fracture zone for example. In fact it would be an exaggeration to say that all magmas erupted along a single ridge segment have a common mantle source: only that the first-order major-element variability is consistent with this. Another point is that an instability-driven process only requires a negative contrast in viscosity and density between a partially molten zone and the overlying mantle. Such a zone does not have to be stationary, and could even be a migratory melt wave propagating upward through a continuously melting mantle column. It could also be caused by the deceleration and overturn of a mantle diapir as it approaches the lithosphere boundary during two-phase flow of melt and solid (*eg* Nicolas *et al.* 1987). Any scenario creating such a contrast in density and viscosity along a linear melting zone would be adequate. The essential point is not the particular analogue used for lateral mantle flow. Rather, it is the substantial evidence at very slow spreading ridges for punctuated magmatism above an essentially linear zone of melt generation, which requires a process involving non-uniform flow and segregation of melt in the mantle focused to the midpoint of spreading centre segments. This is apparently a fundamental facet of ocean ridge magmatism that dictates much of the structure and evolution of the ocean crust, particularly at slower spreading rates.

ACKNOWLEDGMENTS: This paper is dedicated to Hatten S. Yoder and an early version was presented at the International Conference and Field Study on the Physical Chemical Principles of Magmatic Processes, which was held on 1986 June 16–22 in his honour in Hawaii. Hatten Yoder not only made a remarkable scientific contribution to understanding of the evolution of magmas over his long and productive career but also to the education and advancement of a new generation of scientists from which this individual benefited among many.

This research was supported by the National Science Foundation Grants OCE84-16634, DPP83-16490 and DPP87-20002, as well as by the Woods Hole Oceanographic Institution Center for Analysis of Marine Systems Geodynamics Program. The basic model presented here for melt segregation from the oceanic mantle was the product of discussions with Dr Hans Schouten and Dr Jack Whitehead. In addition, Dr Peter Meyer, Dr Jack Casey, Mr Kevin Johnson and Mr Jon Snow gleefully provided ample criticism. Mr Beecher Wooding provided major technical assistance and supervised much of the collection, curation, preparation and analysis of rocks from along the SW Indian Ridge. Dr Robert L. Fisher is acknowledged for providing numerous samples from the collections of the Scripps Institute of Oceanography and for improving the author's literary style, and Dr Anton P. Le Roex and Dr Hugh Berg for assistance in gathering the samples and surveying the regions in which they were collected. The author gratefully acknowledges useful reviews from Professor Dan Mckenzie, Dr Mike Perfit and Dr Andy Saunders.

References

AHERN, J. L. & TURCOTTE, D. L. 1979. Magma migration beneath an ocean ridge. *Earth and Planetary Science Letters* **45**, 115–122.

ANDERSON, R. N. & NISHIMORI, R. K. 1979. Gabbro, serpentinite and mafic breccia from the east Pacific. *Journal of Physics of the Earth* 27, 467–480.

VAN ANDEL, TJ. H., VON HERZEN, R. P. & PHILLIPS, J. D. 1971. The Vema Fracture Zone and the tectonics of transverse shear zones in oceanic crustal plates. *Marine Geophysical Researches* **1**, 261–283.

AUGE, T. 1983. *Etude Minéralogique et Pétrographique de Roches Basiques et Ultrabasiques du Complexe Ophiolotique du Nord Oman.* Thèse de doctorat de troisième cycle, Universite de Orléans, Série Documents du B.R.G.M. **65**, 263 pp.

AUMENTO, F. & LOUBAT, H. 1971. The Mid-Atlantic Ridge near 45°N. XVI. Serpentinized ultramafic intrusions. *Canadian Journal of Earth Sciences* **8**, 631–663.

BALLARD, R. D. & FRANCHETEAU, J. 1982. The relationship between active sulfide deposition and the axial process of the mid-ocean ridge. *Marine Technology Society Journal* **16**, 8–22.

BATIZA, R. & VANKO, D. 1984. Petrology of young Pacific seamounts. *Journal of Geophysical Research* **89**, 11 235–11 260.

BONATTI, E. 1976. Serpentinite protrusions in the oceanic crust. *Earth and Planetary Science Letters* **32**, 107–113.

—— & HAMLYN, P. R. 1978. Mantle uplifted block in the Western Indian Ocean. *Science* **201**, 249–252.

—— & —— 1980. Petrology of mantle-derived ultramafics from the Owen Fracture Zone, northwest

Indian Ocean: implications for the nature of the oceanic upper mantle. *Earth and Planetary Science Letters* **48**, 65–79.
—— & Honnorez, J. 1976. Sections of the earth's crust in the equatorial Atlantic. *Journal of Geophysical Research* **81**, 4104–4116.
Boudier, F. & Coleman, R. G. 1981. Cross section through the peridotite in the Samail ophiolite, southeastern Oman mountains. *Journal of Geophysical Research* **86**, 2573–2592.
—— & Nicolas, A. 1986. Harzburgite and lherzolite subtypes in ophiolitic and oceanic environments. *Earth and Planetary Science Letters* **76**, 84–92.
—— & —— 1977. Structural controls on partial melting in the Lanzo peridotites. *In*: Dick H. J. B. (ed.) *Magma Genesis, Proceedings of the American Geophysical Union Chapman Conference on Partial Melting in the Earth's Upper Mantle*. Oregon, Department of Geology and Mineral Industries, Bulletin **96**, 63–78.
Bryan, W. B. 1979. Regional variation and petrogenesis of basalt glasses from the FAMOUS area, Mid-Atlantic Ridge. *Journal of Petrology* **20**, 293–325.
—— & Dick, H. J. B. 1982. Contrasted abyssal basalt liquidus trends: evidence for mantle major element heterogeneity. *Earth and Planetary Science Letters* **58**, 15–26.
Casey, J. F. 1986. Ultramafic rocks from the MAR at 23°N: evidence for high-temperature alteration and high-temperature, low to moderate stress deformation of mantle tectonites beneath the median valley. *Eos (Transactions of the American Geophysical Union)* **67**, 1214.
Cassard, D., Nicolas, A., Rabinovitch, M., Moutte, J., Leblanc, M. & Prinzhofer, A. 1981. Structural classification of chromite pods in southern New Caledonia. *Economic Geology* **76**, 805–831.
Caytrough, 1979. Geological and geophysical investigation of the Mid-Cayman Rise Spreading Center: initial results and observations. *In*: Talwani, M., Harrison, C. J. & Hays, D. E. (eds) *Deep Drilling Results in the Atlantic Ocean: Ocean Crust*. American Geophysical Union Maurice Ewing Series 2, 66–93.
Church, W. R. & Stevens, R. K. 1971. Early Paleozoic ophiolite complexes of the Newfoundland Appalachians as mantle–oceanic crust sequences. *Journal of Geophysical Research* **76**, 1460–1466.
Crane, K. 1985. The spacing of rift axis highs: dependence upon diapiric processes in the underlying athenosphere. *Earth and Planetary Science Letters* **72**, 405–414.
Detrick, R. S., Buhl, P., Vera, E., Mutter, J., Orcutt, J., Madsen, J. & Brocher, T. 1987. Multi-channel seismic imaging of a crustal magma chamber along the East Pacific Rise. *Nature* **326**, 35–41.
Dick, H. J. B. 1977. Evidence of partial melting in the Josephine Peridotite. *In*: Dick, H. J. B. (ed.) *Magma Genesis, Proceedings of the American Geophysical Union Chapman Conference on Partial Melting in the Earth's Upper Mantle*. Oregon, Department of Geology and Mineral Industries, Bulletin **96**, 59–62.
—— 1979. Alteration and metamorphism of peridotite at the Islas Orcadas Fracture Zone. *Eos (Transactions of the American Geophysical Union)* **60**, 973.
——, Bryan, W. B. & Thompson, G. 1981. Low-angle faulting and steady-state emplacement of plutonic rocks at ridge–transform intersections. *Eos (Transactions of the American Geophysical Union)* **62**, 406.
—— & Bullen, T. B. 1984. Chromian spinel as a petrogenetic indicator in abyssal and alpine-type peridotites and spatially associated lavas. *Contributions to Mineralogy and Petrology* **86**, 54–76.
—— & Fisher, R. L. 1984. Mineralogic studies of the residues of mantle melting: abyssal and alpine-type peridotites. *In*: Kornprobst, J. (ed.) *Kimberlites II: The Mantle and Crust–Mantle Relationships*. Elsevier, Amsterdam, 295–308.
—— & Bryan, W. B. 1984. Mineralogic variability of the uppermost mantle along mid-ocean ridges. *Earth and Planetary Science Letters* **69**, 88–106.
——, Meyer, P. S. & Gallo, D. G. 1987. Crustal variability at the Atlantis II F. Z. *Eos (Transactions of the American Geophysical Union)* **68**, 408.
——, Thompson, G. & Le Roex, A. P. 1982. Ferrogabbros and the evolution of ferrobasalts. *Eos (Transactions of the American Geophysical Union)* **63**, 475.
Engel, C. G. & Fisher, R. L. 1975. Granitic to ultramafic rock complexes of the Indian Ocean Ridge system, western Indian Ocean. *Geological Society of American Bulletin* **96**, 1553–1578.
Fisher, R. L., Dick, H. J. B., Natland, J. H. & Meyer, P. S. 1986. Mafic/ultramafic suites of the slowly spreading Southwest Indian Ridge: Protea exploration of the Antarctic Plate Boundary, 24° E–47°E, 1984. *Ophiditi* **11**, 147–178.
Fox, P. J., Schreiber, E., Rowlett, H. & McCamy, K. 1976. The geology of the Oceanographer Fracture Zone: a model for fracture zones. *Journal of Geophysical Research* **81**, 4117–4128.
Francheteau, J. & Ballard, R. D. 1983. The East Pacific Rise near 21°N, 13°N and 20°S; inferences for along strike variability of axial processes. *Earth and Planetary Science Letters* **64**, 93–116.
——, Choukroune, P., Hekinian, R., Le Pichon, X. & Needham, H.D. 1976. Oceanic fractures do not provide deep sections in the crust. *Canadian Journal of Earth Sciences* **13**, 1223–1235.
Greenbaum, D. 1972. *The Geology and Evolution of the Troodos Plutonic Complex and Associated Chromite Deposits, Cyprus*. Unpublished PhD dissertation. Department of Earth Sciences, University of Leeds, 141 pp.
Gronvold, K. 1988. Krafla lavas and lava composition within a fault swarm. *Symposium on Geologic and Geochemical Evidence for Segmentation of Continental and Oceanic Rifts*. Woods Hole Oceanographic Institution, Woods Hole, MA, 17.
Hamlyn, P. R. & Bonatti, E. 1980. Petrology of

mantle-derived ultramafics from the Owen Fracture Zone, northwest Indian Ocean: implications for the nature of the oceanic upper mantle. *Earth and Planetary Science Letters* **48**, 65–79.

HERBERT, A., BIDEAU, R. D. & HEKINIAN, R. 1983. Ultramafics and mafic rocks from the Garret transform fault near 13°30′S on the East Pacific Rise: igneous petrology. *Earth and Planetary Science Letters* **65**, 107–125.

HESS, H. H. 1955. Serpentinites, orogeny and epeirogeny. *Geological Society of America, Special Paper* **62**, 391–408.

—— 1962. History of the ocean basins. *In*: ENGEL, A. E. J., JAMES, H. L. & LEONARD, B. F. (eds) *Petrologic Studies: a Volume in Honor of A. F. Buddington*. Geological Society of America, 599–620.

HOPSON, C. A., COLEMAN, R. G., GREGORY, R. T., PALLISTER, J. S. & BAILEY, E. H. 1981. Geologic section through the Samail ophiolite and associated rocks along a Muscat Ibra transect. *Journal of Geophysical Research* **86**, 2527–2544.

KARSON, J. & DICK, H. J. B. 1983. Tectonics of ridge–transform intersections at the Kane Fracture Zone. *Marine Geophysical Researches* **6**, 51–98.

——, J., THOMPSON, G., HUMPHRIS, S. E., EDMOND, J. M., BRYAN, W. B., BROWN, J. R., WINTERS, A. T., POCKALNY, R. A., CASEY, J. F., CAMPBELL, A. C., KLINKHAMMER, G., PALMER, M. R., KINZLER, R. J. & SULANOWSKA, M. M. 1987. Along-axis variations in seafloor spreading in the MARK area. *Nature* **328**, 681–685.

KIMBALL, K. L., SPEAR, F. S. & DICK, H. J. B. 1985. High temperature alteration of abyssal ultramafics from the Islas Orcadas Fracture Zone, South Atlantic. *Contributions to Mineralogy and Petrology* **91**, 307–320.

KLEIN, E. & LANGMUIR, C. H. 1987. Global correlations of ocean ridge basalt chemistry with axial depth and crustal thickness. *Journal of Geophysical Research* **92**, 8089–8115.

KOMOR, S. C., ELTHON, D. & CASEY, J. F. 1985. Mineralogic variation in a layered ultramafic cumulate sequence at North Arm Mountain Massif, Bay of Islands ophiolite, Newfoundland. *Journal of Geophysical Research* **90**, 7705–7736.

KURAT G., PALME, H. SPETTEL, B. BADDENHAUSEN, H. HETMEISLER, H. PALME, C. & WANKE, H. 1980. Geochemistry of ultramafic xenoliths from Kapfenstein, Austria: evidence for a variety of upper mantle processes. *Geochimica, Cosmochim, Acta* **44**, 45–60.

KUZNIR, N. J. 1980. Thermal evolution of the oceanic crust; its dependence on spreading rate and effect on crystal structure. *Geophysical Journal of the Royal Astronomical Society* **61**, 167–181.

LANGMUIR, C. H., BENDER, J. F., BENCE, A. E. & HANSON, G. N. 1977. Petrogenesis of basalts from the FAMOUS area: Mid-Atlantic Ridge. *Earth and Planetary Science Letters* **36**, 133–156.

LE ROEX, A. P., DICK, H. J. B., ERLANK, A. J., REID, A. M., FREY F. A. & HART, S. R. 1983. Geochemistry, mineralogy and petrogenesis of lavas erupted along the Southwest Indian Ridge between the Bouvet Triple Junction and 11 Degrees East. *Journal of Petrology* **24**, 267–318.

——, ——, REID, A. M., FREY, F. A., ERLANK, A. J. & HART, S. R. 1985 Petrology and geochemistry of basalts from the American–Antarctic Ridge, southern ocean: implications for the westward influence of the Bouvet mantle plume. *Contributions to Mineralogy and Petrology* **90**, 367–380.

MACDONALD, K. C., CASTILLO, D. A., MILLER, S. P., FOX, P. J., KASTENS, K. A. & BONATTI, E. 1986. Deep-tow studies of the Vema Fracture Zone 1. Tectonics of a major slow slipping transform fault and its intersection with the Mid-Atlantic Ridge. *Journal of Geophysical Research* **91**, 3334–3354.

——, SEMPERE, J.-C., FOX, P. J. & TYCE, R. 1987. Tectonic evolution of ridge-axis discontinuities by the meeting, linking or self-decapitation of neighboring ridge segments. *Geology* **15**, 993–997.

MAALOE, S. & AOKI, K. 1977. The major element composition of the upper mantle estimated from the composition of lherzolites. *Contributions to Mineralogy and Petrology* **63**, 161–173.

—— & SCHEIE, A. 1984. The permeability controlled accumulation of primary magma. *Contributions to Mineralogy and Petrology* **81**, 350–357.

MARSH, B. D. 1979. Island arc development: some observations, experiments, and speculations. *Journal of Geology* **87**, 687–714.

MCKENZIE, D. 1984. The generation and compaction of partially molten rock. *Journal of Petrology* **25**, 713–765.

MELSON, W. G., HART, S. R. & THOMPSON, G. 1972. St Paul's Rocks, equatorial Atlantic: petrogenesis, radiometric ages and implications on seafloor spreading. *Geological Society of America, Memoir* **132**, 241–272.

MENARD, H. W. & ATWATER, T. 1968. Changes in the direction of sea-floor spreading. *Nature* **219**, 463–467.

MENZIES, M. 1973. Mineralogy and partial melt textures within an ultramafic–mafic body, Greece. *Contributions to Mineralogy and Petrology* **42**, 273–285.

—— & ALLEN, C. 1974. Plagioclase lherzolite–residual mantle relationships within two eastern Mediterranean ophiolites. *Contributions to Mineralogy and Petrology* **45**, 197–213.

MICHAEL, P. J. & BONATTI, E. 1985. Peridotite composition from the North Atlantic: regional and tectonic variations and implications for partial melting. *Earth and Planetary Science Letters* **73**, 91–104.

MIYASHIRO, A., SHIDO, F., & EWING, M. 1969. Composition and origin of serpentinites from the Mid-Atlantic Ridge near 24° and 30°N latitude. *Contributions to Mineralogy and Petrology* **23**, 38–52.

MOHR, P. A. & WOOD, C. A. 1974. Volcano spacings and lithospheric attenuation in the eastern rift of Africa. *Earth and Planetary Science Letters* **33**, 126–144.

MUTTER, J. C. & NORTH ATLANTIC TRANSECT STUDY GROUP 1985. Multichannel seismic images of the oceanic crust's internal structure: evidence for a magma chamber beneath the Mesozoic Mid-Atlantic Ridge. *Geology* **13**, 629–632.

NICOLAS, A. & DUPUY, C. 1984. Origin of ophiolitic and oceanic lherzolites. *Tectonophysics* **110**, 177–187.

—— & VIOLETTE, J. F. 1982. Mantle flow at oceanic spreading centers: models derived from ophiolites. *Tectonophysics* **81**, 319–339.

OTTER SCIENTIFIC TEAM 1985. The geology of the Oceanographer Transform; the transform domain. *Marine Geophysical Researches* **7**, 329–358.

PEDERSEN, R. B. 1986. The nature and significance of magma chamber margins in ophiolites: examples from the Norwegian Caledonides. *Earth and Planetary Science Letters* **77**, 100–112.

POCKALNY, R., DETRICK, R. S. & FOX, P. J. 1988. The morphology and tectonics of the Kane Transform from Seabeam bathymetry data *Journal of Geophysical Research* **93**, 3179–3194.

PRESNALL, D. C., DIXON, S. A., DIXON, J. R., O'DONNEL, T. H., BREENNER, N. L., SCHROCK, R. L. & DYCUS, D. W. 1978. Liquidus phase relations on the join diopside-forsteriteanorthite from 1 atm to 20 kbar: their bearing on the generation and crystallization of basaltic magma. *Contributions to Mineralogy and Petrology* **66**, 203–220.

QUICK, J. E. 1981*a*. The origin and significance of large, tabular dunite bodies in the Trinity Peridotite, Northern California. *Contributions to Mineralogy and Petrology* **78**, 413–422.

—— 1981*b*. Petrology and petrogenesis of the Trinity Peridotite, an upper mantle diapir in the eastern Klamath Mountains, northern California. *Journal of Geophysical Research* **92**, 11 837–11 863.

RABINOWICZ, M., CEULENEER,, G. & NICOLAS, A. 1987. Melt segregation and flow in mantle diapirs below spreading centers: evidence from the Oman ophiolite. *Journal of Geophysical Research* **92**, 3475–3486.

RYAN, M. P. 1987. The elasticity and contractancy of Hawaiian olivine tholeiite and its role in the stability and structural evolution of subcaldera magma reservoirs and rift systems. *US Geological Survey Professional Paper 1350*, **2**, 1395–1447.

SCHILLING, J. G. 1973. Iceland mantle plume, Geochemical evidence along the Reykjanes Ridge. *Nature* **242**, 565–571.

SCHOUTEN, H. & KLITGORD, K. D. 1982. The memory of the accreting plate boundary and the continuity of fracture zones. *Earth and Planetary Science Letters* **59**, 255–266.

—— & WHITE, R. S. 1980. Zero offset fracture zones. *Geology* **8**, 175–179.

——, KLITGORD, K. D. & WHITEHEAD, J. A. 1985. Segmentation of mid-ocean ridges. *Nature* **317**, 225–229.

SCLATER, J. G., DICK, H. J. B., NORTON, I. & WOODROFFE, D. 1978. Tectonic structure and petrology of the Antarctic Plate Boundary near the Bouvet Triple Junction. *Earth and Planetary Science Letters* **36**, 393–400.

SHAND, S. J. 1949. Rocks of the Mid-Atlantic Ridge. *Journal of Geology* **57**, 89–92.

SIGURDSSON, H. & SPARKS, S. R. J. 1978. Lateral magma flow within rifted Icelandic crust. *Nature* **274**, 126–130.

SLEEP, N. 1975. Formation of the oceanic crust: some thermal constraints. *Journal of Geophysical Research* **80**, 4037–4042.

SPIEGELMAN, M. & MCKENZIE, D. 1987. Simple 2-D models for melt extraction at mid-ocean ridges and island arcs. *Earth and Planetary Science Letters* **83**, 137–152.

THOMPSON, G. & MELSON, W. G. 1972. The petrology of oceanic crust across fracture zones in the Atlantic ocean: evidence of a new kind of sea floor spreading. *Journal of Geology* **80**, 526–538.

——, BRYAN, W. B., BALLARD, R., HAMURO, K. & MELSON, W. G. 1985. Axial processes along a segment of the East Pacific Rise, 10°–12°N. *Nature* **318**, 429–433.

TURCOTTE, D. L. 1982. Magma migration. *Annual Review of Earth and Planetary Science* **10**, 392–408.

VANKO, D. A. & BATIZA, R. 1982. Gabbroic rocks from the Mathemetician Ridge failed rift. *Nature* **300**, 742–744.

WHITEHEAD, J. A., DICK, H. J. B. & SCHOUTEN, H. 1984. A mechanism for magmatic accretion under spreading centers. *Nature* **312**, 146–148.

WOLFE, E. W. 1988. The Puu Oo eruption of Kilauea: magma transport, storage, differentiation, and eruption in a Hawaiian shield volcano. *Symposium on Geologic and Geochemical Evidence for Segmentation of Continental and Oceanic Rifts*. Woods Hole Oceanographic Institution, Woods Hole, MA, 40.

WRIGHT, T. L., KINOSHITA, W. T. & PECK, D. L. 1968. March 1965 eruption of Kilauea Volcano and the formation of Makaopuhi Lava Lake. *Journal of Geophysical Research* **73**, 3181.

ZINDLER, A., STAUDIGEL, H. & BATIZA, R. 1984. Isotope and trace element geochemistry of young Pacific seamounts: implications for the scale of upper mantle heterogeneity. *Earth and Planetary Science Letters* **70**, 175–195.

HENRY J. B. DICK, Department of Geology and Geophysics, Woods Hole Oceanographic Institution, Woods Hole, MA 02543, USA.

Appendix 1
Modal compositions of SW Indian and American–Antarctic Ridge peridotites

Sample number	OL	OPX	CPX	SP	Plag	Total
American–Antarctic Ridge: Bullard Fracture Zone						
VULC 5: 34–37	76.40	18.00	5.23	0.38	0.00	100.01
VULC 5: 34–42	75.80	20.70	2.89	0.67	0.00	100.06
VULC 5: 34–43	75.40	19.10	4.90	0.60	0.00	100.00
VULC 5: 34–48	79.00	18.80	2.00	0.30	0.00	100.10
VULC 5: 34–51	69.80	21.90	7.47	0.85	0.00	100.02
VULC 5: 34–56	74.80	17.10	7.50	0.60	0.00	100.00
VULC 5: 35–1	74.40	19.20	5.97	0.44	0.00	100.01
VULC 5: 35–3	83.60	15.10	0.00	0.29	1.02	100.01
VULC 5: 35–15	76.80	15.70	6.97	0.50	0.04	100.01
VULC 5: 35–19	70.00	22.10	6.89	1.09	0.00	100.08
VULC 5: 35–22	77.40	18.90	2.96	0.73	0.04	100.03
VULC 5: 35–30	74.70	20.10	4.27	1.00	0.00	100.07
VULC 5: 35–36	74.30	22.20	2.71	0.88	0.00	100.09
VULC 5: 35–37	72.70	22.60	4.00	0.67	0.00	99.97
VULC 5: 35–40	68.40	25.80	5.33	0.40	0.00	99.93
VULC 5: 35–47	71.60	22.40	5.70	0.40	0.00	100.10
Average	74.69	19.98	4.67	0.61	0.07	100.03
Standard deviation	3.65	2.72	2.08	0.24	0.25	0.05
American–Antarctic Ridge: Vulcan Fracture Zone						
VULC 5: 41–13	71.80	18.10	8.94	1.21	0.00	100.05
VULC 5: 41–14	76.40	19.10	3.91	0.38	0.15	99.94
VULC 5: 41–15	63.90	22.70	12.00	1.44	0.00	100.04
VULC 5: 41–29	74.70	16.60	8.10	0.57	0.00	99.97
VULC 5: 41–30	65.70	27.20	5.79	1.33	0.00	100.02
VULC 5: 41–33	68.10	21.00	9.93	0.96	0.04	100.03
VULC 5: 41–41	73.40	18.80	7.04	0.69	0.04	99.97
VULC 5: 41–45	68.10	22.30	8.45	1.14	0.04	100.03
VULC 5: 41–52	71.30	20.00	7.59	1.06	0.04	99.99
VULC 5: 41–55	75.70	19.00	4.57	0.72	0.00	99.99
VULC 5: 41–63	73.20	21.90	3.47	1.44	0.00	100.01
VULC 5: 41–67	71.30	20.00	8.11	0.61	0.00	100.02
Average	71.13	20.56	7.33	0.96	0.03	100.01
Standard deviation	3.78	2.64	2.42	0.35	0.04	0.03
SW Indian Ridge: Bouvet Fracture Zone						
AII 107: 35–4	91.90	7.80	0.04	0.30	0.00	100.04
AII 107: 35–6	87.80	12.00	0.00	0.10	0.00	99.90
AII 107: 39–4	86.10	12.10	0.37	0.55	0.88	100.00
AII 107: 39–13	79.10	18.20	0.36	0.87	1.45	99.98
AII 107: 39–14	80.90	17.50	0.97	0.61	0.00	99.98
AII 107: 39–15	77.50	21.20	0.51	0.62	0.11	99.94
AII 107: 39–21	85.60	13.30	0.36	0.36	0.40	100.02
AII 107: 40–2	81.90	16.90	0.80	0.40	0.00	100.00
AII 107: 40–4	76.00	19.20	3.50	1.30	0.00	100.00
AII 107: 40–6	78.20	18.10	3.40	0.30	0.00	100.00
AII 107: 40–8	77.20	19.60	3.10	0.10	0.00	100.00
AII 107: 40–11	81.10	15.50	2.90	0.40	0.00	99.90
AII 107: 40–13	78.70	18.80	2.00	0.50	0.00	100.00
AII 107: 40–24	82.30	15.00	2.10	0.60	0.00	100.00

Appendix 1. (*cont.*)

Sample number	OL	OPX	CPX	SP	Plag	Total
AII 107: 40–27	80.70	15.50	3.30	0.40	0.00	99.90
	79.80	nc	nc	0.30	0.00	80.10
Average	81.55	16.05	1.58	0.48	0.18	99.84
Standard deviation	4.18	3.44	1.31	0.29	0.40	4.81
SW Indian Ridge: Islas Orcadas Fracture Zone						
10 11/76: 56–10	73.40	18.70	7.18	0.67	0.00	99.95
10 11/76: 56–29	73.70	22.40	2.92	0.52	0.42	99.96
10 11/76: 56–50	72.00	22.90	4.54	0.64	0.00	100.08
10 11/76: 56–54	69.00	21.70	8.20	1.14	0.00	100.04
10 11/76: 56–54–1	72.10	21.70	5.31	0.80	0.00	99.91
10 11/76: 56–57	69.60	24.50	4.32	1.66	0.00	100.08
10 11/76: 58–8	69.10	24.10	6.18	0.57	0.04	99.99
10 11/76: 58–10	65.50	26.90	7.18	0.47	0.00	100.05
10 11/76: 58–12	65.90	27.90	4.45	1.29	0.52	100.06
10 11/76: 58–18	62.50	26.00	11.00	0.59	0.00	100.09
10 11/76: 58–23	76.60	20.50	1.95	0.92	0.00	99.97
10 11/76: 58–30	67.10	24.00	7.06	0.55	1.29	100.00
10 11/76: 58–34	70.00	25.90	3.51	0.62	0.00	100.03
10 11/76: 59–21	76.40	19.30	4.05	0.25	0.00	100.00
10 11/76: 59–49	72.80	20.40	6.06	0.71	0.00	99.97
10 11/76: 59–65	78.10	15.60	5.20	1.09	0.00	99.99
10 11/76: 59–72	67.20	28.20	3.76	0.76	0.05	99.97
10 11/76: 59–78	74.70	22.40	2.27	0.58	0.00	99.95
10 11/76: 59–95	72.00	23.20	4.08	0.76	0.00	100.04
10 11/76: 60–20	79.50	17.30	2.25	0.90	0.00	99.95
10 11/76: 60–21	78.60	16.70	4.21	0.42	0.00	99.93
10 11/76: 60–27	71.60	23.00	4.39	0.69	0.23	99.91
10 11/76: 60–34	79.00	17.70	2.99	0.30	0.00	99.99
10 11/76: 60–41	76.00	20.60	2.63	0.66	0.09	99.98
10 11/76: 60–51	70.80	23.10	5.59	0.50	0.00	99.99
10 11/76: 60–52a	74.40	21.70	3.01	0.96	0.00	100.07
10 11/76: 60–56	73.30	19.70	6.59	0.43	0.00	100.02
10 11/76: 60–61	81.00	16.70	2.09	0.18	0.00	99.97
10 11/76: 60–103	64.00	27.80	7.60	0.53	0.00	99.93
10 11/76: 60–126	82.40	16.00	0.64	0.50	0.45	99.99
10 11/76: 60–143	79.70	17.50	1.81	1.01	0.00	100.02
10 11/76: 60–151	72.70	22.50	4.15	0.71	0.00	100.06
Average	72.83	21.77	4.60	0.70	0.10	100.00
Standard deviation	5.00	3.56	2.20	0.30	0.25	0.05

Appendix 2.
Electron microprobe analyses of SW Indian and American–Antarctic Ridge peridotites

Sample number	Olivine							
	SiO_2	FeO	MnO	MgO	CaO	NiO	Sum	Mg#
American–Antarctic Ridge: Bullard Fracture Zone								
VULC 5: 35–1	40.97	9.37	0.11	49.33	0.04	0.32	100.14	0.904
VULC 5: 35–7	40.49	9.58	0.17	49.19	0.03	0.32	99.78	0.901
VULC 5: 35–15	40.80	9.21	0.16	49.67	0.06	0.32	100.22	0.906
VULC 5: 35–19	40.68	9.86	0.18	49.19	0.05	0.31	100.27	0.899
VULC 5: 35–22	40.75	9.00	0.16	49.42	0.02	0.34	99.69	0.907
VULC 5: 35–30	40.79	8.77	0.14	49.91	0.03	0.34	99.98	0.910
VULC 5: 35–36	40.70	8.76	0.17	49.91	0.03	0.35	99.92	0.910
VULC 5: 35–37	40.80	9.13	0.17	49.37	0.06	0.33	99.86	0.906
VULC 5: 35–47	40.06	9.53	0.13	49.22	0.00	0.36	99.30	0.902
VULC 5: 35–71	40.54	9.10	0.15	49.32	0.03	0.34	99.48	0.906
Average	40.66	9.23	0.15	49.45	0.04	0.33	99.86	0.905
Standard deviation	0.24	0.34	0.02	0.26	0.02	0.01	0.30	0.004
American–Antarctic Ridge: Vulcan Fracture Zone								
VULC 5: 41–13	40.38	9.89	0.18	48.67	0.03	0.32	99.47	0.898
VULC 5: 41–14	40.68	9.31	0.16	49.53	0.06	0.31	100.05	0.905
VULC 5: 41–15	40.70	9.78	0.17	49.23	0.04	0.31	100.23	0.900
VULC 5: 41–29	40.42	9.40	0.16	49.13	0.05	0.28	99.44	0.903
VULC 5: 41–30	40.66	9.39	0.18	49.66	0.04	0.34	100.27	0.904
VULC 5: 41–33	40.63	9.40	0.20	49.74	0.03	0.30	100.30	0.904
VULC 5: 41–41	40.38	9.76	0.15	48.76	0.06	0.33	99.44	0.899
VULC 5: 41–45	40.56	10.00	0.24	49.33	0.04	0.26	100.43	0.898
VULC 5: 41–52	40.49	9.50	0.16	48.84	0.05	0.33	99.37	0.902
VULC 5: 41–55	40.87	9.33	0.12	49.60	0.02	0.29	100.23	0.905
VULC 5: 41–63	40.72	9.79	0.12	49.32	0.06	0.30	100.31	0.900
VULC 5: 41–71	39.84	10.32	0.13	48.33	0.04	0.34	99.00	0.893
Average	40.53	9.66	0.16	49.18	0.04	0.31	99.88	0.901
Standard deviation	0.25	0.30	0.03	0.42	0.01	0.02	0.47	0.003
Indian Ridge: Bouvet Fracture Zone								
AII 107: 39–7	40.63	8.71	0.09	48.90	0.04	0.30	98.67	0.909
AII 107: 40–6	40.36	8.98	0.12	49.10	0.04	0.29	98.89	0.907
AII 107: 40–60	40.47	8.91	0.16	49.26	0.01	0.32	99.13	0.908
Average	40.49	8.87	0.12	49.09	0.03	0.30	98.90	0.908
Standard deviation	0.11	0.11	0.03	0.15	0.01	0.01	0.19	0.001
SW Indian Ridge: Islas Orcadas Fracture Zone								
10 11/76: 56–10	40.42	9.24	0.26	49.22	0.01	0.40	99.55	0.905
10 11/76: 56–29	40.79	9.62	0.14	49.10	0.04	0.22	99.91	0.901
10 11/76: 56–50	40.60	9.71	0.16	48.90	0.01	0.23	99.61	0.900
10 11/76: 56–57	39.84	9.52	0.13	48.99	0.02	0.27	98.77	0.902
10 11/76: 58–8	40.54	9.52	0.18	49.02	0.05	0.36	99.67	0.902
10 11/76: 58–10	40.40	9.59	0.19	49.08	0.02	0.32	99.60	0.901
10 11/76: 58–12	40.63	9.28	0.14	49.19	0.04	0.23	99.51	0.904
10 11/76: 58–30	40.68	9.51	0.13	49.37	0.03	0.26	99.98	0.902
10 11/76: 58–34	40.40	9.49	0.08	49.09	0.02	0.26	99.34	0.902
10 11/76: 58–61	40.39	9.28	0.21	49.56	0.02	0.33	99.79	0.905
10 11/76: 59–72	40.68	9.56	0.10	49.27	0.04	0.25	99.90	0.902

Appendix 2. (*cont.*)

Sample number	Olivine							
	SiO_2	FeO	MnO	MgO	CaO	NiO	Sum	Mg#
10 11/76: 59–78	40.76	9.21	0.12	49.63	0.02	0.24	99.98	0.906
10 11/76: 59–95	40.64	9.12	0.07	49.40	0.02	0.27	99.52	0.906
10 11/76: 60–56	40.62	9.57	0.13	49.26	0.04	0.28	99.90	0.902
10 11/76: 60–61	40.60	8.96	0.07	49.65	0.02	0.25	99.55	0.908
10 11/76: 60–126	40.37	9.27	0.15	48.97	0.00	0.23	98.99	0.904
Average	40.52	9.40	0.14	49.23	0.03	0.28	99.60	0.903
Standard deviation	0.22	0.20	0.05	0.23	0.01	0.05	0.33	0.002

American–Antarctic Ridge: Bullard Fracture Zone														
VULC 5: 34–37	54.03	0.00	4.72	5.63	0.15	31.88	2.75	0.02	0.00	1.06	100.24	0.910	0.861	0.053
VULC 5: 34–48	54.65	0.00	3.64	5.53	0.13	32.54	2.91	0.04	0.01	0.78	100.23	0.913	0.862	0.055
VULC 5: 34–56	54.22	0.00	4.38	5.41	0.14	31.99	2.55	0.02	0.00	1.02	99.73	0.913	0.868	0.050
VULC 5: 35–1	54.21	0.02	4.03	6.09	0.25	31.36	1.93	0.05	0.02	0.64	98.60	0.902	0.867	0.038
VULC 5: 35–3	55.91	0.00	2.26	6.69	0.17	32.32	2.10	0.04	0.00	0.68	100.17	0.896	0.860	0.040
VULC 5: 35–15	54.62	0.00	4.31	5.53	0.15	32.42	2.58	0.02	0.00	0.89	100.52	0.913	0.867	0.050
VULC 5: 35–22	55.09	0.00	3,74	5.10	0.15	32.77	2.36	0.03	0.00	0.60	99.84	0.920	0.878	0.045
VULC 5: 35–30	54.76	0.00	4.37	5.24	0.14	32.34	2.60	0.01	0.01	0.51	99.98	0.917	0.871	0.050
VULC 5: 35–36	54.54	0.06	4.24	5.43	0.10	31.18	2.20	0.01	0.02	0.70	98.48	0.911	0.871	0.044
VULC 5: 35–37	54.63	0.00	4.14	5.43	0.13	32.62	2.35	0.02	0.00	0.75	100.07	0.915	0.873	0.045
VULC 5: 35–40	54.56	0.00	4.12	5.63	0.14	32.25	2.41	0.01	0.01	0.95	100.08	0.911	0.868	0.047
VULC 5: 35–47	53.77	0.08	4.67	6.29	0.12	31.99	1.82	0.05	0.04	0.57	99.40	0.901	0.869	0.036
VULC 5: 35–71	53.23	0.00	5.08	5.75	0.16	31.91	2.76	0.02	0.00	1.19	100.10	0.908	0.860	0.053
Average	54.48	0.01	4.13	5.67	0.15	32.12	2.41	0.03	0.01	0.80	99.80	0.910	0.867	0.047
Standard deviation	0.62	0.03	0.66	0.42	0.03	0.45	0.32	0.01	0.01	0.20	0.60	0.006	0.005	0.006
American–Antarctic Ridge: Vulcan Fracture Zone														
VULC 5: 41–14	53.51	0.00	5.07	5.39	0.15	32.32	2.16	0.10	0.00	0.78	99.48	0.914	0.876	0.042
VULC 5: 41–29	53.88	0.00	5.10	5.49	0.16	31.84	2.40	0.04	0.00	0.77	99.68	0.912	0.869	0.047
VULC 5: 41–30	53.66	0.10	5.35	5.90	0.12	30.75	2.69	0.06	0.03	0.67	99.33	0.903	0.854	0.054
VULC 5: 41–33	52.74	0.09	5.48	6.07	0.04	30.06	2.86	0.05	0.01	0.38	97.78	0.898	0.846	0.058
VULC 5: 41–41	53.73	0.00	5.32	5.76	0.15	31.79	2.35	0.05	0.00	0.74	99.89	0.908	0.866	0.046
VULC 5: 41–45	53.74	0.11	5.56	6.38	0.12	30.45	2.02	0.05	0.04	0.65	99.12	0.895	0.858	0.041
VULC 5: 41–52	53.72	0.00	5.10	5.78	0.15	31.86	2.37	0.05	0.00	0.73	99.76	0.908	0.866	0.046
VULC 5: 41–55	54.34	0.09	4.94	5.97	0.11	31.26	2.04	0.04	0.02	0.72	99.53	0.903	0.866	0.041
VULC 5: 41–63	53.02	0.10	5.83	6.13	0.08	30.06	2.63	0.06	0.00	0.42	98.33	0.897	0.849	0.053
VULC 5: 41–71	54.00	0.00	3.85	5.86	0.13	32.26	2.78	0.01	0.00	0.85	99.74	0.907	0.859	0.053
Average	53.63	0.05	5.16	5.87	0.12	31.27	2.43	0.05	0.01	0.67	99.26	0.905	0.861	0.048
Standard deviation	0.44	0.05	0.51	0.28	0.04	0.83	0.29	0.02	0.01	0.15	0.65	0.006	0.009	0.006
Indian Ridge: Bouvet Fracture Zone														
AII 107: 35–4	55.55	0.02	3.18	5.70	0.13	33.64	2.05	0.00	0.01	0.64	100.92	0.913	0.878	0.038
AII 107: 35–6	55.80	0.01	2.91	5.63	0.15	33.92	1.95	0.00	0.01	0.68	101.06	0.915	0.881	0.036
AII 107: 39–3	55.78	0.00	2.70	6.21	0.11	33.37	1.85	0.00	0.01	0.60	100.63	0.905	0.874	0.035
AII 107: 39–7	55.94	0.07	2.95	5.62	0.12	33.31	1.81	0.00	0.01	0.72	100.55	0.914	0.882	0.034
AII 107: 39–14	55.62	0.01	3.06	5.71	0.14	33.71	2.11	0.00	0.01	0.68	101.05	0.913	0.877	0.039
AII 107: 40–6	55.19	0.00	3.82	5.58	0.16	33.21	2.23	0.00	0.00	0.66	100.85	0.914	0.875	0.042
AII 107: 40–60	54.38	0.02	3.60	5.54	0.13	33.75	2.95	0.05	0.00	0.53	100.95	0.916	0.866	0.054
Average	55.47	0.02	3.17	5.71	0.13	33.56	2.14	0.01	0.01	0.64	100.86	0.913	0.876	0.040
Standard deviation	0.50	0.02	0.37	0.21	0.02	0.24	0.36	0.02	0.00	0.06	0.18	0.003	0.005	0.006

Appendix 2. (*cont.*)

Sample number	Enstatite												
	SiO_2	TiO_2	Al_2O_3	FeO	MnO	MgO	CaO	Na_2O	K_2O	Cr_2O_3	Sum	Mg#	En	Wo
SW Indian Ridge: Islas Orcadas Fracture Zone														
10 11/76: 56–10	54.27	0.07	5.28	5.54	0.13	31.33	1.88	0.10	0.01	0.34	98.95	0.910	0.875	0.038
10 11/76: 56–29	53.76	0.10	5.33	6.12	0.10	31.51	1.87	0.09	0.02	0.58	99.48	0.902	0.868	0.037
10 11/76: 56–50	54.09	0.10	5.69	6.08	0.06	31.63	1.91	0.07	0.01	0.71	100.35	0.903	0.869	0.038
10 11/76: 56–54	54.41	0.09	5.44	5.89	0.10	31.23	1.87	0.04	0.01	0.73	99.81	0.904	0.870	0.037
10 11/76: 56–57	54.21	0.10	5.21	6.03	0.13	31.20	1.91	0.04	0.02	0.73	99.58	0.902	0.868	0.038
10 11/76: 58–8	53.82	0.06	5.13	5.21	0.08	30.89	2.15	0.10	0.02	0.68	98.14	0.914	0.874	0.044
10 11/76: 58–10	53.92	0.05	5.15	5.84	0.12	30.90	2.04	0.09	0.01	0.82	98.94	0.904	0.867	0.041
10 11/76: 58–12	54.22	0.07	5.13	5.85	0.09	31.26	1.94	0.02	0.01	0.78	99.37	0.905	0.870	0.039
10 11/76: 58–18	53.93	0.05	5.21	5.94	0.13	31.58	2.07	0.05	0.02	0.51	99.49	0.905	0.868	0.041
10 11/76: 58–30	53.96	0.14	4.97	5.90	0.08	31.46	1.88	0.01	0.01	1.07	99.48	0.905	0.871	0.037
10 11/76: 58–34	54.60	0.05	5.21	5.74	0.06	31.71	1.95	0.08	0.00	0.41	99.81	0.908	0.873	0.039
10 11/76: 58–61	54.77	0.03	4.51	5.58	0.17	31.85	1.61	0.14	0.02	0.56	99.24	0.910	0.881	0.032
10 11/76: 59–72	53.68	0.05	4.76	5.68	0.20	32.42	2.19	0.07	0.00	0.54	99.59	0.910	0.872	0.042
10 11/76: 59–78	53.79	0.08	5.03	5.37	0.16	31.59	2.33	0.03	0.01	0.84	99.23	0.913	0.871	0.046
10 11/76: 59–95	53.59	0.05	4.98	5.74	0.23	31.46	2.26	0.11	0.01	1.17	99.60	0.907	0.867	0.045
10 11/76: 60–27	53.99	0.04	5.07	5.94	0.13	31.71	2.10	0.05	0.01	0.38	99.42	0.905	0.868	0.041
10 11/76: 60–41	54.60	0.02	4.50	5.64	0.10	32.18	1.82	0.05	0.02	0.41	99.34	0.910	0.878	0.036
11 11/76: 60–45	54.83	0.05	4.21	5.63	0.07	32.25	1.80	0.09	0.01	0.46	99.40	0.911	0.879	0.035
10 11/76: 60–51	54.14	0.04	4.85	5.83	0.11	31.33	2.01	0.07	0.02	0.68	99.08	0.905	0.869	0.040
10 11/76: 60–56	54.10	0.03	5.20	5.98	0.14	31.32	2.00	0.06	0.01	0.72	99.56	0.903	0.867	0.040
10 11/76: 60–61	54.60	0.13	3.93	5.50	0.14	31.80	1.82	0.11	0.05	0.97	99.05	0.912	0.879	0.036
10 11/76: 60–87	54.39	0.02	4.97	5.70	0.16	31.37	1.98	0.10	0.02	0.69	99.40	0.907	0.872	0.040
10 11/76: 60–88	54.04	0.04	4.99	5.86	0.13	31.19	1.97	0.03	0.02	0.53	98.80	0.905	0.869	0.039
10 11/76: 60–97	54.27	0.05	5.08	5.99	0.13	31.54	2.20	0.06	0.01	0.56	99.89	0.904	0.865	0.043
10 11/76: 60–103	54.60	0.03	7.61	5.69	0.11	32.21	1.93	0.05	0.00	0.47	99.70	0.910	0.875	0.038
10 11/76: 60–126	54.39	0.04	5.08	5.91	0.08	31.42	2.07	0.05	0.01	0.40	99.45	0.905	0.867	0.041
Average	54.19	0.06	4.98	5.78	0.12	31.55	1.98	0.07	0.01	0.64	99.39	0.907	0.871	0.039
Standard deviation	0.34	0.03	0.37	0.22	0.04	0.38	0.16	0.03	0.01	0.21	0.41	0.003	0.004	0.003

Sample number	Diopside												
	SiO_2	TiO_2	Al_2O_3	FeO	MnO	MgO	CaO	Na_2O	K_2O	Cr_2O_3	Sum	Mg#	En	Wo
American–Antarctic Ridge: Bullard Fracture Zone														
VULC 5: 34–37	50.71	0.09	5.46	3.07	0.03	18.74	19.61	0.15		1.46	99.32	0.916	0.542	0.408
VULC 5: 34–48	50.94	0.13	4.76	3.47	0.10	18.96	19.22	0.40		1.42	99.40	0.907	0.546	0.398
VULC 5: 34–56	50.97	0.09	5.27	3.76	0.12	19.49	18.06	0.17		1.24	99.17	0.902	0.564	0.375
VULC 5: 35–1	51.24	0.07	5.32	3.54	0.08	17.86	19.84	0.35	0.03	1.28	99.61	0.900	0.524	0.418
VULC 5: 35–15	50.46	0.09	5.30	3.49	0.15	19.40	18.92	0.18		1.19	99.18	0.908	0.555	0.389
VULC 5: 35–22	51.34	0.10	4.56	2.98	0.10	18.69	20.02	0.24		1.19	99.22	0.918	0.538	0.414
VULC 5: 35–30	50.83	0.04	5.05	4.09	0.05	21.56	16.55	0.11	0.02	0.90	99.20	0.904	0.603	0.333
VULC 5: 35–36	50.91	0.14	5.12	3.03	0.11	17.51	20.49	0.34	0.13	1.16	98.94	0.911	0.516	0.434
VULC 5: 35–37	50.63	0.11	5.20	2.58	0.12	18.75	19.73	0.20		1.25	98.57	0.928	0.545	0.413
VULC 5: 35–40	50.65	0.08	5.24	3.10	0.14	18.93	19.03	0.10		1.27	98.54	0.916	0.551	0.398
VULC 5: 35–47	50.35	0.09	5.79	3.42	0.04	18.06	19.53	0.32	0.02	0.81	98.43	0.904	0.531	0.413
Average	50.82	0.09	5.19	3.32	0.09	18.90	19.18	0.23	0.05	1.20	99.05	0.910	0.547	0.399
Standard deviation	0.29	0.03	0.31	0.40	0.04	1.02	1.03	0.10	0.05	0.18	0.37	0.008	0.022	0.026
American–Antarctic Ridge: Vulcan Fracture Zone														
VULC 5: 41–14	50.55	0.45	6.68	3.41	0.09	18.54	17.67	1.12	0.01	1.31	99.83	0.906	0.559	0.383
VULC 5: 41–29	50.40	0.25	5.95	3.26	0.09	18.76	19.46	0.50	0.00	1.27	99.94	0.911	0.543	0.405
VULC 5: 41–30	49.82	0.23	6.42	3.45	0.09	18.93	18.99	0.41	0.00	1.24	99.58	0.907	0.548	0.396
VULC 5: 41–33	50.18	0.23	6.83	3.22	0.03	17.57	19.08	0.45	0.00	1.58	99.17	0.907	0.531	0.415
VULC 5: 41–41	49.89	0.21	6.59	3.56	0.09	18.53	18.85	0.48	0.00	1.02	99.22	0.903	0.544	0.398
VULC 5: 41–45	50.36	0.24	6.37	3.49	0.07	17.42	20.51	0.41	0.02	1.11	100.00	0.899	0.510	0.432
VULC 5: 41–52	50.38	0.19	6.33	3.50	0.07	18.75	18.33	0.53	0.00	1.07	99.15	0.905	0.553	0.389
VULC 5: 41–55	49.37	0.26	6.10	3.14	0.01	17.30	21.93	0.57	0.01	0.67	99.36	0.908	0.497	0.453
VULC 5: 41–63	49.34	0.26	7.14	3.65	0.01	17.63	19.99	0.51	0.00	0.53	99.06	0.896	0.518	0.422
VULC 5: 41–71	50.83	0.23	5.32	3.37	0.12	18.66	19.15	0.57	0.00	1.29	99.54	0.908	0.544	0.401
Average	50.11	0.26	6.37	3.41	0.07	18.21	19.40	0.56	0.00	1.11	99.49	0.905	0.535	0.409
Standard deviation	0.47	0.07	0.48	0.15	0.04	0.61	1.13	0.20	0.01	0.30	0.33	0.004	0.019	0.020

Appendix 2. (*cont.*)

Sample number	Spinel													
	SiO_2	TiO_2	Al_2O_3	FeO	MnO	MgO	CaO	Na_2O	K_2O	Cr_2O_3	Sum	Mg#	En	Wo
SW Indian Ridge: Bouvet Fracture Zone														
AII 107: 40–6	51.52	0.00	4.70	3.01	0.11	18.66	21.11	0.05	0.02	1.20	100.38	0.917	0.525	0.427
SW Indian Ridge: Islas Orcadas Fracture Zone														
10 11/76: 56–10	51.26	0.18	6.79	2.74	0.06	16.86	19.73	0.79	0.00	1.01	99.42	0.916	0.517	0.435
10 11/76: 56–29	50.90	0.21	6.71	2.86	0.06	17.93	19.34	0.74	0.00	1.07	99.82	0.918	0.536	0.416
10 11/76: 56–50	50.66	0.23	6.55	2.65	0.07	17.49	20.11	0.74	0.00	1.30	99.80	0.922	0.523	0.432
10 11/76: 56–54	50.45	0.16	6.35	2.60	0.14	17.74	19.74	0.72	0.03	1.30	99.23	0.924	0.531	0.425
10 11/76: 56–57	50.99	0.28	6.46	2.67	0.07	17.01	20.33	0.77	0.01	1.14	99.73	0.919	0.514	0.441
10 11/76: 58–8	51.15	0.14	6.27	3.06	0.03	17.30	19.62	0.56	0.00	1.15	99.28	0.910	0.522	0.426
10 11/76: 58–10	51.14	0.12	6.62	3.25	0.05	17.68	18.34	0.57	0.00	1.82	99.59	0.906	0.541	0.403
10 11/76: 58–12	50.90	0.19	6.23	2.96	0.08	17.28	20.28	0.55	0.01	1.22	99.70	0.912	0.515	0.435
10 11/76: 58–18	50.50	0.16	6.58	3.25	0.15	17.67	19.42	0.59	0.01	1.58	99.91	0.906	0.528	0.417
10 11/76: 58–30	50.98	0.30	6.25	3.18	0.12	18.16	18.61	0.53	0.02	1.46	99.61	0.911	0.545	0.402
10 11/76: 58–34	50.36	0.09	6.48	3.14	0.11	17.38	19.53	0.71	0.01	1.46	99.27	0.908	0.524	0.423
10 11/76: 59–72	50.73	0.24	5.58	2.74	0.14	19.38	19.73	0.40	0.03	0.94	99.91	0.926	0.552	0.404
10 11/76: 59–78	50.30	0.28	5.81	2.78	0.12	19.39	19.12	0.51	0.02	1.49	99.82	0.926	0.559	0.396
10 11/76: 59–95	50.57	0.10	5.66	2.74	0.17	19.06	19.57	0.35	0.01	1.35	99.58	0.925	0.550	0.406
10 11/76: 60–45	51.04	0.15	5.50	2.87	0.11	16.76	20.26	0.87	0.02	1.70	99.28	0.912	0.509	0.442
10 11/76: 60–51	51.07	0.06	5.87	3.21	0.04	17.20	19.92	0.47	0.00	1.68	99.52	0.905	0.516	0.430
10 11/76: 60–56	50.57	0.11	6.26	3.40	0.14	17.44	20.19	0.49	0.04	1.40	100.04	0.901	0.515	0.429
10 11/76: 60–61	51.19	0.20	5.09	2.70	0.09	16.74	20.14	0.84	0.05	1.70	98.74	0.917	0.511	0.442
10 11/76: 60–87	50.46	0.09	5.94	3.17	0.12	17.60	19.30	0.53	0.03	1.51	98.75	0.908	0.529	0.417
10 11/76: 60–88	50.65	0.07	6.05	3.17	0.18	17.27	20.22	0.43	0.02	1.36	99.42	0.907	0.514	0.433
10 11/76: 60–103	51.00	0.11	5.66	2.95	0.09	17.28	20.10	0.72	0.02	1.71	99.64	0.913	0.518	0.433
Average	50.80	0.17	6.13	2.96	0.10	17.65	19.70	0.61	0.02	1.40	99.53	0.914	0.527	0.423
Standard deviation	0.29	0.07	0.45	0.24	0.04	0.75	0.53	0.15	0.01	0.24	0.34	0.007	0.014	0.014

Sample number	Spinel											
	TiO_2	Al_2O_3	FeO	Fe_2O_3	MnO	MgO	Cr_2O_3	NiO	Sum	Mg#	Cr#	Fe^{3+}#
American–Antarctic Ridge: Bullard Fracture Zone												
VULC 5: 34–37	0.06	48.26	10.61	3.33	0.11	18.52	17.96	0.35	99.20	0.757	0.200	0.034
VULC 5: 34–42	0.39	37.85	12.49	5.42	0.15	16.66	27.17	0.27	100.40	0.704	0.325	0.058
VULC 5: 34–43	0.12	42.29	10.88	4.47	0.17	17.83	23.54	0.29	99.59	0.745	0.272	0.047
VULC 5: 34–48	0.41	32.39	12.62	5.24	0.24	15.97	33.37	0.33	100.57	0.693	0.409	0.058
VULC 5: 34–56	0.05	44.64	11.18	5.21	0.13	18.11	21.31	0.34	100.97	0.743	0.242	0.053
VULC 5: 34–88	0.26	29.76	18.07	4.77	0.21	12.17	34.91	0.06	100.21	0.545	0.440	0.054
VULC 5: 35–1	0.00	44.19	11.01	3.65	0.09	17.98	22.43	0.19	99.54	0.744	0.254	0.038
VULC 5: 35–7	0.10	47.73	10.58	3.55	0.16	18.71	19.12	0.33	100.28	0.759	0.212	0.036
VULC 5: 35–15	0.09	45.94	10.42	3.23	0.17	18.65	21.59	0.34	100.43	0.761	0.240	0.033
VULC 5: 35–19	0.06	54.85	9.13	3.28	0.11	20.23	11.54	0.39	99.59	0.798	0.124	0.032
VULC 5: 35–22	0.08	42.99	12.14	3.58	0.11	17.32	24.06	0.25	100.53	0.718	0.273	0.037
VULC 5: 35–30	0.14	46.32	10.32	3.22	0.11	18.79	20.54	0.10	99.54	0.764	0.229	0.033
VULC 5: 35–36	0.00	42.25	10.23	3.71	0.11	18.22	24.40	0.13	99.05	0.760	0.279	0.039
VULC 5: 35–37	0.05	46.44	10.69	2.75	0.13	18.61	21.75	0.31	100.73	0.756	0.239	0.028
VULC 5: 35–40	0.05	46.53	11.09	3.35	0.12	18.31	20.80	0.33	100.58	0.746	0.231	0.034
VULC 5: 35–47	0.00	48.13	9.60	3.58	0.05	19.29	18.34	0.13	99.12	0.782	0.204	0.036
VULC 5: 35–60	0.00	40.61	10.89	5.65	0.11	17.66	24.32	0.13	99.37	0.743	0.287	0.060
VULC 5: 35–71	0.07	47.44	10.18	4.64	0.22	18.75	18.40	0.49	100.19	0.766	0.206	0.047
Average	0.11	43.81	11.23	4.04	0.14	17.88	22.53	0.26	99.99	0.74	0.26	0.042
Standard deviation	0.12	5.74	1.87	0.88	0.05	1.66	5.27	0.11	0.60	0.05	0.07	0.010

Reanalysis of some internal standards at the time the proofs of this paper were corrected suggested that there may be a problem with the standard values used for Mg and Fe during pyroxene analysis. Unfortunately, the specific errors arising from this could not be determined prior to the printing of this volume. Readers are advised that the errors in the determination of Mg and Fe in pyroxene could lead to an error of as much as 1.5% in the computed value of the Mg/(Mg + Fe) ratio. The error, if it exists, is believed to be systematic and should not change the conclusions drawn, except to produce a minor change in the computed bulk rock values for Mg and Fe.

Appendix 2. (*cont.*)

Sample number	Spinel											
	TiO_2	Al_2O_3	FeO	Fe_2O_3	MnO	MgO	Cr_2O_3	NiO	Sum	Mg#	Cr#	Fe^{3+}#
American–Antarctic Ridge: Vulcan Fracture Zone												
VULC 5: 41–13	0.08	55.44	8.76	3.46	0.14	20.41	10.55	0.46	99.30	0.806	0.113	0.034
VULC 5: 41–14	0.00	48.35	8.79	5.13	0.07	19.67	16.57	0.26	98.84	0.800	0.187	0.052
VULC 5: 41–15	0.06	54.27	7.82	4.44	0.15	20.69	10.40	0.42	98.25	0.825	0.114	0.044
VULC 5: 41–28	0.32	40.23	12.71	5.62	0.25	16.73	24.78	0.35	100.99	0.701	0.292	0.059
VULC 5: 41–29	0.12	47.98	9.60	3.76	0.19	19.29	18.84	0.42	100.20	0.782	0.208	0.038
VULC 5: 41–30	0.00	49.68	9.50	3.92	0.04	19.42	16.19	0.20	98.95	0.785	0.179	0.040
VULC 5: 41–33	0.00	51.11	9.48	2.99	0.06	19.59	15.43	0.18	98.84	0.786	0.168	0.030
VULC 5: 41–41	0.44	52.35	9.26	3.29	0.11	19.94	14.10	0.92	100.41	0.793	0.153	0.033
VULC 5: 41–45	0.00	51.27	10.23	3.25	0.05	19.06	14.57	0.18	98.61	0.769	0.160	0.033
VULC 5: 41–52	0.29	49.22	8.84	4.59	0.16	19.82	16.83	0.88	100.63	0.800	0.187	0.046
VULC 5: 41–55	0.00	48.69	9.70	3.13	0.06	19.19	17.80	0.15	98.72	0.779	0.197	0.032
VULC 5: 41–63	0.00	53.10	9.23	2.97	0.05	19.91	13.30	0.24	98.80	0.794	0.144	0.030
VULC 5: 41–67	0.79	43.69	11.89	4.66	0.23	17.65	21.36	0.79	101.06	0.726	0.247	0.049
VULC 5: 41–71	0.00	41.54	10.77	5.83	0.16	17.62	22.74	0.26	98.92	0.745	0.269	0.062
Average	0.15	49.07	9.76	4.07	0.12	19.21	16.68	0.41	99.47	0.78	0.19	0.042
Standard deviation	0.23	4.39	1.25	0.94	0.07	1.09	4.07	0.26	0.93	0.03	0.05	0.010
SW Indian Ridge: Bouvet Fracture Zone												
AII 107: 35–4	0.00	33.19	14.65	4.22	0.13	14.70	33.62	0.13	100.64	0.641	0.404	0.046
AII 107: 35–6	0.08	32.05	14.88	3.65	0.20	14.27	34.59	0.15	99.87	0.631	0.420	0.040
AII 107: 39–3	0.50	22.65	17.64	5.47	0.26	11.73	41.71	0.16	100.12	0.542	0.553	0.065
AII 107: 39–4	0.02	33.00	12.33	3.85	0.16	16.00	34.10	0.10	99.56	0.698	0.409	0.042
AII 107: 39–14	0.01	33.25	13.10	4.93	0.17	15.59	32.91	0.12	100.08	0.680	0.399	0.054
AII 107: 40–6	0.04	38.66	11.20	3.34	0.10	17.39	29.09	0.22	100.04	0.735	0.335	0.035
AII 107: 40–60	0.00	35.08	12.67	3.99	0.14	16.09	31.92	0.06	99.95	0.694	0.379	0.043
Average	0.09	32.55	13.78	4.21	0.17	15.11	33.99	0.13	100.04	0.66	0.41	0.047
Standard deviation	0.17	4.52	1.98	0.69	0.05	1.67	3.58	0.05	0.30	0.06	0.06	0.009
SW Indian Ridge: Islas Orcadas Fracture Zone												
10 11/76: 56–10	0.14	52.21	8.70	1.95	0.04	20.32	14.94	0.08	98.38	0.806	0.161	0.020
10 11/76: 56–29	0.01	52.50	9.78	1.69	0.09	19.47	15.02	0.33	98.89	0.780	0.161	0.017
10 11/76: 56–50	0.03	54.14	9.86	1.50	0.06	19.74	13.64	0.32	99.29	0.781	0.145	0.015
10 11/76: 56–54	0.02	53.63	9.23	1.64	0.07	20.05	14.35	0.32	99.31	0.795	0.152	0.016
10 11/76: 56–57	0.03	53.18	9.80	1.31	0.06	19.63	14.58	0.30	98.89	0.781	0.155	0.013
10 11/76: 58–8	0.14	49.08	9.73	2.84	0.06	19.37	17.43	0.09	98.74	0.780	0.192	0.029
10 11/76: 58–10	0.14	49.69	9.71	2.60	0.04	19.39	16.73	0.09	98.39	0.781	0.184	0.027
10 11/76: 58–12	0.03	50.59	10.12	2.74	0.08	19.33	16.71	0.30	99.90	0.773	0.181	0.028
10 11/76: 58–18	0.11	48.43	9.57	3.56	0.09	19.47	18.40	0.32	99.95	0.784	0.203	0.036
10 11/76: 58–30	0.12	47.50	10.00	3.66	0.09	19.07	18.95	0.28	99.67	0.773	0.211	0.037
10 11/76: 58–34	0.00	50.79	10.04	2.51	0.04	19.22	16.14	0.34	99.08	0.773	0.176	0.025
10 11/76: 58–61	0.14	46.08	11.13	2.79	0.06	18.13	20.11	0.04	98.48	0.744	0.226	0.029
10 11/76: 59–72	0.00	48.05	9.78	2.23	0.09	19.13	19.69	0.27	99.24	0.777	0.216	0.023
10 11/76: 59–78	0.00	47.61	9.78	2.71	0.09	19.01	19.50	0.31	99.01	0.776	0.215	0.028
10 11/76: 59–95	0.00	47.49	9.89	2.51	0.06	18.81	19.38	0.34	98.48	0.772	0.215	0.026
10 11/76: 60–27	0.03	49.37	9.70	2.88	0.11	19.01	16.99	0.58	98.67	0.777	0.187	0.029
10 11/76: 60–41	0.03	44.04	10.89	3.84	0.17	17.88	21.99	0.40	99.24	0.745	0.251	0.040
10 11/76: 60–45	0.10	41.41	11.38	3.12	0.18	17.28	25.19	0.37	99.03	0.730	0.290	0.033
10 11/76: 60–51	0.14	47.87	9.40	3.71	0.05	19.26	17.33	0.12	97.88	0.785	0.195	0.038
10 11/76: 60–56	0.02	49.36	9.80	3.22	0.09	19.12	17.02	0.37	99.00	0.777	0.188	0.033
10 11/76: 60–61	0.09	41.95	10.85	3.03	0.17	17.88	25.43	0.31	99.71	0.746	0.289	0.032
10 11/76: 60–87	0.03	47.36	9.98	3.10	0.09	18.96	19.63	0.33	99.48	0.772	0.217	0.032
10 11/76: 60–88	0.02	50.67	10.12	3.27	0.07	19.29	16.22	0.39	100.05	0.773	0.177	0.033
10 11/76: 60–97	0.07	48.57	10.03	2.59	0.10	19.17	19.18	0.30	100.01	0.773	0.209	0.026
10 11/76: 60–103	0.02	48.11	9.75	2.74	0.08	19.26	19.42	0.23	99.61	0.779	0.213	0.028
10 11/76: 60–126	0.08	49.14	10.45	2.22	0.08	18.95	18.40	0.21	99.53	0.764	0.201	0.023
Average	0.06	48.80	9.98	2.69	0.09	19.08	18.17	0.28	99.15	0.773	0.200	0.027
Standard deviation	0.05	3.10	0.57	0.67	0.04	0.65	2.88	0.12	0.56	0.016	0.036	0.007

Mineral relationships in gabbroic rocks from fracture zones of Indian Ocean ridges: evidence for extensive fractionation, parental diversity and boundary-layer recrystallization

Sherman H. Bloomer, James H. Natland & Robert L. Fisher

SUMMARY: Evolved two-pyroxene gabbros and ferrogabbros predominate in gabbroic suites dredged from five fracture zones on the SW and Central Indian Ridges. Compositions of olivines ($Fo_{56}-Fo_{85}$) and plagioclases ($An_{10}-An_{80}$) and the generally low magnesium numbers of orthopyroxene (0.58–0.84) and clinopyroxene (0.50–0.89) indicate that most gabbros crystallized from liquids more fractionated than those represented by basalts from adjacent ridge segments and the fracture zones themselves. This disparity, the paucity of diabases and gabbros, the absence of more magnesian gabbros and olivine–spinel cumulates, and the abundance of serpentinite in the dredge collections suggest that the fracture-zone gabbros crystallized in small magma bodies such as dykes or sills. These were emplaced laterally from large central magma chambers along ridge axes towards the fracture zones beneath a carapace of less fractionated basalt.

Rare olivine gabbros, however, contain magnesian clinopyroxene (magnesium number >0.84) and orthopyroxene (magnesium number >0.70), compositions which are more magnesian than those of pyroxenes that have been crystallized experimentally from certain ocean-ridge basalts at 1 atm. It has been suggested that such high-magnesium pyroxenes result from moderate- to high-pressure fractionation. However, the magnesian pyroxenes in the Indian Ocean samples follow olivine and plagioclase in the crystallization sequence and have compositions appropriate for crystallization at low pressure. For example, they match compositions of phenocrysts in refractory (low-Na_2O, low-TiO_2) siliceous (51–53% SiO_2) basalts from Deep Sea Drilling Project sites in Eocene–Cretaceous portions of the Indian Ocean. Such refractory basalts have not been studied experimentally, but the early crystallization of bronzite at low pressure in one of them is consistent with the siliceous and magnesian composition of the host glass.

Alternatively, the more magnesian pyroxenes in the gabbros may have resulted from a process of *in situ* boundary-layer recrystallization (resorption and recrystallization) of minerals on the walls of small magma bodies as high-temperature magmas were repeatedly injected along them. This mechanism is suggested by phase relationships in simplified basaltic systems at low pressure and by the compositions of rounded (resorbed) mega-crystal minerals which occur in many tholeiites from the Indian Ocean and elsewhere.

Gabbroic rocks are an integral part of ocean crust. They are much less studied than their basaltic counterparts, yet their textures and mineral compositions are crucial to understanding the dynamics of crystal–liquid fractionation and the evolution of ocean-ridge basalt magmas. The majority of the gabbroic rocks recovered from the ocean basins have come by dredging in fracture zones. The mechanism of exposure of such rocks in these environments is not known, but it appears to be a combination of the existence of thin crust near fracture zones (Detrick & Purdy 1980; Stroup & Fox 1981), a diminished magmatic budget (Stroup & Fox 1981) and intense faulting. Ocean crust formed in or near a fracture zone is likely to be different from that formed at typical spreading-centre segments, but at slowly spreading ridges such as the SW Central Indian Ridge, fracture zones are very closely spaced (Fig.1). Thus much of the crust in this region formed close to fracture zones.

Plutonic rocks comprise an important proportion of the crust exposed at fracture zones of slowly spreading ridges (Engel & Fisher 1975; Fisher *et al.* 1986). Few gabbroic suites obtained from fracture zones have been studied in detail, in part because they are often extensively altered but primarily because ample and diverse suites have rarely been obtained. Textural and mineralogical data exist for limited suites from the Mid-Atlantic Ridge (Hodges & Papike 1976; Prinz *et al.* 1976; Clarke & Loubat 1977; Miyashiro & Shido 1980; Tiezzi & Scott 1980; Walker & DeLong 1984), the Pacific (Vanko & Batiza 1982; Hébert *et al.* 1983) and the Mid-Cayman Rise (Elthon 1982, 1987).

Recently, a history of high-pressure crystallization of ocean-ridge basalts has been inferred

From Saunders, A. D. & Norry, M. J. (eds), 1989, *Magmatism in the Ocean Basins*, Geological Society Special Publication No. 42, pp. 107–124

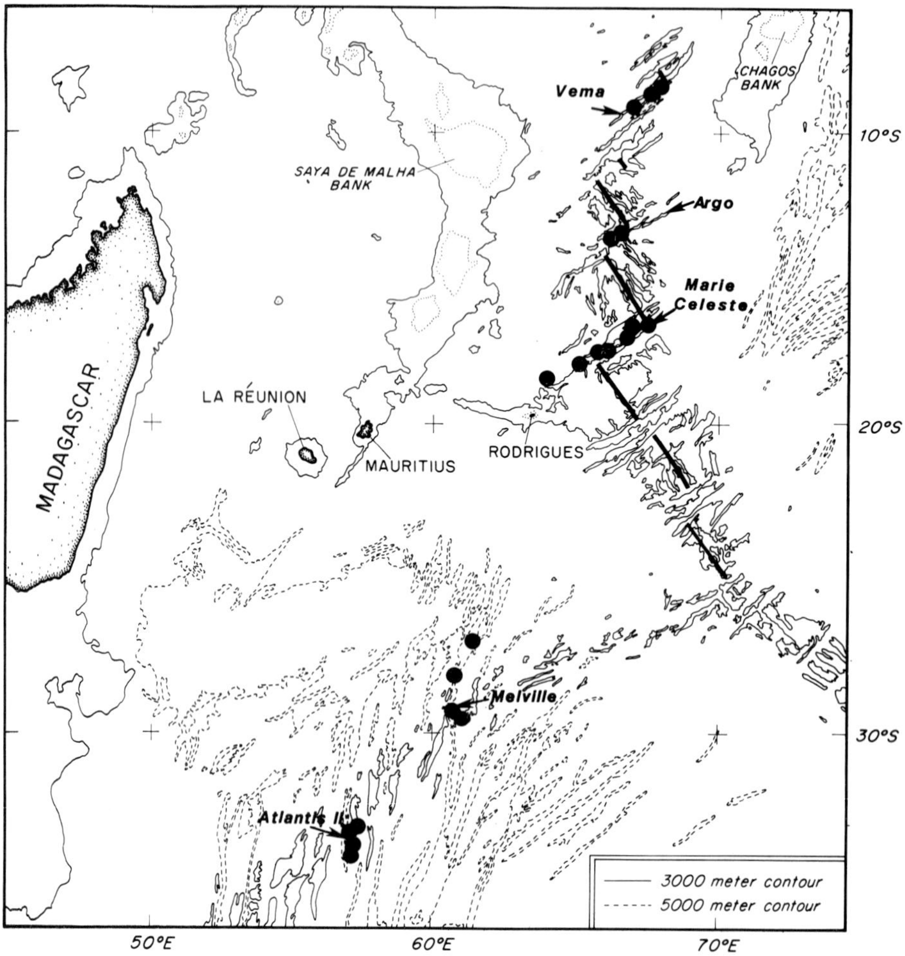

FIG. 1. Location of fracture zones from which the gabbroic samples discussed here were obtained: ●, approximate dredge locations. Detailed charts and sample locations (except for Atlantis II Fracture Zone) are available in Engel & Fisher (1975).

from pyroxene compositions in ultramafic cumulates in the Bay of Islands ophiolite (Elthon *et al.* 1982) and gabbros from the Mid-Cayman Rise (Elthon 1982, 1987). These rocks contain pyroxenes which are more magnesian than those that have been crystallized experimentally at 1 atm from certain abyssal tholeiites. Somewhat similar magnesian pyroxenes are found in several gabbros we have studied from Indian Ocean fracture zones, but both textures and mineral compositions imply that crystallization occurred at low pressure in these rocks. We suggest that neither the full range of compositions nor all the physical conditions relevant to the crystallization of such minerals in the ocean crust has been considered in experimental work, and we offer evidence and arguments in support of both hypotheses.

Source and selection of samples

We summarize mineral data from a detailed study of an extensive suite of gabbroic samples obtained in 21 dredge hauls from five fracture zones on the Central Indian Ridge (CIR) and SW Indian Ridge (SWIR). These materials were obtained by dredging during two expeditions of the Scripps Institution of Oceanography in 1968 and 1970–1971 (Engel & Fisher 1975) and a sampling programme directed by Fisher and performed aboard Woods Hole's *Atlantis II* in

1976. We discuss the implications of the mineralogical data for magmatic processes near fracture zones, and the occurrence and possible origin of high-magnesium orthopyroxene and clinopyroxene in some of these gabbros.

Rocks from the Vema, Argo and Marie Celeste Fracture Zones of the CIR and the Melville and Atlantis II Fracture Zones of the SWIR (Fig. 1) were studied. The CIR is spreading at 2–3 cm a^{-1} and the SWIR is spreading at 0.8–1.1 cm a^{-1} (Fisher *et al.* 1971; Fisher & Sclater 1983). Ridge segments are very deeply rifted, to depths in excess of 3.5 km, especially between the Melville Fracture Zone and the triple junction (Fig. 1). The offsets separating ridge segments adjoining the northern fracture zones are large (300 km for the Vema Fracture Zone, 110 km for the Argo Fracture Zone and 210 km for the Marie Celeste Fracture Zone (Fisher *et al.* 1971)). Relief exceeds 4000 m, 2800 m and 4600 m in the Vema, Argo and Marie Celeste Fracture Zones respectively (Engel & Fisher 1975). The location of individual hauls and detailed topographic maps of each fracture zone, except Atlantis II, are presented in Engel & Fisher (1975). The latter fracture zone has recently been extensively charted at high resolution using Sea Beam (Dick *et al.* 1986).

All mineral analyses on gabbros were performed using an automated nine-channel ARL electron probe at Virginia Polytechnic Institute and State University, Blacksburg, VA. The beam size varied from 2 to 20 μm. Natural mineral standards were used for calibration. All gabbros from 31 dredge hauls (about 150 samples) were examined in thin section. Forty-seven of the least-deformed and least-altered gabbros, including samples representative of all textural types in the hauls, were

TABLE 1. *Representative analyses of olivine and orthopyroxene in magnesian gabbros from the Central and SW Indian Ridges*

Sample	AII 1–50		ANTP 126–42		ANTP 126–43			ANTP 86–12	
Location	Atlantis II		Argo		Argo			Marie	Celeste
SiO_2	38.05	39.11	39.37	40.03	40.35	39.53	39.72	39.75	40.03
TiO_2	0.07	0.07	0.08	0.07	0.08	0.03	0.24	0.04	0.06
Al_2O_3	0.0	0.0	0.0	0.0	0.0	0.0	0.01	0.01	0.03
FeO	22.76	22.4	16.74	16.69	16.47	15.06	15.9	17.58	18.34
MnO	0.33	0.38	0.31	0.31	0.31	0.25	0.31	0.31	0.27
MgO	40.12	40.73	44.75	44.33	43.97	45.42	44.55	43.11	42.85
CaO	0.05	0.05	0.08	0.07	0.07	0.09	0.1	0.09	0.1
Na_2O	0.0	0.0	0.0	0.0	0.01	0.02	0.01	0.01	0.01
K_2O	0.01	0.01	0.0	0.0	0.01	0.01	0.01	0.01	0.01
NiO	ND	ND	ND	ND	ND	0.17	0.18	0.15	0.17
Cr_2O_3	ND	ND	ND	ND	ND	0.03	0.05	0.08	0.03
Sum	101.39	102.75	101.33	101.5	101.27	100.61	101.08	101.14	101.9
Mg/(Mg+Fe)	0.759	0.764	0.827	0.826	0.826	0.843	0.833	0.814	0.806

Sample	ANTP 130–57		AII 1–50		ANTP 126–42	ANTP 130–57
Location	Vema		Atlantis II		Argo	Vema
SiO_2	39.39	40.1	54.51	54.45	54.98	54.53
TiO_2	0.05	0.07	0.34	0.34	0.18	0.43
Al_2O_3	0.04	0.01	1.29	1.46	1.25	1.31
FeO	19.45	19.25	13.75	13.86	10.45	11.77
MnO	0.28	0.31	0.34	0.35	0.29	0.28
MgO	42.16	42.2	29.48	29.17	31.31	30.31
CaO	0.09	0.05	1.12	1.23	0.89	0.99
Na_2O	0.0	0.0	0.03	0.02	0.01	0.0
K_2O	0.02	0.01	0.0	0.01	0.01	0.0
Sum	101.48	102	100.86	100.89	99.37	99.62
Mg/(Mg+Fe)	0.794	0.796	0.793	0.790	0.842	0.821

Analyses were performed on a nine-channel automated ARL electron microprobe at Virginia Polytechnic Institute and State University. ND, not determined.

TABLE 2. *Representative analyses of clinopyroxene in magnesian gabbros from the Central and SW Indian Ridges*

Sample	All 1–50				
Location	Atlantis II				
SiO_2	51.73	51.04	52.86	51.25	51.88
TiO_2	0.65	0.66	0.65	0.67	0.64
Al_2O_3	2.66	3.18	2.31	2.46	2.36
FeO	6.76	6.57	6.05	5.66	5.96
MnO	0.25	0.21	0.26	0.2	0.23
MgO	16.88	17.06	16.73	16.56	16.03
CaO	20.59	20.55	21.41	22.44	21.76
Na_2O	0.37	0.36	0.37	0.36	0.38
K_2O	0.01	0.01	0.01	0.02	0.01
NiO	ND	ND	ND	ND	ND
Cr_2O_3	ND	ND	ND	ND	ND
Sum	99.9	99.64	100.65	99.62	99.25
Mg/(Mg+Fe)	0.817	0.822	0.831	0.839	0.827

Sample	ANTP 126–43			
Location	Argo			
SiO_2	52.6	51.49	52.61	51.88
TiO_2	0.54	0.76	0.49	0.45
Al_2O_3	3.14	3.25	3.17	3.43
FeO	4.62	4.72	4.64	3.84
MnO	0.16	0.17	0.16	0.14
MgO	16.33	16.26	17.7	17.13
CaO	21.64	21.87	21.34	21.21
Na_2O	0.36	0.46	0.35	0.33
K_2O	0.02	0.02	0.02	0.01
NiO	ND	ND	ND	ND
Cr_2O_3	ND	ND	ND	ND
Sum	99.41	99	99.48	98.42
Mg/(Mg+Fe)	0.863	0.860	0.872	0.888

Sample	ANTP 126–42				
Location	Argo				
SiO_2	53.05	51.86	52.41	52.87	52.63
TiO_2	0.35	0.39	0.53	0.4	0.47
Al_2O_3	2.49	2.75	2.27	2.37	2.69
FeO	5.13	4.46	5.34	5.22	5.38
MnO	0.17	0.17	0.21	0.18	0.19
MgO	18.56	17.24	17.17	17.25	18.67
CaO	19.89	22.33	21.33	21.25	20.39
Na_2O	0.27	0.3	0.3	0.29	0.26
K_2O	0.02	0.01	0.02	0.02	0.01
NiO	ND	ND	ND	ND	ND
Cr_2O_3	ND	ND	ND	ND	ND
Sum	99.93	99.51	99.58	99.85	100.69
Mg/(Mg+Fe)	0.866	0.873	0.851	0.855	0.861

Sample	ANTP 86–12					
Location	Marie Celeste					
SiO_2	53	53.62	53.09	53.46	52.13	52.82
TiO_2	0.38	0.37	0.38	0.4	0.83	0.62
Al_2O_3	2.98	3	2.99	2.66	2.9	2.68
FeO	5.71	4.74	5.2	4.78	5.76	6.02
MnO	0.19	0.13	0.15	0.2	0.25	0.18
MgO	17.34	18.04	17.42	16.58	17.68	17.55
CaO	20.65	20.92	21.04	22.87	20.69	20.17
Na_2O	0.31	0.28	0.33	0.31	0.28	0.29
K_2O	0.02	0.01	0.01	0.02	0.02	0.01
NiO	0.09	0.04	0.09	0.11	0.06	0.07
Cr_2O_3	0.51	0.63	0.61	0.38	0.38	0.32
Sum	101.18	101.78	101.31	101.77	100.98	100.73
Mg/(Mg+Fe)	0.844	0.872	0.857	0.861	0.845	0.839

Sample	ANTP 130–57			
Location	Vema			
SiO_2	52.82	52.57	52.82	52.58
TiO_2	0.64	0.93	0.54	0.53
Al_2O_3	2.32	2.63	3.85	2.33
FeO	6.16	6.41	7.09	5.9
MnO	0.19	0.19	0.22	0.19
MgO	16.93	16.55	18.83	16.42
CaO	21.19	20.45	16.97	20.93
Na_2O	0.27	0.31	0.23	0.23
K_2O	0.01	0.02	0.01	0.02
NiO	ND	ND	ND	ND
Cr_2O_3	ND	ND	ND	ND
Sum	100.53	100.06	100.56	99.13
Mg/(Mg+Fe)	0.830	0.821	0.826	0.832

Analyses were performed on a nine-channel automated ARL electron microprobe at Virginia Polytechnic Institute and State University. ND, not determined.

TABLE 3. *Representative analyses of plagioclase in magnesian gabbros from the Central and SW Indian Ridges*

Sample	AII 1–50		ANTP 126–42				ANTP 126–43			
Location	Atlantis II		Argo				Argo			
SiO_2	52.13	53.07	50.15	49.82	50.21	49.07	51.51	49.7	49.46	48.39
TiO_2	0.06	0.09	0.06	0.06	0.06	0.06	0.09	0.06	0.05	0.07
Al_2O_3	30.77	30.7	31.79	32.99	31.94	32.45	30.97	33.2	32.99	33.1
FeO	0.4	0.18	0.29	0.31	0.28	0.27	0.42	0.34	0.43	0.37
MnO	0.03	0.01	0	0.01	0.01	0	0.01	0	0.01	0.01
MgO	0.4	0.1	0.09	0.18	0.16	0.15	0.13	0.08	0.12	0.13
CaO	12.11	11.99	13.55	14.44	14.5	14.66	13.7	14.93	15.5	16.15
Na_2O	4.28	4.59	3.52	3.11	2.96	2.95	3.56	3.09	2.7	2.23
K_2O	0.04	0.05	0.04	0.04	0.04	0.04	0.1	0.08	0.07	0.04
Sum	100.22	100.78	99.49	100.96	100.16	99.65	100.49	101.48	101.33	100.49
An	0.608	0.588	0.678	0.717	0.728	0.731	0.676	0.724	0.757	0.798

Sample	ANTP 86–12						ANTP 130–57		
Location	Marie	Celeste					Vema		
SiO_2	48.32	46.98	50.47	48.88	46.02	50.88	52.75	52.37	50.05
TiO_2	0.05	0.03	0.08	0.03	0.04	0.06	0.11	0.1	0.07
Al_2O_3	33.47	35.65	32.38	33.35	36.49	31.86	29.92	31.26	31.82
FeO	0.38	0.37	0.39	0.42	0.41	0.56	0.32	0.46	0.41
MnO	0.01	0.01	0.01	0.0	0.01	0.02	0.01	0.0	0.0
MgO	0.14	0.08	0.17	0.14	0.10	0.28	0.16	0.0	0.05
CaO	16.19	18.55	14.66	16.29	19.04	13.96	13.16	13.71	14.69
Na_2O	2.37	1.18	3.0	2.19	0.94	3.35	3.64	3.22	2.9
K_2O	0.04	0.03	0.05	0.04	0.03	0.07	0.04	0.03	0.04
Sum	100.97	102.94	101.21	101.34	103.16	101.04	100.11	101.15	100.03
An	0.788	0.895	0.727	0.802	0.916	0.694	0.664	0.700	0.734

Analyses were performed on a nine-channel automated ARL electron microprobe at Virginia Polytechnic Institute and State University.

analysed in reconnaissance (Table 1–3) with 4–10 analyses of all major phases in each sample. Nineteen samples were examined in detail to study the extent of zoning, minor-element variation and sample heterogeneity. We also summarize mineral data in two basalts which were obtained at the Scripps Institution using procedures described by Natland *et al.* (1983). The detailed mineralogical results and the results of a companion study on the basalts from the fracture zones and adjacent ridge segments will be presented elsewhere.

Petrology and mineral chemistry

General characteristics

The full plutonic mafic rock suite from all five fracture zones is diverse, including olivine gabbro, olivine–two-pyroxene gabbro, gabbro, two-pyroxene gabbro, norite, ferro-gabbro, amphibolitized gabbro and pyroxenites (orthopyroxenite and websterite) in association with basalts and ultramafic rocks. Bulk rock and trace-element analyses of some of these samples are presented by Engel & Fisher (1975). Of the gabbros in which primary phases can clearly be identified, olivine gabbro accounts for 27% of the total number of specimens, two-pyroxene gabbro for 31.5% and gabbro for 21%. Ferro-gabbro and two-pyroxene gabbro account for over 50% of these samples. Pyroxenites compose less than 5% of all samples. Four samples of altered, iron-rich gabbroic veins in serpentinites were also recovered. Percentages by weight have not been determined. Based on mineral occurrences, abundances and compositions through the sample set as a whole, the general crystallization sequence was olivine (OL), plagioclase (PLAG), clinopyroxene (CPX), orthopyroxene (OPX) and opaque oxides, usually ilmenite and titanomagnetite. However, there are many individual samples in which textural relations show the sequence to be different, including those exhibiting simultaneous crystallization of PLAG and OL, or of PLAG and CPX, and the appearance of CPX as a primocryst shortly before PLAG. For the most part, then, samples display the consequences of multiple saturation of magmas with two or more silicate phases and can be described as mesocumulates or heteradcumulates. Primary textures, however, are almost always modified by partial recrystallization and granulation due to high-temperature deformation.

The samples are texturally and mineralogically heterogeneous, even on a thin-section scale. Grain size may vary from 2–3 mm to nearly 1 cm across a section. Within single thin sections, OL may vary by 6% Fo, PLAG by 20% An, OPX by a magnesium number of 0.15 (magnesium number=$Mg/(Mg+Fe^{2+})$) and CPX by a magnesium number of 0.18. In the sample set as a whole, OL ranges from Fo_{56} to Fo_{85}, PLAG ranges from An_{10} to An_{80}, the magnesium number of OPX ranges from 0.58 to 0.84 and the magnesium number of CPX ranges from 0.50 to 0.89. The samples tend to be most heterogeneous in the more iron-rich end-members (those with magnesium numbers of CPX less than 0.75). As a rule the individual grains are not strongly zoned; most of the variation is between grains within single samples.

The iron and titanium contents of liquids calculated to have been in equilibrium with CPX in gabbro were significantly higher, *ie* more fractionated, than most of the basalts sampled from the fracture zones and adjacent ridge segments (Fig. 2). There is overlap at the magnesian end, but over 60% of the samples have lower magnesium numbers and higher TiO_2 contents than the CIR and SWIR basaltic glasses. The trend of increasing titanium with decreasing magnesium is like that found in iron enriched basalts and ferrobasalts from the fast-spreading ridges of the eastern Pacific (Fig. 2) (Rhodes *et al.* 1976; Natland 1980; Perfit & Fornari 1983). The scatter in calculated liquid compositions probably results from imprecision in analysis of the low TiO_2 concentrations in most of the CPX. Liquid compositions based on coexisting OL in these gabbros (Roeder & Emslie 1970) yield a very similar range of magnesium numbers.

The magnesium numbers of OPX and CPX in the CIR and SWIR gabbros vary sympathetically (Fig. 3); An and Fo contents also correlate positively with CPX magnesium numbers (Bloomer *et al.* 1985). The variation is similar to that of other mafic and ultramafic plutonic rocks

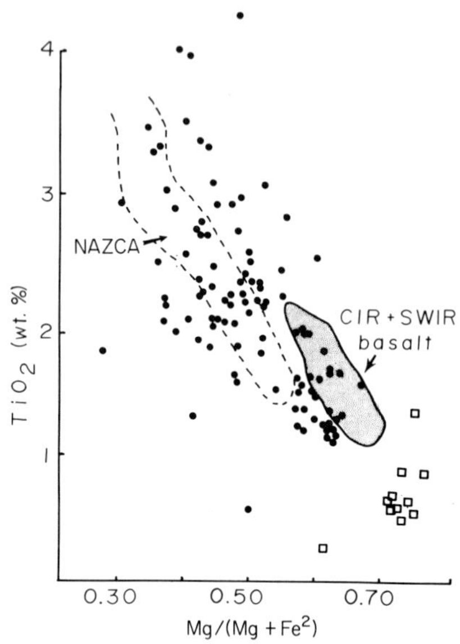

FIG. 2. TiO_2 versus $Mg/(Mg + Fe^2)$ for basalt glasses (shaded field) from the CIR and SWIR and for liquids calculated to have been in equilibrium with CPX in gabbros (●) and CPX in serpentinites (□). Liquid compositions were calculated using a modification of the equation due to Duke (1976) and partition coefficients from Pearce & Norry (1979). The field for Nazca plate basalts is from Rhodes *et al.* (1976).

from oceanic ridges, although the CIR collection includes far more fractionated pyroxenes than are normally found in the Atlantic or Pacific (Fig. 3).

The rare gabbroic veins found in serpentinites have among the most iron-enriched CPX of any samples (Fig. 3; magnesium number, 0.54). The veins are extensively altered, and the CPX is the only primary phase remaining; in Fig. 3 the CPX composition in the veins is plotted against the OPX composition of the host serpentinite (within which no unaltered CPX remains). These veins are clearly not gabbros 'sweated' from the peridotites, but are iron-rich compositions injected into the serpentinized peridotites.

Magnesian pyroxene compositions

Although the gabbros have predominantly iron-rich compositions, some do include a number carrying CPX with magnesium numbers greater than 0.84 and OPX with magnesium numbers over 0.76 (Fig. 4). Similar magnesium-rich pyroxenes have been described from gabbros in the

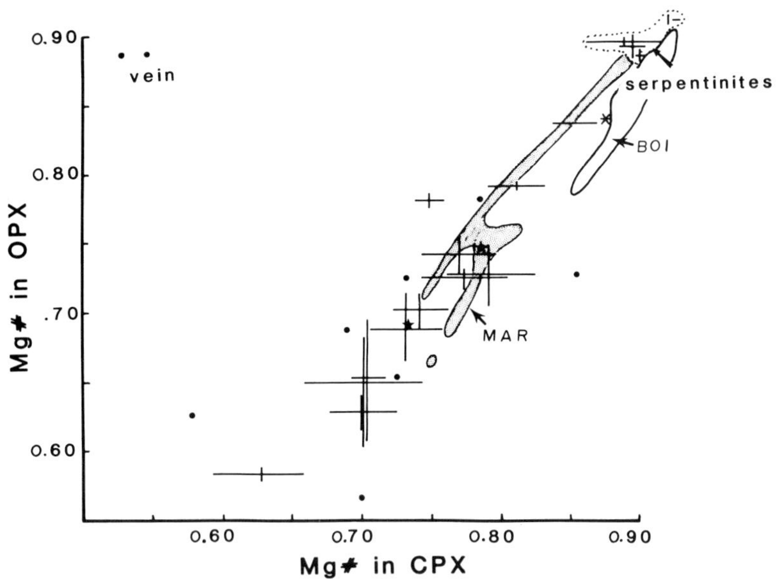

FIG. 3. Variation in coexisting OPX and CPX in mafic and ultramafic samples from the CIR and the SWIR (bars indicate ranges within individual samples): MAR, Mid-Atlantic Ridge field from Tiezzi & Scott (1980) and Hodges & Papike (1976); BOI, Bay of Islands field from Elthon *et al.* (1982); ★, compositions from Vanko & Batiza (1982) for Mathematician Ridge gabbros; points labelled 'vein', iron-rich CPX in a gabbroic vein in a serpentinite with magnesium-rich OPX; ∗, compositions of coexisting pyroxenes in glomerocrysts in a DSDP Site 236 basalt (Table 4).

Mid-Cayman Rise (Elthon 1982, 1987) and ultramafic cumulates from the Bay of Islands ophiolite (Elthon *et al.* 1982, 1984; Komor *et al.* 1985). The first clinopyroxenes to crystallize in experiments on certain 'normal' abyssal tholeiites at 1 atm have magnesium numbers less than 0.84 (Walker *et al.* 1979; Grove & Bryan 1983). The first low-calcium pyroxenes to crystallize at 1 atm have magnesium numbers less than 0.55, although at low pressures compositions with magnesium numbers as high as 0.81 may occur (Elthon *et al.* 1982; Elthon 1987). It is the occurrence of pyroxenes with higher magnesium numbers than these which have been the basis for the suggestion of moderate- to high-pressure crystallization.

Both OPX and CPX compositions in the Indian ocean gabbroic samples overlap the range of magnesium-rich pyroxenes in the Bay of Islands ultramafic rocks (Fig. 3) and approach the most magnesian CPX described from the Mid-Cayman Rise (Fig. 5), for both of which a moderate to high-pressure crystallization has been inferred. The magnesium-rich pyroxenes in the Indian Ridge samples, however, show no textural or mineralogical evidence of crystallization at elevated pressures. The CPX in the Indian Ocean samples is clearly interstitial to OL and PLAG, and in general was the third phase to crystallize. The magnesium-rich OPX in the OL-gabbros is interstitial to the other phases, and usually partially rims olivine. In the most magnesian gabbro without olivine (Fig. 5) and with two pyroxenes, the CPX and OPX are clearly crystallizing after PLAG. All these samples crystallized at pressures low enough for both OL and PLAG to have preceded CPX on the liquidus.

These gabbroic magnesian pyroxenes show neither the high titanium nor high aluminium contents that might be expected in high-pressure pyroxenes, although the low TiO_2 may also depend on liquid composition. Although there is a significant range within samples, there is a consistent trend from low Ti-low Fs CPX in serpentinites to pyroxenes with higher titanium in magnesian gabbros (Fig. 6). Even in the most magnesian CPX (magnesium number, 0.84), are low TiO_2 contents but at low Fs (Fig. 6) are nevertheless higher than in serpentinites. Abundances of Cr_2O_3 (1.5% to less than 0.1%) and Al_2O_3 (4% to less than 1%) both decrease smoothly, from serpentinites to gabbros to iron-enriched gabbros. Only a single pyroxenite sample deviates from these trends. These are all characteristics of low-pressure crystallization

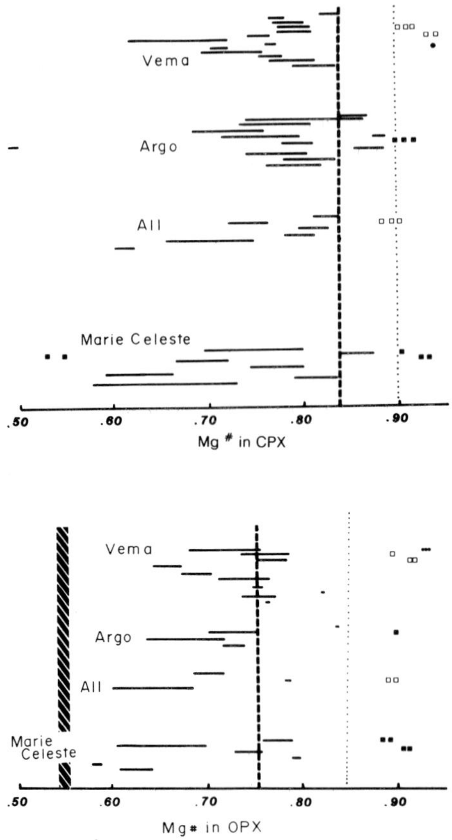

FIG. 4. Range of pyroxene compositions in individual gabbroic samples from all five fracture zones (bars indicate ranges of composition in single gabbro samples): □, compositions in peridotites and pyroxenites; ———, approximate limit of 2 kbar OPX crystallization (magnesium number, 0.76) and 1 atm CPX crystallization (magnesium number, 0.84); bold line beginning of 1 atm OPX crystallization;. . . , upper limit of DSDP 236 megacryst and phenocryst compositions. OPX crystallization at 2 kbar may occur at magnesium numbers as high as 0.81 in some ocean-ridge basalt compositions (see discussion by Elthon 1987).

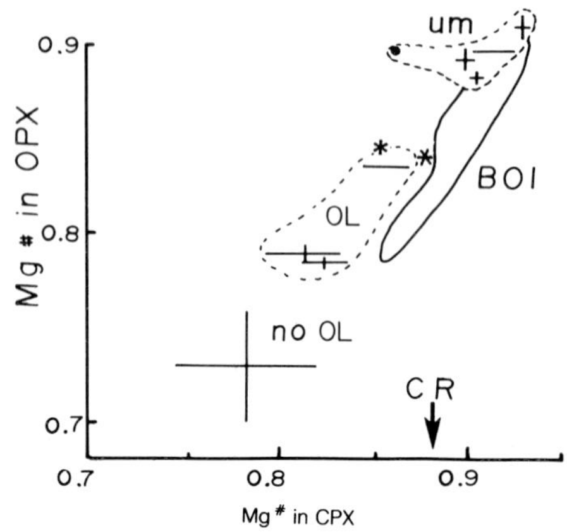

FIG. 5. Coexisting OPX and CPX in the most magnesian mafic and ultramafic (um) samples: OL, olivine gabbros; ∗, DSDP Site 236 phenocrysts; BOI, Bay of Islands field (Elthon *et al.* 1982); CR, sample with highest magnesium number from Cayman Rise gabbros (Elthon 1987).

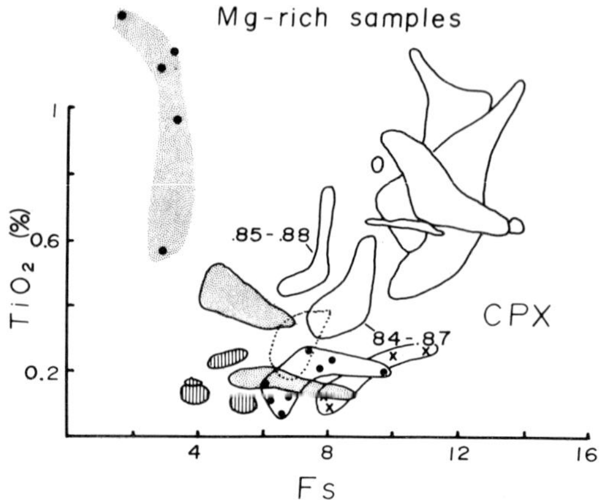

FIG. 6. TiO_2 versus Fs in CPX in magnesium-rich gabbroic and ultramafic samples. Each field encloses compositions within single samples. Shaded fields are for pyroxenites, vertically lined fields are for serpentinites and open fields are for gabbros. The dotted field encloses phenocryst compositions in magnesian basalts from the CIR. Those from DSDP Site 212 (field with full points) and Site 236 (field with crosses) are also shown. The numbers indicate the magnesium numbers of CPX in the indicated OL gabbro samples, and are the most magnesian found in any of the gabbro sample set.

(Grove & Bryan 1983), and *in toto* suggest that something has been overlooked or left untried in the experimental studies of crystallization of abyssal tholeiites.

Discussion

Fracture-zone structure

Several restrictions on the distribution of rock types in the CIR and SWIR fracture zones and on the processes that produce that distribution can be deduced from the mineralogical data from the gabbroic rocks. The most striking characteristic of the gabbros examined here is that a large percentage of them (over 50%) have crystallized from liquids which are more iron-enriched than any represented by basalt glasses from the fracture zones and adjacent ridge crests. These liquids are as fractionated as many basaltic glasses sampled from the Pacific and Nazca Plates (Fig. 2) but are confined to the plutonic portion of the crust. Although there is some overlap of the more magnesian calculated liquid compositions with the range of liquids represented by basalt glasses, there are no basalts equivalent in composition to the more iron-enriched gabbros. Those more iron-rich compositions are in some way isolated in the ridge system and are prevented from erupting.

The covariation of major and minor element compositions in OL, PLAG, CPX and OPX suggests that crystal–liquid fractionation was a principal control in the evolution of the liquids from which the Indian Ridge gabbros crystallized. In a suite of fractionated volcanic rocks from the Galapagos Rift, fractionation from a magnesian basalt (magnesium number 0.61) to an andesitic liquid with a magnesium number comparable with that of the most fractionated Indian Ridge gabbros requires nearly 75%–80% crystallization of the original liquid (Perfit & Fornari 1983); 51% of the resulting solid should be an OL-bearing gabbro. The most magnesian basalt from the CIR (magnesium number, 0.70) is significantly more magnesian than the parental basalt (magnesium number, 0.61) of Perfit & Fornari (1983) and would produce an even larger volume of OL-cumulates if fractionated to an andesitic composition. If a similar high-magnesian basalt is parental to most of the Indian Ridge basalts and gabbros, these OL-rich cumulates must exist somewhere in the ridge system. They are under-represented in the fracture zones, relative to their expected abundance in the crustal section as a whole.

Diabase is rare in the dredge hauls described here and from fracture zones farther SW on the SWIR (Fisher *et al.* 1986). Ultramafic rocks (largely serpentinized) and pillow basalts together account for most (about 85%) of the recovery from those SWIR fracture zones. Ultramafic rocks in the collection from fracture zones of the CIR, although not as abundant as in the SWIR fracture zones, are common and relatively fresh (Engel & Fisher 1975). These patterns of rock distribution have been used to infer a thinned gabbroic section, an absence of diabase dykes and the presence of serpentinite as a principal crustal component in the fracture zones of the western Indian Ocean (Fisher *et al.* 1986).

The isolation of iron-rich liquids in the plutonic section from erupted liquids, the abundance of serpentine in the fracture zones, the paucity of OL-gabbro and the thinned gabbroic sections in the fracture zones, place constraints on models for magmatic systems at slow-spreading ridges. One model which satisfies these observations is that in which ridge segments between fracture zones or offsets are fed by large upwelling diapirs, and where a central magma chamber develops beneath the middle portion of the ridge segment (Whitehead *et al.* 1984; Crane 1985). Crustal growth in the more distal parts of the ridge segment is by lateral flow of magma along the ridge towards the fracture zone, as has been observed in Iceland (Sigurdsson & Sparks 1978). Both basalt flows and gabbroic bodies will be fed from this central chamber, the gabbros to be injected as dykes and sills longitudinally down the ridge towards the fracture zone (Fig. 7). Much of the fractionation will occur in the wings and distal ends of the chamber, and in the small dykes and sills. Gabbros further from the central chamber will be even more fractionated; the bulk of the OL-cumulates will be restricted to the region around the central magma chamber. Magnesian eruptives from the central chamber may form a carapace over the gabbroic dykes. This carapace, and the increased density of the iron-enriched liquids (Stolper & Walker 1980), may keep these iron-enriched compositions from erupting. We envisage that much of the later-stage gabbroic crystallization and emplacement occurs in a network of dykes and sills injected along-axis, thinning towards the fracture zone. Thus the scarcity of dykes and the thinned gabbroic sections in the fracture zones can be explained; magma supply to the fracture zone is low because the source for the magmas is a considerable distance away. The material deficiency within the fracture zone resulting from that low magma supply may be made up by plastic flow of hot and partially serpentinized peridotite up into the fracture zone (Whitehead *et al.* 1984; Fisher *et al.* 1986). This mechanism may therefore explain the thinning of ocean crust detected seismically near some fracture zones (Detrick & Purdy 1980; Stroup & Fox 1981; Cormier *et al.* 1984).

Lateral injection of fractionated magmas also provides an explanation for the iron-rich gabbroic veins found cutting serpentinites. Normally such fractionated gabbros are considered to be 'high level', *ie* they crystallized in the central parts of a nearly solidified magma chamber. However, if fractionated gabbroic dykes are injected laterally down the ridge and if serpentinites comprise a significant part of the fracture zone crust, those gabbroic dykes may be injected into serpentinites or partially serpentinized peridotites at shallow depths (Fig. 7).

A lateral injection model does not explain all the chemical variations seen near some fracture zones and at small offsets, *eg* in the Pacific, where there is evidence of various sources and degrees of melting across and near ridge discontinuities (Langmuir *et al.* 1986). Individual ridge segments at any spreading centre may also be fed by large central diapirs, several small diapirs or a combination of the two along their length, depending on the spreading rate, the lengths of fracture zone offsets and other tectonic characteristics. These offer ample opportunity for the production of diverse parental magma types, which may be displayed in gabbroic rocks. We now provide evidence that this is the case at CIR and SWIR fracture zones.

Magnesian pyroxenes: high-pressure crystallization or evidence for unusual parentage?

The occurrence in mafic–ultramafic rocks from the Cayman Trough and Bay of Islands ophiolite of pyroxenes more magnesian than those which have been produced in 1 atm experiments on certain abyssal tholeiites has been taken as evidence for high-pressure crystallization (Elthon 1982, 1987; Elthon *et al.* 1982). Theoretically, as pressure increases, the clinopyroxene primary phase volume in the OL–OPX–

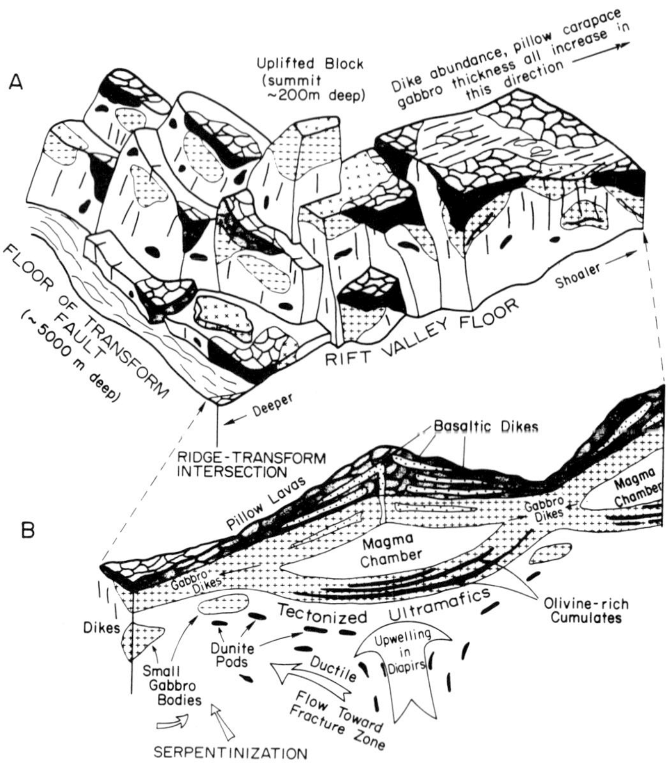

FIG. 7. Possible relationship between outcrop relations deduced from dredged collections from the Indian Ocean ridges at and near the intersection of a spreading-ridge section and a transform fault (section A) and a schematic longitudinal cross-section through the crust and upper mantle along the axial magma chamber nearest to the transform fault (section B). Section A is a view with sediment and talus removed. Both ridge-parallel and transform-parallel faults are shown delineating fault blocks. Gabbros are shown as underlying (intruding the base of) extrusive pillows with few, if any, intervening dykes and as intruding ultramafic rocks. Ol-rich cumulates precipitating beneath principal magma chambers are not shown as extending toward the fracture zone, although evolved gabbros (including ferrogabbros) are injected towards the fracture zone from the tip of the magma chamber.

CPX–PLAG system expands and the amount of OL crystallization that occurs before the CPX liquidus is reached is less, so that the first pyroxene to crystallize will be more magnesian than a pyroxene that would form from the same magma at 1 atm. At sufficiently high pressures, CPX can replace OL as the liquidus phase in normal abyssal tholeiites (Elthon & Scarfe 1984). Such a mechanism might account for occurrences of very magnesian pyroxenes in plutonic rocks with CPX or OPX as the first crystallizing phase, as in the Bay of Islands ultramafic cumulates. Some pyroxenites from the Indian Ocean ridge system may have a similar origin (Fig. 6). This shift in phase volumes may also be important in producing more magnesian phases at low to moderate pressures (2–5 kbar), as suggested for gabbros from the Mid-Cayman Rise (Elthon 1982) and phenocryst assemblages in basalts from Deep Sea Drilling Project (DSDP) Sites 504 and 505 (Costa Rica Rift) (Natland *et al.* 1983) and the Mid-Atlantic Ridge (Tormey *et al.* 1987).

Some of the magnesian pyroxenes in the Indian ocean fracture zone suite, although not having compositions as extreme as pyroxenes in some Bay of Islands samples, clearly fall outside the known limits of 1 atm pyroxene compositions produced experimentally from abyssal tholeiites (Fig. 3). Whereas part of this shift in composition may result from a moderate-pressure crystallization as discussed above, there is no other textural or mineralogical data which supports such a contention.

We do not wish to contest the theoretical basis of the high-pressure hypothesis. However, this argument has always been based on the demonstrated impossibility of producing magnesian pyroxenes from abyssal tholeiites experimentally at low pressure. This assumes, however, that all relevant abyssal tholeiite compositions have been studied (or, conversely, that the few which have been studied are relevant to all). The Indian Ocean provides evidence to the contrary. Moreover, the experiments have not necessarily recreated the physical conditions in which the gabbros crystallized. We discuss these alternatives in turn.

Concerning parental compositions, several DSDP sites have been drilled into Eocene and older crust produced at ridge segments trending E–W that existed in the Indian Ocean prior to a major change in the direction and reduction in the rate of spreading about 40 Ma ago (Fisher *et al.* 1971). These basalts (from DSDP Sites 212, 220, 221 and 236) have generally refractory compositions, with low Na_2O and TiO_2 contents for their state of fractionation and unusually high SiO_2 contents (Frey *et al.* 1980; Mahoney & Natland 1985). Basalts from two of these sites (212 and 236) (Table 4) are porphyritic and bear explicitly on the question here. Basalt glass from DSDP Site 212 has a high magnesium number of 0.703, but is multiply saturated in very calcic plagioclase (An_{87-89}), olivine ($Fo_{89.6-90.4}$), spinel and magnesian clinopyroxene (Fs_{6-7}). Glass from DSDP Site 236 is somewhat less magnesian (magnesium number, 0.62) and lacks spinel. However, it is also multiply saturated in plagioclase (An_{77-82}), olivine ($Fo_{85.1}$), clinopyroxene (Fs_{8-9}) and bronzite ($Wo_5En_{80}Fs_{15}$). As can be seen in Table 4, both these basalts are siliceous in comparison with so-called 'normal' abyssal tholeiites (*eg*. Elthon, this volume) and their abundances of Na_2O and TiO_2 are among the lowest in the ocean basins. Indeed, Frey *et al.* (1980) compare the DSDP Site 236 basalt with island-arc basalts in these aspects of its composition and because of its extraordinarily depleted abundances of rare earth elements. There is no indication that the minerals in these rocks are far out of equilibrium with the present host glasses (indeed, olivines and plagioclases have almost precisely the expected compositions). Since many of the pyroxenes occur in glomerocrysts with at least plagioclase present and form small tabular crystals in the glass, there is every reason to suppose that these are low-pressure near-equilibrium minerals in the basalts. They also have low contents of TiO_2 (Fig. 6) and Al_2O_3 (1%–4%), similar to abundances in the magnesian pyroxenes in the gabbro suites.

We thus interpret gabbroic rocks having similar pyroxenes in CIR and SWIR fracture zone suites as plausibly having precipitated from parents with compositions similar to those listed in Table 4. Such compositions not only have high SiO_2 contents compared with typical abyssal tholeiites but they also have high CaO/Al_2O_3 ratios. Both factors will lead to precipitation of high-temperature (hence magnesian) pyroxenes at low pressure, on the basis of long-standing interpretations of experimental data (*eg* Bowen & Schairer 1935; Lindsley & Anderson 1983).

The question is, what is the role of such refractory and siliceous basalts in the present CIR and SWIR? There are no such basalts out of more than 80 compositional types published or in our files of natural glass analyses from spreading segments adjacent to the five fracture zones discussed here, nor from the fracture zones themselves. However, in all fracture zones, a range of parental basalt compositions exists, or can be inferred, which has variably

TABLE 4. *Refractory basalt compositions from the Indian Ocean*

		DSDP 212		DSDP 236
SiO_2		51.33		53.38
Al_2O_3		15.60		14.43
FeO^*		7.86		9.93
MgO		8.97		7.80
CaO		13.48		11.38
Na_2O		1.42		1.65
K_2O		0.05		0.06
TiO_2		0.61		0.61
P_2O_5		0.03		0.04
Sum		99.35		100.28
Mg number		0.703		0.619
Mineral compositions[a]				
Olivine	(P)	$Fo_{89.6-90.4}$	(P)	$Fo_{85.1}$
Plagioclase	(P)	An_{87-89}	(P)	An_{77-82}
	(G)	An_{81}		
Clinopyroxene	(P)	$Wo_{40}En_{54}Fs_6$	(P)	$Wo_{37}En_{55}Fs_8$
	(G)	$Wo_{40}En_{53}Fs_7$		
Orthopyroxene		—	(P)	$Wo_5En_{80}Fs_{15}$

Electron microprobe analyses by T. O'Hearn, Smithsonian Institution, Washington, DC.
[a](P), phenocryst; (G), groundmass.

refractory bulk compositions (expressed in terms of Na_2O and TiO_2 contents at particular magnesium numbers). Moreover, on the basis of the common occurrence of very calcic plagioclase phenocrysts (An_{85-90}) in many of the basalts and of refractory glass inclusions in these and associated minerals (Price *et al.* 1986; J. Natland, unpublished data), it is clear that the appropriate refractory magma types are present in the region and coalesce with others to make the more normal abyssal tholeiites that have been analysed. The situation is comparable with that of the Siqueiros Fracture Zone, where refractory melt compositions are deduced from crystallization histories of basalts but where no such lava type has actually been sampled (Natland, this volume).

Thus we do not suppose that the gabbroic rocks dredged from any of these fracture zones are entirely consanguineous, *ie* related to a succession of liquids representing a single line of descent. Instead, to the extent that a spectrum of magma types is typically supplied to adjacent ridge segments, so too are the gabbros a consequence of crystallization and mixing of diverse parental magma strains. It is important in this respect that the magnesian pyroxenes do not occur as isolated minerals in individual cumulates; rather, they are the typical ferromagnesian silicates of individual samples. Therefore these samples are evidence for crystallization of isolated, and perhaps unusually refractory, magma stems. These did not enter any long-lived axial magma chamber to mix with more typical parental and hybrid compositions. Thus either the schematic magma chamber shown in Fig. 7 should be viewed as an ephemeral feature, or the samples are evidence for a more dispersed type of magma supply system from the mantle between such chambers and the fracture zones.

The mineral compositions in the magnesian gabbros and in two basalt samples reported here compare well with those in DSDP Site 334 near the FAMOUS area, Mid-Atlantic Ridge (Hodges & Papike 1976). Basalt glass compositions at this site are among the more siliceous in the North Atlantic (Aumento *et al.* 1977), suggesting that the gabbros there are another example of low-pressure crystallization of particular melt compositions, thus far not studied experimentally.

Now, concerning the physical conditions of crystallization, we propose that the pyroxene compositions may, alternatively or in addition, be a consequence of a process we call *in situ* or boundary-layer recrystallization. McBirney & Noyes (1979) suggested that many of the textures seen in layered intrusions could be explained by a process of crystal growth along the boundaries of magma chambers rather than by the more traditional concept of crystal settling and accumulation. Irvine (1980) also invoked double-diffusive fractional crystallization along the boundaries of magma chambers, wherein the mat of crystals adhering to the boundaries is modified by both continued contact with magma in the chamber and reaction with evolved melt squeezed out from within the mat on the floor of the chamber by the weight of accumulating crystals. Weight of crystals may not be necessary to produce such expulsion; continued crystal growth within the mat can expel interstitial melt very efficiently (Walker *et al.* 1985). Langmuir (1986) has shown that the effects of such *in situ* crystallization on both trace and major-element compositions of basaltic liquids can be very different from those predicted by simple crystal–liquid separation. Additionally, the compositions of accumulated olivines and pyroxenes in contact with the melt in a steadily fractionating magma chamber will become more iron rich as they react with interstitial melts.

Consider now the situation where a small magma body is repeatedly injected with primitive (high-magnesium) magma which passes through it *en route* to a larger body or eruption on the seafloor. The volume of each pulse is large with respect to that of the magma body and to the crystals adhering to its walls and sides. Such a body might be a dyke, a sill or some other small intrusion. The heterogeneous textures and mineralogy and the distribution of gabbros in the Indian Ocean fracture zones indicate that many of them formed in small dykes or sills rather than in large open magma chambers. Pulses of magma passing through such small dykes or sills might represent the ascent of one primitive magma batch produced in the mantle but introduced over a period of time, or the ascent of several primitive batches (*cf* Kilauea E rift; Wright & Fiske 1970). Probably, as at Kilauea, somewhat fractionated magmas would make their way from principal magma chambers down-rift towards fracture zones in the ocean crust (Fig. 7).

In a larger magma chamber repeated injections of primitive magma might result in numerous reversals in cryptic variation in layered cumulates such as those observed, for example, at Rhum (Brown 1956; Dunham & Wadsworth 1978). Usselman & Hodge (1978) considered what would happen to compositions of melts in a magma chamber as a consequence of mixing between evolved compositions in the chamber and primitive injected magmas, and then predicted the relevant changes in compositions of cumulus minerals (specifically, the reversals in cryptic variation). They did not consider, however, that both the injected high-temperature melts and, for a time, the resulting hybrids in the chamber would have the capacity to dissolve a fraction of those cumulates with which they came in contact. To the extent that the melts which produced the cumulates had a lower temperature than either the injected or hybrid magmas, the latter could resorb the minerals in the cumulates. Scavenging of mineral-melt aggregates (the mat of crystals) from magma-chamber walls has been described from the Juan de Fuca Ridge (Eaby *et al.* 1986), and scavenging of aggregates with gabbroic texture from coarse-grained portions of dykes has been described at DSDP Hole 504B, Costa Rica Rift (Kempton *et al.* 1985). This is evidence for mechanical interaction between migrating magmas and the walls of conduits. Partial resorption of gabbroic clots also has been described from Hole 504B basalts (Natland *et al.* 1983).

Because of the open-system nature of the process, however, this would not be a case of equilibrium dissolution of crystals. Indeed, in the extreme case we are considering, we might imagine that all the liquid originally in the small magma body could be expelled (pushed out) ahead of the injected magma before much mineral resorption could occur and that the remaining crystals on the walls of the chamber would suddenly come into contact with primitive melt (Fig. 8). Those crystals would tend to become resorbed, and interstitial melts in the cumulate mat would mix with the injected melt at the boundary layer. In such circumstances, what would be the consequences for the injected melt compositions and for cumulates produced from them?

Walker *et al.* (1979) demonstrated that one of the consequences of mixing between primitive and evolved abyssal tholeiites lying on the plagioclase–clinopyroxene cotectic surface in the system Ol–Di–Plag–SiO_2 would be to produce a hybrid within the clinopyroxene phase volume; the hybrids become supersaturated with respect to clinopyroxene, and hence might be expected for a time to precipitate that mineral alone, ahead of plagioclase, until the cotectic surface is reattained. Repeated mixing events of this type could conceivably produce a sequence of olivine–spinel and olivine–spinel–clinopyroxene (dunite–wehrlite) cumulates. Such a cumulate association is not yet known from fracture zones in the ocean crust (Dick *et al.* 1986; Fisher *et al.* 1986) but is fairly common in ophiolites and may be represented by the rare pyroxenites in our Indian Ocean collections.

Our example, however, would result in hybrids having compositions primarily between

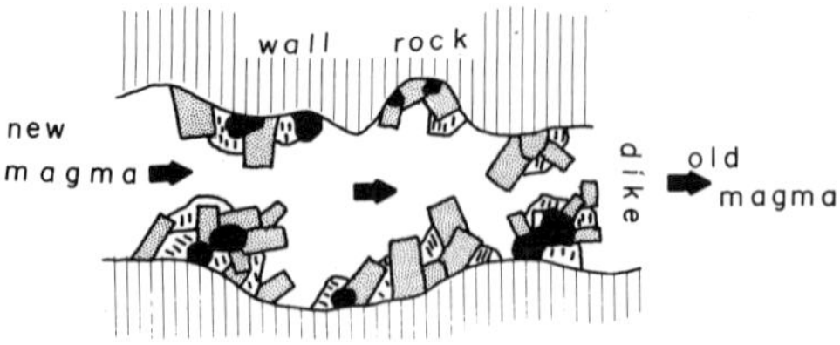

FIG. 8. *In situ* boundary-layer crystallization of gabbro on the walls of a small dyke or sill with periodic magma flow which has purged the intrusion of all liquid previously in contact with the crystals: black areas, OL; shaded PLAG; lined areas, pyroxenes.

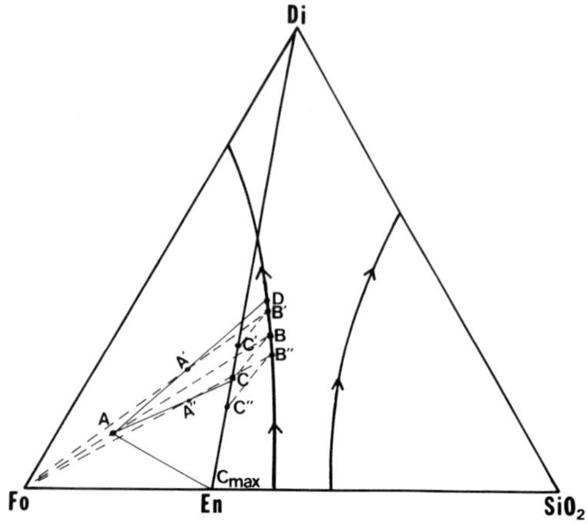

FIG. 9. Diagram (after Bowen 1927) illustrating courses of crystallization in the Di–Fo–SiO_2 system. Phase boundaries are distorted to show relationships. Olivine(Fo)-controlled fractionation pathways through liquid compositions A, A′ and A″ to liquids on the boundary curve B, B′ and B″ and lines linking B, B′ and B″ to the compositions of pyroxenes C, C′ and C″ in equilibrium with those liquids are broken. Mixing lines between liquid compositions A and D, A and C, and A and C_{max} are full.

those of injected primitive melts and of particular mineral compositions—olivines, to some extent, and more probably plagioclases and clinopyroxenes. The clinopyroxenes could well be more magnesian and probably more siliceous than the engulfing melts, on the basis of analyses of typical partially resorbed megacrysts in basalts from the Indian Ocean (J. Natland, unpublished data) and elsewhere (*eg* Wilkinson 1982; Natland *et al* 1983). Associated partly resorbed plagiolases in the glomerocrysts of such basalts are less siliceous, more calcic and less sodic than the host melts and most injected melts that plausibly might be inferred for our example. The precise path of liquid modification assuming resorption of calcic plagioclase and magnesian clinopyroxene cannot be specified, but the hybrids would probably become more magnesian and less sodic, more calcic and perhaps also more siliceous. In short, the hybrid compositions would tend to approach those in Table 4, with the result that both clinopyroxene and orthopyroxene would crystallize earlier and have more magnesian compositions than the corresponding equilibrium minerals cyrstallized from the injected primitive melt alone.

The question of early crystallization of orthopyroxene can now be considered, and this general hypothesis illustrated, using the ternary diopside –forsterite–silica (Di–Fo–SiO_2) (Bowen 1914) in the schematic form employed by Bowen (1927). This ternary (Fig. 9) is a well-known example of a system with a series of binary solid solutions that melt incongruently. A simplified basaltic liquid of composition A in Fig. 9 will, upon equilibrium or fractional crystallization, move away from the olivine corner until the boundary curve is reached at point B. Here, reaction will ensue between forsterite and melt, with pyroxene of composition C replacing the olivine. The liquid remaining after this process is complete, or additional liquid released by fractional crystallization of forsterite, will proceed at equilibrium along the boundary curve to some point D in equilibrium with pyroxene C.

If another magma of composition A were to mix with D to produce, say, a hybrid A′ along the mixing line shown, forsterite would then precipitate until the differentiating liquid reached composition B′, whence reaction to pyroxene of composition C′ would ensue. This pyroxene would form at a lower temperature than pyroxene C (and thus would be formed later in a comparative crystallization sequence).

In our example, however, we assume that a newly injected magma of composition A first displaces all liquid in contact with pyroxene C and then simply causes pyroxene C to dissolve; there is no mixing with any liquid having the range of compositions B–D. The resultant hybrid between the melt and the resorbed minerals would lie along a line between A and C, for example A″. Crystallization of forsterite from A″ would reach the boundary curve at B″, whence reaction to produce pyroxene of composition C″ would follow. C″ crystallizes at a higher temperature than C, and thus is formed earlier in a comparative crystallization sequence and is more magnesian.

This process can be imagined to continue, with yet another injected magma of composition

A interacting with pyroxenes, this time of composition C″, to produced a hybrid even closer to the forsterite–silica sideline. The temperature of the first pyroxene to crystallize from the new hybrid will be even higher than that of C″, and the mineral itself will be even more magnesian (closer to the olivine–silica sideline). An iterative process of re-injection with primitive magma may ultimately result in interaction with quite magnesian pyroxene, to the following limiting cases. If A is crystallizing olivine (is not superheated), the limiting temperature of pyroxene produced from hybrids reaching the boundary curve will be that of the liquidus temperature at A in the ternary or the temperature of the reaction to produce enstatite along the olivine–silica sideline (at C_{max}), whichever is reached first. The ratio of injected melt to resorbed crystals will determine where along the mixing lines (such as A–C) the magma hybrids will lie. If the melt proportion is high, many episodes of melt injection may be required to drive pyroxenes to magnesian compositions. Furthermore, higher ratios or resorbed crystals to melt at each successive contact with new melt will accelerate the process. Theoretically, even extraordinarily magnesian pyroxenes could be produced by such a mechanism. None of the hybrid melt compositions will ever be more forsteritic (magnesian) than composition A; all will be richer in silica and successively poorer in diopside. The mechanism need not be envisaged to occur solely between melts and cumulus minerals; reaction between wandering melts and pyroxenes of appropriate composition in peridotites could produce the same effects.

Phase relations in natural basaltic systems will of course be more complex than those shown in Fig. 9. This general mechanism could explain, for example, the occurrence of enstatites and bronzites in some basalts near Iceland (Sigurdsson & Brown 1970; Sigurdsson 1981), the very magnesian pyroxenes in gabbroic rocks at DSDP Site 334 (Hodges & Papike 1976), and the less (but still quite) magnesian pyroxenes reported in gabbros here and from the Cayman Trough (Elthon 1987).

Whether such mineral relationships in abyssal gabbros result from the compositions of parental magmas or depend on boundary-layer recrystallization as just described remains to be determined, primarily by experimental work and integrated mineralogical–geochemical studies of the gabbros. We conclude that high-pressure crystallization is not the only way to produce such relationships, although in cases of substantial thicknesses of cumulates it may have been operative.

Conclusions

(1) Plutonic rocks from five fracture zones in the Indian Ocean include a diverse suite of gabbroic rocks dominated by two-pyroxene gabbro and ferrogabbro. Olivine gabbros are much less abundant than expected. The gabbros are highly variable in grain size and composition, even on a thin-section scale.

(2) Over half the gabbroic samples have crystallized from liquids which are more iron rich than any represented by basalts from the fracture zones or adjacent ridge segments. Some of the calculated liquid compositions that produced the gabbros were as fractionated as those found on fast-spreading ridges in the eastern Pacific.

(3) The gabbroic compositions, their textures and the distribution of rock types are consistent with the fracture zones being fed by longitudinal injection of magmas from central magma chambers. Primitive basalts form carapaces along the ridge crests, beneath which small bodies of melt are injected which crystallize to form the iron- rich gabbros. Much of the crustal section in the fracture zones may comprise serpentinite.

(4) There are some gabbroic samples which have pyroxenes as magnesian as those which have been cited as evidence for moderate to high-pressure crystallization. These magnesian pyroxenes can also be derived by low-pressure crystallization of refractory siliceous magmas such as those sampled at DSDP 236, which may periodically be supplied to the spreading ridge, or by low-pressure boundary-layer crystallization of magmas in small dykes, sills or chambers.

Acknowledgments: The laboratory work was supported by National Science Foundation grants OCE–83–20589 to S.H. Bloomer and OCE–83–08696 to J.H. Natland. The work benefited from discussions with H.J.B. Dick, D. Elthon and J. Karson. T. Sohlburg of Virginia Polytechnic Institute and State University assisted with the microprobe analyses. Basalt glass analyses were performed by T. O'Hearn at the Smithsonian Institution, Washington, DC. P. Browning and D. Elthon provided thoughtful reviews.

References

Aumento, F. *et al.* (eds) 1977. *Initial Reports of the Deep Sea Drilling Project* **37**.

Bloomer, S. H., Natland, J. H. & Fisher, R. L. 1985. Mineralogic studies of gabbros from the Central and Southwest Indian Ridges. *EOS (Transactions of the American Geophysical Union)* **66**, 405 (abstract only).

Bowen, N. L. 1914. The Ternary system: diopside–forsterite–silica. *American Journal of Science* **38**, 207–264.

——1927. *The Evolution of the Igneous Rocks*. Princeton University Press, Princeton, NJ.

——& Schairer, J. F. 1935. The system, MgO–FeO–SiO_2 *American Journal of Science* **29**, 151–217.

Brown, G. M. 1956. The layered ultrabasic rocks of Rhum, Inner Hebrides. *Philosophical Transactions of the Royal Society of London, Series B* **240**, 1–53.

Clarke, D. B. & Loubat, H. 1977. Mineral analyses from the peridotite–gabbro–basalt complex at Site 334, DSDP Leg 37. *In*: Aumento, F. *et al.* (eds) *Initial Reports of the Deep Sea Drilling Project* **37**, 847–856.

Cormier, M.-H., Detrick, R. S. & Purdy, G. M. 1984. Anomalously thin crust in oceanic fracture zone; new seismic constraints from the Kane Fracture Zone. *Journal of Geophysical Research* **89**, 10249–10266.

Crane, K. 1985. The spacing of rift axis highs: dependence on diapiric processes in the underlying lithosphere? *Earth and Planetary Science Letters* **85**, 405–414.

Detrick, R. S. Jr & Purdy, G. M. 1980. The crustal structure of the Kane fracture zone from seismic refraction studies. *Journal of Geophysical Research* **85**, 3759–3777.

Dick, H. J. B., Gallo, D. G., Meyer, P. S., Schouten, H., Natland, J., Berg, H. & Patriat, P. 1986. The geology of the Atlantis II F. Z., SW Indian Ridge, and the potential for mantle drilling. *EOS* (*Transactions of the American Geophysical Union*) **67**, 1231 (abstract only).

Duke, J. M. 1976. Distribution of the Period Four transition elements among olivine, calcic, clinopyroxene and mafic silicate liquid: experimental results. *Journal of Petrology* **17**, 499–521.

Dunham, A. C. & Wadsworth, W. J. 1978. Cryptic variation in the Rhum layered intrusion. *Mineralogical Magazine* **42**, 347–356.

Eaby, J. S., Clague, D. A. & Eissen, J.-P. 1986. Gabbroic xenoliths from the southern Juan de Fuca Ridge: implications for magma evolution at midocean ridge spreading centers. *Journal of Geophysical Research* **91**, 3795–3820.

Elthon, D. 1982. Mineral chemistry of gabbroic rocks from the Mid-Cayman Rise *EOS* (*Transactions of the American Geophysical Union)* **63**, 1215 (abstract only).

—— 1987. Mineral chemistry of gabbroic rocks from the Mid-Cayman Rise spreading center. *Journal of Geophysical Research* **92**, 658–682.

—— & Scarfe, C. M. 1984. High-pressure phase equilibria of a high-magnesia basalt and the genesis of primary oceanic basalts. *American Mineralogist* **69**, 1–15.

—— Casey, J. F. & Komor, S. 1982. Mineral chemistry of ultramafic cumulates from the North Arm Mountain massif of the Bay of Islands ophiolite: evidence for high pressure crystal fractionation of oceanic basalts. *Journal of Geophysical Research* **87**, 8717–8734.

——, —— & —— 1984. Cryptic mineral chemistry variations in a detailed traverse through the cumulative ultramafic rocks of the North Arm Mountain massif of the Bay of Islands ophiolite, Newfoundland. *In*: Gass, I. G., Lippard, S. J. & Shelton, A. W. (eds) *Ophiolites and Oceanic Lithosphere*. Blackwell Scientific Publications Oxford, 83–100.

Engel, C. G. & Fisher, R. L. 1975. Granitic to ultramafic rock complexes of the Indian Ocean ridge system, western Indian Ocean. *Bulletin of the Geological Society of America* **86**, 1553–1578.

Fisher, R. L. & Sclater, J. G. 1983. Tectonic evolution of the Southwest Indian Ocean since the Mid-Cretaceous: plate motions and stability of the pole of Antarctica/Africa for at least 80 Myr. *Geophysical Journal of the Royal Astronomical Society* **73**, 553–576.

——, Dick, H. J. B., Natland, J. H. & Meyer, P. S. 1986. Mafic/ultramafic suites of the slowly spreading Southwest Indian Ridge: PROTEA exploration of the Antarctic plate boundary, 24°E–47°E, 1984. *Ofioliti* **11**, 147–178.

——, Sclater, J. G. & McKenzie, D. P. 1971. Evolution of the Central Indian Ridge, Western Indian Ocean. *Bulletin of the Geological Society of America* **82**, 553 562.

Frey, F. A., Dickey, J. S. Jr, Thompson, G., Bryan, W. B. & Davies, H. L. 1980. Evidence for heterogeneous primary MORB and mantle sources. *Contributions to Mineralogy and Petrology* **74**, 387–402.

Grove, T. L. & Bryan, W. B. 1983. Fractionation of pyroxene-phyric MORB at low pressure: an experimental study. *Contributions to Mineralogy and Petrology* **69**, 75–82.

Hébert, R., Bideau, D. & Hekinian, R. 1983. Ultramafic and mafic rocks from the Garrett Transform Fault near 13°30′S on the East Pacific Rise: igneous petrology. *Earth and Planetary Science Letters* **65**, 107–125.

Hodges, F. N. & Papike, J. J., 1976. DSDP Site 334: magmatic cumulates from oceanic layer 3. *Journal of Geophysical Research* **81**, 4135–4151.

Irvine, T. N. 1980. Magmatic density currents and cumulus processes. *American Journal of Science* **280A**, 1–58.

Kempton, P. D., Autio, L. K., Rhodes, J. M.,

HOLDAWAY, M. J., DUNGAN, M. A. & JOHNSON, P. 1985. Petrology of basalts from Hole 504B, Deep Sea Drilling Project. *In*: ANDERSON, R. N. *et al.* (eds) *Initial Reports of the Deep Sea Drilling Project* **83**, 129–164.

KOMOR, S. C., ELTHON, D. & CASEY, J. F. 1985. Mineralogic variation in a layered ultramafic cumulate sequence at the North Arm Mountain massif, Bay of Islands ophiolite, Newfoundland. *Journal of Geophysical Research* **90**, 7705–7736.

LANGMUIR, C. H. 1986. Some possible geochemical consequences of *in situ* fractionation in a boundary layer. *EOS (Transactions of the American Geophysical Union)* **67**, 385 (abstract only).

——, BENDER, J. F. & BATIZA, R. 1986. Petrological and tectonic segmentation of the East Pacific Rise, 5°30′S–14°30′N. *Nature* **322**, 422–429.

LINDSLEY, D. H. & ANDERSON, D. J. 1983. A two-pyroxene geothermometer. *Journal of Geophysical Research* **88**, A887–A906.

MAHONEY, J. & NATLAND, J. H. 1985. Domains of isotopic and basalt-glass compositions at Indian Ocean ridges. *EOS (Transactions of the American Geophysical Union)* **66**, 1108 (abstract only).

MCBIRNEY, A. R. & NOYES, R. M. 1979. Crystallization and layering of the Skaergaard Intrusion. *Journal of Petrology* **20**, 487–554.

MIYASHIRO, A. & SHIDO, F. 1980. Differentiation of gabbros in the mid-Atlantic Ridge near 24°N. *Geochemical Journal* **14**, 145–154.

NATLAND, J. 1980. Effect of axial magma chambers beneath spreading centers on the compositions of basaltic rocks. *In*: ROSENDAHL, B. R. *et al.* (eds) *Initial Reports of the Deep Sea Drilling Project* **54**, 605–633.

——, ADAMSON, A. C., LAVERNE, C., MELSON, W. G. & O'HEARN, T. 1983. A compositionally nearly steady-state magma chamber at the Costa Rica rift: evidence from basalt glass and mineral data, Deep Sea Drilling Project sites 501, 504 and 505. *In*: CANN, J. R. *et al.* (eds) *Initial Reports of the Deep Sea Drilling Project* **69**, 811–858.

PEARCE, J. A. & NORRY, M. J. 1979. Petrogenetic implications of Ti, Zr, Y, and Nb variations in volcanic rocks. *Contributions to Mineralogy and Petrology* **69**, 33–47.

PERFIT, M. R. & FORNARI, D. J. 1983. Geochemical studies of abyssal lavas recovered by DSRV Alvin from Eastern Galapagos Rift, Inca Transform and Equador Rift 2. Phase chemistry and crystallization history. *Journal of Geophysical Research* **88**, 10530–10550.

PRICE, R. C., KENNEDY, A. K., RIGGS-SNEERINGER, M. & FREY, F. A. 1986. Geochemistry of basalts from the Indian Ocean triple junction: implications for the generation and evolution of Indian Ocean basalts. *Earth and Planetary Science Letters* **78**, 379–396.

PRINZ, M., KEIL, K., GREEN, J. A., REID, A. M., BONATTI, E. & HONNOREZ, J. 1976. Mineralogy and petrology of some ultramafic and mafic dredge samples from the equatorial Mid-Atlantic Ridge and fracture zones. *Journal of Geophysical Research* **81**, 4087–4103.

RHODES J. M., BLANCHARD, D. P., RODGERS, K. V., JACOBS, J. W. & BRANNON, J. C. 1976. Petrology and chemistry of basalts from the Nazca plate: Pt. 2—Major and trace element chemistry. *In*: YEATS, R. S. *et al.* (eds) *Initial Reports of the Deep Sea Drilling Project* **34**, 239–244.

ROEDER, P. L. & EMSLIE, R. F. 1970. Olivine–liquid equilibrium. *Contributions to Mineralogy and Petrology* **29**, 275–289.

SIGURDSSON, H. 1981. First-order major element variation in basalt glasses from the Mid-Atlantic Ridge: 29°N to 73°N. *Journal of Geophysical Research* **86**, 9483–9502.

—— & BROWN, G. M. 1970. An unusual enstatite–forsterite basalt from Kolbeinsey Island, north of Iceland. *Journal of Petrology* **11**, 205–220.

—— & SPARKS, S. R. J. 1978. Lateral magma flow within rifted Icelandic crust. *Nature* **274**, 126–130.

STOLPER, E. & WALKER, D. 1980. Melt density and the aveıage composition of basalt. *Contributions to Mineralogy and Petrology* **74**, 7–12.

STROUP, J. B. & FOX, P. J. 1981. Geologic investigations in the Cayman Trough: evidence for thin oceanic crust along the Mid-Cayman Rise. *Journal of Geology* **89**, 395–420.

TIEZZI, L. J. & SCOTT, R. B. 1980. Crystal fractionation in a cumulate gabbro, Mid-Atlantic Ridge, 26°N. *Journal of Geophysical Research* **85**, 5438–5454.

TORMEY, D. R., GROVE, T. L., & BRYAN, W. B. 1987. Experimental petrology of normal MORB near the Kane Fracture Zone: 22°–25°N, Mid-Atlantic Ridge. *Contributions to Mineralogy and Petrology* **96**, 121–139.

USSELMAN, T. M. & HODGE, D. S. 1978. Thermal control of low-pressure fractionation processes. *Journal of Volcanology and Geothermal Research* **4**, 265–281.

VANKO, D. A. & BATIZA, R. 1982. Gabbroic rocks from the Mathematician Ridge failed rift. *Nature* **300**, 742–744.

WALKER, D. & DELONG, S. E. 1984. A small Soret effect in spreading center gabbros. *Contributions to Mineralogy and Petrology* **85**, 203–208.

—— JUREWICZ, S. & WATSON, E. B. 1985. Experimental observation of an isothermal transition from orthocumulus to adcumulus texture. *EOS (Transactions of the American Geophysical Union)* **66**, 362 (abstract only).

—— SHIBATA, T. & DELONG, S. E. 1979. Abyssal tholeiites from the Oceanographer Fracture Zone: II, phase equilibria and mixing. *Contributions to Mineralogy and Petrology* **70**, 111–125.

WHITEHEAD, J., DICK, H. J. B. & SCHOUTEN, H. 1984. A mechanism for magmatic accretion under spreading centres. *Nature* **312**, 146–148.

WILKINSON, J. G. F. 1982. The genesis of mid-ocean ridge basalts. *Earth Science Reviews* **18**, 1–57.

WRIGHT, T. L. & FISKE, R. S. 1970. Origin of the differentiated and hybrid lavas of Kilauea volcano, Hawaii. *Journal of Petrology* **12**, 1–65.

S. H. BLOOMER, Department of Geology, Duke University, Box 6729, College Station, Durham, NC 27708, USA.

J. H. NATLAND & R. L. FISHER, Scripps Institution of Oceanography, University of California, San Diego, La Jolla, CA 92093, USA.

Pressure of origin of primary mid-ocean ridge basalts

Don Elthon

SUMMARY: High-pressure phase equilibrium studies of mid-ocean ridge basalts (MORBs) have shown that some samples are in equilibrium with olivine + orthopyroxene (OPX) + clinopyroxene ± plagioclase at about 10 kbar, whereas others are not in equilibrium with OPX at this pressure. Samples that are OPX-saturated at about 10 kbar are characterized by higher SiO_2 (>49.7 wt%) than most (about >75%) MORB glasses with more than 9.5 wt% MgO, suggesting that only about 25% of these glasses could be derived from primary magmas generated at about 10 kbars. Orthopyroxenes in abyssal peridotites are characterized by Al_2O_3 contents that range from 2.0 to 6.0 wt% (probably about 2.5–6.5 wt% before subsolidus equilibration), but orthopyroxenes in peridotite partial-melting experiments and liquidus orthopyroxenes for MORBs at about 10 kbar have only 3.5–4.3 wt% Al_2O_3. Liquidus orthopyroxenes for high-MgO basalts at 25 kbar and 20 kbar have about 6.6 wt% and 5.5 wt% Al_2O_3 respectively. It appears that the least depleted (Al_2O_3-rich end) of abyssal peridotite orthopyroxenes indicate the dominance of melting and primary MORB genesis at pressures substantially greater than 10 kbar, and probably at 20–25 kbar.

The compositions of primary mid-ocean ridge basalt (MORB) magmas and their pressures of origin are two of the most fundamental questions in discussions of the evolution of MORBs and the upper mantle. Two major schools of thought have developed over the years. One school (*eg* Kushiro & Thompson 1972; Fujii & Kushiro 1977; Hart & Davis 1978, 1979; Presnall *et al.* 1979; Fujii & Bougault 1983) interprets the data to indicate that the most MgO-rich MORBs (9.5–10.5 wt% MgO) are primary or near-primary magmas produced at about 10 kbar. The other school (*eg* O'Hara 1968; Clarke 1970; Clarke & O'Hara 1979; Green *et al.* 1979; Elthon & Scarfe 1980, 1984; Stolper 1980) suggests that MORBs are derived from high-MgO basalt (16–18 wt% MgO) primary magmas produced at higher pressures (generally 20–30 kbar).

The term 'primary magma' is used here to describe a partial melt of the mantle that has not undergone crystallization or other differentiation processes, whereas the term 'most primitive' is used to describe basaltic glasses with the highest (>9.5 wt%) MgO contents which appear to have undergone the smallest amounts of crystallization. Although tightly constrained criteria for recognizing primary MORBs are not uniformly agreed upon by workers in the field, certain features are recognized as being essential. Primary MORB magmas, by definition, must be in equilibrium with the residual mantle, which is widely believed to be represented by abyssal peridotites. Primary MORBs must therefore be in equilibrium with olivine and orthopyroxene (and often clinopyroxene and/or spinel (or garnet)) at the pressure of melting. The experimental test for such equilibrium for a given magma composition is the presence of liquidus olivine and orthopyroxene (± clinopyroxene, spinel or garnet) at the inferred pressure of melting. In addition, the compositions of these liquidus minerals must match the compositions of minerals in the residual mantle under the conditions of melting.

It seems that the major debate in discussions of primary MORB genesis focuses on the relative importance of high-MgO basalts in the generation of the range of MORB compositions. In other words, are high-MgO basalts petrological oddities of limited significance or are most MORBs ultimately derived from such basalts? The critical evidence bearing on the pressure of origin of primary MORBs is briefly reviewed in this paper. Pressure and composition are necessarily related in this discussion because moderate amounts of partial melting at about 10 kbar can produce primary basalts that are broadly similar to some oceanic basalts, but partial melting at 25 kbar will produce high-MgO basalts. Constraints placed on the pressure of melting therefore place general constraints on the compositions of primary magmas.

Constraints from pseudo-liquidus phase diagrams

O'Hara (1965, 1968), following Coombs (1963), developed the conceptual framework of the high-pressure pseudo-liquidus phase diagram in

From Saunders, A. D. & Norry, M. J. (eds), 1989, *Magmatism in the Ocean Basins*, Geological Society Special Publication No. 42, pp. 125–136.

which most subsequent discussions of the high-pressure phase equilibria of basalts have taken place. One of the most powerful of O'Hara's early arguments that MORBs are not primary magmas was the observation that they are not in equilibrium with orthopyroxene at any pressure. Subsequent studies of MORBs, however, have shown that some are in equilibrium with orthopyroxene at 7.5–11 kbar (Fujii & Kushiro 1977; Fujii & Bougault 1983). In the light of these studies, it is clear that *some* MORBs are in equilibrium with orthopyroxene at about 10 kbar and that these basalts might be suitable candidates for primary magmas.

Green *et al.* (1979), Stolper (1980) and Elthon & Scarfe (1980, 1984), however, have proposed that MORB liquids that are multiply saturated with olivine + orthopyroxene + clinopyroxene ± plagioclase ± spinel at about 10 kbar are unlikely to be parental to most primitive MORBs. They proposed that the orthopyroxene-saturated primary phase volume is sufficiently displaced from primitive MORBs to preclude the derivation of most primitive MORBs from magmas generated at about 10 kbar.

The basis for this interpretation is shown in Figs 1*a* and 1*b*, in which the compositions of MORBs and related phase equilibria studies are projected onto the olivine–clinopyroxene–silica plane (Elthon & Scarfe 1984). The only modification from Elthon & Scarfe (1984) is that experimental run 333 at 10 kbar is not used to constrain the locations of multiply saturated phase equilibria boundaries. Fujii & Bougault (1983) have criticized Elthon & Scarfe's use of this datum point because the liquid's Fe/Mg ratio is too high to be in equilibrium with mantle olivines. C. Ford (pers. comm.) has shown that the olivine and liquid compositions do not agree with the equilibrium distribution coefficients obtained by Ford *et al.* (1983), possibly indicating problems with either the experiment or the analyses. This experimental run also has a large abundance of crystals (68%), which makes analysis of the glass more difficult than in near-liquidus runs and probably makes this experiment less reliable than near-liquidus studies such as those by Fujii & Kushiro (1977) and Fujii & Bougault (1983). The inclusion or exclusion of this experiment has little effect on the locations of multiply saturated phase equilibria (Fig. 1*a*); it is excluded in the following discussion simply to remove unnecessary controversy. The remaining data (Kushiro & Thompson 1972; Fujii & Kushiro 1977; Bender *et al.* 1978; Green *et al.* 1979; Fujii & Bougault 1983) indicate that liquids in equilibrium with orthopyroxene at a pressure of about 10 kbar lie within or adjacent to the shaded region indicated in Fig. 1*a*.

There are other data, particularly from studies of the melting of peridotites and from peridotite–basalt sandwich experiments, that are potentially relevant in constraining the locations of the high-pressure phase equilibria shown in Fig. 1a. High-pressure melting studies of peridotites, however, have generally been hampered by either substantial iron loss to platinum containers or modification of the glass composition by quench crystal overgrowth. Loss of iron from the peridotite sample to the enclosing platinum capsule was a major problem in the studies by Mysen & Kushiro (1977) and Jaques & Green (1979, 1980). Jaques & Green (1980) estimated the compositions of liquids produced during melting of two peridotite samples using mass balance. The residual minerals, however, were often variable in composition (particularly clinopyroxene), and so estimates of liquid compositions in these mass balance calculations were dependent on the assumed equilibrium mineral compositions. The only study of peridotite melting in which direct analysis of the glass compositions was possible and in which no iron loss occurred is that by Sen (1982), who conducted his experiments in graphite capsules. The data of Sen for liquids in equilibrium with olivine, orthopyroxene, clinopyroxene and spinel at 9 kbar are shown in Fig. 1*a*, where they agree with the phase equilibria constructed on the basis of basalt melting experiments.

The problems encountered in determining liquid compositions during partial melting of peridotites has led to a series of experimental investigations in which a basalt is encapsulated within peridotite minerals (Stolper 1980; Takahashi & Kushiro 1983; Fujii & Scarfe 1985). In these 'basalt–peridotite sandwich' experiments, the liquid (glass) is anticipated to change compositionally until it is in equilibrium with the surrounding peridotite minerals.

A major problem with the direct application of basalt–peridotite sandwich experiments to the petrogenesis of primary MORBs is that the compositions of the liquids produced in these experiments depend upon the compositions of both the peridotite minerals and the basaltic glasses used as starting materials. Takahashi & Kushiro (1983), for example, equilibrated a spinel lherzolite assemblage (HK66) with various basaltic compositions at 1300 °C and 15 kbar; there is a substantial range in the compositions of the final glasses (SiO_2, 47.5–51.3 wt%; TiO_2, 0.5–1.7 wt%; Al_2O_3, 15.3–16.6 wt%; FeO* 9.2–10.3 wt%; MgO, 9.8–10.9 wt%; CaO,

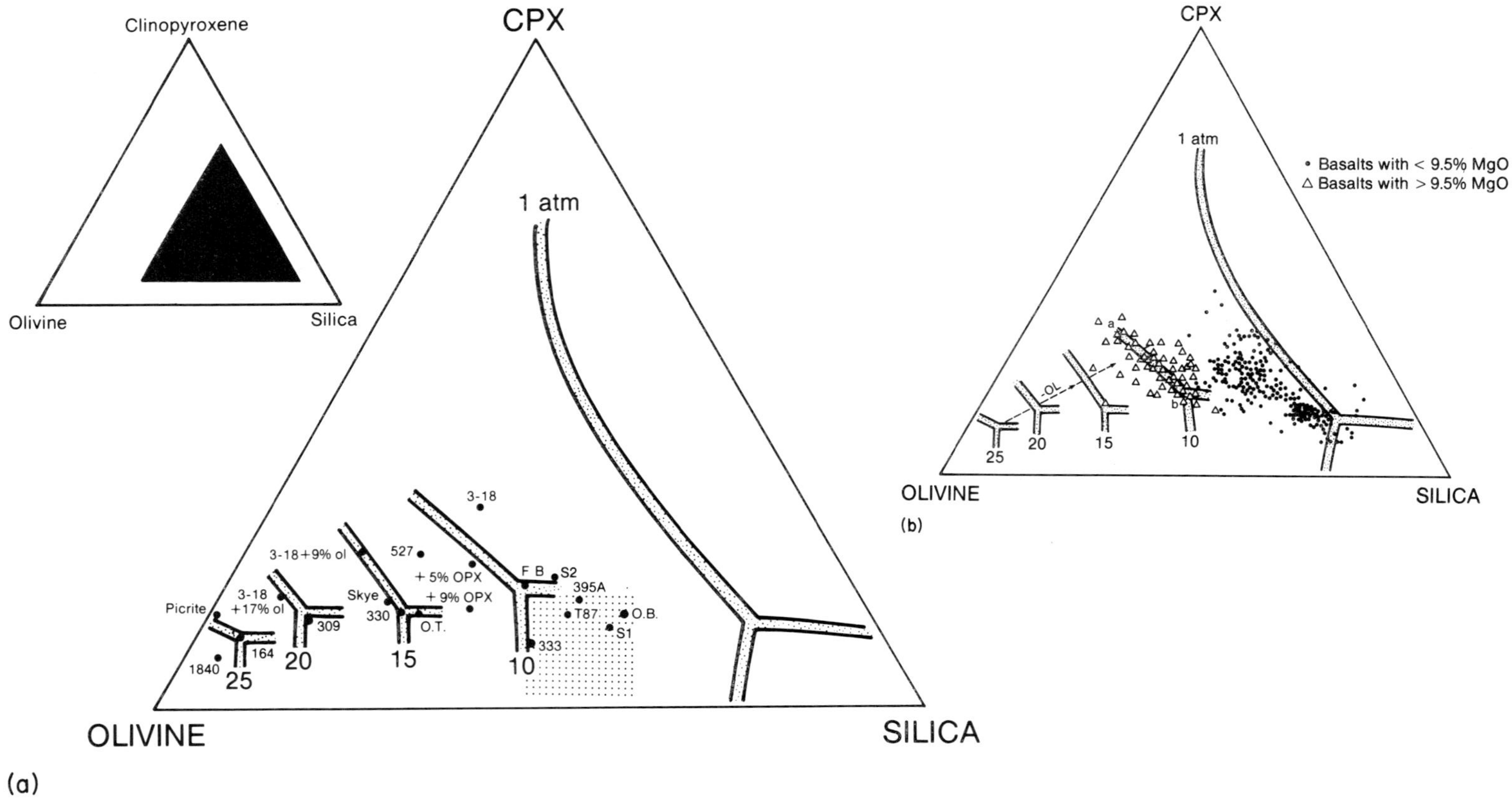

FIG. 1. Phase equilibria of (*a*) basalts and (*b*) compositions of mid-ocean ridge basalts projected from anorthite onto a portion of the olivine–clinopyroxene–silica plane (modified slightly from Elthon & Scarfe (1984), with S1 and S2 from Sen (1982)). The apices of the triangle are (top) $CPX_{60}OL_{10}SIL_{30}$, (left) $CPX_{10}OL_{60}SIL_{30}$ and (right) $CPX_{10}OL_{10}SIL_{80}$. Sources of experimental data are listed in Elthon & Scarfe (1984). Pressures indicated are in kilobars. Sources of data on MORB glasses with more than 9.5 wt% MgO are listed in the caption to Fig. 2. Line ab is the 10 kbar olivine + clinopyroxene + spinel + liquid 'cotectic'.

9.3–10.8 wt%, Na_2O; 2.3–3.2 wt %; K_2O, 0.1–1.0 wt%), indicating that the composition of the final glass is influenced by the composition of the starting glass. Furthermore, Fujii & Scarfe (1985) have shown that the compositions of the minerals in the basalt–peridotite sandwich experiments influence the compositions of the resulting final liquids. They show that the composition of the spinel has a substantial effect on the composition of the coexisting liquid. It is probable, although not discussed by them, that the compositions of pyroxenes also strongly influence the compositions of basaltic liquids in equilibrium with these peridotite mixtures.

Another problem with basalt–peridotite sandwich experiments is whether the basaltic liquid changes in composition to be in equilibrium with the mantle minerals (the intended result) or whether there are large changes in the compositions of minerals. The compositions of minerals *after* equilibration are not reported in much detail, but the data of Takahashi & Kushiro (1983) clearly show that orthopyroxene rims have changed to approach equilibrium with the basaltic liquid, while cores of very different composition are preserved. In this case at least (and probably with the other experiments as well because they were similar) the compositions of the rims of minerals do not match the starting mineral compositions, and therefore these experiments do not determine the compositions of liquids in equilibrium with residual mantle mineral compositions. For these reasons the results from basalt–peridotite sandwich experiments are not used to constrain the locations of high-pressure phase equilibria in Fig. 1*a*.

The compositions of MORB glasses (and some whole-rock data) are projected onto this pseudo-liquidus phase diagram in Fig. 1*b*. The MORB glasses with more than 9.5 wt% MgO are considered to be the most primitive magmas, probably having undergone the smallest amount of differentiation because MgO progressively decrease with increasing differentiation in MORBs. These glasses generally cluster near the 10 kbar olivine + clinopyroxene + liquid 'cotectic' (line ab in Fig. 1*b*), with some samples projecting near the region of orthopyroxene-saturated magmas (near B in Fig. 1*b*). Elthon & Scarfe (1984) suggested that, because most MORB glasses with more than 9.5 wt% MgO do not lie near orthopyroxene saturation at 10 kbar, most primitive MORBs are not produced by partial melting at about 10 kbar but are derived from primary magmas generated at higher pressures (about 20–25 kbar). This conclusion was previously reached by Green *et al.* (1979), Stolper (1980) and Elthon & Scarfe (1980) for similar reasons.

Presnall & Hoover (1984, 1986) have criticized the conclusions of these investigators by arguing that the analytical uncertainties of microprobe analyses are so large that most primitive MORBs are actually within analytical uncertainties of orthopyroxene saturation. These conclusions have been disputed by Elthon (1986) and Tormey *et al.* (1987), who argued that the analytical uncertainties of Presnall & Hoover are substantially overstated and that the well-defined trends in experimental data argue against the magnitude of the uncertainties proposed by these workers. None the less, Presnall & Hoover's work raises the fundamental question of what compositional differences distinguish MORBs with more than 9.5 wt% MgO that do not show orthopyroxene saturation at about 10 kbar (Bender *et al.* 1978; Green *et al.* 1979) from those that do (Fujii & Kushiro 1977; Fujii & Bougault 1983). One of these differences, the SiO_2 content, is addressed below.

Silica contents of primitive mid-ocean ridge basalt glasses

Experimental studies of the partial melting of peridotites (Jaques & Green 1980; Takahashi & Kushiro 1983) have shown that the compositions of liquids in equilibrium with a lherzolite residuum vary as a function of pressure, particularly showing a decrease in normative silica with increasing pressure (*eg* Fig. 1*a*). This decrease in normative silica is coupled with a decrease in the SiO_2 contents of primary liquids (*eg* Jaques & Green 1980 Tables 4 and 5). The SiO_2 contents of *primary* MORBs might therefore be anticipated to vary as a function of the pressure of melting, decreasing with an increase in pressure.

The SiO_2 contents of MORB glasses with more than 9.5 wt% MgO are shown in Fig. 2. This discussion is restricted to glasses because they represent liquid compositions unmodified by alteration or accumulation processes. The SiO_2 contents of MORB glasses with more than 9.5 wt% MgO range from 46.8 to 50.7 wt% (Fig. 2). Those MORBs that show orthopyroxene (or pigeonite) saturation at about 10 kbar (Kushiro & Thompson 1972; Fujii & Kushiro 1977; Fujii & Bougault 1983) contain 49.7–50.0 wt% SiO_2 (Fig. 2). The liquids produced by melting of peridotite at 9 kbar contain 50.6–51.6 wt% SiO_2 (Sen 1982). Those MORB-type liquids

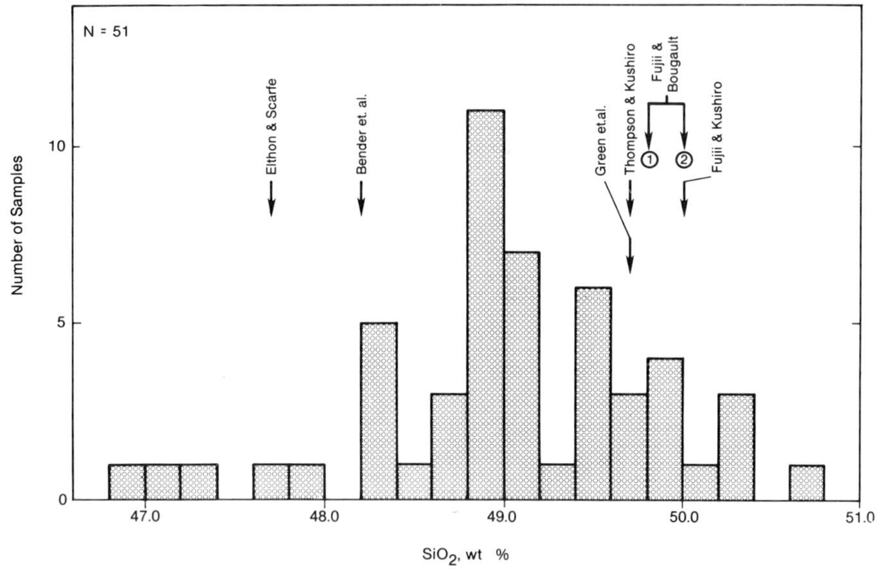

FIG. 2. Histogram of the SiO_2 contents of MORB glasses with more than 9.5 wt% MgO. SiO_2 contents of experimentally studied basalts are shown; note that Fujii & Bougault (1983) present three analyses, one with 48.8 wt% SiO_2 and two with 50.0 wt% SiO_2. Sources of the data on MORB glasses are Hekinian *et al.* (1976), Bryan & Moore (1977), Langmuir *et al.* (1977), Melson *et al.* (1977), Bryan (1979), Dungan *et al.* (1979), Melson & O'Hearn (1979), Natland & Melson (1979), Wood *et al.* (1979), Sigurdsson (1981), Byers *et al.* (1984), Eaby *et al.* (1984) and Humphris *et al.* (1985).

that do not exhibit orthopyroxene saturation at about 10 kbar (Bender *et al.* 1978; Green *et al.* 1979; Elthon & Scarfe (1984) have SiO_2 contents of 47.7–49.7 wt% (Fig. 2).

Differentiation of primitive MORBs results in a gradual increase in SiO_2 when olivine crystallization occurs. The SiO_2 contents of the most primitive MORB glasses (more than 9.5 wt% MgO) must therefore be very similar to those of the primary magma from which they were derived if melting occurred at about 10 kbar, because only a small amount (about 5%) of olivine crystallization could have occurred prior to eruption. For this reason, those MORB glasses with more than 49.7 wt% SiO_2 (about 25% of primitive MORB glasses (Fig. 2)) could develop from MORBs that show orthopyroxene saturation at about 10 kbar, but MORBs with lower SiO_2 contents (about 75% of primitive MORB glasses (Fig. 2)) would not. Most primitive MORB glasses must be derived from primary liquids with SiO_2 contents lower than 49.7 wt%. It is most likely that the lower SiO_2 contents of primitive MORB glasses directly reflect their origin from primary magmas produced by melting at substantially higher pressures (suggested below to be 20–25 kbar).

Mineral compositions in the residual mantle

As noted in the introduction, the compositions of liquidus minerals of a primary MORB must match the compositions of minerals in the residual mantle. The compositions of olivine in residual mantle rocks range from about Fo_{89} to $Fo_{91.8}$ and average $Fo_{90.5}$ (Fig. 3*a*), comparable with Dick & Fisher's (1983) estimate ($Fo_{90.7}$). The NiO contents of residual mantle olivines range from 0.32 to 0.52 wt% and average 0.41 wt% (Fig. 3*b*). Dick & Fisher's estimate of the average NiO contents of mantle olivines (0.28 wt%) is substantially lower than all published data and is not considered further. The liquidus olivine composition in the 10 kbar experiment by Fujii & Bougault (1983) was $Fo_{88.5}$, whereas the liquidus olivine composition at 25 kbar for the high-MgO basalt of Elthon & Scarfe (1984) was $Fo_{89.5}$. These liquidus olivine compositions overlap those found in the residual mantle and it appears that the forsterite content of mantle olivines does not constrain the composition or pressure of origin for primary MORBs; both primitive MORB glasses and high-MgO basalts

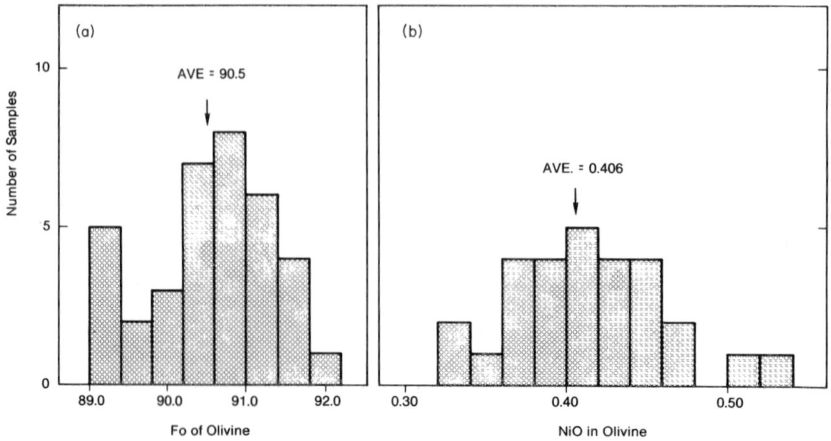

FIG. 3. Histograms of the Fo and NiO contents of olivines in abyssal peridotites. Sources of the data are Prinz *et al.* (1976), Arai & Fujii (1978), Sinton (1978), Hamlyn & Bonatti (1980), Hébert *et al.* (1983) and Shibata & Thompson (1986).

could be in equilibrium with mantle olivines at high pressures.

One of the commonly cited arguments against high-MgO basalts as primary MORBs is that the abundance of nickel in MORBs is too high to have been derived from high-MgO basalts (*eg* Sato 1977; Hart & Davis 1978). The NiO in abyssal peridotite olivines averages 0.41 wt% (about 3200 ppm), with most values between 0.36 wt% (about 2850 ppm) and 0.47 wt% (about 3750 ppm) (Fig. 3*b*). The nickel contents of liquids in equilibrium with these olivines can be calculated from the following 1 atm distribution coefficient (from Elthon 1987):

$$D_{\mathrm{Ni}} = \frac{\text{NiO in olivine}}{\text{NiO in liquid}} = \frac{122.16}{\text{wt\% MgO (liq)}} - 2.27 \quad (r = 0.97)$$

The compositions of liquids in equilibrium with residual mantle olivines and MORB glasses are shown in Fig. 4. The liquids in equilibrium with residual mantle olivines are calculated assuming that there is no pressure effect on D_{Ni} between olivine and liquid (Takahashi 1980).

The nickel versus MgO variations (Fig. 4) are consistent with either of two end-member models. On the one hand, olivine crystallization from a primary magma with 11.0 wt% MgO and 360 ppm nickel is capable of producing the Ni–MgO variations of MORBs. It is also shown that olivine crystallization starting from a high-MgO basalt composition similar to NT-23 (Table 1 and Elthon & Scarfe 1984) results in derivative liquids that closely resemble MORBs (Fig. 4). These results suggest that the nickel abundances of MORBs are consistent with either the generation of primary MORBs at about 10 kbar followed by a small amount (less than 5%) of olivine crystallization or the generation of high-MgO basalts at 20–25 kbar followed by about 20% olivine crystallization prior to eruption.

One of the minerals whose composition changes substantially during partial melting of the mantle and is dependent on the pressure of melting is orthopyroxene. Jaques & Green (1980) have shown that the Al_2O_3 content of orthopyroxene decreases as the extent of melting increases and increases as the pressure of melting increases (Fig. 5, inset). Because the Al_2O_3 contents of orthopyroxenes in abyssal peridotites are well documented, the potential exists to constrain the pressure of melting in the suboceanic mantle from these compositions.

The magnesium number of orthopyroxene versus the Al_2O_3 content of orthopyroxene for various abyssal peridotites is shown in Fig. 6. The overall trend of increasing magnesium number with decreasing Al_2O_3 is consistent with partial melting studies of peridotites (Jaques & Green 1980; Dick & Fisher 1983; Michael & Bonatti 1985). Within the suites from the Owen Fracture Zone and the Lewis Hills massif, where numerous samples are available from the same localities, the trends run almost perpendicular to the inferred partial melting

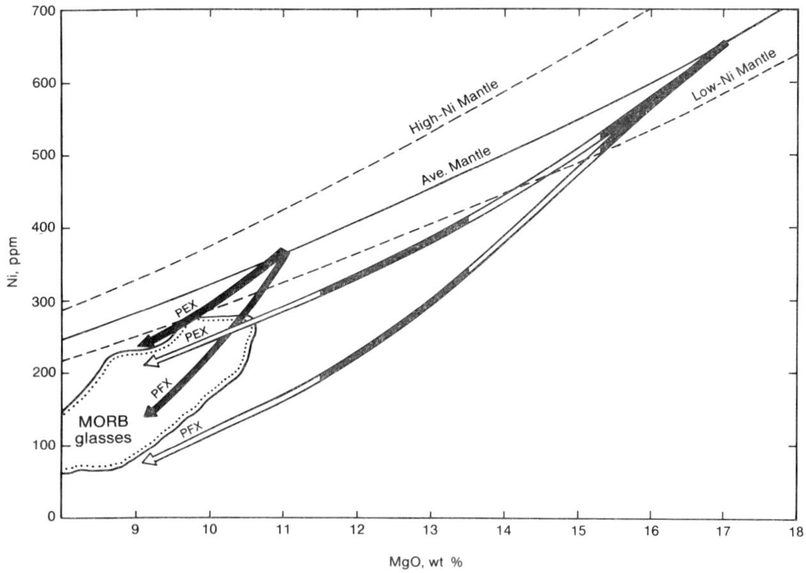

FIG. 4. MgO versus nickel for MORB glasses and potential primary magmas. The lines for high-nickel mantle, average mantle and low-nickel mantle correspond to calculated primary liquids assuming 3750 ppm Ni, 3200 ppm Ni and 2850 ppm Ni respectively in residual mantle olivines. The nickel values for residual mantle olivines come from data in Fig. 3. Paths for olivine crystallization from primary magmas with 11 and 17 wt% MgO are shown for perfect equilibrium crystallization (PEX) and perfect fractional crystallization (PFX).

TABLE 1. *Geochemical data on sample NT−23*

Major elements[a] (wt%)		XRF[b] (ppm)		INAA[c] (ppm)	
SiO_2	47.73	Y	12	La	1.01
TiO_2	0.85	Zr	38	Ce	2.75
Al_2O_3	13.42	Sr	30	Sm	1.34
FeO^*	9.58	Rb	<5	Eu	0.67
MnO	0.20	Zn	51	Yb	2.18
MgO	16.89	Ni	615	Lu	0.30
CaO	9.94	Cr	1657	Co	64.1
Na_2O	1.07	Nb	2.4	Cr	1597
K_2O	0.05	Cu	62	Hf	1.32
				Sc	33.6
				Ta	0.10

[a]Microprobe data of fused sample from Elthon & Scarfe (1984).
[b]X-ray fluorescence data on pressed powder pellets obtained at Lamont−Doherty Geological Observatory, 1980.
[c]Instrumental neutron activation analysis data from laboratory at Johnson Space Center, 1986.

trend. The Lewis Hills and Owen Fracture Zone trends are interpreted as being subsolidus equilibration trends (Smith & Elthon 1988). Studies of abyssal and ophiolitic residual mantle rocks suggest that these samples have undergone substantial subsolidus re-equilibration (generally to 700−900 °C) and that corrections of the mineral compositions to magmatic temperatures (1300−1400 °C) are indicated.

These data suggest that the least-depleted end of the abyssal peridotite orthopyroxene cluster would have 6.0−6.5 wt% Al_2O_3 under magmatic temperatures and the most-depleted end would have about 2.5−3.0 wt% Al_2O_3 (Fig. 6). The liquidus orthopyroxene in Fujii & Bougault's (1983) experiment at 10.5 kbar had 4.1 wt% Al_2O_3, the liquidus orthopyroxene in Green *et al.*'s (1979) experiment at 20 kbar had 5.5 wt% Al_2O_3 and the liquidus orthopyroxene in Elthon & Scarfe's (1984) 25 kbar experiment had 6.7 wt % Al_2O_3 (Fig. 5). These results are also in agreement with melting studies of peridotites by Jaques & Green (1980) and Takahashi & Kushiro (1983) which show that the Al_2O_3 contents of orthopyroxenes in a lherzolite residuum increase from 3.4−4.0 wt% at 8−10 kbar to 4.7 wt% at 20 kbar (Fig. 5).

The Al_2O_3 contents of orthopyroxenes at the least-depleted end of the abyssal peridotite trend (6.0−6.5 wt% Al_2O_3) suggest that

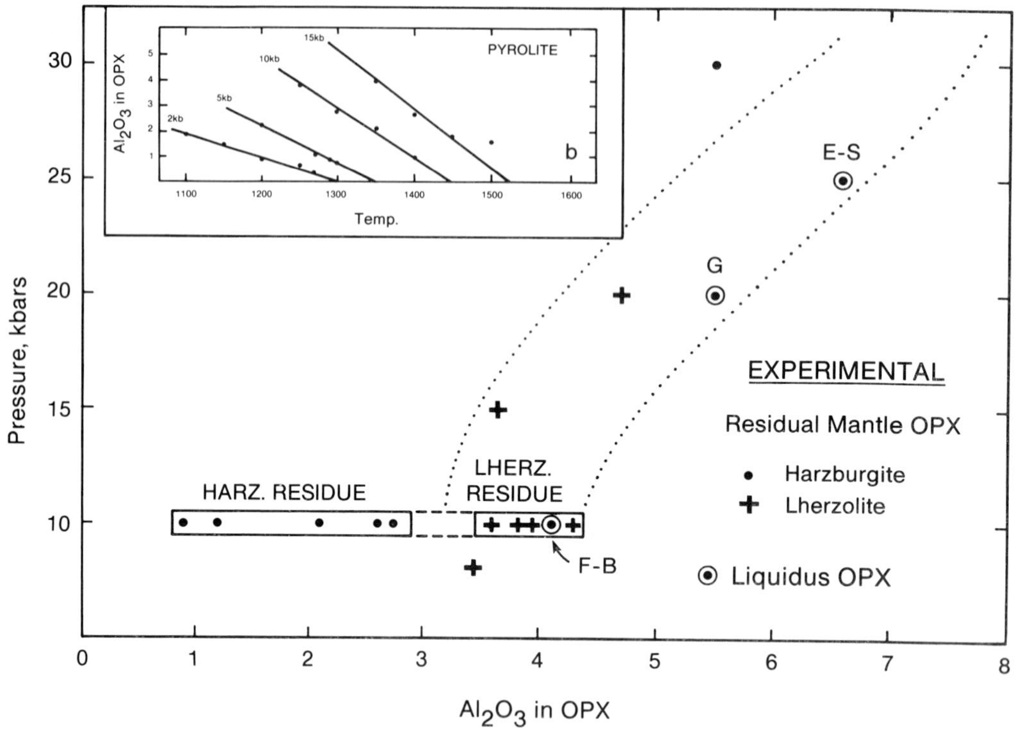

FIG. 5. The Al_2O_3 contents of orthopyroxene (OPX) versus the pressure of equilibration for experimental melting of peridotites (Jaques & Green 1980; Takahashi & Kushiro 1983). Also shown are the compositions of liquidus (or near-liquidus) orthopyroxenes from high-pressure experimental studies of MORB-type liquids (F-B, Fujii & Bougault 1983; G, Green *et al.* 1979; E-S, Elthon & Scarfe 1984). The inset (from Jaques & Green 1980) shows how Al_2O_3 in OPX varies as a function of pressure with increasing melting (temperature).

partial melting must have occurred at about 20–25 kbar because this is the pressure at which the Al_2O_3 contents of liquidus orthopyroxene and residual mantle orthopyroxene match (compare Figs 5 and 6). In fact, the entire field of abyssal peridotite orthopyroxenes (Fig. 6) could be produced by partial melting at 20–25 kbar, with the low-Al_2O_3 and high-magnesium-number end corresponding to the largest extent of partial melting (see Fig. 5, inset) The liquidus orthopyroxenes at 10 kbar contain 3.6–4.3 wt% Al_2O_3(Fig. 5), suggesting that partial melting at 10 kbar could produce only the third of the abyssal peridotite field with the highest magnesium numbers and the lowest Al_2O_3 contents. For approximately two-thirds of the abyssal peridotite field, melting at about 10 kbar does not appear to be capable of producing the higher Al_2O_3 contents in orthopyroxenes.

With these data strongly suggesting that the Al_2O_3 contents of orthopyroxenes constrain the majority of abyssal peridotites to have produced primary magmas at high pressures (20–25 kbar), what other possibilities exist for producing their high Al_2O_3 contents? There is the distinct possibility that some abyssal peridotites do not represent 'clean' residues, but that they might have a small amount of trapped basaltic liquid resulting from either infiltration by precursor melts (McKenzie 1985) or liquid that was not removed during separation of the primary magma. Either process would result in higher Al_2O_3 contents in the peridotite compared with that of a 'clean' residue but would not substantially increase the Al_2O_3 contents in orthopyroxene. This is concluded from the fact that the peridotite compositions chosen for melting studies by Jaques & Green (1980) had 5.4–5.9 wt% Al_2O_3 in the *bulk rock* (much higher than the 2–4 wt% Al_2O_3 in residual mantle samples), which would make the Jaques and Green samples compositionally equivalent to massively infiltrated peridotites with large amounts of trapped liquid. It is therefore un-

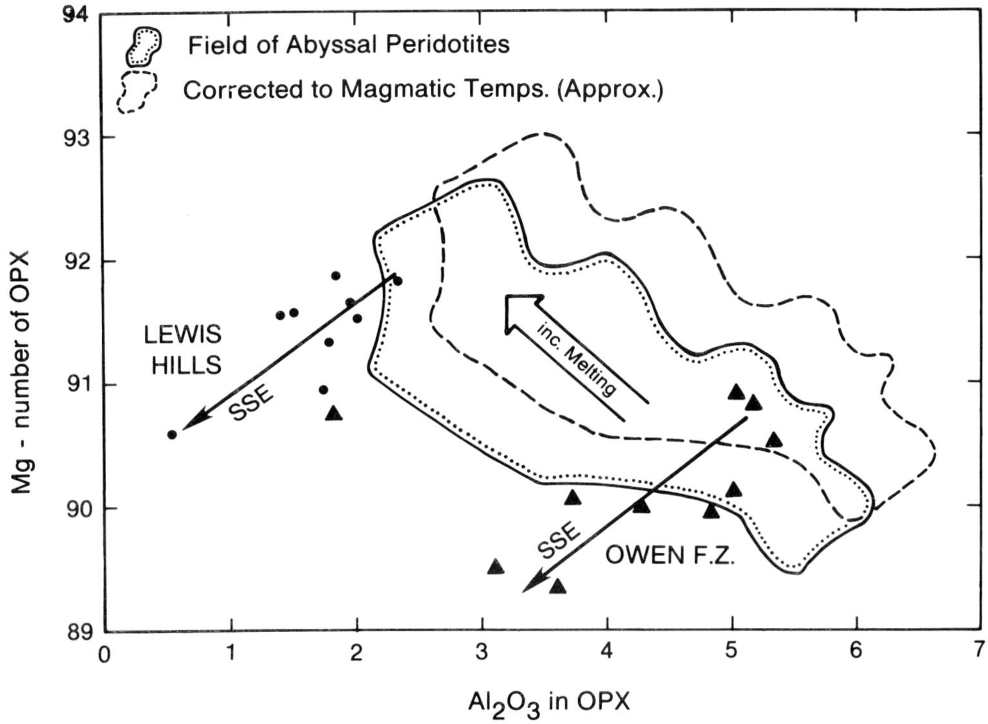

FIG. 6. The magnesium number (100×Mg/(Mg+Fe)) versus Al_2O_3 content of orthopyroxene (OPX). The field of abyssal peridotites includes data from Prinz *et al.* (1976), Nicholls *et al.* (1981), Dick & Fisher (1983) and Michael & Bonatti (1985). The large arrow in the centre of this field indicates the direction for increasing partial melting. The Owen Fracture Zone data are from Hamlyn & Bonatti (1980) and the Lewis Hills data are from Smith & Elthon (1988). The arrows for these two data sets correspond to estimated subsolidus equilibration (SSE) trends (see Smith & Elthon 1988). The field of abyssal peridotites is shifted slightly to indicate what the compositions were before subsolidus equilibration. Note that this shift corresponds to an average abyssal peridotite composition but that individual samples will move different distances along different trajectories. The magnitude of this shift is therefore only approximate.

likely that the higher Al_2O_3 contents in abyssal peridotite orthopyroxenes compared with liquidus orthopyroxenes from primitive MORBs at 10 kbar can be explained with an appeal to a mantle richer in Al_2O_3.

Concluding remarks

On the basis of high-pressure experimental results on the melting of peridotites and basalts, it is concluded that the majority (probably about 75%) of primitive MORB glasses are derived from primary MORBs that were produced at pressures substantially greater than 10 kbar. The SiO_2 abundances, for example, suggest that only about 25% of primitive MORB glasses could be derived from primary magmas generated at approximately 10 kbar; the remaining primitive glasses appear to be derived from primary magmas generated at substantially higher pressures. The Al_2O_3 contents of orthopyroxenes in abyssal peridotites also do not support models that involve the production of primary MORBs at about 10 kbar; instead a pressure of 20–25 kbar appears to fit the data much better. It is noted, however, that these conclusions are based on limited experimental data and that additional studies will be required before they can be considered to be well established.

Melting of the mantle at 20–25 kbar, near the transition from the spinel lherzolite to the garnet lherzolite facies, will produce primary high-MgO basalt liquids (O'Hara 1968; Elthon & Scarfe 1984). Examples of high-MgO basalts that are potential primary magmas generated at about 20–25 kbar are sample NT–23 of Elthon & Scarfe (1984) and the Gorgona Island komatiities (Echeverria 1980). Francis (1985) has shown that the Baffin Bay picrites previously considered by Clarke (1970) to be potential primary magmas are olivine-enriched accumulative magmas, but that very forsterite-rich olivine megacrysts (Fo_{93}) suggest that high-MgO basalts were present at earlier stages of the evolution of these magmas.

With such poor sampling of high-MgO basalts at present, the full range of their compositions is not well established. For this reason, it seems that the most reasonable course of action in the continuing assessment of the role of high-MgO basalts in MORB petrogenesis is to evaluate the compositions of samples known at present without deriving great satisfaction when these samples fit simple models and without rejecting their role when they do not seem to fit. It is important to keep in mind that advocates of the primary high-MgO basalt hypothesis have generally argued that the compositions of *erupted* basalts indicate that they have been derived from primary high-MgO basalts at 20–30 kbar (*eg* Fig. 1*a*), even in the absence of samples of aphyric high-MgO basalts (*eg* O'Hara 1968). Present samples provide an important perspective on the nature of high-MgO basalts, but the view is limited by the small number of samples available and the uncertainty of how representative they are of primary high-MgO basalts that never reach the Earth's surface.

ACKNOWLEDGMENTS: I thank A. Reid, A. D. Saunders and R. N. Thompson for their constructive comments. This research was supported by National Science Foundation Grant OCE81–21232 and by the Exxon Education Foundation. This work was done while D. E. was a visiting scientist at the Lunar and Planetary Institute, which is operated by the Universities Space Research Association under Contract NASW–4066 from NASA. This paper is LPI Contribution No. 651.

References

ARAI, S. & FUJII, T. 1978. Petrology of ultramafic rocks from Site 395. *Initial Reports of the Deep Sea Drilling Project* **45**, 587–594.

BENDER, J. F., HODGES, F. N. & BENCE, A. E. 1978. Petrogenesis of basalts from the project FAMOUS area: Experimental study from 0 to 15 kbars. *Earth and Planetary Science Letters* **41**, 277–302.

BRYAN, W. B. 1979. Regional variation and petrogenesis of basalt glasses from the FAMOUS area, Mid-Atlantic Ridge. *Journal of Petrology* **20**, 293–325.

—— & MOORE, J. G. 1977. Compositional variation of young basalts in the Mid-Atlantic Ridge rift valley near 36°49′N. *Bulletin of the Geological Society of America* **88**, 556–570.

BYERS, C. D., CHRISTIE, D. M., MUENOW, D. W. & SINTON, J. 1984. Volatile contents and ferric–ferrous ratios of basalt, ferrobasalt, andesite and rhyodacite glasses from the Galapagos 95.5°W propagating rift. *Geochimica et Cosmochimica Acta* **48**, 2239–2245.

CLARKE, D. B. 1970. Tertiary basalts of Baffin Bay: possible primary magma from the mantle. *Contributions to Mineralogy and Petrology* **25**, 203–224.

—— & O'HARA, M. J. 1979. Nickel and the existence of high-MgO liquids in nature. *Earth and Planetary Science Letters* **44**, 153–158.

COOMBS, D. S. 1963. Trends and affinities of basaltic magmas and pyroxenes as illustrated on the diopside–olivine–silica diagram. *Mineralogical Society of America, Special Paper* **1**, 227–250.

DICK, H. J. B. & FISHER, R. L. 1983. Mineralogic studies of the residues of mantle melting: abyssal and alpine-type peridotites. *In*: KORNPROBST, J. (ed.) *Kimberlites II. The Mantle and Crust–Mantle Relationships*. Elsevier, Amsterdam, 295–308.

DUNGAN, M. A., RHODES, J. M., LONG, P. E., BLANCHARD, D. P., BRANNON, J. C. & RODGERS, K. V. 1979. The petrology and geochemistry of basalts from Site 396, Legs 45 and 46 of the Deep Sea Drilling Project. *Initial Reports of the Deep Sea Drilling Project* **46**, 89–113.

EABY, J. S., CLAGUE, D. A. & DELANEY, J. R. 1984. Sr isotopic variations along the Juan de Fuca Ridge. *Journal of Geophysical Research* **89**, 7883–7890.

ECHEVERRIA, L. M. 1980. Tertiary or Mesozoic komatiites from Gorgona Island, Colombia: field relations and geochemistry. *Contributions to Mineralogy and Petrology* **73**, 253–266.

ELTHON, D. 1979. High magnesia liquids as the parental magma for ocean floor basalts. *Nature* **278**, 514–518.

—— 1986. Comments on 'Composition and depth of origin of primary mid-ocean ridge basalts' by D. C. Presnall and J. D. Hoover. *Contributions to Mineralogy and Petrology* **94**, 253–256.

—— 1987. Olivine–liquid partitioning in high–MgO basalts and komatiites. *Abstracts of the 18th*

Lunar and Planetary Science Conference, pp. 258–259.

—— & SCARFE, C. M. 1980. High-pressure phase equilibria of a high-magnesia basalt: implications for the origin of mid-ocean ridge basalts. *Carnegie Institution of Washington Yearbook* **79**, 277–281.

—— & —— 1984. High-pressure phase equilibria of a high-magnesia basalt and the genesis of primary oceanic basalts. *American Mineralogist* **69**, 1–15.

FORD, C. E., RUSSELL, D. G., CRAVEN, J. A. & FISK, M. R. 1983. Olivine–liquid equilibria: temperature, pressure and composition dependence on the crystal/liquid cation partition coefficients for Mg, Fe^{2+}, Ca and Mn. *Journal of Petrology* **24**, 256–265.

FRANCIS, D. 1985. The Baffin Bay lavas and the value of picrites as analogues of primary magmas. *Contributions to Mineralogy and Petrology* **89**, 144–154.

FUJII, T. & BOUGAULT, H. 1983. Melting relations of a magnesian abyssal tholeiite and the origin of MORBs. *Earth and Planetary Science Letters* **62**, 283–295.

—— & KUSHIRO, I. 1977. Melting relations and viscosity of an abyssal tholeiite. *Carnegie Institution of Washington Yearbook* **76**, 461–465.

—— SCARFE, C. M. 1985. Composition of liquids coexisting with spinel lherzolite at 10 kbar and the genesis of MORBs. *Contributions to Mineralogy and Petrology* **90**, 18–28.

GREEN, D. H., HIBBERSON, W. O. & JAQUES, A. L. 1979. Petrogenesis of mid-ocean ridge basalts. *In*: MCELHINNEY, M. W. (ed.) *The Earth: Its Origin, Structure, and Evolution*. Academic Press, London, 265–299.

HAMLYN, P. R. & BONATTI, E. 1980. Petrology of mantle-derived ultramafics from the Owen Fracture Zone, northwest Indian Ocean: implications for the nature of the oceanic upper mantle. *Earth and Planetary Science Letters* **48**, 65–79.

HART, S. R. & DAVIS K. E. 1978. Nickel partitioning between olivine and silicate melt. *Earth and Planetary Science Letters* **40**, 203–219.

—— & —— 1979. Reply to D. B. Clarke and M. J. O'Hara, 'Nickel and the existence of high-MgO basalts in nature'. *Earth and Planetary Science Letters* **44**, 159–161.

HEBERT, R., BIDEAU, D. & HEKINIAN, R. 1983. Ultramafic and mafic rocks from the Garrett transform fault near 13°30'S on the East Pacific Rise: igneous petrology. *Earth and Planetary Science Letters* **65**, 107–125.

HEKINIAN, R., MOORE, J. G. & BRYAN, W. B. 1976. Volcanic rocks and processes of the Mid-Atlantic Ridge rift valley near 36°49'N. *Contributions to Mineralogy and Petrology* **58**, 83–110.

HUMPHRIS, S. E., THOMPSON, G., SCHILLING, J. -G. & KINGSLEY, R. H. 1985. Petrological and geochemical variations along the Mid-Atlantic Ridge between 46°S and 32°S: influence of the Tristan da Cunha mantle plume. *Geochimica et Cosmochimica Acta* **49**, 1445–1464.

JAQUES, A. L. & GREEN, D. H. 1979. Determination of liquid compositions in experimental high-pressure melting of peridotite. *American Mineralogist* **64**, 1312–1321.

—— & —— 1980. Anhydrous melting of peridotite at 0–15 kb pressure and the genesis of tholeiitic basalts. *Contributions to Mineralogy and Petrology* **73**, 278–310.

KUSHIRO, I. & THOMPSON, R. N. 1972. Origin of some abyssal tholeiites from the Mid-Atlantic Ridge. *Carnegie Institution of Washington Yearbook* **71**, 403–406.

LANGMUIR, C. H., BENDER, J. F., BENCE, A. E., HANSON, G. N. & TAYLOR, S. R. 1977. Petrogenesis of basalts from the FAMOUS area: Mid-Atlantic Ridge. *Earth and Planetary Science Letters* **36**, 133–156.

MCKENZIE D. 1985. ^{230}Th–^{238}U disequilibrium and the melting processes beneath ridge axes. *Earth and Planetary Science Letters* **72**, 149–157.

MELSON, W. G. & O'HEARN, T. 1979. Basaltic glasses erupted along the Mid-Atlantic Ridge between 0–37°N: relationship between composition and latitude. *In*: TALWANI, M., HARRISON, C. G. & HAYES, D. E. (eds) *Deep Drilling Results in the Atlantic Ocean: Oceanic Crust*. American Geophysical Union, Washington, DC, 249–261.

——, BYERLY, G. R., NELEN, J. A., O'HEARN, T., WRIGHT, T. L. & VALLIER, T. 1977. A catalog of the major element chemistry of abyssal volcanic glasses. *Mineral Sciences Investigations, Smithsonian Contributions to the Earth Sciences* **19**, 31–60.

MICHAEL, P. J. & BONATTI, E. 1985. Peridotite composition from the North Atlantic: regional and tectonic variations and implications for partial melting. *Earth and Planetary Science Letters* **73**, 91–104.

MYSEN, B. O. & KUSHIRO, I. 1977. Compositional variations of coexisting phases with degree of melting of peridotite in the upper mantle. *American Mineralogist* **62**, 843–865.

NATLAND, J. H. & MELSON, W. G. 1979. Compositions of basaltic glasses from the East Pacific Rise and the Siqueiros Fracture Zone, near 9°N. *Initial Reports of the Deep Sea Drilling Project* **54**, 705–724.

NICHOLLS. I. A., FERGUSON, J., JONES, H., MARKS, G. P. & MUTTER, J. C. 1981. Ultramafic blocks from the ocean floor southwest of Australia. *Earth and Planetary Science Letters* **56**, 362–374.

O'HARA, M. J. 1965. Primary magmas and the origin of basalts. *Scottish Journal of Geology* **1**, 19–40.

—— 1968. The bearing of phase equilibria studies on the origin and evolution of basic and ultrabasic rocks. *Earth Science Reviews* **4**, 69–133.

PRESNALL, D. C. & HOOVER, J. D. 1984. Composition and depth of origin of primary mid-ocean ridge basalts. *Contributions to Mineralogy and Petrology* **87**, 170–178.

—— & —— 1986. Composition and depth of origin of primary mid-ocean ridge basalts — reply to D. Elthon. *Contributions to Mineralogy and Petrology* **94**, 257–261.

——, Dixon, J. R., O'Donnell, T. H. & Dixon, S. A. 1979. Generation of mid-ocean ridge tholeiites. *Journal of Petrology* **20**, 3–35.

Prinz, M., Keil, K., Green, J. A., Reid, A. M., Bonatti, E. & Honnorez, J. 1976. Mineralogy and petrology of some ultramafic and mafic dredge samples from the equatorial Mid-Atlantic Ridge and fracture zones. *Journal of Geophysical Research* **81**, 4087–4103.

Sato, H. 1977. Nickel content of basaltic magmas: identification of primary magmas and a measure of the degree of olivine fractionation. *Lithos* **10**, 113–120.

Sen, G. 1982. Composition of basaltic liquids generated from a partially depleted lherzolite at 9 kbars pressure. *Nature* **299**, 336–338.

Shibata, T. & Thompson, G. 1986. Peridotites from the Mid-Atlantic Ridge at 43°N and their petrogenetic relation to abyssal tholeiites. *Contributions to Mineralogy and Petrology* **93**, 144–159.

Sigurdsson, H. 1981. First-order major element variation in basalt glasses from the Mid-Atlantic Ridge: 29°N to 73°N. *Journal of Geophysical Research* **86**, 9483–9502.

Sinton, J. M. 1978. Petrology of (alpine-type) peridotites from Site 395, DSDP Leg 45. *Initial Reports of the Deep Sea Drilling Project* **45**, 595–601.

Smith, S. E. & Elthon, D. 1988. Mineral compositions of plutonic rocks from the Lewis Hills massif, Bay of Islands ophiolite. *Journal of Geophysical Research* **93**, 3450–3468.

Stolper, E. 1980. A phase diagram for mid-ocean ridge basalts: preliminary results and implications for petrogenesis. *Contributions to Mineralogy and Petrology* **74**, 13–27.

Takahashi, E. 1980. Olivine/liquid nickel partitioning at high pressures: experiments with an olivine capsule. *EOS (Transactions of the American Geophysical Union)* **61**, 397.

—— & Kushiro, I. 1983. Melting of a dry peridotite at high pressures and basalt magma genesis. *American Mineralogist* **68**, 859–879.

Tormey, D. R., Grove, T. L. & Bryan, W. B. 1987. Experimental petrology of normal MORB near the Kane Fracture Zone: 22°–25°N, mid-Atlantic Ridge. *Contributions to Mineralogy and Petrology* **96**, 121–139.

Wood, D. A., Varet, J., Bougault, H., Corre, O., Joron, J. L., Treuil, M., Bizouard, H., Norry, M. J., Hawkesworth, C. J. & Roddick, J. C. 1979. The petrology, geochemistry, and mineralogy of North Atlantic basalts: a discussion based on IPOD Leg 49. *Initial Reports of the Deep Sea Drilling Project* **49**, 597–655.

D. Elthon, Lunar and Planetary Institute, 3303 NASA Road One, Houston, TX 77058, USA, and Department of Geosciences, University of Houston, Houston, TX 77004, USA.

Genesis of mid-ocean ridge basalts

Toshitsugu Fujii

SUMMARY: Analyses of the compositions of liquids from various melting experiments indicate that the liquids coexisting with magnesian olivine, orthopyroxene, clinopyroxene and spinel at 10 kbar occupy a large volume in the normative projection OL–CPX–PL–SIL and that the application of the invariant melting concept analogous to the simple system CMAS is not warranted in the natural peridotite system. The compositions of magnesian mid-ocean ridge basalts (MORBs) are similar to those of liquids coexisting with olivine, orthopyroxene, clinopyroxene and spinel at 10 kbar. The relationships between Cr/(Cr+Al) in spinel and Mg/(Mg+Fe) in olivine or orthopyroxene in both abyssal peridotites and the solid phases of the melting experiments of peridotites indicate that abyssal peridotites may represent the residual materials after the extraction of liquids formed at pressures around 10 kbar. Reconstructed modal compositions of abyssal peridotites indicate that most of them included calcium-rich clinopyroxene at their solidus temperatures. It is suggested that many magnesian MORBs represent primary magmas formed at pressures around 10 kbar, leaving residual materials composed of olivine, orthopyroxene, calcium-rich clinopyroxene and spinel.

In the last 15 years a large amount of data on the nature of mid-ocean ridge basalts (MORBs) has been accumulated. Several melting experiments on MORBs at 1 atm and at high pressures have been carried out (*eg* Kushiro 1973; Bender *et al.* 1978; Fujii *et al.* 1978; Green *et al* 1979; Walker *et al.* 1979; Fujii & Bougault 1983). Compositions of melts which can coexist with mantle minerals have also been determined by high-pressure experiments to clarify the genesis of MORB magmas (Jaques & Green 1980; Stolper 1980; Takahashi & Kushiro 1983; Elthon & Scarfe 1984; Fujii & Scarfe 1985). However, there is some controversy over the genesis of MORBs and the nature of the residual mantle materials.

The debate on the nature of primary magmas of MORBs has divided into two distinct groups. One group favours a picrite magma as the parental primary magma for MORBs and believes that most MORBs are not primary magmas but evolved magmas (O'Hara 1965; Elthon & Scarfe 1980, 1984; Jaques & Green 1980; Stolper 1980). The other group favours the proposal that some magnesian MORBs could be primary magmas (Kushiro 1973; Fujii *et al.* 1978; Presnall *et al.* 1979; Fujii & Bougault, 1983; Takahashi & Kushiro 1983; Presnall & Hoover 1984; Fujii & Scarfe 1985). These two groups also have different views on the depth of magma segregation from the mantle materials and on the compositions of residual materials; the latter group prefers relatively low pressures ranging from 8 to 11 kbar leaving lherzolite or calcium-rich pyroxene-bearing harzburgite, whereas the former group envisages pressures higher than 15 kbar, and mostly more than 20 kbar, leaving a calcium-rich pyroxene-free harzburgite at the conditions of melt segregation. An exception to the latter model is that of Elthon & Scarfe (1980, 1984) who discussed the role of lherzolitic residual mantle.

After Stolper (1980) and Elthon & Scarfe (1980) published pseudo-liquidus high-pressure phase diagrams, which were conceptually developed by O'Hara (1965), to discuss the possibility of picrite primary magma, most discussions of the genesis of MORBs have been based on this kind of phase diagram. The main argument by O'Hara (1965), Stolper (1980) and Elthon & Scarfe (1980, 1984) against the primary MORB concept is that MORBs, when projected in OL–CPX–PL–SIL normative space, are displaced from the orthopyroxene liquidus field at all pressures and therefore cannot coexist with mantle peridotite.

Against this argument, Fujii & Bougault (1983) showed that one of the most primitive MORBs from the FAMOUS area is saturated in olivine, orthopyroxene, clinopyroxene and spinel–plagioclase. Presnall & Hoover (1984) argued that projected positions of natural basalts within this kind of normative diagram could be scattered because of analytical uncertainty and that most primitive MORBs are within analytical uncertainty of orthopyroxene saturation at about 10 kbar. Tormey *et al.* (1987) argued, however, that the scattering of data as a result of analytical uncertainty is not as large as suggested by Presnall & Hoover (1984).

In this paper I discuss the genesis of MORBs using data mainly based on high-pressure

From Saunders, A. D. & Norry, M. J. (eds), 1989, *Magmatism in the Ocean Basins*, Geological Society Special Publication No. 42, pp. 137–146.

melting experiments. It is assumed that the silica-poor nature of some MORBs is real, as argued by Stolper (1980), Elthon & Scarfe (1980, 1984) and Tormey *et al.* (1987).

Normative silica contents of mid-ocean ridge basalts

Stolper (1980) noted that primitive MORBs plot at the silica-poor end of the basalt cluster in the OL−CPX−PL−SIL normative projection (Walker *et al.* 1979) and that more evolved MORBs plot at the silica-rich end. Therefore he argued that, although silica-rich MORBs are close to the orthopyroxene-saturated field at slightly less than 10 kbar, they cannot be primary magmas because they are evolved. It is clear from Fig. 1, however, that his observation is not entirely correct. Magnesian MORBs which can coexist with magnesian olivine ($Fo_{>88}$) cannot be discriminated from the mildly evolved MORBs coexisting with olivine ranging from Fo_{88} to Fo_{85}. Although many of the most evolved MORBs, which can coexist with iron-rich olivine ($Fo_{<85}$), do indeed plot toward the silica-rich end, some highly evolved MORBs plot at the silica-poor end. This distribution can be seen more clearly in Fig. 2, where no positive relationship between the Mg/(Mg+Fe) ratio and

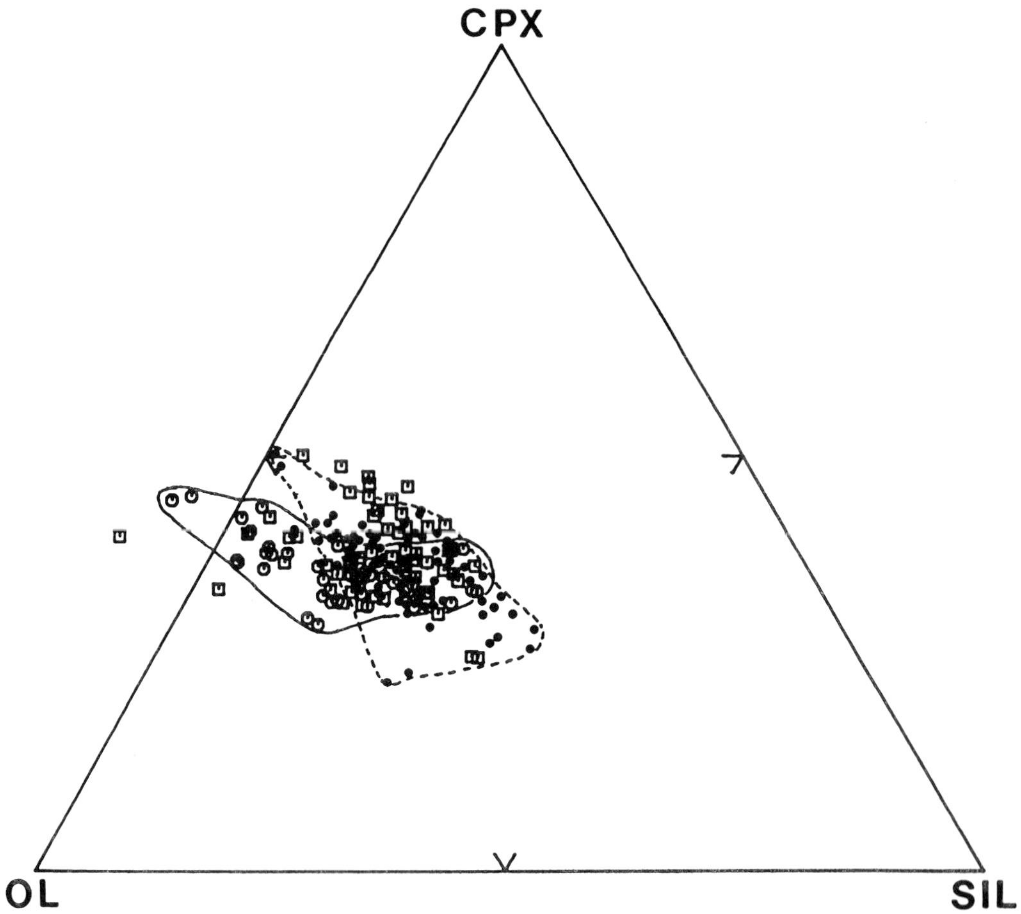

FIG. 1. Normative projection of MORBs from the Atlantic region (the projection method is that of Walker *et al.* 1979): ○, primitive MORBs with FeO/MgO ratios less than 0.94 which can coexist with magnesian olivine ($Fo_{>88}$); □, MORBs with FeO/MgO ratios between 0.94 and 1.2; ●, evolved MORBs with FeO/MgO ratios larger than 1.2 which can coexist with olivine more iron rich than Fo_{85}. The primitive MORBs and the most evolved MORBs are enclosed in full and broken lines respectively. Data sources: Bryan & Moore (1977), Melson *et al.* (1977), Bryan (1979), O'Donnell & Presnall (1980).

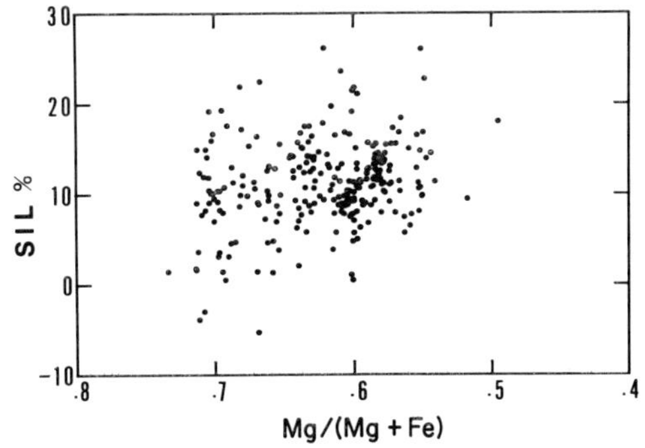

FIG. 2. SIL% (100×SIL/(OL+CPX+PL+SIL) in the molar normative projection (Walker *et al.* 1979) versus Mg/(Mg+Fe) ratio for MORBs from the Atlantic region. Data sources as for Fig. 1.

SIL% (100×SIL/(OL+CPX+PL+SIL)) can be observed. This suggests that the degree of silica saturation cannot be used as an indicator of the degree of differentiation in this normative projection. Similar arguments have been made by Elthon (1983), who proposed a different method of projection which relates the degree of differentiation more directly to that of SIL enrichment.

The original reason for using the normative projection was to obtain a comprehensive view of the partial melting of mantle materials, which more simplistic projection could not readily provide; and not just to pursue the differentiation trend of the magmas. It is not yet clear, however, which of the two projection methods is better suited for this purpose. Therefore the Walker-type projection is adopted in this paper simply because it is much more commonly used.

Partial melts of peridotites in normative projection

Several attempts have been made to determine the liquid compositions which can coexist with mantle materials. Stolper (1980) determined the compositions of melt coexisting with olivine and orthopyroxene at pressures of 10–20 kbar using a harzburgite capsule as a container for the basalt melt. More detailed experiments were performed by Takahashi & Kushiro (1983) at pressures ranging from 8 to 30 kbar using the peridotite–basalt sandwich technique similar to that described by Stolper (1980). Jaques & Green (1980) used a more direct approach to determine the composition of partial melts of mantle peridotite, which raised some suspicions about the accuracy of the liquid composition (Takahashi & Kushiro 1983) because the volumetric correction for quench crystals cannot be so accurate. Fujii & Scarfe (1985) used a method similar to that of Takahashi & Kushiro (1983) to examine the compositions of liquids coexisting with olivine, orthopyroxene, clinopyroxene and spinel at 10 kbar at various temperatures and bulk compositions. Their results are shown in Fig. 3.

The compositions of liquids that coexist with four phases of mantle minerals occupy a large

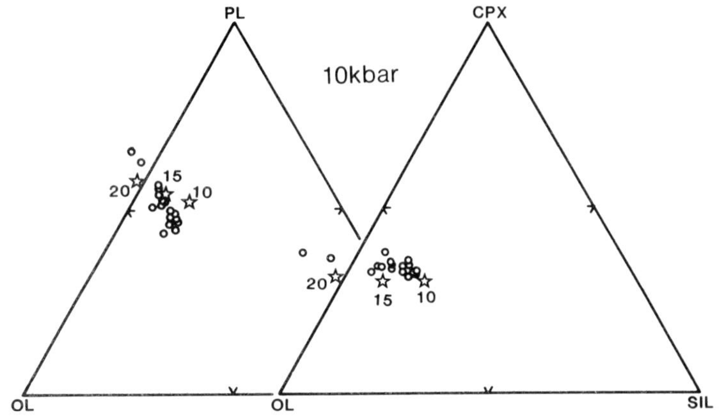

FIG. 3. Normative projections of melts coexisting with olivine, orthopyroxene, clinopyroxene and spinel at 10 kbar (Fujii & Scarfe 1985). The asterisks indicate melt compositions coexisting with mantle olivine, orthopyroxene and clinopyroxene, which were thought to be isobaric invariant melting points by Stolper (1980), and the numbers indicate their pressure in kilobars.

area in the normative projection, even at a fixed pressure. This is probably explained by the effect of the solid solution of minerals and the multicomponent nature of the natural system. On the basis of this observation Fujii & Scarfe (1985) argued that the partial melting of natural peridotite cannot be simulated by invariant eutectic melting. They claimed that the proposals made by Stolper (1980) and Elthon & Scarfe (1984) might not be appropriate because the composition of the liquid cannot be determined uniquely even if it coexists with four phases of mantle minerals at a certain pressure.

Falloon & Green (1987) recently suggested that, because the compositions of the peridotites used by Fujii & Scarfe (1985) were different from the natural peridotite compositional array, their results are not applicable to MORB genesis. As clearly stated by Falloon & Green (1987), however, bulk equilibrium through the run charge cannot be obtained in the sandwich experiments, but local equilibrium can be achieved between the glass composition and the residual crystals near the glass layer and the rims of crystals away from the glass layer. Similar observations were made by Takahashi & Kushiro (1983) and Fujii & Scarfe (1985). This clearly indicates that the liquid composition of the sandwich experiments does not represent the composition of the equilibrium partial melt

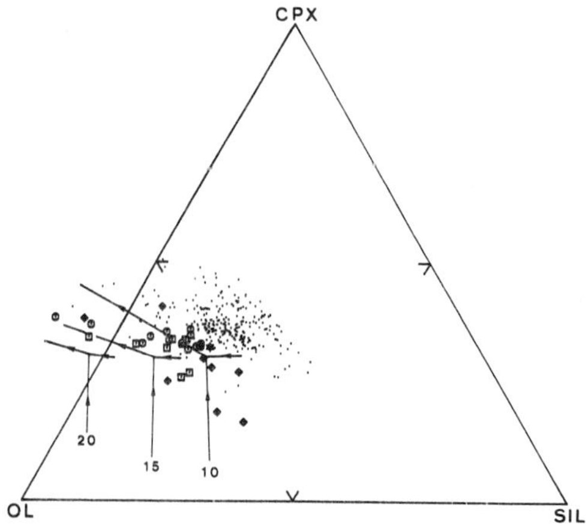

FIG. 4. Normative projections (CPX−OL−SIL) of melts coexisting with mantle minerals (olivine + orthopyroxene + clinopyroxene ± spinel) at 10 kbar and MORBs from the Atlantic region: ○ Fujii & Scarfe (1985); □, Takahashi & Kushiro (1983); ◆, Jaques & Green (1980); ●, MORB glasses from the Atlantic region; ★, ARP-74−10−16 (Fujii & Bougault 1983). The liquidus boundaries of Stolper (1980) at 10, 15 and 20 kbar are also shown for comparison. MORB data sources as for Fig. 1.

of the starting bulk composition but *does* represent the composition of a liquid which can coexist with mantle minerals. As far as the disequilibrium nature of the sandwich experiment is concerned, the evaluation of any sandwich experiment should not be based on the composition of the peridotite capsule but rather on the compositions of the glass layer and the residual minerals. The compositions of liquids obtained from the sandwich experiments of Fujii & Scarfe (1985) are therefore candidates for primary liquids at 10 kbar, even if their peridotite compositions are removed from those of the natural peridotite array, because their liquids were co-existing with four phases of mantle minerals.

The compositions of liquids which could coexist with natural mantle minerals at 10 kbar (Jaques & Green 1980; Takahashi & Kushiro 1983; Fujii & Scarfe 1985) are projected onto the Walker diagram in Fig. 4. As shown below, not all of these may apply directly to MORB genesis, because of their quite different compositions from MORBs; however, all the data are reproduced in the figure to show the possible composition range of liquids coexisting with mantle materials (olivine + orthopyroxene + clinopyroxene ± spinel). The scattering of the compositional range is much larger than that proposed by Fujii & Scarfe (1985) and overlaps the field of magnesian MORB glasses from the Atlantic region.

The magnesian MORB (ARP-74−10−16) which is multiply saturated with mantle minerals at 10 kbar (Fujii & Bougault 1983) is projected at the silica-rich end of the possible primary magma region (Fig. 4), supporting Fujii & Bougault's conclusion that such a magnesian MORB could represent a primary magma. Falloon & Green (1987), however, suspected the multiply saturated nature of ARP-74−10−16, and argued that Fujii & Bougault (1983) did not give the electron probe microanalysis (EPMA) data for the glass coexisting with olivine, orthopyroxene and clinopyroxene and that the multiple saturation occurred well below the liquidus temperature at 10 kbar. Careful examination of their results, however, indicates that the suspicions of Falloon & Green are not valid and confirms the conclusions of Fujii & Bougault. ARP-74−10−16 is saturated at least in olivine and clinopyroxene at 10 kbar and less than 10 °C below the liquidus temperature, and is saturated in orthopyroxene and clinopyroxene at 10.5 kbar and less than 10 °C below the liquidus temperature (Fujii & Bougault 1983, Table 2). These two runs indicate that there must be multiple saturation in olivine, orthopyroxene and clinopyroxene between 10 and 10.5 kbar at or close to the liquidus temperature (within 10 °C). The similarity of compositions between glasses from these two runs and the starting material (Fujii & Bougault 1983, Table 3) indicates that the multiply saturated liquid must not have deviated from the starting composition. It is of little value to publish the composition of glass which is multiply saturated in olivine, orthopyroxene and clinopyroxene at temperatures more than 25 °C below the liquidus temperature because the composition must be fractionated.

It is noteworthy that even silica-poor magnesian MORBs, which are projected near the plane of CPX–OL–PL, are also within or close to the region of possible primary magmas at 10 kbar (Fig. 4). This seems to indicate that most magnesian MORBs, ranging from silica poor to silica rich, could be formed by the partial melting of peridotite near 10 kbar.

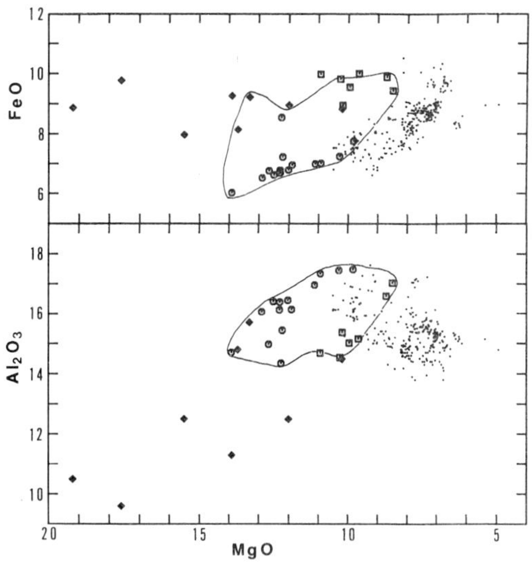

FIG. 6. Total FeO and Al_2O_3 versus MgO for possible primary magmas formed at 10 kbar together with MORBs from the Atlantic region. All the compositions are in weight per cent. The symbols are defined in Fig. 4. The compositions of primary magmas which can coexist with olivine, orthopyroxene, clinopyroxene and/or spinel are enclosed by a full line.

Partial melts of peridotite in oxide variation diagrams

As discussed by many workers (Presnall *et al.* 1979; Elthon 1983; Fujii & Bougault 1983; Presnall & Hoover 1984; Fujii & Scarfe 1985), the normative projection described above has some limitations. Small analytical errors in SiO_2 and alkalis may cause some scattering in normative silica contents (Elthon 1983; Presnall & Hoover 1984), and some distortion on the liquidus phase boundaries is possible because the reduction of the natural multicomponent system to a four-component system is based on a number of assumptions (Fujii & Bougault 1983; Fujii & Scarfe 1985). Because of these limitations, this projection is inadequate for a detailed discussion of the difference or similarity of the projected positions between natural basalt compositions and the pseudo-liquidus phase boundaries. It is therefore necessary to compare the compositions in the usual oxide variation diagram in order to discuss the similarities or differences between MORBs and the possible primary melts which were determined by experiments.

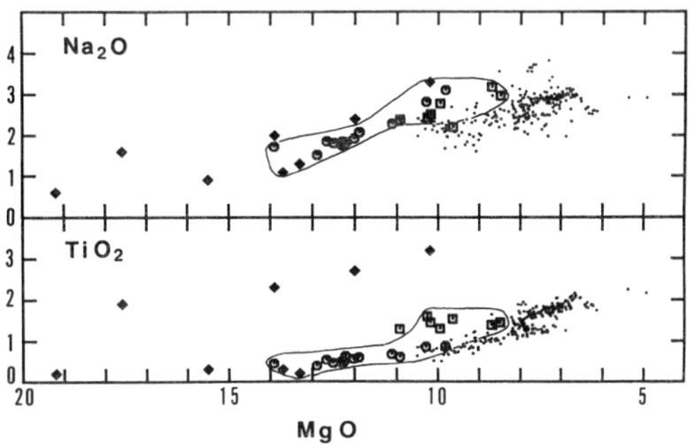

FIG. 5. Na_2O and TiO_2 versus MgO for possible primary magmas formed at 10 kbar together with MORBs from the Atlantic region. All the compositions are in weight per cent. The symbols are as defined in Fig. 4. The compositions of possible primary magmas which can coexist with olivine, orthopyroxene, clinopyroxene and/or spinel are enclosed by a full line.

In Figs 5, 6 and 7, oxide compositions of MORB glasses are compared with the compositions of possible primary liquids at 10 kbar determined by Jaques & Green (1980), Takahashi & Kushiro (1983) and Fujii & Scarfe (1985). Because some melts determined by Jaques & Green (1980) and Takahashi & Kushiro (1983) coexist only with olivine and orthopyroxene, the composition of liquids coexisting with olivine, orthopyroxene, clinopyroxene and/or spinel are enclosed within a line. In most of the diagrams, magnesian MORBs overlap or plot close to the field of the four-phase saturated primary melts. This

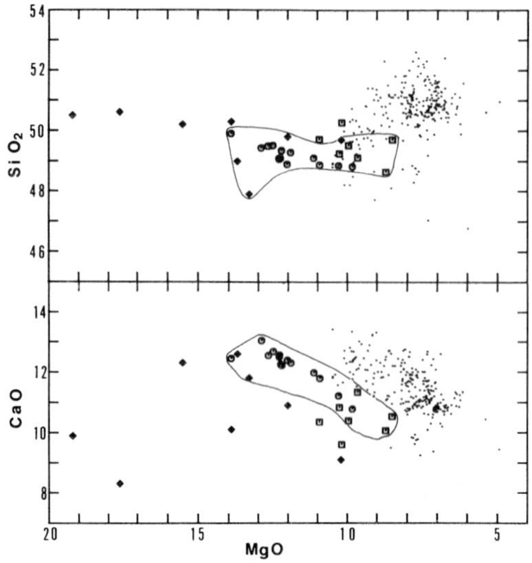

FIG. 7. SiO_2 and CaO versus MgO for possible primary magmas formed at 10 kbar together with MORBs from the Atlantic region. All the compositions are in weight per cent. The symbols are defined in Fig. 4. The compositions of primary magmas which can coexist with olivine, orthopyroxene, clinopyroxene and/or spinel are enclosed by a full line.

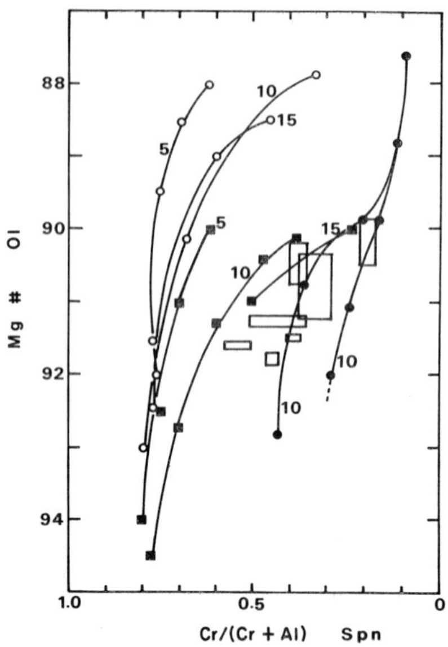

FIG. 8. Relationship between the Cr/(Cr+Al) ratio of spinel and the magnesium number of olivine coexisting with basaltic liquids: ○, pyrolite composition (Jaques & Green 1980); ■ Tinaquillo peridotite (Jaques & Green 1980); ●, basalt–peridotite sandwich experiments (Fujii & Scarfe 1985 and in preparation). Numbers adjacent to the lines connecting the symbols indicate the pressure in kilobars. The rectangles represent olivine–spinel pairs in abyssal peridotites (Arai & Fujii 1978; Dick *et al.* 1984; Shibata & Thompson 1986).

suggests that the magnesian MORBs could be primary magmas formed by the partial melting of spinel lherzolite, leaving residual minerals of olivine, orthopyroxene, clinopyroxene and spinel. However, the liquids formed by the partial melting of some specific peridotites are distinctly different from MORBs. In the diagram of TiO_2 versus MgO (Fig. 5), for example, the liquids formed by the direct melting of pyrolite are distinctly enriched in TiO_2 relative to the MORBs, indicating that the pyrolite composition adopted by Jaques & Green (1980) may not be adequate for the MORB source material.

Jaques & Green (1980) found that, although some liquid compositions formed by the partial melting of pyrolite and Tinaquillo peridotite at 10 kbar plot close to MORBs in the normative projection, their Al_2O_3 contents are distinctly lower than those of MORBs. Therefore they argued that depletion in Al_2O_3 in the partial melts at 10 kbar is essential, and any MORBs which are high-Al_2O_3 basalts cannot be formed by the direct melting of mantle peridotite. As discussed by Fujii & Scarfe (1985), however, the Al_2O_3 content in the partial melt is controlled by the initial Al_2O_3 content of the source materials and is reflected in the composition of the residual spinel. The liquids coexisting with aluminous spinel at 10 kbar have similar Al_2O_3 contents to MORBs (Fig. 6; Fujii & Scarfe 1985). The arguments put forward by Jaques & Green (1980) cannot therefore be justified.

Residual materials and abyssal peridotite

Ishiwatari (1985) and Ozawa (1986) suggested that the relationship between the forsterite content of olivine and the Cr/(Cr+Al) ratio of spinel in peridotite is a useful criterion for estimating the degree of melting, even for a peridotite which has undergone subsolidus re-equilibration after the melting event. Their suggestion may be justified because the diffusion rate of trivalent cations within spinel and pyroxene is low, and consequently the redistribution of Cr/(Cr+Al) would be unlikely. Furthermore, the forsterite content of olivine would show little variation during subsolidus re-equilibration because olivine is the predominant phase within peridotite and buffers the Mg/(Mg+Fe) ratio.

The compositions of spinel and coexisting olivine from the partial melting experiments of Jaques & Green (1980) and Fujii & Scarfe (1985 and in preparation) are shown in Fig. 8 together with those of peridotite recovered from the oceanic regions (Arai & Fujii 1978; Dick *et al.* 1984; Shibata & Thompson 1986). The Cr/(Cr +Al) ratios of spinels formed during the partial melting of pyrolite are too high at any pressure, and they are completely different from those of abyssal peridotites. This indicates that pyrolite is not adequate for a MORB source material, at least in terms of aluminium content, and that the experimental results based on the melting of pyrolite are not applicable to the genesis of MORBs. The results on Tinaquillo peridotite are more relevant for the genesis of abyssal peridotites. Spinels, however, have a higher Cr/(Cr+Al) ratio than the abyssal peridotites at a given coexisting olivine composition. This indicates that the low Al_2O_3 content of melts formed by the partial melting of Tinaquillo peridotite (Jaques & Green 1980) was due to Al_2O_3 depletion in this peridotite. The spinel and olivine formed by the equilibration experiments of basalt and peridotite at 10 kbar (Fujii & Scarfe 1985), however, resemble those of abyssal peridotites. This indicates that the high-Al_2O_3 nature of MORBs and the existence of aluminium-rich spinel in the residual materials (abyssal peridotites) can be explained for certain source peridotite compositions at 10 kbar. Ozawa (1986) showed that the vast majority of spinel peridotite nodules in alkaline basalts and kimberlites, which are believed to be residual materials, have undergone various degrees of melting (*eg* Zindler & Hart 1986) and contain aluminous spinel with a Cr/(Cr+Al) ratio between 0.1 and 0.3. This indicates that the mantle materials must have contained more aluminous spinel before undergoing melting and that the aluminous nature of the abyssal mantle source materials is justified.

The compositions of coexisting olivine and spinel within abyssal peridotites are not sufficiently known because olivine is often altered to serpentine in abyssal peridotites. In order to extend the above discussion, orthopyroxene was selected instead of olivine because it often survives even when coexisting olivine has been completely replaced by serpentine. In this case a similar argument to that put forward for the olivine–spinel pair is possible because the Fe–Mg partitioning between olivine and orthopyroxene is insensitive to temperature (*eg* Matsui

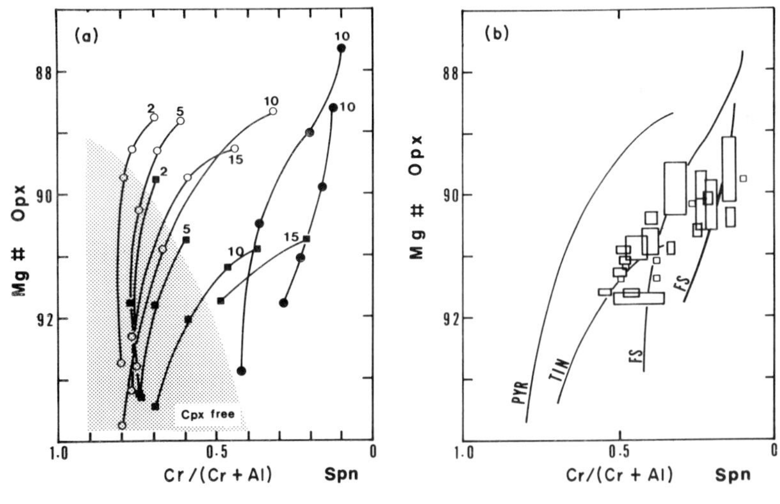

FIG. 9. Relationship between the Cr/(Cr+Al) ratio of spinel and the magnesium number of orthopyroxene. (**a**) Orthopyroxene–spinel pairs from melting experiments of peridotites: ■, Tinaquillo peridotites (Jaques & Green 1980): ○ pyrolite (Jaques & Green, 1980); •, basalt–peridotite sandwich experiments (Fujii & Scarfe 1985 and in preparation). The numbers beside the lines are pressures in kilobars and the shaded region represents the pairs which do not coexist with calcium-rich clinopyroxene. (*b*) Orthopyroxene–spinel pairs of abyssal peridotites: PYR, pyrolite; TIN, Tinaquillo peridotite; FS, basalt–peridotite sandwich experiments. The lines are pairs from melting experiments at 10 kbar (see Fig. 9*a*). Sources of data for abyssal peridotites are Arai & Fujii (1978), Dick *et al.* (1984), Michael & Bonatti (1985) and Shibata & Thompson (1986).

& Nishizawa 1974) and the Fe/Mg ratio of orthopyroxene can be buffered by that of olivine even in the subsolidus re-equilibration process. Figure 9 suggests that most abyssal peridotites are in the residual array of the melting of Al_2O_3-rich peridotite at 10 kbar and that only highly depleted abyssal peridotites (Shibata & Thompson 1986) show a similar relationship with the residual materials formed by the partial melting of Tinaquillo peridotite. Pyrolite residues are again far removed from the abyssal peridotite trend. This confirms the above conclusions derived from the relationship between the magnesium number of olivine and the Cr/(Cr+Al) ratio of spinel.

Other information on the nature of residual materials after the partial melting of peridotite can be obtained from Fig. 9. Because the residual spinel becomes chromium-enriched and olivine and orthopyroxene become more magnesian with increasing degree of melting, residual materials which lack clinopyroxene may contain more chromium-rich spinel and more magnesian olivine and orthopyroxene than is the case for clinopyroxene-bearing residuals. As is shown in Fig. 9*a*, the orthopyroxene–spinel pairs co-existing only with olivine are in the shaded region. The orthopyroxene–spinel pairs outside that region coexist with olivine and also with calcium-rich clinopyroxene. Comparison of this figure with Fig. 9*b* suggests that most abyssal peridotites are saturated with clinopyroxene when the partial melts for MORBs are produced. This, in turn, indicates that MORBs coexisted with olivine, orthopyroxene, clinopyroxene and spinel when they were formed, if it is assumed that the abyssal peridotites represent the residual materials after the MORB magmas were extracted.

Fujii & Scarfe (1985) suggested that the residual peridotite must be lherzolite and not harzburgite. Their terminology, however, was not strictly petrographical. Their term 'lherzolite' simply indicated the coexistence of clinopyroxene with olivine, orthopyroxene and spinel at the solidus temperature. Green *et al.* (1979) argued that the parental picrite magmas for MORBs coexist only with olivine and orthopyroxene and not with calcium-rich clinopyroxene. In addition, they argued that the abundance of abyssal harzburgite is strong supporting evidence for their conclusions. It is well known, however, that the solubility of calcium-rich clinopyroxene in orthopyroxene decreases with decreasing temperature (*eg* Lindsley 1983). The clinopyroxene-free assemblage at the peridotite solidus temperature may form a clinopyroxene-bearing assemblage when it is re-

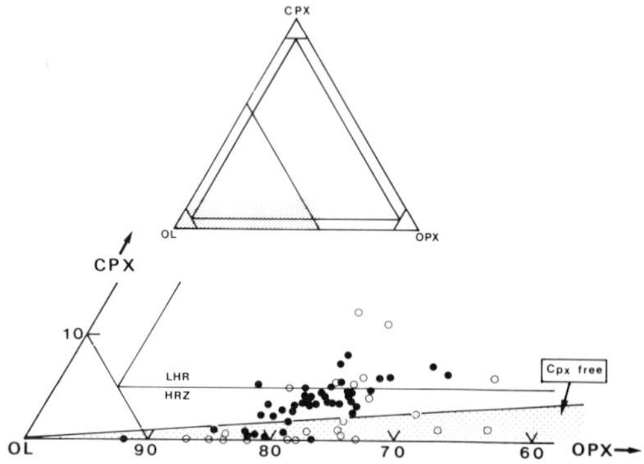

FIG. 10. Modal composition of abyssal peridotites: ○, Michael & Bonatti (1982); ●, Dick *et al.* (1984). The shaded region indicates the composition of peridotites which lack calcium-rich clinopyroxene at the solidus temperature.

equilibrated at lower temperature. Therefore the arguments concerning the residual phases must be based on the reconstructed assemblage at the solidus temperatures because most natural abyssal peridotites are equilibrated at subsolidus temperatures (*eg* Arai & Fujii 1978) and may not represent the assemblage at solidus temperature.

Modal ratios of abyssal peridotites (Dick *et al.* 1984; Michael & Bonatti, 1985) are shown in Fig. 10. The peridotite in the shaded region will not contain calcium-rich clinopyroxene at its solidus temperature. Many of the abyssal peridotites are within the region where coexistence of orthopyroxene and clinopyroxene is expected at the solidus temperature. This tendency is much clearer when we restrict the data to those carefully counted by Dick *et al.* (1984). Only a few abyssal peridotites are plotted within the clinopyroxene-free region; most peridotites are within the region where coexistence of orthopyroxene and clinopyroxene is expected at the solidus temperature. This indicates that primary magmas for MORBs coexist with clinopyroxene-bearing peridotite rather than with clinopyroxene-free peridotite. Thus the possibility of picrite primary magmas is rare if they coexist with clinopyroxene-free peridotites at the condition of melt segregation, as discussed by Green *et al.* (1979).

Concluding remarks

It is concluded on the basis of the analysis of the results of high-pressure experiments on the melting of primitive MORBs and peridotites that many primitive (high-magnesian) MORBs could be primary magmas produced at approximately 10 kbar. The comparison of mineral compositions which are not modified by subsolidus re-equilibration between experimentally determined residual phases and the abyssal peridotites indicates that a range of primary magmas could be formed by the different degrees of partial melting, leaving olivine, orthopyroxene, clinopyroxene and spinel as residual phases. These clinopyroxene-saturated residual rocks may recrystallize to lherzolite and clinopyroxene-bearing harzburgite at subsolidus temperatures.

The above discussions do not preclude picritic magma as a possible primary magma for MORBs; however, it can be shown on the basis of analyses of abyssal peridotites that the existence of primary magmas leaving clinopyroxene-free peridotite as a residue is rare. This indicates that picritic magmas (Green *et al.* 1979), which require a higher degree of melting and leave clinopyroxene-free residues at the site of melt segregation, are not primary magmas for MORBs. However, this is not yet conclusive because the petrographical and mineralogical data sets for abyssal peridotites are still inadequate. Further sampling and examination of abyssal peridotites is necessary.

ACKNOWLEDGMENTS: The author thanks T. L. Grove, K. Ozawa, C. M. Scarfe and R. N. Thompson for constructive comments and stimulating discussions on problem of MORB genesis. A. D. Saunders and D. Elthon are thanked for their critical reviews and helpful comments. D. H. Green is thanked for his review and for supplying a copy of an article in the press.

References

ARAI, S. & FUJII, T. 1978. Petrology of ultramafic rocks from Site 395. *Initial Reports of the Deep Sea Drilling Project* **45**, 587–594.

BENDER, J. F., HODGES, F. N. & BENCE, A. E. 1978. Petrogenesis of basalts from the project FAMOUS area: experimental study from 0 to 15 kbars. *Earth and Planetary Science Letters* **41**, 277–302.

BRYAN, W. B. 1979. Regional variation and petrogenesis of basalt glasses from the FAMOUS area, Mid-Atlantic Ridge. *Journal of Petrology* **20**, 293–325.

—— & MOORE, J. G. 1977. Compositional variation of young basalts in the Mid-Atlantic Ridge rift valley near 360°49′N. *Bulletin of the Geological Society of America* **88**, 556–570.

DICK, H. J. B., FISHER, R. L. & BRYAN, W. B. 1984. Mineralogic variability of the uppermost mantle along mid-ocean ridges. *Earth and Planetary Science Letters* **69**, 88–106.

ELTHON, D. 1983. Isomolar and isostructural pseudo-liquidus phase diagrams for oceanic basalts. *American Mineralogist* **68**, 506–511.

—— & SCARFE, C. M. 1980. High-pressure phase equilibria of a high-magnesia basalt: implication for the origin of mid-ocean ridge basalts. *Carnegie Institution of Washington Yearbook* **79**, 277–281.

—— & —— 1984. High-pressure phase equilibria of a high-magnesia basalt and the genesis of primary oceanic basalts. *American Mineralogist* **69**, 1–15.

FALLOON, T. J. & GREEN, D. H. 1987. Anhydrous partial melting of MORB pyrolite and other peridotite compositions at 10 kbar: implications for the origin of primitive MORB glasses. *Mineralogy and Petrology* **37**, 181–219.

FUJII, T. & BOUGAULT, H. 1983. Melting relations of a magnesian abyssal tholeiite and the origin of MORBs. *Earth and Planetary Science Letters* **62**, 283–295.

—— & SCARFE, C. M. 1985. Composition of liquids coexisting with spinel lherzolite at 10 kbar and the genesis of MORBs. *Contributions to Mineralogy and Petrology* **90**, 18–28.

—— KUSHIRO, I. & HAMURO, K. 1978. Melting relations and viscosity of an olivine tholeiite. *Initial Reports of the Deep Sea Drilling Project* **45**, 513–517.

GREEN, D. H., HIBBERSON, W. O. & JAQUES, A. L. 1979. Petrogenesis of mid-ocean ridge basalts. *In*: MCELHINNEY, M. W. (ed.) *The Earth: Its Origin, Structure, and Evolution*. Academic Press, London, 265–299.

ISHIWATARI, A. 1985. Igneous petrogenesis of Yakuno Ophiolite (Japan) in the context of the diversity of ophiolites. *Contributions to Mineralogy and Petrology* **89**, 155–167.

JAQUES, A. L. & GREEN, D. H. 1980. Anhydrous melting of peridotite at 0–15 kb pressure and the genesis of tholeiitic basalts. *Contributions to Mineralogy and Petrology* **73**, 278–310.

KUSHIRO, I. 1973. Origin of some magmas in oceanic and circum oceanic regions. *Tectonophysics* **17**, 211–222.

LINDSLEY, D. H. 1983. Pyroxene thermometry. *American Mineralogist* **68**, 477–493.

MATSUI, Y. & NISHIZAWA, O. 1974. Iron(II)–

magnesium exchange equilibrium between olivine and calcium-free pyroxene over a temperature range 800 °C to 1300 °C. *Bulletin of the Mineralogical and Crystallographical Society of France* **97**, 122–130.

MELSON, W. G., BYERLY, G. R., NELEN, J. A., O'HEARN, T., WRIGHT, T. L. & VALLIER, T. 1977. A catalog of major element chemistry of abyssal volcanic glasses. *Mineral Sciences Investigations, Smithsonian Contributions to the Earth Sciences* **19**, 31–60.

MICHAEL, P. J. & BONATTI, E. 1985. Peridotite composition from the North Atlantic: regional and tectonic variations and implications for partial melting. *Earth and Planetary Science Letters* **73**, 91–104.

O'DONNELL, T. H. & PRESNALL, D. C. 1980. Chemical variations of the glass and mineral phases in basalts dredged from 25°–30°N along the Mid-Atlantic Ridge. *American Journal of Science* **280**, 845–868.

O'HARA, M. J. 1965. Primary magmas and the origin of basalts. *Scottish Journal of Geology* **1**, 19–40.

OZAWA, K. 1986. Partitioning of elements between constituent minerals in peridotites from Miyamori Ultramafic Complex, Kitakami Mountains, Northeast Japan: estimation of *P–T* condition and igneous composition of minerals. *Journal of the Faculty of Science, University of Tokyo, Section II* **21**, 115–137.

PRESNALL, D. C., DIXON, J. R., O'DONNELL, T. H. & DIXON, S. A. 1979. Generation of mid-ocean ridge tholeiites. *Journal of Petrology* **20**, 3–35.

—— & HOOVER, J. D. 1984. Composition and depth of origin of primary mid-ocean ridge basalts. *Contributions of Mineralogy and Petrology* **87**, 170–178.

SHIBATA, T. & THOMPSON, G. 1986. Peridotites from the Mid-Atlantic Ridge at 43°N and their petrogenetic relation to abyssal tholeiites. *Contributions to Mineralogy and Petrology* **93**, 144–159.

STOLPER, E. 1980. A phase diagram for mid-ocean ridge basalts: preliminary results and implications for petrogenesis. *Contributions to Mineralogy and Petrology* **74**, 13–27.

TAKAHASHI, E. & KUSHIRO, I. 1983. Melting of a dry peridotite at high pressures and basalt magma genesis. *American Mineralogist* **68**, 859–879.

TORMEY, D. R., GROVE, T. L. & BRYAN, W. B. 1987. Experimental petrology of normal MORB near the Kane Fracture Zone: 22°–25°N, mid-Atlantic ridge. *Contributions to Mineralogy and Petrology* **96**, 121–139.

WALKER, D., SHIBATA, T. & DELONG, S. E. 1979. Abyssal tholeiites from the Oceanographer Fracture Zone. II. Phase equilibria and mixing. *Contributions to Mineralogy and Petrology* **70**, 111–125.

ZINDLER, A. & Hart, S. 1986. Chemical geodynamics. *Annual Review of Earth and Planetary Science* **14**, 493–571.

T. FUJII, Earthquake Research Institute, University of Tokyo, Yayoi, Tokyo 113, Japan.

Strontium, neodymium and lead isotope characteristics of the Granadilla Pumice, Tenerife: a study of the causes of strontium isotope disequilibrium in felsic pyroclastic deposits

Z. A. Palacz & J. A. Wolff

SUMMARY: The Granadilla Pumice deposit, the product of a single explosive eruption of phonolitic magma, exhibits stratigraphically systematic internal compositional variations interpreted as the result of fractionation of the observed phenocryst assemblage in the pre-eruptive magma body. Whole-rock $^{87}Sr/^{86}Sr$ data exhibit a consistent profile, with the most differentiated compositions having the most radiogenic strontium. The resultant $^{87}Sr/^{86}Sr$ versus 1/Sr mixing line has a positive slope, suggesting the addition of a relatively radiogenic strontium component to the existing strontium concentration profile. There is no significant systematic variation, however, in neodymium or lead isotopes. Separated alkali feldspar phenocrysts have $^{87}Sr/^{86}Sr$ values of 0.7031–0.7032, identical with Tenerife mafic magmas; the strontium isotopic variation is restricted to the complementary glasses. The more radiogenic component in the glasses is only partially removed by acid leaching and has a strontium isotope ratio of about 0.707, deduced from analysis of the leachant. Strong acid leaching, accompanied by up to 25% sample dissolution, results in greater, but not complete, loss of the more radiogenic component, with a good correlation between pre- and post-leaching glass isotopic compositions. These relationships cannot be easily explained by either post-eruptive alteration or contamination of the magma chamber with high-$^{87}Sr/^{86}Sr$ country rock. The data are instead consistent with a model of interaction between vesiculating magma and hydrothermal fluids with high $^{87}Sr/^{86}Sr$, immediately before and/or during explosive eruption, involving limited diffusion of strontium from the fluid into bubble walls before quenching of the magma to form glassy pumice. This process may be common during large explosive eruptions, and we suggest that whole-rock strontium isotope variations found in strontium-poor felsic pyroclastic deposits may not reflect purely magmatic processes, even though they closely mimic plausible magmatic variations.

Variations in radiogenic isotope ratios within the rocks of a single igneous province or centre are usually taken to indicate mantle source heterogeneity and/or the involvement of continental crust in their petrogenesis. In the latter case, the crustal component is usually regarded as the contaminant of a volume of mantle-derived magma; the actual physical mechanism may be simple melting of wall rock and mixing of the two liquids, combined assimilation–fractional crystallization or dissolution of the crustal rock in the magma. The essence of all these processes is the involvement of liquid magma, and it follows that isotopic variations should be detectable in single igneous bodies that have not internally rehomogenized prior to cooling. Such variations have been documented from layered intrusions (DePaolo 1985; Palacz & Tait 1985) but could represent temporal rather than spatial isotopic heterogeneity within a single magma chamber. In contrast, the pyroclastic deposits formed during large explosive eruptions represent voluminous samples of magma chambers quenched while in a largely liquid condition, and the internal compositional variations they exhibit (Hildreth 1979; Wolff & Storey 1984; Wörner & Schmincke 1984) therefore represent real heterogeneities, in the form of vertical compositional zonation, that existed within individual magma bodies at single points in time. Furthermore, the low strontium concentrations (attributable to an extensive history of feldspar fractionation) of the leucocratic magmas that typically erupt in this fashion should render their strontium isotope compositions particularly sensitive to contamination and/or magmatic recharge, while high Rb/Sr ratios (often more than 20) allow the possibility of *in situ* ageing and elevation of $^{87}Sr/^{86}Sr$ in the case of a long-lived felsic magma chamber (Christensen & DePaolo 1987). Recycling of high-Rb/Sr felsic rocks existing within the volcanic edifice is also conceivable. Such processes will not greatly affect the neodymium isotope composition of the magma because of the generally high neodymium contents in evolved magmas.

Figure 1 is a compilation of Sr–Nd isotope

From Saunders, A. D. & Norry, M. J. (eds), 1989, *Magmatism in the Ocean Basins*, Geological Society Special Publication No. 42, pp. 147–159

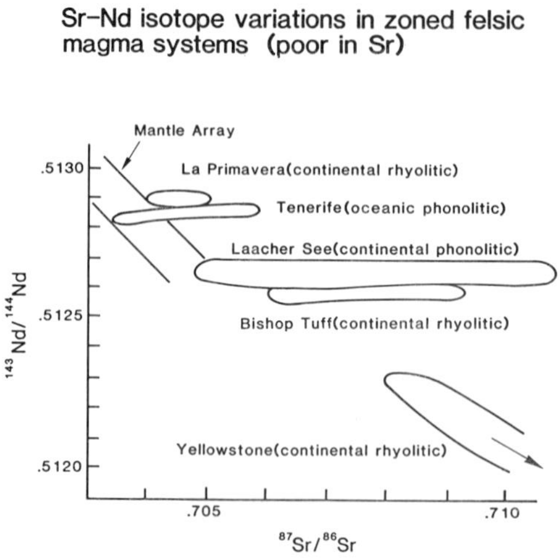

FIG. 1. Compilation of Sr–Nd isotope data for zoned felsic magma systems. The four systems showing horizontal arrays are all zoned to highly evolved strontium-depleted compositions. Fields for Bishop, Laacher See and Tenerife are for the products of individual eruptions from zoned magma chambers. For these three systems, the strontium isotope compositions of feldspar phenocrysts plot at the low-$^{87}Sr/^{86}Sr$ end of each array. Note the contrasting Yellowstone trend to high $^{87}Sr/^{86}Sr$ and low $^{143}Nd/^{144}Nd$, which must result from mixing of isotopically distinct components before fractionation to strongly strontium-depleted compositions. Data sources: Bishop tuff, Halliday *et al.* (1984); La Primavera and Yellowstone, Halliday *et al.* (1985); Laacher See, Wörner *et al.* (1985); Tenerife, this study.

data from well-studied zoned felsic magma systems, most of which exhibit substantial variations in $^{87}Sr/^{86}Sr$ while $^{143}Nd/^{144}Nd$ effectively remains constant. This is consistent with the low strontium contents of rhyolitic and phonolitic liquids, and hence the relative ease with which crustal strontium can perturb their strontium isotope ratios. Both the Bishop tuff (Halliday *et al.* 1984) and Laacher See tephra (Wörner *et al.* 1985, 1986) exhibit phenocryst–matrix isotopic disequilibrium, in that the $^{87}Sr/^{86}Sr$ values in separated feldspars show little variation and always plot close to the least radiogenic end of the range; the introduction of relatively radiogenic strontium thus post-dates phenocryst formation. Isotopic variability in glasses has been attributed to post-eruptive alteration (Halliday *et al.* 1984) or exchange between magma and hydrothermal fluids shortly before eruption (Wörner *et al.* 1986).

Prior to eruption, the Bishop and Laacher See magma bodies were located within continental crust. However, similar strontium and neodymium isotope behaviour is seen in felsic pyroclastic rocks from Tenerife (Fig. 1), an ocean island where continental crust, or fluids which may have equilibrated with the crust, does not exist. We have therefore carried out a detailed study of the compositionally zoned Granadilla Pumice of Tenerife in order to understand more fully the nature and cause of isotopic variations in pyroclastic formations.

Background

The Canary Islands lie in the eastern Atlantic Ocean off the continental slope of NW Africa. Tenerife, the largest island, consists of a late Tertiary alkali basalt shield capped by younger differentiated volcanics and later basalts. Within this young complex is developed the Las Cañadas caldera, a multiple structure of two or three overlapping collapse features modified by late northward landslip to give a horseshoe-shaped depression (Borley 1974; Wolff 1983; B. Booth & G. P. L. Walker, unpublished data); the twin stratovolcanoes of Teide and Viejo stand within the caldera. Lavas of the young complex are invariably silica-undersaturated and show the simple evolutionary trend alkali basalt/basanite–trachybasalt–trachyandesite–phonolite (Ridley 1970; Borley 1974).

An extensive late Quaternary pyroclastic series of dominantly phonolitic composition is prominent in the S and E of the island. Booth & Walker (unpublished data) distinguish the products of some 40 individual eruptions from vents located within Las Cañadas. The deposits are mostly non-welded plinian pumice falls and ignimbrites, and at least 12 have bulk volumes greater than 0.5 km^3. The total volume of the series is very roughly 70 km^3 of equivalent dense rock.

In thin section, phonolitic pumices consist of a few per cent phenocrysts set in highly vesicular glass. Alkali feldspar (sodic sanidine or anorthoclase) is always the dominant phenocryst, with variable lesser amounts of clinopyroxene salite or aegirine-augite), magnetite and sphene, commonly accompanied by biotite, ilmenite and sodalite, and occasionally by nepheline and/or an alkali amphibole. Apatite and pyrrhotite are commonly included in other phases.

Where magnetite and ilmenite occur in equilibrium, they yield temperatures between 800 and 900 °C and oxygen fugacities close to the Ni–NiO buffer (Wolff & Storey 1983). At least 15 deposits are petrographically heterogeneous in that they bear both mafic and felsic pumice with respective contrasting phenocryst populations and bulk compositions. In many cases phonolitic and mafic magmas have been quenched by eruption in the act of physically mixing, to give petrographically-complex banded pumices. These relations demonstrate that phonolitic magma was generated as a capping layer to a magma system which was zoned downwards to mafic–intermediate compositions and subject to recharge by more primitive basaltic magma (Wolff 1983, 1985*a*; Wolff & Storey 1984).

Nepheline-syenite blocks occur as lithics in some deposits and provide solidified representatives of phonolitic magma. The mineralogy of these blocks indicates that they crystallized from the most strongly differentiated liquids represented among pumice compositions and probably formed within the roof zone of the system (Wolff 1987). Some 80% of syenites are hydrothermally altered, indicating the presence of a large hydrothermal system that was active prior to or during formation of the pumice series. Present-day warm springs attest to the continued existence of this system.

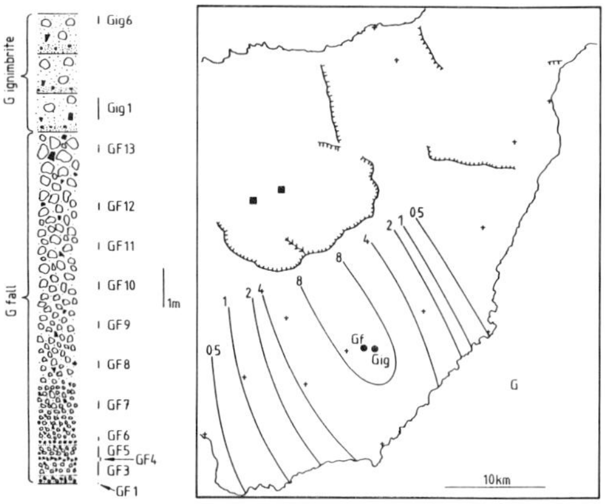

FIG. 2. Isopach map of the Granadilla Pumice fall deposit, contoured in metres, together with a section through the deposit. The hatched lines indicate prominent scarps. The partly enclosed central area is the Las Cañadas caldera containing the twin post-caldera stratovolcanoes Teide and Viejo (■). The crosses indicate the principal towns. Sampling localities for the fall deposit (Gf) and ignimbrite (Gig) are indicated by solid circles. The stratigraphic section is a composite of the two localities: open symbols, juvenile pumice clasts qualitatively showing clast size variation: solid symbols, lithics: dots, ash-size material. The numbers refer to the stratigraphic position of whole-rock samples.

The Granadilla Pumice deposit

The field characteristics of the Granadilla Pumice have been described by Booth (1973) and Wolff (1983); a summary is given here. The Granadilla deposit is distributed over a wide area of southern Tenerife around the town of Granadilla (Fig. 2) and consists of a plinian fall deposit overlain by a non-welded ignimbrite. The transition from plinian to ignimbrite-forming activity during the course of a single explosive eruption is a widely recognized phenomenon, and can be attributed to gravitational collapse of the plinian eruption column (Sparks *et al.* 1973; Sparks & Wilson 1976) in response to changing conditions at the vent. Booth (1973) divides the plinian deposit into four fall units, the lowest of which (fall unit 1) rests on a palaeosol at the sampling locality (Fig. 2). Partially disaggregated crystalline mafic trachyandesite inclusions are common in pumice clasts from this fall unit and decrease in abundance upwards into fall unit 2. These are interpreted as fragments of wall rock incorporated in the first-erupted magma as it broke through to the surface; they are virtually absent from fall units 3 and 4. The lowermost 2 m of the deposit (fall units 1, 2 and most of 3) are well stratified and contain two prominent fine ash bands (Fig. 2), attesting to a rather discontinuous initial eruptive phase. Fall unit 4, which makes up the bulk of the fall deposit, is very coarse grained (pumices up to 30 cm in diameter, 20 km from source), reversely graded and poorly stratified, and is the product of a powerful sustained plinian eruption. The capping ignimbrite is best preserved in palaeotopographic lows; at the sampling locality it consists of three main flow units with intercalated small laterally non-persistent flow units. The ignimbrite everywhere rests conformably on the top of fall unit 4; it consists of poorly to moderately rounded pumice clasts dispersed in a matrix of fine ash and loose crystals.

The age of the Granadilla deposit is uncertain; the only reliable age determination for the entire pumice series is from a unit younger than the Granadilla which contains carbonized wood with a ^{14}C age of $28\,500^{+1580}_{-1320}$ years BP (Shotton *et al.* 1974; B. Booth & G. P. L. Walker, unpublished data). On the basis of palaeosol thicknesses, the Granadilla Pumice may be some 10 000–30 000 years older than this. Carbonized wood from the Granadilla ignimbrite itself has yielded a ^{14}C age of more than 31 000 years BP (Shotton & Williams 1971).

Duration of the Granadilla eruption

It is of some interest to attempt to determine the duration of the plinian phase of the Granadilla eruption. This can be approached indirectly as follows. Carey & Sparks (1986) have recently presented a model for the fallout of clasts from a plinian eruption column which permits quantitative estimation of column height from analysis of the dispersal pattern of tephra in the resultant deposit. The eruption column height is a function of the mass discharge rate at the vent (Wilson *et al.* 1978; Sparks 1986). If the total erupted volume is known, it is thus possible to arrive at a figure for eruption duration. The procedure is fraught with uncertainties, which arise largely from ways in which field data on tephra dispersal are collected and presented (Carey & Sparks 1986) and from total erupted volume estimates based on field data, but tends to yield minimum estimates for eruption duration. Application of the Carey & Sparks model to the lithic and pumice size distribution data of Booth (1973) gives a range of values between 29 and 33 km for the eruption column height, with a mean value of 31.7 km. These values are approximate for the above reasons, and also because the clast size data of Booth (1973) are somewhat sparse for this purpose. From the relationship between column height and mass discharge rate given by Sparks (1986), this height range corresponds to discharge rates between 1×10^8 and 1.5×10^8 kg s^{-1} for a magmatic temperature of 800 °C. Walker (1981) has calculated the total erupted volume of the plinian deposit, including the unseen distal portion, as approximately 45 km^3 of vesiculated material. Using a mean value of 650 kg m^{-3} for tephra density gives a calculated eruption duration of 2–3.5 days. As noted above, this is unlikely to be an accurate estimate, particularly as the lower part of the deposit records fluctuations and breaks in the eruption, but may be a reasonable minimum value. It is not possible to estimate the additional length of time during which the ignimbrite was erupted.

Geochemistry of the Granadilla Pumice

Various aspects of Granadilla Pumice geochemistry have been mentioned by Wolff & Storey (1984) and Wolff (1984, 1985*b*). Major- and trace-element whole-rock data, including all whole-rock strontium, neodymium and lead isotope determinations made for this study, are given for eight representative samples in Table 1. Each whole-rock sample represents several of the largest individual juvenile pumice clasts, within a narrow size range, taken from a small vertical height interval within the deposit (Fig. 2); this procedure tends to minimize spurious compositional variations induced by pyroclast formation and transport (Wolff 1985*b*). Samples from fall units 1 and 2 perforce consist of pumice clasts around 1–2 cm in diameter, and these may have experienced some crystal loss during clast formation (Wolff 1985*b*).

The behaviour of the major and trace elements is consistent with fractionation of the observed phenocryst assemblage (alkali feldspar, sodalite, salite, alkali amphibole, biotite, sphene and Fe–Ti oxides). Noteworthy features are the compatibility of middle rare earths and fractionation of niobium from tantalum, both attributable to sphene (Wolff 1984; Wolff & Storey 1984). Strontium decreases sharply with fractionation, whereas neodymium behaves as a moderately compatible element. With the exception of sodium and potassium, which are very susceptible to low-temperature ion-exchange processes, elements potentially mobile during post-eruptive hydration covary systematically with immobile elements such as titanium and zirconium. From about 30% thickness to the top of the plinian, the deposit is normally zoned in that abundances of compatible elements (titanium, calcium, phosphorus, strontium, barium, vanadium and middle rare earths) increase upwards while incompatible elements (rubidium, zirconium, hafnium, thorium, uranium, niobium, lanthanum and ytterbium) progressively decrease. The opposite trend is found below 30% thickness (Fig. 3). This reverse zoning is due to a combination of contamination with more mafic country rock, as noted above, and the sampling of genuinely less evolved magma during the opening stages of the eruption (Wolff & Storey 1984). The same less evolved chemical signature is recorded in an anomalously 'mafic' composition from 6 m height (Fig. 3). It should be noted that the whole-rock compositional variation is not simply a function of phenocryst content; separated glasses display an identical trend shifted to lower abundances of compatible elements (Wolff 1985*b*). Furthermore, samples from full units 1 and 2, which from grain-size considerations should, if anything, be depleted in phenocrysts, and have less differentiated compositions which define the basal reverse-zoning trend.

TABLE 1. *Whole-rock major- and trace-element abundances and strontium, neodymium and lead isotope compositions of Granadilla pumice samples.*

	Gig6 (12.5m)[a]	Gig1 (8.5m)[a]	Gf12 (7m)[a]	Gf11 (6m)[a]	Gf10 (4.9m)[a]	Gf8 (2.9m)[a]	Gf6 (1.15m)[a]	Gf5 (0.8m)[a]
Major elements (wt%)[b]								
SiO_2	61.94	61.70	61.57	61.62	61.18	61.40	61.04	60.69
TiO_2	0.66	0.63	0.55	0.60	0.52	0.47	0.47	0.50
Al_2O_3	19.37	19.33	19.84	19.35	19.91	20.03	19.94	19.78
Fe_2O_3	3.31	3.29	3.45	3.34	3.49	3.49	3.38	3.44
MnO	0.19	0.24	0.24	0.23	0.25	0.27	0.27	0.25
MgO	0.46	0.53	0.32	0.39	0.27	0.28	0.42	0.42
CaO	0.77	0.73	0.64	0.78	0.62	0.56	0.59	0.65
Na_2O	6.78	6.67	7.08	7.56	7.73	7.50	7.72	8.26
K_2O	6.43	6.81	6.24	6.06	5.97	5.94	6.11	5.95
P_2O_5	0.08	0.08	0.06	0.06	0.05	0.05	0.05	0.05
Trace elements (ppm)								
Rb	163	141	166	155	184	203	188	175
Sr	16	16	11	30	12	11	14	21
Ba	45	60	25	215	30	—	35	102
Cu	29	23	27	23	34	9	8	7
Ni	5	4	2	3	3	5	4	3
Zn	134	118	142	132	148	148	156	155
V	34	42	37	38	34	31	30	32
Y	50	50	53	54	57	59	58	57
La	136	132	144	131	150	153	nd	nd
Ce	226	245	237	213	235	233	219	220
Nd	61	73	62	59	58	53	58	56
Sm	8.3	10.8	8.4	8.6	8.2	7.4	7.5	7.8
Eu	1.8	2.2	1.7	1.8	1.5	1.5	1.5	1.6
Tb	1.4	1.5	1.4	1.1	1.4	1.8	1.2	1.1
Yb	4.3	3.8	4.5	4	4.7	5.1	5.3	4.9
Zr	960	790	1120	1000	1210	1220	1280	1230
Hf	nd	nd	23	nd	23	27	27	27
Th	28.6	23.8	29.9	24.8	35.2	36	33.4	33.5
U	8.3	6.9	7.9	5.7	8.5	9.6	8.4	8.5
Nb	306	293	369	334	384	368	389	381
Ta	19	22	21	nd	21	20	21	21
Isotope ratios								
$^{87}Sr/^{86}Sr$	0.70413	0.70458	0.70506	0.70356	0.70513	0.70571	0.70567	0.70483
$^{143}Nd/^{144}Nd$			0.512863	0.512825	0.512876	0.512871	0.512892	0.512862
$^{206}Pb/^{204}Pb$	19.694	19.621	19.704		19.706	19.734	19.704	19.704
$^{207}Pb/^{204}Pb$	15.596	15.604	15.592		15.594	15.608	15.600	15.607
$^{208}Pb/^{204}Pb$	39.54	39.48	39.54		39.547	39.547	39.547	39.565

[a]Sample height above base at sampling locality.
[b]Oxide weight percentages normalized to a total of 100% volatile-free.

Isotope results

Whole rocks

The isotopic ratios of strontium, neodymium and lead were measured on whole-rock pumice samples using a VG 54E mass spectrometer. Strontium ratios are normalized to $^{86}Sr/^{88}Sr$= 0.1194 and neodymium ratios to $^{146}Nd/^{144}Nd$= 0.7219; NBS 981 gave $^{206}Pb/^{204}Pb$=16.907, $^{207}Pb/^{204}Pb$=15.451 and $^{208}Pb/^{204}Pb$=36.566. Blanks gave 2 ng strontium, 1 ng neodymium and 0.3 ng lead, and can be considered negligible for this study.

The whole-rock $^{87}Sr/^{86}Sr$ ratio varies between 0.70356 and 0.70571, significantly higher than for basalts on Tenerife and other Canary Islands which are remarkably uniform at 0.7031–0.7033. However, the $^{143}Nd/^{144}Nd$ ratio is effectively constant around 0.51286, identical with values obtained from basalts. With the exception of one sample collected from the ignimbrite, the lead isotope ratios are also constant within the Granadilla Pumice, although Tenerife

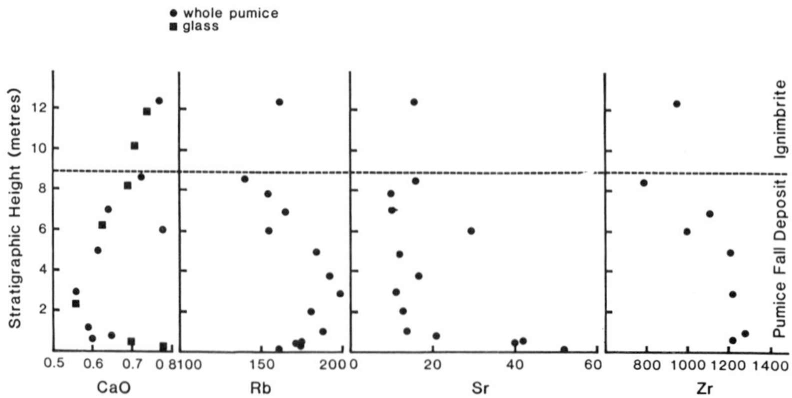

FIG. 3. Variation in CaO, zirconium, rubidium and strontium through the Granadilla fall deposit and overlying ignimbrite. The whole-pumice samples correspond to the sample numbers given in Fig. 2. The glass samples correspond to the sample numbers and vertical height above base given in Table 2.

basalts have variable lead isotope compositions (Sun 1980). Vertical profiles for the $^{87}Sr/^{86}Sr$, $^{143}Nd/^{144}Nd$ and $^{206}Pb/^{204}Pb$ ratios for the Granadilla Pumice, together with three samples from two similar plinian deposits (D and E) that lie directly under the Granadilla deposit, are shown in Fig. 4. Deposits D and E are also more radiogenic in strontium than the Tenerife basalts, and have different lead isotope compositions from the Granadilla Pumice. All samples have lead isotope ratios that lie within the range described for Tenerife basalts (Sun 1980), and the distinct lead signatures of the three deposits

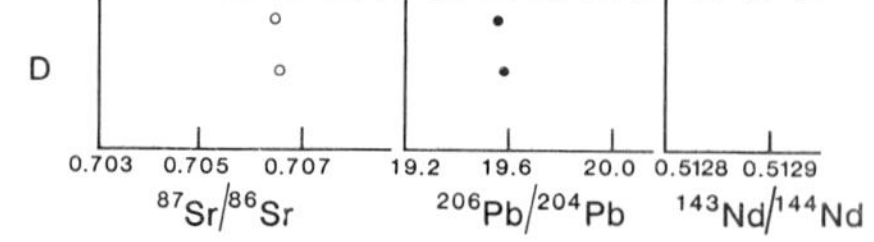

FIG. 4. Strontium, lead and neodymium isotope compositions versus stratigraphic height for the Granadilla Pumice (G) and two underlying fall deposits (E and D) believed to be the products of earlier eruptions from the same magma system. The maximum 2σ errors are 0.00004 for $^{87}Sr/^{86}Sr$, 0.00002 for $^{143}Nd/^{144}Nd$ and 0.01 for $^{206}Pb/^{204}Pb$.

can therefore be plausibly ascribed to different mixtures of mantle lead components in three separate erupted batches of fractionated phonolitic magma (Wolff & Palacz 1989). This is not the case for the strontium ratios, which are not only decoupled from the lead and neodymium ratios, but are invariably more radiogenic than the local mantle.

Whole-rock $^{87}Sr/^{86}Sr$ ratios are clearly related to major- and trace-element composition, with the most differentiated samples having the most radiogenic strontium (Fig. 5). This is expressed on a plot of a $^{87}Sr/^{86}Sr$ versus 1/Sr (Fig. 6), where the positive correlation implies the addition of a low-strontium high-$^{87}Sr/^{86}Sr$ component to a higher-strontium low-$^{87}Sr/^{86}Sr$ magma. Constant neodymium and lead isotope ratios can be explained as the result of the buffering by high neodymium and lead contents of the magma against the effects of any addition of a component of a different isotopic composition.

Glass and feldspar strontium isotope compositions

The whole-rock isotope composition for any rock is determined by the isotope composition of its constituent minerals, and in volcanic rocks it will be affected by possible isotope disequilibrium between the phenocrysts and the matrix. Strontium isotope disequilibrium has been recognized in the Bishop tuff (Halliday *et al.* 1984) and the Laacher See tephra (Wörner *et al.* 1985), both of which are located in areas of continental volcanism. In order to test for isotopic disequilibrium, phenocrysts and glasses

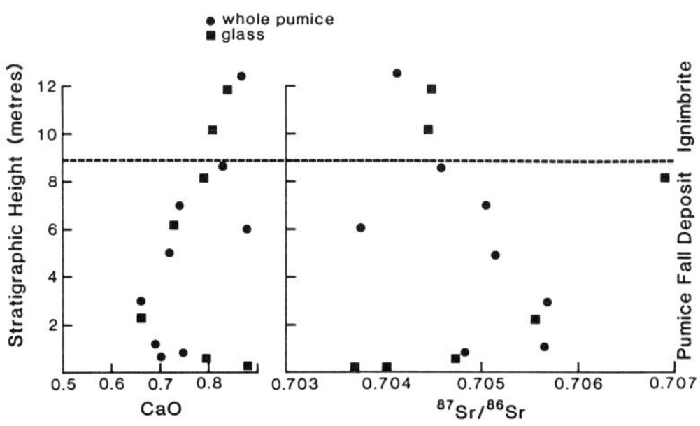

FIG. 5. Comparison of variations in strontium isotope composition and CaO content (as an indicator of fractionation) of whole rock and glass with the stratigraphic height through the Granadilla fall deposit and ignimbrite.

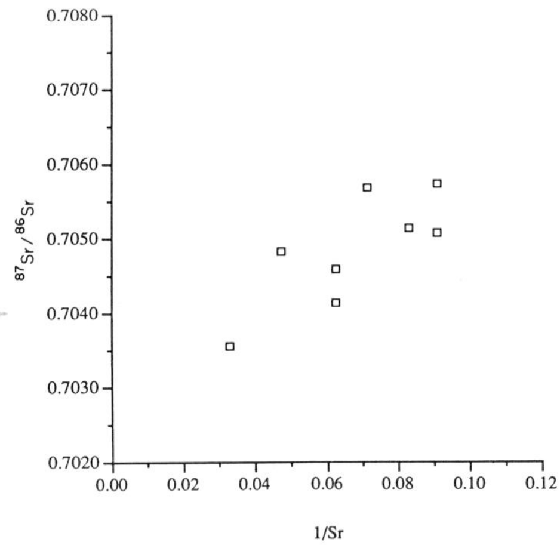

FIG. 6. 1/Sr versus $^{87}Sr/^{86}Sr$ in whole-rock samples of the Granadilla Pumice.

were separated from pumices and analysed. Feldspar strontium contents are expected to be much higher than those of the glass, and thus could affect pumice isotope compositions if feldspar and matrix glass are not in isotope equilibrium.

The results are presented in Table 2. Glass and whole-rock data are plotted on the same diagram in Fig. 5 and are broadly similar throughout the unit. The radiogenic component appears to be restricted to the glass which, owing to its low crystal content (less than 5%), volumetrically dominates the whole rock. In contrast, the feldspar separates have effectively constant $^{87}Sr/^{86}Sr$ ratios at 0.7031–0.7032, identical with values obtained for Tenerife basaltic rocks. The Granadilla Pumice therefore exhibits significant phenocryst–matrix strontium isotope disequilibrium (Fig. 7 and Table 2).

Leaching experiments

Previous workers have suggested three mechanisms for isotopic disequilibrium in glassy pyroclastics. First, high $^{87}Sr/^{86}Sr$ ratios in the low-strontium glasses, but not in the high-strontium feldspars, could be a response to post-eruptive alteration by high-$^{87}Sr/^{86}Sr$ groundwater or other fluids (*cf* Halliday *et al.* 1984). This would be consistent with the neodymium and lead isotope data because such fluids would have negligible neodymium and lead contents. Second, contamination of the magma chamber with relatively radiogenic strontium *after* feldspar crystallization could be responsible (Wörner *et al.* 1985); this could be achieved by interaction of magma with hydrothermal fluids as the magma chamber carapace fractures prior to eruption (Wörner *et al.* 1986). A possible third mechanism is *in situ* aging of magma with very high Rb/Sr ratios, *after* crystallization of feldspar with lower Rb/Sr ratios (Christensen & DePaolo 1987). The last mechanism can be ruled out for the Granadilla Pumice because its Rb/Sr values (≤ 16) are too low to cause significant elevation of $^{87}Sr/^{86}Sr$ during the lifetime of phonolitic magma batches on Tenerife (estimated at 10^3–10^4 years by Wolff & Storey (1984)). To test for alteration, leaching experiments were carried out on separated glasses on the assumption that a surficial strontium component introduced during alteration should be fully leachable. Whole-rock powders were not leached because of the possible complication of differential dissolution of feldspar.

Glasses were initially leached in cold 2.5 M HCl for 10 min and rinsed thoroughly in pure water; the leachants and residual glasses were then analysed. No effervescence was observed during acid leaching. Strontium isotope data for this series of determinations were obtained using a Finnegan MAT 261 mass spectrometer in the static mode; the results are given in Table 2 and Fig. 7. In all cases, the $^{87}Sr/^{86}Sr$ ratios of the leached glasses are lower than the pre-leach values. However, the leached glasses are still significantly richer in radiogenic strontium than the feldspar separates from the same samples. The leachable strontium has a high but near-constant isotope ratio around 0.707. In order to try and completely remove the high-$^{87}Sr/^{86}Sr$

TABLE 2. *Major-element compositions and strontium isotope compositions of leached and non-leached glasses with isotopic compositions of separated feldspars*

	7814 (11.8m)[a]	7812 (10.3m)[a]	789 (8.3m)[a]	786 (2.2m)[a]	784 (0.65m)[a]	782 (0.3m)[a]	781 (0.15m)[a]
Major elements (wt%)[b]							
SiO_2	62.34	61.69	62.67	60.9	60.91	61.09	61.53
TiO_2	0.66	0.64	0.59	0.47	0.48	0.53	0.59
Al_2O_3	19.53	18.98	19.29	20.08	20.08	19.92	19.8
Fe_2O_3	3.17	3.17	3.31	3.54	3.5	3.45	3.42
MnO	0.19	0.23	0.21	0.26	0.24	0.23	0.23
MgO	0.41	0.42	0.68	0.31	0.36	0.4	0.39
CaO	0.74	0.71	0.69	0.56	0.6	0.69	0.78
Na_O	6.33	7.55	5.82	7.78	7.94	7.96	7.52
K_2O	6.54	6.54	6.67	6.04	5.81	5.67	5.65
P_2O_5	0.08	0.07	0.07	0.05	0.07	0.06	0.07
$^{87}Sr/^{86}Sr$ (non-leached)	0.70453	0.70447	0.706881	0.70555	0.704758	0.70401	0.70371
$^{87}Sr/^{86}Sr$ (2.5 M HCl)			0.705783	0.70482		0.703572	
$^{87}Sr/^{86}Sr$ (6 M HCl)		0.703732	0.705198	0.703526		0.703336	0.703236
Leachant	0.707232		0.70719				0.70691
Feldspar	0.703184	0.703151	0.703175	0.703135			0.703146

[a]Sample height above base at sampling locality.
[b]Oxide weight percentages normalized to a total of 100% volatile-free.

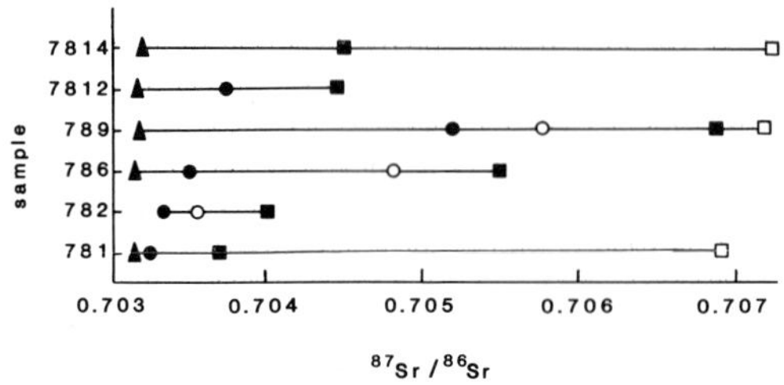

FIG. 7. Strontium isotopic composition of feldspars, glasses and leached glasses from the Granadilla Pumice: ▲, feldspars; ● severely leached (6 M HCl) glasses; ○, mildly leached (2.5 M HCl) glasses; ■, non-leached glasses; □, leachable strontium. Sample numbers on the vertical axis are in stratigraphic order.

component, leached and non-leached glasses were subjected to a severe leach using cold 6 M HCl for 3 h. Phonolitic glass is soluble in strong acid, and up to 20% of the sample dissolved during the severe leach. Isotope ratios of severely leached glasses are lower than those obtained after the 2.5 M HCl leach, but invariably remain in excess of the ratios in co-existing feldspars (Fig. 7 and Table 2). No correlation was found between the degree of sample dissolution and the magnitude of isotope composition shift. The high $^{87}Sr/^{86}Sr$ component in the glasses is thus remarkably tenacious inasmuch as it is only partly removed during severe acid leaching leading to partial sample dissolution.

Discussion

A progressive increase in $^{87}Sr/^{86}Sr$ ratio with fractionation in a comagmatic suite is usually interpreted in terms of combined assimilation and fractional crystallization (AFC; DePaolo 1981). Although at first glance no suitable contaminant would be expected in the ocean island environment, it could be argued that material with an elevated $^{87}Sr/^{86}Sr$ ratio exists in the Tenerife volcanic edifice. Firstly, marine sediments may be intercalated within the volcanic pile. Secondly, the hydrothermal system is likely to be dominated by seawater (especially as the upper part of the subaqueous pile will consist largely of permeable phreatomagmatic tuffs) with $^{87}Sr/^{86}Sr \approx 0.709$. Thus hydrothermally altered rocks, particularly low-strontium syenites, could have strontium isotope ratios between 0.703 and 0.709. Thirdly, Tenerife is a long-lived volcanic centre which has been producing felsic magmas for several million years; for example, an early trachyte has been dated at 5.35 Ma (Sun 1980). The high Rb/Sr ratio in the felsic rocks analysed (2–40; Wolff 1983) could easily raise the $^{87}Sr/^{86}Sr$ ratio to the required values over this time period. Thus, AFC is a potentially viable mechanism, albeit difficult to reconcile with phenocryst-matrix isotopic disequilibrium, together with post-eruptive alteration and interaction of the magma with fluids immediately prior to eruption. We shall now evaluate each of these mechanisms in turn, in the light of the feldspar, glass and leaching data.

Post-eruptive alteration

We reject post-eruptive alteration as a means of introducing *all* the radiogenic strontium component on the following grounds.

(1) Sampling localities are well away from the caldera region (Fig. 2) and the hot arid coastal zone; alteration is commonly moderate to intense in both these areas. The pumice is loose, non-welded and non-cemented and retains its primary physical characteristics such as grain size and density (Booth 1973). Although the matrix material is variably hydrated, in thin section it consists entirely of isotropic glass with no evidence of devitrification or clay mineral growth.
(2) Rubidium, strontium and barium are sometimes mobile during hydration of volcanic glasses (*eg* Zielinski *et al.* 1977). In the Granadilla Pumice variations in the concentrations of these elements mirror those of immobile elements such as zirconium, and they are therefore themselves immobile in our sample suite.
(3) The Na_2O/K_2O ratio is a very sensitive indicator of alteration in volcanic glasses since these two elements readily undergo ion exchange (Scott 1971). Although these elements appear to have been slightly affected by hydration, inasmuch as they only partly follow immobile-element profiles through the deposit (Wolff 1983), there is no correlation with $^{87}Sr/^{86}Sr$.
(4) Most significantly, acid leaching does not remove all the radiogenic strontium component, as would be expected for a surficial component introduced at low temperature. This is true even for the severely leached glasses that experienced substantial dissolution in acid.

Contamination of magma by country rock

The following points are inconsistent with the introduction of radiogenic strontium by a contamination or AFC process in the magma chamber:

(1) If the feldspar strontium isotope composition, which is identical with that of parental basalts, reflects that of the magma prior to introduction of the more radiogenic component, contamination must have been a single-stage event (not continuous AFC) post-dating feldspar crystallization. While this is possible, kinetic considerations limit the time-scale of the process to less than 1000 years (*cf* Wörner *et al.* 1985).
(2) Samples from fall units 1 and 2 were indeed visibly contaminated shortly before eruption, as noted previously. However, these samples have the least radiogenic compositions.
(3) The partly leachable nature of the more radiogenic component with a weak acid leach, further (but incomplete) removal during the strong leach accompanied by partial sample dissolution and the correlation between pre- and post-leaching $^{87}Sr/^{86}Sr$ ratios can only be explained if the analysed sample powder is isotopically heterogeneous *both between and within the individual glassy particles that make up the powder*. We cannot envisage how this could result from any process involving country-rock assimilation in the magma chamber.

It is conceivable that a two-stage contamination process, where the effects of surficial alteration are superimposed on a variably contaminated magma composition, could give rise to the observed strontium isotope variations. In this case the contaminant is unlikely to be rock, since possible high-$^{87}Sr/^{86}Sr$ lithologies within the Tenerife edifice will have lead concentrations similar to those of the magma. Unless the contaminant has a lead isotopic composition almost identical with that of the magma (an unlikely event given the range in Tenerife lead isotopes reported by Sun (1980), Granadilla lead should show some degree of sympathetic variation with geochemistry and strontium isotopes. This constraint does not apply to contamination of magma by fluids as envisaged by Wörner *et al.* (1986). As a modification of Wörner *et al.*'s mechanism, we shall now consider single-stage magma–fluid interaction, without post-eruptive alteration, as a possible means of generating the strontium isotope variations in Granadilla feldspar, glasses and leached glasses.

Syneruptive interaction between vesiculating magma and hydrothermal fluid

The radiogenic isotope characteristics of the Granadilla Pumice are similar to those of the phonolitic Laacher See tephra (Wörner *et al.* 1985, 1986) in that relatively radiogenic strontium has been introduced only into the groundmass glass. The effect is most noticeable in strongly fractionated, strontium-poor phonolite in the cupola of the Laacher See magma chamber. Wörner *et al.* (1985, 1986) attribute this to interaction of magma with hydrothermal fluids which had previously equilibrated with upper continental crust.

Although continental crust is absent from Tenerife, seawater provides a suitable source of radiogenic strontium ($^{87}Sr/^{86}Sr \approx 0.709$). The phonolitic magma chamber that supplied the Granadilla and associated pumice fall deposits is believed to have possessed a carapace of partly to fully crystallized nepheline-syenite (Wolff 1987). Some 80% of these blocks have suffered intense hydrothermal alteration, indicating that the hydrothermal system of the volcano was partly located in older felsic rocks away from the immediate thermal aureole of the active magma chamber, within which fresh and incompletely crystallized syenite existed. Nepheline-syenites crystallized from the most strongly differentiated phonolitic magmas (Wolff 1987) and have similar compositions. Equilibration of seawater with strongly fractionated low-strontium nepheline-syenite could thus easily produce a fluid strontium isotope composition of *c.* 0.707, the minimum ratio of the contaminant, as indicated by the leachable component.

FIG. 8. Scanning electron micrograph of powder separated from a Granadilla Pumice sample after grinding in a Tema mill. Note the original bubble-wall surfaces bounding the large fragment.

Results of leaching experiments require that the $^{87}Sr/^{86}Sr$ ratios of margins of individual particles in the analysed powders are higher than those of the cores. Scanning electron microscope examination of powders reveals that, even after crushing, particles are commonly partly bounded by bubble-wall surfaces which have survived from the internal morphology of the original pumice (Fig. 8). This is because the range of particle sizes (about 2–30 μm) in the powder is on the same scale as the vesicle sizes in pumice (Witham & Sparks 1986). Thus the leaching data are consistent with a preferential concentration of ^{87}Sr close to the bubble walls in pumice. We have already argued that this can only be due to post-eruptive alteration if some other process supplied a relatively radiogenic *non*-leachable component prior to alteration. Here we wish to draw attention to a possible alternative mechanism—the interaction of hydrothermal fluid with *vesicular* magma resulting in limited diffusional strontium isotope exchange between the fluid and the pumice bubble walls.

Eichelberger *et al.* (1986) and Witham & Sparks (1986) have emphasized that the vesicularity of erupting felsic magmas reaches the point where the vesicles become interconnected and gas flow through the magma framework becomes possible. Whether or not the magma degasses in the conduit to erupt ultimately as lava or fragments explosively depends on such factors as the ease with which magmatic gas can escape from the system and the distribution of interconnected versus non-connected vesicles (Eichelberger *et al.* 1986; Witham & Sparks 1986). If hydrothermal fluids can gain access to the vesiculating magma, they may flush through the magma froth. If fluids and bubble walls remain in contact for a sufficient length of time at elevated temperatures, diffusional exchange between low-$^{87}Sr/^{86}Sr$ glass and high-$^{87}Sr/^{86}Sr$ fluid will occur. The data of Magaritz & Hofmann (1978) for strontium diffusion in obsidian yield characteristic diffusional lengths of the order of microns for times of 10–100 s at magmatic temperatures. This is on the same time-scale as the uprise of magma and pyroclasts during explosive eruptions (see the eruption models of Wilson (1976) and Sparks (1986)). In other words, diffusional exchange of strontium to the required depth in pumice bubble walls could occur in the time between vesiculation and formation of a pumice clast through magmatic fragmentation in the conduit and quenching of the clast in the eruption column. In fact, the diffusion data suggest that significant diffusional exchange of strontium can occur even if incorporation of hydrothermal fluids into an eruption is restricted to levels above the magma fragmentation surface in the conduit. Invasion of a rising vesiculated magma column by fluids prior to magma fragmentation will increase the contact time between magma and fluids, and enhance the effect.

Is it reasonable to postulate invasion of an explosively erupting magma by hydrothermal fluids from the surrounding country rock? Available data and models of eruption processes and hydrothermal systems suggest that the answer is in the affirmative. The eruption conduit will pierce the hydrothermal system as it propagates toward the surface. Wilson *et al.* (1980) have argued that the pressure above the fragmentation level of the erupting magma in the conduit is less than lithostatic. Hence if the fluid pressure in the hydrothermal system is equal to or greater than the lithostatic pressure, the resultant pressure gradient will draw fluid into the erupting mixtures. Nelson & Giles (1985) state that maximum fluid pressures in over-pressured vapour-dominated hydrothermal systems may exceed the lithostatic load to depths of roughly 1 km. The same workers note that overpressured fluids may develop at much greater depths if the system receives energy from adjacent magmatic intrusion, such as would occur during the initial uprise of magma towards the surface. The common field occurrence of hydrofractures in and around igneous intrusions, fossil hydrothermal systems and associated ore bodies is evidence that fluid pressures can at least temporarily exceed the lithostatic pressure (Henley & McNabb 1978; Nelson & Giles 1985). Also, Delaney (1982) has modelled the hydraulic consequences of rapid intrusion of magma into groundwater systems and found that, under certain circumstances, water can invade a permeable magma in the short term. There is thus a reasonable theoretical basis for the idea that the participation of hydrothermal fluids in explosive eruptions is possible. We emphasize that the mechanism is different from that of phreatomagmatic (hydroclastic) eruptions, where the driving energy is derived largely from explosive interaction of magma with relatively cold liquid water and the products have distinctive field and microscopic characteristics. Rather, we envisage fluids being drawn into the conduit of an essentially magmatic eruption, where the explosive energy is provided by the intrinsic magmatic volatile content. One remaining question is whether fluid input can be sustained throughout the 2–3.5 day period calculated for the duration of the Granadilla eruption; this cannot be answered at present.

Given that most or all volcanic centres that produce pumice-rich eruptives also possess hydrothermal systems, strontium isotope disequilibrium between phenocrysts and matrix in strontium-poor felsic pyroclastic rocks may be an artefact of the explosive eruption process itself. The effect will not be noticeable in less fractionated compositions richer in strontium; Fig. 6 indicates that for the Granadilla deposit, roughly 50 ppm strontium in pumice would be sufficient to buffer magmatic $^{87}Sr/^{86}Sr$ against the contaminant. None the less it seems clear that models for the petrogenesis of highly evolved strontium-poor magmas should be based on isotopic data from phenocrysts rather than whole-rock samples. This applies equally to small phonolitic and trachytic pumice deposits on oceanic islands and to large-volume silicic ignimbrites in continental settings.

Finally, it is worth noting that felsic strontium-poor lavas may not be immune from such processes. Two recent models for the formation of lava domes have emphasized the role of subsurface gas release through a temporary permeable magma foam (Eichelberger *et al.* 1986) or actual fragmentation of the magma followed by recoalescence of particles in the conduit after the manner of a welded tuff (Wolff 1986). In each case, fluids could penetrate and contaminate the magma whilst it is temporarily vesicular.

Acknowledgments: Field work on Tenerife and the major- and trace-element analysis were carried out (by JAW) during tenure of a Natural Environment Research Council (NERC) research studentship at Imperial College. G. P. L. Walker provided access to valuable unpublished field data on the Tenerife pumice series. Isotope work was carried out at the Open University. ZAP acknowledges receipt of a NERC research fellowship. Robin Parker and Nick Rogers are thanked for assistance with analytical work, James Talbot for the scanning electron microscopy, and Gerhard Wörner and Mick Storey for helpful reviews.

References

BOOTH, B. 1973. The Granadilla Pumice deposit of southern Tenerife, Canary Islands. *Proceedings of the Geologists' Association* **84**, 353–370.

BORLEY, G. D. 1974. Aspects of the volcanic history and petrology of the island of Tenerife, Canary Islands. *Proceedings of the Geologists' Association* **85**, 259–279.

CAREY, S. & SPARKS, R. S. J. 1986. Quantitative models of the fallout and dispersal of tephra from volcanic eruption columns. *Bulletin of Volcanology* **48**, 109–125.

CHRISTENSEN, J. N. & DEPAOLO, D. J. 1987. Isotopic study of processes in a silicic magma chamber: the Bishop Tuff, Long Valley, California. *EOS (Transactions of the American Geophysical Union)* **68**, 434.

DELANEY, P. T. 1982. Rapid intrusion of magma into wet rock: groundwater flow due to pore pressure increases. *Journal of Geophysical Research* **87**, 7739–7756.

DEPAOLO, D. J. 1981. Trace element and isotopic effects of combined wallrock assimilation and fractional crystallisation. *Earth and Planetary Science Letters* **52**, 177–184.

—— 1985. Isotopic studies of processes in mafic magma chambers: I. The Kiglapait Intrusion, Labrador. *Journal of Petrology* **26**, 925–951.

EICHELBERGER, J. C., CARRIGAN, C. R., WESTRICH, H. R. & PRICE, R. H. 1986. Nonexplosive silicic volcanism. *Nature* **323**, 598–602.

HENLEY, R. W. & MCNABB, A. 1978. Magmatic vapor plumes and ground-water interaction in porphyry copper emplacement. *Economic Geology* **73**, 1–20.

HALLIDAY, A. N., FALLICK, A. E., HUTCHINSON, J. & HILDRETH, W. 1984. A Nd, Sr and O isotopic investigation into the causes of chemical and isotopic zonation in the Bishop Tuff, California. *Earth and Planetary Science Letters* **68**, 379–391.

—— MAHOOD, G. & HILDRETH, W. 1985. Mantle signature for high-silica rhyolites with <1 ppm Sr: Sierra La Primavera, Mexican Volcanic Belt (MVB). *EOS (Transactions of the American Geophysical Union)* **66**, 1136.

HILDRETH, W. 1979. The Bishop Tuff: evidence for the origin of compositional zonation in silicic magma chambers. *Geological Society of America Special Paper* **180**, 43–75.

MAGARITZ, M. & HOFMANN, A. W. 1978. Diffusion of Sr, Ba and Na in obsidian. *Geochimica et Cosmochimica Acta* **42**, 595–606.

NELSON, C. E. & GILES, D. L. 1985. Hydrothermal eruption mechanisms and hot spring gold deposits. *Economic Geology* **80**, 1633–1639.

PALACZ, Z. A. & TAIT, S. R. 1985. Isotopic and geochemical investigation of unit 10 from the Eastern Layered Series of the Rhum intrusion, north-west Scotland. *Geological Magazine* **122**, 485–490.

RIDLEY, W. I. 1970. The petrology of the Las Cañadas volcanoes, Tenerife, Canary Islands. *Contributions to Mineralogy and Petrology* **26**, 124–160.

SCOTT, R. B. 1971. Chemical variations in glass shards and interstitial dust of ignimbrite cooling units. *American Journal of Science* **270**, 116–173.

SHOTTON, F. W. & WILLIAMS, R. E. G. 1971. Birmingham University radiocarbon dates V. *Radiocarbon* **13**, 150.

——, —— & JOHNSON, A. S. 1974. Birmingham University radiocarbon dates VIII. *Radiocarbon* **16**, 293.

SPARKS, R. S. J. 1986. The dimensions and dynamics of volcanic eruption columns. *Bulletin of Volcanology* **48**, 3–15.

—— & WILSON, L. 1976. Model for ignimbrite formation by gravitational column collapse. *Journal of the Geological Society, London* **132**, 441–451.

——, SELF, S. & WALKER, G. P. L. 1973. Products of ignimbrite eruptions. *Geology* **1**, 115–118.

SUN, S.-S 1980. Lead isotopic study of young volcanic rocks from mid-ocean ridges, ocean islands and island arcs. *Philosophical Transactions of the Royal Society of London, Series A* **297**, 409–445.

WALKER, G. P. L. 1981. Plinian eruptions and their products. *Bulletin Volcanologique* **44**, 223–240.

WILSON, L. 1976. Explosive volcanic eruptions — III Plinian eruption columns. *Geophysical Journal of the Royal Astronomical Society* **45**, 543–556.

WILSON, L., SPARKS, R. S. J., HUANG, T. C. & WATKINS, N. D. 1978. The control of volcanic column heights by eruption energetics and dynamics. *Journal of Geophysical Research* **83**, 1829–1836.

——, SPARKS, R. S. J. & WALKER, G. P. L. 1980. Explosive volcanic eruptions — IV. The control of magma properties and conduit geometry on eruption column behaviour. *Geophysical Journal of the Royal Astronomical Society* **63**, 117–148.

WITHAM, A. G. & SPARKS, R. S. J. 1986. Pumice. *Bulletin of Volcanology* **48** 209–223.

WOLFF, J. A. 1983. *Petrology of Quaternary Pyroclastic Deposits from Tenerife, Canary Islands*. PhD thesis, University of London.

—— 1984. Variation in Nb/Ta during differentiation of phonolitic magma, Tenerife, Canary Islands. *Geochimica et Cosmochimica Acta* **48**, 1345–1348.

—— 1985*a*. Zonation, mixing and eruption of silica-undersaturated alkaline magma: a case study from Tenerife, Canary Islands. *Geological Magazine* **122**, 623–640.

—— 1985*b*. The effect of explosive eruption processes on geochemical patterns within pyroclastic deposits. *Journal of Volcanology and Geothermal Research* **26**, 189–201.

—— 1986. Welded-tuff dykes, conduit closure, and lava dome growth at the end of explosive eruptions. *Journal of Volcanology and Geothermal Research* **28**, 379–384.

—— 1987. Crystallisation of nepheline syenite in a subvolcanic magma system: Tenerife, Canary Islands. *Lithos* **20**, 207–223.

—— & PALACZ, Z. A. Pb isotope and trace-element

variation in the phonolitic Granadilla Series, Tenerife: evidence for recycling within an ocean island volcano. *Mineralogical Magazine*, in press.

—— & STOREY, M. 1983. The volatile component of some pumice-forming alkaline magmas from the Azores and Canary Islands. *Contributions to Mineralogy and Petrology* **82**, 66–74.

—— & —— 1984. Zoning in highly alkaline magma bodies. *Geological Magazine* **121**, 563–575.

WÖRNER, G. & SCHMINCKE, H.-U. 1984. Mineralogical and chemical zonation of the Laacher See tephra sequence (East Eifel, W. Germany). *Journal of Petrology* **25**, 805–835.

——, HARMON, R. S. & HOEFS, J. 1986. Stable isotope relations in an open magma system, Laacher See, Eifel (F. R. G.) *Contributions to Mineralogy and Petrology* **95**, 343–349.

——, STAUDIGEL, H. & ZINDLER, A. 1985. Isotopic constraints on open system evolution of the Laacher See magma chamber (Eifel, West Germany). *Earth and Planetary Science Letters* **75**, 37–49.

ZEILINSKI, R. A. LIPMAN, P. W. & MILLARD, M. T. 1977. Minor-element abundances in obsidian, perlite, and felsite of calc-alkaline rhyolites *American Mineralogist* **62**, 426–437.

Z. A. PALACZ, Department of Earth Sciences, The Open University, Walton Hall, Milton Keynes MK7 6AA, Buckinghamshire, UK. *Present address*: VG Isotopes Ltd, Ion Path Road Three, Winsford, Cheshire CW7 3BX, UK.

J. A. WOLFF, Department of Geology, University of Texas at Arlington, UTA Box 19049, Arlington, TX 76019, USA.

Origin of hybrid lavas from Agua de Pau volcano, Sao Miguel, Azores

M. Storey, J. A. Wolff, M. J. Norry & G. F. Marriner

SUMMARY: The island of Sao Miguel in the Azores exhibits bimodal volcanism: eruptives from Agua de Pau volcano and the 'Waist' consist almost entirely of basalt or trachyte with rare intermediate products. Although some intermediate lavas are fractionates from basalt, the majority are mixed-magma hybrids anomalously enriched in europium, barium and potassium. These features clearly indicate that they have also accumulated alkali feldspar. The source of the alkali feldspar contaminant must be the least evolved trachyte in the Agua de Pau system, because barium, europium and strontium are rapidly depleted in liquids and crystals with differentiation to more evolved trachytes. Many less evolved trachytes are also accumulative in alkali feldspar, suggesting a genetic association with the hybrid lavas. Geochemical modelling indicates that the hybrid lavas are the result of contamination of basalt by accumulative trachyte containing up to 75% alkali feldspar.

Field and petrological evidence points to the episodic injection and ponding of basalt beneath trachytic magma at Agua de Pau volcano. One trachytic lava, with alkali feldspar-contaminated basaltic inclusions, represents a sample of this interface. Density calculations indicate that highly accumulative trachyte can have a neutral or negative buoyancy with respect to basalt and could descend through the interface. Resorption of alkali feldspar will lower the bulk density of the contaminated basalt, with the effect of favouring even greater contamination by overlying accumulative trachyte. The most attractive mechanism for achieving the accumulative trachyte compositions is the slumping of side-wall cumulates and the formation of magmatic density currents which flowed along the compositional interface between trachyte and basalt. Flowage differentiation and crystal settling would result in local 'parcels' of denser highly accumulative trachytic magma. The process is illustrated with results from a simple laboratory experiment.

A common feature of many oceanic island volcanoes is the strong basic–felsic bimodality of the compositions of erupted rocks (Daly 1925; Walker 1966; Sigurdsson 1977; Booth *et al.* 1978). Some workers have considered that the apparent over-representation of felsic magmas and the scarcity of intermediate compositions in the eruptive sequences raise questions about the viability of fractional crystallization models. Therefore arguments against the derivation of felsic magmas from basalt via fractional cystallization are essentially based on these volumetric, rather than petrological, inconsistencies. Hence a number of alternative mechanisms have been proposed for the origin of felsic magmas in these environments. These include primary mantle-derived melts (Yoder 1973), melting of a plagiogranite layer within the oceanic crust (Sigurdsson 1977), or hydrous partial melting of basalt (O'Nions & Gronvold 1973). However, the 'Daly gap' is to some extent dependent on the geochemical parameters chosen to express it, and an apparent gap in the range of silica contents may be unsupported by incompatible-trace-element abundances (Clague 1978).

Another aspect of bimodal volcanism is that physical interaction of basalt and felsic magmas might well be expected to be a common phenomenon; for example, evidence for magma mixing has been documented from pyroclastic rocks for the bimodal suites of Askja in Iceland (Sigurdsson & Sparks 1981) and Tenerife in the Canary Islands (Wolff 1985). Theoretical and experimental studies suggest that several important processes may result from the intrusion of basalt into felsic magma; for example, the resulting increase in fluid pressure may act as a 'trigger' for explosive pumice-forming eruptions (Sparks *et al.* 1977). Experiments by Campbell & Turner (1986) indicate that, owing to the large density and viscosity differences between basaltic and felsic magmas, little physical comingling is expected; if basalt is intruded into felsic magma it would be expected to pond at the base of the chamber as a discrete layer. This, of course, poses the problem of how such liquids can actually mix to produce the hybrid magmas witnessed in, for example, the 1875 eruption of Askja (Sigurdsson & Sparks 1981). Several mechanisms have been suggested.

(1) Basaltic magma pooled beneath more evolved magma crystallizes to the point

From SAUNDERS, A. D. & NORRY, M. J. (eds), 1989, *Magmatism in the Ocean Basins*, Geological Society Special Publication No. 42, pp. 161–180.

where the density of the residual melt is less than that of the overlying magma (*cf* Huppert & Sparks 1980).

(2) Basaltic magma pooled beneath felsic magma crystallizes to the point where the residual liquid is volatile saturated and begins to vesiculate, causing an exsolution-driven overturn (Eichelberger 1980; Huppert *et al.* 1982).

(3) The basaltic and felsic magmas mix during actual eruption (Koyaguchi 1985; Blake & Campbell 1986).

The island of Sao Miguel in the Azores (Fig. 1) contains an example of this bimodal style of volcanism and therefore provides a good opportunity for examining some of these petrological and volcanological problems. The volcanology of Sao Miguel was the subject of a detailed study by Booth *et al.* (1978), who demonstrated that recent eruptives mostly consist of basalt and trachyte, with rarer examples of lavas with silica contents intermediate between these two end-members. Basalt and trachyte occur in comparable amounts, with this style of activity probably operating for at least the past 50 000 years on Sao Miguel. The trachytic eruptions have largely been confined to the central calderas of the three active stratovolcanoes on the island (Fig. 1). Basalts, in contrast, have been erupted from flank vents on the stratovolcanoes and from a WNW–ESE-trending fissure zone in the low-lying tract of land between Agua de Pau and Sete Cidades volcanoes known as the 'Waist'.

In this paper we describe the petrology of the alkali basalt, intermediate, trachyte and hybrid lavas from Agua de Pau and the Waist. Despite the reservations expressed by other workers about fractional crystallization processes (because of the bimodal volume relationships), data for Agua de Pau and the Waist define geochemical trends which are wholly consistent with this mechanism, with the exception of the hybrid lavas, some of which have been described by Marriner *et al.* (1982) and which form the main subject of this paper.

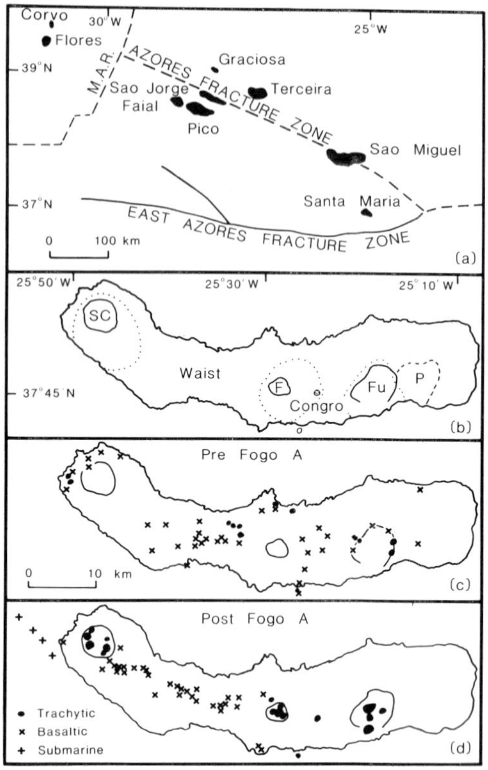

FIG. 1. (*a*) The Azores archipelago: – – –, active plate boundary (after Laughton & Whitmarsh 1975; Flower *et al.* 1976). (*b*) Sao Miguel, showing the location of calderas associated with the three active stratovolcanoes Sete Cidades (SC), Fogo (F) caldera of Agua de Pau volcano and Furnas (Fu); . . . ., isopachs for trachytic pumice deposits formed in the past 5000 years from each of the three centres. Also shown is the extinct centre of Povoacao (P). (*c*) Location of basaltic and trachytic vents which erupted during the few millenia preceding Fogo A. (*d*) Location of vents on Sao Miguel which have erupted during the past 5000 years. (*b–d*; all from Booth *et al.* 1978.)

Agua de Pau volcano

Agua de Pau is one of three active Quaternary stratovolcanoes on Sao Miguel, the largest of the nine islands forming the Azores (Fig. 1). Volcanic rocks from Agua de Pau belong to the alkali basalt–trachyte suite and are amongst the most potassic in the Azores (White *et al.* 1979). The nomenclature used to describe the suite is based on the silica content (Table 1).

Basalts erupted from the flank vents of Agua de Pau and the fissure zone in the Waist occur as lava flows and scoria cones. Rocks of trachytic composition occur mainly as air-fall pumice deposits produced by subplinian to plinian eruptions with subordinate lavas and pyroclastic flow deposits. A large plinian eruption 4500 years ago (Fogo A) was largely responsible for the formation of the

TABLE 1. *Phenocryst assemblages*

Rock type	SiO_2 (wt%)	Mineralogy
Basalts	<50	Olivine + titaniferous augite ± plagioclase ± Cr-spinel + Fe–Ti oxides
Intermediates	50 – 59	Plagioclase ± augite + kaersutite + Fe–Ti oxides + apatite
Trachytes	>59	Alkali feldspar + clinopyroxene (salite) + biotite ± kaersutite + Fe–Ti oxides + apatite
Hybrids ('intermediates') Basaltic assemblage	51–59	Titaniferous augite + olivine ± plagioclase ± Cr-spinel + Fe–Ti oxides
Trachytic assemblage		Alkali feldspar + clinopyroxene + Fe–Ti oxides ± biotite

Fogo caldera of Agua de Pau (Walker & Croasdale 1971). There have been five post-Fogo A trachytic eruptions from a single continuously fractionating magma batch (Storey 1981). The most recent trachytic eruption (1563 AD) was followed within a short time by the extrusion of two small basaltic flows from a scoria cone on the NW flanks of Agua de Pau, near Queimado. All but one (Congro Ash, Fig. 1*b*) of these explosive trachytic eruptions had vents located within or marginal to the present-day caldera walls (Booth *et al.* 1978). The majority of intermediate lavas are hybrids which to a first approximation can be considered to have been formed by mixing of basaltic and trachytic magmas (Table 1). They are subdivided on the basis of several criteria into two groups (types 1 and 2) as discussed in subsequent sections; however, it is worth noting that type 1 hybrids have largely been erupted from vents on the NW flanks of Agua de Pau, whereas type 2 hybrids are mainly restricted to the NE flanks of the volcano. Intermediate lavas with equilibrium phenocryst assemblages and which plot on liquid lines of descent have been erupted from vents on the southern flanks of Agua de Pau. Syenite xenoliths (Cann 1967) occur in several basalt lavas and trachytic pyroclastic deposits from Agua de Pau.

The magmatic system of Agua de Pau volcano

The idea that subvolcanic felsic magma bodies in both the oceanic and the continental environment may be compositionally and thermally zoned has gained wide acceptance in recent years (*eg* Hildreth 1979; Smith 1979; Leat *et al.* 1984; Wolff & Storey 1984; Worner & Schmincke 1984). Typically, such systems consist of a differentiated low-density volatile-rich capping layer (rhyodacite, rhyolite, trachyte or phonolite magma) underlain by more mafic magma. Capping layers and their internal chemical gradients are largely the result of fractional crystallization (Michael 1983*a*, *b*; Wolff & Storey 1984). Theoretical and laboratory analogue studies have emphasized the importance of the physical properties of melts (density, viscosity) and dynamic effects (*eg* convection) in controlling the formation and evolution of magma bodies (Sparks *et al.* 1984). One mechanism for the generation of the felsic caps of zoned systems, suggested by such experiments, is the formation and uprise of buoyant melt due to crystallization at a chamber wall. Sparks *et al.* (1984) used the term 'convective fractionation' to describe this process. Gravitational settling of individual crystals is unlikely to be a viable fractional crystallization mechanism for felsic magma bodies as crystal-settling velocities are greatly exceeded by convective velocities (Sparks *et al.* 1984).

The Fogo A deposit provides direct evidence for the existence of a compositionally zoned trachytic magma body beneath Agua de Pau volcano. The deposit grades upwards from buff-coloured pumice into first a streaky and then a more mafic variety (Walker & Croasdale 1971). On the basis of a small number of samples, Wolff & Storey (1984) showed that there is a significant compositional difference between pumice from different parts of the deposit, particularly in the abundances of the compatible elements strontium, europium and barium. Data published recently (Widom *et al.* 1987) show that the compositional variation in Fogo A is considerably greater than that reported by Wolff & Storey (1984), with weakly peralkaline trachytic pumice as the initial eruption product. The substantial evolution of trachytic magma at Agua de Pau over 4000 years (Storey 1981) suggests rapid generation of such highly evolved liquids.

There is a curious dichotomy between the distribution of basaltic vents at Agua de Pau and its environs before and after the eruption of the 4500-year-old Fogo A deposit (Fig. 1*c*, *d*). Post-Fogo A basaltic eruptions are largely restricted to the fissure zone in the Waist, whereas pre-Fogo A basaltic vents show a much wider geographical occurrence. Booth *et al.* (1978) suggested that a large body of trachytic magma had spread out laterally between Agua de Pau volcano and Furnas volcano following the Fogo A plinian eruption, thereby restricting the ability of basaltic magma to reach the surface. The lack of basaltic vents in the caldera similarly argues for a 'shadow zone' caused by the uprise and ponding of basalt beneath more differentiated, less dense magmas. Explosive trachytic eruptions from vents within or close to the Fogo caldera have by their very nature mostly sampled the uppermost volatile-rich part of the magma chamber. However, the occurrence of basalt-derived phenocrysts in some of the trachytic pumice deposits erupted from Agua de Pau (*eg* Fogo A, B and 1563 (Storey 1981)) may suggest some contamination by and due to the presence of basalt at a deeper level. The near-contemporaneous eruption of trachytic pumice and basalt lava from Agua de Pau in 1563 AD also demonstrates the simultaneous availability of both magma types. There is therefore excellent field and petrological evidence for the presence of basalt beneath trachytic magma at Agua de Pau volcano.

Mineralogy and petrography of Agua de Pau eruptives

Basalts

Basalts vary from aphyric types to highly porphyritic MgO-, nickel- and chromium-rich ankaramites, with phenocrysts of titanaugite (up to 3.9 wt% TiO_2), olivine (Fo_{75-90}), Fe–Ti oxides, Cr-spinel and plagioclase (An_{60-86}). Some of the basalts contain aggregates of forsteritic olivine and clinopyroxene.

Lavas of intermediate composition

For those intermediate lavas which fall on liquid lines of descent, the phenocryst assemblage consists of kaersutite, plagioclase and Fe–Ti oxides; apatite microphenocrysts are an important accessory phase. The hybrid lavas, which have similar silica contents but very different mineralogical and trace-element characteristics, are described below.

Trachytes

The phenocryst content of glassy pumice does not usually exceed 10%, although the trachyte lavas are more porphyritic (up to 30% phenocrysts). The phenocryst assemblage consists of large euhedral sanidines with smaller amounts of anorthoclase, sometimes showing resorption and polygonization textures. Anorthoclase ($An_2Ab_{61}Or_{36}$–$An_6Ab_{46}Or_{48}$) is seen only as phenocrysts in the least evolved trachytes (LETs), which are characterized by low to moderate zirconium contents. Sanidines show a trend towards progressively more albite-rich and anorthite-poor compositions with increasing chemical evolution of the trachyte (Storey 1981). Small amounts of green salite, biotite and Fe–Ti oxides and apatite microphenocrysts occur, with rare kaersutite in some LET samples. Basaltic xenocrysts occur in a number of the trachytes. These include labradorite feldspar, titanaugite and forsteritic olivine (Fo_{78-80}) in the trachytic pumice deposits Fogo A and B (Storey 1981). Marriner *et al.* (1982) report olivines with Fo_{90} compositions from one trachyte lava from Agua de Pau.

Syenites

The syenite xenoliths are composed mostly of sanidine (or a lower-temperature form) with quartz, arfvedsonite and aegirine. Accessory phases include Fe–Ti oxides, fayalite, apatite, biotite, zircon, dalyite and astrophyllite (Cann 1967). Cumulate textures are absent.

Hybrid lavas from Agua de Pau

Many lavas of intermediate silica content erupted from Agua de Pau volcano possess xenocrysts which must have originally crystallized from either a trachytic or basaltic melt. Their most prominent feature is the presence of large

TABLE 2. *Selected feldspar analyses (wt%) of hybrid lavas*

	AZ1721			AZ1705			MA239	
SiO_2	59.75	62.28	60.48	53.91	64.72	62.00	65.99	60.61
Al_2O_3	24.93	22.28	23.12	28.55	19.60	19.99	19.43	23.77
FeO	0.61	0.51	0.45	0.58	0.27	0.62	0.19	0.64
CaO	6.70	3.69	4.93	11.48	1.02	3.14	0.77	6.55
Na_2O	7.01	6.40	6.26	4.32	3.52	5.45	5.86	6.55
K_2O	1.38	4.36	3.10	0.62	10.97	6.60	8.15	2.26
BaO	0.48	1.58	1.13	NA	NA	NA	NA	NA
Total	100.86	101.10	99.47	99.46	100.10	97.80	100.39	100.38
Recalculated on the basis of $8O^{2-}$								
Si	2.663	2.800	2.748	2.452	2.942	2.862	2.960	2.726
Al	1.310	1.181	1.238	1.531	1.051	1.089	1.027	1.260
Fe	0.023	0.019	0.017	0.022	0.010	0.024	0.007	0.024
Ca	0.320	0.178	0.240	0.560	0.050	0.156	0.037	0.264
Na	0.606	0.558	0.551	0.381	0.310	0.488	0.510	0.571
K	0.079	0.250	0.180	0.036	0.636	0.389	0.467	0.130
Ba	0.008	0.028	0.020	—	—	—	—	—

NA, not analysed.

anhedral carlsbad-twinned sanidines and cross-hatched twinned anorthoclases (Fig. 2 *a*, *b*). These occur as individual grains or aggregates, and often have overgrowths and show evidence for polygonization into smaller grains. Feldspathic patches in the groundmasses of these lavas are suggestive of either remnants of alkali feldspars which have been completely resorbed or inclusions of incompletely mixed trachytic magma (Fig. 2 *c*). The anorthoclases are notable for their large and variable BaO contents (Table 2).

Also observed in the hybrid lavas are trachyte-derived xenocrysts of biotite (usually oxidized), salite and low-Al–Mg magnetite, all of which can be found in association with aggregate grains of alkali feldspar. Basaltic xenocrysts consist of forsteritic olivine, titanaugite or diopside, Cr-spinel and bytownite plagioclase up to An_{88}. One criterion for the subdivision of the hybrid lavas is the composition of olivine xenocrysts. Typically type 1 hybrids have compositions around Fo_{80} whereas those of type 2 are near Fo_{90}. These observations are consistent with an origin for the type 1 and type 2 hybrid lavas which involved mixing of trachyte with low-MgO and more primitive high-MgO basaltic magmas respectively.

Geochemistry

Representative analyses are shown in Table 3. Whole-rock compositional variation trends are illustrated in Fig. 3. Rapid depletion of nickel and chromium from the MgO-rich to MgO-poor basalts is consistent with olivine and clinopyroxene fractionation. Three ankaramite lavas erupted from Agua de Pau which contain up to 40% olivine and clinopyroxene and which have anomalously high MgO (14–17 wt%), nickel (434–827 ppm) and chromium (1414–1812 ppm) contents are considered to be accumulative. We have no evidence for the eruption of similar highly accumulative basalts from the fissure zone in the Waist. Several basalts plot on the back-projection of the liquid lines of descent for titanium, phosphorus and strontium, which suggests that they are accumulative in amphibole–magnetite, plagioclase and apatite. Inflections or maxima in the variation trends in the more differentiated rocks correspond to either a change in the relative modal abundances of the separating phases or a change in the phenocryst mineralogy (Fig. 3).

Elemental abundances in the trachytes provide convincing evidence for a fractional crystallization origin; the mildly peralkaline types show extreme depletion of strontium, europium, barium, phosphorus, titanium, vanadium and magnesium coupled with strong enrichment of incompatible elements such as zirconium, niobium, thorium and the rare-earth elements (Figs 3 and 4). The level of enrichment of zirconium, for example, indicates that the most evolved trachytes could be related to primitive basalts by *c.* 80%–90% crystallization, assuming simple Rayleigh fractionation.

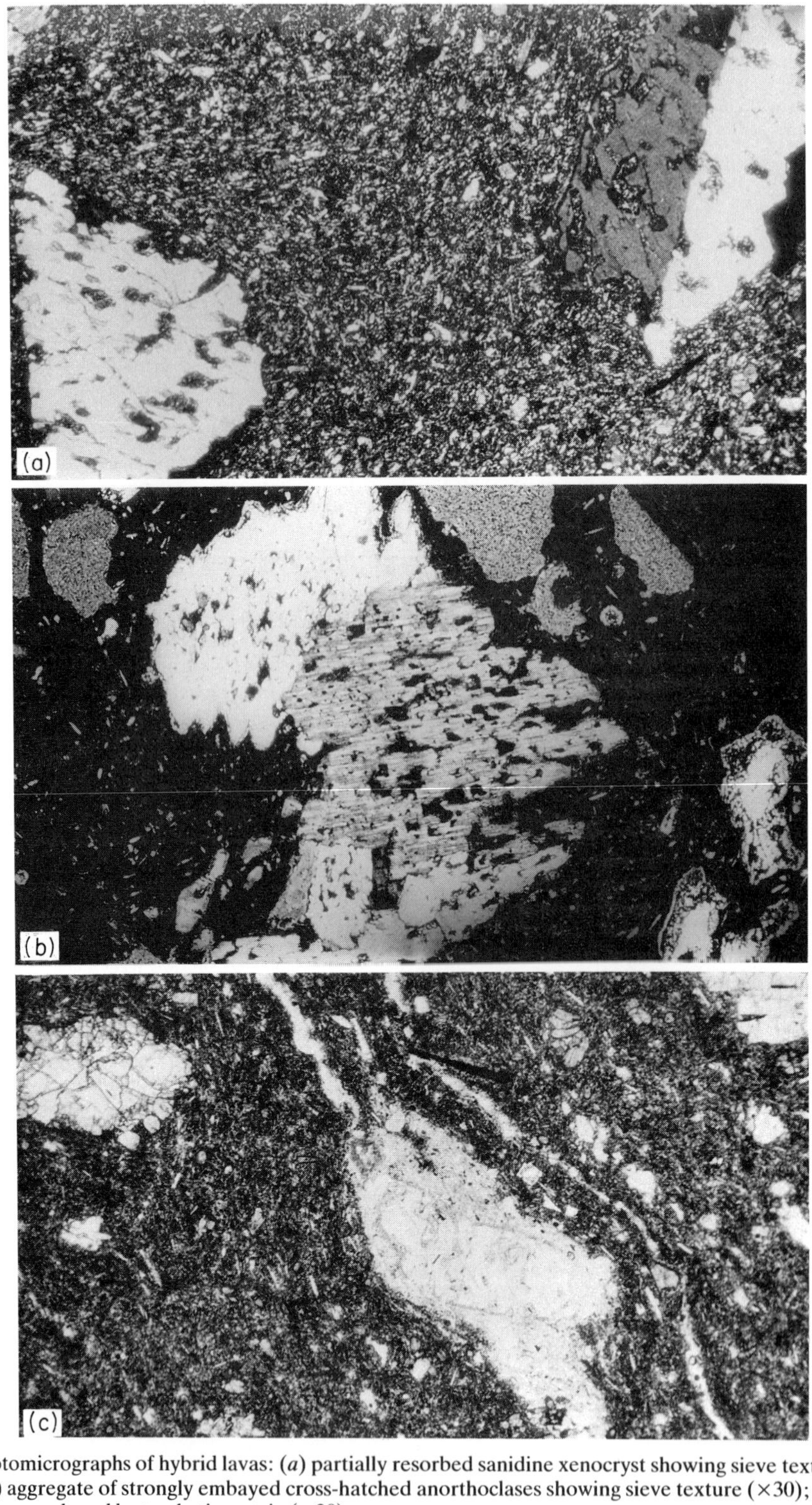

FIG 2. Photomicrographs of hybrid lavas: (*a*) partially resorbed sanidine xenocryst showing sieve texture (×30); (*b*) aggregate of strongly embayed cross-hatched anorthoclases showing sieve texture (×30); (*c*) sanidine enveloped by trachytic matrix (×30).

TABLE 3. *Representative whole-rock major- and trace-element analyses*

	Basalts														
	Waist						Agua de Pau								
	AZ1039	SM30	AZ1324	AZ1880	AZ1874	AZ1872	MA23	SM62	SM28	SM18	36AP	SM49	SM23	SM27	SM11
Major elements (wt%)															
SiO_2	45.5	45.8	46.9	47.4	46.3	45.3	45.6	46.0	44.8	47.2	45.9	48.2	45.0	45.1	46.1
TiO_2	4.00	3.95	3.88	2.52	3.80	3.44	2.91	2.98	3.86	3.42	4.06	3.57	4.99	4.69	4.45
Al_2O_3	13.3	14.0	17.1	13.0	14.1	12.0	8.9	10.5	11.9	13.7	14.6	14.6	14.1	14.2	14.8
Fe_2O_3	12.7	12.3	11.4	11.1	13.0	13.2	12.1	11.4	12.9	11.4	12.5	11.1	13.9	13.7	13.2
MnO	0.17	0.16	0.19	0.17	0.18	0.17	0.17	0.15	0.16	0.16	0.16	0.14	0.16	0.16	0.15
MgO	9.0	7.2	5.2	10.6	8.6	11.6	16.9	13.9	10.5	8.5	6.6	6.5	5.6	5.6	5.1
CaO	11.3	10.1	8.7	11.9	9.5	11.8	10.7	9.5	10.3	9.2	10.0	9.2	11.0	10.7	9.0
Na_2O	2.9	3.3	4.3	2.7	3.0	2.5	1.5	2.5	2.3	3.3	3.2	3.2	2.4	2.4	2.8
K_2O	1.63	1.93	2.07	1.15	1.87	1.25	1.28	1.43	1.71	1.98	2.49	2.65	1.77	1.73	2.28
P_2O_5	0.53	0.64	0.76	0.36	0.59	0.42	0.39	0.63	0.87	0.65	0.66	0.69	1.06	1.05	0.87
Trace elements (ppm)															
Cr	425	268	13	719	469	648	1812	1414	797	422	224	221	20	18	56
Ni	106	83	10	222	168	217	563	434	202	173	199	132	26	24	45
V	294	273	217	257	280	296	247	222	238	234	295	245	337	320	295
La	44.0	47.7	55.0	29	51	37	33.9	40.0	52.4	47.8	53.0	53.0			
Ce	90.0	98.9	116.0				70.8	85.7	110.8	102.7	110.0	112.6	104.8	118.7	104.4
Nd	46.0	55.7	65.0				34.2	45.2	61.9	54.4	55.0	55.1	59.8	62.5	59.7
Sm	9.4	11.8	13.0				8.7	9.6	12.9	11.3	11.3	10.1	12.0	13.7	11.7
Eu	3.0	3.7	4.1				2.0	2.8	4.2	3.3	3.1	3.8	4.3	4.6	3.8
Gd	8.2		11.3									9.1	10.4	11.8	8.8
Tb		1.4					0.9	1.0	1.4	1.3	1.1	1.2	1.3	1.5	1.1
Tm		0.4						0.3	0.4	0.4		0.4	0.4	0.4	0.3
Yb	2.0	2.7	3.0				1.8	1.8	2.2	2.4	2.5	2.1	2.1	2.5	2.3
Lu		0.4					0.2	0.2	0.3	0.4	0.3	0.3	0.3	0.3	0.3
Y	21.0	35	34	25	26	21	20	35	33	30	27	36	35	36	37
Nb	65.0	67	77	49	78	44	50	59	57	68	70	63	58	61	63
Zr	255.0	275	299	149	277	199	204	251	269	284	297	304	249	257	283
Hf		8.2						6.5	8.2	7.9		8.3	6.6	7.5	7.5
Ta		5.4						4.1	4.6	5.1		4.9	4.6	5.0	4.7
Th		5.8						4.9	6.2	5.6		6.8		6.1	7.0
Sr	687	791	945	469	742	520	455	607	789	716	781	794	1001	951	749
Rb	38	43	47	49	39	28	29	37	33	44	49	59	38	38	55
Ba	565	584	653	428	574	404	392	400	502	618	417	784	624	609	738

TABLE 3 (*cont.*)

	Basalts												
	Type 1 hybrids					Type 2 hybrids					Intermediates		
	SM13	AZ1705	SM25	SM8	SM4	SM21	SM29	SM20	MA239	SM56	SM71	SM61	AZ1396
Major elements (wt%)													
SiO_2	52.8	53.7	55.4	55.8	56.3	53.8	57.8	58.3	58.4	54.4	52.2	56.3	57.5
TiO_2	2.87	2.86	2.14	1.86	1.78	2.12	1.35	1.49	1.60	2.02	2.36	1.55	1.61
Al_2O_3	16.2	16.9	16.5	17.8	17.5	14.7	16.5	16.2	16.5	14.8	17.3	17.5	18.4
Fe_2O_3	8.6	8.7	6.8	6.3	6.0	7.6	5.5	5.6	6.2	7.5	8.4	5.9	6.3
MnO	0.12	0.13	0.11	0.11	0.11	0.13	0.13	0.12	0.15	0.14	0.16	0.16	0.17
MgO	3.7	3.6	2.8	2.4	2.3	6.2	3.0	3.6	3.4	6.3	3.3	2.1	2.0
CaO	7.0	6.6	4.9	4.7	4.7	5.6	3.8	4.3	3.9	5.4	6.1	4.1	4.2
Na_2O	3.7	4.3	4.4	4.5	4.7	4.3	5.4	4.7	5.1	4.1	4.5	5.5	5.6
K_2O	4.28	4.56	5.45	4.95	5.08	4.42	4.98	5.30	5.10	4.45	3.77	4.58	4.60
P_2O_5	0.51	0.38	0.45	0.35	0.36	0.45	0.32	0.32	0.30	0.44	0.95	0.51	0.42
Trace elements (ppm)													
Cr	47	32	37	21	21	440	146	265	165	457		3	
Ni	30	40	17	16	15	157	40	66	60	170	4		2
V	183	179	125	105	105	120	72	84	75	118	119	52	49
La		51.0	57.3		59.4	62.4	69.8		80	71.8		95.6	89
Ce	95.7	108.0	112.7	92.3	105.7	70.8	136.5	116.2	142	121.3	185.3	187.0	
Nd	45.8	42.0	55.1	41.1	43.4	31.1	60.8	52.2	58	55.1	87.8	81.3	
Sm	9.0	9.9	10.5	7.8	9.6		11.7	9.6	10.2	11.0		14.7	
Eu	4.1	3.6	4.9	3.6	4.2	2.1	5.8	3.2	2.3	3.6	5.1	4.3	
Gd	7.5			5.5				7.9			14.7		
Tb	0.9	1.0	1.2	0.7	1.0	0.7	1.3	0.9	1.1	1.1	1.8	1.5	
Tm	0.3		0.4	0.2	0.3	0.2	0.4	0.3		0.4	0.6	0.5	
Yb	1.9	2.0	2.1	1.8	1.9	1.3	2.3	2.1	2.6	2.0	3.3	3.5	
Lu	0.3	0.3	0.3	0.3		0.2	0.4	0.3		0.2	0.3	0.5	
Y	31	26	31	24	32	31	34	31	33	33	54	56	47
Nb	53	55	60	61	79	72	75	74	111	77	109	128	131
Zr	258	265	259	286	288	345	359	318	464	351	505	613	661
Hf	6.8		7.6	6.4	7.1		9.6	7.3		9.0	12.4	14.7	
Ta	4.0		4.9	3.9	4.9	3.2	5.5	4.9		5.5	7.6	9.0	
Th	6.2		6.5	6.5	7.4	4.7	9.3	8.3		8.4	11.8	16.5	
Sr	685	679	485	614	597	480	435	313	357	463	1010	886	926
Rb	63	58	67	75	71	74	75	79	94	74	92	126	130
Ba	1536	1616	957	1304	1326	828	845	818	701	794	1057	1255	1286

	Trachytes														Mixed lava		Syenite xenoliths	
	SM5	SM19	SM12	SM22	SM53	AZ1672	AZ1146	SM64	SM68	AZ1377	AZ1378	AZ1394	AZ1188	AZ1149	AZ1721	BAZ1721T	AZ1204	AZ1219
Major elements (wt%)																		
SiO_2	63.0	60.9	62.2	63.9	64.9	63.1	60.4	62.2	62.9	61.7	62.2	63.2	65.1	64.8	53.6	61.8	64.8	64.5
TiO_2	0.59	0.88	0.82	0.60	0.72	0.69	1.34	0.76	0.62	1.02	0.92	0.58	0.33	0.45	2.53	1.06	0.50	0.32
Al_2O_3	17.0	17.1	17.6	16.1	17.1	17.9	17.3	18.2	18.1	17.6	18.2	18.8	16.5	16.8	17.0	17.3	15.9	16.7
Fe_2O_3	4.1	4.2	4.0	4.1	3.2	3.0	4.9	3.6	3.3	4.2	4.1	3.5	4.8	3.5	8.9	5.2	4.6	3.5
MnO	0.19	0.13	0.14	0.18	0.18	0.10	0.14	0.12	0.13	0.14	0.15	0.15	0.29	0.22	0.17	0.20	0.26	0.22
MgO	0.5	2.2	1.0	0.5	0.5	0.6	1.7	0.6	0.4	1.1	1.0	0.3	0.2	0.3	3.4	1.0	0.4	0.2
CaO	0.8	1.8	1.3	0.7	0.7	1.4	2.9	1.1	1.1	2.2	1.7	1.0	0.6	0.9	5.2	1.6	0.2	0.5
Na_2O	6.7	6.2	6.2	7.0	6.9	5.8	5.6	5.9	6.4	6.0	6.0	6.7	7.0	7.6	4.7	6.3	6.3	7.2
K_2O	5.79	5.80	6.11	5.71	6.00	6.52	5.66	6.49	6.24	6.07	5.88	5.89	5.28	5.46	4.00	5.54	4.96	5.42
P_2O_5	0.11	0.17	0.15	0.11	0.10	0.11	0.21	0.14	0.10	0.19	0.14	0.08	0.08	0.07	0.42	0.18	0.04	0.10
Trace elements (ppm)																		
Cr		138					38								132	3		
Ni		67			2		15								31	7	7	16
V		7	16		6	73	59								150	26		
La		59.0		131.2	133.7	165.0	75			80.7	112.7	136.2	189.7	197.1			194.0	242
Ce	231.3	166.1	182.5	234.8	156.6	54.0		190.7	222.1	172.8	230.8	290.6	439.9	362.3	152.5	223.9	382.0	469.0
Nd	98.6	76.0	117.1	104.6	97.5	11.0		86.9	82.1	58.8	87.1	99.8	151.0	129.7	85.8	93.9	155.0	156.0
Sm	17.3	13.1	18.7	19.1	17.8			12.7	13.1	12.0	16.7	17.5	30.1	27.3	15.8	15.5	31.0	27.0
Eu	1.5	2.4	2.2	1.8	2.7	2.6		1.8	1.5	2.4	1.7	1.2	0.4	0.6	5.8	3.3	1.2	0.7
Gd	13.1	10.4	15.1					9.5	13.8									
Tb	1.8	1.4	1.9	1.9	2.0	1.0		1.3	1.6	1.3	1.9	2.1	3.1	2.7	1.1	1.5	3.8	2.6
Tm	0.6	0.5	0.6	0.6	0.8	2.9		0.4	0.7									
Yb	4.0	3.1	3.6	4.9	5.8			2.4	4.6	3.6	5.4	6.1	9.8	8.1	2.5	3.1	8.9	8.5
Lu	0.3	0.2	0.3	0.6	0.8	33			0.4	0.5	0.6	0.9	1.2	1.1			1.2	1.2
Y	55	46	60	51	66	127	34	31	56	37	53	67	107	100	39	37	111	76
Nb	167	99	115	167	186	753	94	95	173	118	176	222	339	442	72	158	308	272
Zr	777	466	563	929	894		435	476	1066	681	974	1274	1779	1906	328	812	1473	907
Hf	19.0	11.5	13.3	22.1	21.6			11.3	24.3						9.2	20.9		
Ta	10.6	6.8	7.6	11.6	13.1			6.6	12.5						4.3	10.2		
Th	18.9	11.9	14.2	20.1	22.9			12.3	27.7	15.2	23.5	34.3	41.3	48	5.5	15.6	34.0	65
Sr	3	113	54			202	235	44	26	278	115	14	6	14	777	257	6	76
Rb	146	117	131	162	161.0	150	96	122	190	132	160	189	281	348	58	115	231	250
Ba	12	398	186	35	10	582	734	132	56	531	200	74	15	63	3905	1258	47	38

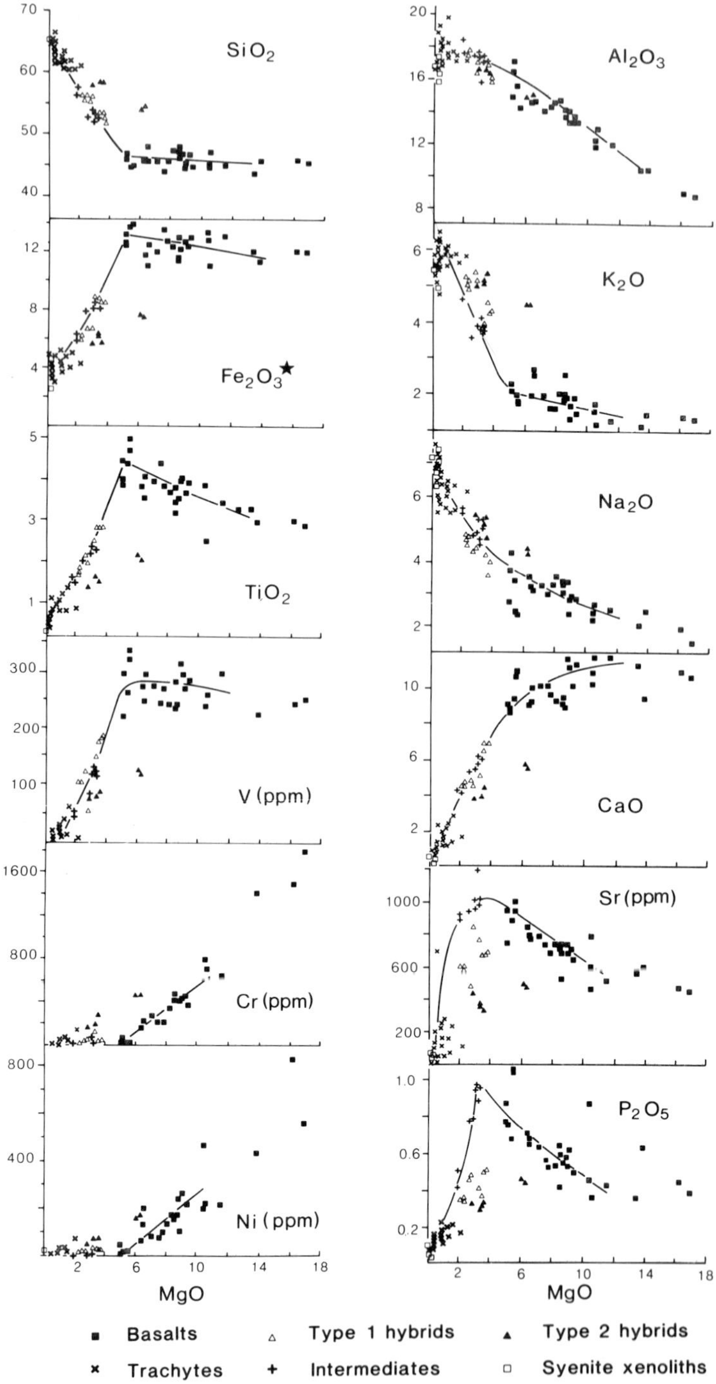

FIG 3. Major and selected trace elements versus MgO for volcanic rocks from Agua de Pau and the Waist: ———, inferred liquid line of descent. Inflections or maxima in the variation trends in the more differentiated rocks correspond to either a change in the relative modal abundances of the separating phases or a change in the phenocryst mineralogy; for example, the strontium and P_2O_5 peaks at approximately 3% MgO are consistent with feldspar fractionation, while the inflections in SiO_2, Fe_2O_3, TiO_2 and vanadium at 5% MgO suggest a larger proportion of Fe–Ti oxides in the separating assemblage. Total iron is expressed as Fe_2O_3.

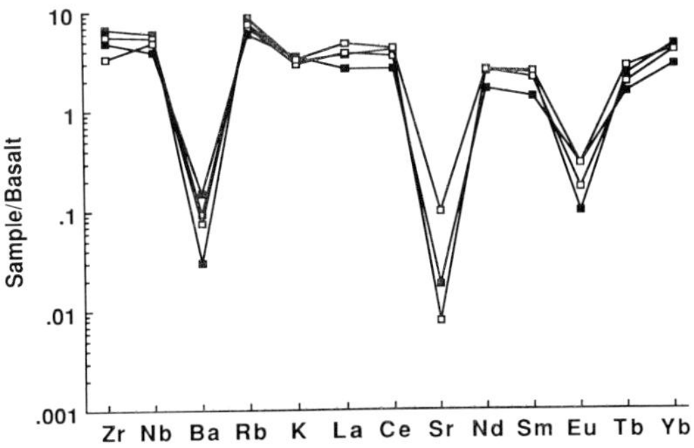

FIG 4. Element abundances in syenites (AZ1204, AZ1209) and most evolved trachytes (AZ1188, AZ1394) from Agua de Pau normalized to a high-MgO basalt (SM18) (Table 3): □, syenites; ■ trachytes.

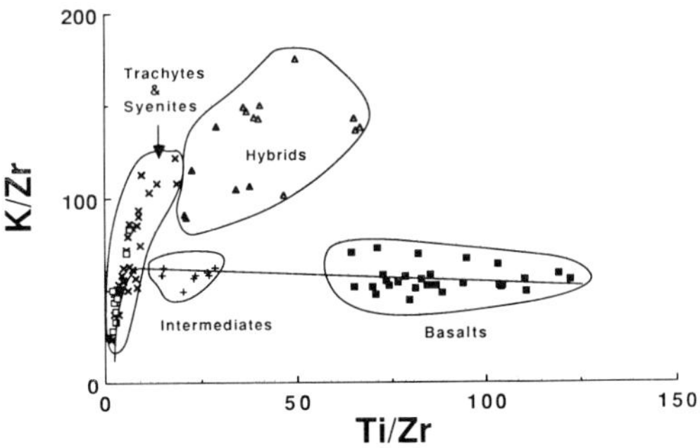

FIG 5. K/Zr versus Ti/Zr for volcanic rocks from Agua de Pau and the Waist: ——, inferred liquid line of descent. Symbols as in Fig. 3.

The syenite xenoliths have compositions very similar to the most evolved trachytes, suggesting that they are 'frozen' liquids (Fig. 4).

Chemical trends across the whole compositional spectrum are consistent with progressive fractional crystallization of, ultimately, primitive alkali basalt. This is not the case for the hybrid lavas, which plot off liquid lines of descent of elements which show maxima or inflections at intermediate MgO contents (Fig. 3). This is particularly well seen in the variation in strontium, P_2O_5 and TiO_2, where the hybrid lavas show two distinct mixing trends between basaltic and trachytic compositions. This forms the basis for their division into type 1 and type 2 hybrids, the former suggesting the involvement of a low-MgO basaltic component and the latter the involvement of a high-MgO basaltic component.

Figure 5 is a plot of K/Zr against Ti/Zr and illustrates several important features of volcanic rocks from Agua de Pau which are not readily apparent from other chemical variation diagrams. With differentiation from basalt to intermediate compositions the K/Zr ratio increases slightly, suggesting that potassium is more incompatible than zirconium (ascribed to clinopyroxene fractionation). The trachytes show two opposing trends, with about half having K/Zr ratios lower than intermediate compositions; this is consistent with alkali feldspar fractionation. However, it is also apparent that the hybrid lavas and a number of trachytes from Agua de Pau are accumulative in alkali feldspar as they have anomalously high K/Zr ratios (and high Eu/Eu* ratios (Marriner *et al.* 1982)). Thus the hybrids are not simple basalt–trachyte mixtures. The occurrence of accumulative trachyte lavas suggests a possible genetic relationship between them and the hybrid lavas.

Type 1 and type 2 hybrid lavas are distinguished from each other by a plot of K/Zr versus Ba/Zr (Fig. 6). Basalts and intermediate lavas define a tight cluster with similar K/Zr and Ba/Zr ratios, and this forms a good reference point with which to compare other compositions. Trachytes have generally lower Ba/Zr ratios but variable K/Zr ratios, consistent with some being fractionates and others accumulative. Type 1 hybrids are distinguished from type 2 hybrids by having significantly higher Ba/Zr ratios. In type 1 hybrids the Ba/Zr ratio also shows a small negative correlation with

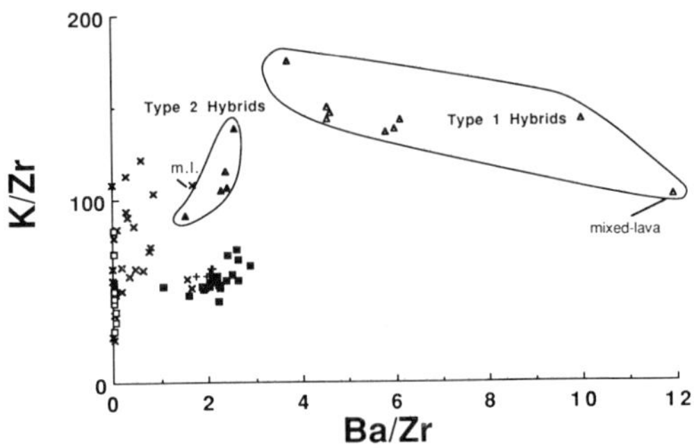

FIG 6. K/Zr versus Ba/Zr for volcanic rocks from Agua de Pau and the Waist. Also indicated are mafic and trachytic components of mixed lava (m.l.; AZ1721). Symbols as in Fig. 3.

K/Zr, consistent with the positive correlation between the albite and barium contents of anorthoclases from these rocks (Table 2). Also plotted in Fig. 6 is a trachyte lava from Agua de Pau which is unusual because of the presence of more mafic magmatic inclusions which have a mineralogy and composition similar to the type 1 hybrids. The lower Ba/Zr ratio of the enclosing trachyte relative to the mafic inclusions demonstrates that the trachyte, although accumulative, cannot be the high-barium, high-europium end-member required by mixing models to produce the type 1 hybrid compositions. Some of the type 2 hybrids could possibly be due to mixing of basalt with the most accumulative trachyte compositions observed in Fig. 5.

Geochemical constraints on the origin of the hybrid lavas

Selective potassium diffusion

The europium-, barium- and potassium enriched nature of the hybrid lavas is a feature which suggests that they do not simply represent mixtures of basaltic and trachytic magmas but are actually accumulative in alkali feldspar. However, an alternative mechanism for the observed potassium enrichment is that it is diffusion controlled. Watson (1982) has demonstrated that rapid selective potassium diffusion from solid felsic rocks into more basic magmas may be an important mechanism of potassium enrichment for mafic liquids exposed to a large surface area of solid felsic rock, *eg* a basaltic rock carrying syenite xenoliths. Some of the potassium enrichment observed in the contaminated intermediates from Agua de Pau volcano might conceivably be due to this mechanism; syenite xenoliths are observed in some trachytes and basalts from Sao Miguel. However, this mechanism cannot explain the positive europium anomalies of the hybrids, or their relative lack of rubidium enrichment, nor does it explain the relative potassium enrichment in some trachytes where solid–melt potassium activity gradients would be negligible.

Nature of the alkali feldspar contaminant

At Agua de Pau alkali feldspar only precipitates as a phenocryst phase in magmas of trachytic composition, and therefore the anorthoclase and sanidine xenocrysts in the hybrid lavas must be derived from a trachyte magma, associated cumulates or a solid rock of trachytic composition. The elements which are most sensitive to fractionation in the Agua de Pau trachytes are barium, europium and strontium, all of which partition strongly into alkali feldspar. Alkali feldspar first starts to precipitate as a phenocryst phase in metaluminous LET which has high barium, europium and strontium contents and a moderate zirconium content. With crystal fractionation, principally involving alkali feldspar (more than 80% of the separating assemblage), LET evolves to mildly peralkaline most evolved trachyte (MET) which is strongly enriched in incompatible trace elements (*eg* zirconium) and highly depleted in barium, europium and strontium (Storey 1981). A large section of this compositional spectrum, from LET to MET, is shown by the zoned Fogo A deposit (Widom *et al.* 1987). The favoured differentiation mechanism for producing MET from LET is side-wall crystallization and boundary-layer uprise of MET liquid (Wolff & Storey 1984). Clearly, the trace-element contents of any cumulate crystals which are forming on the walls of the chamber would be a function of the composition of the crystallizing magma and the element bulk distribution coefficents *D*. Figure 7 is a schematic illustration of the compositional gradient of

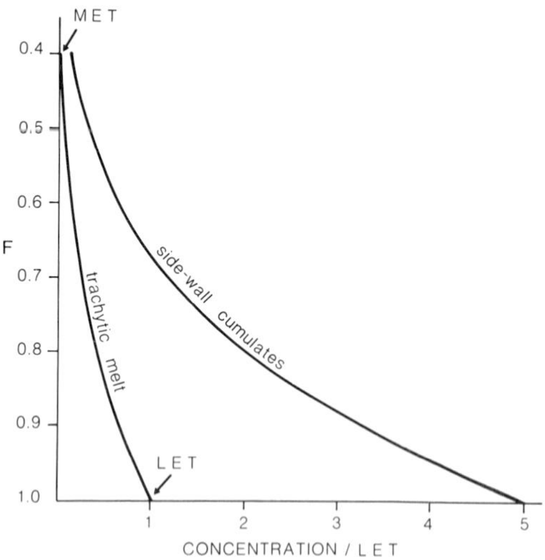

FIG 7. Schematic illustration of the variation in barium concentration, assuming Rayleigh fractionation, in a compositionally zoned trachytic magma body with coexisting side-wall cumulates. *F* is the weight fraction of melt remaining. It is assumed that $D_{Ba} = 5$ (Hanson 1978).

barium, assuming simple Rayleigh fractionation, in a trachyte melt with coexisting sidewall cumulates in a zoned system. Barium (and europium and strontium) concentrations are rapidly depleted in both the trachyte and sidewall cumulates by fractional crystallization. It is evident that only alkali feldspar crystallizing from LET will have sufficient barium to produce the observed enrichment in type 1 hybrids. Type 2 hybrid lavas, although clearly alkali feldspar accumulative, are not as strongly enriched in barium and europium, which suggests that they have been contaminated by feldspars which have precipitated from a slightly more evolved trachyte.

The occurrence of large cognate syenite blocks in several Agua de Pau eruptives suggests a possible contamination mechanism involving settling through trachyte magma into underlying basalt. Dynamic objections to gravitational settling are overcome by invoking sufficiently large blocks (Storey *et al.* 1984). However, the syenites are incapable of supplying the required chemical signature. Analysed syenites (Table 3) generally have the major-and trace-element compositions of MET liquids, and are therefore non-accumulative and strongly *depleted* in europium and barium. Furthermore, anorthoclase, which is observed to accompany sanidine as xenocrysts in the hybrid lavas, is absent from these syenites.

Identification of end-members for hybrids

The basaltic end-member

The hybrid lavas define two separate mixing trends. Back-projection of these mixing trends onto the liquid line of descent gives an estimate of the MgO content of the basalt mixing component. This indicates 5–6 wt% MgO (evolved basalt (EB)) for type 1 hybrids and 9–12 wt% MgO (primitive basalt (PB)) for type 2 hybrids, consistent with the higher nickel and chromium contents of the latter (Table 3). Both hybrid types contain olivine xenocrysts, and therefore the Mg/Fe^{2+} ratio of the liquid from which they precipitated can be estimated (Roeder & Emslie 1970). If the smallest measured value for the Fe_2O_3/FeO ratio in Sao Miguel basalts (0.27) is taken to approximate magmatic values and it is assumed that the total iron is equivalent to 12 wt% Fe_2O_3, it is possible to estimate the MgO content of the basalt from which the olivine grew. This indicates that the olivines (Fo_{81}) in two type 1 hybrids (AZ1705, AZ1721) are derived from a basalt with 6.5 wt% MgO. The olivines (Fo_{88}) in MA239 (Type 2 hybrid) would have been in equilibrium with a basalt with 11.4 wt% MgO. This is in good agreement with the basaltic end-members deduced from the mixing trends on the MgO variation diagrams.

The felsic end-member

The anomalously high K/Zr, Ba/Zr and Eu/Eu* ratios of the hybrid lavas show that the trachytic mixing component is accumulative in alkali feldspar. However, it is not easy to identify the relative proportions of alkali feldspar and LET on the basis of major-element variations because of the close compositional similarity. This is demonstrated by least-squares modelling of the hybrids (Table 4); either alkali feldspar or LET as the felsic mixing component can give reasonably low residuals and both solutions involve similar proportions of basalt.

A more profitable approach for estimating the relative proportion of alkali feldspar in the LET component (degree of accumulation) is to compare the concentration of incompatible elements (*eg* zirconium) in the hybrid lavas with those predicted by mixing various proportions of phenocrysts, LET and EB–PB (Fig. 8). Type 1 and type 2 hybrids from Agua de Pau are plotted in Fig. 8 using the respective mixing proportions of EB and PB obtained from major-element mass balance calculations (Table 4). The observed dilution of zirconium relative to the abundances expected if simple binary mixing of basalt and trachyte had operated suggests that the LET component was 55%–75% accumulative for type 1 hybrids and 30%–65% accumulative for type 2 hybrids. Using these mixing proportions it is possible to obtain a rough estimate of what the 'average' barium content of the alkali feldspar contaminant should be in order to produce the observed concentrations in the hybrids. This has been done for AZ1721 and indicates an average BaO content of 1.3 wt% in the alkali feldspar. This agrees with the measured range of BaO contents in alkali feldspars from this sample (Table 2).

Physical mechanism of alkali feldspar accumulation and contamination

There are several mechanisms by which LET may have accumulated alkali feldspar.

(1) *In situ* growth of alkali feldspar from LET,

TABLE 4. *Petrological mixing models for hybrid magmas from Agua de Pau volcano*

Mixing components	Type 1 hybrids							Type 2 hybrids			
	AZ1705			SM8			SM25	SM21	SM29	MA239	SM56
LET (AZ1377)	57.4		17.9	67.5		44.6	37.6	57.4	82.6	80.2	58.0
EB (AZ1324)	42.6	51.3	55.1	32.5	48.2	36.0	36.5				
PB1 (AZ1872									17.4	19.8	
PB2 (SM62)								42.6			42.0
Alkali feldspar		40.5	21.4		49.4	17.7	21.4				
Clinopyroxene		6.4	5.0		2.4	1.7	4.5				
Titanomagnetite		1.8	0.7								
Σr^2	2.064	0.428	0.226	0.981	2.932	0.099	0.408	0.705	0.118	0.163	0.510

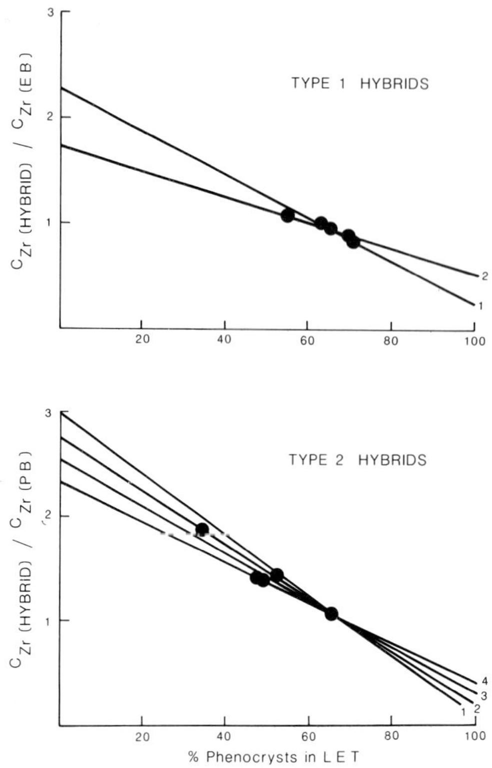

FIG 8. (*a*) Model concentration of zirconium in type 1 hybrid magmas normalized to EB as a function of percentage phenocrysts in the LET mixing component: line 1, hybrid = 25% EB + 75% LET; line 2, 55% EB + 45% LET. (*b*) Model concentration of zirconium in type 2 hybrid magmas normalized to PB as a function of percentage phenocrysts in the LET mixing component: line 1, hybrid = 10% PB + 90% LET; line 2, 20% PB + 80% LET; line 3, 30% PB + 70% LET; line 4, 40% PB + 60% LET. Assumed zirconium contents for LET, EB and PB are 800 ppm, 300 ppm and 250 ppm respectively.

on the walls and floor of the trachyte layer, with convective uprise of less dense fractionate. This would give the highly accumulative LET compositions required to produce the observed contamination of the hybrids. Although this mechanism cannot be rejected outright, it is difficult to envisage how it could account for the occurrence of erupted LETs which are only moderately accumulative. If the LET is underlain by hotter basaltic magma, it seems unlikely that the interface itself could act as a nucleation zone for alkali feldspar precipitation. A more attractive possibility is that the crystals have been derived from elsewhere in the chamber and concentrated at the base of the trachytic layer by essentially sedimentary processes.

(2) Settling of individual crystals can be rejected as a viable process in felsic magma chambers on the basis of dynamic arguments (Sparks *et al.* 1984).

(3) Sinking of crystal-rich plumes from the magma chamber roof. Brandeis & Jaupart (1986) have shown that thermal convection is weak in felsic magmas and consequently crystal-rich plumes descending from the roof are unlikely. In this case, however, trachytic magma is underlain by basaltic magma and so convection may be stronger. Even so, crystallizing trachytic magma at the chamber roof in the zoned Agua de Pau system has been shown to be an unlikely contaminant for producing the high europium and barium contents of the hybrid lavas.

(4) Density flows. Evidence for crystal deposition from flowing magma is common in layered intrusions, and has been recorded from a number of granitic and syenitic plutons (Smith 1975; Barriere 1981; Parsons

& Butterfield 1981). Sedimentary structures in these examples include cross-bedding, channel scours and load structures. Brandeis & Jaupart (1986) suggest that the occurrence of these structures in layered intrusions is dependent on side-wall proximity and that the effects of the currents that produce them are most strongly felt in small magma chambers that have a high surface area-to-volume ratio. One probable way of forming sedimented layers is by the slumping of crystal mush from the walls of the chamber (Irvine 1980). Theoretical studies (Spera *et al.* 1982; Thompson & McBirney 1985) predict that a crystal mush zone should occur close to the walls of the crystallizing magma chamber. The greater bulk density of the crystal mush zone, relative to that of the ambient magma, will render it gravitationally unstable and prone to form slumps. These may develop into magmatic density currents when they encounter a chamber floor or a density barrier (*ie* a compositional interface) which channels the crystal–liquid suspension into horizontal flow.

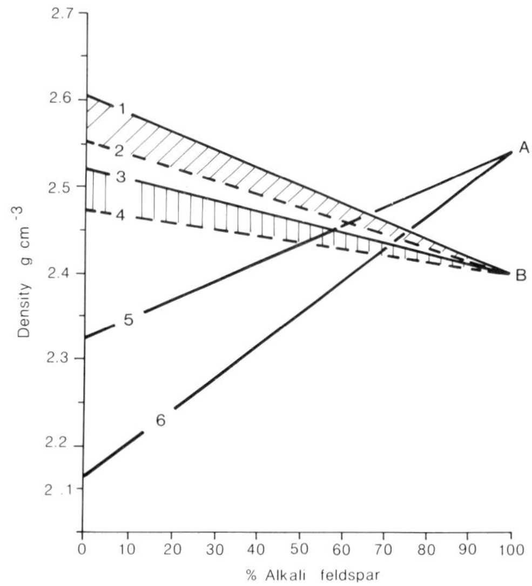

FIG 9. Density variation of LET (AZ1378) and basalt (AZ1324) with added solid (A) and liquid (B) alkali feldspar respectively: line 1, EB with 1 wt% H_2O plus liquid alkali feldspar (Or_{30}); line 2, as line 1 but shifted by -0.05 g/cm^{-3}, the maximum discrepancy between the measured values of Murase & McBirney (1973) and values obtained by the calculation method of Mo *et al.* (1982); lines 3 and 4, as lines 1 and 2 respectively but with 2 wt% H_2O in MEB; line 5, LET with 2 wt% H_2O plus solid alkali feldspar; line 6, as line 5 but with 5 wt% H_2O. The measured density of liquid alkali feldspar is from Carmichael *et al.* (1977).

Mechanisms for contaminating basalt with accumulative least evolved trachyte

The density relationships between basalt, LET and solid and liquid alkali feldspar are important in that they will determine the relative buoyancy of crystal-rich pods of LET if stratified above or intruded by basalt. Densities have been calculated for LET and basalt for various water contents (Fig. 9) using the method of Mo *et al.* (1982). The accuracy of this procedure has been estimated for volcanic rocks with a wide compositional range by comparing our calculated densities with laboratory measurements given by Murase & McBirney (1973). The discrepancy between the two values is never more than about ± 0.05 g cm^{-3}. LET densities have been calculated for magmatic water contents of 2 and 5 wt% H_2O, the latter being in the range given for trachytic magmas from Agua de Pau by Wolff & Storey (1983*a*, *b*). It is not unreasonable to suggest that basalts from Sao Miguel would contain between 1 and 2 wt% H_2O (Aoki *et al.* 1981), and for these magmatic water contents the density relationships in Fig. 9 indicate that LET magma is always less dense than basalt. Alkali feldspar is more dense than both LET and basalt with 2 wt% H_2O.

It is therefore conceivable that, if basalt pooled below accumulative LET magma which contained a sufficiently high proportion of suspended alkali feldspar, then it might be possible for it to be contaminated by denser crystal-rich plumes of LET sinking through the interface. An alternative, but not necessarily mutually exclusive, mechanism would be the injection and ponding of basalt at the base of accumulative LET magma. The neutral or negative buoyancy of highly accumulative LET would allow efficient mixing with the basalt. Resorption of alkali feldspar into basalt will lower the melt density owing to the lower density of alkali feldspar liquid (Fig. 9). This would aid the first mechanism which suggests that, if it is initiated, the sinking process may rapidly become irreversible and lead to massive contamination of the basaltic magma.

A simple experiment to illustrate the behaviour of a sediment-laden density current in a zoned system

The behaviour of crystal-laden density currents or plumes descending through a zoned magma column will depend on the changing density contrast between plume and ambient liquid. In order to illustrate a possible mode of behaviour of a suspension pooled at a density step in a zoned system, the following simple laboratory experiments were carried out.

A layered system was created in a Plexiglass tank measuring 35 cm × 30 cm × 10 cm by carefully emplacing tapwater above a concentrated Epsom salt ($MgSO_4{\cdot}7H_2O$) solution with a density of approximately 1.4 g cm^{-3}. Slight mixing of the two liquids resulted in a transitional zone less than 0.5 cm deep. A suspension of finely ground glass powder in water was then prepared such that the density was intermediate between the densities of the two liquid layers in the tank; the solids content of the suspension was approximately 20%.

Roughly 100 cm^3 of suspension was injected at the interface between the upper and lower layers (Fig.10*a*). During injection of the suspension some entrainment of both upper and lower layers occurred. After the flow was shut off, the suspension became evenly distributed as a discrete layer of uniform depth between the upper and lower liquids (Fig.10*b*) and remained pooled for a significant time with settling of only a few anomalously large suspended particles (Fig.10*c*). After 1.5 min plumes were observed to begin falling from the suspension into the lower layer (Fig.10*d*). The number of plumes rapidly increased until massive penetration of the lower layer by falling plumes occurred across the entire width of the tank (Fig.10*e,f*), making further observation by means of the shadowgraph technique impossible.

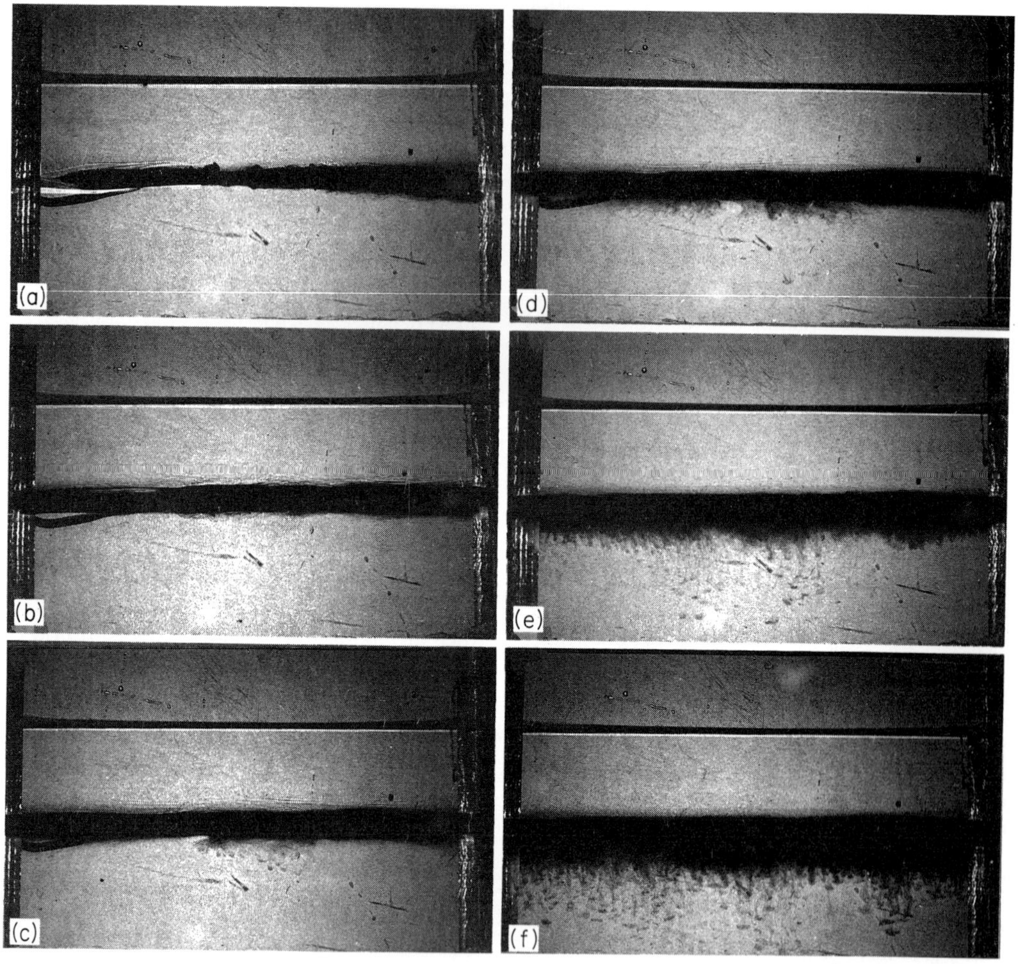

FIG 10. Shadowgraphs of tank experiments: (*a*) after 13 s, still injecting slurry; (*b*) after 33 s, injection of slurry stopped; (*c*) after 1 min 10 s; (*d*) after 1 min 33 s; (*e*) after 2 min 12 s; (*f*) after 2 min 41 s.

This experiment illustrates a number of important features of the case where a solids-laden flow pools above a denser fluid as, for example, may also occur in oceans when a turbidity current encounters a bottom layer of dense brine in a semiclosed marine basin (K. T. Pickering, pers. comm.). In the Plexiglass tank the suspension initially behaves as a homogeneous fluid with uniform density intermediate between the upper and lower ambient layers, as indicated by the pooling behaviour at the interface observed during the first 1.5 min of the experiment. The transition to plumes of suspension falling through the interface can only have been possible when the density of the suspension locally exceeded that of the underlying layer. As the transition occurred significantly after the flow came to rest, this heterogeneous density distribution within the suspension must have been due to gravitational settling of solid particles. Clearly, the solids concentration in the descending plumes must exceed that of the suspension that remains at the original level of the interface. It is this relationship between the separation of high-concentration and low-concentration suspensions from an initially homogeneous dispersion that may be relevant to the case studied here.

Discussion

We find that the most attractive explanation for the association of the hybrid and accumulative LET lavas at Agua de Pau is that alkali feldspar and the other (minor) trachytic phenocrysts, clinopyroxene, Fe–Ti oxides and biotite were preferentially concentrated on the floor of a trachyte magma chamber or, alternatively, at a trachyte–basalt compositional interface by crystal-rich slumps and density currents (Fig. 11). A trachytic density current carrying suspended alkali feldspar crystals will have a higher density than the ambient trachyte, although for

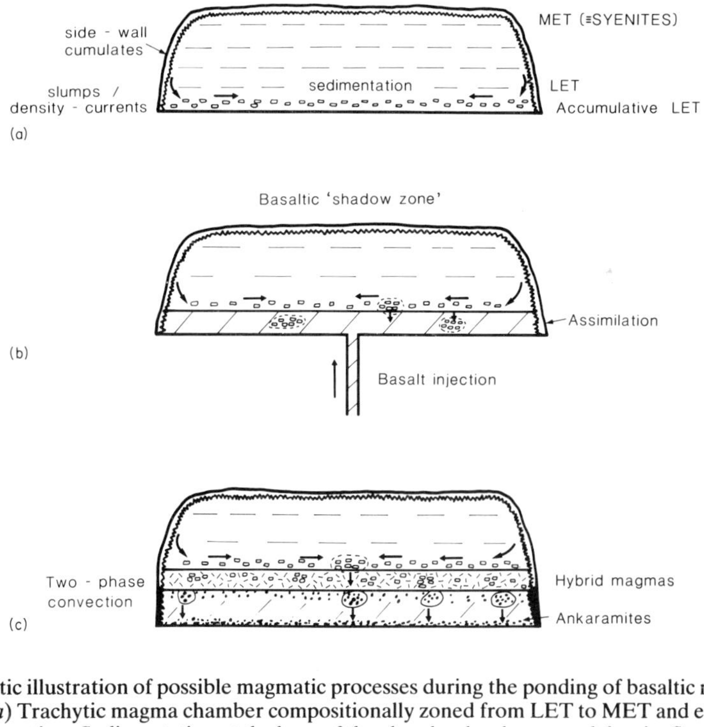

FIG. 11. Schematic illustration of possible magmatic processes during the ponding of basaltic magma beneath felsic magmas. (*a*) Trachytic magma chamber compositionally zoned from LET to MET and evolving by convective fractionation. Sedimentation at the base of the chamber by slumps and density flows forms an accumulative 'carpet'. (*b*) Episodic injection and ponding of basalt at the the base of LET. Assimilation of LET side-wall cumulates and accumulative LET layer at base. Highly accumulative LET sinks through the interface. (*c*) As hybridization proceeds, the density of the contaminated magma is lowered, aiding further contamination by sinking plumes of highly accumulative LET. Undercooling of ponded basalt results in phenocryst nucleation below the interface and two-phase convection, forming ankaramites at the base of the chamber.

crystal contents that still permit the current to behave as a fluid (less than 50%–60% crystals (Marsh 1981)) it will probably be less dense than the basaltic magma (Fig. 9). The crystal-rich current could then flow and eventually pool above the interface between the higher-density basaltic magma and the lower-density LET. Slumping of crystal mush from the walls of the trachytic portion of the chamber provides a simple mechanism for generating crystal-rich accumulative trachyte compositions either on the LET floor or in the neighbourhood of a compositional interface with basalt. Sedimentation within the dense crystal-rich layer then generates highly accumulative negatively buoyant 'packets' which fall into the underlying mafic magma. This mechanism thus both produces the highly accumulative 'contaminant' required by geochemical modelling of the hybrids and accounts for associated moderately accumulative LET magma (*ie* the low concentration suspension which remains pooled above the interface). However, segregation of crystals may also occur within the moving density current. Irvine (1980) carried out a series of laboratory experiments using viscous materials (glycerine and silicone) to investigate the dynamics of magmatic density currents flowing across the floor of a mafic magma chamber. Irvine found that the glass shards suspended in his density flows were transferred from the fastest part of the current, located at the top, to the slowest part, located at the bottom. This process, akin to flowage differentiation and operating in a magmatic density current, would produce local parcels of highly accumulative magma at the base of the moving flow. This mechanism is also attractive because it duplicates the solids segregation features of our experiments but does not depend on particle settling through a stationary felsic magma which may well have possessed a finite yield strength (Murase & McBirney 1973). Whichever mechanism operates, the end result is a crystal 'carpet' located at the interface beneath accumulative trachyte magma and above basaltic magma. In this regard the following points deserve mention.

(1) Geochemical modelling indicates that the trachytic component of the contaminated basalts may contain up to 75% alkali feldspar. At this phenocryst content any 'magma' effectively behaves as a solid (Marsh 1981). The contaminant may therefore enter the basaltic magma in the form of rafts and clumps of crystals with trapped liquid rather than as a crystal mush which is still fluid. The feldspar component of these clumps will remelt almost immediately upon entry into the hotter basaltic magma, both by conductive heating and by flushing of basaltic liquid through the porous block.

(2) As we have already indicated, at low water contents (1–2 wt%) basaltic magma is denser than liquid alkali feldspar and therefore assimilation of solid alkali feldspar will reduce its density (Fig. 9), particularly if a proportion of hydrous trachytic liquid is added simultaneously. This will aid the contamination process envisaged.

(3) Magma-mixing mechanisms proposed previously typically involve some form of catastrophic overturn of a compositionally zoned magma column as a result of eruption or the development of less dense magma at the base of the chamber by fractional crystallization or volatile exsolution (Eichelberger 1980; Huppert & Sparks 1980; Huppert *et al.* 1982; Koyaguchi 1985; Blake & Campbell 1986). However, the fact that the trachyte lava and the contaminated basaltic inclusions (mixed lava AZ1721) have both preserved their individual geochemical identities and the crucial involvement of accumulative magmas in the hybridization suggest that the basalt must have *passively* received highly accumulative LET while it was stably ponded at the base of the magma chamber.

(4) Basalt pooled beneath trachytic magma will experience a significant amount of undercooling next to the interface. If this situation prevails for any length of time, large amounts of crystal nucleation should be promoted, leading to the possibility of rapid differentiation of the basalt by two-phase convection (Morse 1986). This could account for the eruption of some highly accumulative ankaramite lavas from flank vents on Agua de Pau volcano, and the apparent absence of such rocks from fissure zones in the Waist (Fig. 11).

(5) The general applicability of the hybridization mechanism proposed here awaits further investigation of the densities of accumulative packets of felsic magmas in relation to less evolved melts. However, the occurrence of geochemically similar hybrid lavas from Furnas volcano on Sao Miguel and from the island of Faial (unpublished data) suggests that similar processes may operate at other centres of bimodal volcanism.

ACKNOWLEDGMENTS: We are indebted to Professor G. P. L. Walker, Dr I. L. Gibson and Dr R. Croasdale for supplying the samples from Sao Miguel. Laboratory assistance during the tank experiments was provided by James Talbot, Jay Miller and Lee Stiborek. We appreciate the thoughtful and careful reviews of this paper by P. E. Baker, R. M. Briggs, R. Macdonald and A. D. Saunders.

References

AOKI, K., ISHIWAKA, K. & KANISAWAS, S. 1981. Fluorine geochemistry of basaltic rocks from continental and oceanic regions and petrogenetic application. *Contributions to Mineralogy and Petrology* **76**, 53–59.

BARRIERE, M. 1981. On curved laminae, graded layers, convection currents and dynamic crystal sorting in the Ploumanc'h (Brittany) subalkaline granite. *Contributions to Mineralogy and Petrology* **77**, 214–224.

BLAKE, S. & CAMPBELL, I. H. 1986. The dynamics of magma-mixing during flow in volcanic conduits. *Contributions to Mineralogy and Petrology* **94**, 72–81.

BOOTH, B., CROASDALE, R. & WALKER, G. P. L. 1978. A quantitative study of five thousand years of volcanism on Sao Miguel, Azores. *Philosophical Transactions of the Royal Society of London, Seres A* **288**, 271–319.

BRANDEIS, G. & JAUPART, C. 1986. On the interaction between convection and crystallization in cooling magma chambers. *Earth and Planetary Science Letters* **77**, 345–361.

CAMPBELL, I. H. & TURNER, J. S. 1986. The influence of viscosity on fountains in magma chambers. *Journal of Petrology* **27**, 1–30.

CANN, J. R. 1967. A second occurrence of dalyite and the petrology of some ejected syenite blocks from Sao Miguel, Azores. *Mineralogical Magazine* **36**, 227–232.

CARMICHAEL, I. S. E., NICHOLLS, J., SPERA, F. J., WOOD, B. J. & NELSON, S. A. 1977. High-temperature properties of silicate liquids: applications to the equilibration and ascent of basic magma. *Philosophical Transactions of the Royal Society of London, Series A* **286**, 373–431.

CLAGUE, D. A. 1978. The oceanic basalt–trachyte association: an explanation of the Daly Gap. *Journal of Geology* **86**, 739–743.

DALY, R. A. 1925. The geology of Ascension Island. *Proceedings of the American Academy of Arts and Sciences* **5**, 1–80.

EICHELBERGER, J. C. 1980. Vesiculation of mafic magma during replenishment of silicic magma reservoirs. *Nature* **288**, 446–450.

FLOWER, M. F. J., SCHMINCKE, H.-U. & BOWMAN, H. 1976. Rare-earth and other trace elements in historic Azorean lavas. *Journal of Volcanology and Geothermal Research* **1**, 127–147.

HANSON, G. N. 1978. The application of trace elements to the petrogenesis of igneous rocks of granitic composition. *Earth and Planetary Science Letters* **38**, 26–43.

HILDRETH, W. 1979. The Bishop Tuff: evidence for the origin of compositional zonation in silicic magma chambers. In: CHAPIN, C. E. & ELSTON, W. E. (eds) *Ash-flow Tuffs*. Geological Society of America Special Publication **180**, 43–75.

HUPPERT, H. E. & SPARKS, R. S. J. 1980. The fluid dynamics of a basaltic magma chamber replenished by influx of hot, dense ultrabasic magma. *Contributions to Mineralogy and Petrology* **75**, 279–289.

——, —— & TURNER, J. S. 1982. Effects of volatiles on mixing in calc-alkaline magma systems. *Nature* **297**, 554–557.

IRVINE, T. N. 1980. Magmatic density currents and cumulus processes. *American Journal of Science* **280-A**, 1–58.

KOYAGUCHI, T. 1985. Magma mixing in a conduit. *Journal of Volcanology and Geothermal Research* **25**, 365–369.

LAUGHTON, A. S. & WHITMARSH, R. B. 1975. The Azores–Gibraltar plate boundary. *In*: KRISTJANSSON, (ed.) *Geodynamics of Iceland and the North Atlantic Area*. Reidel, Dordrecht, 63–81.

LEAT, P. T. MACDONALD, R. & SMITH, R. L. 1984. Geochemical evolution of the Menengai caldera volcano, Kenya. *Journal of Geophysical Research* **89**, 8571–8592.

MARRINER, G. F., NORRY, M. J. & GIBSON, I. L. 1982. The petrology and geochemistry of Agua de Pau volcano, Sao Miguel, Azores. *Proceedings of 1980 International Symposium on the Activity of Oceanic Volcanoes (IAVCEI)*, 159–173.

MARSH, B. D. 1981. On the crystallinity, probability of occurrence, and rheology of lava and magma. *Contributions to Mineralogy and Petrology* **78**, 85–98.

MICHAEL, P. J. 1983*a*. Chemical differentiation of the Bishop Tuff and other high-silica magmas through differentiation processes. *Geology* **11**, 31–34.

—— 1983*b*. Chemical differentiation of the Bishop Tuff and other high-silica magmas through differentiation processes: reply. *Geology* **11**, 623–624.

MO, X., CARMICHAEL, I. S. E., RIVERS, M. & STEBBINS, J. 1982. The partial molar volume of Fe_2O_3 in multicomponent silicate liquids and the pressure dependence of oxygen fugacity in magmas. *Mineralogical Magazine* **45**, 237–245.

MORSE, S. A. 1986. Thermal structure of crystallizing magma with two-phase convection. *Geological Magazine* **123**, 205–214.

MURASE, T. & MCBIRNEY, A. R. 1973. Properties of some common igneous rocks and their melts at high temperatures. *Bulletin of the Geological Society of America* **84**, 3563–3592.

O'NIONS, A. K. & GRONVOLD, K. 1973. Petrogenetic relationships of acid and basic rocks in Iceland: Sr-isotope and rare earth elements in late and post-glacial volcanics. *Earth and Planetary Science Letters* **19**, 397–409.

PARSONS, I. & BUTTERFIELD, A. W. 1981. Sedimentary features of the Nunarssuit and Klokken syenites, S. Greenland. *Journal of the Geological Society, London* **138**, 289–306.

ROEDER, P. L. & EMSLIE, R. F. 1970. Olivine–liquid equilibrium. *Contributions to Mineralogy and Petrology* **26**, 275–289.

SIGURDSSON, H. 1977. Generation of Icelandic rhyolites by melting of plagiogranites in the oceanic layer. *Nature* **269**, 25–28.

—— & SPARKS, R. S. J. 1981. Petrology of rhyolitic and mixed-magma ejecta from the 1875 eruption of Askja, Iceland. *Journal of Petrology* **22**, 41–84.

SMITH. R. L. 1979. Ash-flow magmatism. *In*: CHAPIN. C. E. & ELSTON, W. E. (eds) *Ash-flow Tuffs*. Geological Society of America Special Paper **180**, 5–72.

SMITH, T. E. 1975. Layered granitic rock at Chebucto Head, Halifax County, Nova Scotia. *Canadian Journal of Earth Sciences* **12**, 456–463.

SPARKS, R. S. J., HUPPERT, H. E. & TURNER, J. S. 1984. The fluid dynamics of evolving magma chambers. *Philosophical Transactions of the Royal Society of London, Series A* **310**, 511–534.

——, SIGURDSSON, H. & WILSON, L. 1977. Magma mixing: a mechanism for triggering acid explosive eruptions. *Nature* **267**, 315–318.

SPERA, F. J., YUEN, D. A. & KIRSCHVINK, S. J. 1982. Thermal boundary layer convection in silicic magma chambers: effects of temperature-dependent rheology and implications for thermogravitational chemical fractionation. *Journal of Geophysical Research* **87**, 8755–8767.

STOREY, M. 1981. Trachytic pyroclastics from Agua de Pau volcano, Sao Miguel, Azores: evolution of a magma body over 4,000 years. *Contributions to Mineralogy and Petrology* **78**, 423–432.

——, WOLFF, J. A. & BALSLEY, S. D. 1984. Mixed-magma lavas from the zoned magma chambers of Agua de Pau and Furnas volcanoes, Sao Miguel, Azores: evidence supporting crystal settling. *EOS (Transactions of the American Geophysical Union)* **65**, 1122.

THOMPSON, M. E. & MCBIRNEY, A. R. 1985. Redistribution of phenocrysts by convective flow in a viscous boundary layer. *Journal of Volcanology and Geothermal Research* **24**, 83–94.

WALKER, G. P. L. 1966. Acid volcanic rock in Iceland. *Bulletin Volcanologique* **29**, 375–402.

—— & Croasdale, R. 1971. Two plinian-type eruptions in the Azores. *Journal of the Geological Society, London* **127**, 17–55.

WATSON, E. B. 1982. Basalt contamination by continental crust: some experiments and models. *Contributions to Mineralogy and Petrology* **80**, 73–87.

WHITE, W. M., TAPIO, M. D. M. & SCHILLING, J.-G. 1979. The petrology and geochemistry of the Azores Islands. *Contributions to Mineralogy and Petrology* **69**, 201–213.

WIDOM, E., SCHMINCKE, H.-U. & GILL, J. 1987. Chemical zonation and U-series disequilibria in Fogo A, Sao Miguel, Azores. *EOS (Transactions of the American Geophysical Union)* **68**, 1522.

WOLFF, J. A. 1985. Zonation, mixing and eruption of silica-undersaturated alkaline magma: a case study from Tenerife, Canary Islands. *Geological Magazine* **122**, 623–640.

—— & STOREY, M. 1983*a*. The volatile component of some pumice-forming alkaline magmas from the Azores and Canary Islands. *Contributions to Mineralogy and Petrology* **82**, 66–74.

—— & —— 1983*b*. The use of tephra studies to determine volatile release during large explosive volcanic eruptions. *In*: PEPIN, R.O. & O'CONNELL, R. (eds) *Conference on Planetary Volatiles*. Lunar and Planetary Institute Technical Report **83–01**, 194–198.

—— & —— 1984. Zoning in highly alkaline magma bodies. *Geological Magazine* **121**, 563–575.

WÖRNER, G. & SCHMINCKE, H.-U. 1984. Mineralogical and chemical zonation of the Laacher See tephra sequence. *Journal of Petrology* **25**, 805–835.

YODER, H. S. 1973. Contemporaneous basaltic and rhyolitic magmas. *American Mineralogist* **58**, 153–171.

M. STOREY, Department of Earth Sciences, University of Waikato, Private Bag, Hamilton, New Zealand. *Present address*: Department of Geology, University of Leicester, Leicester LE1 7RH, UK.

J. A. WOLFF, Department of Geology, University of Texas at Arlington, UTA Box No 19049, Arlington, TX 76019, USA.

M. J. NORRY, Department of Geology, University of Leicester, Leicester LE1 7RH, UK.

G. F. MARRINER, Department of Geology, Royal Holloway and Bedford New College, Egham Hill, Surrey TW20 0EX, UK.

Axial volcanism on the East Pacific Rise, 10–12°N

G. Thompson, W. B. Bryan & S. E. Humphris

SUMMARY: The East Pacific Rise axis, 10–12°N, consists of three discrete topographic or tectonic units each of which has erupted compositionally distinct magmas. The two major units have an axial bathymetric high, are bounded by transform or overlapping spreading centres and are shown to be discrete volcanoes. One of the units shows systematic variation in magma composition along the rift consistent with injection of lava down-rift and concomitant low-pressure fractional crystallization with distance from the topographic high. Compositional variation within and between magmatic units is largely controlled by fractional crystallization at low and intermediate pressures. The eastern and western limbs of a large overlapping spreading centre at about 11°45′N have compositionally distinct lavas, apparently from separate magma batches. Axial eruptions in the vicinity of a large transform fault (Clipperton Fracture Zone) show a large variation in composition and at least one independent magma injection. Highly fractionated lavas with compositions up to 59% silica are included in these eruptions. One of the axial tectonic units has, in addition to normal tholeiitic eruptions, some lavas highly enriched in incompatible elements such as potassium, titanium, strontium, barium, zirconium and lanthanum. These lavas are also isotopically enriched with respect to $^{87}Sr/^{86}Sr$, suggesting that the mantle beneath some portions of the axis is compositionally heterogeneous. The presence of compositionally distinct magma batches, often in close proximity, suggests that magma chambers underlying the East Pacific Rise axis are small and probably intermittent.

The segmentation of oceanic spreading centres into discrete units bounded by transform faults has been recognized for some time (Francheteau & Ballard 1983; Schouten & Klitgord 1983). Segments along fast spreading ridges such as the East Pacific Rise (EPR) have been shown to be bounded by transform faults and by overlapping spreading centres (OSCs), which have also been termed non-transform offsets (Lonsdale 1983; MacDonald *et al.* 1984). These and other workers (*eg* Ballard *et al.* 1981; Whitehead *et al.* 1984; Crane 1985) further showed that the individual segments often have bathymetric highs between their boundaries and have suggested that they result from thermal uplift and relatively high rates of extrusion from underlying magma chambers. Francheteau & Ballard (1983) showed that these highs are the locii of hydrothermal and magmatic activity, with a high ratio of sheet flows to pillow lavas. They suggested that the ridge axes away from the topographic highs would show decreasing hydrothermal activity, decreasing volcanic activity and decreasing ratios of sheet flows to pillows.

Based on preliminary petrological and geochemical studies of lavas along a segment of the EPR 10°–12°N, Thompson *et al.* (1985) first demonstrated that the ridge segments had distinct magma compositions and represented discrete volcanoes. Langmuir *et al.* (1986), in a wider ranging study of the EPR axis from 5°N to 14°N, showed that compositionally diverse lavas occurred within even smaller discrete tectonic segments, and that petrological boundaries often coincided with small offsets of the ridge axis within larger segments. They also confirmed that the ridge axis was made up of discrete volcanoes or magmatic units.

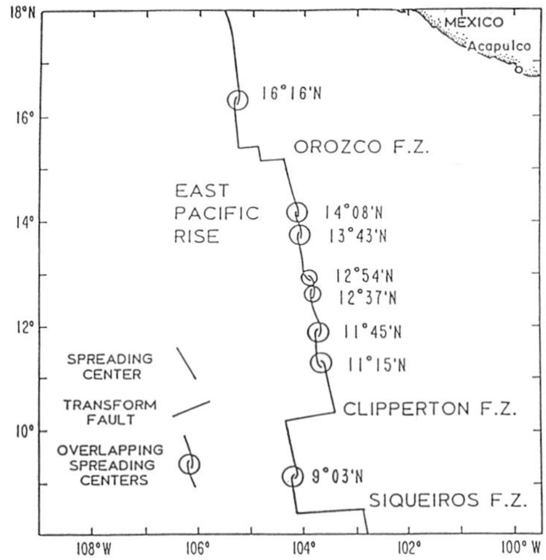

FIG. 1. Major tectonic units and segmentation of the EPR axis, 8°–18°N (from MacDonald *et al.* 1984).

The EPR from 10°N to 12°N (Fig. 1) has been

From SAUNDERS, A. D. & NORRY, M. J. (eds), 1989, *Magmatism in the Ocean Basins*, Geological Society Special Publication No. 42, pp. 181–200.

the focus of Seabeam and Seamarc studies (MacDonald *et al.* 1984; Crane 1985). In 1984, we made ANGUS (Ballard 1980) photographic studies and dredge hauls in selected portions of this part of the EPR (Thompson *et al.* 1985). In 1985 the same region was the focus of detailed ARGO (Harris *et al.* 1986) photographic coverage. The resulting geological study from these two expeditions and an assessment of seafloor topography and accreting plate fabric are discussed elsewhere (Argo Rise 1988). In this paper we examine the geochemistry and petrology of basalts erupted over the 180 km stretch of EPR axis between 10°N and 12°N. Preliminary data, based on a few selected samples (Thompson *et al.* 1985), indicated magmatic segmentation along this portion of the EPR. Here we summarize data from over 300 samples and discuss the variation in composition within and between magmatic units, their relation to tectonic segments and their possible petrogenesis.

Geological setting

The axial segment of the EPR being studied starts at a major transform, the Clipperton Fracture Zone at 10°15′N, where the neovolcanic zone is at a depth of 3100 m. The zero-age axial depth then shows a progressive shallowing northward to a first topographic high of 2500 m at about 10°55′N, and deepens slightly to a topographic low at 11°15′N, thought to be a small OSC (MacDonald *et al.* 1984) (Figs 1 and 2). The axis then rises to a second topographic high at 11°30′N and from there it deepens to a large OSC at 11°45′N; a deep nodal basin separates the overlapping ridge segments (Fig. 3*a*, *b*). In terms of ridge segmentation and the concept of 'spreading cells' (Schouten & Klitgord 1983), we appear to have two major cells or tectonic units: the first is bounded by the Clipperton Fracture Zone and the small topographic depression at 11°15′N and includes the first topographic high at 10°55′N; the second is bounded by the OSCs at 11°15′N and 11°45′N and includes the second topographic high at 11°30′N. The eastern limb of the large OSC at 11°45′N is presumably the distal end of a third cell or tectonic unit. Previous studies in this region showed that the second topographic high at 11°30′N was the locus of intense hydrothermal activity (Hekinian *et al.* 1983; Hekinian *et al.* 1985). Our ANGUS photography showed similar hydrothermal activity on the first topographic high at 10°55′N (Thompson *et al.* 1985; McConachy *et al.* 1986). Both topographic

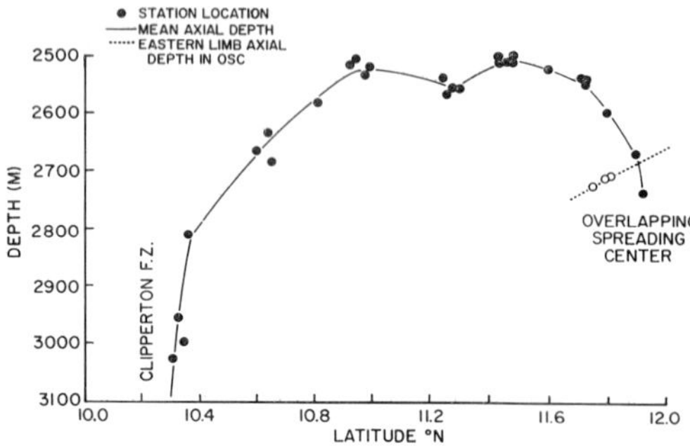

FIG. 2. Station locations along the axis of the EPR, 10–12°N. Mean axial depth along this segment is also shown.

highs have distinct axial grabens and a high ratio of sheet flows to pillow lavas (Argo Rise 1988).

Along this 180 km of EPR axis we positioned transponder nets at seven discrete localities within the three principal tectonic units where we did photography and transponder-guided dredging. We also dredged between nets, using Seabeam bathymetry and 12 kHz echo-sounding to guide the sampling. Samples were recovered on the ANGUS camera frame in some lowerings, and from an ALVIN dive at a hydrothermal vent (McConachy *et al.* 1986). Sample locations are shown in Table 1. In Fig. 2, station locations and depths are plotted as a function of latitude together with the mean axial depth.

Analytical results

Selected thin sections of samples from each site were studied petrographically for modal mineralogy mineral composition and petrogenesis. Glass fragments from individual samples were analysed for major elements using the Smithsonian microprobe (Melson 1977). Whole-rock fragments from these samples and in some instances glass fragments were also analysed for major and trace elements by X-ray fluorescence (XRF) spectroscopy (Schroeder *et al.* 1980). In all, we described and numbered over 1000 samples from which we microprobed 345 glasses and analysed 244 by XRF spectroscopy. For purposes of discussion and plotting we have taken this extensive data set and compiled the compositional mean and standard deviation at each site. In certain instances where a large variation in composition exists at a single site, the mean and standard deviation of specific chemical groupings within that section are

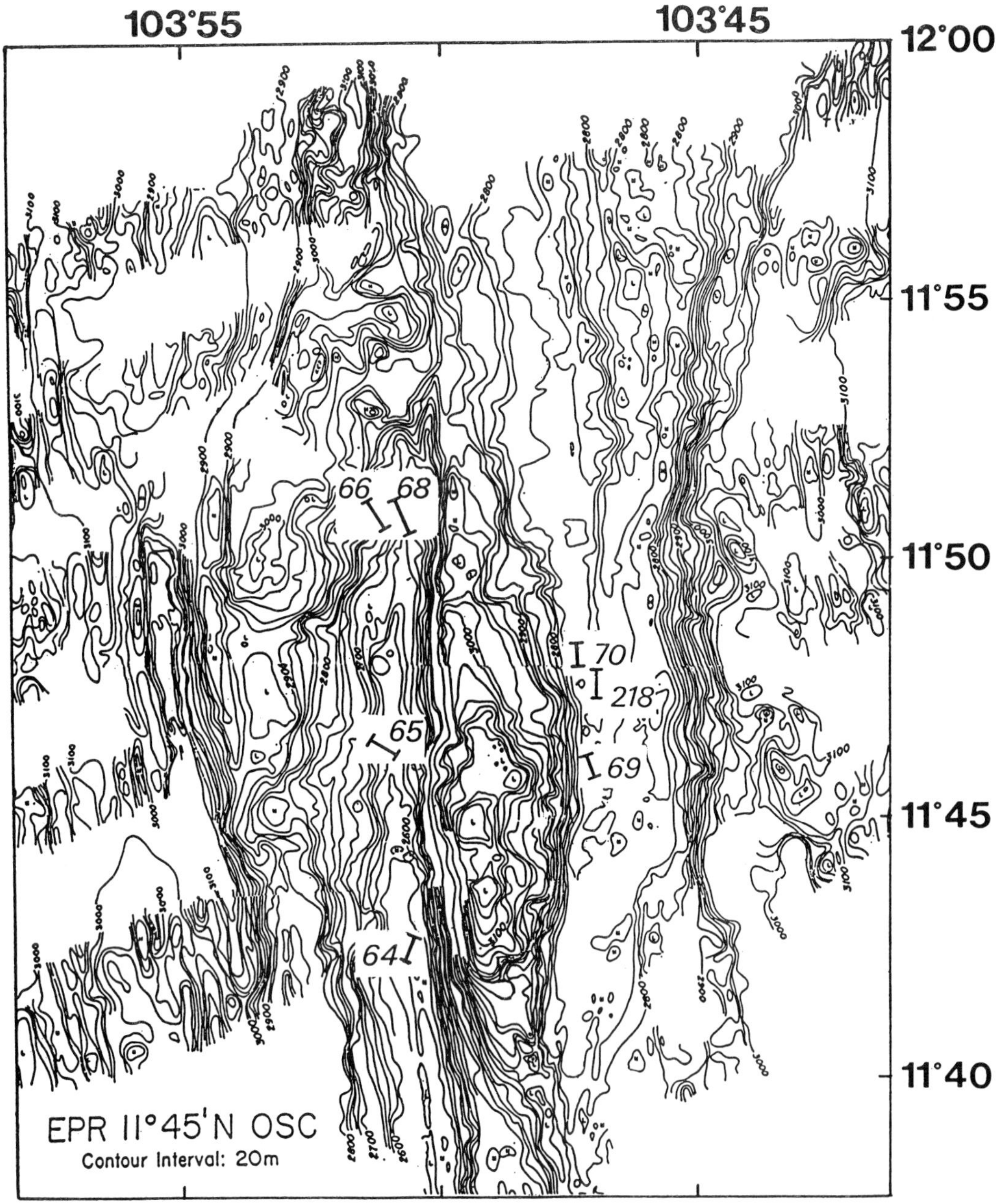

(a)

FIG. 3. (*a*) Bathymetry and (*b*) topography and station locations in the OSC at 11°45′N. (*c*) Bathymetry and station locations on the EPR axis near the Clipperton Fracture Zone.

shown as subgroups. These data are presented in Table 2. It is apparent from the small deviation about the mean at each site or subgroup that this approach effectively characterizes and discriminates between the various magma groups. For aphyric samples the whole-rock XRF and the microprobe data are in good agreement; similarly for these same samples the XRF analyses of the whole rock and the XRF analyses of the glass are essentially identical.

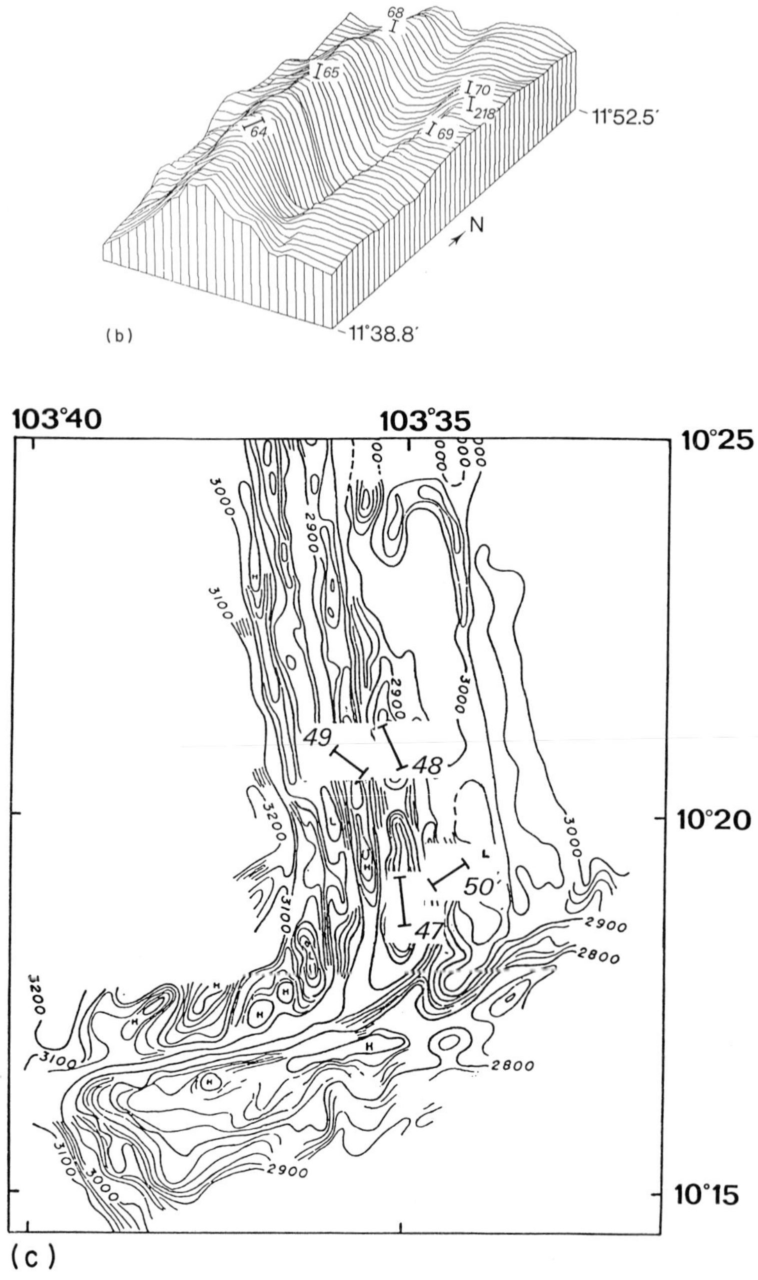

FIG. 3. (*cont.*)

The more porphyritic samples show differences in composition between the whole-rock and the glass (microprobe) analyses in the direction expected from the most abundant phenocryst phases. For plotting purposes in this paper and for most petrological discussion and modelling, we have utilized the microprobe glass and mineral data to study major-element variation. Rare earth element (REE) analyses are presented in Table 3.

Within the three main tectonic units, we recognize five compositional groupings:

TABLE 1. *Station data*

Station no.	Latitude		Longitude	Depth	Tectonic units	Locality
Dr 47	10°18′N	(10.31)	103°35′W	3025	Southern axis and high	Area 1, low near Clipperton Fracture Zone
Dr 50	10°19′N	(10.33)	103°35′W	2954		Area 1, eastern ridge near Clipperton Fracture Zone
Dr 48	10°21′N	(10.35)	103°35′W	2995		Area 1, low, flank of ridge axis, near Clipperton Fracture Zone
Dr 49	10°21′N	(10.36)	103°36′W	2807		Area 1, western ridge axis, near Clipperton Fracture Zone
Dr 51	10°36′N	(10.80)	103°37.7′W	2663		Area 2, ridge axis
Dr 52	10°38.4′N	(10.64)	103°38′W	2633		Area 2, ridge axis
Dr 53	10°38.9′N	(10.66)	103°38.5′W	2682		Area 2, ridge axis
Dr 54	10°48.8′N	(10.82)	103°39.3′W	2583		Mid-area 2 and 3, axis
Dr 57	10°55.6′N	(10.93)	103°41.2′W	2514		Area 3, 1st topographic high, axis
Alv 1384	10°55.6′N	(10.93)	103°40.6′W	2510		Area 3, 1st topographic high, axis
Ang. 209	10°58.6′N	(10.98)	103°42.0′W	2530		Area 3, 1st topographic high, axis
Dr 55	10°59.2′N	(10.99)	103°42.3′W	2524		Area 3, 1st topographic high, axis
Dr 59	11°14.8′N	(11.25)	103°45.2′W	2534	Northern axis and high	Area 4, 1st topographic low, axis
Dr 58	11°15.8′N	(11.27)	103°45.6′W	2564		Area 4, 1st topographic low, axis
Ang 212	11°16.5′N	(11.27)	103°45.3′W	2560		Area 4, 1st topographic low, axis
Dr 60	11°17.7′N	(11.30)	103°45.9′W	2554		Area 4, 1st topographic low, axis
CLDr 7–1	11°26′N	(11.43)	103°47′W	2500		Area 5, 2nd topographic high, axis
Dr 62	11°26.1′N	(11.44)	103°47.5′W	2504		Area 5, 2nd topographic high, axis
Dr 61	11°26.9′N	(11.46)	103°47.7′W	2504		Area 5, 2nd topographic high, axis
Dr 63	11°27.9′N	(11.47)	103°47.9′W	2504		Area 5, 2nd topographic high, axis
CLRAIE-8	11°28′N	(11.47)	103°48′W	2500		Area 5, 2nd topographic high, axis
Dr 67	11°34.9′N	(11.59)	103°49.9′W	2524		Mid area 5 and 6, axis
Ang 215	11°42.6′N	(11.71)	103°49.9′W	2535		Area 6, W OSC, axis
Dr 64	11°43.0′N	(11.72)	103°49.9′W	2534		Area 6, W OSC, axis
Ang 217	11°43.4′N	(11.72)	103°50′W	2550		Area 6, W OSC, axis
Dr 65	11°46.7′N	(11.79)	103°50.9′W	2593		Area 6, W OSC, axis
Dr 66	11°50′N	(11.85)	103°50′W	2663		Area 6, W OSC, axis
Dr 68	11°51′N	(11.85)	103°50.1′W	2732		Area 6, W OSC, axis
Dr 69	11°45.0′N	(11.75)	103°47.2′W	2722	Eastern OSC	Area 7, E OSC, axis
Ang 218	11°48.2′N	(11.79)	103°46.4′W	2709		Area 7, E OSC, axis
Dr 70	11°48.5′N	(11.80)	103°47.1′W	2707		Area 7, E OSC, axis

(1) Near the Clipperton Fracture Zone (area 1)—the region about 10°20′N at axial depths of 2800–3000 m where four dredges were sited (Fig. 3*c*).
(2) The southern axis and first topographic high (areas 2 and 3)—the region from 10°30′N to 10°55′N where eight sites were sampled but are clustered in separate areas, the topographic high and a distal point deeper and along the axis to the S with one site midway between.
(3) The northern axis and second topographic high (areas 4, 5 and 6)—the region from 11°15′N (the topographic low thought to be a small OSC) to 11°45′N, which marks the northernmost end of this segment of the ridge axis and is also the western limb of the large OSC at 11°45′N. We have sampled 16 sites along this unit.
(4) The eastern limb of the OSC at 11°45′N (area 7)—three sites were sampled in this unit.
(5) E-type MORB—several samples of basalt compositionally enriched in incompatible elements and having alkaline affinities were sampled along the northern axis and high.

Discussion

Compositional variation

In Fig. 4 we show the magnesium number (Mg/(Mg+Fe*)) versus latitude for the samples

TABLE 2. *Mean and standard deviation composition of individual stations or groups within a station, East Pacific Rise 10°–12°N*

	Area 1 Lat. 10.31 Depth 3025 Dr 47	Area 1 Lat. 10.33 Depth 2954 Dr 50	Area 1 Lat. 10.35 Depth 2995 Dr 48			
			Group A	Group B	48–24	48–25
	n = 7:4	n = 5:5	n = 17:6	n = 6:3	n = 1:1	n = 1:1
SiO_2	51.19±0.24	50.45±0.22	51.31±0.92	50.79±0.20	59.39	53.70
TiO_2	2.30±0.03	2.55±0.06	2.42±0.12	1.56±0.02	1.37	2.28
Al_2O_3	13.67±0.09	13.21±0.07	13.40±0.27	14.20±0.10	13.02	13.29
FeO*	12.66±0.13	13.56±0.15	12.99±0.48	10.98±0.10	9.64	12.27
MnO						
MgO	5.66±0.05	5.63±0.02	5.58±0.48	7.23±0.03	2.87	4.58
CaO	10.30±0.09	10.07±0.21	10.05±0.50	11.90±0.07	6.42	8.90
Na_2O	3.11±0.03	3.10±0.01	3.13±0.17	2.66±0.01	4.37	3.52
K_2O	0.24±0.02	0.20±0.01	0.23±0.06	0.07±0.01	0.74	0.34
P_2O_5	0.21±0.01	0.24±0.02	0.21±0.01	0.15±0.01	0.22	0.20
Sum	99.23±0.29	99.02±0.23	99.33±0.49	99.54±0.13	98.04	99.08
FeO/MgO	2.23±0.03	2.41±0.03	2.35±0.24	1.52±0.01	3.36	2.68
Mg#	0.44	0.43	0.43	0.54	0.35	0.40
Rb	4±0.5	4±1	4±0.5	3± 1	11	5
V	346± 6	421±9	405±9	335± 2	144	350
Cr	140±18	40±5	49±17	165±16	26	40
Co	44± 2	46±1	46± 1	47± 1	27	40
Ni	64± 3	43±1	45± 7	68± 2	24	34
Cu	66± 5	60±5	56± 6	74± 2	35	55
Zn	93± 1	113±2	110± 6	85± 3	110	109
Sr	137±0.5	128±1	126± 4	76± 2	101	118
Ba					88	
Y	48±1.5	58±1	59± 6	41± 0	138	75
Zr	146± 5	171±2	184±23	97± 1	551	253
Nb	5± 0	5.5±0.5	5.5±0.5	2± 0	9	6

	Area 1 Lat. 10.36 Depth 2807 Dr 49	Area 2 Lat. 10.60 Depth 2663 Dr 51	Area 2 Lat. 10.64 Depth 2633 Dr 52	Area 2 Lat. 10.66 Depth 2682 Dr 53	Mid-area 2–3 Lat. 10.82 Depth 2583 Dr 54	Area 3 Lat. 10.93 Depth 2514 Dr 57
	n = 4:4	n = 5:6	n = 32:9	n = 20:8	n = 8:8	n = 23:14
SiO_2	51.50±0.09	50.65±0.13	50.77±0.23	50.80±0.25	50.77±0.33	50.83±0.39
TiO_2	2.21±0.02	2.62±0.02	2.31±0.06	2.21±0.16	1.94±0.06	1.88±0.13
Al_2O_3	13.36±0.09	13.14±0.13	13.65±0.10	13.72±0.16	14.16±0.11	14.36±0.20
FeO*	13.14±0.05	13.79±0.13	12.61±0.28	12.50±0.42	11.24±0.22	10.87±0.40
MnO						
MgO	5.72±0.09	5.40±0.05	5.97±0.12	6.07±0.29	6.58±0.15	6.78±0.27
CaO	9.83±0.06	9.92±0.11	10.45±0.23	10.63±0.25	11.41±0.11	11.43±0.38
Na_2O	3.01±0.07	3.13±0.11	3.14±0.06	3.02±0.16	2.92±0.06	2.93±0.09
K_2O	0.19±0.01	0.23±0.02	0.20±0.01	0.18±0.02	0.18±0.02	0.17±0.03
P_2O_5	0.20±0.01	0.25±0.01	0.21±0.02	0.19±0.02	0.19±0.01	0.18±0.02
Sum	99.16±0.18	99.13±0.27	99.31±0.27	99.36±0.30	99.38±0.44	99.43±0.33
FeO/MgO	2.30±0.04	2.56±0.03	2.11±0.08	2.07±0.16	1.71±0.06	1.61±0.12
Mg#	0.44	0.41	0.46	0.46	0.51	0.53

	Area 1 Lat. 10.36 Depth 2807 Dr 49	Area 2 Lat. 10.60 Depth 2663 Dr 51	Area 2 Lat. 10.64 Depth 2633 Dr 52	Area 2 Lat. 10.66 Depth 2682 Dr 53	Mid-area 2–3 Lat. 10.82 Depth 2583 Dr 54	Area 3 Lat. 10.93 Depth 2514 Dr 57
	n = 4:4	n = 5:6	n = 32:9	n = 20:8	n = 8:8	n = 23:14
Rb	4±0.5	4±5	4±0.5	3±5	3±1	3±0.5
V	399±4	402±11	485±21	374±7	337±5	334±20
Cr	68±4	52±5	47±13	65±15	162±31	200±42
Co	46±1	46±1	47±2	46±1	44±1	44±2
Ni	55±2	49±2	47±4	52±5	69±6	72±7
Cu	58±3	61±4	62±4	62±5	70±3	71±7
Zn	110±1	108±3	103±6	97±4	85±1	85±6
Sr	106±1	134±1	136±1	128±6	138±4	139±4
Ba						
Y	54±0.5	54±1	50±2	47±3	41±1	43±3
Zr	155±1	162±2	153±7	139±10	123±4	129±13
Nb	4±0.5	6±0.5	5±0.5	5±1	5±0.5	5±1

	Area 3 Lat. 10.93 Depth 2510 Alv 1384–1	Area 3 Lat. 10.98 Depth 2530 Ang 209	Area 3 Lat. 10.99 Depth 2524 Dr 55	Area 4 Lat. 11.25 Depth 2534 Dr 59	Area 4 Lat. 11.27 Depth 2564 Dr 58	Area 4 Lat. 11.27 Depth 2560 Ang 212	Area 4 Lat. 11.30 Depth 2554 Dr 60
	n = 1:1	n = 9:6	n = 36:21	n = 15:9	n = 18:7	n = 3:3	n = 8:10
SiO_2	50.90	50.97±0.28	50.76±0.27	50.52±0.36	50.59±0.21	50.44±0.15	50.28±0.17
TiO_2	2.12	1.85±0.04	1.87±0.07	1.78±0.04	2.17±0.03	1.90±0.01	1.90±0.03
Al_2O_3	13.87	14.38±0.16	14.43±0.17	14.99±0.11	13.97±0.09	14.72±0.06	14.70±0.20
FeO*	11.98	10.89±0.12	11.00±0.30	10.22±0.10	11.74±0.06	10.48±0.01	10.50±0.08
MnO	0.22						
MgO	6.76	6.81±0.08	6.88±0.13	7.09±0.08	6.42±0.08	7.03±0.03	6.94±0.09
CaO	11.06	11.31±0.11	11.57±0.17	11.68±0.13	10.97±0.21	11.39±0.18	11.56±0.17
Na_2O	2.40	2.94±0.05	2.88±0.08	2.85±0.06	3.01±0.06	2.86±0.03	2.94±0.04
K_2O	0.15	0.15±0.02	0.16±0.01	0.23±0.02	0.21±0.01	0.20±0.01	0.20±0.01
P_2O_5	0.17	0.20±0.02	0.18±0.01	0.20±0.01	0.20±0.02	0.21±0.02	0.20±0.01
Sum	99.63	99.51±0.46	99.72±0.39	99.56±0.33	99.29±0.21	99.25±0.37	99.23±0.22
FeO/MgO	1.77	1.60±0.03	1.60±0.07	1.44±0.03	1.83±0.03	1.49±0.01	1.51±0.03
$Mg^{\#}$	0.50	0.53	0.53	0.55	0.49	0.54	0.54
Rb	3	3±1	4±1	3±0.5	4±1	3±0.5	4±0.5
V	370	329±4	333±16	307±7	380±7	325±2	328±3
Cr	124	209±5	207±32	241±12	100±3	280±5	276±8
Co	48	44±1	45±2	44±2	46±1	43±1	43±2
Ni	64	73±2	72±6	96±3	57±1	83±1	82±1
Cu	61	77±4	73±6	67±2	62±4	68±3	69±2
Zn	98	83±3	84±6	79±2	98±1	84±1	84±2
Sr	133	133±1	135±4	163±5	143±2	154±1	155±2
Ba							
Y	46	45±1	41±1	38±1	48±1	41±1	41±0.5
Zr	142	136±3	123±4	120±1	147±1	132±1	131±1
Nb	5	4±0.5	5±0.5	6±0.5	6±0.5	6±0	6±0.5

TABLE 2. (*cont.*)

	Area 5 Lat. 11.43 Depth 2500 CLDR 7–1	Area 5 Lat. 11.44 Depth 2504 Dr 62	Area 5 Lat. 11.46 Depth 2504 Dr 61	Area 5 Lat. 11.47 Depth 2504 Dr 63	Area 5 Lat. 11.47 Depth 2500 CL RAIE-8
	n = 1:1	*n* = 13:10	*n* = 22:10	*n* = 9:7	*n* = 1:1
SiO_2	50.06	50.68±0.29	50.46±0.31	50.52±0.28	50.14
TiO_2	2.15	1.85±0.07	1.84±0.04	1.80±0.02	2.07
Al_2O_3	15.36	14.68±0.20	14.73±0.12	14.67±0.10	15.35
FeO*	10.05	10.45±0.31	10.36±0.17	10.35±0.07	9.84
MnO					
MgO	6.34	6.93±0.14	7.01±0.09	7.06±0.10	6.33
CaO	11.01	11.50±0.22	11.82±0.17	11.61±0.04	10.92
Na_2O	3.24	2.90±0.05	2.87±0.07	2.90±0.04	3.18
K_2O	0.69	0.18±0.02	0.19±0.01	0.17±0.01	0.64
P_2O_5	0.24	0.18±0.02	0.18±0.02	0.20±0.02	0.23
Sum	99.14	99.35±0.34	99.45±0.40	99.30±0.34	98.70
FeO/MgO	1.59	1.51±0.08	1.48±0.04	1.47±0.02	1.55
Mg#	0.53	0.54	0.55	0.55	0.54
Rb	10	4±0.5	4±0.5	4±0.5	10
V	272	323±26	333±5	316±6	276
Cr	210	250±57	271±28	274±8	201
Co	41	43±2	44±1	43±1	42
Ni	103	78±10	77±3	79±1	98
Cu	63	69±5	70±3	70±3	62
Zn	78	81±6	81±1	80±2	79
Sr	287	147±3	145±2	148±2	277
Ba	134				135
Y	37	41±3	41±0.5	41±1	37
Zr	155	127±8	127±2	130±2	153
Nb	16	5±0.5	6±0.5	5.5±0.5	15

	Mid Area 5–6 Lat. 11.59 Depth 2524 Dr 67		Area 6 Lat. 11.71 Depth 2535 Ang 215	Area 6 Lat. 11.72 Depth 2534 Dr 64	Area 6 Lat. 11.73 Depth 2550 Ang 217	Area 6 Lat. 11.79 Depth 2593 Dr 65
	Group A *n* = 10:6	Group B *n* = 5:4	*n* = 2:4	*n* = 16:13	*n* = 3:4	*n* = 4:4
SiO_2	49.89±0.21	50.41±0.31	49.63±0.01	50.36±0.35	50.30±0.17	50.74±0.24
TiO_2	1.60±0.02	1.93±0.02	2.17±0.01	1.53±0.06	1.53±0.24	1.63±0.01
Al_2O_3	15.84±0.07	15.00±0.20	16.28±0.04	15.59±0.37	15.37±0.20	15.10±0.07
FeO*	9.46±0.07	10.55±0.08	9.65±0.02	9.43±0.25	9.60±0.19	9.62±0.04
MnO						
MgO	7.70±0.09	6.87±0.08	6.59±0.02	7.80±0.22	7.53±0.26	7.40±0.06
CaO	11.74±0.12	11.24±0.11	10.41±0.02	11.90±0.09	13.18±0.51	12.09±0.07
Na_2O	2.93±0.03	2.91±0.04	3.24±0.01	2.65±0.12	2.71±0.34	2.71±0.11
K_2O	0.16±0.01	0.23±0.01	0.73±0.00	0.15±0.01	0.17±0.08	0.16±0.01
P_2O_5	0.17±0.01	0.21±0.01	0.26±0.00	0.17±0.01	0.18±0.02	0.16±0.01
Sum	99.49±0.21	99.37±0.44	99.51±0.40	99.58±0.27	99.60±0.04	99.60±0.22
FeO/MgO	1.23±0.02	1.53±0.03	1.46±0.01	1.21±0.07	1.28±0.01	1.30±0.02
Mg#	0.59	0.54	0.55	0.60	0.58	0.58

	Mid Area 5–6 Lat. 11.59 Depth 2524 Dr 67		Area 6 Lat. 11.71 Depth 2535 Ang 215	Area 6 Lat. 11.72 Depth 2534 Dr 64	Area 6 Lat. 11.73 Depth 2550 Ang 217	Area 6 Lat. 11.79 Depth 2593 Dr 65
	Group A n = 10:6	Group B n = 5:4	n = 2:4	n = 16:13	n = 3:4	n = 4:4
Rb	3±0.5	4± 0	11±1	4±0.5	4±0	3±0.5
V	254±10	317±13	270±9	273±16	266±2	271±11
Cr	283±10	222±12	229±8	299±11	291±6	304± 8
Co	43± 1	43± 1	41±1	44± 1	43±1	43± 1
Ni	122± 2	97± 2	123±4	101±15	98±3	98± 4
Cu	72± 3	63± 6	56±2	76± 3	74±2	77± 4
Zn	70± 2	86± 3	80±3	73± 3	71±1	70± 1
Sr	171± 1	165± 2	320±8	150± 3	196±1	152± 1
Ba			180±13			
Y	35± 1	42± 1	36±1	34± 1	34±1	34± 1
Zr	115± 1	138± 1	166±1	107± 5	124±1	106± 1
Nb	4±0.5	6±0.5	19±1	5±0.5	6±0.5	4± 0

	Area 6 Lat. 11.85 Depth 2663 Dr 66	Area 6 Lat. 11.85 Depth 2732 Dr 68		Area 7 Lat. 11.75 Depth 2722 Dr 69	Area 7 Lat. 11.79 Depth 2709 Ang 218	Area 7 Lat. 11.80 Depth 2707 Dr 70	
	n = 1	Group A n = 9:9	Group B n = 3:3	n = 12:10	n = 1	Group A n = 6:4	Group B n = 11:10
SiO_2	50.60	50.63±0.18	50.66±0.29	50.47±0.36	49.67	49.80±0.36	50.75±0.21
TiO_2	2.06	2.09±0.03	2.05±0.01	2.47±0.13	2.12	1.29±0.12	2.12±0.13
Al_2O_3	14.21	14.25±0.16	13.87±0.12	13.23±0.08	13.63	15.97±0.41	13.70±0.19
FeO*	10.61	11.17±0.17	11.62±0.02	13.67±0.31	13.43	9.24±0.42	12.35±0.33
MnO							
MgO	6.70	6.50±0.06	6.53±0.08	6.00±0.12	6.09	8.15±0.25	6.63±0.19
CaO	11.38	11.25±0.07	11.21±0.16	10.38±0.23	11.02	12.30±0.11	11.12±0.20
Na_2O	2.64	2.94±0.06	2.77±0.03	2.68±0.06	2.87	2.50±0.06	2.48±0.07
K_2O	0.23	0.20±0.03	0.13±0.01	0.15±0.03	0.16	0.10±0.01	0.11±0.02
P_2O_5	0.24	0.20±0.02	0.19±0.01	0.22±0.02	0.21	0.13±0.01	0.18±0.01
Sum	98.71	99.23±0.34	99.04±0.32	99.29±0.37	99.20	99.38±0.31	99.45±0.29
FeO/MgO	1.59	1.72±0.02	1.78±0.03	2.28±0.09	2.21	1.14±0.10	1.87±0.10
$Mg^{\#}$	0.53	0.51	0.50	0.44	0.45	0.61	0.49
Rb		4±0.5	3±0.5	2.5±0.6		2.5±0.5	2±0.5
V		319±6	361±7	468±13		245±6	426±19
Cr		265±12	199±2	70±4		323±4	119±36
Co		43±2	44±1	50±3		48±1	48±2
Ni		85±3	75±1	56±1		137±2	64±5
Cu		64±4	63±4	59±4		81±4	64±4
Zn		83±3	92±2	118±6		69±1	107±3
Sr		161±1	124±1	97±3		125±2	96±4
Ba							
Y		41±1	45±0	58±2		29±0.5	52±3
Zr		133±1	131±2	159±9		90±1	139±10
Nb		6±0.5	4±0.5	4±1		3±0.5	3±0.5

Major element data (wt%) are from probe analyses; trace elements (ppm) are from XRF analyses.
n, number of analyses — major: trace. $Mg^{\#} = Mg/(Mg + Fe^{2+})$

TABLE 3. *REE analyses (ppm) of selected samples, East Pacific Rise 10°N–12°N*

	Samples from near Clipperton Fracture Zone							Samples from first topographic high and axis						
	47–1	48–1	48–18	48–24	48–25	49–3	50–2	51–2	52–4	53–16	54–1	55–13	209–2	57–4
La	5.7	5.8	3.2	21.9	9.4	5.8	6.0	5.4	4.7	5.2	3.9	4.2	—	3.1
Ce	17.0	17.7	8.2	66.6	29.2	17.6	20.4	19.7	16.8	17.5	13.9	14.0	14.4	12.4
Nd	15.7	17.1	10.1	55.8	28.0	17.8	20.1	18.9	15.9	16.7	13.1	14.3	14.7	12.7
Sm	4.63	5.38	3.69	15.97	8.47	5.57	6.24	5.70	4.96	5.05	4.22	4.25	4.62	4.26
Eu	1.81	2.01	1.49	3.71	2.52	2.03	2.28	1.96	1.68	1.75	1.45	1.51	1.56	1.37
Gd	6.77	7.25	5.32	19.20	10.49	7.38	8.28	7.76	6.57	6.82	5.77	5.91	6.30	5.53
Dy	7.48	8.36	6.23	22.54	12.55	8.86	9.66	8.97	7.61	8.05	6.70	6.77	7.13	6.00
Er	4.62	5.23	4.16	15.06	7.93	5.88	6.13	5.51	4.65	5.01	4.14	4.21	4.42	3.86
Yb	4.62	5.10	4.03	14.46	7.73	5.56	5.83	5.20	4.44	4.63	3.83	3.94	4.25	3.54
Lu	0.70	0.81	0.59	2.17	1.19	0.82	0.86	0.84	0.72	0.76	0.64	0.63	0.83	0.57

	Samples from second topographic high and axis													
	58–1	59–1	60–1	61–1	62–14	63–1	64–2	65–1	67–1	67–2	68–1	68–4	212–2	217–4
La	5.4	4.8	4.5	4.5	6.4	5.8	3.9	3.4	3.7	5.7	5.0	3.7	5.7	5.1
Ce	17.9	14.6	15.5	15.3	15.4	14.8	11.2	12.2	12.9	17.4	16.3	13.6	15.6	15.8
Nd	17.3	13.2	14.5	14.4	14.8	14.5	11.6	11.5	12.5	15.5	14.8	14.5	14.9	13.4
Sm	5.13	4.01	4.36	4.62	4.61	4.70	3.44	3.71	3.49	4.75	4.77	4.63	4.38	3.76
Eu	1.78	1.69	1.82	1.77	1.77	1.77	1.52	1.54	1.59	1.85	1.79	1.79	1.79	1.65
Gd	6.80	5.36	6.22	7.63	5.78	5.98	4.79	4.83	4.81	6.15	6.00	6.22	5.95	5.12
Dy	7.78	6.26	6.97	6.83	6.64	6.87	5.35	5.44	5.64	6.97	6.68	7.35	6.68	5.57
Er	4.88	3.90	4.19	4.30	4.22	4.14	3.34	3.50	3.60	4.27	4.21	4.58	4.12	3.37
Yb	4.55	3.54	3.91	3.89	3.86	3.95	3.04	3.26	3.15	3.85	3.87	4.32	3.84	3.03
Lu	0.73	0.55	0.62	0.62	0.60	0.60	0.48	0.49	0.50	0.62	0.60	0.68	0.60	0.47

	Samples from eastern OSC			Samples of E-type MORB		
	69–1	70–8	70–20	215–2	CL7–1	R8–1
La	4.6	3.5	2.4	11.7	11.0	9.9
Ce	16.8	14.0	9.0	30.9	28.5	25.6
Nd	17.8	14.6	9.2	21.6	20.6	18.8
Sm	5.62	4.96	3.12	5.89	5.29	5.06
Eu	2.09	1.84	1.29	2.14	2.15	2.06
Gd	7.71	6.72	3.92	6.39	6.45	6.04
Dy	9.37	8.18	4.66	6.36	6.57	6.23
Er	6.09	5.32	2.91	3.71	4.03	3.82
Yb	5.66	4.90	2.81	3.45	3.54	3.51
Lu	0.92	0.79	0.44	0.41	0.45	0.40

of Table 2. It is clear that the areas near the Clipperton Fracture Zone and in the eastern OSC have a wide range in magnesium number, although these represent small regions of the EPR and the samples are relatively closely spaced. The southern axis and first topographic high samples also show a range in magnesium number (0.4–0.54) but this is distributed in a systematic manner. The highest magnesium number at the topographic high decreases systematically with distance (and axial depth) from the topographic high. The northern axis and second topographic high also show a range in magnesium number but in a less systematic manner. Samples with the highest magnesium number are offset to the N of the topographic high; although there is considerable scatter, the values do decrease gradually to the S, and much more rapidly to the N. A plot of FeO* versus MgO (Fig. 5) indicates differences and systematic variation — in these major components between and within each of the three main tectonic units. Compositional groups represented by the samples near Clipperton and the E-MORB do not display systematic variations but do plot as distinct groups. Plots of Ti/K (Fig. 6), strontium

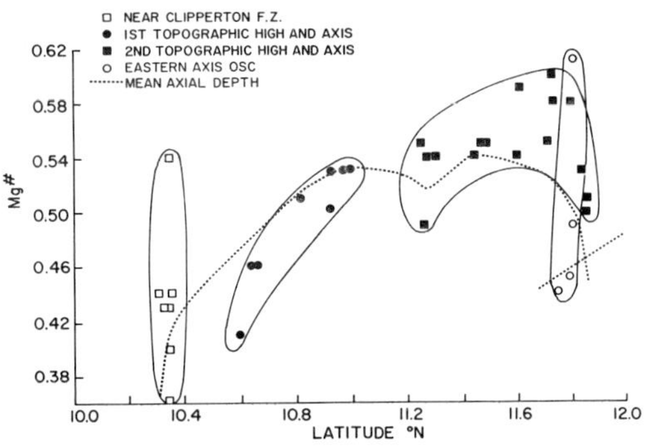

FIG. 4. Magnesium number versus latitude for each station mean composition or magma group mean within a station. Each axial region is shown by a different symbol and ringed. The mean axial depth is superimposed on the diagram.

(Fig. 7), Zr/Nb (Fig. 8) and La/Sm (Fig. 9) versus latitude all confirm the differences between the compositional groups.

Near Clipperton Fracture Zone

In this region of the EPR, the neovolcanic zone is ill defined and is a region of sparse volcanic activity; tectonism clearly dominates and characterizes the region. Eruptive units of different ages are in close proximity. Dredge 49 is from the top of the axial ridge that defines the neovolcanic zone, dredges 47 and 48 are from the side of that same ridge, and dredge 50 is from an older ridge to the E (Fig. 3*c*). Thus, variation in composition in this region reflects variation in time as well as space. In all the

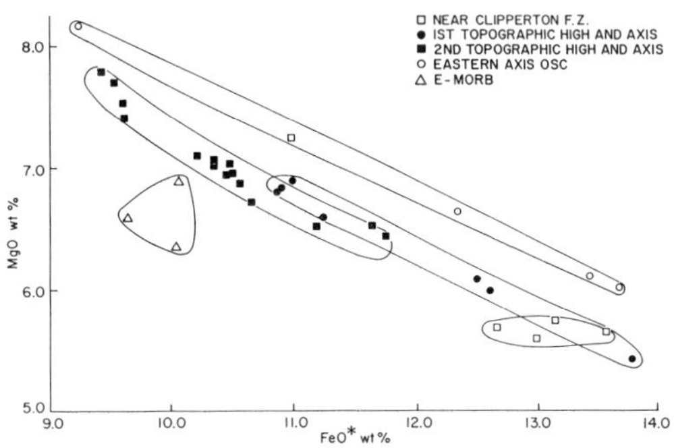

FIG. 5. MgO versus FeO* for each station mean or magma group mean within a station. Symbols and groupings as in Fig. 4. Note that magma group B from Station 48 near the Clipperton Fracture Zone (see Table 2) plots in the same field as the eastern OSC magmas.

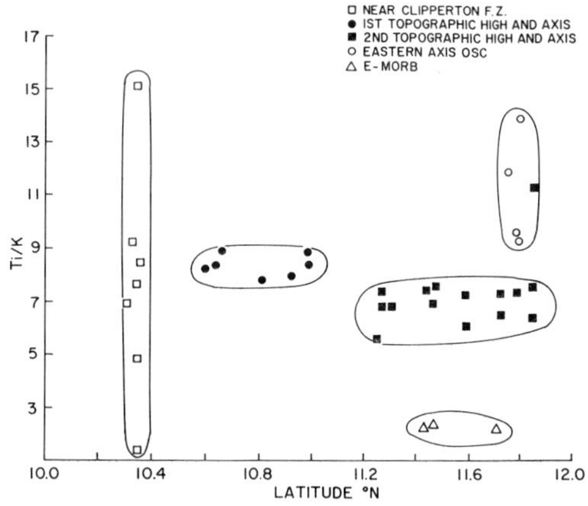

FIG. 6. Ti/K versus latitude for each station mean or magma group mean within a station. Symbols and groupings as in Fig. 4. Note that magma group B from Station 68 in the western OSC (see Table 2) plots in the same field as the eastern OSC magmas.

other tectonic units our dredges were placed only in the neovolcanic zone. The lavas from the Clipperton region show a range in composition as noted in Table 2 and in Figs 4–9. The variations in FeO/MgO, Ti/K, strontium, Zr/Nb, Sr/Y, Zr/Y and La/Sm are not compatible with simple fractionation from a common parent. The samples which comprise subgroup

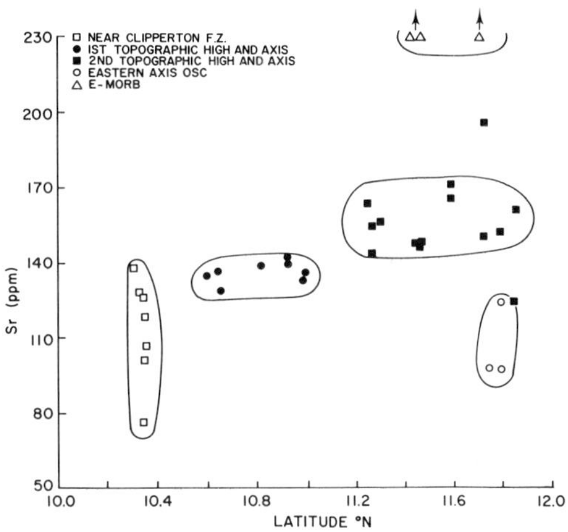

FIG. 7. Strontium versus latitude for each station mean or magma group mean within a station. Symbols and groupings as in Fig. 4. Note that magma group B from Station 68 in the western OSC plots in the same field as the eastern OSC magmas. Also note that magmas from Station ANGUS 217 in the second topographic high and axis region plots midway between that region and the E-type MORB field.

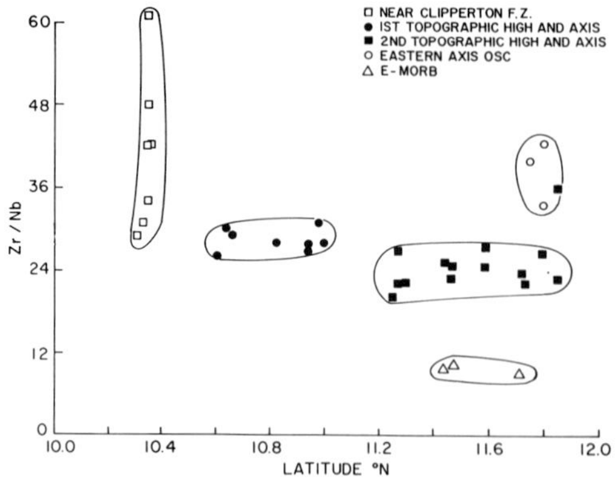

FIG. 8. Zr/Nb versus latitude for each station mean or magma group mean within a station. Symbols and groupings as in Fig. 4. Note that magma group B from Station 68 in the western OSC lies in the field of the eastern OSC magmas.

B of dredge 48 are the least fractionated but have distinctly higher magnesium number, Ti/K and Zr/Nb and lower strontium and Zr/Y than all other samples from this region, and even than subgroup A sample from the same dredge.

Also of interest in dredge 48 are highly silicic andesite and rhyodacite (Table 2, samples 48–25 and 48–24 respectively). Although these were recovered from the side of the ridge comprising the neovolcanic zone in this region, they are very fresh glassy samples and apparently were recently erupted. Similar high-silica varieties have been reported from other regions of the EPR (Bender *et al.* 1986; Langmuir *et al.* 1986; Natland *et al.* 1986). It is tempting to surmise that if they result from magmatic fractionation, which requires about 80% crystallization, then this takes place in isolated magma pockets in segments of the ridge where cooling is more rapid, such as close to transform

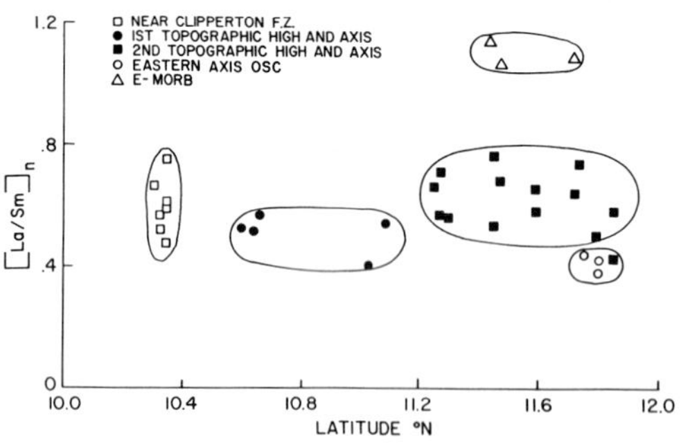

FIG. 9. $(La/Sm)_n$ versus latitude for each station mean or magma group mean within a station. Symbols and groupings as in Fig. 4.

faults or near tips of propagating ridge segments (Christie & Sinton 1981). During the period of cooling, such a magma pocket may be carried off axis and interrupted by local tectonism to allow the last stages of liquid magma to be injected on the side of, rather than in, the actual axial graben.

The overall variability in composition of all the samples recovered near the Clipperton Fracture Zone may indicate a rift tip close to a cold lithospheric edge. In such regions variability in magma compositions have been noted by other workers (Christie & Sinton 1981; Sinton *et al.* 1983; Bender *et al.* 1984; Langmuir & Bender 1984). These 'edge effects' may result from low magma supply, faster cooling rates and variation in extent of melting because of cooler temperatures.

Southern axis and first topographic high

Our original interpretation (Thompson *et al.* 1985) that this tectonic unit was a good example of an axial volcano with a unique magma composition appears to be confirmed. The lavas from this region show a systematic variation in composition with distance to the S along the axis from the topographic high. Magnesium number decreases (Fig. 4), indicating a wide range in fractionation, whilst elemental ratios such as Ti/K and Zr/Nb are similar for all samples independent of changes in degree of fractionation, indicative of a similar source. Although these samples and the samples from the second topographic high and northern axis are normal mid-ocean ridge basalts (N-type MORBs), there are subtle differences in both chemistry and petrography between the two units. Mineralogically, the magmas from the southern axis and first topographic high are characterized by plagioclase and clinopyroxene with little or no olivine; those from the northern axis, by plagioclase and olivine with rare clinopyroxene. Compositionally, the magmas from the southern axis and first topographic high have higher iron and lower titanium for a given magnesium content than the basalts from the northern unit. They also have lower silicon contents and Zr/Y and La/Sm ratios and higher Zr/Nb ratios than the basalts from the northern axis and second topographic high. They are compositionally distinct from the lavas from the eastern OSC and from the E-type MORB, as noted in Figs 5–9. A few of the lavas from the region near the Clipperton Fracture Zone are compositionally similar to the samples (although clearly not all) from the southern axis and first topographic high, suggesting that the southern axis magma unit may have erupted down into that region. Analyses for REEs (Table 3) confirm the compositionally distinct groupings based on other elements. Basalts from the southern axis and first topographic high have a mean $(La/Sm)_n$ of 0.49, those from the northern axis 0.61, those from the eastern OSC 0.41 and the Clipperton Fracture Zone region basalts have a range in $(La/Sm)_n$ from 0.48 to 0.75 (Fig. 9). The E-type MORBs have a mean $(La/Sm)_n$ of 1.07 and a distinctly different REE pattern from the N-type MORBs of the other units, as shown in Fig. 10.

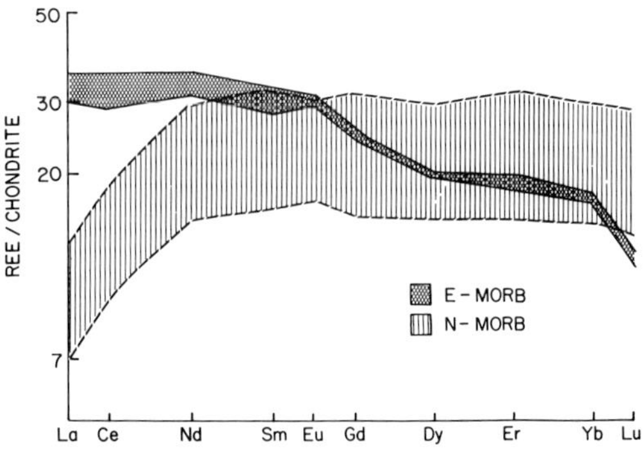

FIG. 10. REE distributions relative to chondrite REE abundances in E-type MORB compared with the range of distributions in N-type MORB from all three segments of the EPR axis between 10°N and 12°N.

Northern axis and second topographic high

Most of the lavas from this second tectonic unit we grouped as a distinct magma type in our original study (Thompson *et al.* 1985). We did note, however, that a few samples were very different in composition and were markedly enriched in incompatible elements such as potassium, titanium, strontium, barium, zirconium, niobium and light REEs. These we called E-MORB and placed in a separate magma group. Plots in Figs 5–9 here confirm our original observations but do suggest some complexity to the region not previously noted. Although the samples with the lowest magnesium number are found at the distal portions of the axis in the unit, samples with the highest values appear to be located slightly N of the second topographic high and are quite variable. A central magma supply with injection down-axis does not appear to be a viable model in this case. Basalts from dredges 58, 59 and 60 and ANGUS 212 located in the topographic low about 11°15′N, previously thought to be a small OSC, appear to have the chemical characteristics of the samples from the northern axis and second topographic high and are included in that unit in our grouping in Figs 4–9. C. Langmuir (pers. comm.) confirms that the petrological boundary between the northern axis and second topographic high and the southern axis and first topographic high is not in the topographic low at 11°15′N but is somewhere between 11°05′N and 11°10′N on the basis of their dredge data. This supports our suggestion that, if the topographic low at 11°15′N was once an OSC, then the northern segment appears to have propagated southward through the region.

The northern tip of the axis of this unit is also the western limb of the large OSC at 11°45′N. This limb is marked by older fractionated lavas, and tectonism has broken up the region. We infer that this limb of the OSC has been inactive for some time. Dredge 68 at the distal portion of this limb contains mostly basalts with the chemical and mineralogical characteristics of the northern axis and second topographic high magma group. However, within this dredge a few basalt samples, subgroup 68B, are chemically distinct and clearly must represent a different magma batch. These subgroup B lavas have strontium contents and Ti/K, Zr/Y, strontium, Zr/Nb and $(La/Sm)_n$ ratios much more akin to the lavas from the eastern limb of the OSC and plot with that group in Figs 4–9. Dredge 68 is at a point on the western limb where it curves towards and is in closest proximity to the eastern limb and is past the deepest point of the nodal basin separating the limbs (Fig. 3*b*). We suggest that some of the eastern limb magma may have been tapped and injected into the western limb at this point.

Eastern limb of the overlapping spreading centre at 11°45′N

Our previous ANGUS photography and the more recent ARGO studies clearly show that the axial portion of the EPR in the eastern limb of this OSC is morphologically well defined. It is volcanically active, with recent eruptions showing a high ratio of sheet flows to pillows and with ongoing hydrothermal activity even near the propagating tip. The lavas from this region are compositionally distinct from most lavas on the western limb or from the other axial units (Figs 5–9). They have higher magnesium contents for a given iron content and lower strontium contents, as well as higher Ti/K and Zr/Nb ratios and lower Sr/Y and $(La/Sm)_n$ ratios. Only a few samples from subgroup B of dredge 68 in the western OSC limb have similar compositions, as noted earlier.

The lavas from the eastern limb of the OSC have a wide range in composition (magnesium number from 0.42 to 0.6). These lavas come from only a limited axial length and this large range in composition is represented in basalts from a single dredge haul, dredge 70; subgroup A basalts are highly fractionated whilst subgroup B lavas are quite primitive in composition (Table 2; Figs 4 and 5). This region is a propagating rift tip and we might expect some variation in composition and degree of melting as the tip propagates through colder lithosphere. Clearly, the same mantle source is involved under this eastern segment because all the lavas here have similar chemical properties.

E-type mid-ocean ridge basalt

In the northern axis and second topographic high region E-type MORBs were recovered from three separate sites in the axis. Two sites at the topographic high, CLDr7−1 and CLRAEI, were recovered by the French (Hekinian *et al.* 1983) and made available for this study. The third (ANGUS 215) was from further N along the axis and slightly into the western limb of the OSC. The basalts at these sites are characterized by relatively high concentrations (compared with the N-type MORB from this area of the EPR) of K_2O (>0.6%), strontium (>250 ppm), barium (>100 ppm), zirconium (>150 ppm) and niobium (>15 ppm). REE distribution patterns of the E-type MORB, relative to chondrite abundances, show light REE enrichment (Fig. 10). They also differ from the N-type MORB in having lower Ti/K and Zr/Nb ratios and higher Sr/Y, Zr/Y and $(La/Sm)_n$ ratios. The E-type basalts are not alkaline lavas (*Ne* normative) or as enriched in the incompatible elements as alkalic lavas reported from seamounts in the East Pacific at this latitude (Batiza & Johnson 1980), but appear to be transitional between N-type MORB and alkalic basalts. Samples from ANGUS 217 site on the second topographic high region are N-type MORB in most respects but do have slightly elevated concentrations of elements such as strontium and zirconium so that they plot between the N-type MORB and E-type MORB compositions (*eg* Figs 6 and 7). Apart from their enrichment in incompatible elements, these E-type MORB lavas are also isotopically distinct with $^{87}Sr/^{86}Sr$ values of 0.70278, compared with 0.70250 for the N-type MORB from this region (D. Graham, pers. comm.). It is clear that these more radiogenic and enriched lavas must arise from a different mantle source than the other basalts.

Petrogenesis

Consideration of the petrogenesis of the basalts recovered in each of the three spreading centre cells or tectonic units points to fractional crystallization within magma batches as a major factor governing the observed compositional variation and mineralogy. Because most samples are plagioclase saturated, we have used the projection from plagioclase onto the pyroxene−olivine−silica face in the CMAS tetrahedron to consider such variations. In Fig. 11*a* we have plotted the variation in CMAS coordinates for glass and whole-rock analyses from the first tectonic unit, the southern rift and the first topographic high. Basalts from the near Clipperton Fracture Zone compositional group are also plotted (Fig. 11*b*). There is a striking correspondence of the plotted data points to the experimental 1 atm pyroxene-saturated phase boundary (Walker *et al.* 1979). This is entirely consistent with the prominent appearance of pyroxene even in the least fractionated basalts

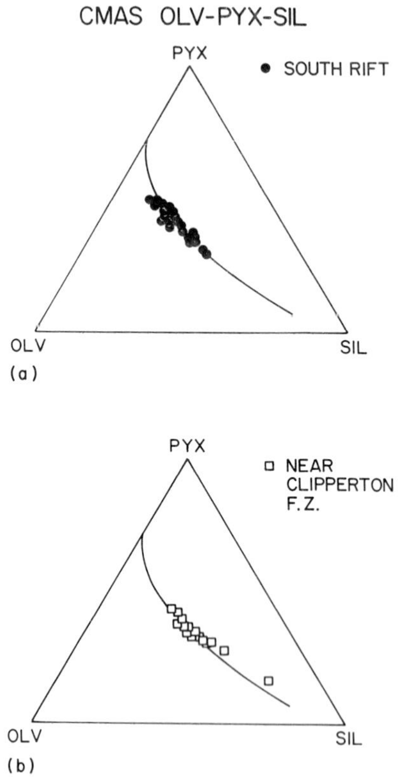

FIG. 11. Variation in CMAS coordinates for glass and whole-rock analyses from (*a*) the southern topographic high and axis and (*b*) stations near the Clipperton Fracture Zone.

from this region and suggests a major role for low-pressure crystal melt equilibria in the genesis of these basalts.

We have previously noted that many (but not all) of the basalts from the near Clipperton Fracture Zone region have similar chemical compositions and ratios of incompatible elements to the basalts from the first topographic high and southern axis. The simplest and most coherent compositional relationships appear among the basalts extending down the southern rift region from the first topographic high. It is possible to model successive stages of equilibrium crystallization starting with the 'most primitive' basalt from dredge 57 on the high at 11°N as the parent. The increments in crystallization can be nicely extended to the ferrobasalt of dredge 51 in this region and then to the andesitic basalt 48–15 in the near Clipperton Fracture Zone region with the rhyodacite 48–24 as the final residual liquid. In Fig. 12*a* we show the variation in composition of the southern rift unit basalts leading to generation of dacite as the residual end liquid. The plotted points are the actual measured analytical values; the 'per cent crystallized' value associated with each analysis is simply the sum of the weight per cents of all crystal phases required to generate that liquid in the materials balance calculation. The curves are best-fit second-to fourth-degree polynomials. We note a strong similarity between our fractionation model using the site 57 parent to produce the dacite and the fractionation model to explain the basalt to dacite relationship documented on the Galapagos rift (Clague *et al.* 1981). Details of our fractionation model are discussed in a separate paper (Bryan *et al.* 1988). In the course of this modelling we have examined the alternative possibility that the very low-K basalts found in subgroup B of dredge 48, for example, could be parental to the andesite and dacite. Solutions involving this

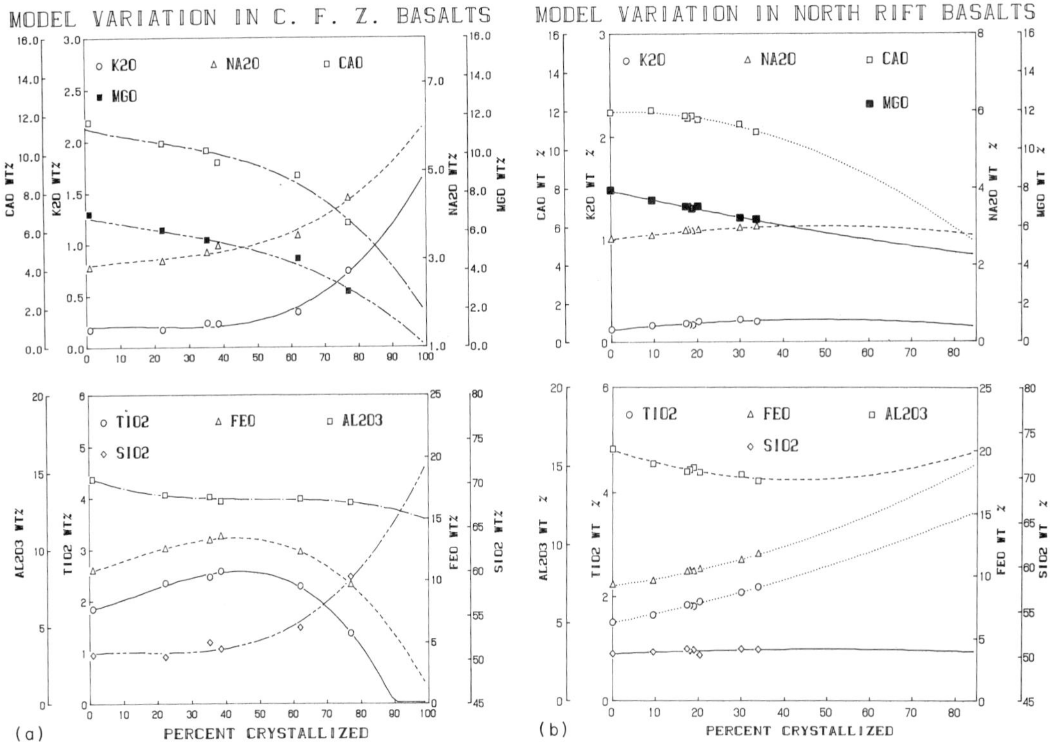

FIG. 12. (*a*) Compositional variation in selected basalt glass compositions from the S rift and aphyric bulk rock 'andesite' and 'dacite' from adjacent to the Clipperton Fracture Zone. 'Per cent crystallized' is the sum of the solid phase weight per cents derived from materials balance calculations. Variation curves are best-fit polynomials. (*b*) Compositional variation in selected basalt glass compositions from the N rift system. The figure was constructed in the same way as (*a*), with trends extrapolated beyond 80% crystallization. These curves are undefined past 35% crystallization and probably do not correctly predict compositional variation beyond that point. See text for discussion.

possible parent contain large residuals for potassium and titanium and do not account for other enrichments in trace elements such as zirconium, barium and lanthanum. Thus we conclude that some of the basalts from the Clipperton Fracture Zone region must represent discrete and compositionally distinct magma batches. They may derive their compositional characteristics from different degrees of partial melting of the mantle source or melting at different depths in the mantle.

The calculated phase assemblages for the basalts from this first tectonic unit closely match the observed modal phase assemblages. This along with the excellent fits of major and trace elements in the modelling of equilibrium crystallization strongly supports the view that simple low-pressure crystal–liquid fractionation is the major control on compositional variability in this unit. All the basalts with the exception of a few near the Clipperton Fracture Zone evolve from a single parent liquid. This does not necessarily mean that there is a continuous magma chamber underlying this rift segment; on the contrary such a chamber might be expected to buffer compositional changes and obscure relations along the rift. Indeed, Detrick *et al.* (1987), based on seismic profiles in this area of the EPR axis, could not find any reflector indicative of a magma chamber from 10°20′N to 11°N. Although a small magma chamber may exist under the topographic high at 11°N where volcanism is most active, most of the zero-age eruptives down-rift towards the Clipperton Fracture Zone may represent leakage from narrow rootless dyke injections analogous to the rift eruptions in Iceland and Hawaii (Sigurdsson & Sparks 1978; Ryan 1987). This is supported by the observed systematic variation in composition of the lavas from the topographic high towards the Clipperton Fracture Zone. This segment of the ridge axis shows the most pronounced change in topography from the high to the southern boundary. Clearly not all segments of the EPR show marked compositional or topographic changes along strike (Langmuir *et al.* 1986) and this model may not be applicable to every segment.

The second tectonic unit, the northern axis and topographic high, does not show the simple relationship amongst the erupted basalts that we have noted for the first tectonic unit. In Fig. 13*a* the CMAS projection for glass and whole-rock data is shown. The plotted points are consistent with the petrographic data which indicate plagioclase–olivine assemblages in most of these samples and only a little pyroxene in the most evolved samples. Clearly the E-type

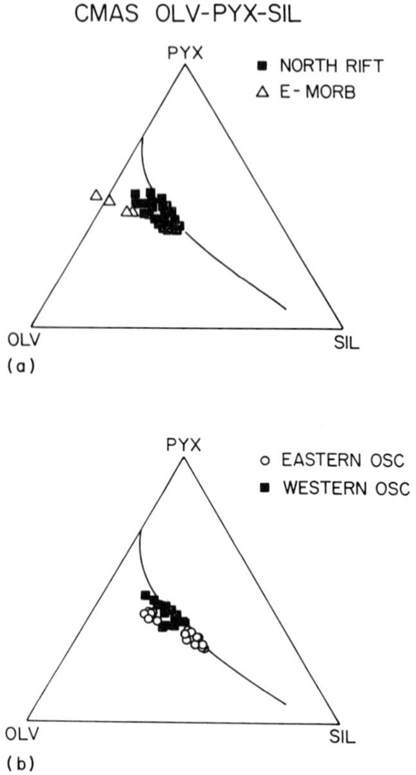

FIG. 13. Variation in CMAS coordinates for glass and whole-rock analyses from (*a*) the northern topographic high and axis and (*b*) stations in the OSC at 11°45′N.

MORBs are a compositionally distinct unit and are discussed separately. The fit to the 1 atm phase boundary is not as perfect as in the southern rift, and there is the possibility of multiple liquid lines of descent originating from any of the potential parent liquids offset towards olivine. We suggest that this offset may be due to fractionation at elevated pressure (Tormey *et al.* 1987).

Much of the compositional variation along the northern axis and second topographic high nevertheless can be explained by low-pressure equilibrium crystallization as observed for the southern rift. Fractionation models starting with the least evolved basalts of dredge 64 as parent yield good solutions for most intermediate basalts with the ferrobasalt of dredge 58 being the most evolved product. In Fig. 12*b* we show the model fits for the compositional variation in these basalts from the northern rift unit. To facilitate comparison with the southern unit basalts (Fig. 12*a*) we have extrapolated the N rift trends beyond 80% crystallization, but the trends are really undefined beyond 35%. These differences reflect the more primitive starting composition of the N rift parental liquid, as well as the much more limited range of crystallization. Comparison of the variation trends of the S and N rift units suggests that the S rift parent is very similar to the N rift basalt at about 20% crystallization. However, MgO and CaO start at distinctly lower values in the S rift basalt. These differences, along with important trace-element differences noted elsewhere, suggest that these trends represent two distinct 'liquid lines of descent' and probably the N rift magma would evolve differently even at higher degrees of crystallization. A similar set of solutions accounts for the relatively evolved basalts of subgroup A, dredge 68, at the northern end of the rift where it terminates at the western limb of the OSC at 11°45′N. However, some of the basalts from dredge 67 and dredge 59 do not fit well into this scheme. Thus for this tectonic unit we suggest that some fractionation of the magma may have occurred parallel to the pyroxene-saturated phase boundary, with simultaneous precipitation of plagioclase, olivine and pyroxene at 2–9 kbar. The array of pyroxene-saturated liquids at low pressure may then have been derived by dominantly olivine–plagioclase fractionation from the correspondingly diverse array of parental basalts generated at higher pressure.

The compositional complexity and distribution of lavas along the northern axis and second topographic high may be a function of changes in magmatic activity with time along this axial segment, *ie* there is some periodicity to magmatic injection. We have noted that the eastern limb of the OSC at 11°45'N is volcanically active and apparently propagating southward whilst the western limb is volcanically inactive, tectonically degraded and apparently 'retreating'. The 'magmatic centre' of the northern axial segment as identified by the least evolved basalts lies to the N of the topographic high. The topographic high has the most recent eruptions and is the focus of hydrothermal activity, suggesting that, as the eastern limb of the OSC has propagated S, so has the magmatic activity in the northern rift migrated from its earlier position near dredge site 64 to its present location on the topographic high. These observations, along with the compositional complexity and possible multiple magma batches in this unit, suggest that magma chambers may be small and/or relatively short lived. Detrick *et al.* (1987) reported the presence of a seismic reflector thought to be indicative of a magma chamber underneath the axis at the topographic high of this region.

Basalts from the eastern limb of the OSC at 11°45′N are compositionally distinct from those of the northern rift and the western limb of the OSC. They are part of a third tectonic unit or segment of the EPR axis. In Fig. 13*b* it can be seen that not all the basalts plot along the 1 atm phase boundary and similar crystallization histories may have operated within this magma batch as are deduced for the basalts of the northern rift and second topographic high. Basalts of subgroup B, dredge 68, in the western limb of the OSC have compositional characteristics of the basalts from the eastern limb; we suggest that the magma from which the eastern limb basalts are derived may have broken through into the western limb whilst tectonic fracturing was degrading that limb.

Both highly fractionated and primitive basalts are erupted in close proximity in the eastern limb of the OSC. Fractionation modelling and compositional correlation (similar Zr/Nb and $(La/Sm)_n$ ratios) suggest that they are derived from the same magma source. However, the juxtaposition of two compositionally discrete units is not compatible with derivation by injection along axis from a remote topographic high to the N. Rather, it suggests that in the region of a propagating ridge tip, the balance between magma supply and temperature variation may lead to high degrees of fractionation in some magma batches. Similar observations have been noted at other propagating rift tips (Christie & Sinton 1981; Sinton *et al.* 1983). It is also interesting to note that this eastern limb of the OSC is the locus of high magnetic intensity measurements. Similar high intensities are noted at the propagating tip of other OSCs along the EPR (Sempere & MacDonald 1986). Each of these other propagating ridge axes is characterized by eruptions of high iron basalts (12–14 wt% FeO) similar to the basalts of dredge 69 and subgroup B, dredge 70 (Table 2), suggesting a common source for the high magnetic intensities.

The E-type MORBs recovered from the second tectonic unit in the northern axis are clearly compositionally distinct. There are too few samples to infer anything about processes that might govern their compositional variability. We have noted that they are not as alkalic as lavas erupted from seamounts off axis at this latitude (Batiza & Johnson 1980). If they are derived from a similar source in the mantle as those alkalic lavas, then clearly they have been compositionally affected before eruption on the seafloor. The very slight enrichment in incompatible elements noted in the lavas of ANGUS 217 compared with those in the E-type MORBs, and the fact that the E-type MORBs are themselves transitional in composition between the N-type MORB and alkalic lavas, suggest that these basalts may result from mixing of magmas from an N-type and an alkalic mantle source. Our preliminary models of mixing of such end-members suggest that they can indeed give rise to the observed compositions.

Other workers have noted eruptions of similar E-type MORB along other segments of the EPR (Hekinian & Fouquet 1985; Langmuir *et al.* 1986). Just as we have noted for the EPR axis 10–12°N, these workers also found that the E-type MORB eruptions do not occur in every tectonic unit. The presence of these enriched basalts implies that the mantle beneath the ridge at that site must be compositionally heterogeneous. Zindler *et al.* (1984) similarly implied such heterogeneity for the mantle beneath seamounts that had erupted radiogenically enriched alkalic lavas. The question is whether this heterogeneity is uniformly distributed under the ridge axis or is found only in discrete separated sections. It is possible that magma batches arising from a ubiquitous radiogenically and incompatible-element-enriched source are small and are often masked by the more abundant N-type MORB magmas (or never reach the surface); thus they would not be seen on all segments. Their presence along the northern ridge axis and second topographic high could be explained by noting that normal magmatic activity is waning in the northern part of the section and reactivating to the S, thus allowing passage of the magmas from the enriched source to reach the surface only slightly modified *ie* mixed with N-type MORB). However, it is not clear that this is true for all EPR segments where isotopically enriched lavas have been recovered. Moreover, if this hypothesis were true, one might expect to have recovered such lavas in the other regions of sparse magmatic activity such as in the first tectonic unit along the southern axis. We thus favour the notion that mantle heterogeneity occurs in discrete 'pockets' under the EPR axis.

Conclusions

(1) A large part of the compositional variation observed within distinct magma batches in each of three tectonic units or segments of the EPR can be attributed to low-pressure fractional crystallization.

(2) Compositional differences between the principal magma batches of the three main units can be attributed, in part, to variable degrees of high- or intermediate-pressure fractional crystallization.

(3) Compositional differences between some magma batches, particularly in the region of the ridge axis close to a cold lithospheric plate, may be attributed to different degrees of melting or to melting at greater depths in the mantle in a region of steep thermal gradients.

(4) A model of magma upwelling beneath a bathymetric high, with lateral injection accompanied by increasing fractionation along rift, appears to be viable only for one of the tectonic segments we have studied—the S axis and first topographic high. This segment also shows the steepest topographic gradient along axis.

(5) There is a strong asymmetry to the volcanic processes along the EPR axis between 10°N and 12°N. Most of the lateral injection of lava in the southern segment appears to be to the S, down-rift toward the Clipperton Fracture Zone. There is chemical and petrographic evidence that the second segment may have propagated southward through the small topographic depression at 11°15′N. Certainly the focus of magmatic injection in this northern segment appears to have moved southward away from the western limb of the OSC at 11°45′N. The eastern limb of this OSC also is propagating southward. Thus there is a southerly bias both to rift propagation and magma injection. It is not clear whether injection actively drives fracture propagation or passively follows it.

(6) The opposing limbs of OSCs derive their lavas from separate and compositionally different magma batches. The eastern limb is volcanically active and propagating southward, whereas the other limb is magmatically inactive. Compositional variation in magmas erupted in the propagating tip may be attributed to variation in magma supply, with some batches undergoing high-level fractionation and others reaching the seafloor with little or no fractionation.

(7) One of the segments of the EPR between 10°N and 12°N erupts enriched lavas compositionally and isotopically distinct from the N-type MORB typically erupted in this and all other segments. These E-type MORBs require a different mantle source than the N-type MORBs, and the mantle beneath this segment must be compositionally heterogeneous. The magmas de-

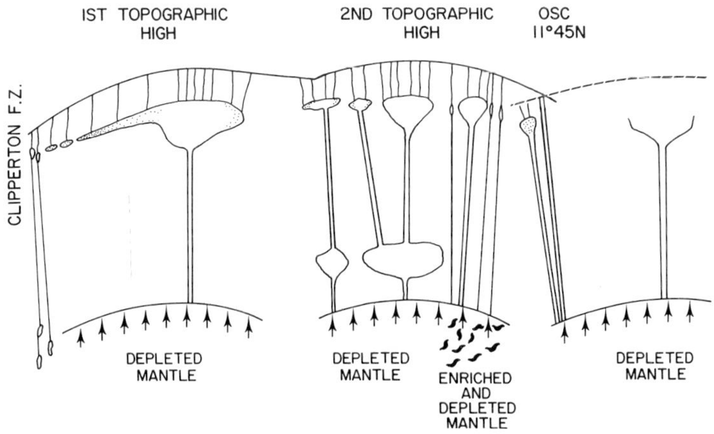

FIG. 14. Schematic interpretation of magmatic evolution beneath the EPR axis at 10°N to 12°N. Note that these features are not necessarily drawn to scale, only to illustrate the concept.

rived from the more radiogenic component in the heterogeneous mantle undergo some mixing with the more common N-type MORB magmas before eruption. The zones of mantle heterogeneity must occur in separate and discrete localities separated by typically homogeneous depleted mantle.

(8) The occurrence in the EPR axis between 10°N and 12°N of compositionally distinct magma batches, often in close proximity, taken with the seismic data of Detrick *et al.* (1987) and the observations of Langmuir *et al.* (1986) of small-scale magmatic segmentation in other regions of the EPR, suggests that very large long-lasting or steady-state magma chambers do not underlie the EPR axis. Rather, one must infer the presence of small intermittent magma chambers, particularly underlying topographic highs. Magma batches may pass from the mantle to the seafloor maintaining to a large degree their distinct and different compositional integrity.

A suggested schematic representation of what may be occurring under the EPR axis from 10°N to 12°N is shown in Fig. 14.

ACKNOWLEDGMENTS: We thank the crew and officers of the R/V *Melville* and colleagues R. Ballard, K. Hamuro, W.G. Melson, R. Zierenberg, M. Sulanowska and B. Schroeder for help in obtaining and analysing the samples. This work was supported by NSF Grant OCE83–09977. This is Contribution No. 6729 of the Woods Hole Oceanographic Institution.

References

ARGO RISE 1988. Generation of accreting plate fabric at fast spreading centers. *Canadian Mineralogist*, in press.

BALLARD, R. D. 1980. Mapping the mid-ocean ridge. *In: Proceedings of the 12th Annual Meeting of the Offshore Technical Conference OTC–3682*, 55–64.

FRANCHETEAU, J., JUTEAU, T., RANGIN, C. & NORMARK, W. 1981. East Pacific Rise near 21°N: the volcanic, tectonic and hydrothermal processes of the central axis. *Earth and Planetary Science Letters* **55**, 1–10.

BATIZA, R. & JOHNSON, J. R. 1980. Trace element and isotopic evidence for magma mixing in alkalic and transitional basalts near the East Pacific Rise. *In*: ROSENDAHL *et al.* (eds) *Initial Reports of the Deep Sea Drilling Project* **54**, 63–69.

BENDER, J. F., LANGMUIR, C. H. & HANSON, G. N. 1984. Basalt glasses from the Tamayo region, East Pacific Rise. *Journal of Petrology* **25**, 213–254.

——, ——, NATLAND, J. & BATIZA, R. 1986. Petrogenesis of East Pacific Rise high-silica glasses. *Eos (Transactions of the American Geophysical Union)*

BRYAN, W. B., HAMURO, K. & THOMPSON, G. 1988. Along-rift fractionation in zero-age contrasting magma groups, East Pacific Rise at 10–12°N. In preparation.

CHRISTIE, D. M. & SINTON, J. M. 1981. Evolution of abyssal lavas along propagating segments of the Galapagos spreading center. *Earth and Planetary Science Letters* **56**, 321–335.

CLAGUE, D. A., FREY, F. A., THOMPSON, G. &

Rindge, S. 1981. Minor and trace element geochemistry of volcanic rocks dredged from the Galapagos spreading center: role of crystal fractionation and mantle heterogeneity. *Journal of Geophysical Research* **86**, 9469–9482.

Crane, K. 1985. The spacing of rift axis highs: dependence upon diapiric processes in the underlying asthenosphere. *Earth and Planetary Science Letters* **72**, 405–414.

Detrick, R. S., Buhl, P., Vera, E., Mutter, J., Orcutt, J., Madsen, J. & Brocher, T. 1987. Multichannel seismic imaging of crustal magma chamber along the East Pacific Rise. *Nature* **326**, 35–41.

Francheteau, J. & Ballard, R. D. 1983. The East Pacific Rise near 21°N, 13°N and 20°S: inferences for along strike variability of axial processes of the mid-ocean ridge. *Earth and Planetary Science Letters* **64**, 93–116.

Harris, S. E., Marquet, W. M. & Ballard, R. D. 1984. Development status: ARGO deep ocean instrument platform. *ROV '84 Proceedings, May 14–17, San Diego, CA*.

Hekinian, R. & Fouquet, Y. 1985. Volcanism and metallogenesis of axial and off-axial structures on the East Pacific Rise near 13°N. *Economic Geology* **80**, 221–249.

——, Fevrier, M. *et al.* 1983. East Pacific Rise near 13°N: geology of new hydrothermal fields. *Science* **219**, 1321–1324.

——, Francheteau, J. & Ballard, R. D. 1985. Structural evolution of hydrothermal deposits at the axis of the East Pacific Rise. *Oceanologica Acta* **8**, 1–9.

Langmuir, C. H. & Bender, J. F. 1984. The geochemistry of oceanic basalts in the vicinity of transform faults: observations and implications. *Earth and Planetary Science Letters* **69**, 107–127.

——, Bender, J. F. & Batiza, R. 1986. Petrologic and tectonic segmentation of the East Pacific Rise, 5°30′N–14°30′N, *Nature* **322**, 422–427.

Lonsdale, P. 1983. Overlapping rift zones at the 5.5°S offset of the East Pacific Rise. *Journal of Geophysical Research* **88**, 9393–9406.

MacDonald, K., Sempere, J. C. & Fox, P. J. 1984. East Pacific Rise from Siqueiros to Orozco fracture zones: along strike continuity of axial neovolcanic zone and structure and evolution of overlapping spreading centers. *Journal of Geophysical Research* **89**, 6049–6069.

McConachy, T., Ballard, R. D., Mottl, M. J. & Von Herzen, R. P. 1986. Geologic form and setting of a hydrothermal vent field at Lat. 10°56′N East Pacific Rise: a detailed study using ANGUS and ALVIN. *Geology* **14**, 295–298.

Melson, W. G. 1977. Microprobe analysis of basalt glass from the world's ocean. *Smithsonian Contributions to the Earth Sciences* **19**, 31–60.

Natland, J., Langmuir, C. H., Bender, J. F., Batiza, R. & Hopson, C. 1986. Petrologic systematics in the vicinity of the 9°N non-transform offset, East Pacific Rise. *Eos (Transactions of the American Geophysical Union)* **67**, 1254.

Ryan, M. P. 1987. Neutral buoyancy and the mechanical evolution of magmatic systems. *In*: Mysen, B. O. (ed.) *Magmatic Processes: Physicochemical Principles*. Geochemical Society Special Publication **1**, 259–287.

Schouten, H. & Klitgord, K. 1983. Overlapping spreading centers on the East Pacific Rise: comment. *Nature* **303**, 549–550.

Schroeder, B., Thompson, G., Sulanowska, M. & Ludden, J. N. 1980. Analysis of geologic materials using an automated X-ray fluorescence system. *X-Ray Spectrometry* **9**, 198–205.

Sempere, J. C. & Macdonald, K. C. 1986. Magnetic and petrologic discontinuities and the surficial boundaries of spreading cells beneath mid-ocean ridges. *Eos (Transactions of the American Geophysical Union)* **67**, 359.

Sigurdsson, H. & Sparks, S. R. J. 1978. Lateral magma flow within rifted Icelandic crust. *Nature* **274**, 126–130.

Sinton, J. M., Wilson, D. S., Christie, D. M., Hey, R. N. & Delaney, J. R. 1983. Petrologic consequences of rift propagation on oceanic spreading centers. *Earth and Planetary Science Letters* **62**, 193–207.

Thompson, G., Bryan, W. B., Ballard, R. D., Hamuro, K. & Melson, W. G. 1985. Axial processes along a segment of the East Pacific Rise, 10–12°N. *Nature* **318**, 429–433.

Tormey, D. R., Grove, T. L. & Bryan, W. B. 1987. Experimental petrology of normal MORB near the Kane Fracture Zone, 22–25°N, Mid Atlantic Ridge. *Contributions to Mineralogy and Petrology* **96**, 121–139.

Walker, D., Shibata, T. & Delong, S. E. 1979. Abyssal tholeiites from the Oceanographic Fracture Zone: phase equilibria and mixing. *Contributions to Mineralogy and Petrology* **70**, 111–125.

Whitehead, J. A., Dick, H. J. B. & Schouten, H. 1984. A mechanism for magmatic accretion under spreading centers. *Nature* **312**, 146–148.

Zindler, A., Staudigel, H. & Batiza, R. 1984. Isotope and trace element geochemistry of young Pacific seamounts: implications for the scale of upper mantle heterogeneity. *Earth and Planetary Science Letters* **70**, 175–195.

Geoffrey Thompson, Wilfred B. Bryan & Susan E. Humphris, Woods Hole Oceanographic Institution, Woods Hole, MA 02543, USA.

The Fjallgardar volcanic ridge in NE Iceland: an aborted early stage plate boundary or a volcanically dormant zone?

Johann Helgason

SUMMARY: The extinct Fjallgardar volcanic ridge (FVR) in NE Iceland is approximately 15–20 km wide and extends for about 190 km from the Vatnajökull glacier toward the Slétta peninsula in the far N. The FVR was mainly formed subglacially and strikingly parallels the curvature of the present median line (Askja to Theistareykir centres) of the neovolcanic zone (NVZ) which lies about 40 km to the W.

The basalts from the Fjallgardar and Slétta regions are dominantly evolved tholeiites as shown by major- and trace-element compositions of 42 samples. The average magnesium number is 47 (range 37–58). 90 per cent of the samples can be grouped into either 'low-' potassium (0.1%–0.4% K_2O) or 'high-' potassium (0.5%–0.7% K_2O) units. The high-potassium lavas in the FVR are chemically similar to Tertiary tholeiites of eastern Iceland.

A test for temporal geochemical variations for three stratigraphic formations was carried out for the Mödrudalur area of the FVR. The oldest of the three formations (M3) consists of low-potassium basalts only, whereas the two younger formations (M4 and M5) consist predominantly of more evolved high-potassium basalts. Formation M5 is most probably fed into Mödrudalur from the Kverkfjöll volcanic system some 75 km further S.

Volcanism in the FVR S of Slétta was short-lived and lasted for perhaps 200 000–300 000 years. Subsurface lateral flow of magma from the NVZ is probably responsible for the Recent subaerial volcanism in Slétta. In the Slétta area the older subglacially formed ridges have only a low-potassium basalt composition.

The FVR may be viewed as 'abortive' and as the precursor to the NVZ in northeastern Iceland. During the Brunhes magnetic polarity chron (less than 0.7 Ma) a rift jump is postulated to have occurred from the FVR to the present NVZ, approximately 40 km to the W. E–W lateral rift jumps (20–40 km) may have been common in the eastern half of Iceland in the vicinity of the Iceland hot-spot located 150–200 km E of the Reykjanes and Kolbeinsey mid-ocean ridges.

Geological setting

In the neovolcanic zone of NE Iceland, volcanic centres coincide with crustal accretion maxima (Fig. 1). Collectively, the centres and associated dyke/fissure swarms constitute the plate boundary. In this part of Iceland the neovolcanic zone joins with the main arm of the Tjörnes transcurrent fault zone by Husavik (Einarsson 1976; McMaster *et al.* 1977; Mamula & Voight 1985). S, towards the centre of Iceland, the neovolcanic zone merges with the Kverkfjöll complex close to an assumed mantle plume maximum (Sigvaldason *et al.* 1974).

The first crustal growth model to propose that the volcanic centres in the Iceland volcanic regime would 'jump', after a life span of 0.5–1.0 Ma, was that of Gibson & Piper (1972). Large-scale rift jumps, incorporating shifts of the volcanic zone over lateral distances of 100–200 km, have been proposed for the Iceland region by several workers (*eg* Ward 1971; Saemundsson 1974). More recently, it has been postulated that the volcanic zone, consisting of several volcanic centres, may *frequently* jump to either side of a dying crustal axis, each jump being of the order of 20–40 km (Helgason 1984). The frequency of such shifts is thought to control the life span of volcanic centres in the accreting rift zones of Iceland. On a time-average basis the early dynamic rift models generate a symmetrical crustal structure across the rift axis. In contrast, a model incorporating small-scale rift jumping predicts a complex crustal structure that is asymmetrical or 'random' across the rift axis with regard, for instance, to dipping stratigraphic reflectors (Helgason 1984, 1985).

A subglacially formed complex lies 40 km E of and parallel to the neovolcanic zone in NE Iceland, henceforth referred to as the Fjallgardar volcanic ridge (FVR). It has been suggested (Walker 1975) that this ridge complex may have been an independent volcanic region that can be regarded either as 'abortive' or

From Saunders, A. D. & Norry, M. J. (eds), 1989, *Magmatism in the Ocean Basins*, Geological Society Special Publication No. 42, pp. 201–213.

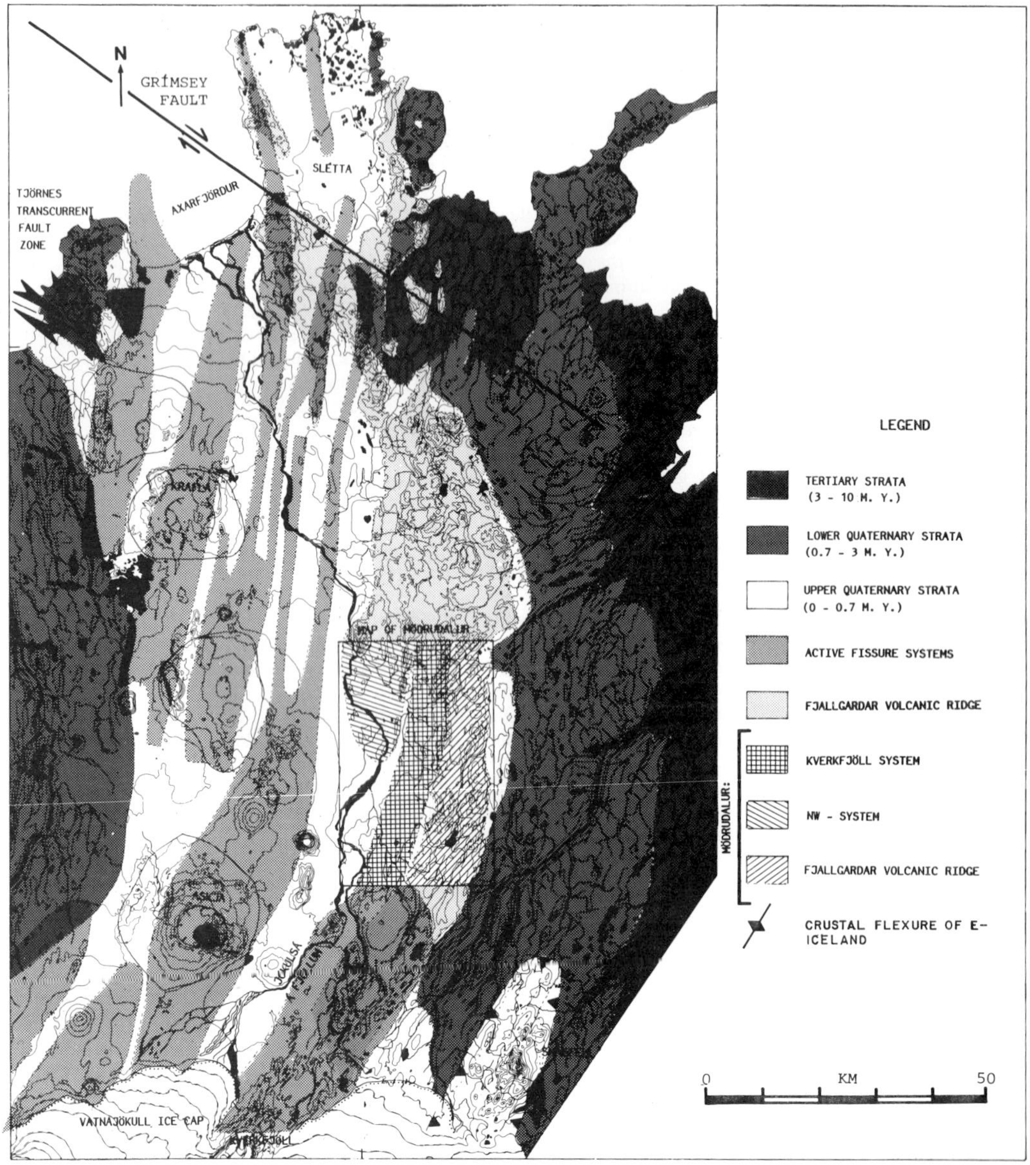

FIG. 1. Main division of strata in northeastern Iceland (after Saemundsson 1974): Tertiary (3–10 Ma), Lower Quaternary (0.7–3 Ma) and Upper Quaternary (0–0.7 Ma). Volcanic centres (*eg* Askja and Krafla) and fissure swarms of the neovolcanic zone occur on the western side of the area formed during the Upper Quaternary. The mapped area of Mödrudalur is shown on the eastern half of the Upper Quaternary region.

volcanically dormant. Work in progress for this region is reported here and an attempt is made to establish some structural and geochemical characteristics of the FVR that may eventually lead to a better understanding of how rift jumping proceeds in Iceland.

The Fjallgardar volcanic ridge

As a supergroup the FVR can be traced for some 190 km, *ie* approximately from Vatnajökull ice cap (S) to the sea in the Slétta region (N, Fig. 1). The Fjallgardar supergroup consists of numerous ridges of various sizes that rise some 50–500 m above the surrounding plateau. From S to N the plateau altitude grades from 500 m down to sea level on Slétta in the far N.

Volcanism in the Fjallgardar region has been predominantly subglacial. Mainly hyaloclastite and pillow breccia accumulated on eruptive fissures with only minimal lateral flow underneath the ice sheet. The ridges formed entirely during the Brunhes geomagnetic polarity chron and strictly delineated a region of crustal accretion and volcanism.

In order to demonstrate relief in the Fjallgardar region, four sections, with an E–W orientation, are shown in Fig. 2. The sections are centrally located on the FVR and cover an interval of 70 km from Lake Thrihyrningsvatn in the S (section a at 65°09′N) to Bunga in the N (section d at 65°47′N). Not indicated in Fig. 2 is the gentle dip of the Fjallgardar ridge substratum, *ie* W toward the neovolcanic zone. The central segment of the FVR (enhanced in grey) has a maximum width of only 15 km, although collectively the area of Brunhes subglacial volcanism by Mödrudalur is up to 30 km in width. This is seen in section c in Fig. 2 where subglacially formed ridges extend from Ytri-Grimsstadanupur in the W to the eastern side of Thjodfell. The cross-sections clearly show that, in the S by Lake Thrihyrningsvatn, the Fjallgardar central ridge segment consists of two broad ridges that merge together at Bunga. Also, the overall ridge elevation is clearly higher at Bunga than at Lake Thrihyrningsvatn, suggesting that the volcanic output increased toward the N with an overall maximum at approximately 65°20′N Fig. 2, section c.

Geological mapping in Mödrudalur

The subglacially formed ridges were mapped in detail in an area of 1200 km², extending from approximately 65°06′N to 65°30′N, by Mödrudalur (Fig. 1) (Helgason 1987). The purpose of this mapping was to characterize a cross-section of the FVR and in particular to reveal the different formations that have built

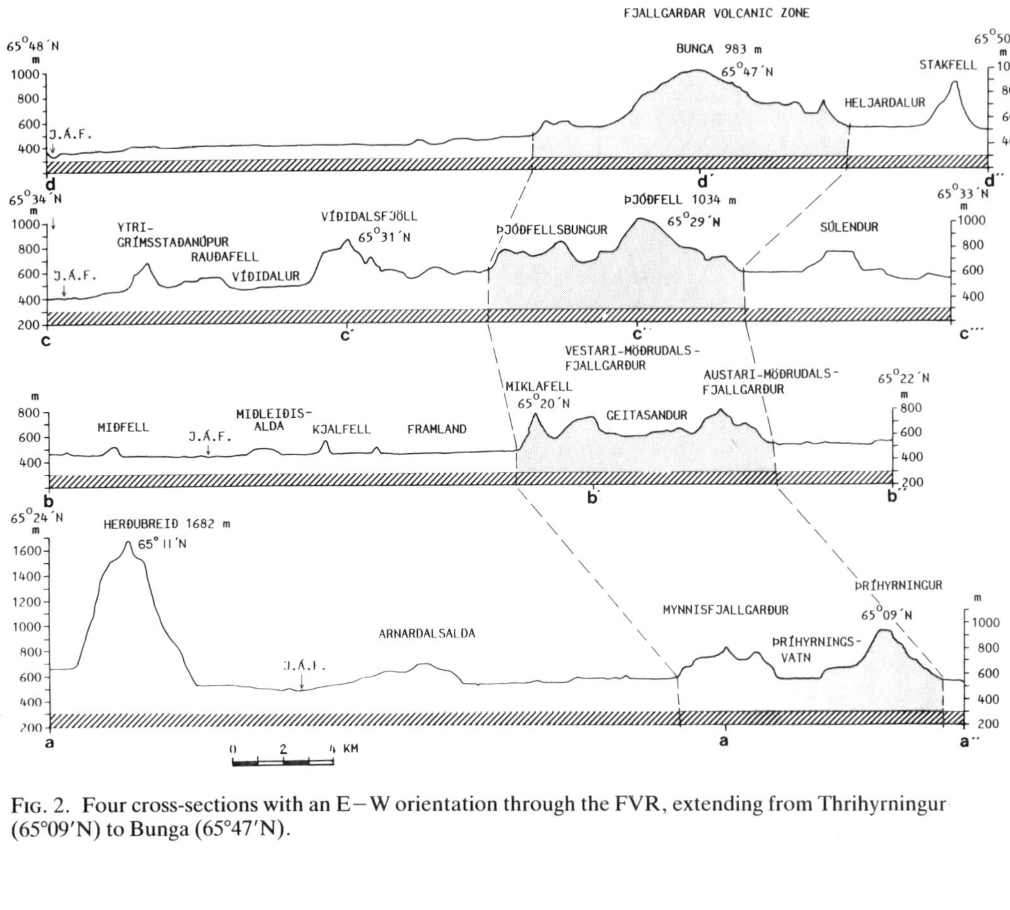

FIG. 2. Four cross-sections with an E–W orientation through the FVR, extending from Thrihyrningur (65°09′N) to Bunga (65°47′N).

this ridge complex. A geological map of Mödrudalur is presented in Fig. 3. On the basis of morphological character, fissure orientation and petrography, the ridges and other volcanic strata were provisionally divided into six 'subglacial' formations in addition to two lava formations that erupted under subaerial conditions, one being interglacial and the other postglacial. These eight formations provide a good E–W cross-section for the Mödrudalur strata. A name is assigned to each formation and, for convenience, an abbreviation (M1–M6, H1, N1) is used on the map (Fig. 3).

Ferjufjall–Lambafjöll (M1, subglacial basalt) and Raudinupur (M2, subglacial dacite) formations

The two oldest formations, located in the NW part of the area, are referred to here as the

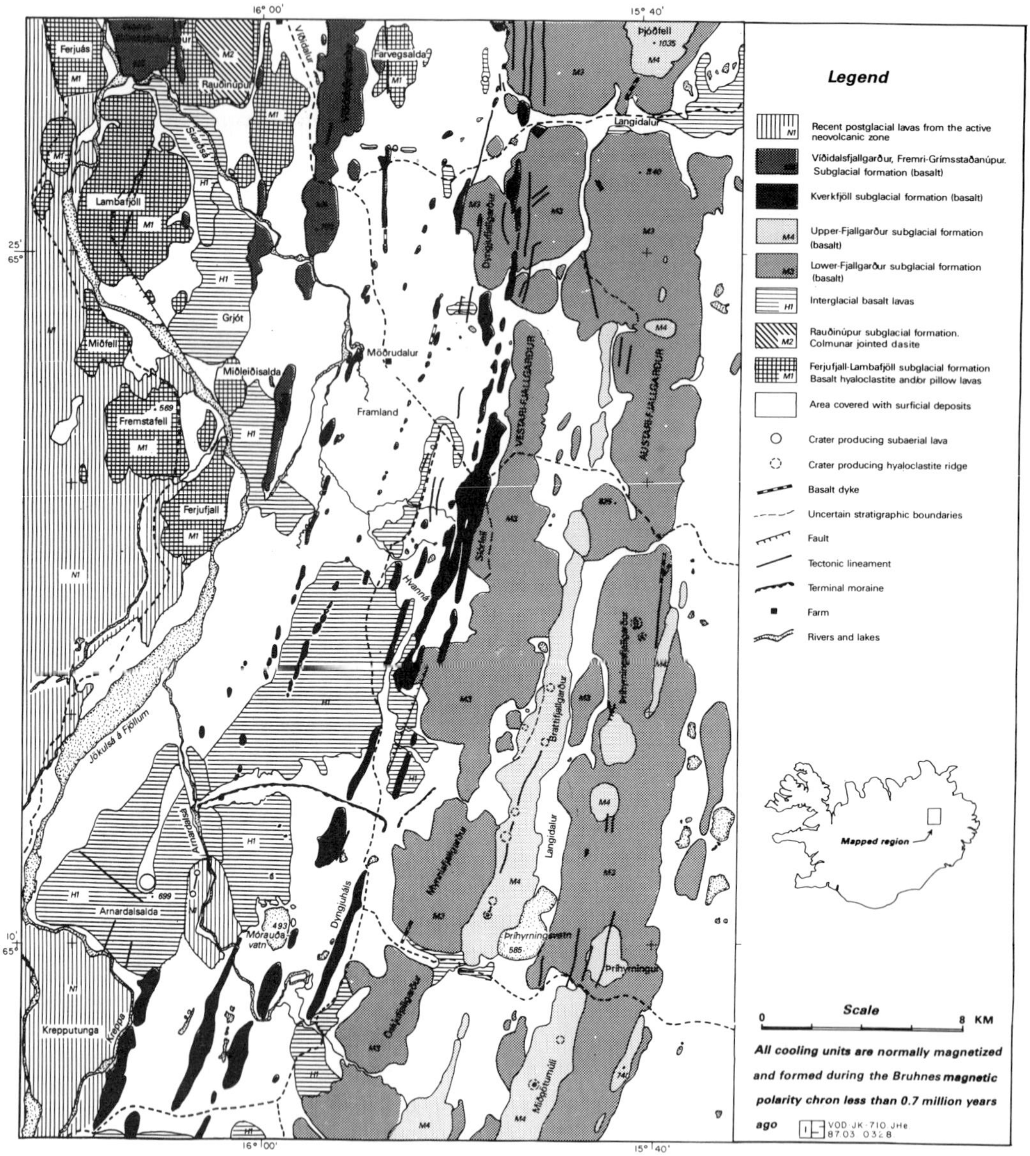

FIG. 3. Geological map of Mödrudalur, mid-east Iceland. The area has been divided into six subglacial formations (M1–M6), one interglacial lava formation (H1) and one postglacial lava formation (N1).

Ferjufjall—Lambafjöll (or M1) and Raudinupur (M2) formations. Both erupted under subglacial conditions. In contrast with other formations in the region, individual hills have been flattened substantially by erosion. Formation M1 consists of columnar-jointed basalt, hyaloclastite and pillow breccia. Chemical analysis has demonstrated that the Raudinupur hill (8 km^2) is of dacitic composition (Karl Grönvold, pers. comm.). Dacitic rocks have not been found elsewhere in the FVR and are therefore considered to be of local distribution.

Formation H1: subaerially erupted interglacial lavas

At this stage the ice sheet receded and fissure-type volcanism began under subaerial conditions to the E (Fig. 3). However, the main volcanic phase proceeded at this new location under subglacial conditions. Although a rare event for this region, lava shields may have erupted during this interglacial, *eg* at Arnardalsalda (SW).

Lower Fjallgardar (M3, subglacial basalt) and Upper Fjallgardar (M4, subglacial basalt) formations

Climatic conditions deteriorated and volcanism now proceeded under subglacial conditions. A major volcanic episode followed in the FVR, producing two major ridges, 5–15 km wide and 300–400 m high, that merge together in the N at Thjodfell. Individual units within M3 are voluminous in comparison with other strata, suggesting that this formation was produced in large volcanic eruptions.

The few faults observed are minor and parallel the ridge trend, approximately N–S. No graben structures were noted within formation M3 or M4.

The Upper Fjallgardar formation (M4) is partly superimposed on the Lower Fjallgardar formation (M3) and locally it fills the valley between the two ridges of the Lower Fjallgardar formation. Numerous craters occur within the hyaloclastite rock of M4 that are absent from the lower formation (M3) as a result of erosion. Formation M4 is typically plagioclase–porphyritic (10% by volume).

Kverkfjöll (M5, subglacial basalt) formation

This formation consists of numerous cone-shaped hills, 200–500 m long and no more than 50 m high, that align on a fissure swarm. The swarm was mainly volcanically active during the Weichselian glaciation. Volcanism within this swarm was less voluminous and clearly orders of magnitude smaller than when formations M3 and M4 accumulated. The fissure swarm has been traced southward from Mödrudalur (75 km) into the active Kverkfjöll volcanic complex. It is strongly suggested that the ridges of this formation were fed by subsurface lateral magma flow from a shallow magma chamber of the Kverkfjöll volcanic system. The mechanism of subsurface magma flow across great distances has been established for the Krafla volcanic system in NE Iceland (*eg* Einarsson & Brandsdóttir 1980).

Vididalsfjallgardur–Fremri–Grimsstadanupur (M6, subglacial basalt) formation

This subglacial formation also erupted during the Weichselian glaciation. It consists of ridges, 500–1000 m in width, that rise some 300 m over the surrounding plateau. The location of the formation coincides roughly with formations M1 and M2. It is postulated that these three formations may collectively constitute a single volcanic system, with a centre in Raudinupur, which is of dacitic composition.

Formation N1: postglacial basalt lavas

When the ice sheet receded to its present location, some 10 000 years ago, lavas began to flow into the area from the neovolcanic zone to the W. Only one lava is known to have erupted in Mödrudalur in postglacial times, by Arnardalsalda. This location is on a fissure that is probably part of the Kverkfjöll fissure swarm and thus originated in the neovolcanic zone.

The three volcanic systems of Mödrudalur

The eight stratigraphic formations described above all erupted at different times although some overlapping cannot be excluded. The six subglacial formations in Mödrudalur can be related to three volcanic systems.

The Fjallgardar volcanic ridge

With regard to origin and distribution the two most voluminous subglacial formations (M3 and M4) were probably erupted from a deep-seated regional magma source underneath the general Fjallgardar region. This volcanic stage produced only basalts and probably pre-dated the development of shallow magma chambers or local volcanic systems of the Krafla type that was to follow. In the Mödrudalur area these two formations are referred to as the *Fjallgardar volcanic ridge* (Fig. 1). This ridge is by far the most voluminous system in Mödrudalur. The volcanism that produced M3 and M4 must have extended at least from Vatnajökull in the S to Hólaheidi in the N (at 66°15′N), *ie* over a distance of some 140 km. The volcanism in the Slétta region N of 66°15′N is probably also part of the FVR; thus the total ridge length might be approximately 190 km.

The NW system

Formations M1, M2 and M6 are confined to the NW sector and the occurrence of dacite there (M2) suggests that a volcanic complex may have developed locally. Henceforth, these formations are informally referred to as the *NW system* (Fig. 1).

The Kverkfjöll system

The subglacially formed ridges of formation M5 (Fig. 3) have the same trend as the fissure swarm from which a surface lava of formation N1 has erupted in postglacial times in Mödrudalur. There, this fissure swarm of the present neovolcanic zone cuts into the FVR at an angle of approximately 15°. Because the fissure swarm can be traced into the Kverkfjöll volcanic centre to the S (75 km) it is strongly indicated that these volcanic products of Mödrudalur were fed to the eruptive site by subsurface lateral magma flow from the same source. Therefore formations M5 (subglacially formed ridges) and N1 (a surface lava flow of Arnardalsalda) are regarded as part of the *Kverkfjöll volcanic system*. Being younger and still active this system is not regarded as part of the FVR proper.

Geochemistry: sampling and analytical methods

A detailed stratigraphic subdivision of the Mödrudalur area was presented (Fig. 3) in a previous section. In that area the FVR is 45 km long, and 17 representative samples were collected from it. In addition 25 samples were collected within the FVR, *ie* further S and N of Mödrudalur. The purpose of the sampling was firstly to study regional geochemical variations in the FVR and secondly to correlate the geochemistry with the more detailed stratigraphic work in the Mödrudalur area.

Geochemical analyses for major and trace elements were carried out on 42 samples: 34 subaqueously cooled units (mainly pillow lavas or para-pillow lavas*) (Jones 1970), three dyke units and five subaerially erupted lavas. Sample localities are shown in Fig. 4.

Approximately 15 cm^3 of rock were dryed at 110–120°C for two days to release H_2O^-. Fusion discs were prepared and analysed for major and trace elements using a Philips PW 1400 X-ray fluorescence spectrometer at Ruhr University (FRG). A total of 29 reference standards was routinely included with the analysed samples. An estimate of analytical precision was readily obtained from duplicate analysis; this proved to be excellent (Table 1).

For Fe^{2+} analysis, each sample was dissolved in hydrofluoric acid–silver perchlorate and a semi-automatic potentiometric titration was carried out using standard potassium bromide solution. CO_2 and H_2O^+ were determined by closed-system colorimetric titration: the CO_2 of a barium perchlorate solution with which the gases produced by passing oxygen over the heated sample were mixed, and the H_2O^+ by a Karl–Fisher reagent charged with the inert carrier gas (N_2) loaded with the water removed from the sample at 1300°C.

Geochemical results

Selected representative analyses (eight out of 42) that span the chemical range observed in the study region are presented in Table 2. The silica values of the 42 samples analysed range from 48.2 to 50.6 wt%. All samples fall well within the basalt tholeiite range on the silica versus combined potassium and sodium diagram of Macdonald & Katsura (1964). The CIPW norm was calculated assuming an Fe_2O_3/FeO ratio of 0.15 (Brooks 1976; Jakobsson 1979) and normalizing for major and trace elements

* Jones (1970, p. 128): 'The term "para-pillow lava" has been coined for lava consisting of structural units similar to pillows in their dimensions, fabric, and internal structure, but with typically ragged outlines and separated by a breccia of glass. The surface of these units is remarkably jagged, and the breccia which encloses them commonly contains lava nodules, some no larger than a pea, with a hackly glass crust. Voluminous irregular masses of closely jointed lava up to tens of metres across are an integral part of the total structure of para-pillow lava.'

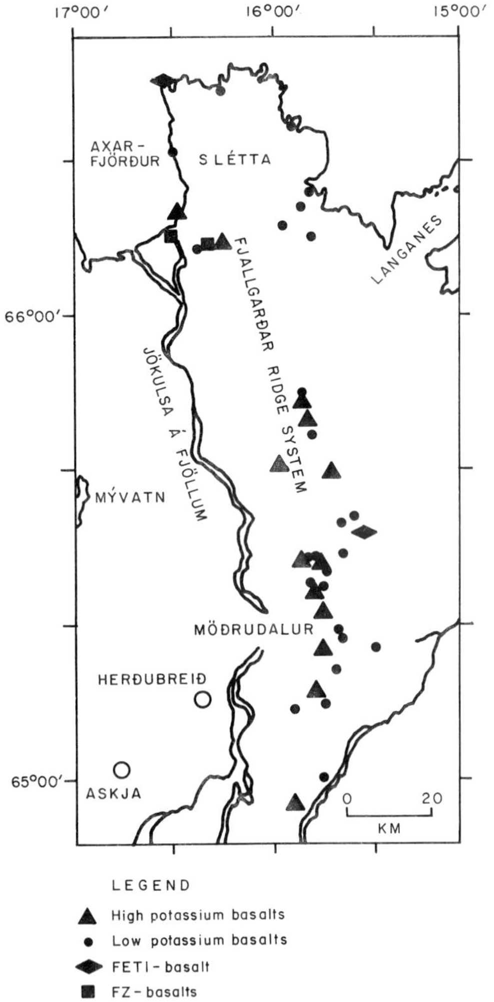

FIG. 4. Map showing sample locations for geochemistry and the main geochemical grouping (explained in the text): ●, LPB; ▲, HPB; ◆, FETI; ■, FZ. The same symbols are used in subsequent figures.

(H_2O-, S- and Cl-free). The norm calculations for the Fjallgardar basalts show that 32 units are *Qz* normative whereas 10 units are *Ol* normative. Maximum *Qz*- and *Ol*-normative values are 3.11% and 3.71% respectively.

In order to test for possible chemical grouping the analysed units were plotted on an yttrium versus zirconium diagram (Fig. 5). Two main chemical trends were obtained that subsequently were found to have different potassium ranges. Thus, in Fig. 5, 38 (90%) of the 42 samples fall on two differentiation trends: 13 are of a high-potassium and 25 of a low-potassium type, henceforth referred to as the HPB and LPB units. A plot of the K_2O contents of the Fjallgardar basalts against latitude (Fig. 6) shows a non-overlapping grouping of the LPB and HPB units. The 'low' values range between approximately 0.1 and 0.4 wt% K_2O and the 'high' values between approximately 0.5 and 0.7 wt% K_2O. The basalt grouping is partly confirmed by the norm calculations. Thus LPB units range from *Ol* normative (3.71% *Ol*) to *Qz* normative (1.65% *Qz*) whereas all the HPB units are *Qz* normative only (0.69%–3.11% *Qz*).

The two groupings are further confirmed by plotting K_2O against zirconium (Fig. 7) and the Y/Nb ratio against zirconium (Fig. 8). Whereas the LPB units show a rather uniform increase in K_2O and zirconium (Fig. 7) with a slightly greater dispersion at the upper bound, the HPB units tend to be further internally grouped. In Fig. 8 the HPB units have close to uniform Y/Nb ratios (1.6–2.0) whereas the LPB units have a Y/Nb range three times greater (2.2–3.4). Thus the source material for the LPB units appears to have been less homogeneous than that for the HPB units, notwithstanding that a greater error in the determination of both yttrium and niobium could be involved at the lower concentration levels.

Two units are rich in iron and titanium and defined as FETI basalts. A further two units were possibly mixed with gabbroic xenoliths and are provisionally termed 'fracture zone basalts' (FZ) because their sampling location coincided with one branch of the Tjörnes transcurrent fault zone at approximately 66°15′N (Fig. 1)—the Grimsey fault zone of Einarsson (1976). Subsequently, Mamula & Voight (1985) referred to this belt as the Grimsey–Vopnafjördur lineament system.

Regional stratigraphic and temporal geochemical variations

The distribution of each of the four basalt groups (*ie* LPB, HPB, FETI and FZ) is indicated in Fig. 4. The LPB units are found within the entire region of study whereas the HPB units occur only S of the Grimsey fault located at latitude 66°15′N.

Of the 42 samples analysed for chemistry 17 were collected in the mapped area of Mödrudalur (Fig. 3). Nine of these 17 samples were collected from the Lower Fjallgardar subglacial formation (M3), four from the Upper

TABLE 1. *Results of duplicate chemical analyses from the Fjallgardar ridge segment, NE Iceland*

Field number grouping	84–68	84–68d	84–75	84–75d
Major elements (wt%)				
SiO_2	49.60	49.40	49.60	49.70
TiO_2	2.06	2.05	2.31	2.32
Al_2O_3	16.91	16.89	13.32	13.32
Fe_2O_3	1.98	1.38	5.73	5.78
FeO	9.02[a]	—	7.98[a]	—
MnO	0.18	0.18	0.23	0.22
MgO	5.21	5.17	6.49	6.55
CaO	11.64	11.57	11.01	11.06
Na_2O	2.27	2.21	2.33	2.29
K_2O	0.47	0.46	0.30	0.30
P_2O_5	0.23	0.24	0.24	0.24
Total	99.86	—	99.97	—
Volatiles (wt%)				
H_2O+	0.25[a]	—	0.28[a]	—
CO_2	0.04[a]	—	0.08[a]	—
S	0.01[a]	—	0.06[a]	—
Cl	0.01[a]	—	0.01[a]	—
Trace elements (ppm)				
V	319	307	387	386
Cr	69	65	115	113
Co	35	34	44	42
Ni	59	59	78	80
Cu	163	160	183	171
Zn	90	94	121	118
Rb	11	10	9	10
Sr	256	254	191	184
Y	27	26	35	35
Zr	137	136	131	131
Nb	15	16	13	13
Ba	134	153	123	131

[a] Not analysed in duplicate.

Fjallgardar subglacial formation (M4) and four from the Upper Fjallgardar subglacial formation (M5). The question whether a relationship exists between geochemistry and these three stratigraphic formations can therefore be addressed. It is relevant that the eight stratigraphic formations in the Mödrudalur area belong essentially to three volcanic systems and the 17 samples, either LPB or HPB units, belong to only two of these, *ie* the FVR and the Kverkfjöll system (see Fig. 1). The results show that all the samples collected from the Lower Fjallgardar formation (older) are low-potassium basalts. For each of the Upper Fjallgardar formation and the Kverkfjöll formation (younger) three of four samples are high-potassium basalts.

First this demonstrates a clear temporal relationship between stratigraphy and chemistry. Thus the oldest of these three formations, *ie* the Lower Fjallgardar formation, consists entirely of LPB units whereas the younger stratigraphic formations, *ie* the Upper Fjallgardar formation and the Kverkfjöll formation, are predominantly HPB units. That two LPB units fall within the Kverkfjöll and Upper Fjallgardar formations is thought to be due to slight overlapping in the volcanic extrusion of the basalt types rather than inaccuracy in the stratigraphic mapping.

Discussion

It is interesting that the temporal change in geochemistry in the FVR is from LPB units in the Lower Fjallgardar formation to HPB units in both the Upper Fjallgardar formation and the Kverkfjöll volcanic system. This change is

TABLE 2. *Chemical composition of selected analyses from the Fjallgardar region, NE Iceland*

Field number grouping	84–28 LPB	84–46 LPB	83–38 LPB	84–43 LPB	84–49 HPB	84–44 HPB	84–73 HPB	84–36 FETI
Major elements (wt%)								
SiO_2	49.78	49.83	49.46	48.79	50.13	50.58	49.76	50.12
TiO_2	2.62	1.93	1.59	1.31	3.11	2.79	1.99	3.30
Al_2O_3	13.08	14.07	14.50	17.17	12.94	13.61	16.67	13.31
Fe_2O_3	4.25	4.04	2.60	3.51	4.47	4.13	2.83	3.85
FeO	10.40	8.50	9.11	6.33	11.03	9.86	8.22	10.86
MnO	0.23	0.21	0.20	0.16	0.24	0.22	0.18	0.23
MgO	6.10	6.98	7.78	7.33	5.10	5.46	5.72	5.19
CaO	10.59	11.87	12 .37	13.25	9.49	9.90	11.81	9.61
Na_2O	2.38	2.13	2.06	1.77	2.46	2.50	2.18	2.66
K_2O	0.32	0.26	0.17	0.12	0.65	0.62	0.43	0.51
P_2O_5	0.25	0.18	0.15	0.11	0.36	0.33	0.21	0.35
Total	100.21	100.07	100.09	100.40	99.19	99.04	99.94	99.47
Volatiles (wt%)								
H_2O_+	0.46	0.59	0.13	0.61	0.71	0.68	0.52	0.54
CO_2	0.05	0.06	0.04	0.05	0.06	0.04	0.06	0.07
S	0.01	0.01	0.01	0.01	0.03	0.05	0.03	0.01
Cl	0.01	0.01	0.01	0.01	0.01	0.01	0.01	0.01
Trace elements (ppm)								
V	439	368	336	270	468	420	304	486
Cr	56	131	271	292	37	45	91	32
Co	47	49	46	41	39	47	41	38
Ni	64	82	91	111	42	45	60	50
Cu	157	169	142	138	258	170	148	202
Zn	129	100	95	72	135	120	91	132
Rb	9	7	6	3	15	16	13	12
Sr	195	172	158	179	230	233	252	218
Y	37	30	25	19	41	35	26	46
Zr	140	103	83	64	204	184	129	201
Nb	12	10	8	7	23	22	14	19
Ba	139	94	85	38	233	188	138	213
Magnesium number	46.5	53.8	57.9	61.0	40.7	44.9	51.8	42.3

from primitive type basalts to more evolved *Qz*-normative tholeiites. It appears that the two trends represent two different basalt sources but a more thorough investigation is needed to confirm or reject this idea. The temporal change in chemistry towards more evolved basalts (or perhaps a different source) is thought to be tectonically controlled. Thus the Lower Fjallgardar formation was erupted in the absence of 'mature' volcanic centres and during this period magma transport may have been fast toward the surface and essentially without prolonged residence in shallow crustal magma chambers. This led to only slight differentiation of LPB material, which has a potassium range of only 0.1%–0.4%. High-potassium basalts are first introduced in Mödrudalur with the build-up of the Upper Fjallgardar formation. Because high-potassium basalts also dominate the Kverkfjöll formation which can be traced S toward Kverkfjöll, it appears likely that the HPB material is somehow related to volcanic activity in the Kverkfjöll volcanic system.

Schilling (1973), Brooks & Jakobsson (1974) and Sigvaldason *et al.* (1974) have shown that the potassium range in tholeiitic basalts increases to approximately 0.6% K_2O towards central Iceland, *ie* Kverkfjöll. The two combined K_2O ranges observed in the FVR span exactly the total potassium range which earlier studies reported for basaltic tholeiites in the Iceland region. The non-overlapping of the two

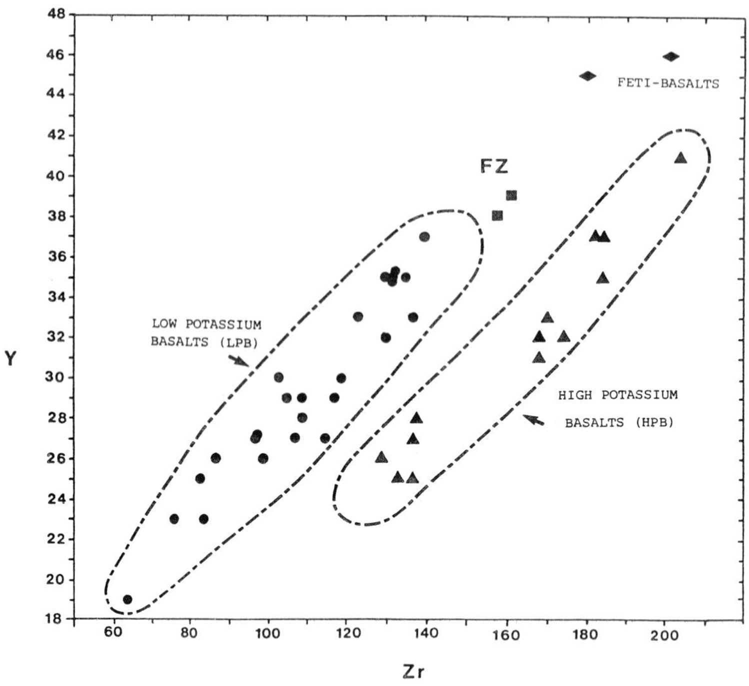

FIG. 5. Plot of yttrium versus zirconium showing that the high-and low-potassium units (of Fig. 6) have separate differentiation trends. The FETI and FZ basalts appear to be derivatives of the LPB group.

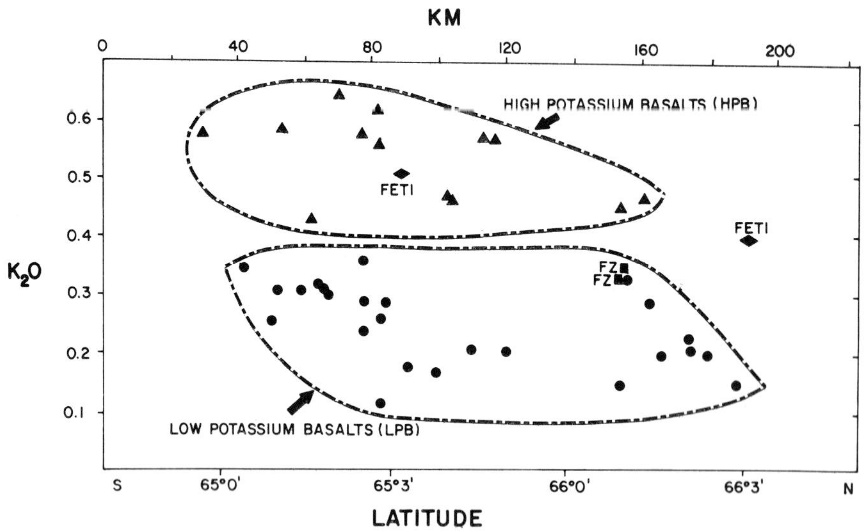

FIG. 6. Potassium (K_2O) plotted against latitude and approximate distance from the Kverkfjöll region. The two main groups are either 'high' (0.5%–0.7%) or 'low' (0.1%–0.4%) in potassium. Also included are the FETI and FZ units.

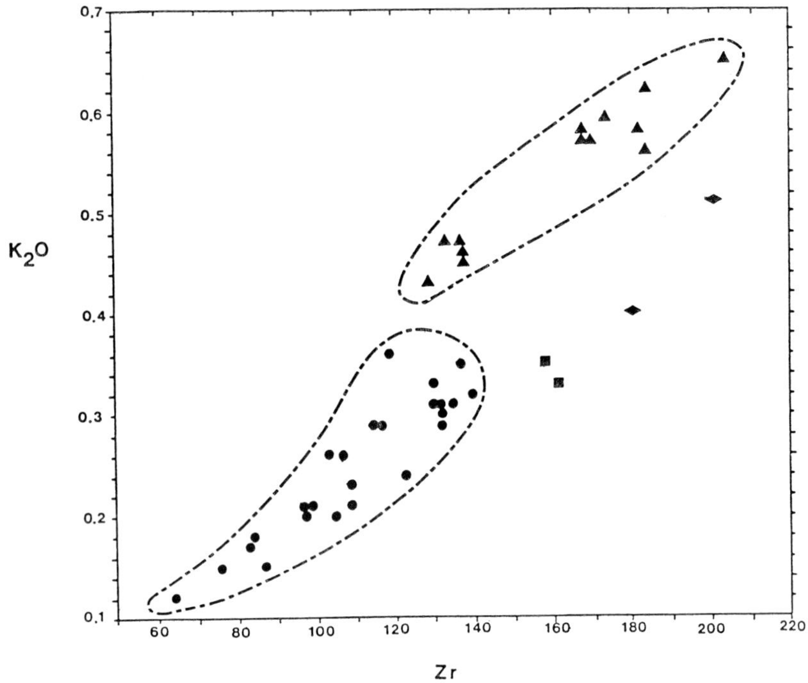

FIG. 7. Potassium versus zirconium. The same grouping is observed as for Fig. 5 (same explanation).

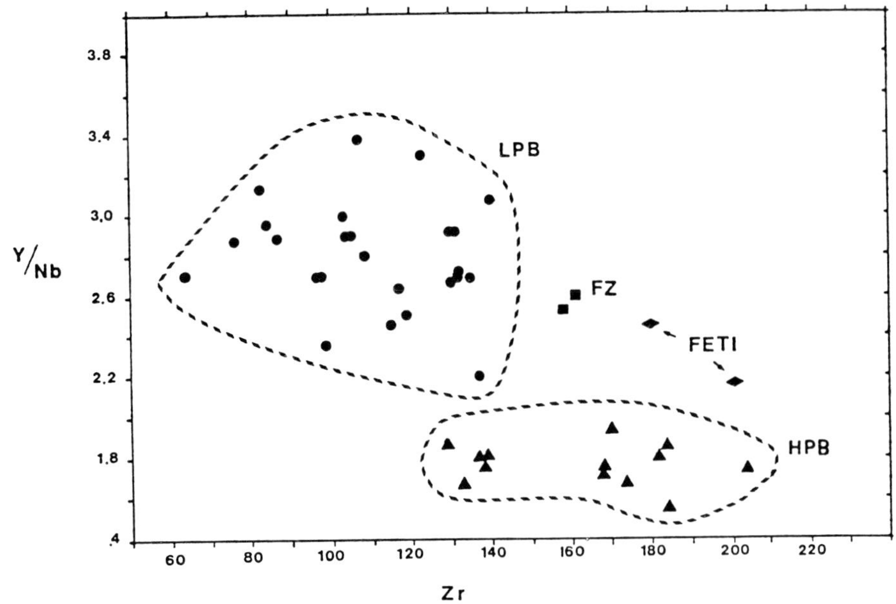

FIG. 8. Y/Nb versus zirconium. The high-potassium units have a restricted Y/Nb scatter (1.6–2.0) unaffected by the zirconium concentration, whereas for the low-potassium units the scatter is much greater (2.2–3.4) and appears to increase with zirconium concentration.

basalt types in the FVR suggests that the apparently wide K_2O range in Iceland can be subdivided on the basis of distinct basalt types/sources and that differentiation of individual basalt types increases the K_2O value by of the order of 0.2%−0.4% K_2O by weight. The minimum K_2O values of the LPB and HPB units are approximately 0.1% and 0.45% respectively, and all higher values are considered to have resulted from differentiation. Accordingly, 'undifferentiated' tholeiite in the FVR has a maximum potassium value of only 0.45% K_2O.

Basalts enriched in iron and titanium are not uncommon in the Iceland region and have been reported for the Skagi peninsula, N Iceland (Sigurdsson *et al.* 1978), and eastern Iceland (Wood 1976). From Table 2 the FETI unit (84−36) has a strikingly similar trace-element pattern to that of the most evolved HPB unit (84−49). Thus the two FETI units found in the FVR may have evolved from the HPB material. However, examination of Fig. 5 suggests that they, as well as the FZ units, might be derivatives of the LPB units through extended differentiation.

At approximately 66°10′N, fissure swarms of the neovolcanic zone extend into the Slétta region in a similar manner to the Kverkfjöll system further S in Mödrudalur. Volcanism was probably renewed in postglacial times in Slétta but exact timing of events there remains poorly understood. Subsurface lateral flow of magma from the neovolcanic zone may thus be responsible for most of the Recent Slétta volcanism.

It can be speculated that the HPB units represent a basalt source that originates S of the Grimsey fault, possibly as far S as Kverkfjöll (140 km), and that this fault boundary prevents flow of magma to the N. The LPB eruptive units may represent a separate source, the distribution of which is not influenced in the N by the Grimsey fault. This source is petrologically more primitive and resembles ocean floor basalts as well as some lavas of the neovolcanic zone.

It is relevant that volcanism in the Fjallgardar volcanic zone was greatest at approximately 66°30′N or some 85 km N of the Kverkfjöll region, which has often been regarded as the centre of the Iceland hot-spot. This indicates that the FVR may have been mostly independent of volcanic production in the Kverkfjöll region. Alternatively, crustal spreading may have been at a maximum at 66°30′N which in turn may have caused greater volcanism there.

The question whether the Fjallgardar volcanic zone was 'aborted' or remains 'dormant' has not been answered here. The six subglacial ridge formations defined for the Mödrudalur area were not active contemporaneously although some overlapping may have occurred there. One of these formations (M5) clearly extends from the active Kverkjföll system of the neovolcanic zone into the Fjallgardar region. Volcanism continued in Fjallgardar during the Weichselian glaciation as well as during earlier glaciations. It is therefore quite likely that a 'punctuated' mode of crustal construction still exists between the Fjallgardar region and the neovolcanic zone. The median lines of the neovolcanic zone and the FVR are parallel and are 40 km apart but the fissure systems of the neovolcanic zone are at an angle of some 25° to the median line and thus are capable of feeding lavas into the Fjallgardar region by subsurface lateral magma flow. Predominantly, however, volcanism in the FVR was independent of magma flow from the neovolcanic zone. It can further be speculated that the 'punctuated' mode of volcanism that is responsible for the different stratigraphic formations in the FVR may be related to punctuated activity or shifting of activity between separate branches of the Tjörnes transcurrent fault zone. The proposed rift jump between the neovolcanic zone and the FVR has similarities to mid-ocean ridge rift jumps although it is more complicated in that the specific plate boundary in NE Iceland grades in the N into a transcurrent fault zone (made up of three branches) and in the S to a mantle plume maximum in the Kverkfjöll region.

Conclusions

For the FVR this study infers that the increased K_2O range towards central Iceland is partly clarified in that the FVR units can be subdivided into low- and high-potassium basalts. In other words a single potassium range does not appear to be present: instead two basalt groups with separate non-overlapping potassium ranges characterize the tholeiitic basalt that erupted in the FVR region. Separate differentiation trends for the basalt groups indicate two sources of origin.

The FVR probably represents an area where volcanism began as the result of a small-scale rift jump. It is concluded that small-scale lateral rift jumps, of the order of 20−40 km, can lead to drastic geochemical changes over extended periods of time during which stratigraphic formations of different geochemical compositions may accumulate. The present study may help towards an understanding of long-term geochemical evolution in the Iceland region, *eg* the Tertiary basalt lava succession of eastern Iceland. Thus geochemical variations involving evolution from primitive to more evolved tholeiitic-type basalts may be caused by rift jumping: in the initial stages of rift opening, shallow crustal chambers have not yet developed, which prevents magma residence, mixing and evolution within major volcanic centres.

ACKNOWLEDGMENTS: Financial support from the Icelandic Science Foundation to carry out fieldwork in NE Iceland is gratefully acknowledged. I thank the Alexander von Humboldt Foundation for a Humboldt Fellowship to work on the field data at the Department of Mineralogy, Ruhr-Universität Bochum, FRG.

I thank members of the Department of Mineralogy, Ruhr-Universität Bochum, for their pleasant company, and in particular Mrs Henny Niephaus for carrying out chemical analysis and Dr Hans-Ulrich Schmincke for suggestions whilst I was preparing an early version of this manuscript. Also, I thank Dr Ulrich Bednarz and Dr Gerhard Wörner of the same institution for many fruitful discussions. I thank the reviewers, Dr Ian L. Gibson (University of Waterloo, Ontario, Canada) and Dr Andrew Saunders (University of Leicester), for helpful suggestions to improve the text. Finally, I thank Ingunn Sigurdardóttir and Helga Sveinbjörnsdóttir of the National Energy Authority of Iceland for skillful drafting of figures.

References

BROOKS, C. K. 1976. The Fe_2O_3/FeO ratio of basalt analyses: an appeal for a standardized procedure. *Bulletin of the Geological Society of Denmark* **25**, 117–120.

—— & JAKOBSSON, S. P. 1974. Petrochemistry of the volcanic rocks of the North Atlantic Ridge system. *In*: KRISTJANSSON, L. (ed.) *Geodynamics of Iceland and the North Atlantic Area*. Reidel, Amsterdam, 139–154.

EINARSSON, P. 1976. Relative location of earthquakes in the Tjörnes fracture zone. *Societas Scientiarum Islandica* **5**, 45–60.

—— & BRANDSDÓTTIR, B. 1980. Seismological evidence for lateral magma intrusion during the July 1978 deflation of the Krafla volcano in NE Iceland. *Journal of Geophysics* **47**, 160–165.

GIBSON, I. L. & PIPER, J. D. A. 1972. Structure of the Iceland basalt plateau and the process of drift. *Philosophical Transactions of the Royal Society of London, Series A* **271**, 141–150.

HELGASON, J. 1984. Frequent shifts of the volcanic zone in Iceland. *Geology* **12**, 212–216.

—— 1985. Shifts of the plate boundary in Iceland: some aspects of Tertiary volcanism. *Journal of Geophysics Research* **80**, 10 084–10 092.

—— 1987. *Geological Map of Mödrudalur, Mid-east Iceland*. National Energy Authority (Hydro-Power Division) and National Power Company of Iceland.

JAKOBSSON, S. P. 1979. Petrology of Recent basalts of the Eastern Volcanic Zone, Iceland. *Acta Naturalia Islandica* **26**, 103.

JONES, J. G. 1970. Intraglacial volcanoes of the Laugarvatn region, southwest Iceland—II. *Journal of Geology* **78**, 127–140.

MACDONALD, G. A. & KATSURA, T. 1964. Chemical composition of Hawaiian lavas. *Journal of Petrology* **5**, 82–133.

MAMULA, N. & VOIGHT, B. 1985. The tectonic analysis of lineaments near a spreading axis, north-eastern Iceland. *Tectonophysics* **116**, 63–93.

MCMASTER, R. L., SCHILLING, J.-G. & PINET, P. R. 1977. Plate boundary within the Tjörnes Fracture Zone on northern Iceland's insular margin. *Nature* **269**, 663–668.

SAEMUNDSSON, K. 1974. Evolution of the axial rifting zone in northern Iceland and the Tjörnes Fracture Zone. *Bulletin of the American Geological Society* **85**, 495–504.

SCHILLING, J.-G. 1973. Iceland mantle plume: geochemical study of the Reykjanes Ridge. *Nature* **244**, 565–571.

SIGURDSSON, H., SCHILLING, J.-G. & MEYER, P. S. 1978. Skagi and Langjökull volcanic zones in Iceland: 1. Petrology and structure. *Journal of Geophysical Research* **83**, 3971–3982.

SIGVALDASON, G. E., STEINTHORSSON, S., OSKARSSON, N. & IMSLAND, P. 1974. Compositional variation in recent Icelandic tholeiites and the Kverkfjöll hot spot. *Nature* **251**, 579–582.

WALKER, G. P. L. 1975. Excess spreading axis and spreading rate in Iceland. *Nature* **255**, 468–470.

WARD, P. L. 1971. New interpretation of the geology of Iceland. *Geological Society of America Bulletin* **82**, 2991–3012.

WOOD, D. A. 1976. Spatial and temporal variations in the trace element geochemistry of the eastern Iceland flood basalt succession. *Journal of Geophysical Research* **81**, 4353–4360.

J. HELGASON, Institut für Mineralogie, Ruhr-Universität Bochum, FRG, *Present address*: Science Institute, University of Iceland, Dunhaga 3, 101 Reykjavik, Iceland.

Geochemical features of intraplate oceanic plateau basalts

P.A. Floyd

SUMMARY: Coeval with the Cretaceous thermal uplift of a large region of the 'old Pacific' oceanic lithosphere was the widespread development of two expressions of intraplate volcanism: edifice building (ocean islands, seamounts) and submarine sheet-flows (oceanic plateaux). The geochemical features of the development of oceanic plateaux are reviewed using the Nauru Basin complex as a model. Ocean plateau basalts (OPBs) are mildly altered, poorly phyric, *hy*- or *qz*-normative tholeiites, rapidly extruded submarine sheet-flow lavas with a relatively uniform chemistry. Although their eruptive setting is apparently intraplate they have a chondrite-normalized incompatible-element-depleted pattern similar to normal-type mid-ocean ridge basalt (MORB) formed at ocean spreading centres. They are chemically distinguished from normal-type MORB by being enriched in selected incompatible elements and can be matched by various enriched or 'transitional' MORB types. Isotopically they have similarities to some types of ocean island basalt (OIB), such as Hawaii. They were probably derived by high degrees of melting from an upper mantle MORB-type source with an enriched OIB-type component; melts being mixtures derived from the two mantle components. In terms of crustal development they are characteristic of anomalously thickened oceanic crust and were formed in an off-axis (or near-axis) eruptive setting shortly after the development of axis-generated oceanic basement over which they flowed. OPB appears to be restricted to the widespread Cretaceous magmatic event.

Throughout much of the present-day western and central Pacific Ocean basin are seen the products of regionally extensive and voluminous mid-Cretaceous intraplate volcanism (Winterer 1973; Watts *et al.* 1980; Schlanger *et al.* 1981). Major activity took place between about 110 and 70 Ma (Larson & Schlanger 1981; Schlanger *et al.* 1981) and is now represented by linear island chains, groups of seamounts and submarine plateaux (Fig. 1). Widespread mantle thermal disturbances and related volcanism produced lithospheric uplift such that parts of the 'old Pacific' floor were elevated above their 'normal' subsidence level for crust of Cretaceous age (Schlanger *et al.* 1981; Schlanger & Moberly 1986). In particular this produced anomalous shallow areas over thickened oceanic segments, such as the Nauru Basin, Ontong-Java and Manihiki plateaux (Fig. 1).

Cretaceous intraplate volcanism in the Pacific basin is represented by two different morphological expressions that exhibit chemically contrasting basalt types: (i) edifice structures with ocean island basalt (OIB) features, and (ii) oceanic plateaux composed of extensive sheet-flows with a chemistry akin to that of mid-ocean ridge basalts (MORBs). This paper summarizes the main characteristics of the plateau sheet-flows that constitute a chemically significant and volumetrically important addition to intraplate volcanism supported on an oceanic basement of Mesozoic age.

Not only are plateau sheet-flows clearly different for the abundant ocean islands and seamounts, but their tectonic significance is indicated by (i) their generation in an intraplate environment but with MORB-like chemical features (Batiza *et al.* 1980) and (ii) the fact that in areal and volumetric terms they could be the oceanic analogues of continental plateau basalt provinces (Tokuyama & Batiza 1981; Saunders 1986).

It is considered that these volcanics represent a special group and they are therefore referred to here as oceanic plateau basalts (OPBs) (*cf* Tokuyama & Batiza 1981).

Oceanic plateaux

The best known oceanic sheet-flows, in terms of volcanic stratigraphy, petrology and geochemistry, are represented by the Nauru Basin complex (Fig. 1) investigated during Deep Sea Drilling Project (DSDP) Legs 61 (Larson *et al.* 1981) and 89 (Moberly *et al.* 1986). Today the Nauru complex is situated in an oceanic basin and does not have the positive bathymetry of other submarine plateaux, such as the Ontong-Java plateau, in the western-central Pacific. However, when it was formed during the Cretaceous thermal uplift it probably stood 0.5–1.5 km above the surrounding Jurassic ocean floor before subsequent regional subsidence and part

From Saunders, A. D. & Norry, M. J. (eds), 1989, *Magmatism in the Ocean Basins.* Geological Society Special Publication No. 42, pp. 215–230.

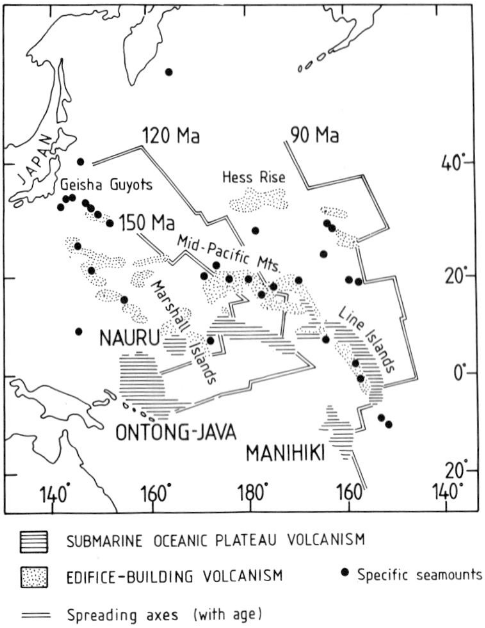

FIG. 1. Map of the western-central Pacific showing the distribution of Cretaceous intraplate volcanism (after Winterer 1973; Schlanger *et al.* 1981). Specific seamounts of Cretaceous age from Haggerty *et al.* (1982) and isochrons from Watts *et al.* (1980).

burial by penecontemporaneous seamount-derived volcaniclastic debris.

The present work draws on the published chemical data base and views expressed in the above references, as well as the writer's own data (Floyd 1986), and presents a summary of OPBs. The Nauru complex will be used as a model against which the volcanics of other plateaux of about the same age can be compared, although the chemical data for other areas are sparse. On the basis of seismic reflection interpretations plateaux similar to Nauru are considered to exist between the Nauru Basin and the Line Islands (Winterer 1973; Schlanger *et al.* 1981), in addition to the Ontong-Java and Manihiki plateaux to the S (Fig. 1) which, on the basis of plate reconstructions of the early Pacific, are thought to have been generated off-axis at 105–110 Ma (Winterer 1976). Parts of the Caribbean crust may also have a similar construction, since the sea is anomalously shallow (Burke *et al.* 1978) and according to Cretaceous plate configurations for about 120 Ma the crust formed part of the Pacific-generated Farallon plate (Larson & Pitman 1972); not far removed from the Nauru complex in time and space. Table 1 summarizes the generation ages and suggested origins for oceanic plateaux and anomalous crust of Cretaceous age.

Nauru Basin complex

DSDP Hole 462A is near the centre of the seismically defined Nauru Basin complex and reached a depth of 1209 m sub-bottom. The basal section is represented by about 650 m of basic volcanics with minor sedimentary interlayers and is composed of a lower unit of sheet-flow basalts and an upper unit of (intrusive?) dolerite sheets (Moberly *et al.* 1986). The intraplate character of the Cretaceous volcanic complex is based on undisturbed local magnetic lineation patterns that show that it rests on Jurassic ocean crust of about 150 Ma (Larson & Schlanger 1981). Faunal evidence in the few sediment intercalations is sparse and radiometric dating of the eruptive age of the complex is made difficult owing to alteration and the very low Rb/Sr ratios and potassium contents of the basalts that make Rb–Sr and Ar–Ar dating imprecise. Schlanger & Moberly (1986), however, recognize three Cretaceous volcanic events in the Pacific, the earliest of which is the Nauru Basin sheet-flow complex, based on 'acceptable' $^{40}Ar/^{39}Ar$ plateau ages (Ozima *et al.* 1981; Takigami *et al.* 1986):

(1) 130 Ma, development of the Nauru complex sheet-flows;
(2) 100 Ma, intrusion of the higher-level Nauru sills; this was probably coeval with local edifice-building volcanism during the Aptian–Albian (Moberly & Jenkyns 1981; Rea & Vallier 1983);
(3) 75 Ma, major edifice-building volcanic event during the Campanian throughout much of the western and central Pacific (Watts *et al.* 1980; Schlanger *et al.* 1981).

If an extrusion age for the earliest sheet-flows of about 130 Ma is correct, then they were emplaced some 20 Ma after the development of the local Jurassic ocean crust. However, uncertainty in dating the two events allows for a greater overlap in time or a closer age relationship than is implied by the 'acceptable' generation ages. This has implications for ideas concerning the generation of anomalously thickened crustal segments if they are of about the same age as nearby spreading axes.

TABLE 1. *General features and suggested origins for some Pacific oceanic plateaux*[a]

Plateau	Approximate area (km^2)	Mean water depth (km)	Mean relief above ocean floor (km)	Basement age (and method)	General data reference (DSDP volume)
Hess Rise	760 000	2.0	3.0	Late Albian (oldest, sediment): no basement reached; related to seamount activity(?)	32
Magellan	50 000	3.0	2.5	132–140 Ma (oldest sediment)	17
Manihiki	650 000	2.5	3.0	107 Ma (oldest sediment) 110–120 Ma (K–Ar)	33
Nauru	360 000	3.5 (max. 5.1)	(none)	> 112 Ma (oldest sediment) 127–131 Ma (Ar–Ar)	61, 89
Ontong-Java	1 500 000	1.7	3.0	104–109 Ma (oldest sediment) >121 Ma (estimate)	7, 30
Shatsky Rise	680 000	2.0	3.5	Early Berriasian (oldest sediment); no basement reached	6, 32
Caribbean	1 000 000	2.5	(none)	*c.*110 Ma, max. of 120 Ma (Ar–Ar)	51, 52, 53

Tectonic setting	Origin	Reference
1. Anomalous oceanic crust	Off-axis (intraplate) non-hotspot volcanism (*eg* Nauru)	Batiza *et al.* 1980
2. Anomalous oceanic crust	Genetically associated with anomalous (plume-enhanced?) axis activity (*eg* Nauru)	Saunders 1986
3. Thick ocean crust	Thermal-anomaly-induced volcanism at oceanic triple junction (*eg* Manihiki)	Winterer 1973
4. Large seamount-type structure on ocean crust	Activity caused by realignment of spreading ridge due to change in spreading direction (*eg* Manihiki)	Winterer 1973
5. Oceanic spreading axis	Build-up of axis volcanism due to slow spreading rates (*eg* Ontong-Java)	Coleman & Kroenke 1981
6. Originally part of continental crust	Fragmented and drifted old continental block (*eg* part of Ontong-Java)	Carlson *et al.* 1980 Nur & Ben-Avraham 1982

[a] A review of world-wide oceanic plateaux, rises and ridges has been presented by Nur & Ben-Avraham (1982), from which some of the above data have been taken.

Petrography

Only a brief outline will be given here of basalt petrography, as flow unit details are documented by Larson *et al.* (1981) and Moberly *et al.* (1986). The sheet-flows represent a series of rapidly extruded, alternating hypocrystalline to holocrystalline aphyric and poorly phyric non-vesicular tholeiitic basalts. Coarser-grained subophitic textures are found in the few thicker (up to 25 m) flow units, although quench textures are common throughout most thinner units. The dolerite sills are generally coarser grained and texturally variable with pyroxene-rich and granophyric patches. Proportions of phenocrysts are generally low (predominantly less than 5%) throughout the complex and indicate that the flows can be considered as liquid compositions. The following phenocryst assemblages are exhibited: olivine–plagioclase–clinopyroxene, clinopyroxene–plagioclase, and clinopyroxene–olivine, with clinopyroxene and plagioclase often forming glomerocrystic clusters and olivine as individual microphenocrysts. Phenocrystic olivine decreases down the lava flow sequence, being common throughout much of the top portion of the sheet-flows. Basalt matrix consists of plagioclase–clinopyroxene–titanomagnetite both with and without glass. Electron microprobe data for the phenocrystic and matrix pyroxene and plagioclase are given by Tokuyama & Batiza (1981), Floyd & Rowbotham (1986) and Castillo *et al.* (1986).

Alteration

Secondary minerals are seen throughout the basalts and dolerites, although the degree of alteration is generally minor and low grade. Palagonitization of glass, various smectites and zeolites are common, whereas celadonite, carbonate, K-feldspar and iron oxides are rare. Post-consolidation alteration was governed by low-temperature, non-oxidative seawater hydration processes in the zeolite facies. Although the top of the volcanic complex is slightly more hydrated (and carbonated) than deeper down (Floyd 1986), no systematic chemical or mineralogical changes are seen with depth.

Chemically, alteration in the lower sheet-flows is indicated by slightly higher than usual values for H_2O contents and Fe_2O_3/FeO ratios compared with fresh oceanic basalts, highly variable K/Rb ratios (100–700) and non-magmatic distributions of potassium, rubidium, barium, strontium, uranium and to a lesser extent iron, magnesium and nickel (Floyd 1986). One factor that may be general for the Nauru and other plateau lavas compared with altered ocean-ridge-generated basalts is the lack of extensive oxidation and carbonation. This may be a consequence of the rapid extrusion rates of sheet-flows, together with the lack of chaotic pillow lavas which provide fluid channelways, such that the lava pile is effectively sealed and blanketed from later seawater penetration.

Chemostratigraphy

On the basis of incompatible-element abundances and trends throughout the sampled sequence, the Nauru OPB complex has been divided into three chemical groups (Castillo *et al.* 1986; Saunders 1986). These comprise (i) the upper sheet-flows from 730 to 995 m depth (type 1 basalts), (ii) the lower sheet-flows from 997 to 1209 m depth (type 2 basalts), which are separated by a thin volcaniclastic siltstone horizon, and (iii) the 'sills' which are divided by Castillo *et al.* (1986) into a minor, stratigraphically lower group with type 2 basalt chemistry, which is separated by about 30 m of sediment from a thick higher sequence (termed type 3).

The chemical groups and down-hole variation for selected elements are illustrated in Fig. 2 (based on compilations of data from Seifert (1981), Tokuyama & Batiza (1981), Castillo *et al.* (1986), Floyd (1986) and Saunders (1986). A number of features are exhibited.

(1) The chemical break between the type 1 and type 2 sheet-flows is marked by an increase in compatible-element abundance (*eg* chromium) and a corresponding decrease in all incompatible elements.
(2) In general, the sills revert to higher abundances of incompatible elements but display greater internal variation than the sheet-flows.
(3) The sheet-flow basalts all have relatively uniform or restricted elemental ranges. However, there is a difference between the sheet-flow groups, with the type 2 basalts showing little or possibly minor 'sawtooth' variation (*eg* zirconium), whereas type 1 basalts exhibit decreasing trends of incompatible-element abundances with stratigraphic height.

The chemical uniformity of tholeiitic OPB (Fig. 2) relative to MORB and continental basalts appears to be a characteristic feature and implies short residence times in magma chambers such that chemical evolution was limited (as witnessed by the poorly phyric nature and lack of cumulates) and/or continuous and

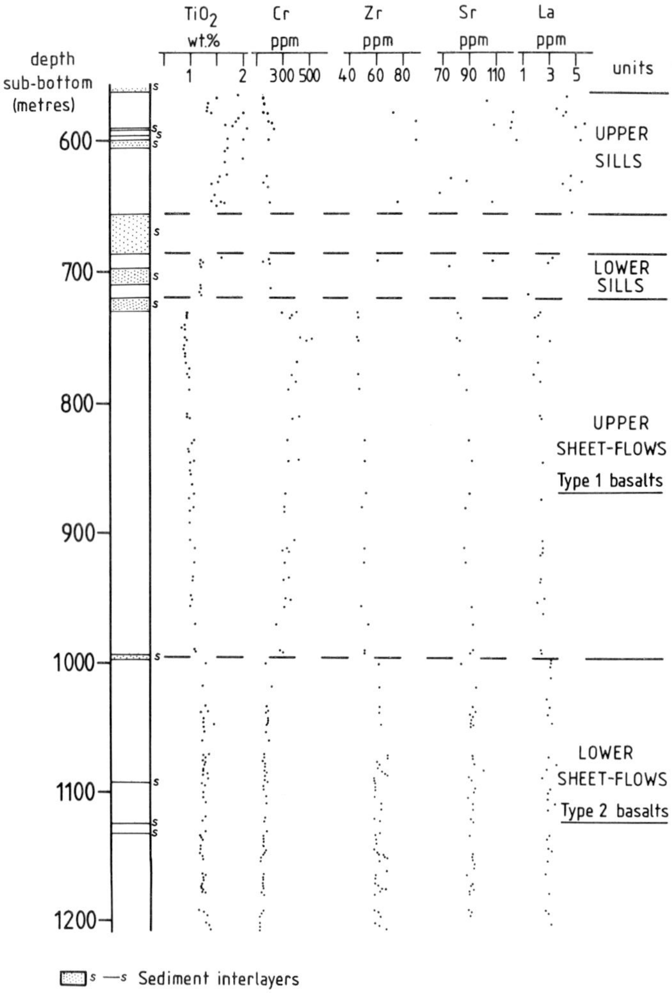

FIG. 2. Chemostratigraphy of DSDP Hole 462A in the Nauru ocean plateau complex. Chemical data selected from papers in Larson *et al.* (1981) and Moberly *et al.* (1986).

efficient magma mixing to near 'steady-state' compositions.

Chemical variations and magmagenesis

Selected chemical analyses for the basaltic sheet-flows are given in Table 2. As a model for OPB, the Nauru tholeiites exhibit the following general chemical characteristics.

(1) They are chemically uniform throughout with only limited internal magmatic differentiation.

(2) They have low abundances of large-ion lithophile (LIL) elements, such as potassium, rubidium, barium, strontium, uranium, thorium and light rare earth elements (REEs), which show chondrite-normalized depleted patterns similar to those of normal-type (N-type) MORBs. Although the most incompatible of the LIL elements are mobile and systematically enhanced by low-grade submarine alteration (Hart *et al.* 1974; Floyd & Fuge 1982), all the Nauru tholeiites are characterized by low potassium and rubidium, irrespective of the degree of alteration.

(3) Chondrite-normalized REE patterns vary slightly from mildly light REE depleted ($(Ce/Yb)_n \approx 0.85$ for type 1 basalts) to flat (approximately × 10–18 chondritic) for type 3 sills.

TABLE 2. *Representative chemical analyses of Nauru complex oceanic plateau basalt type 1 and type 2 sheet-flows (samples 1–5 and 6–10 respectively)*

	Samples									
	1	2	3	4	5	6	7	8	9	10
SiO_2	49.6	49.6	49.78	49.22	49.35	49.8	49.42	49.60	50.18	49.87
TiO_2	0.95	1.08	—	—	—	1.23	1.31	1.23	1.26	1.33
Al_2O_3	14.9	14.20	13.74	13.86	13.81	13.6	13.83	13.61	14.06	13.88
Fe_2O_3	1.42	1.42	—	—	—	4.11	4.11	3.72	4.27	4.46
FeO	9.44	9.46	11.67	11.67	11.57	8.63	8.67	9.04	8.34	8.04
MnO	0.22	0.20	0.19	0.20	0.21	0.20	0.22	0.46	0.23	0.20
MgO	7.56	8.02	7.76	7.71	7.81	6.97	7.02	6.74	6.75	6.80
CaO	12.74	12.44	11.83	12.31	12.20	11.40	11.57	11.66	11.43	10.52
Na_2O	2.00	2.01	2.11	2.08	2.15	2.12	2.43	2.22	2.55	3.51
K_2O	0.05	0.06	0.06	0.01	0.02	0.03	0.12	0.17	0.13	0.13
P_2O_5	0.07	0.10	—	—	—	0.12	0.10	0.10	0.11	0.11
H_2O_+	—	—	—	—	—	0.29	1.26	0.83	1.10	1.25
Ba	7	2	—	—	—	—	11	5	5	29
Co	—	—	49	50	50	49	—	—	—	—
Cr	484	342	313	325	325	181	162	170	143	119
Hf	1.52	1.55	1.70	1.71	1.77	2.05	1.84	1.93	1.89	2.26
Nb	4	2	—	—	—	—	6	5	6	5
Ni	168	125	—	—	—	—	88	87	77	79
Rb	1	1	—	—	—	—	3	6	2	6
Sc	—	—	47	49	48	49	44	46	48	49
Sr	83	88	—	—	—	—	111	109	109	107
Ta	0.20	0.21	0.17	0.21	0.16	0.23	0.50	0.86	0.42	0.79
Th	0.26	0.35	0.12	—	0.25	0.15	0.28	0.27	0.29	0.34
U	—	—	—	—	—	—	0.07	0.02	0.08	0.10
Y	23	21	—	—	—	—	31	28	29	28
Zr	47	52	—	—	—	—	62	65	62	73
La	3.11	2.71	2.83	2.66	2.73	3.29	3.69	3.54	3.64	3.88
Ce	8.85	8.11	8.12	8.31	8.49	9.21	9.38	9.28	9.45	10.14
Nd	6.16	6.47	—	—	—	—	7.97	7.74	7.75	8.24
Sm	2.19	2.30	2.25	2.26	2.25	2.62	2.61	2.60	2.58	2.72
Eu	0.86	0.91	0.86	0.92	0.88	0.98	0.98	0.97	0.98	1.02
Gd	3.21	3.31	—	—	—	—	3.39	3.61	3.47	3.67
Tb	0.57	0.56	0.59	0.56	0.62	0.82	—	—	—	—
Dy	—	—	—	—	—	—	4.32	4.35	4.31	4.41
Ho	—	—	—	—	—	—	0.91	0.90	0.90	0.90
Er	—	—	—	—	—	—	2.74	2.76	2.71	2.76
Yb	2.64	2.37	2.24	2.19	2.43	2.71	2.62	2.64	2.68	2.63
Lu	0.43	0.38	0.34	0.37	0.36	0.43	0.42	0.43	0.42	0.42

Data from Castillo *et al.* (1986), Floyd (1986 and unpublished REE data), Saunders (1986) and Seifert (1981). Major oxides in weight per cent and trace elements in parts per million; —, no data; total Fe as FeO for samples 3–5.

(4) Relative to primary mantle melts, OPBs have evolved compositions with low magnesium numbers (100 Mg/Mg+Fe) (maximum about 66), low chromium and nickel contents and high FeO^*/MgO ratios, indicative of high-pressure olivine + pyroxene (and/or spinel) fractionation prior to any subsequent high-level ponding in magma chambers.

(5) Pairs of highly incompatible elements have ratios that are approximately, or slightly lower than, chondritic.

Variations between and within the recognized chemical groups (especially the types 1 and 2 sheet-flows which are well documented) are illustrated in Figs 3 and 4. The main chemical feature shown is that the type 1 basalts are the most primitive group, type 2 having more evolved compositions with high FeO^*/MgO

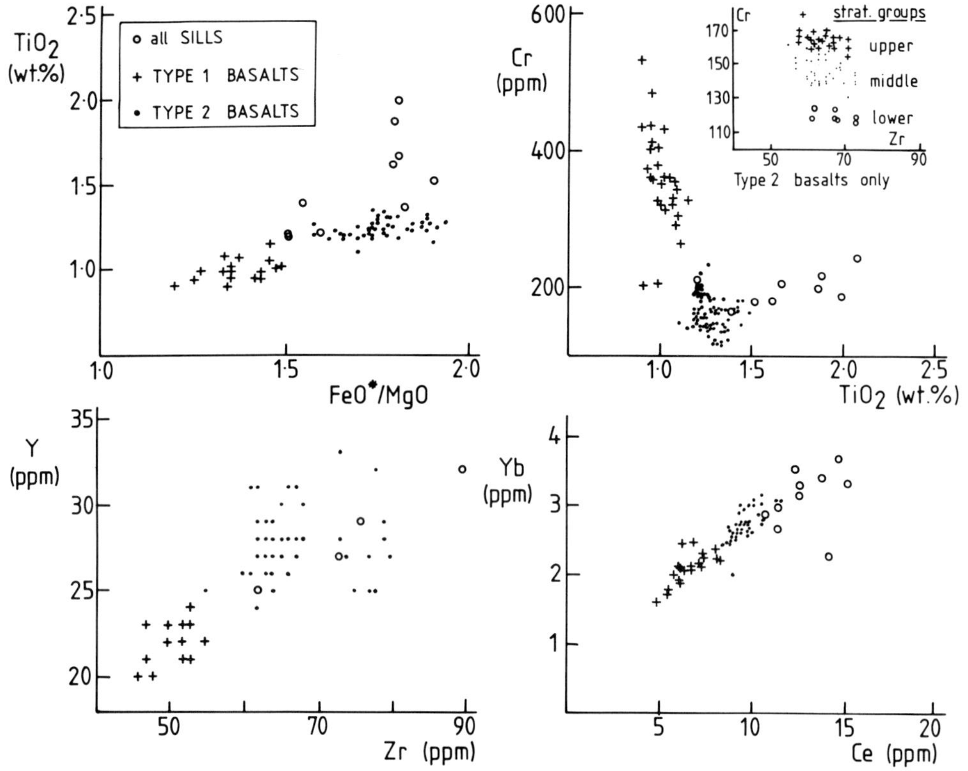

FIG. 3. Chemical features of Nauru complex OPB. Cr–TiO_2 inset from Floyd (1986).

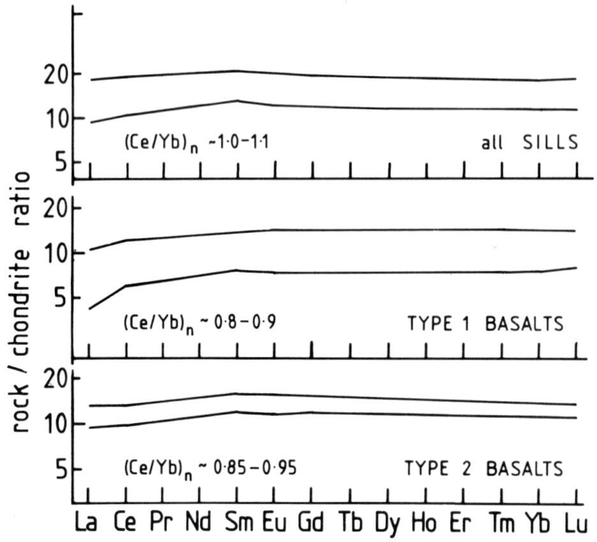

FIG. 4. Envelopes enclosing chondrite-normalized REE patterns for Nauru complex sheet-flows (types 1 and 2) and upper sills. Data from Tokuyama & Batiza (1981), Castillo *et al.* (1986), Saunders (1986) and Floyd (unpublished type 2 basalt data).

ratios and incompatible-element abundances and corresponding lower chromium and nickel contents. Systematic covariance between pairs of incompatible elements, such as Ce–Yb and Zr–Y (Fig. 3), and parallel REE patterns (Fig. 4) imply that either progressive fractional crystallization or variable partial-melting processes could cause all the incompatible-element variation displayed and link types 1 and 2 basalts into one petrogenetic scheme. However, progressive and marked decreases in chromium (and nickel) contents and variations in the mafic differentiation index FeO*/MgO indicate that both clinopyroxene and olivine fractionation were important factors in the evolution of the sheet-flow basalts. Parallel REE patterns (with constant $(Ce/Yb)_n$ ratios for different samples) can also be interpreted as indicative of olivine and/or plagioclase fractionation, whereas minor changes in the light-to-heavy REE ratio can be caused by clinopyroxene fractionation (which selectively incorporates the heavy REEs). Plagioclase is a major phenocryst phase, although its fractionation is not easy to illustrate chemically as the irregular distribution of strontium is due to alteration and negative europium anomalies are not readily observed.

Although major-element modelling by Castillo *et al.* (1986) demonstrated that type 2 basalts could be derived from type 1 by fractional crystallization, the marked chemical gap with lack of content overlap for many elements and differences in the distribution of transition elements argues against a single comagmatic series. For example, in the $Cr-TiO_2$ plot (Fig. 3) the type 1 basalt fractional variation is defined by a well-developed decreasing trend, whereas type 2 basalts form a cluster within which Floyd (1986) demonstrated the existence of three separate series (Fig. 3, inset) whose limited variation might be due to mixing of melts of broadly similar composition. It seems likely that type 1 and type 2 basalts are separate chemical series with more extensive mafic fractionation governing the variation in the former, whilst minor fractionation plus magma mixing governed the latter.

On the basis of REE modelling Castillo *et al.* (1986) demonstrated that type 1 and 2 basalts could be generated by respectively 20%–35% and 17%–20% batch partial melting of a common spinel peridotite source. These high levels of melting correspond to that considered possible for N-type MORB (Sun *et al.* 1979) but are higher than the 10%–15% suggested by Green (1971) for the production of tholeiites from a spinel peridotite mantle. Although useful there are difficulties in this type of calculation, as the actual source composition is not known with any certainty and thus the degree of melting cannot be adequately constrained. The respective geneses of the types 1 and 2 basalt parental melts are best approximated in relative terms until trace-element mantle compositions are adequately documented for this area. For example, the generally incompatible-element and total REE abundances seen in the type 2 basalts, coupled with only minor light REE depletion, could be accounted for by a relatively lower degree of melting compared with the type 1 basalts. If the melting ranges suggested by Castillo *et al.* (1986) are taken at face value then transition-trace-element abundances and magnesium numbers should be similar for the 'parental melts' of the two chemical series of sheet-flows. However, this is not the case, with the least evolved compositions observed in the two groups having markedly different MgO, chromium and nickel contents. This suggests that, although the two series could have been formed by varying degrees of melting, type 2 melts underwent a further stage of evolution with (high-pressure) olivine and pyroxene fractionation prior to minor higher-level fractionation.

Outline of source characteristics

Chondrite-normalized multi-element patterns (Floyd 1986; Saunders 1986) together with uniform strontium and neodymium isotopic ratios (Fujii *et al.* 1981; Castillo *et al.* 1986; Notsu *et al.* 1986) indicate that the Nauru complex OPB source was relatively homogeneous and similar for all three chemical groups. To a first approximation the mantle source had a depleted chemistry but was selectively enriched in some incompatible elements relative to an N-type MORB source. It is clearly distinct from a typical continental basalt source (Saunders 1986) which generally shows high relative enrichment in most LIL elements (Thompson 1982; Thompson *et al.* 1983).

Comparison with mid-ocean ridge basalt

Although the Nauru complex OPBs have depleted chemical characteristics broadly similar to those of N-type MORB, they exhibit some distinctive incompatible-element and isotopic differences (Batiza *et al.* 1980; Castillo *et al.* 1986; Floyd 1986; Saunders 1986).

Incompatible-element patterns

Various incompatible-element ratios for OPB deviate considerably from N-type MORB values (*eg* lower P/Ce, Zr/Nb, Zr/Ta and La/Ta, but higher La/Sm and Ce/Yb). The differences between the two basaltic types can best be illustrated using an N-type MORB normalized multi-incompatible-element plot (Fig. 5). This diagram is double normalized, however, first to N-type MORB (values from Saunders & Tarney 1984) and then to a constant zirconium content (set arbitrarily at 10 ppm) which allows comparison at the same level of chemical evolution and removes the parallel-pattern effect of variable fractional crystallization between samples. Samples from the same source should have the same (superimposed) incompatible-element pattern. As seen in Fig. 5 basalts from the three chemical groups have almost identical closely coherent, patterns and reflect the homogeneity of the Nauru complex OPB source for all lavas sampled to date. Note the marked difference in compatible elements chromium and nickel, which as stated above probably indicates near-source fractionation prior to higher-level magma chamber ponding.

The major elemental difference shown by OPB (Fig. 5*a*) is the relative enrichment in selected incompatible elements: (i) LIL elements rubidium, thorium, lanthanum, cerium, neodymium and europium and (ii) high field strength (HFS) elements niobium, tantalum and titanium. Some type 2 basalts also have very high FeO* contents (more than 13 wt %) which, in conjunction with the enriched HFS element contents, could imply the melting of iron oxide phases in the mantle source. Note that the 'enriched pattern' is almost exactly matched by the more enriched types of MORB (especially 'transitional' types (T-type MORB), Fig. 5*b*) found at the Reykjanes Ridge (Tarney *et al.*

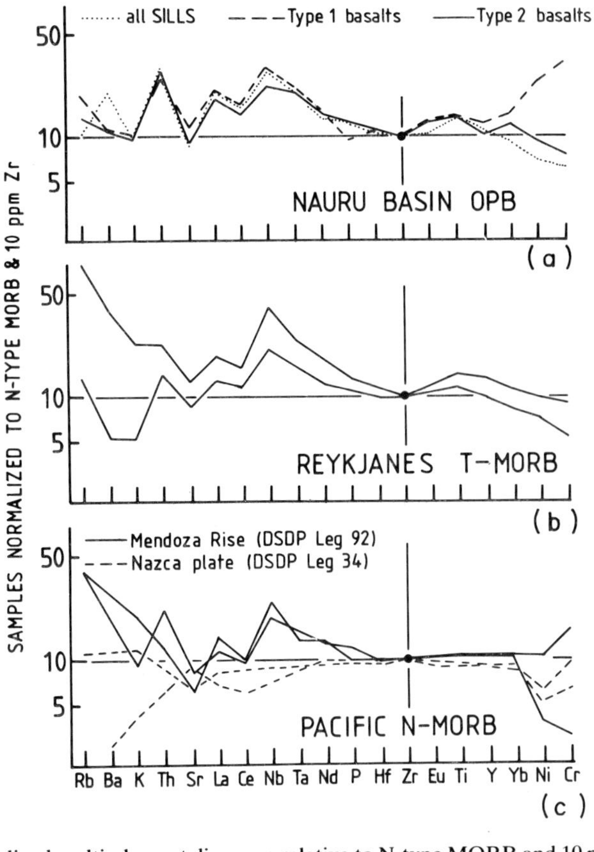

FIG. 5. Double normalized multi-element diagrams relative to N-type MORB and 10 ppm Zr. (a) Representative examples of the Nauru complex sheet-flows and sills; (*b* Reykjanes Ridge T-type MORB (enveloped data from Tarney *et al.* 1979); (*c*) variable N-type MORB data from the eastern Pacific (Rhodes *et al.* 1976; Pearce *et al.* 1986).

1979) and FAMOUS segments (Flower *et al.* 1977; Wood *et al.* 1979) of the Mid-Atlantic Ridge. This chemical comparison has also been commented on by Batiza *et al.* (1980) and Saunders (1986), although as seen in Fig. 5*c* there are parts of the Pacific oceanic crust which are similarly enriched in thorium, light REEs, niobium and tantalum. Much of the relatively young Pacific crust (*eg* the Nazca plate) has a typical uniform N-type MORB chemistry, whereas recent drilling in the eastern Pacific (Leg 92; Pearce *et al.* (1986)) has revealed some enriched features (Fig. 5*c*) that can be matched by the Nauru OPB. Also, transform-fault-bounded segments of the East Pacific Rise are enriched in highly incompatible elements such as barium (Langmuir *et al.* 1986). In summary it is clear that the Nauru complex OPB can be matched with special types of MORB with mildly enriched characteristics.

Isotopic features

Strontium and neodymium isotope data are limited for the Nauru complex (Fujii *et al.* 1981; Castillo *et al.* 1986; Notsu *et al.* 1986). $^{87}Sr/^{86}Sr$ ratios are relatively uniform, falling within the range 0.7036 − 0.7039, and are systematically higher than the average of 0.7025 for MORB (Hofmann & Hart 1978). $^{143}Nd/^{144}Nd$ ratios ($\approx$0.5130) are marginally lower than those for typical MORB. As seen in the Sr−Nd isotope diagram (Fig. 6) the data (corrected for an age of 100 Ma) fall within the mantle array but outside the MORB field defined by White & Hofmann (1982). The apparent OIB character of the isotope data (Fig. 6) with marginally enriched ^{87}Sr and depleted ^{143}Nd reflects the selected trace-element enrichments (rubidium and samarium respectively) noted above for the mantle source of OPB. The sheet-flow basalts have undergone some low-grade submarine alteration, but it seems unlikely that any exchange with seawater (which would enhance the $^{87}Sr/^{86}Sr$ ratio) has affected the strontium isotopic value to any marked degree at the level of alteration observed (*cf* Fujii *et al.* 1981).

Ocean plateau basalt source features

To a first approximation the mantle source for the Nauru complex OPB had a depleted MORB-like character. However, as shown by incompatible-element patterns and isotopic correlations it is more enriched than this with higher levels of rubidium, thorium, light REEs and ^{87}Sr, as well as titanium, niobium and tantalum, similar to T-type MORB and some tholeiitic OIBs such as those from Hawaii. It seems likely that the sheet-flow basalts were derived from an upper mantle MORB source with an enriched component (*cf* Outong-Java plateau, Bielski-Zyskind *et al.* 1984), either in the form of incompatible-element-rich veins (*eg* Tarney *et al.* 1980) or OIB-composition 'plums' (*eg* Morris & Hart 1983).

On the assumption of a two-component source for OPB comprising a dominant MORB-type matrix interspersed with a minor OIB-type element, trace-element mixing relationships were investigated using the method of Langmuir *et al.* (1978). Utilizing pairs of incompatible elements whose ratio reflects the source composition, any variation can be interpreted in terms

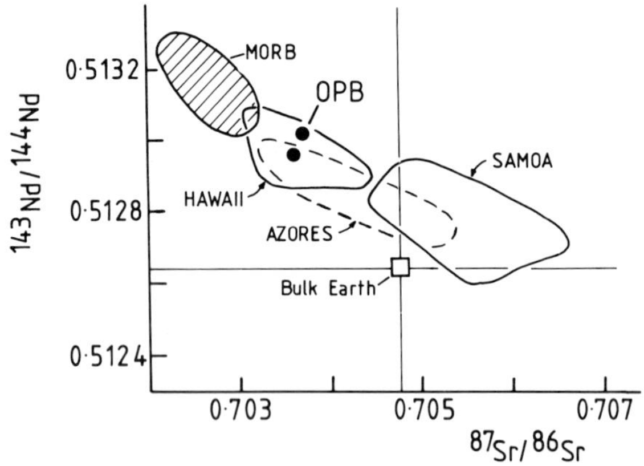

FIG. 6. Sr−Nd isotope correlation diagram showing Nauru complex data (Castillo *et al.* 1986) relative to the MORB and OIB segments of the mantle array (White & Hofmann 1982).

of source heterogeneity. As seen in Fig. 7 calculated mixing curves between idealized 'end-member' MORB and OIB source compositions for Zr/Nb–Ce/Y and its companion plot Zr/Nb–Y/Nb could suggest that N-type and T-type MORB variations reflect the mixing of varying proportions of melt derived from the two sources. That is, T-type MORB magmas represent the melting of mantle with a higher OIB component than N-type MORB (*cf* Tarney *et al.* 1980). On the basis of this simple mixing model, the Nauru complex OPB magmas could have been obtained by the melting of a dominant N-type MORB source that mixed with melts derived from a sparsely distributed OIB source (Fig. 7). As data for all three chemical groups from the Nauru complex completely overlap, their initial chemical differences (relative incompatible-element abundances) cannot be attributed to the mixing of varying proportions of melt from the more enriched OIB source component. It seems likely that, at the high levels of melting envisaged by Castillo *et al.* (1986), all the minor OIB component would have disappeared relatively early during melting.

Other oceanic plateau provinces

As indicated in an earlier section, a number of other oceanic plateaux may have been generated during the Cretaceous volcanic event in the 'old Pacific' region (Fig. 1). Some of these have been investigated by the DSDP and include the Ontong-Java plateau (Andrews *et al.* 1975), Manihiki plateau (Schlanger *et al.* 1976) and the Caribbean (Donnelly *et al.* 1979). However, the

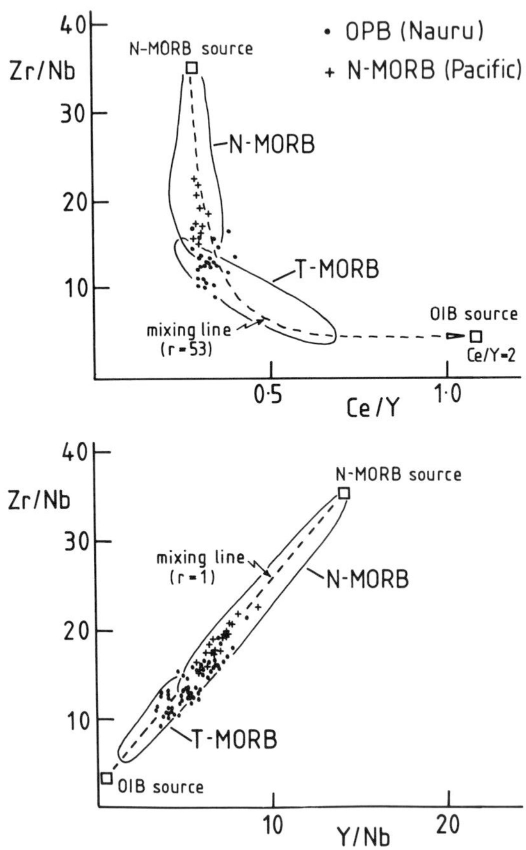

FIG. 7. Chemical diagrams depicting possible mixing relationships between MORB and OIB source end-members. Mixing lines calculated by the method of Langmuir *et al.* (1978). The position of the Nauru complex basalt data implies a complement of enriched OIB within melts dominated by the MORB component. End-member ratios for N-type MORB are Zr/Nb=35, Y/Nb=14, Ce/Y=0.29, and for OIB Zr/Nb=3.2, Y/Nb=0.26, Ce/Y=2.0.

paucity or selectivity of analytical data precludes a normalized multi-element chemical comparison with the Nauru complex OPB, although a number of relationships can be illustrated (Fig. 8). In each case the plateau basalts show significant differences from N-type MORB, especially in the low and restricted absolute abundances of HFS incompatible elements such as hafnium and zirconium. Also, recent work by Marriner (pers. comm.) has demonstrated that the Ta/Zr ratios for these plateau basalts are higher than those of N-type MORB and identical with those of Nauru OPBs. Previous chemical work on the Ontong-Java and Manihiki plateaux suggested that the basalts obtained from them had typical MORB characteristics (Stoeser 1975; Jackson *et al.* 1976). The available data for these provinces mirror the distribution and trends for the Nauru complex (Fig. 8), in particular the steep chromium trend and the variable Zr/Y ratios, and suggest that they are also characteristic of the OPB type. There are petrographic similarities between the Nauru basalts and the Ontong-Java and Manihiki plateau basalts. They are characteristically quenched, poorly phyric (less than a few per cent phenocrysts) and aphyric tholeiitic sheet-flows with phenocrystic plagioclase and clinopyroxene forming glomerocrystic clusters. It appears that the depth of water into which they were extruded can vary, as the Manihiki basalts are vesicular (approximately 10%), whereas the Nauru and Ontong-Java basalts are essentially non-vesicular.

The basalts of the anomalously thick Caribbean crust often show greater chemical variation than the other OPBs, being divisible into many different chemical groups of limited compositional range (Flower *et al.* 1979). Although most basalts are light REE depleted (more so than the Nauru OPBs), some have enriched normalized patterns (Bence *et al.* 1975; Rice *et al.* 1979). Possible primary $^{87}Sr/^{86}Sr$ ratios are generally greater than 0.7036 (*cf* Nauru OPB) and can be as high as 0.7068 in leached samples (Staudigel *et al.* 1979). The Caribbean basalts are petrographically similar to other OPBs, but again exhibit greater variation from quenched and subophitic aphyric

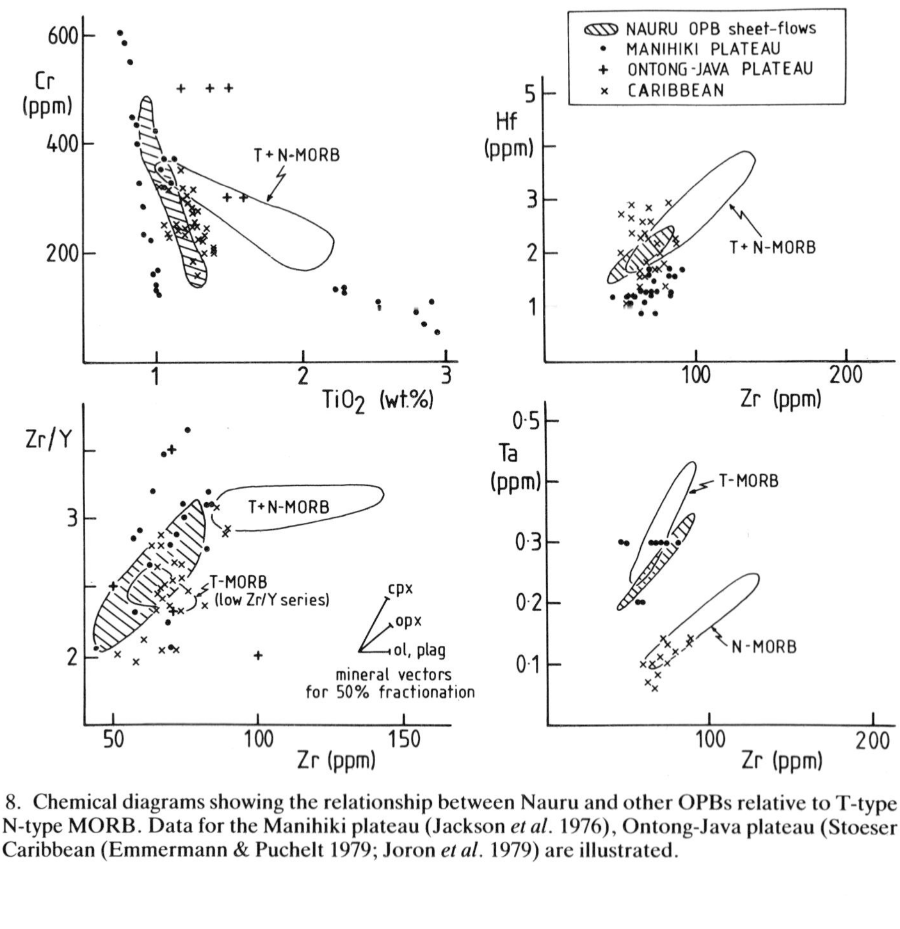

FIG. 8. Chemical diagrams showing the relationship between Nauru and other OPBs relative to T-type MORB and N-type MORB. Data for the Manihiki plateau (Jackson *et al.* 1976), Ontong-Java plateau (Stoeser 1975) and Caribbean (Emmermann & Puchelt 1979; Joron *et al.* 1979) are illustrated.

basalts to plagioclase–olivine–clinopyroxene and plagioclase–olivine–phyric basalts with glomerophyric clots (Bence *et al.* 1975; Rice *et al.* 1979). Unlike the other OPBs, pillow lavas are common, but sheet-flows, as well as high-level sills, make up much of the basaltic pile.

The greater chemical–petrographical variation and cyclicity of the Caribbean crust is not dissimilar to N-type MORB, although there are many features which suggest that it is anomalous and more akin to OPBs (Fig. 8). Also, Joron *et al.* (1979) showed that many incompatible-element ratios (Zr/Hf, Y/Tb, Ta/Th etc.) exactly overlap T-type MORB compositions from the Mid-Atlantic Ridge.

It is not possible at this stage to gauge the importance of OPB development throughout the geological column, or whether this form of intraplate volcanism was restricted to the special tectonic conditions during the Cretaceous magmatic event with its attendant rapid spreading rates, changes in plate motion and widespread lithospheric thermal disturbance. One feature common to all oceanic plateaux is that they represent a thickening of the oceanic crust shortly after the formation of the underlying basement at a nearby spreading centre. The Nauru complex was formed any time up to about 20 Ma after the Jurassic oceanic crust, and the Ontong-Java and Manihiki plateaux only about 5–15 Ma after the local Cretaceous basement (Lanphere & Dalrymple 1976), whereas the Caribbean basalts have an Ar/Ar isochron age of 120 Ma (Ozima *et al.* 1979), apparently 10 Ma older than the MO magnetic anomaly age! Possibly this reflects a very short time interval between, or merging of, axis and off-axis crustal development in the Caribbean. In this connection, Saunders (1986) has suggested that thickened ocean crust could result from volcanism in both axis and off-axis environments producing an Iceland-type ridge segment. However, in the case of the Cretaceous plateaux, the sheet-flows were probably fed via a diffuse complex of numerous short-lived fractures, none of which acted as spreading centres, otherwise the underlying crustal magnetic lineations would have been disturbed. Also, OPBs do not apparently show any chemical evidence of having an Iceland-type plume-derived component, although this could be a function of the limited sampling of these extensive submarine provinces.

Relative to their size and volume, oceanic plateaux are very poorly known. Data are limited to a small collection of basalts from a few, often not very deep, DSDP holes in the plateaux of the western-central Pacific. Oceanic plateau provinces require further study, not only because they represent a unique form of intraplate volcanism, but because they provide evidence of the magmatic processes involved in the anomalous thickening of oceanic crust near to spreading centres and the nature of the mantle sampled relative to the adjacent ridge.

Recently, thick Cretaceous basaltic sequences within segments of the Western Cordillera of Colombia and Ecuador have been shown to have OPB chemical features (Millward *et al.* 1984). They were probably formed in part of the 'old Pacific' marginal to the South American continent prior to eastward transport and eventual collision. Recognition of these basaltic piles as obducted OPB fragments is clearly important in the reconstruction of local plate tectonics, especially as various units have been considered to represent either MORBs (Bourgois *et al.* 1982), island arc basalts (Henderson 1979; McCourt *et al.* 1984) or marginal basin basalts (Marriner & Millward 1984). Thus they may be important factors in the orogenic development of continental margins with the capacity of blocking subduction zones owing to the increased crustal thickness involved. There is little doubt that the Ocean Drilling Program is a suitable vehicle for the study of the structure, composition and genesis of oceanic plateaux, especially adjacent to a coeval spreading axis or aseimic ridge.

Conclusions

(1) One expression of the extensive intraplate Cretaceous volcanic episode in the 'old Pacific' was the development of tabular oceanic plateaux composed of thick piles of sheet-flow poorly phyric tholeiitic basalts. Although intraplate in character they may be temporally and areally related to nearby ridge spreading centres and produce a local thickening of relatively young oceanic crust over which they were erupted via short-lived fractures.

(2) In gross terms they exhibit the chondrite-normalized incompatible-element-depleted characteristics of N-type MORB, but are sufficiently different to be recognized as a specific type of intraplate basalt — OPB. Relative to N-type MORB (at the same degree of chemical evolution) they are enriched in rubidium, thorium, light REEs, titanium, niobium, tantalum and ^{87}Sr, and depleted in ^{143}Nd. On average they have higher $^{87}Sr/^{86}Sr$ (0.7036–0.7039), $(La/Sm)_n$ (0.6–0.7) and $(Ce/Yb)_n$ (0.8–0.9) ratios,

but lower P/Ce (70–90), Zr/Nb (10–16), Zr/Ta (260–275) and La/Ta (15–17) ratios and marginally lower $^{143}Nd/^{144}Nd$ (0.5130) ratios. OPBs can be matched with mildly enriched types of MORB (*eg* Reykjanes Ridge T-type MORB), although isotopically they have an ocean island character, falling on the OIB Sr–Nd mantle array.

(3) Overall OPBs are chemically and petrographically uniform with only minor differences between chemical groups internally related via limited mafic fractionation. Uniformity probably results from a combination of the rapid expulsion of melts from short-lived magma chambers and frequent mixing with new pulses of more primitive magma.

(4) A two-component upper mantle source region is envisaged for the generation of OPB via relatively high degrees of melting. OPB could represent the mixing of melts (in variable proportions) derived from a dominant MORB-type depleted matrix, together with a minor contribution from an entrained OIB-type enriched element.

Acknowledgments: I gratefully acknowledge the work and ideas of scientists on DSDP Legs 61 and 89, especially Dr R.I. Larson, Dr R. Moberly and Dr S.O. Schlanger, on the regional implications of Cretaceous volcanism and tectonics. I have also freely drawn on the comments and chemical data base provided by Dr R. Batiza (and various co-workers) and Dr A. Saunders for the Nauru complex, for which thanks are due. My own data input has benefited from the analytical expertise of D. Emley and M. Aiken (X-ray fluorescence analysis, University of Keele), Dr N. Walsh (ICP analysis of REE, Royal Holloway and Bedford New College, University of London) and Dr G. Gilmore (INAA analysis of hafnium, tantalum, thorium and uranium, Universities Research Reactor, Risley, UK).

References

Andrews, J. E., Packham, G. *et al.* (eds) 1975. *Initial Reports of the Deep Sea Drilling Project* **30**.

Batiza, R., Larson, R. L., Schlanger, S. O., Shcheka, S. A. & Tokuyama, H. 1980. Trace element abundances in basalts of the Nauru Basin. *Nature* **286**, 476–478.

Bence, A. E., Papike, J. J. & Ayuso, R. A. 1975. Petrology of submarine basalts from the Central Caribbean: DSDP Leg 15. *Journal of Geophysical Research* **80**, 4775–4804.

Bielski-Zyskind, M., Wasserburg, G. J. & Nixon, P. H. 1984. Sm–Nd and Rb–Sr systematics in volcanics and ultramafic xenoliths from Malaita, Soloman Islands, and the nature of the Ontong-Java Plateau. *Journal of Geophysical Research* **89**, 2415–2424.

Bourgois, J., Calle, B., Tournon, J. & Toussaint, J. F. 1982. The Andean ophiolitic megastructure in the Buga-Buenaventura traverse (western Cordillera-Valle, Columbia). *Tectonophysics* **82**, 207–229.

Burke, K., Fox, P. J. & Sengor, A. M. C. 1978. Buoyant ocean floor and the evolution of the Caribbean. *Journal of Geophysical Research* **83**, 3949–3954.

Carlson, R. L., Christensen, N. I. & Moore, R. P. 1980. Anomalous crustal structures in ocean basins: continental fragments and oceanic plateaus. *Earth and Planetary Science Letters* **51**, 171–180.

Castillo, P., Batiza, R. & Stern, R. J. 1986. Petrology and geochemistry of Nauru Basin igneous complex: large-volume, off-ridge eruptions of MORB-like basalt during the Cretaceous. *In*: Moberly, R., Schlanger, S. O. *et al.* (eds) *Initial Reports of the Deep Sea Drilling Project* **89**, 555–576.

Coleman, P. J. & Kroenke, L. W. 1981. Subduction without volcanism in the Soloman Islands arc. *GeoMarine Letters* **1**, 129–134.

Donnelly, T., Francheteau, J. *et al.* (eds) 1979. *Initial Reports of the Deep Sea Drilling Project* **51**, **52**, **53**.

Emmermann, R. & Puchelt, H. 1979. Major and trace element chemistry of basalts from Holes 417D and 418A, Deep Sea Drilling Project Legs 51–53. *In*: Donnelly, T., Francheteau, J. *et al.* (eds). *Initial Reports of the Deep Sea Drilling Project* **51**, **52**, **53**, 987–1000.

Flower, M. F. J., Ohnmacht, W., Robinson, P. T., Marriner, G. & Schmincke, H.-U. 1979. Lithologic and chemical stratigraphy at Deep Sea Drilling Project Sites 417 and 418. *In*: Donnelly, T., Francheteau, J. *et al.* (eds) *Initial Reports of the Deep Sea Drilling Project* **51**, **52**, **53**, 939–956.

——, Robinson, P. T., Schmincke, H.-U. & Ohnmacht, W. 1977. Petrology and geochemistry of igneous rocks, DSDP Leg 37. *In*: Aumento, F., Melson, W. G. *et al.* (eds) *Initial Reports of the Deep Sea Drilling Project* **37**, 653–679.

Floyd, P. A. 1986. Petrology and geochemistry of oceanic intraplate sheet-flow basalts, Nauru Basin, Deep Sea Drilling Project, Leg 89. *In*: Moberly, R., Schlanger, S. O. *et al.* (eds) *Initial Reports of the Deep Sea Drilling Project* **89**, 471–497.

—— & Fuge, R. 1982. Primary and secondary alkali and halogen element distribution in Iceland Research Drilling Project basalts from eastern Iceland. *Journal of Geophysical Research* **87**, 6477–6488.

—— & ROWBOTHAM, G. 1986. Chemistry of primary and secondary phases in intraplate basalts and volcaniclastic sediments, Deep Sea Drilling Project Leg 89. *In*: MOBERLY, R., SCHLANGER, S. O. *et al. Initial Reports of the Deep Sea Drilling Project* **89**, 459–470.

FUJII, N., NOTSU, K. & ONUMA, N. 1981. Chemical composition and Sr isotopes of Deep Sea Drilling Project Leg 61 basalts. *In*: LARSON, R. L., SCHLANGER, S. O. *et al.* (eds) *Initial Reports of the Deep Sea Drilling Project* **61**, 697–700.

GREEN, D. H. 1971. Composition of basaltic magmas as indicators of conditions of origin: application to oceanic volcanism. *Philosophical Transactions of the Royal Society of London, Series A* **268**, 707–725.

HAGGERTY, J. A., SCHLANGER, S. O. & PREMOLI-SILVA, I. 1982. Late Cretaceous and Eocene volcanism in the southern Line Islands and implications for hotspot theory. *Geology* **10**, 433–437.

HART, S. R., ERLANK, A. J. & KABLE, E. J. D. 1974. Sea floor basalt alteration: some chemical and Sr isotopic effects. *Contributions to Mineralogy and Petrology* **44**, 219–230.

HENDERSON, W. G. 1979. Cretaceous to Eocene volcanic arc activity in the Andes of northern Ecuador. *Journal of the Geological Society, London* **141**, 367–378.

HOFMANN, A. W. & HART, S. R. 1978. An assessment of local and regional isotopic equilibrium in the mantle. *Earth and Planetary Science Letters* **38**, 44–62.

JACKSON, E. D., BARGAR, K. E., FABBI, B. P. & HEROPOULOS. C. 1976. Petrology of the basaltic rocks drilled on Leg 33 of the Deep Sea Drilling Project. *In*: SCHLANGER, S. O., JACKSON, E. D. *et al.* (eds) *Initial Reports of the Deep Sea Drilling Project* **33**, 571–630.

JORON, J. L., BOLLINGER, C., QUISEFIT, J. P., BOUGAULT, H. & TREUIL, M. 1979. Trace elements in Cretaceous basalts at 25°N in the Atlantic Ocean: alteration, mantle compositions and magmatic processes. *In*: DONNELLY, T., FRANCHETEAU, J. *et al.* (eds) *Initial Reports of the Deep Sea Drilling Project* **51, 52, 53**, 1087–1098.

LANGMUIR, C. H., BENDER, J. F. & BATIZA, R. 1986. Petrological and tectonic segmentation of the East Pacific Rise, 5°30′–14°30′N. *Nature* **322**. 422–429.

——, VOCKE, R. D., HANSON, G. N. & HART, S. R. 1978. A general mixing equation with application to Icelandic basalts. *Earth and Planetary Science Letters* **37**, 380–392.

LANPHERE, M. A. & DALRYMPLE, G. B. 1976. K–Ar ages of basalts from DSDP Leg 33: Sites 315 (Line Islands) and 317 (Manihiki plateau). *In*: SCHLANGER, S. O., JACKSON, E. D. *et al.* (eds) *Initial Reports of the Deep Sea Drilling Project* **33**, 649–654.

LARSON, R. L. & PITMAN, W. C. 1972. World-wide correlations of Mesozoic magnetic anomalies and its implications. *Bulletin of the Geological Society of America* **83**, 3645–3661.

—— & SCHLANGER, S. O. 1981. Geological evolution of the Nauru Basin and regional implications. *In*: LARSON, R. L., SCHLANGER, S. O. *et al.* (eds) *Initial Reports of the Deep Sea Drilling Project* **61**, 841–862.

——, SCHLANGER, S. O. *et al.* (eds) 1981. *Initial Reports of the Deep Sea Drilling Project* **61**.

MARRINER, G. F. & MILLWARD, D. 1984. The petrology and geochemistry of Cretaceous to Recent volcanism in Colombia: the magmatic history of an accretionary plate margin. *Journal of the Geological Society, London* **141**, 473–486.

MCCOURT, W. J., ASPDEN, J. A. & BROOK, M. 1984. New geological and geochronological data from the Colombian Andes: continental growth by multiple accretion. *Journal of the Geological Society, London* **141**, 831–845.

MILLWARD, D., MARRINER, G. F. & SAUNDERS, A. D. 1984. Cretaceous tholeiitic volcanic rocks from the Western Cordillera of Colombia. *Journal of the Geological Society, London* **141**, 847–860.

MOBERLY, R. & JENKYNS, H. C. 1981. Cretaceous volcanogenic sediments of the Nauru Basin, Deep Sea Drilling Project Leg 61. *In*: LARSON, R. L., SCHLANGER, S. O. *et al.* (eds) *Initial Reports of the Deep Sea Drilling Project* **61**, 533–548.

——, SCHLANGER, S. O. *et al.* (eds) 1986. *Initial Reports of the Deep Sea Drilling Project* **89**.

MORRIS, J. D. & HART, S. R. 1983. Isotopic and incompatible element constraints on the genesis of island arc volcanics from Cold Bay and Amak Islands, Aleutians, and implications for mantle structure. *Geochimica et Cosmochimica Acta* **47**, 2015–2030.

NOTSU, K., ONUMA, N. & FUJII, N. 1986. Rb–Sr isotope systematics and Sr/Ca–Ba/Ca ratios of Nauru Basin basalts, Deep Sea Drilling Project Leg 89. *In*: MOBERLY, R., SCHLANGER, S. O. *et al.* (eds) *Initial Reports of the Deep Sea Drilling Project* **89**, 523–527.

NUR, A. & BEN-AVRAHAM, Z. 1982. Oceanic plateaus, the fragmentation of continents and mountain building. *Journal of Geophysical Research* **87**, 3644–3661.

OZIMA, M., KANEOKA, I. & YANAGISAWA, M. 1979. $^{40}Ar-^{39}Ar$ geochronological studies of drilled basalts from Leg 51 and Leg 52. *In*: DONNELLY, T., FRANCHETEAU, J. *et al.* (eds) *Initial Reports of the Deep Sea Drilling Project* **51**, **52**, **53**, 1127–1128.

——, SAITO, K. & TAKIGAMI, Y. 1981. $^{40}Ar/^{39}Ar$ geochronological studies on rocks drilled at Holes 462 and 462A, Deep Sea Drilling Project, Leg 61. *In*: LARSON, R. L., SCHLANGER, S. O. *et al.* (eds) *Initial Reports of the Deep Sea Drilling Project* **61**, 701–703.

PEARCE, J. A., ROGERS, N., TINDLE, A. J. & WATSON, J. S. 1986. Geochemistry and petrogenesis of basalts from Deep Sea Drilling Project Leg 92, Eastern Pacific. *In*: LEINEN, M., REA, D. K. *et al.* (eds) *Initial Reports of the Deep Sea Drilling Project* **92**, 435–457.

REA, D. K. & VALLIER, T. L. 1983. Two Cretaceous

volcanic episodes in the western Pacific Ocean. *Bulletin of the Geological Society of America* **94**, 1430–1437.

RHODES, J. M., BLANCHARD, D. P., RODGERS, K. V., JACOBS, J. W. & BRANNON, J. C. 1976. Petrology and chemistry of basalts from the Nazca Plare. Part 2: Major and trace element chemistry. *In*: HART, S. R., YEATS, R. S. *et al.* (eds) *Initial Reports of the Deep Sea Drilling Project* **34**, 239–244.

RICE, S., LANGMUIR, C. H., BENDER, J. F., HANSON, G. N., BENCE, A. E. & TAYLOR, S. R. 1979. Basalts from Deep Sea Drilling Project Holes 417A and 417D, fractionated melts of a light rare earth depleted source. *In*: DONNELLY, T., FRANCHETEAU, J., *et al.* (eds) *Initial Reports of the Deep Sea Drilling Project* **51**, **52**, **53**, 1099–1111.

SAUNDERS, A. D. 1986. Geochemistry of basalts from the Nauru Basin, Deep Sea Drilling Project Legs 61 and 89: implications for the origin of oceanic flood basalts. *In*: MOBERLY, R., SCHLANGER, S.O. *et al.* (eds) *Initial Reports of the Deep Sea Drilling Project* **89**, 499–517.

—— & TARNEY, J. 1984. Geochemical characteristics of basaltic volcanism within back-arc basins. *In*: KOLELAAR, B. P. & HOWELLS, M. F. (eds) *Marginal Basin Geology*. Geological Society of London Special Publication **16**, 59–76.

SCHLANGER, S. O. & MOBERLY, R. 1986. Sedimentary and volcanic history: East Mariana Basin and Nauru Basin. *In*: MOBERLY, R., SCHLANGER, S. O. *et al.* (eds) *Initial Reports of the Deep Sea Drilling Project* **89**, 653–678.

——, JACKSON, E. D. *et al.* (eds) 1976. *Initial Reports of the Deep Sea Drilling Project* **33**.

——, JENKYNS, H. & PREMOLI-Slva, I. 1981. Volcanism and vertical tectonics in the Pacific basin related to global Cretaceous transgressions. *Earth and Planetary Science Letters* **52**, 435–449.

SEIFERT, K. E. 1981. Geochemistry of Nauru Basin basalts from the lower portion of Hole 462A, Deep Sea Drilling Project, Leg 61. *In*: LARSON, R. L., SCHLANGER, S. O. *et al.* (eds) *Initial Reports of the Deep Sea Drilling Project* **61**, 705–708.

STAUDIGEL, H., FREY, F. A. & HART, S. R. 1979. Incompatible trace element geochemistry and $^{87}Sr/^{86}Sr$ in basalts and corresponding glasses and palagonites. *In:* DONNELLY, T., FRANCHETEAU, J. *et al.* (eds) *Initial Reports of the Deep Sea Drilling Project* **51**, **52**, **53**, 1137–1144.

STOESER, D. B. 1975. Igneous rocks from Leg 30 of the Deep Sea Drilling Project. *In*: ANDREWS, J. E., PACKHAM, G. *et al.* (eds) *Initial Reports of the Deep Sea Drilling Project* **30**, 401–414.

SUN, S.-S., NESBITT, R. W. & SHARASKIN, A. Y. 1979. Geochemical characteristics of mid-ocean ridge basalts. *Earth and Planetary Science Letters* **44**, 119–138.

TAKIGAMI, Y., AMARI, S., OZIMA, M. & MOBERLY, R. 1986. $^{40}Ar/^{39}Ar$ geochronological studies of basalts from Hole 462A, Nauru Basin, Deep Sea Drilling Project Leg 89. *In*: MOBERLY, R., SCHLANGER, S. O. *et al.* (eds) *Initial Reports of the Deep Sea Drilling Project* **89**, 519–527.

TARNEY, J., SAUNDERS, A. D., WEAVER, S. D., DONNELLAN, N. C. B. & HENDRY, G. L. 1979. Minor element geochemistry of basalts from Leg 49, North Atlantic Ocean. *In*: LUYENDYK, B. P., CANN, J. R. *et al.* (eds) *Initial Reports of the Deep Sea Drilling Project* **49**, 657–691.

——, WOOD, D. A., SAUNDERS, A. D., CANN, J. R. & VARET, J. 1980. Nature of mantle heterogeneity in the North Atlantic: evidence from Deep Sea Drilling. *Philosophical Transactions of the Royal Society of London, Series A* **297**, 179–202.

THOMPSON, R. N. 1982. Magmatism of the British Tertiary volcanic province. *Scottish Journal of Geology* **18**, 49–107.

——, MORRISON, M. A., DICKIN, A. P. & HENDRY, G. D. 1983. Continental flood basalts... arachnids rule OK? *In*: HAWKESWORTH, C. J. & NORRY, M. J. (eds) *Continental Basalts and Mantle Xenoliths*. Shiva Publishing, Nantwich, 158–185.

TOKUYAMA, H. & BATIZA, R. 1981. Chemical composition of igneous rocks and origin of the sill and pillow basalt complex of Nauru Basin, southwest Pacific, *In*: LARSON, R. L., SCHLANGER, S. O., *et al.* (eds) *Initial Reports of the Deep Sea Drilling Project,* **61**, 673–688.

WATTS, A. B., BODINE, J. H. & RIBE, N. M. 1980. Observations of flexure and the geological evolution of the Pacific ocean basin. *Nature* **283**, 522–537.

WHITE, W. M. & HOFMANN, A. W. 1982. Sr and Nd isotope geochemistry of oceanic basalts and mantle evolution. *Nature* **296**, 821–825.

WINTERER, E. L. 1973. Regional problems. In: WINTERER, E. L., EWING, J. I. *et al.* (eds) *Initial Reports of the Deep Sea Drilling Project* **17**, 911–922.

—— 1976. Anomalies in the tectonic evolution of the Pacific. *In*: SUTTON, G. H., MANGHAMI, M. & MOBERLY, R. (eds) *Geophysics of the Pacific Ocean Basin and its Margins*. American Geophysical Union Monograph **19**, 269–278.

WOOD, D. A., TARNEY, J., VARET, J., SAUNDERS, A. D., BOUGAULT, H., JORON, J. L., TREUIL, M. & CANN, J. R. 1979. Geochemistry of basalts drilled in the North Atlantic by IPOD Leg 49: implications for mantle heterogeneity. *Earth and Planetary Science Letters* **42**, 77–97.

P. A. FLOYD, Department of Geology, University of Keele, Staffordshire ST5 5BG, UK.

A combined chemical and Pb–Sr–Nd isotope study of the Azores and Cape Verde hot-spots: the geodynamic implications

G. R. Davies, M. J. Norry, D. C. Gerlach & R. A. Cliff

SUMMARY: The mafic lavas of the Azores and Cape Verde Islands are of a highly varied composition, reflecting a complex history of magma genesis and a variety of source compositions. The lavas of the Cape Verde Islands are characteristically highly silica undersaturated, with alkali-rich ankaramites, larnite-normative melilitites and carbonatites. In contrast, the lavas of the Azores vary from strongly nepheline- to hypersthene-normative types. Isotopic ratios and trace elements also show considerable variation, consistent with derivation from multicomponent mantle sources. Three distinct groupings of lava compositions can be seen in Pb–Sr–Nd isotope space, and each is characterized by its own trace-element enrichment. (i) The majority of the lavas from the Azores and northern Cape Verdes Islands have isotope and trace-element systematics similar to basalts recovered from elevated segments of the mid-Atlantic Ridge (*eg* during Deep Sea Drilling Project Leg 82) (Ba/La <12, $^{206}Pb/^{204}Pb=19.3-20.0$). (ii) The islands of Sao Miguel and Faial in the Azores are characterized by radiogenic strontium and lead isotope ratios and marked fractionation of the most incompatible trace elements (*eg* $^{87}Sr/^{86}Sr=0.70522$, $^{206}Pb/^{204}Pb=19.88$, $^{207}Pb/^{204}Pb=15.75$, Ba/Nb=12, La/Nb=0.84). (iii) In contrast, the southern Cape Verde Islands have relatively unradiogenic Pb–Sr–Nd isotope ratios (*eg* $^{87}Sr/^{86}Sr=0.70393$, $^{206}Pb/^{204}Pb=18.74$, $^{207}Pb/^{204}Pb=15.53$, Ba/La=22, Ba/Nb=15, La/Nb=0.67).

Mixing relationships between isotope and trace-element ratios indicate the involvement of at least four, and possibly five, chemically distinct source regions in the petrogenesis of mafic lavas. It is shown that these different source regions represent combinations of the depleted upper mantle (mid-ocean ridge basalt source), recycled oceanic lithosphere and two components of the subcontinental lithospheric mantle. No direct unequivocal evidence is found for a fifth component, namely undepleted primitive mantle reservoir, although helium isotope data suggest influxes of material and heat from the lower mantle. Similarly, it is suggested that continental crust has no direct contribution to the petrogenesis of these ocean island basalts and its role in the mantle is limited to the production/modification of the subcontinental lithosphere above subduction zones.

Chemical and isotope variations within mantle-derived peridotites and mafic volcanic rocks demonstrate that the mantle retains chemical and isotopic heterogeneities on scales from less than 0.01 m (Hawkesworth *et al.* 1983; Menzies *et al.* 1985) to greater than 10^5 m (Hart 1984). The extent to which these heterogeneities represent partial-melting processes within the mantle or the recycling of oceanic and continental lithosphere remains unclear (see Zindler & Hart (1986) for a recent review). The major chemical fractionation occurring within the Earth happens during the extraction of the continental crust from the mantle. This is a two-stage process: formation of oceanic crust and the production of continental crust upon its subsequent subduction (Allegre 1987). Recycling of crustal material should therefore have a significant influence on the composition of the mantle.

The suggestion that continental crust was recycled into the mantle was first proposed by Armstrong (1968). In order to explain the trace-element and isotope characteristics of young oceanic volcanic rocks, recycling models have since invoked combinations of oceanic lithosphere and continental crust (Hofmann & White 1982, Ringwood 1982) and subcontinental lithosphere (McKenzie & O'Nions 1983). Seawater contains chemical components that are derived from the continents (Martin & Meybeck 1979; Piepgrass & Wasserburg 1980) such that, during subduction, crustally derived material is probably recycled into the mantle in the form of hydrothermally altered oceanic crust with or without sediment. One of the questions we wish to address is the ultimate destination of the different portions of the recycled oceanic lithosphere. For instance, subduction-related volcanic rocks record helium, beryllium and lead isotope evidence of sediment involvement in their petrogenesis (Cohen & O'Nions 1982*b*; Poreda 1985; Tera *et al.* 1986). Chemical models devised to explain the generation of island arc magmas generally invoke some melting of the down-going lithosphere (Anderson *et al.* 1978; Wyllie 1984) implying that any sediments present would

From SAUNDERS, A. D. & NORRY, M. J. (eds), 1989. *Magmatism in the Ocean Basins.* Geological Society Special Publication No. 42, pp. 231–255.

be greatly above their solidus temperature. The extent to which sediments could survive subduction and hence be recycled into the convecting upper mantle is therefore open to question. (We use the term upper mantle as that above the 670 km seismic discontinuity.) Equally the extent to which subducted oceanic lithosphere mixes with the convecting upper mantle is unclear. Recent geophysical and experimental investigations have concluded that the oceanic lithosphere may either pass through (Creager & Jordan 1984; Grand 1987) or be trapped at (Anderson 1987; Ringwood & Irifune 1988) the 670 km discontinuity. If subduction of oceanic lithosphere has continued at present-day rates over the last 3.5 Ga (a conservative estimate) then 4.5×10^{10} km^3 of ocean crust (the top 7 km of the oceanic lithosphere) have been recycled. This represents approximately 5% of the whole mantle and 17% of the upper mantle. Establishing the ultimate destination of the recycled oceanic lithosphere is therefore of great importance, as it has the potential for being a major chemical reservoir within the Earth comparable in size with the subcontinental lithosphere.

Despite the complex processes involved in the extraction of partial melts from the mantle the resultant lavas sampled at the surface will retain certain characteristics of their source regions. For instance Sr–Nd–Pb isotope ratios are unchanged by partial melting and indicate that the mantle is extremely heterogeneous. In addition, although partial melting will produce fractionation between trace elements, source characteristics such as a high ratio of large-ion lithophile (LIL) to high field strength (HFS) elements will generally be retained. The existence of combined trace-element and isotope relationships strongly suggest that melt extraction has not destroyed the chemical signature of the mantle source. Using such relationships we may be able to constrain the petrogenesis of mantle sources and hence assess the possible role of recycled material.

The major chemical characteristics of the continental crust, and sediments therefrom, are well known (*eg* high LIL to HFS element ratios; Taylor & McLennan 1981, 1985; Gormet *et al.* 1984) and distinct from bulk Earth and present-day mantle compositions. The effects of recycling unmodified sediment into the mantle can therefore be predicted. Weaver *et al.* (1987) argue that recycled sediments are responsible for the chemical characteristics in the islands of Tristan da Cunha and Gough. Equally the relatively radiogenic strontium and $^{208}Pb/^{204}Pb$ isotope ratios in these and other ocean island basalts (OIBs) that form the 'Dupal' anomaly (Hart 1984) may also be used to argue for sediment recycling (*eg* Hofmann & White 1982). It is therefore apparent that OIBs show certain chemical and isotope characteristics that may be used to argue for the recycling of continental crust.

In this paper we review the comprehensive combined chemical and isotopic data from the Cape Verde and Azores hot-spots. In particular we assess the scale and origin of the chemical heterogeneities in order to test the validity of recycling models.

General geological setting

The Azores hot-spot has developed a shallow oceanic platform in the central Atlantic that straddles the mid-Atlantic Ridge (MAR) between 37°N and 40°N (Fig. 1). Nine emergent volcanic islands form a broad WNW–ESE lineament between 31°W and 24°W (Ridley *et al.* 1974). The MAR in this area has been extensively sampled by dredging (Schilling 1975, 1985), drilling (Deep Sea Drilling Project (DSDP) Legs 37 and 82) and submersible operations (Shirey *et al.* 1987). The islands are dominantly composed of alkali basalts with minor differentiation products, comendite and pantellerite (White *et al.* 1979). Alkali basalt compositions vary from approximately 20% normative nepheline on Santa Maria to less than 5% on the majority of the islands. A few hypersthene-normative basalts occur on Terceira (Self & Gunn 1976) and Sao Miguel (Storey *et al.*, this volume)

The Cape Verde Islands are situated approximately 500 km W of the Senegalese coast of Africa and 1500 km E of the MAR between latitudes 14.6°–17°N and longitudes 22°–25°W (Fig. 1). There are nine major islands in the group which are formed on a topographic anomaly 1000 km in diameter that rises approximately 1.5 km from the surrounding seafloor. The islands rise a further 2 km from the ocean floor, which is Jurassic to Cretaceous in age (Hayes & Rabinowitz 1975). All the islands include highly undersaturated volcanic rocks ranging from larnite-normative melilitites and carbonatites to alkali basalts (Gunn & Watkins 1976; Silva *et al.* 1981). Representative analyses are given in Table 1, which includes the degree of silica saturation expressed as the amount of normative nepheline (Ne) + leucite (Lc + larnite (Cs). Evolved volcanic rocks (phonolites) representing products of extensive igneous differentiation are rare and are not considered

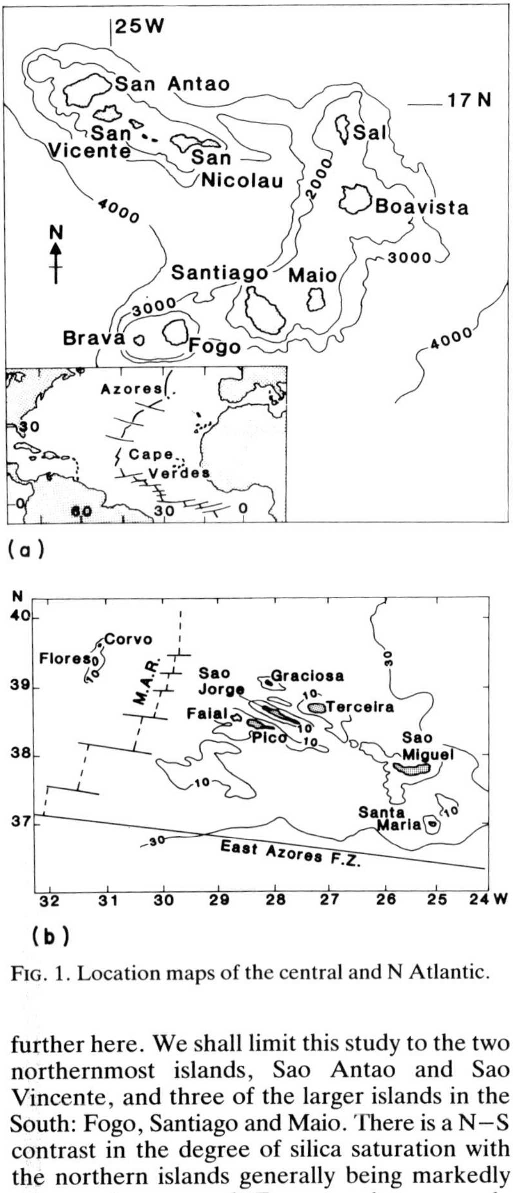

FIG. 1. Location maps of the central and N Atlantic.

further here. We shall limit this study to the two northernmost islands, Sao Antao and Sao Vincente, and three of the larger islands in the South: Fogo, Santiago and Maio. There is a N–S contrast in the degree of silica saturation with the northern islands generally being markedly more undersaturated. For example, extremely silica-undersaturated samples from Sao Antao have up to 45% normative Ne + Lc + Cs.

Altered Jurassic tholeiitic pillow basalts locally form the basement to recent volcanism on Maio and Santiago. The basalts are overlain by thin pelagic sediments and together are interpreted as uplifted ocean floor (Stillman *et al.* 1982). Alkalic magmatism commenced at least 20 Ma ago. Above the basement each Cape Verde island records a distinct magmatic history. The oldest rocks on Santiago, for example, are carbonatites and syenites. Subsequent volcanism (Flamengos Formation, 15 Ma; Pico de Antonia Formation, 7 Ma; Assomada Formation, 6 Ma and Quaternary) is basanitic. In contrast, early Tertiary magmatism on Maio is basaltic and evolves through basanitic (Penoso Formation) to highly larnite-normative volcanic rocks (Malhadra Pedra Formation). The islands are located near the pole of rotation of the African plate and thus have remained almost stationary in relation to the hot-spot reference frame (Duncan 1980). The Cape Verde Islands therefore provide an opportunity to study the chemical and isotope variations of a diverse rock assemblage derived from a constant section of the mantle over a 20 Ma period.

Both the Azores and Cape Verde ocean platforms are covered by minor thicknesses of sediment and hence the recent volcanic rocks are unlikely to have been contaminated by sediments at a shallow level.

Sr–Nd–Pb isotope variations

Ocean floor

In Table 1 and Figs 2–5 we present representative Sr–Nd–Pb isotope data for basaltic rocks from the Azores and Cape Verde platforms (see Davies & Lloyd (in press) for analytical techniques). Alteration effects are probably responsible for a large proportion of the strontium isotope variations in the MAR basalts and account for the radiogenic $^{87}Sr/^{86}Sr$ ratios of the Jurassic tholeiitic basement rocks in the Cape Verde group (Gerlach *et al.*, in press). Glass samples from DSDP Leg 82, to the W and SW of the Azores platform, define a correlation trending to relatively low $^{143}Nd/^{144}Nd$ ratios typical of some OIBs. These samples and glasses from Site 558, along strike from the FAMOUS region, show marked trace-element enrichment in addition to their low $^{143}Nd/^{144}Nd$ ratios (Jenner *et al.* 1985; Rideout & Schilling 1985).

Tholeiitic basalts formed at the MAR between 37°N and 40°N extend over 60% of MORB and OIB $^{206}Pb/^{204}Pb$ ratios. The majority of these MAR basalts lie close to the Northern Hemisphere Reference Line (NHRL) (Hart 1984) as shown in Figs 3 and 4. The notable exceptions are samples from between 37°N and 38°N (DSDP Leg 37) which have relatively radiogenic $^{207}Pb/^{204}Pb$ and $^{208}Pb/^{204}Pb$ ratios. The samples showing the greatest deviation from the NHRL are from adjacent boreholes (Yamaguchi *et al.* 1977). More recent analyses of samples from the same DSDP leg and the

TABLE 1. *Representative chemical analyses from the Cape Verde and Azores hot-spots*

	Azores				
	Sao Miguel[1]			Terciera[2]	FAMOUS region[3]
	AZ1704	SM30	SM49	T19	527–1–1
SiO_2	46.84	45.85	48.17	46.28	48.3
TiO_2	3.47	3.95	3.57	1.84	0.65
AlO_3	13.58	14.00	14.60	16.04	16.7
FeO_3	13.77	12.06	11.11	3.09	8.7
FeO	—	—		6.62	
MnO	0.17	0.16	0.14	0.16	0.15
MgO	9.23	7.17	6.54	8.75	9.65
CaO	11.32	10.12	9.23	9.48	12.5
Na_2O	2.92	3.28	3.22	2.62	2.04
K_2O	1.35	1.93	2.65	1.11	0.08
P_2O_5	0.52	0.64	0.69	0.40	—
CIPW norms					
Hy	—	—	—	1	8.6
Ne	5.8	6.2	3.6	—	—
Lc	—	—	—	—	—
Cs	—	—	—	—	—
$^{87}Sr/^{86}Sr$	0.70348±1	0.70432±1	0.70522±1	0.70344±8	0.70281±7
$^{143}Nd/^{144}Nd$	0.51292±2	0.51285±2	0.51281±2	—	—
$^{206}Pb/^{204}Pb$	19.333	19.750	19.884	19.61	18.67
$^{207}Pb/^{204}Pb$	15.601	15.699	15.754	15.57	15.50
$^{208}Pb/^{204}Pb$	39.13	39.89	40.17	39.05	38.20
△7/4	+1.4	+6.7	+10.8	−4.7	−1.5
△8/4	+12.9	+38.5	+50.3	−29	0
La	—	47.7	53.0	34.8	2.15
Ce	85.9	103	122	27.8	5.24
Nd	45.2	51.5	54.4	37.4	4.07
Sm	9.5	10.7	10.4	8.3	1.37
Eu	2.88	3.35	3.45	2.6	0.53
Gd	8.01	9.31	8.73	—	2.2
Dy	5.62	6.88	6.45	5.9	3.12
Er	2.58	3.14	2.91	—	2.22
Yb	1.94	2.35	2.18	3.2	2.27
Zr	211	275	304	—	31
Nb	62	67	63	—	3.0
Hf	—	8.2	8.3	4.1	0.8
Y	23	35	36	—	24
Th	—	5.8	6.8	2.3	0.22
U	0.94	1.19	1.46	—	0.07
Ba	478	584	784	412	20
Rb	28	43	59	13	—
Sr	649	791	794	355	—
Pb	2.04	3.06	3.96	—	—
Cr	459	132	221	311	510
Ni	174	83	132	131	252

Sample localitites: 1, see Hawkesworth *et al.* (1979); 2, see White *et al.* (1979); 3, see Langmuir *et al.* (1977); 4, see Gerlach *et al.* (in press).
$(^{207}Pb/^{204}Pb)_{NHRL} = 0.1084(^{206}Pb/^{204}Pb) + 13.491$; $(^{208}Pb/^{204}Pb)_{NHRL} = 1.209(^{206}Pb/^{204}Pb) + 15.627$. $\triangle 7/4 = [(^{207}Pb/^{204}Pb)_{DS} - (^{207}Pb/^{204}Pb)_{NHRL}] \times 100$. $\triangle 8/4 = [(^{208}Pb/^{204}Pb)_{DS} - (^{208}Pb/^{204}Pb)_{NHRL}] \times 100$.
Oxides in wt%; trace elements in ppm.

Cape Verde[4]							
Sao Antao			Santiago		Maio		Fogo
NA60	NA51	N17	ZI50	ZI47	ZM55	ZM60	NF34
35.64	42.02	43.70	42.82	43.48	36.60	41.89	41.23
3.91	4.96	3.30	4.03	3.44	3.25	4.90	3.67
10.34	12.97	12.60	13.78	14.70	9.99	15.17	13.70
13.25	13.94	13.15	13.55	13.72	12.10	13.50	14.30
		—		—			
0.26	0.20	0.18	0.18	0.19	0.19	0.22	0.20
10.25	6.74	11.48	7.78	8.79	14.41	5.27	8.37
16.34	12.50	11.51	13.15	11.32	17.38	12.47	13.42
4.35	4.10	3.38	1.74	3.54	3.16	3.63	2.92
2.1	0.92	0.87	1.60	0.36	0.81	1.56	1.93
2.28	0.72	0.57	0.70	0.55	1.27	0.95	0.70
—	—	—	—	—	—	—	—
20.0	12.9	11.6	12.0	9.0	14.5	13.9	21.3
9.7	—	—	—	—	3.8	—	9.5
15.0	—	—		—	16.6	—	1.2
0.70302±1	0.70317±1	0.70365±2	0.70328±2	0.70325±1	0.70325±1	0.70372±2	0.70342±
0.51297±1	0.51287±2	0.51275±2	0.51286±2	0.51261±2	0.51290±2	0.51281±1	0.51275±
19.68	19.651	19.013	19.118	18.743	19.287	19.200	18.926
15.621	15.623	15.560	15.577	15.532	15.581	15.565	15.551
39.28	39.28	38.86	39.97	38.66	39.06	39.04	38.82
0	0	+0.8	+1.4	+0.9	0	−0.7	+0.3
−14.6	−10.5	+24.6	+22.9	+37.2	+11.5	+20.0	+31.3
115	62.3	39.4	61.2	43.5	59.8	64.8	58.1
241	137.3	87	129	94	130	140	132
121.3	65.6	46	65	46	64	68	68
22.22	12.38	8.4	11.1	7.96	11.20	11.7	12.0
6.77	3.84	2.73	3.50	2.61	3.46	3.55	3.80
17.9	10.5	—	—	—	—	—	—
—	—	—	—	—	—	—	—
—	—	—	—	—	—	—	—
2.91	2.15	1.45	1.92	1.76	1.95	1.92	2.30
444	311	329	328	278	300	436	300
156	104	72	76	65	100	103	65
9.75	8.8	7.1	7.6	6.1	9.0	8.9	8.3
50	31	24	29	24	31	39	28
9.28	6.48	4.1	6.3	4.03	5.7	6.2	5.3
3.29	—	0.85	1.30	0.93	1.2	1.67	0.91
709	649	632	823	982	874	977	694
49	30	20	41	90	25	56	46
1452	915	828	1063	1150	1256	1136	841
5.12	—	2.22	3.89	2.96	2.45	3.9	2.81
266	135	379	161	248	489	230	141
116	108	264	101	168	282	100	72

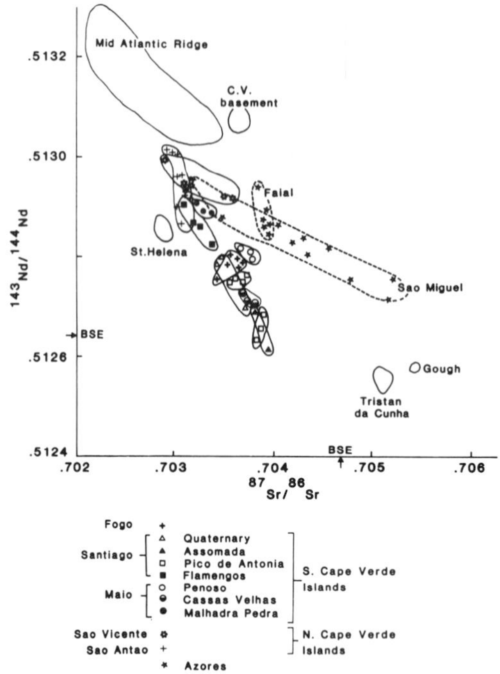

FIG. 2. Nd–Sr isotope variation diagram for selected oceanic basalts. The 'mantle array' is used in this work as a descriptive term for oceanic volcanic rocks that have coupled Nd–Sr isotope ratios that lie between those of MORB and Tristan da Cunha. Data sources: Cohen & O'Nions (1982*a*), Hofmann & White (1982), Jenner *et al.* (1985). Rideout & Schilling (1985), Palacz & Saunders (1986) and Chaffey *et al.* (this volume). All data are normalized to a $^{143}Nd/^{144}Nd$ ratio of 0.512 62 for BCR–1 and an $^{87}Sr/^{86}Sr$ ratio of 0.708 00 for E&A. $^{143}Nd/^{144}Nd$ ratios previously reported by Hawkesworth *et al.* (1979) were not normalized to a BCR–1 value of 0.512 62 which accounts for their being higher than those of Hofmann & White (1982). In this study three of the Hawkesworth *et al.* samples have been duplicated within analytical error (see Table 1).

nearby Oceanographer Fracture Zone have significantly lower $^{207}Pb/^{204}Pb$ ratios (Cohen *et al.* 1980; Dupre & Allegre 1980; Shirey *et al.* 1987) suggesting that the earlier analyses may have suffered from blank contamination (Armstrong, pers. comm., 1987). Only the most recent data are plotted in the lead isotope diagrams. However, glasses from the Oceanographer Fracture Zone have $^{207}Pb/^{204}Pb$ and $^{208}Pb/^{204}Pb$ ratios that are displaced to relatively high values relative to the NHRL (Shirey *et al.* 1987).

The altered Jurassic basement from Cape Verde has radiogenic $^{206}Pb/^{204}Pb$ ratios (19.6–20.3) but notably low $^{208}Pb/^{206}Pb$ ratios (1.91–1.99; Gerlach *et al.* in press). The μ values ($^{238}U/^{204}Pb$) of the altered tholeiites (40–130) are significantly higher than typical values for mid-ocean ridge basalt (MORB), which rarely reach 20 (Cohen & O'Nions 1982*a*). These observations suggest that seafloor alteration and hydrothermal processes have significantly increased the U/Pb and U/Th ratios in the ocean floor beneath the Cape Verdes. This is consistent with trace-element studies of hydrothermal waters at mid-ocean ridges which lose uranium to, and leach lead from, the oceanic crust (Michard *et al.* 1983). With time high U/Pb and U/Th ratios would cause the production of uranogenic lead isotope ratios, *ie* relatively high $^{206}Pb/^{204}Pb$ ratios (*eg* a μ value of 50 increases $^{206}Pb/^{204}Pb$ by 0.7 in 100 Ma). It is therefore apparent that if the Tertiary to Recent volcanism of the Cape Verde Islands had assimilated a significant amount of altered oceanic lithosphere they would have inherited relatively radiogenic $^{87}Sr/^{86}Sr$, $^{206}Pb/^{204}Pb$ and $^{143}Nd/^{144}Nd$ ratios and low $^{208}Pb/^{204}Pb$ ratios. In contrast, the oceanic crust beneath the Azores islands is significantly younger and hence has not had time to develop uranogenic lead isotope ratios. However, Sr strontium exchange between seawater and ocean crust in a mid-ocean ridge hydrothermal system has the potential to produce relatively radiogenic $^{87}Sr/^{86}Sr$ ratios even in young MORB.

Cape Verde Islands

The Cape Verde Islands show marked strontium and neodymium isotope variations (0.70292–0.70393 and 0.51305–0.51261) with a slope (−0.44) steeper than the mantle array (−0.25) that connects MORB to estimates of bulk Earth (Fig. 2). Individual islands record significant isotope variation. The northern island of Sao Antao has the most depleted isotope systematics, comparable with MORB glasses from the Azores platform, but defines a trend below the mantle array to values comparable with St Helena (Hofmann & White 1982, Chaffey *et al.*, this volume). The most undersaturated volcanic rocks have MORB-like Sr–Nd isotope systematics. In contrast, the adjacent island of Sao Vicente displays a significantly shallower slope in Fig. 2, with values ranging from those seen on Sao Antao to slightly right of the mantle array. Volcanic rocks from Sao Vicente include carbonatites and basanites (5%–29% norm Ne+Lc+Cs). The samples with radiogenic $^{87}Sr/^{86}Sr$ ratios are dolomite–bearing carbonatites with $\delta^{18}O$ values above + 20, suggesting that they have suffered extensive interaction with

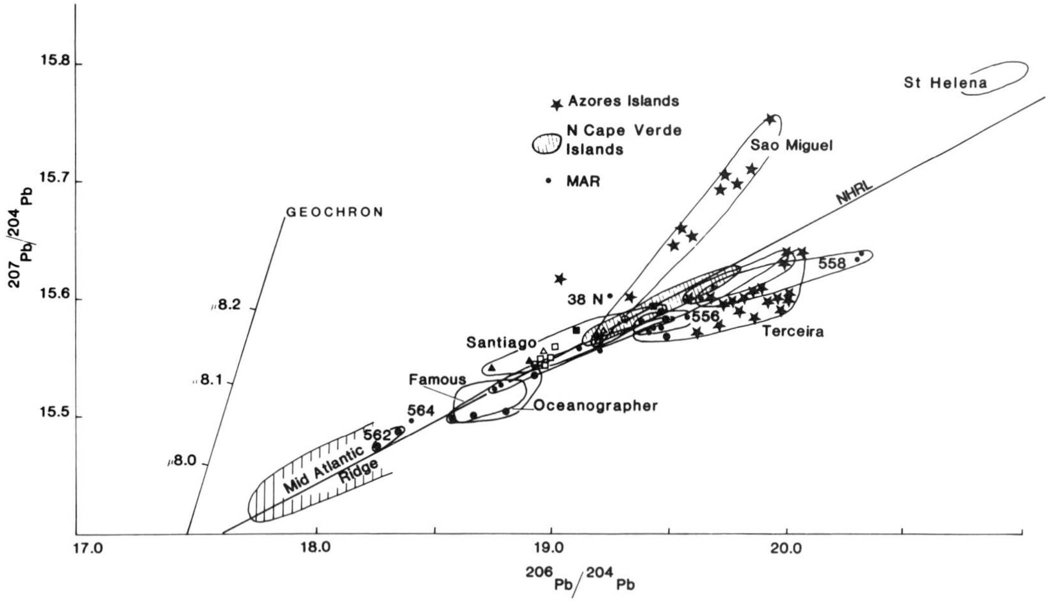

FIG. 3. $^{207}Pb/^{204}Pb$ – $^{206}Pb/^{204}Pb$ isotope variation diagram for selected oceanic basalts. For symbols see Fig. 2. Data sources: Dupre & Allegre (1980, 1983), Sun (1980), Cohen & O'Nions (1982*a*), Dupre *et al.* (1982, & 1985), White *et al.* (1985) and Shirey *et al.* (1987). The 4.55 Ga Pb–Pb geochron is shown with single-stage μ values. The Northern Hemisphere Reference Line (NHRL) of Hart (1984) when interpreted as a secondary isochron has a slope equivalent to 1.77 Ga. Volcanics plotting above the NHRL are said to have a 'Dupal' signature (Hart 1984) (*eg* Sao Miguel).

hydrothermal fluids (Hodgson 1986). The shallow Sr–Nd isotope trend shown by the Sao Vicente rocks is therefore probably an alteration phenomenon.

The southern islands, and Santiago in particular, exhibit relatively high $^{87}Sr/^{86}Sr$ ratios and define the majority of the steep Nd–Sr isotope array. In addition, Fogo and Maio show isotopic evidence of the unradiogenic neodymium isotope component. However, $^{87}Sr/^{86}Sr$ ratios are not as extreme as those of Santiago, suggesting some mixing between the low Nd–Sr isotope component and sources lying on the mantle array. It is perhaps significant to point out at this stage that the Jurassic MORB basement of Santiago has more radiogenic $^{143}Nd/^{144}Nd$ ratios than the recent volcanic rocks and hence does not record evidence of the low Nd–Sr isotope trend, implying that the hot-spot was not active in the Jurassic.

Individual islands do not display consistent temporal isotope variations. For example, on Santiago $^{87}Sr/^{86}Sr$ ratios become more radiogenic from the Flamengos to the Assomada Formation with Quaternary volcanism having intermediate values. The islands show a general N–S difference in Sr–Nd isotope ratios. The less undersaturated rocks of the southern islands have the most radiogenic strontium and least radiogenic neodymium isotope ratios. In addition the most undersaturated volcanic rocks on each island have the most depleted strontium isotope systematics (*eg* Miao, see Table 1). This relationship is also shown within individual stratigraphic sequences, *eg* Pico de Antonia Formation on Santiago (Gerlach *et al.*, in press).

The lead isotope systematics of Cape Verde volcanic rocks are equally heterogeneous ($^{206}Pb/^{204}Pb$=18.74–19.77; $^{207}Pb/^{204}Pb$=15.54–15.63; $^{208}Pb/^{204}Pb$=38.69–39.45; Figs 3 and 4) and vary both with location and major-element composition. The northern islands have the most radiogenic lead isotope ratios, comparable with those of the Canaries (Sun 1980) and lying close to the NHRL (Figs 3 and 4). In contrast, the southern islands have lower $^{206}Pb/^{204}Pb$ ratios with the Assomada Formation on Santiago again being the most extreme. Significantly, the islands that have the greatest displacement from the Nd–Sr mantle array (Fig. 2) are also displaced from the NHRL on the Pb–Pb isotope diagrams (Figs. 3 and 4). The Cape Verde Islands show correlations between Pb–Sr–Nd isotope ratios with the northern and southern islands showing distinct relationships. In a plot of $^{206}Pb/^{204}Pb$ versus strontium

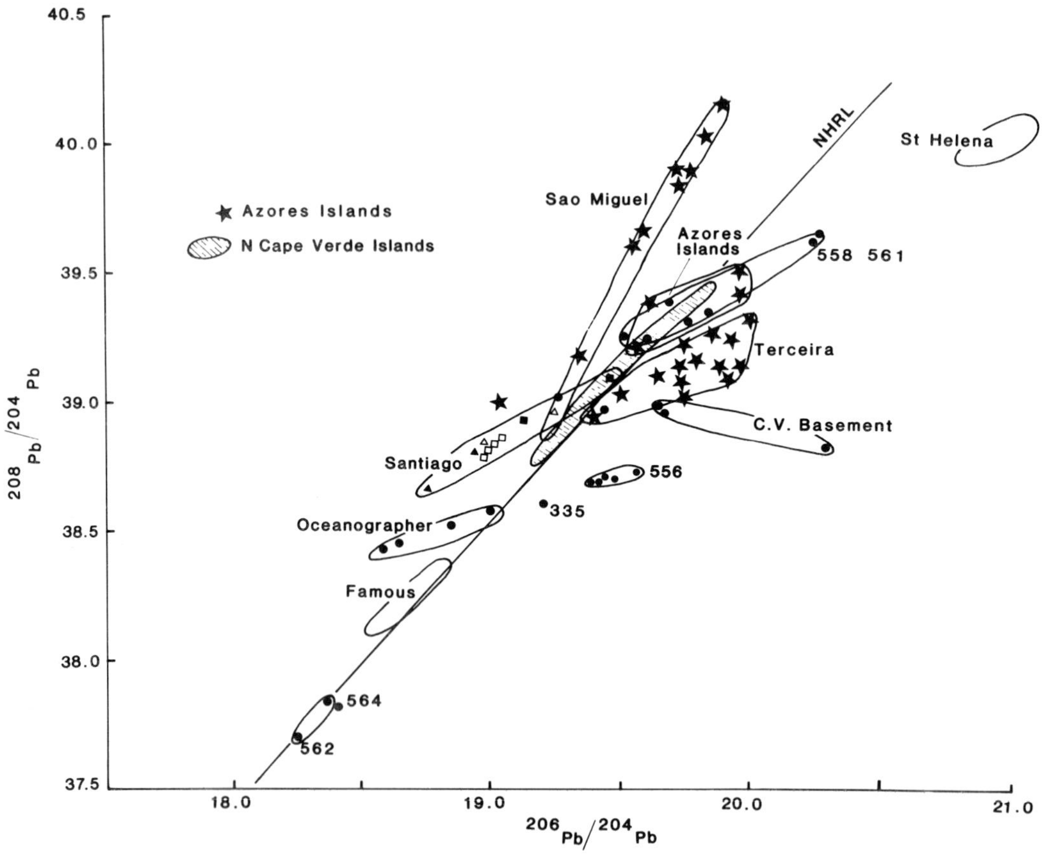

FIG. 4. $^{208}Pb/^{204}Pb - ^{206}Pb/^{204}Pb$ isotope variation diagram for selected oceanic basalts. For symbols and data sources see Figs 2 and 3. The NHRL equates to a second-stage Th/U ratio of 4.0. Note that, unlike Fig. 3, the southern Cape Verde Islands (Fogo, Maio and Santiago) significantly diverge from the NHRL and that samples with the most radiogenic lead isotope ratios (high time-integrated U/Pb, *ie* the high μ (HIMU) of Zindler & Hart (1986) do not lie at the end of the NHRL.

isotopes (Fig. 5) the northern islands show a shallow positive correlation. In contrast, the southern islands have a steep negative correlation that trends towards the field defined by the Walvis Ridge (Richardson *et al.* 1982) and some Hawaiian islands (Stille *et al.* 1983). These isotopic relationships suggest that at least three isotopically distinct sources are involved in the genesis of the Cape Verde volcanic rocks. Following the terminology of Zindler & Hart (1986) these components are (i) a depleted MORB (DMM) source, (ii) a depleted Sr–Nd and uranogenic lead (HIMU) source and (iii) a source with high Sr/Pb isotope ratio (EMI).

Azores

Previous geochemical studies of volcanism from the Azores platform have not provided combined trace-element and Pb–Sr–Nd isotope data (White *et al.* 1979; Sun 1980; Dupre *et al.* 1982; Hofmann & White 1982). We present new lead Pb isotope and trace-element data for the island of Sao Miguel which together with existing Nd–Sr isotope ratios (Hawkesworth *et al.* 1979) now provide a comprehensive data set. The islands of Faial and Sao Miguel are generally displaced to more radiogenic $^{87}Sr/^{86}Sr$ relative to the mantle array (Fig. 2). The Sao Miguel Sr–Nd data array defines a shallow slope (−0.1) that trends away from the MAR of the Azores platform and is in marked contrast to the steep array defined by the Cape Verde Islands.

Sao Miguel and a single Faial sample (Sun 1980) define steep positive slopes in Pb–Pb diagrams and diverge from the NHRL, *ie* record a strong 'Dupal' signature (Figs 3 and 4). Sam-

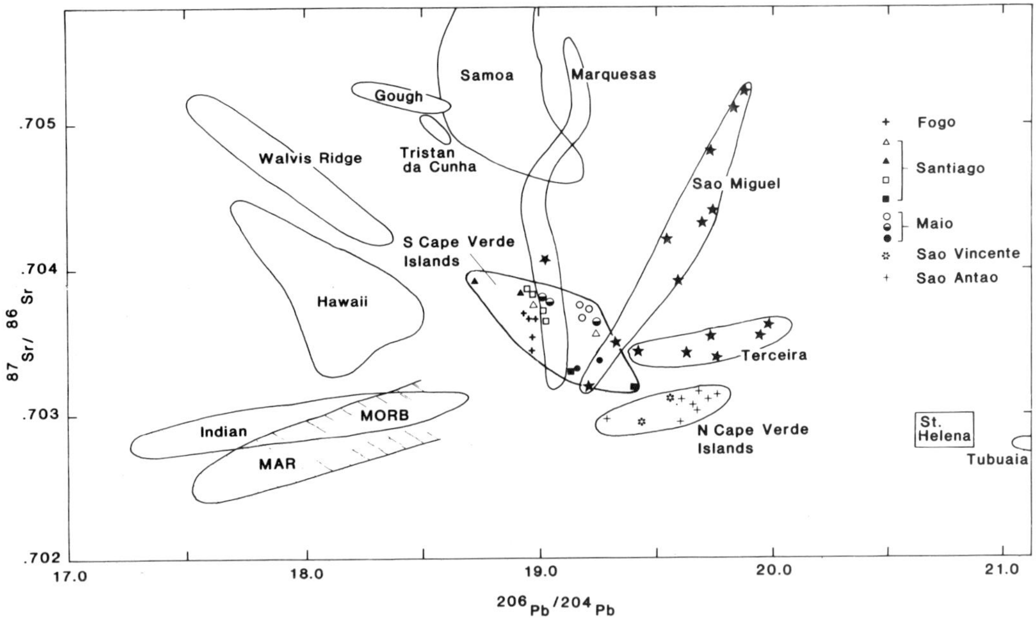

FIG. 5. $^{87}Sr/^{86}SR$ – $^{206}Pb/^{204}Pb$ isotope variation diagram for selected oceanic basalts. For data sources see Figs 2 and 3. In addition to the varied relationships shown by islands with the 'Dupal' signature, the HIMU and NHRL islands have distinct isotope relationships.

ple AZ 1039 from Sao Miguel plots outside the data array displaced towards less radiogenic $^{206}Pb/^{204}Pb$. Data for the island of Terceira (Dupre *et al.* 1982) lie below the NHRL and are more radiogenic than for the northern Cape Verde islands. Available data from Sao Jorge, Flores and Santa Maria (Sun 1980) have lead isotope systematics comparable with those of Terceira. The Azores, as the Cape Verde Islands, also show two distinct Pb–Sr–Nd isotope relationships. On a Pb–Sr diagram (Fig. 5) Terceira data define a shallow positive trend displaced to more radiogenic $^{87}Sr/^{86}Sr$ ratios than the Cape Verde islands of Sao Antao and Sao Vicente. In contrast, Sao Miguel and the single Faial sample record a steep positive correlation. Sao Miguel is the only known ocean island that records this steeply positive Sr–Pb isotope relationship. Overall, the Azores hot-spot records evidence of four components. Following the terminology of Zindler & Hart (1986) these components are (i) a depleted MORB (DMM) source, (ii) a depleted Sr–Nd and uranogenic lead (HIMU) source, (iii) a source with high Sr/Pb isotope ratio (EMI) and (iv) a radiogenic strontium and lead isotope source (EMII).

We can conclude that recent volcanism on the MAR and at the Azores and Cape Verde hot-spots demonstrates marked isotope heterogeneity in their mantle source regions. The isotope data define mixing relationships between typical MORB values and at least three distinct isotopic compositions. Before we can draw any conclusions regarding the origin of these distinct mantle components, it is imperative to consider their possible history. In particular, do the components reflect different source chemistry or are they simply of different ages? That is, have they evolved with the same parent/daughter ratios for different periods of time? In the following section, we show that the coupled trace-element and isotope relationships of the Azores and Cape Verdes represent chemically distinct source materials.

Combined radiogenic isotope and trace-element relationships

Fractional crystallization and partial-melting processes may cause significant changes in the trace-element ratios of volcanic rocks such that trace-element ratios are no longer representative of the mantle source region. The effects of fractional crystallization are well shown by the data for Sao Miguel: there is a general increase in incompatible-trace-element contents as SiO_2 increases and MgO decreases. Rocks of intermediate composition have fractionated magnetite, apatite and plagioclase resulting in a significant increase in Rb/Sr (Storey *et al.*, this volume). In an effort to minimize the effects of crystal fractionation this study is confined to rocks with MgO >5% from both the Azores and Cape Verde hot-spots.

The degree of fractionation between trace elements during partial melting will depend on the difference in their partition coefficients and the effective degree of partial melting. Generally, very incompatible elements can only be fractionated from each other at extremely small degrees of partial melting, *ie* less than 1% (McKenzie 1984, 1985*a*, *b*). Most ocean island basalts show minor fractionation between the most incompatible elements on trace-element-normalized diagrams recording a steady increase in normalized values from ytterbium to niobium and then a flattening or steady decrease to rubidium. The latter case, St Helena for example (Fig. 6), implies derivation from a previously melted source region. This conclusion is compatible with the Nd–Sr isotope ratios of this island (Chaffey *et al.*, this volume). Comparable trace-element patterns are shown by the nepheline-normative basalts from Cape Verde (*eg* Sao Vicente, Fig. 6). However, markedly undersaturated volcanics (*eg* larnite-normative rocks on Sao Antao) show distinct fractionation among the elements more incompatible than niobium, *eg*, rubidium and potassium 'troughs' (Fig. 6). In addition strontium, hafnium and zirconium are relatively depleted compared with rare earth elements (REEs) having similar degrees of incompatibility. These features may not be characteristic of the original mantle source but are most probably due to the influence of residual phases at small degrees of partial melting (Clague & Frey 1982). Retention in the source of potassium, rubidium, strontium, zirconium and hafnium but not barium, REEs and titanium precludes residual phlogopite and titanites. We consider a low-titanium amphibole, potassic richterite for example, as the most likely residual phase.

Melilitites represent the most SiO_2 undersaturated silicate volcanism in an oceanic environment. Hawaiian examples show depletion in

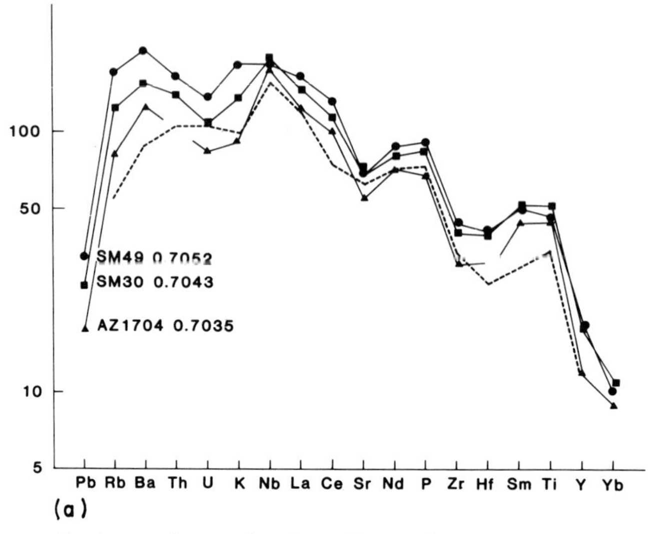

FIG. 6. Chondrite-normalized trace-element abundance diagram (normalization values from Thompson *et al.* (1982) except barium and lead which are from Sun (1980). A typical St Helena basalt is shown in all figures. (*a*) Sao Miguel basalts. Increases in $^{87}Sr/^{86}Sr$ ratios are accompanied by a distinct change in the style of trace-element enrichment. Samples with low $^{87}Sr/^{86}Sr$ ratios have a niobium-rich enrichment style comparable with St Helena samples. As $^{87}Sr/^{86}Sr$ increases the niobium enrichment remains approximately constant whereas other highly incompatible elements become more enriched. (*b*) Sao Antao melilitites (northern Cape Verde). The general style of trace-element enrichment is compatible with derivation from a St Helena type source by a low degree of partial melting which produces greater enrichment in the most incompatible trace elements, and, owing to the presence of minor phases, may result in the retention of potassium in the source, *eg* NA60. (*c*) Santiago basalts (southern Cape Verde). The less undersaturated rocks (43% SiO_2) show increases in $^{87}Sr/^{86}Sr$ accompanied by greater barium, rubidium and lower uranium, thorium and potassium contents, implying variations in both source composition and mineralogy.

niobium in addition to potassium and rubidium (see Figs 6 and 7), resulting in increased La/Nb ratios (1–3 compared with less than 1 for most OIBs) (Clague & Frey 1982). A possible explanation is that the magmas are produced by very small degree partial melts from a depleted MORB source which has a relatively high La/Nb ratio (Norry & Saunders, in press). However, the most undersaturated lavas from the Cape Verde have very variable Nb/Ta ratios. This characteristic suggests that an Nb–Ta–rich phase, probably a critonite series titanate, was retained in their source region and caused the high La/Nb ratios.

For clarity we shall initially confine the following trace-element discussion to the basaltic

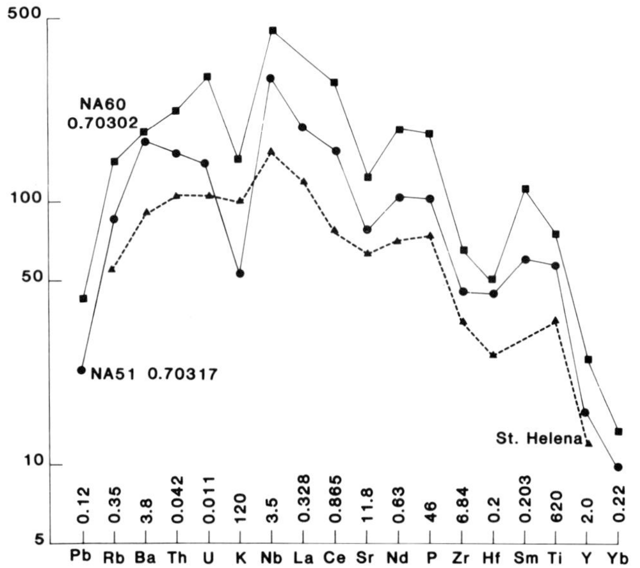

(b)

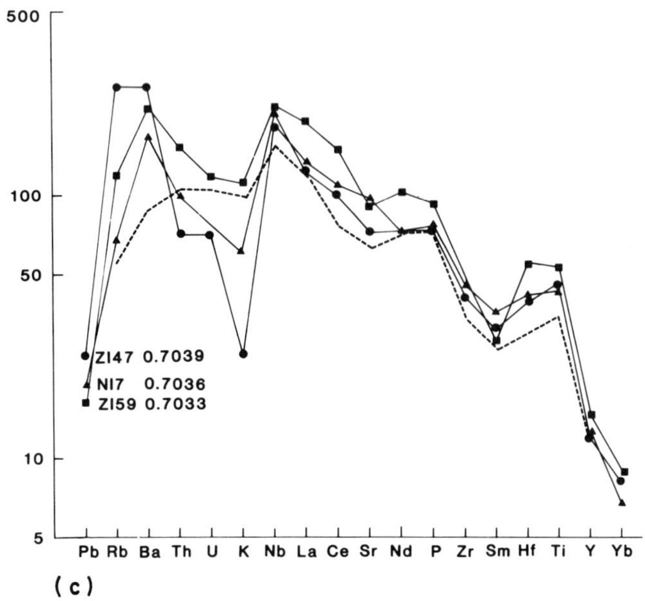

(c)

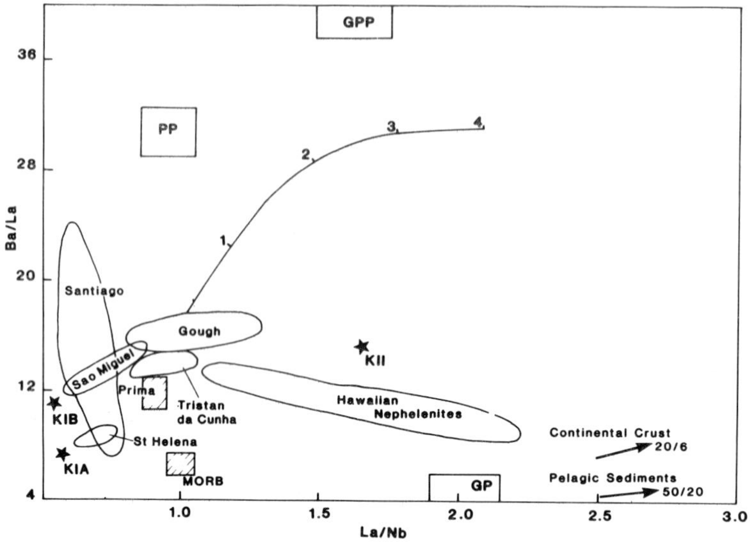

FIG. 7. La/Nb − Ba/La variation diagram for representative oceanic basaltic volcanic rocks. Data sources: Clague & Frey (1982), LeRoex (1985), Weaver *et al.* (1987) and Chaffey *et al.* (this volume). Material believed to be at least in part from the subcontinental lithosphere shows a wide variation in trace-element ratios. Average xenolith populations from southern Africa are plotted (GP average garnet peridotite; GPP, average garnet phlogopite peridotite; PP, average phlogopite peridotite) (Erlank *et al.* 1987). Pyroxenite xenolith assemblages from Uganda (Davies & Lloyd, in press), southern Australia (Menzies & Wass 1983) and NW Britain (Menzies *et al.* 1987) encompass the entire diagram but are notable in that they are the only known mantle- or crustal-derived materials with very low La/Nb ratios. Lamproites and group II kimberlites (KIA, KIB, KII; average group IA, IB and II kimberlites) (Smith *et al.* 1985) have high but variable La/Nb and Ba/La ratios (see text). The retention of minor phases in the source of undersaturated volcanic rocks may generate high La/Nb ratios, *eg* Hawaiian nephelenites (Clague & Frey 1982). The mixing line depicts the addition of pelagic sediment (Hole *et al.* 1984; White *et al.* 1985) to an undifferentiated mantle source. The low La/Nb ratios in Santiago volcanic rocks preclude a role for recycled sediments in the source of these basalts.

volcanic rocks from the islands of Santiago and Sao Miguel as they show the most variable and extreme isotope systematics. The basaltic volcanic rocks of the Azores islands as a whole show a significant range in terms of major-element composition with a general compositional dependence on the degree of light REE enrichment (White *et al.* 1979). Basaltic volcanic rocks on Sao Miguel are mildly hypersthene-normative suggesting that the degree of partial melting in their source region was sufficiently large to leave no minor phases and hence cause only minor fractionation between the most incompatible elements. Although these basalts represent some of the least undersaturated rocks from the Azores island group they record among the greatest degree of light REE enrichment. In addition there is a marked REE variation, with the degree of light REE enrichment, $(Ce/Yb)_n$ ranging from 9.6 to 14.3 and showing a positive correlation with strontium and lead isotope ratios (Fig. 8). Ratios between incompatible elements vary markedly for Sao Miguel basalts (*eg* Ba/Nb=6.3−12.5, see Table 2) and the ratios generally show good correlations with each other and Sr−Nd−Pb isotope ratios (Figs. 7−9). The Rb/Ba ratio is relatively constant throughout the suite. Although the increase in ratios such as Ba/Nb and K/La is consistent with fractionation during small degrees of partial melting, the variations are significantly greater than predicted for even extremely small degrees of partial melting less than (0.01%) (McKenzie 1984, 1985*a*, *b*). In addition, rocks with the highest Ba/Nb ratios also have greater enrichment in potassium than in the more incompatible elements thorium, rubidium and barium (Fig. 6). These trace-element relationships are unlikely to be solely the result of a single partial-melting event and strongly suggest source heterogeneities. The samples with the most radiogenic strontium and lead isotope ratios consistently show the most extreme trace-element enrichment. Relative trace-element enrichment is in the following order: K>Ba=Rb>Th>La>Pb>U>Nb. In

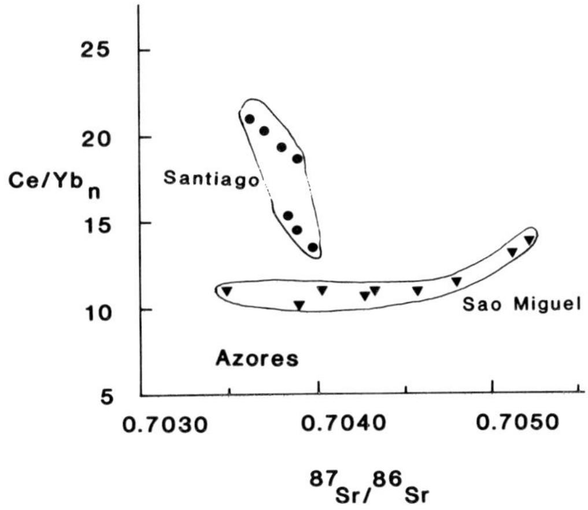

FIG. 8. $(Ce/Yb)_n - {}^{87}Sr/{}^{86}Sr$ variation diagram for basalts from the islands of Santiago and Sao Miguel. Field for Azores basalts from White *et al.* (1979).

the 'enriched' rocks increased Th/U ratios are associated with a noticeable increase in radiogenic ^{208}Pb and ^{206}Pb ($^{208}Pb^*/^{206}Pb^*$ ratios of 0.963–1.011 compared with 0.93 for Terceira). However, there is an apparent decoupling between the U/Pb and $^{206}Pb/^{204}Pb$ ratios: a decrease in values (30–25) is associated with more radiogenic $^{206}Pb/^{204}Pb$ ratios. This decoupling implies that the trace-element signature is of relatively recent origin, possibly related to variations in the degree of partial melting (see later discussion).

The basaltic volcanic rocks on Santiago also show marked trace-element variation (*eg* Ba/Nb=6–17, Ba/La=10–24 see Table 2). Potassium and rubidium show complex relationships with other trace elements. For example the Quaternary and Flamengos Formations have positive correlations between rubidium, K_2O, K_2O/TiO_2 and $^{87}Sr/^{86}Sr$. In contrast, the Assomada Formation with the greatest $^{87}Sr/^{86}Sr$ ratio and rubidium content shows no consistent relationship with K_2O. In addition to source variations, we ascribe these relationships to a variable role for potassium-rich minor phases during melt extraction. Phlogopite in particular may play a role in reducing the K/Rb ratio of the Flamengos Formation. The inter-element relationships of the most isotopically extreme samples are significantly different from those for Sao Miguel. For instance the degree of light REE enrichment, which is strongly composition dependent with the most undersaturated rocks having the greatest REE fractionation ($(Ce/Yb)_n$=13.6–20.7), shows negative and positive correlations with strontium (Fig. 8) and lead isotope ratios respectively. Incompatible-trace-element ratios generally show correlations between themselves (*eg* Ba/La versus La/Nb; Fig. 7) and with strontium, lead (Fig. 9) and neodymium isotope ratios. The 'enriched' component has relatively high $^{208}Pb^*/^{206}Pb^*$ ratios (0.953–0.973) although Th/U ratios are lower than in other rocks from the Cape Verdes. μ values and lead isotope ratios of the Cape Verde 'enriched' component are lower than in samples that lie on the NHRL (20 versus 30). Both the Th/U and μ values are significantly higher than estimates of bulk Earth. The style of trace-element enrichment in the 'enriched' component is distinct from that of Sao Miguel (Ba>Pb>Nb>La>U>Th), with lead and niobium being notably enriched compared with uranium and thorium.

TABLE 2. *Chemical characteristics of Santiago and Sao Miguel*

	Sao Miguel	Santiago
$^{87}Sr/^{86}Sr$	0.7035–0.7052	0.7034–0.7039
$^{206}Pb/^{204}Pb$	19.33–19.88	19.19–18.74
$^{207}Pb/^{204}Pb$	15.60–15.75	15.58–15.53
Ce/Yb	9.6–14.3	24.6–13.6
Ba/Nb	6.3–12.5	5.5–15.5
La/Nb	0.58–0.84	0.8–0.63
Ba/La	11–14.8	7.6–24.0
Ba/Th	52–115	57–244
U/Pb	0.46–0.37	0.73–0.31
Th/U	3.7–4.7	5.1–4.33
U/Nb	0.015–0.023	0.019–0.014
Ce/Pb	42–31	43–32

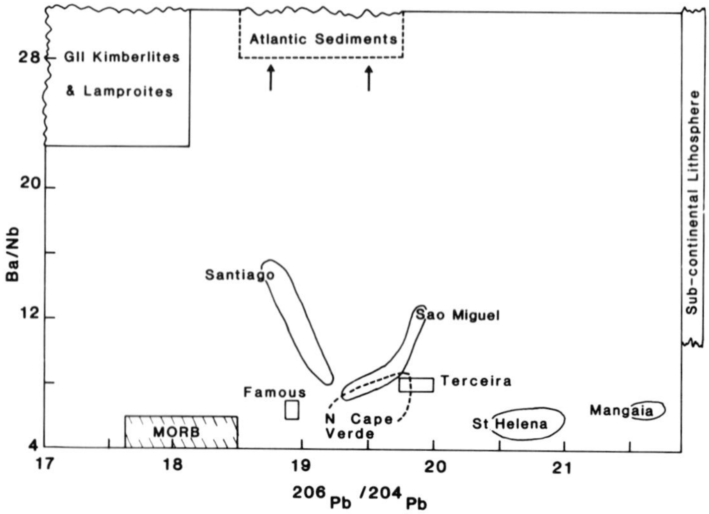

FIG. 9. Ba/Nb$-^{206}$Pb/^{204}Pb isotope variation diagram for selected oceanic basalts and possible recycled lithospheric material. For data sources see Figs 3 and 8. See text for discussion.

Origin of the chemical heterogeneities

We have established that volcanic rocks on the islands of Santiago and Sao Miguel record mixing relationships between mantle sources that, in isotopic terms, lie on the mantle array (Nd–Sr) and NHRL (Pb–Pb) and two distinct 'enriched' components. In detail, the sources of the two enriched components have different trace-element characteristics. It is the origin of these two source regions that we now wish to address.

The 'enriched' components from Cape Verde and Sao Miguel have many characteristics comparable with parts of the continental lithosphere, *eg* high LIL to HFS element ratios. One way to account for these features in the suboceanic mantle is to advocate the involvement of recycled lithospheric material either directly or in the form of the mantle wedge above subduction zones. Establishing the chemical composition of recycled components remains a difficult problem. Although we have good constraints on the present composition of material entering subduction zones, and have a reasonable idea of temporal variations from the geological record, we do not know how the components are modified by dehydration and partial melting during subduction. Below we make reasoned comparisons between the trace-element compositions of the 'enriched' components and possible mantle reservoirs.

Primitive upper mantle

Estimates of bulk Earth compositions are based on chondritic models (see the review by Zindler & Hart (1986)). To date no OIB volcanic rocks have been found that have Pb–Pb and Sr–Nd isotope ratios that could be derived solely from an undifferentiated mantle source. However, some of the islands with Pb–Sr–Nd relationships approaching bulk Earth values (*eg* Hawaii (Stille *et al.* 1983)) have high ^{3}He contents that are thought to imply derivation from a relatively un-degassed source (Kurz *et al.* 1983), implying that they may contain a large primitive mantle component.

Recently Hofmann *et al.* (1986) reported that suboceanic mantle sources have constant Nb/U and Ce/Pb ratios that are greater than primitive mantle estimates and those of the crust (Fig. 10). More data are now available and demonstrate that, although oceanic volcanic rocks are not as constant in terms of these ratios as Hofmann *et al.* reported, their observations remain very pertinent (Table 1; Hofmann 1986; Palacz & Saunders 1986; Chaffey *et al.*, this volume; Sun & McDonough, this volume). The implications of the data are that OIBs and MORBs are not derived from a primitive mantle source or from mixtures of depleted mantle and primitive mantle. Equally the La/Nb ratios of Hawaiian basalts are lower than the bulk Earth value. We therefore conclude that no primitive mantle source is sampled by oceanic volcanism and that the U–Th–He system is effectively

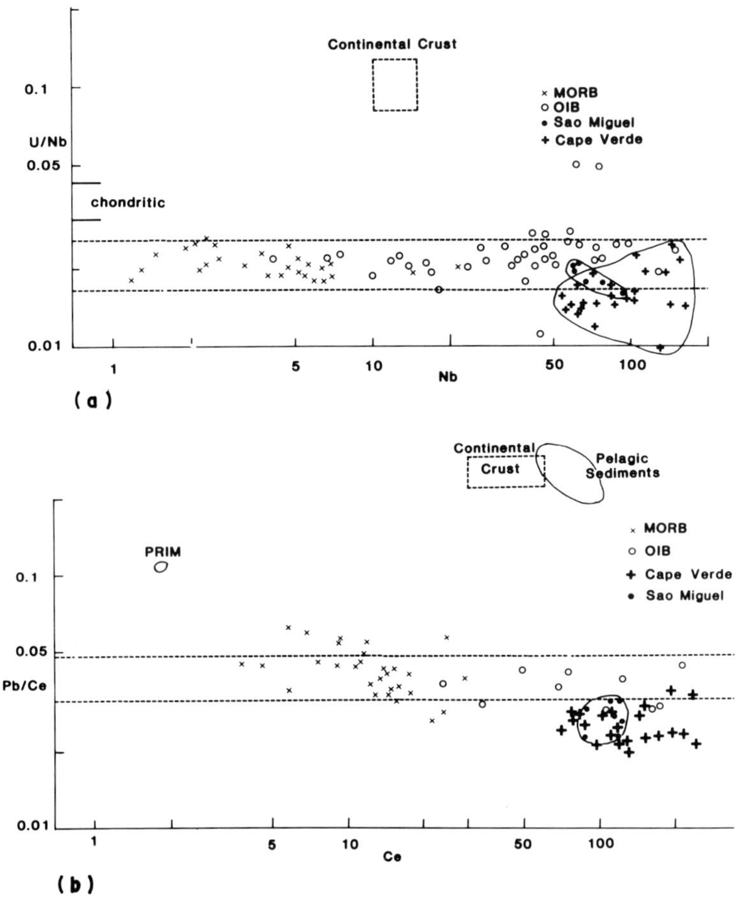

FIG. 10. (*a*) U/Nb–Nb variation diagram after Hofmann *et al.* (1986). Note that the Cape Verde and Sao Miguel volcanic rocks show lower U/Nb ratios than typical OIB and continental crust. Increased $^{87}Sr/^{86}Sr$ ratios in the Sao Miguel samples correlate with greater U/Nb ratios. (*b*) Pb/Ce–Pb variation diagram after Hofmann *et al.* (1986). Note that the Cape Verde and Sao Miguel volcanic rocks show lower Pb/Ce ratios than typical OIB and continental crust. Increased $^{87}Sr/^{86}Sr$ ratios in the Sao Miguel samples correlate with greater Pb/Ce ratios.

decoupled from the lead, strontium and neodymium isotope systems (Anderson 1986).

Sediments

Most pelagic and detrital sediments have major-element compositions that are essentially granitic. Under the hydrous conditions and inferred $P-T$ regimes operating in present-day and ancient subduction zones both pelagic and detrital sediments would melt. In addition, unlike mafic and ultramafic material, the solidus and liquidus temperatures of sediments are similar, so that sediments would undergo large degrees of melting even when only just above their solidus. We predict that under conditions of extensive melting minor phases (*eg* zircon) capable of significantly fractionating the most incompatible trace elements are unlikely to survive in the residue and that the chemical signature of a recycled sedimentary component would be essentially that seen at the surface today (*ie* high Ce/Yb, Ba/Nb, La/Nb, Ba/La, Rb/Sr etc.). Sediment-derived melts would be in marked chemical disequilibrium with the overlying peridotitic mantle and would react rapidly. There are three possible destinations for melts derived from subducted sediments. They may be trapped within the convecting mantle wedge and subsequently recycled into the upper mantle or, in the form of subduction-related magmatism, be stabilized in the overlying subcontinental lithosphere and/or the continental crust.

Basaltic volcanism from the islands of Santiago and Sao Miguel shows correlations between the majority of the trace-element ratios that characterize sediments (Fig. 7) and Pb–Sr–Nd isotope ratios. For example Santiago records a negative correlation between lead isotope ratios and Ba/Nb ratios (Fig. 9). The correlation trends to less radiogenic $^{206}Pb/^{204}Pb$ ratios than present-day pelagic sediments (O'Nions *et al.* 1978; White *et al.* 1985). The low U/Pb ratios of pelagic sediments (Hole *et al.* 1984; White *et al*, 1985) would result in ancient (more than 1 Ga) recycled sedimentary components having less radiogenic lead isotope ratios than present-day examples. Pelagic sediments have generally high incompatible-trace-element contents (*eg* Ba $>$ 300 ppm and Pb $>$ 15 ppm) (Hole *et al.* 1984). Therefore the addition of less than 2% pelagic sediment to the mantle will explain many of the trace-element and isotope mixing relationships defined by the island of Santiago (Fig. 9). However, ancient recycled pelagic sediments would still retain relatively high $^{207}Pb/^{204}Pb$ and $^{208}Pb/^{204}Pb$ ratios inherited from the continental crust. Santiago volcanic rocks are notable for their relatively low $^{207}Pb/^{204}Pb$ ratios, implying that pelagic sediments were not involved in their petrogenesis. In addition, continental crust and pelagic sediments are characterized by relatively high La/Nb, U/Nb and Pb/Ce ratios (Hole *et al.* 1984; White *et al.* 1985; Hofmann *et al.* 1986; Newsom *et al.* 1986). Significantly, samples with the most extreme isotope systematics from Santiago have lower La/Nb (Fig. 7), U/Nb and Pb/Ce ratios (Fig. 10*a*, *b*) than typical OIB volcanism. It is therefore apparent from combined trace-element and Pb–Sr–Nd isotope relationships that pelagic sediments are *not* involved in the genesis of the volcanism on Santiago.

In contrast, Sao Miguel basalts with the greatest trace-element enrichment do show the generally higher La/Nb and U/Nb ratios required for crustal involvement (Figs. 7 and 10). A positive correlation between $^{206}Pb/^{204}Pb$ and Ba/Nb trends towards more radiogenic lead isotope ratios than present-day pelagic sediments (O'Nions *et al.* 1978; White *et al.* 1985). The low U/Pb ratios of pelagic sediments effectively rules out the possible involvement of ancient recycled pelagic sediments. However, clastic sediments represent an additional crustal-derived component that may be recycled into the mantle. Detrital sediments enter deep-sea trenches, and by implication are subducted at mature island arcs and active continental margins (Kay 1984). Chemical and mechanical weathering significantly modify crustal rocks during erosion, transport and deposition (Martin & Meybeck 1979). Although strontium and uranium are particularly susceptible to complexing in water (Martin & Meybeck 1979), clastic sediments generally retain the high U/Pb and Rb/Sr ratios characteristic of continental crust (Gormet *et al.* 1984). Therefore, notwithstanding our reservations regarding the survival of sediments during subduction, the radiogenic lead isotope ratios and many of the trace-element characteristics of Sao Miguel volcanism could be explained by the incorporation of detrital sediments into the mantle. However, an additional characteristic of detrital sediments is that they tend to concentrate heavy minerals compared with their source rocks. Zircon, a common heavy detrital mineral, is notable in that it has high heavy REE and hafnium contents and a low Lu/Hf ratio. Detrital sediments have lower Lu/Hf ratios than average continental crust and estimates for the bulk Earth (Patchett *et al.* 1984). A consequence of the physical separation of zircon into detrital sediments is that authigenic sediments (deep-sea clays) have very high Lu/Hf ratios. Therefore, with time, the two sedimentary types will produce markedly different hafnium isotope ratios that are distinct from mantle values. Oceanic and subduction-related volcanism record a linear relationship between neodymium and hafnium isotopes (Patchett & Tatsumoto 1980; White & Patchett 1984). If either old, pelagic or detrital sediments were a significant component in the petrogenesis of OIBs then the linear Nd–Hf isotope relationship would be disrupted (Patchett *et al.* 1984). Mass balance calculations demonstrate that the addition of 2% sediment, which had been subducted for 0.5 Ga, to a mantle source would be sufficient to disrupt the Nd–Hf isotope correlation. Although hafnium isotope data are not available for Sao Miguel, samples have been analysed from other islands with an EMII component and a comparable style of trace-element enrichment: Samoa and Tristan da Cunha (Patchett & Tatsumoto 1980). These data lie on the Nd–Hf correlation and hence cannot have a large detrital sediment component in their source unless they have been recently subducted (less than 500 Ma). The available hafnium isotope data do not entirely rule out a role for recycled sediment. However, in conjunction with the unlikelihood of sediments surviving subduction without melting, we strongly argue against a role for directly recycled detrital sediments in the mantle sources of OIBs with a 'Dupal' signature.

Subcontinental lithosphere

The subcontinental lithosphere represents a relatively small but significant proportion of the Earth's mantle — less than 20%. Peridotite xenoliths from the subcontinental lithosphere are generally depleted in terms of major elements, implying that they represent residua after partial melting (Nixon *et al.* 1981). However, the majority of these xenoliths have suffered some form of trace-element enrichment event following the depletion event, so that the subcontinental lithosphere must be considered a significant reservoir in terms of the trace-element budget of the Earth (Nixon & Davies 1987). The delamination of subcontinental lithosphere into the convecting upper mantle has been proposed on thermal (Houseman *et al.* 1981) and chemical (McKenzie & O'Nions 1983) grounds. Xenoliths and magmas believed to be derived from the subcontinental lithosphere record variable trace-element and Pb–Sr–Nd isotope characteristics. The majority of highly potassic magmas (lamproites and group II kimberlites), formed away from active subduction zones, have high Ba/Nb, La/Nb, U/Nb, Pb/Ce ratios and low Sm/Nd ratios (*eg* Australia, South Africa and Spain) (Figs 7 and 10) (Smith 1983; Fraser *et al.* 1986; Nelson *et al.* 1986). In addition, these rocks are characterized by relatively unradiogenic neodymium isotope and $^{206}Pb/^{204}Pb$ ratios and radiogenic $^{207}Pb/^{204}Pb$, $^{208}Pb/^{204}Pb$ and $^{87}Sr/^{86}Sr$ ratios. The limited trace-element data from lherzolite and harzburgite xenoliths are also characterized by relatively high, but variable, Ba/Nb, Ba/La and La/Nb ratios (Fig. 7) (Erlank *et al.* 1987; Stolz & Davies 1988). Overall, barium generally shows the greatest degree of relative enrichment in all the above rock types. A similar style of barium-rich trace-element enrichment is displayed in the majority of the continental flood basalt provinces: Columbia River (Carlson 1984); Deccan (Mahoney *et al.* 1982, 1985; Cox & Hawkesworth 1984, 1985); Ethiopia (Norry *et al.* 1987); Karoo (Betton *et al.* 1984; Hawkesworth *et al.* 1984); and Snake River Plain (Doe *et al.* 1982; Menzies *et al.* 1984). This style of trace-element enrichment is again accompanied by high $^{207}Pb/^{204}Pb$, $^{208}Pb/^{204}Pb$ and $^{87}Sr/^{86}Sr$ ratios. All the above examples record high K_2O/TiO_2 ratios and relative niobium depletion, comparable with subduction-related volcanic rocks. In many respects the chemical signature is comparable with that predicted for ancient pelagic sediments. It is therefore a possibility that a large proportion of the subcontinental lithosphere has been modified by subduction-related fluids, probably H_2O rich, that contain a sediment-derived signature. Coupled with the presence of a sedimentary signature in subduction-related volcanism these observations strongly argue against the direct recycling of sediments into the convecting upper mantle.

Alkali-pyroxenite xenolith assemblages record a different style of trace-element enrichment which, although barium rich, is high in TiO_2 and niobium (S Australia (Menzies & Wass 1983) and Uganda (Davies & Lloyd, in press)). Available isotope data again show high $^{207}Pb/^{204}Pb$ and $^{208}Pb/^{204}Pb$ but variable $^{87}Sr/^{86}Sr$ 0.703–0.706), $^{206}Pb/^{204}Pb$ (17–19.5) and unradiogenic neodymium isotope ratios (0.5126–0.5115) (Menzies *et al.* 1987; Davies & Lloyd, in press). In Wyoming and Montana (USA), potassic volcanic rocks are characterized by a similar trace-element and isotope signature (Fraser *et al.* 1986). In addition, several continental flood basalt provinces show a similar chemical signature, *eg* the Karoo, (Hawkesworth *et al.* 1984a) and Deccan (Mahoney *et al.* 1982, 1985; Cox & Hawkesworth 1984, 1985). The overall style of trace-element enrichment outlined above is comparable with the majority of alkaline OIB magmatism, *ie* low K_2O/TiO_2 and high niobium, suggesting that it is related to the migration of small degree silicate melts (Hawkesworth *et al.* 1984b; Nixon & Davies 1987). The presence of apatite and sphene in the pyroxenite xenolith assemblages results in high U/Pb ratios (Davies & Lloyd, in press) that could, with time, produce very radiogenic lead isotope ratios comparable with those from Sao Miguel.

We conclude that the subcontinental lithosphere records at least two styles of trace-element enrichment but always appears to have a notable barium enrichment. In addition, the subcontinental lithosphere records different fractionation of rubidium to strontium and uranium to lead, which coupled with variations in formation age (3.3 Ga to present (Richardson *et al.* 1984)) results in markedly variable lead and strontium isotope systematics. The subcontinental lithosphere appears to represent the only reservoir that can provide the low Sr–Nd isotope array of the Cape Verde Islands that is characterized by high Ba/La and low La/Nb ratios. In addition, the radiogenic strontium and lead isotope ratios of Sao Miguel may also be derived from recycled subcontinental lithosphere characterized by high Ba/Nb and La/Nb ratios which in turn was probably formed by subduction-related fluids. To date, there are insufficient data available to determine the amount of sediment recycled into the subcontinental lithosphere. However, certain peridotite massifs contain carbon-rich garnet pyroxenites with a biogenic carbon isotope signature, $\delta^{13}C=-27$ (up to 15 wt% graphite in the form of octahedral pseudomorphs after diamond) (Nixon *et al.* 1986; Pearson *et al.*, in press), which implies that, at least locally, sediment-derived material forms a significant proportion of the subcontinental lithosphere.

Oceanic crust

Hofmann & White (1982) proposed that the high U/Pb ratios in hydrothermally altered subducted oceanic crust were responsible for the radiogenic lead isotope ratios found in certain OIBs–HIMU (*eg* St Helena (Sun 1980; Chaffey *et al.*, this volume) and Tubuaii (Vidal *et al.* 1984)). Uranium-series disequilibrium studies on subduction-related volcanism demonstrates that uranium is not preferentially released from the mantle during such volcanism (Newman *et al.* 1984). The implication is that uranium is not preferentially lost from the hydrothermally altered oceanic crust during dehydration. Because of the high TiO_2 content, subducted oceanic crust will re-equilibrate to produce sphene- or rutile-bearing eclogite (Hellman & Green 1979). Melting of the subducted basaltic slab will produce a relatively siliceous melt (*eg* dacite) (Sekine & Wyllie 1982). Under such melting conditions titanium is effectively retained in the source (Ryerson & Watson 1987). The formation of titanium-rich phases such as sphene or rutile would retain Nb–Ta in the subducted slab and would explain the relative depletion of HFS elements in all subduction-related volcanism (Pearce 1982). Equally important would be the role of these minor phases in retaining uranium and thorium and causing the subducted lithosphere to produce, with time, uranogenic lead isotope ratios with a low $^{208}Pb^*/^{206}Pb^*$ ratio (*eg* St Helena, 0.92, Chaffey *et al.*, this volume). High U/Pb ratios in sphene, rutile and perovskite enable them to be used for U-Pb dating (Schärer *et al.* 1986). However, the breakdown of rutile at elevated pressures (TiO_2 becomes more soluble in garnet and clinopyroxene with increasing pressure) is unlikely to be accompanied by partial melting under anhydrous conditions and the uranium- and niobium-rich oceanic crust would be recycled into the mantle.

Saunders *et al.* (in press) argue that in order to explain the topology of the La/Nb versus K/Nb diagram, where MORB and continental crust have higher La/Nb and K/Nb ratios than bulk Earth, an additional Nb–Ta-rich component must exist in the mantle to balance the niobium budget of the Earth. We agree with Saunders *et al.* (in press) that OIB sources represent this additional component which comprises recycled oceanic lithosphere (*cf* Sun & McDonough, this volume).

The chemical geodynamics of the mantle

We now wish to examine the topology of various isotope and trace-element ratio diagrams to determine the nature of mixing between the different mantle components discussed above to establish their age and location in the mantle.

The NHRL Pb–Pb isotope correlation represents one of the most impressive mixing relationships shown by ocean basalts. In terms of the Pb–Pb diagram this relationship could represent mixing between MORB and a HIMU component. However, Sr–Nd and Sr–Pb isotope diagrams (Figs 2 and 5) demonstrate that there is little direct mixing between MORB and HIMU. Equally the NHRL does not represent a mixture of MORB and primitive mantle (bulk Earth will lie on the Pb–Pb geochron and have a $^{206}Pb/^{204}Pb$ ratio comparable to that of MORB) (Figs 3 and 5). The most incompatible trace-elements are highly mobile during seawater and hydrothermal alteration processes (Ludden & Thompson 1979) so that in the present trace-element discussion we can only consider MORB glasses and recent subaerial OIB volcanism. Glasses from the FAMOUS area of the MAR have greater trace-element enrichment than typical MORB and HIMU islands such as St Helena (*eg* Ba/Nb <7) (Langmuir *et al.* 1977; Chaffey *et al.*, this volume). Notwithstanding the effect of the smaller degree of partial melting involved in the genesis of OIBs, the Azores (Ba/Nb=6–8 excluding Sao Miguel and Faial) (White *et al.* 1979) and the northern Cape Verde (Ba/Nb=5–9, Table 1) also show greater trace-element enrichment than MORB and HIMU sources. Although from an admittedly sparse data set, samples forming the NHRL define a positive correlation between trace-element ratios (*eg* Ba/Nb) and strontium and lead isotope ratios (Figs 5 and 9). The enrichment style of the most incompatible trace elements is generally compatible with generation as a result of partial-melting processes *ie* increased Ba/Nb, La/Nb and Rb/Sr and low Sm/Nd and K_2O/TiO_2 ratios (Hawkesworth *et al.* 1984). However, niobium shows the greatest relative enrichment (Fig. 6) and we conclude from this observation that the niobium-rich HIMU component is also required. The presence of the HIMU component would also account for the relatively radiogenic lead isotope ratios.

Initially, the above conclusions, *ie* (i) the isotope evidence that the HIMU component is not directly involved in the NHRL and (ii) the trace-element relationships that require a high niobium recycled ocean crust (=HIMU?) component, appear to be contradictory. Before explaining the reason for this apparent paradox we shall outline several other pertinent mixing relationships that will lead us to a chemical geodynamic model.

The basaltic rocks from Cape Verde and the Azores that display the most pronounced 'enriched' isotopic signatures are the least undersaturated in terms of their major elements. The interesting implication is that the most alkalic magmas, which presumably represent the smallest degree of melting, do not record the most extreme trace-element and isotope characteristics. We therefore have another apparent contradiction in that the most fertile sources, in terms of their trace elements, are only sampled by larger degree melts. This implies that the 'enriched' source is inherently less depleted in easily fusable material and will have a lower solidus temperature. In the case of Sao Miguel this relationship could simply be a small-scale regional heterogeneity at any level in the mantle. The 'enriched' signature only occurs along 40 km of the island (Hawkesworth *et al.* 1979). At Cape Verde, however, the case is different. With the possible exception of Sao Vicente, there is evidence of the low Sr–Nd isotope component in *all* the Cape Verde Islands, although it is generally more pronounced in the southern islands. The Cape Verde island chain is approximately stationary in terms of the hot-spot reference frame. An implication is that, should the 'enriched' mantle component be developed at shallow levels in the mantle (*ie* above the ascending plume), it may be expected to become progressively exhausted. The volcanism on Cape Verde does not record consistent temporal isotopic or chemical variations. It is therefore apparent that the 'enriched' isotope component does not reside at a shallow level in the upper mantle. We conclude that the 'enriched' component must be periodically introduced into the source region of the Cape Verde volcanic rocks by the ascending diapir. Advocating greater degrees of partial melting in the source of the 'enriched' volcanism will result in lower degrees of fractionation among incompatible trace elements and may explain the apparent paradox in the lead isotope systematics of the Azores, *ie* that the time-integrated high U/Pb ratios of their source appear less extreme than the more undersaturated rocks.

An additional observation is that none of the mantle components that we believe to be recycled (HIMU, EMI or EMII) mixes directly with isotopically depleted MORB (Figs 2–5). The 'enriched' components always show mixing relationships with a component that lies on the NHRL (Pb–Pb) and mantle array (Sr–Nd). Both the Cape Verde and Sao Miguel are good examples of this relationship. Equally no OIBS record direct mixing between the HIMU component and 'enriched' components. It is therefore apparent that the NHRL Pb–Pb isotope correlation must form prior to mixing with the 'enriched' components and that they all remain isolated from the convecting upper mantle source for MORBs. We conclude that the NHRL relationship and HIMU represent a combination of three processes: firstly the formation of oceanic lithosphere by depletion of the upper mantle; secondly the upper portion of the oceanic crust is modified by hydrothermal alteration and subduction, producing a U–Nb–rich slab which is recycled into the mantle; thirdly diffusion and convection cause some homogenization of the basaltic and peridotitic portions of subducted lithosphere while it remains effectively isolated from the convective upper and lower mantle. Subsequent partial-melt-related trace-element enrichment, derived from the basaltic crust, is superimposed on the peridotitic section of the oceanic lithosphere. With time this enrichment will generate lithosphere with more radiogenic lead and strontium isotope ratios than those of MORBs. Sections of the basaltic ocean crust that remain intact will have the HIMU characteristics.

There are several implications for the above suggestions. Firstly, the isolation of MORB from 'enriched' and HIMU components and the recycled NHRL material implies that a heterogeneous 'plum pudding' model (Zindler *et al.* 1984) for the mantle is unrealistic. Secondly, the lack of mixing between the pure HIMU end-member (recycled oceanic crust) and the 'enriched' components strongly implies that neither EMI and EMII can represent directly subducted sediments trapped within the oceanic crust. This conclusion strengthens our proposal that both EMI and EMII are portions of the subcontinental lithosphere which record different styles of trace-element enrichment. Thirdly, Galer & O'Nions (1985) pointed out that the Th/U ratio in the source of MORB is too low (less than 2.5) to maintain the relatively high $^{208}Pb^{*}/^{206}Pb^{*}$ of MORB. HIMU basalts also record low $^{208}Pb^{*}/^{206}Pb^{*}$ and Th/U ratios (*eg* St Helena, 0.92 and 3 respectively) (see Chaffey *et al.*, this volume) so that recycling of oceanic lithosphere does not balance the Th/U ratio of the Earth. In order to explain this relationship Galer & O'Nions advocate large-scale transfer of undifferentiated mantle (a high Th/U ratio of 3.9) from the lower to the upper mantle. However, we argue above that there was little evidence of primitive undifferentiated mantle in the source of present-day basalts. The subconinental lithosphere represents an additional high Th/U reservoir that we believe produces the EMI and EMII isotope signatures. At present it is not possible to constrain the amount of subcontinental lithosphere in the upper mantle but a significant proportion of OIBs have 'Dupal' signature. The problem remains that on Pb−Pb isotope diagrams MORBs show no mixing relationships with bulk Earth or the 'enriched' components. We suggest that the high Th/U component required to balance the Th/U ratio of the upper mantle represents a mixture of all the recycled components but is only incorporated into the convecting upper mantle reservoir following thorough mixing of all the recycled materials, *ie* following diapiric upwelling and melt extraction.

The combined Pb−Sr−Nd isotope systematics allow certain constraints to be placed upon the ages of the reservoirs and the relative order of their formation/incorporation in the mantle. We pointed out above that the mixing relationships are not random. For example 'Dupal' signatures are not directly mixed with the depleted MORB reservoir but always with a source with greater lead and strontium isotope ratios (Figs 3 and 5). One implication is that the 'Dupal' and MORB reservoirs are effectively isolated from each other whereas the components making up the NHRL correlation are intimately mixed. The NHRL lead isotope correlation equates to an age of 1.7 Ga (Chase 1981), implying that the MORB and HIMU components have been separated for this period of time. The timing of the melt-related trace-element enrichment in certain of the NHRL ocean islands can also be constrained. The Azores and northern Cape Verdes have an $^{87}Sr/^{86}Sr$ ratio close to 0.7035, in contrast with the value of 0.7025 in the MORB and HIMU components. Melt extraction increases the Rb/Sr ratio of the basaltic volcanism by approximately 25% (assuming less than 5% batch melting) which suggests that of the order of 700 Ma are required to increase the $^{87}Sr/^{86}Sr$ ratios from an original MORB source. Assessing the time of recycling of the 'enriched' components is more difficult. They would inherit 'ancient' Pb−Sr−Nd isotope signatures from the continental crust if they represent either subcontinental lithosphere formed above a subduction zone or recycled sediments. Slopes on Pb−Pb isotope diagrams will therefore have no direct age significance. In certain cases, Sao Miguel for example, mixing with a radiogenic portion of the NHRL implies that the component has probably been added to the mantle after the melt-related enrichment had increased the Rb/Sr and Ba/Nb ratios, *ie* relatively recently, less than 700 Ma.

The isotopic signatures in oceanic volcanism are not as extreme as in Archaean continental crust or subcontinental lithosphere, suggesting that the 'enriched' signatures have not remained isolated in the mantle since the Archaean. We therefore conclude that the recycling of large masses of the subcontinental lithosphere (bodies that can retain their chemical signature on a 10−100 km length scale during convection) can only be a recent phenomenon or that evidence for older recycling has been totally destroyed by mantle convection. In addition, the absence of Archaean lithosphere in the sources of OIB indicates that such material has not been recently delaminated or that it has remained isolated from OIB sources. Both these factors may be significant. Firstly, Archaean continental cratons are generally surrounded by younger mobile belts so that they would not be involved in recent continent−continent collision and hence the ancient subcontinental lithosphere would be likely to remain intact. Secondly, if delamination of Archaen material occurred recently, it would be relatively cool and dense and unlikely to act as a magma source prior to achieving thermal equilibrium (McKenzie & O'Nions 1983).

From the above discussion we are able to formulate a model for the likely structure of the Earth's mantle (Fig. 11). Firstly, away from the influences of ocean island hot-spots, the isotope composition of MORB tends to a relatively constant composition. In the proximity of hot-spots MORB develops the chemical and isotopic signatures of the hot-spot volcanism (*eg* S Atlantic)(Hanan *et al.* 1987). It is therefore apparent that the MORB and hot-spot reservoirs are generally isolated from each other. In contrast the NHRL, HIMU and 'Dupal' signatures may all occur in a single hot-spot, the

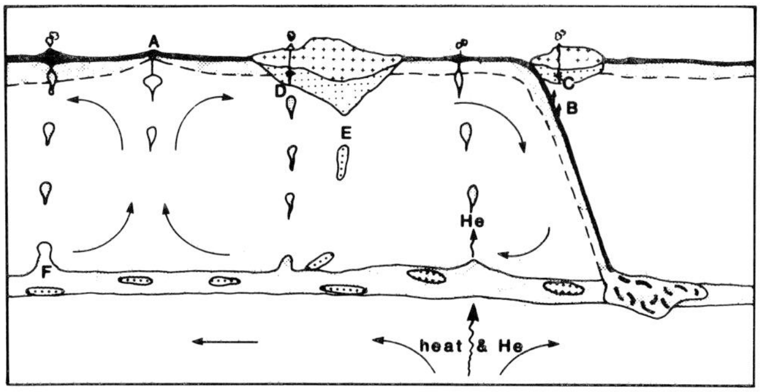

FIG. 11. Schematic diagram depicting the major influences in the generation of ocean island volcanism: (*a*) generation of oceanic crust followed by hydrothermal alteration and enrichment in uranium and rubidium and increased $^{87}Sr/^{86}Sr$; (*b*) dehydration and melting of the subducted oceanic crust, stabilization of rutile eclogite followed by storage of this cold dense material at the 650 km discontinuity; (*c*) modification/formation of the subcontinental lithosphere by subduction-related fluids and magmas; (*d*) modification of subcontinental lithosphere by magmas with an OIB source; (*e*) delamination of subcontinental lithosphere, following crustal thickening, and storage of this cold dense material at the 650 km discontinuity; (*f*) of the order of 1 Ga is required for the stored subcontinental lithosphere and oceanic crust to reach thermal equilibrium and become buoyant. The diapir produces OIB magmas prior to incorporation into the convecting upper mantle. The 650 km discontinuity acts as a thermochemical boundary layer and permits transport from deeper layers, *eg* helium and heat.

Azores for example. This demonstrates that their source region is extremely heterogeneous. To explain the above observations we advocate a layered mantle model (Dupre & Allegre, 1980; Hofmann & White 1982; Ringwood, 1982) in which the convecting upper and lower mantle have limited interaction rather than the heterogeneous 'plum pudding' model of Zindler *et al.* (1984). The cold dense nature of recycled oceanic and subcontinental lithosphere results in their sinking to the 670 km discontinuity between the upper and lower mantle where they form a physical–chemical boundary layer (Ringwood 1982; Anderson 1986; Ringwood & Irifune 1988). This conclusion appears at odds with the large geoid and seismic velocity anomalies associated with subducting slabs which appear to require slab penetration into the lower mantle and hence whole mantle convection (Creager & Jordan 1984; Hager 1984). However, Anderson (1987) and Ringwood & Irifune (1988) have demonstrated that thermally induced phase changes in the subducted cold oceanic lithosphere will greatly increase both seismic velocities and density compared with the 'normal' mantle. Under these conditions whole mantle convection is not required to fit the seismic and geoid data.

The convecting upper mantle is generally well mixed on a local scale (apart from the effects of recent volcanism) but does appear to retain some regional chemical heterogeneities, *eg* between Atlantic and Indian MORBs (Dupre & Allegre 1983). Despite evidence for relatively primitive helium isotope signatures from some ocean islands (Kurz *et al.* 1983), we can find no unequivocal combined Sr–Nd–Pb isotope and trace-element evidence for an undifferentiated primitive mantle component in the source region of OIBs. We conclude that recycled components must act as an effective barrier to mass transport of any possible primitive material that may exist in the lower mantle but does allow limited thermal and gas diffusion. The radiogenic lead isotope ratios and high time-integrated Th/U ratio of the MORB source is maintained by incorporation of mantle plumes that originate from the physical– chemical boundary layer.

ACKNOWLEDGMENTS: We thank P. Guise and R. Green for analytical assistance and J. Davies for producing the diagrams. This paper has benefited from discussions with many colleagues, in particular M. Wilson, N. Rogers, C. Hawkesworth and Z. Palacz. GRD gratefully acknowledges Natural Environment Research Council funding.

References

ALLEGRE, C. J. 1987. Isotope geodynamics. *Earth and Planetary Science Letters* **86**, 175–205.

ANDERSON, D. L. 1986. The transition region of the Earth's upper mantle. *Nature* **320**, 321–328.

—— 1987. Thermally induced phase changes, lateral heterogeneity of the mantle, continental roots, and deep slab anomalies. *Journal of Geophysical Research* **92**, 13 968–13 980.

ANDERSON, R. N., DELONG, S. E. & SCHWARZ. W. H. 1978. Thermal model for subduction with dehydration in the downgoing slab. *Journal of Geology* **86**, 731–739.

ARMSTRONG, R. L. 1968. A model for Sr and Pb isotope evolution in a dynamic Earth. *Reviews of Geophysics* **6**, 175–199.

BETTON, P. J., ARMSTRONG, R. A. & MORTON, W. I. 1984. Variation in the lead isotope composition of Karoo Magmas. *Special Publication of the Geological Society of South Africa* **13**, 331–340.

CARLSON, R. W. 1984. Isotopic constraints on Columbia River flood basalt genesis and the nature of the subcontinental mantle. *Geochimica et Cosmochimica Acta* **48**, 2357–2372.

CHASE, C. G. 1981. Ocean Island Pb: two stage histories and mantle evolution. *Earth and Planetary Science Letters* **52**, 277–284.

CLAGUE, D. A. & FREY, F. A. 1982. Petrology and trace element chemistry of the Honolulu Volcanics, Oahu: implications for the oceanic mantle below Hawaii, *Journal of Petrology* **23**, 447–504.

COHEN, R. A. & O'NIONS, R. K. 1982*a*. The lead, neodymium and strontium isotopic structure of ocean ridge basalts. *Journal of Petrology* **23**, 299–324.

—— & —— 1982*b*. Identification of recycled continental material in the mantle from Sr, Nd and Pb isotope investigations. *Earth and Planetary Science Letters* **61**, 73–84.

——, EVENSEN, N. M., HAMILTON, P. J. & O'NIONS, R. K. 1980. U–Pb, Sm–Nd and Rb–Sr systematics of mid-ocean ridge basalt glasses. *Nature* **283**, 149–153.

COX, K. G. & HAWKESWORTH, C. J. 1984. Relative contribution of crust and mantle to flood basalt magmatism, Mahabaleshwar area, Deccan Traps. *Philosophical Transactions of the Royal Society of London, Series A* **310**, 627–641.

—— & —— 1985. Geochemical stratigraphy of the Deccan Traps at Mahabaleshwar, Western Ghats, India, with implications for open system magmatic processes. *Journal of Petrology* **26**, 355–377.

CREAGER, K. C. & JORDAN, T. H. 1984. Slab penetration in the lower mantle. *Journal of Geophysical Research* **89**, 3031–3049.

DAVIES, G. R. & LLOYD, F. E. Pb–Sr–Nd isotope and trace element data bearing on the origin of the potassic subcontinental lithosphere beneath south west Uganda. *Journal of the Geological Society of Australia*, in press.

DOE, B. R., LEEMAN, M. P., CHRISTIANSEN, R. L. & HEDGE, C. E. 1982. Lead and strontium isotopes and related trace elements as genetic tracers in the Upper Cenozoic Rhyolite–Basalt association of the Yellowstone Plateau Volcanic Field. *Journal of Geophysical Research,* **87B**, 4785–4806.

DUNCAN, R. A. 1980. Hotspots in the Southern Ocean — an absolute frame of reference for motion of the Gondwana continents. *Tectonophysics* **74**, 29–42.

DUPRE, B. & ALLEGRE, C. J. 1980. Pb–Sr–Nd isotopic correlation and the geochemistry of North Atlantic mantle. *Nature* **286**, 17–22.

—— & —— 1983. Pb–Sr–Nd isotope variation in Indian Ocean basalts and mixing phenomena. *Nature* **303**, 142–146.

——, GOPEL, C. & BOUGAULT, H. 1985. Lead isotope variations in old ocean crust near the Azores. *Initial Reports of the Deep Sea Drilling Project* **82**, 497–500.

——, LAMBERT, B. & ALLEGRE, C. J. 1982. Isotopic variations within a single ocean island: the Terceira case. *Nature* **299**, 620–622.

ERLANK, A. J., WATERS, F. G., HAWKESWORTH, C. J., HAGGERTY, S. E., ALLSOPP, H. L., RICKARD, R. S. & MENZIES, M. A. 1987. Evidence for mantle metasomatism in peridotite nodules from the kimberley pipes, South Africa. *In*: MENZIES, M. A. & HAWKESWORTH, C. J. (eds) *Mantle Metasomatism*. Academic Press, New York, 221–312.

FRASER, K. J., HAWKESWORTH, C. J., ERLANK, A. J., MITCHELL, R. H. & SCOTT–SMITH, B. H. 1986. Sr, Nd and Pb isotope and minor element geochemistry of lamproites and kimberlites. *Earth and Planetary Science Letters* **76**, 57–70.

GALER, S. J. G. & O'NIONS, R. K. 1985. Residence time of thorium, uranium and lead in the mantle with implications for mantle convection. *Nature* **316**, 778–782,

GERLACH, D. C., CLIFF, R. A., DAVIES, G. R., NORRY, M. J. & HODGSON, N. Magma sources of the Cape Verdes Archipelagos: isotopic and trace element constraints. *Geochimica et Cosmochimica Acta*, in press.

GORMET, P. L., DYMEK, R. F., HASKIN, L. A. & KOROTER, R. L. 1984. The "North American shale composite": its compilation, major and trace element characteristics. *Geochemica et Geochimica Acta* **48**, 2469–2482.

GRAND, S. P. 1987. Tomographic inversion for shear velocity beneath the North American plate. *Journal of Geophysical Research* **92**, 14 065–14 090.

GUNN, B. M. & WATKINS, N. D. 1976. Geochemistry of the Cape Verde Islands and Fernando de Noronha. *Bulletin of the Geological Society of America* **87**, 1089–1100.

HAGER, B. H. 1984. Subducted slabs and the geoid. *Journal of Geophysical Research* **89**, 6003–6015.

HANAN, B. B., KINGSLEY, R. H. & SCHILLING, J.-G. 1987. Pb isotope evidence in S. Atlantic for migrating ridge hotspot interactions. *Nature* **322**, 137–144.

HART, S. R. 1984. A large-scale isotope anomaly in the Southern Hemisphere mantle. *Nature* **309**, 753–57.

HAWKESWORTH, C. J., ERLANK, A. S., MARSH, J. S., MENZIES, M. A. & VAN CALSTEREN, P. 1983. Evolution of the continental lithosphere: evidence from volcanics and xenoliths in Southern Africa. *In*: HAWKESWORTH, C. J. & NORRY, M. J. (eds) *Continental Basalts and Mantle Xenoliths*. Shiva, 111–138.

——, MARSH, J. S., DUNCAN, A. R., ERLANK, A. S. & NORRY, M. J. 1984*a*. The role of continental lithosphere in the generation of the Karoo volcanic rocks: evidence from combined Nd and Sr isotope studies. *Special Publication of the Geological Society of South Africa* **13**, 341–354.

——, NORRY M. J., RODDICK, J. C. & VOLLMER, R. 1979. $^{143}Nd/^{144}Nd$ and $^{87}Sr/^{86}Sr$ ratios from the Azores and their significance in LIL enriched mantle. *Nature* **280**, 28–31.

——, ROGERS, N. W., VAN CALSTEREN, P. & MENZIES, M. A. 1984*b*. Mantle enrichment processes. *Nature* **311**, 331–335.

HAYES, D. E. & RABINOWITZ, P. D. 1975. Mesozoic magnetic lineations and the magnetic quiet zone off northwest Africa. *Earth and Planetary Science Letters* **28**, 105–115.

HELLMAN, P. L. & GREEN, T. H. 1979. The role of sphene as an accessory phase in the high-pressure partial melting of hydrous mafic compositions. *Earth and Planetary Science Letters* **42**, 191–201.

HODGSON, N. 1986. *Carbonatites and Associated Rocks from the Cape Verde Islands*. Unpublished thesis, University of Leicester.

HOFMANN, A. W. 1986. Nb in Hawaiian magmas: constraints on source composition and evolution. *Chemical Geology* **57**, 17–30.

—— & WHITE, W. M. 1982. Mantle plumes from ancient oceanic crust. *Earth and Planetary Science Letters* **57**, 421–436.

——, JOCHUM, K. P., SENFERT, M. & WHITE, W. M. 1986. Nb and Pb in oceanic basalts: new constraints on mantle evolution. *Earth and Planetary Science Letters* **79**, 33–45.

HOLE, M. J., SAUNDERS, A. D., MARRINER, G. F. & TARNEY, J. 1984. Subduction of pelagic sediments: implications for the origin of Ce-anomalous basalts from the Mariana Islands. *Journal of the Geological Society, London* **141**, 435–472.

HOUSEMAN, G. A., MCKENZIE, D. P., & MOLNAR, P. 1981. Convective instability of a thickened boundary layer and its relevance for the thermal evolution of continental convergent belts. *Journal of Geophysical Research* **86**, 6115–6132.

JENNER, G. A., HERTOGEN, J., SACHTLEBEN, TH. & SCHMINKE, H.-U. 1985. Isotopic and trace element composition of basalts from sites 556–559 and 561–564: constraints on some processes affecting their composition. *Initial Reports of the Deep Sea Drilling Project* **82**, 501–508.

KAY, R. W. 1984. Elemental abundances related to identification of magma sources. *Philosophical Transactions of the Royal Society of London, Series A* **310**, 535–547.

KURZ, M. D., JENKINS, W. J., HART, S. R. & CLAGUE, D. 1983. Helium isotopic variations in volcanic rocks from Loihi Seamount and the Island of Hawaii. *Earth and Planetary Science Letters* **66**, 388–406.

LANGMUIR, C. H., BENDER, J. F., BENCE, A. E., HANSON & TAYLOR, S. R. 1977. Petrogenesis of basalts from the Famous Area: Mid-Atlantic Ridge. *Earth and Planetary Science Letters* **36**, 133–156.

LE ROEX, A. P. 1985. Geochemistry, mineralogy and magmatic evolution of the basaltic and trachytic lavas from Gough Island, South Atlantic. *Journal of Petrology* **26**, 149–186.

LUDDEN, J. N. & THOMPSON, G. 1979. An evaluation of the behavior of rare earth elements during the weathering of sea-floor basalt. *Earth and Planetary Science Letters* **43**, 85–92.

MAHONEY, J. S., MACDOUGALL, J. D., LUGMAIR, G. W., GOPALAN, K. & KRISHNAMURTHY, P. 1985. Origin of contemporaneous tholeiitic and K-rich alkalic lavas: a case study from the northern Deccan Plateau, India. *Earth and Planetary Science Letters* **72**, 39–53.

——, ——, ——, SANELCAR DAS, M., MURALI, A. V. & GOPALAN, K. 1982. Origin of the Deccan Trap flows at Mahabashwar inferred from Nd and Sr isotopic and chemical evidence. *Earth and Planetary Science Letters* **60**, 47–60.

MARTIN, J. M. & MEYBECK, M. 1979. Element mass-balance of material carried by major world rivers. *Marine Chemistry* **7**, 173–206.

MCKENZIE, D. 1984. The generation and compaction of partially molten rock. *Journal of Petrology* **25**, 713–765.

—— 1985*a*. The extraction of magma from the crust and mantle. *Earth and Planetary Science Letters* **74**, 81–91.

—— 1985*b*. $^{230}Th/^{238}U$ disequilibrium and the melting processes beneath ridge axes. *Earth and Planetary Science Letters* **72**, 149–157.

—— & O'NIONS, R. K. 1983. Mantle reservoirs and ocean island basalts. *Nature* **301**, 229–231.

MENZIES, M. A. & WASS, S. 1983. CO_2 rich mantle below eastern Australia: REE, Sr and Nd isotopic study of Cenozoic alkaline magmas and apatite rich xenoliths, Southern Highlands Province, New South Wales, Australia. *Earth and Planetary Science Letters 65*, 287–302.

——, HALIDAY, A. N., PALACZ, Z., HUNTER, R. H., UPTON, B. G. J., ASPEN, P. & HAWKESWORTH, C. J. 1987. Evidence from mantle xenoliths for an enriched lithosphere keel under the Outer Hebrides. *Nature* **325**, 44–46.

——, KEMPTON, P. & DUNGAN, M. 1985. Interaction of continental lithosphere and asthenospheric melts below the Geronimo volcanic field, Arizona, U. S. A. *Journal of Petrology* **26**, 663–693.

——, LEEMAN, W. P. & HAWKESWORTH, C. J. 1984. Geochemical and isotopic evidence for the origin of continental flood basalts with particular reference to the Snake River Plain, Idaho, U. S. A. *Philosophical Transactions of the Royal Society of London, Series A,* **310**, 643–660.

MICHARD, A., ALBAREDE, F., MICHARD, G., MINSTER, J. F. & CHARLOU J. L. 1983. Rare earth elements and uranium in high temperature solutions for East Pacific Rise vent field (13°N). *Nature* **303**, 795–797.

NELSON, D. R., MCCULLOCH, M. T. & SUN, S.-S. 1986. The origin of ultra-potassic rocks as inferred from Sr, Nd and Pb isotopes. *Geochimica et Cosmochimica Acta* **50**, 231–245.

NEWMAN, S., MACDOUGALL, J. D. & FINKEL, R. C. 1984. ^{230}Th–^{238}U disequilibrium in island arcs: evidence from the Aleutians and the Marianas. *Nature* **308**, 268–270.

NEWSOM, H. E., WHITE, W. M., JOCHUM, K. P. & HOFMANN, A. W. 1986. Siderophile and chalcophile element abundances in oceanic basalts, Pb isotope evolution and growth of the Earth's core. *Earth and Planetary Science Letters* **80**, 299–314.

NIXON, P. H. & DAVIES, G. R. 1987. Mantle xenolith perspectives. *In*: NIXON, P. H.(ed.) *Mantle Xenoliths*. Wiley, New York, 741–756.

——, ——, SLODKEVICH, V. V. & BERGMAN, S. G. 1986. Graphite pseudomorphs after diamond in eclogites from Beni Bousera, N. Morocco. *Extended Abstracts, Proceedings of the 4th International Kimberley Conference, Perth, W Australia*, 412–414. Geological Society of Australia.

——, ROGERS, N. W., GIBSON, I. L. & GREY, A. 1981. Depleted and fertile mantle xenoliths from southern African kimberlites. *Annual Review of Earth and Planetary Sciences* **9**, 285–309.

——, DAVIES, G. R., BETTON, P. J. & JONES, P. 1987. Ethiopian flood basalts; evidence for the origin of the sub-continental lithosphere. *Abstr. Terra Cognita* **6**, 622.

O'NIONS, R. K., CARTER, S. R., COHEN, R. S., EVERSEN, N. M. & HAMILTON, P. J. 1978. Pb, Nd and Sr isotopes in oceanic ferromanganese deposits and ocean floor basalts. *Nature* **273**, 435–438.

PALACZ, Z. A. & SAUNDERS, A. D. 1986. Coupled trace element and isotope enrichment in the Cook–Austral islands, S. W. Pacific. *Earth and Planetary Science Letters* **79**, 270–280.

PATCHETT, P. J. & TATSUMOTO, M. 1980. Hafnium isotope variations in oceanic basalts. *Geophysical Research Letters* **7**, 1077–1080.

——, WHITE, W. M., FELDMANN, H., KIELINCZUK, S. & HOFMANN, A. W. 1984. Hafnium rare earth element fractionation in the sedimentary system and crustal recycling into the Earth's mantle. *Earth and Planetary Science Letters* **69**, 365–378.

PEARCE, J. A. 1982. Trace element characteristics of lavas from destructive plate boundaries. *In*: THORPE, R. S. (ed.) *Andesites* Wiley, New York, 525–548.

PEARSON, G. P., NIXON, P. H., DAVIES, G. R. & MILLEDGE, J. J. Graphitised diamonds from Beni Bousera; implications for oceanic lithosphere recycling and the origin of anomalous diamond occurrances. *Nature*, in press.

PIEPGRASS, D. J. & WASSERBURG, G. J. 1980. Neodymium isotopic variations in seawater. *Earth and Planetary Science Letters* **50**, 128–138.

POREDA, R. 1985. Helium–3 and deuterium in back-arc basalts: Lau basin and the Mariana Trough. *Earth and Planetary Science Letters* **73**, 244–254.

RICHARDSON, S. H., ERLANK, A. J., DUNCAN, A. R. & REID, D. L. 1982. Correlated Nd, Sr and Pb isotope variations in Walvis Ridge basalts and implications for the evolution of their mantle source. *Earth and Planetary Science Letters* **34**, 327–342.

——, GURNEY, J. J., ERLANK, A. J. & HARRIS, J. W. 1984. Origin of diamonds in old enriched mantle. *Nature* **310**, 198–202.

RIDEOUT, M. L. & SCHILLING, J.-G. 1985. Rare-earth elements ^{87}Sr/^{86}Sr and ^{143}Nd/^{144}Nd mantle source variations. *Initial Reports of the Deep Sea Drilling Project* **82**, 483–496.

RIDLEY, W. I., WATKINS, N. D. & MACFARLENE, D. J. 1974. The oceanic islands: Azores. *In*: NAIRIN, A. E. M. & STEHLI, F. G. (eds) *The Ocean Basins and Margins*, Vol. 2 Plenum, New York, 443–483.

RINGWOOD, A. E. 1982. Phase transformations and differentiation in subducted lithosphere; implications for mantle dynamics, basalt petrogenesis, and crustal evolution. *Journal of Geology* **90**, 611–643

—— & IRIFUNE, T. 1988. Nature of the 650-km seismic discontinuity: implications for mantle dynamics and differentiation. *Nature* **331**, 131–136.

RYERSON, F. J. & WATSON, E. B. 1987. Rutile saturation in magmas: implications for Ti–Nb–Ta depletion in island-arc basalts. *Earth and Planetary Science Letters* **86**, 225–239.

SAUNDERS, A. D., NORRY, M. J. & TARNEY, J. in press. *In*: MENZIES, M. & COX, K. G. (eds) *Continental and Oceanic Lithosphere. Journal of Petrology* (special volume).

SCHÄRER, U., KROGH, T. E. & GROWER, C. F. 1986. Age and evolution of the Grenville Province in eastern Labrador from U–Pb systematics in accessory minerals. *Contributions to Mineralogy and Petrology* **94** 438–451.

SCHILLING, J.–G. 1975. Azores mantle blob: rare earth evidence. *Earth and Planetary Science Letters* **25**, 103–115.

—— 1985. Upper mantle heterogeneities and dynamics. *Nature* **314**, 62.

SEKINE, T. & WYLLIE, P. J. 1982. Phase relationships in the system $KAlSiO_4$–Mg_2SiO_4–SiO_2–H_2O as a model for hybridization between siliceous melts and peridotite. *Contributions to Mineralogy and Petrology* **79**, 368–374.

SELF, S. & GUNN, B. M. 1976. Petrology, volume and age relations of alkaline and saturated peralkaline volcanics from Terceira, Azores. *Contributions to Mineralogy and Petrology* **54**, 293–314.

SHIREY, S. B., BENDER, J. F. & LANGMUIR, C. H. 1987. Three-component isotopic heterogeneity near the Oceanographer transform, Mid-Atlantic Ridge. *Nature* **325**, 217–223.

SILVA, L. C., LEBAS, M. J. & ROBERSTON, A. H. F.

1981. An oceanic carbonatitic volcano on Santiago, Cape Verde Islands. *Nature* **294**, 644–645.

SMITH, C. B. 1983. Two types of sources for southern African kimberlites. *Nature* **304**, 51–54.

——, GURNEY, J. J., SKINNER, E, M. W., CLEMENT, C. R. & EBRAHIM, N. 1985. Geochemical character of Southern African kimberlites: a new approach based on isotopic constraints. *Transactions of the Geological Society of South Africa* **88**, 267–280.

STILLE, P., UNRUH, D. M. & TATSUMOTO, M. 1983. Pb, Sr, Nd and Hf isotopic evidence of multiple sources for Oahu, Hawaii basalts. *Nature* **304**, 25–29.

STILLMAN, C. J., FURNES, H., LE BAS, M., ROBERTSON, A. H. F. & ZIELONKA, J. 1982. The geological history of Maio, Cape Verde Islands. *Journal of the Geological Society, London,* **134**, 347–361.

STOLZ, A. J. & DAVIES, G. R. Chemical and Sr, Nd and Pb isotopic evidence for the role of fluids with contrasting compositions in the metasomatism of spinel lherzolite xenoliths from south-eastern Australia. *In*: MENZIES, M. & COX, K. G. (eds) *Continental and Oceanic Lithosphere. Journal of Petrology*, Special volume, in press.

SUN, S.-S. 1980. Lead isotopic study of young volcanic rocks from mid-ocean ridges, ocean islands and island arcs. *Philosophical Transactions of the Royal Society of London, Series A* **297**, 409–445.

TAYLOR, S. R. & MCLENNAN, S. M. 1981. The composition of the continental crust: rare earth element evidence from sedimentary rocks. *Philosophical Transactions of the Royal Society of London, Series A* **301**, 381–993.

—— & —— 1985. *The Continental Crust: its Composition and Evolution*. Blackwell Scientific Publications, Oxford.

TERA, F., BROWN, L., MORRIS, J., SACKS, I. S., KLEIN, J. & MIDDLETON, R. 1986. ^{10}Be evidence for sediment incorporation island arc lavas, I: Global coverage. *Terra Cognita* **6**, 199.

THOMPSON, R. N., DICKIN, A. P., GIBSON, I. L. & MORRISON, M. A. 1982. Elemental fingerprints of isotopic contamination of Hebridean Palaeocene mantle-derived magmas by Archaean sial. *Contributions to Mineralogy and Petrology* **79**, 159–168.

VIDAL, PH., CHAUVEL, C. & BROUSSE, R. 1984. Large mantle heterogeneity beneath French Polynesia. *Nature* **307**, 536–538.

——, WOOD, D. A., TARNEY, J. & JORON, J.-L. 1987. Geochemistry of ocean island basalts from the South Atlantic: Ascension, Bouvet, St. Helena, Gough and Tristan da Cunha. *In*: FITTON, F. G. & UPTON, B. G. J. (eds) *Alkaline Igneous Rocks*. Geological Society of London Special Publication **30**, 253–267.

WHITE, W. M. & PATCHETT, J. 1984. Hf–Nd–Sr isotopes and element abundances in island arcs: implications for magma origins and crust mantle evolution. *Earth and Planetary Science Letters* **67**, 167–185.

——, DUPRE, B. & VIDAL, P. 1985. Isotope and trace element geochemistry of sediments from the Barbados Ridge — Demerara Plain region, Atlantic Ocean. *Geochimica et Cosmochimica Acta* **49**, 1875–1886.

——, TAPIA, M. D. M. & SCHILLING, J.-G. 1979. The petrology and geochemistry of the Azores Islands. *Contributions to Mineralogy and Petrology* **69**, 201–213.

WYLLIE, P. J. 1984. Constraints imposed by experimental petrology on possible and impossible magma sources and products. *Philosophical Transactions of the Royal Society of London, Series A* **310**, 439–456.

YAMAGUCHI, M., ARMSTRONG, R. L., RUSSELL, R. D. & SLAUSON, W. F. 1977. Strontium and lead isotopic investigation of igneous rocks from the Mid-Atlantic Ridge. *Initial Reports of the Deep Sea Drilling Project* **37**, 612–616.

ZINDLER, A. & HART, S. R. 1986. Chemical geodynamics. *Annual Review of Earth and Planetary Sciences* **14**, 494–571.

——, STAUDIGEL, H. & BATIZA, R. 1984. Isotope and trace element geochemistry of young Pacific seamounts: implications for the scale of upper mantle heterogeneity. *Earth and Planetary Science Letters* **70**, 175–195.

G. R. DAVIES & R. A. CLIFF, Department of Earth Sciences, University of Leeds, Leeds LS2 9JT, UK.

M. J. NORRY, Department of Geology, University of Leicester, Leicester LE1 7RH, UK.

D. C. GERLACH, Department of Earth Sciences, University of Leeds, Leeds LS2 9JT, UK. *Present address*: Lawrence Livermore Laboratory, Livermore, CA 94550, USA.

Characterization of the St Helena magma source

D. J. Chaffey, R. A. Cliff & B. M. Wilson

SUMMARY: Combined Sr–Nd–Pb isotope and trace-element data for St Helena are interpreted in terms of changing thermal and chemical fluxes impinging on and interacting with the base of the lithosphere over a period of 6 Ma. The data reveal the existence of two geochemically distinct components in the St Helena source region: (i) a HIMU (high U/Pb) component which has extremely radiogenic lead isotopes ($^{206}Pb/^{204}Pb > 20.8$) with $^{87}Sr/^{86}Sr$ and $^{143}Nd/^{144}Nd$ ratios displaced below the mantle array; (ii) a component with less radiogenic lead and strontium isotopic compositions and more radiogenic neodymium compositions. Coupled trace-element and isotope variations are evident during the activity of each volcano. During shield development an increase in incompatible-trace-element enrichment occurs. This is coupled to a decrease in $^{143}Nd/^{144}Nd$, whilst strontium and lead isotope ratios become progressively more radiogenic with time. The time-dependent variations are thought to be consistent with high-level processes occurring at the base of, or within, the lithosphere. A decrease in the signature of the depleted component with time is shown to be the result of a decreasing thermal flux acting on the base of the lithosphere. As a consequence the degree of partial melting of the depleted component (which may reside in the lithosphere or asthenosphere) is reduced, increasing the signature of the HIMU component in the erupted magmas.

St Helena is located at a latitude of 16°S and lies 800 km to the E of the Mid-Atlantic Ridge (Fig. 1). Bathymetric data (Emery & Uchupi 1984) suggest that it forms (together with Bonaparte Seamount) part of a linear volcanic chain related to movement of oceanic lithosphere above a mantle hot-spot (Morgan 1971, 1983). The 4000 m high volcanic pile lies on oceanic crust of magnetic anomaly 10 indicating an age of about 30 Ma (LaBrecque *et al.* 1971). K–Ar age determinations (Baker *et al.* 1967) indicate that subaerial activity occurred between 14 and 7 Ma, although preliminary results on the new sample set indicate that the range of activity may be more restricted (9–7 Ma). This paper presents comprehensive major-element, trace-element and strontium, neodymium and lead isotopic data from a new sample set of 250 rocks spanning the entire subaerial record of activity.

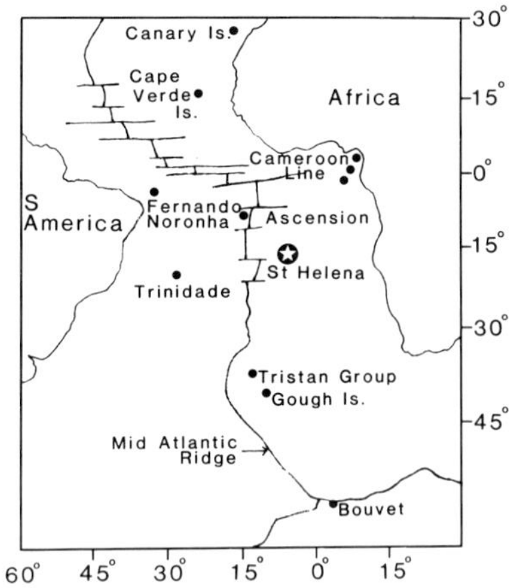

FIG. 1. Location of St Helena.

Previous research

Detailed studies of the volcanic structure and petrology of St Helena have been conducted previously by Melliss (1875), Darwin (1876), Daly (1927) and Baker (1968). Such studies established the mildly alkalic *nepheline*-normative nature of primitive St Helena volcanics and showed the existence of well-defined fractional crystallization trends between parental alkali basalt and hawaiite–mugearite–trachyte–phonolite differentiates. Attention has been focused on St Helena more recently with the discovery that the volcanic rocks possess extreme isotopic characteristics for an ocean island basalt (OIB) source (Sun 1980; Cohen & O'Nions 1982; Hofmann & White 1982). The St Helena magma source has a dominant component, which approaches one of the five end-members which describe the isotopic composition of the majority of oceanic magmas (White 1985; Zindler & Hart 1986). This end-member is variously known as the St Helena component (White 1985) or the HIMU component (Zindler & Hart 1986). The latter term is derived from the characteristic high U/Pb ratios ($^{238}U/^{204}Pb = \mu$) which must be sustained in a source region

From SAUNDERS, A. D. & NORRY, M. J. (eds), 1989, *Magmatism in the Ocean Basins*, Geological Society Special Publication No. 42, pp. 257–276.

discrete from mantle volumes with a lower μ for a sufficiently long period to produce the present-day extreme radiogenic lead compositions. The means by which these isotopic characteristics are produced is a fundamental problem in chemical geodynamics.

St Helena is also important in a regional context because its spatial relation to the other S Atlantic islands places constraints on the scale length of mantle heterogeneity (Hart 1984). Regional trace-element studies (Weaver *et al.* 1987) of Atlantic OIB suites have considered St Helena as a reference for source regions unaffected by sediment addition. Whilst the documentation of mantle heterogeneity from regional studies has yielded valuable results it is now recognized, with the acquisition of data relating to individual volcanoes, that precise constraints may only be placed on mantle heterogeneity with combined regional and detailed studies. This approach has been applied with most success in the Hawaiian chain (*eg* Stille *et al.* 1986). In detailed studies, factors that may modify the magma composition subsequent to its segregation in the source region may be discerned more readily and their effects removed. Such factors include alteration, crystal fractionation and lithospheric interaction. In addition, studies of individual volcanoes provide much better constraints on regionally integrated models. For example, the study of Haleakala by Chen & Frey (1985) has been of great value in the development of models of Hawaiian magmatism.

Stratigraphic relations and sample collection

Two principal effusive centres are present on St Helena (Fig. 2): an older shield volcano in the NE and a younger more complex shield in the SW. The two centres may have been simultaneously active for a limited period, but age determinations and structural relations are inconclusive. The main stratigraphic groups identified by Baker (1968) together with their age relations are shown in Fig. 3. The NE shield unconformably overlies an undated dyke complex (NE (BB)) which was emplaced into breccias of uncertain origin (NEBB). Following initiation of NE shield volcanism, 750 m of lavas were erupted without major unconformity

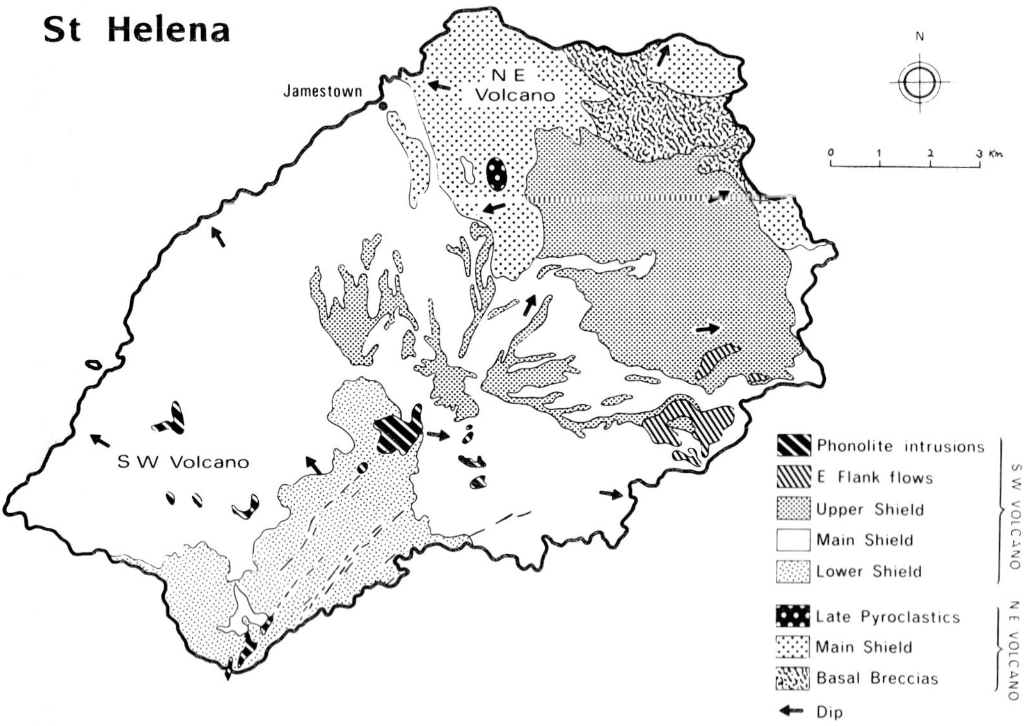

FIG. 2. Geological map of St Helena (after Baker 1968).

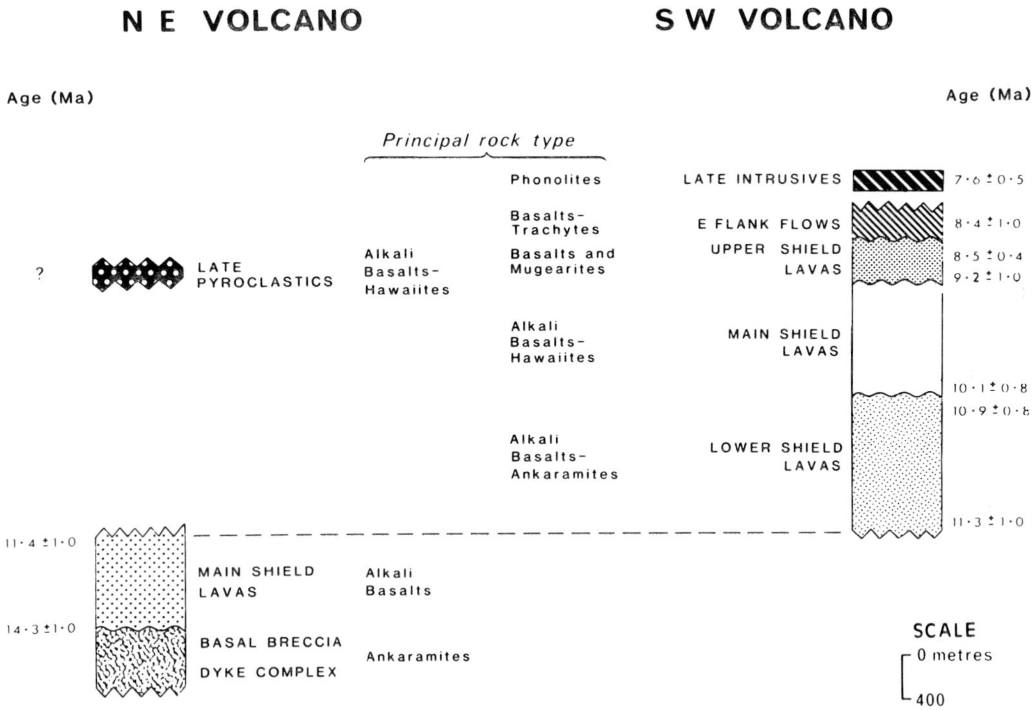

FIG. 3. Principal stratigraphic units and age relations, St Helena. K–Ar ages and unit names from Baker *et al.* (1967).

(NEMS). In contrast, three clear unconformities occur in the SW volcano. These separate the lower shield (SWLS) from the main shield (SWMS), which in turn is overlain by the upper shield (SWUS). Parasitic shield volcanism (SWEF) and intrusion of highly alkaline dykes and arcuate phonolite intrusions occurred during the final stages of activity (SWLI). Although the choice of units made by Baker (1968) to divide the lava pile was found to be generally valid, reassignments of some minor flow units were necessary. In illustrating temporal variations the samples are represented by their stratigraphic position. This was calculated by detailed measurements of type sections representing the maximum thickness of a particular stratigraphic group. Whilst the spatially heterogeneous nature of volcanic activity causes problems in cross-correlation the use of type sections minimizes this problem and produces a good representation of relative positions in the lava pile. Cessation of subaerial activity on St Helena was followed by 7 Ma of intensive erosion. This has resulted in a deeply dissected lava pile which, together with the arid climate and limited vegetation cover, facilitates sample collection. A new large sample set of 250 rocks was collected during a three month field season in 1986 to represent the full range of erupted compositions from primitive to evolved rocks. A bias to more primitive rocks was made to enable identification of source characteristics.

Analytical methods

Hand-picked chips were crushed to a powder (less than 100 μm) in an agate swing mill. Hand picking minimized the content of weathered surfaces and volumes with a high proportion of secondary mineralization (zeolites). Major elements were determined by X-ray fluorescence (XRF) on fusion beads, whilst trace elements were determined on powder briquettes of 15 g of rock bound with 4 ml of Moviol (poly vinylacetate) between polished steel dies at 10 tonnes pressure. Beads were analysed on a Philips PW 1400 XRF spectrometer at Leeds University. Inductively coupled plasma (ICP) techniques were used in the analysis of rare earth elements (REEs). Sample dissolution is described in Walsh & Howie (1980). Operating conditions of the Philips PV 8210 ICP are given in Walsh *et al.* (1981). Selected samples were analysed by instrumental neutron activation analysis (INAA) for La, Ce, Nd, Sm, Eu, Tb, Yb, Lu, Hf, Ta, Th and U. Detailed descriptions of the analytical technique are given in Potts *et al.* (1985). Strontium isotopic ratios were determined on a VG Isomass 54E at Leeds University following separation using standard ion exchange techniques. Neodymium was separated using the method of Richard *et al.* (1976). Lead separation was achieved by the ion exchange techniques of Manhes *et al.* (1978). Analysis for neodymium and lead isotopes was performed on a VG Micromass 30. Age correction for strontium isotopes was made using interpolations of Baker's age data based on stratigraphic height. Such corrections were insignificant for neodymium and lead isotopes. Routine total procedure blanks were Sr < 2 ng, Nd < 0.2 ng and Pb < 2.0 ng.

Major- and trace-element results

Analyses of primitive and selected evolved samples for which isotopic determinations have been made are given in Table 1. The rocks are mildly alkalic (*nepheline* normative) and plot above the line separating tholeiites from alkali basalts in Hawaii on a silica versus total alkalis diagram (MacDonald & Katsura 1964) (Fig. 4). Although all compositions lie above the divide, a range in degree of alkali enrichment is seen with some compositions approaching basanites and others tholeiites. SiO_2 contents vary from 42%–60% which indicate the important role of fractional crystallization in the evolution of the St Helena volcanic suite. Fine-grained basalts and hawaiites dominate, with porphyritic types also important. Ankaramites are uncommon and are found principally at the base of flow units and in large feeder dykes. Mugearites are of limited extent in the NE volcano and only become important late in the development of the SW volcano where extensive low-pressure fractional crystallization also produces less extensive trachytes and phonolites. Petrographic studies together with major-element data indicate that differentiation is controlled by olivine and clinopyroxene fractionation between 13% and 4.5% MgO whereas plagioclase becomes important at a later stage.

Mantle-normalized multi-element plots for a range of rocks from primitive to evolved are shown in Fig. 5. The decrease in relative abundances of elements normally considered to be incompatible (*eg* strontium relative to cerium and neodymium) indicates fractionation of certain phases. For example, strontium becomes compatible in plagioclase, titanium in titanomagnetite and phosphorus in apatite. Samples containing more than 5% MgO will be used in subsequent diagrams to assess source region characteristics. Consideration of variation diagrams shows that fractionation is controlled by olivine and clinopyroxene until this point. These phases do not produce significant differences in incompatible-element ratios.

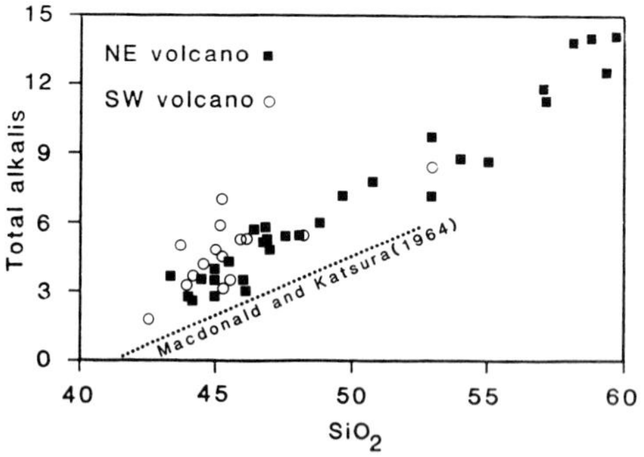

FIG. 4. Total alkalis versus silica diagram for St Helena volcanic rocks. The dotted line separates Hawaiian tholeiites from alkali basalts. (MacDonald & Katsura 1964.)

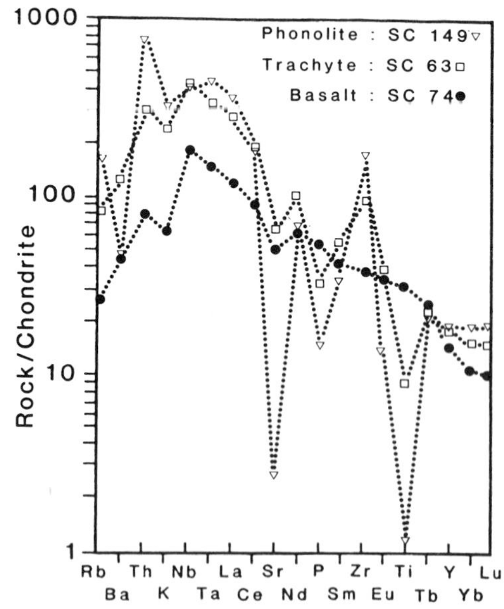

FIG. 5. Modified chondrite-normalized (Thompson 1982) multi-element plot for primitive to evolved compositions, St Helena.

TABLE 1. *Major-element, trace-element and isotope data for igneous rocks from St Helena*

Rock number	37	38	68	69	125	35
Volcano	NE	NE	NE	NE	NE	NE
Height (m)	0	0	50	50	50	150
SiO_2	45.65	42.63	44.14	45.40	44.11	43.85
TiO_2	2.71	1.73	2.31	2.66	2.73	3.64
Al_2O_3	14.99	8.83	11.28	14.30	12.89	16.33
Fe_2O_3	12.55	11.29	12.67	12.19	13.10	12.77
MnO	0.17	0.16	0.19	0.18	0.17	0.19
MgO	6.64	13.69	13.71	8.61	10.78	4.22
CaO	12.51	11.22	10.81	10.99	11.14	10.51
Na_2O	2.71	1.25	1.85	2.31	2.38	3.61
K_2O	0.78	0.43	0.71	0.88	0.96	1.37
P_2O_5	0.38	0.21	0.37	0.46	0.45	0.65
Loss on ignition (wt%)	2.36	2.82	1.83	2.05	1.32	2.62
Cr	194	922	928	359	457	—
Co	460	67	66	53	68	43
Ni	97	420	331	155	243	16
Cu	47	33	62	86	83	37
Zn	88	75	94	102	104	106
Rb	14	10	13	16	17	25
Sr	498	327	372	498	494	746
Y	23	16	20	26	24	29
Zr	173	130	175	217	199	271
Nb	44	29	39	52	52	76
Ba	217	138	182	243	259	382
U	0.97	—	1	—	—	—
Th	2.99	—	2.85	—	—	—
Ta	2.50	—	2.44	—	—	—
V	268	219	216	219	251	184
Sc	39	45	37	37	40	25
Hf	4.51	—	4.32	—	—	—
La	28.20	20.30	26.10	—	31.70	50.60
Ce	57.80	43.80	57.80	—	67.50	107
Nd	31.10	21.10	29.20	—	31.30	48.10
Sm	6.65	4.40	6.33	—	6.50	9.80
Eu	2.19	1.40	2.06	—	2.10	3.16
Gd	—	4.20	—	—	6.10	9.06
Yb	1.90	1.30	1.81	—	1.80	2.45
Lu	0.29	0.18	0.24	—	0.26	0.35
Tb	1.03	—	0.93	—	—	—
$^{87}Sr/^{86}Sr$	0.702854	0.702826	0.702806	0.702818	0.702818	0.702849
$^{143}Nd/^{144}Nd$	0.512965	0.512970	0.512925	0.512923	0.512926	—
$^{208}Pb/^{204}Pb$	39.841	39.736	23.870	39.754	39.820	39.873
$^{207}Pb/^{204}Pb$	15.759	15.708	15.735	15.711	15.724	15.788
$^{206}Pb/^{204}Pb$	20.442	20.401	20.554	20.448	20.440	20.607

The role of alteration

The rocks analysed have been screened for the effects of low-temperature alteration. Samples with a large proportion of secondary minerals were rejected. A chemical screening for altered samples was also conducted by considering elements such as rubidium and potassium which are frequently anomalous in otherwise coherent OIB suites (*eg* Chen & Frey 1985). Largely consistent ratios of mobile-to-immobile elements (*eg* K/La, Rb/Nb) in rocks of a particular unit on St Helena suggest that only limited alteration has occurred. However, plots of niobium versus Ba/Nb (Fig. 6) show large enrichments in barium in five samples. High

TABLE 1 (*cont.*)

Rock number	56	96	80	109	111	28
Volcano	NE	NE	NE	NE	NE	NE
Height (m)	300	500	541	565	618	642
SiO_2	45.21	44.63	45.28	48.37	45.98	46.28
TiO_2	3.68	3.81	3.54	3.49	3.48	3.30
Al_2O_3	16.03	16.87	16.06	17.34	17.56	15.64
Fe_2O_3	12.98	13.47	13.73	12.28	12.26	13.71
MnO	0.20	0.20	0.22	0.17	0.15	0.18
MgO	4.35	4.71	4.16	3.63	3.84	6.86
CaO	9.82	11.18	9.30	9.18	10.12	9.47
Na_2O	4.25	3.00	4.96	3.89	3.63	4.01
K_2O	1.53	1.09	1.93	1.60	1.61	1.28
P_2O_5	0.75	0.63	0.82	0.76	0.66	0.68
Loss on ignition (wt%)	0.70	0.49	0.79	1.19	2.11	0.08
Cr	—	23	13	22	39	79
Co	40	103	39	45	47	43
Ni	19	50	14	15	24	59
Cu	8	61	4	23	30	29
Zn	119	112	126	123	105	107
Rb	30	18	36	32	41	200
Sr	706	748	786	734	621	719
Y	32	49	34	33	29	28
Zr	311	229	344	305	338	260
Nb	84	66	97	86	60	70
Ba	393	544	459	408	343	334
U	2.19	—	—	—	—	1.87
Th	5.83	—	—	—	—	5.04
Ta	4.65	—	—	—	—	4.11
V	—	307	163	237	172	274
Sc	2	35	21	27	26	27
Hf	6.90	—	—	—	—	6.54
La	52.90	50.10	59.70	52.10	—	47.60
Ce	111	116	126.60	110.50	—	99.70
Nd	54.70	50.80	54.50	48.20	—	53.40
Sm	10.70	10.50	10.70	9.60	—	10.10
Eu	3.40	3.40	3.30	3	—	3.14
Gd	—	10.30	9.50	8.60	—	—
Yb	2.69	3.30	2.50	2.50	—	2.31
Lu	0.39	0.47	0.35	0.35	—	0.35
Tb	1.43	—	—	—	—	1.25
$^{87}Sr/^{86}Sr$	0.702840	0.702864	0.702852	0.702875	0.702856	0.702922
$^{143}Nd/^{144}Nd$	0.512895	—	0.512911	—	0.512872	—
$^{208}Pb/^{204}Pb$	39.929	—	39.995	39.846	39.977	39.880
$^{207}Pb/^{204}Pb$	15.753	—	15.755	15.740	15.759	15.740
$^{206}Pb/^{204}Pb$	20.609	—	20.745	20.511	20.617	20.605

Ba/Nb values in evolved samples (*eg* SC 131) may be explained by accumulation of a phase such as K-feldspar, but anomalous enrichments are found across the entire range of compositions. Samples showing enrichment in barium may also possess a markedly different slope on chondrite-normalized REE diagrams compared with the majority of volcanics (Fig. 7). The ankaramite SC 140 has a large enrichment in heavy REEs and a moderate enrichment in light REEs in comparison with a typical ankaramite such as SC 38. SC 96 shows a similar but smaller relative increase in heavy REEs. Because such anomalous features are not specific to particular units and because of their magnitude, they seem unlikely to represent source differences. It is found that the chemical manifestation of alteration is not consistent with high (>>1%) loss on ignition values. Clearly great care must be taken to remove extraneous

TABLE 1 (*cont.*)

Rock number Volcano Height (m)	134 NE 812	8 NE 824	26 NE 1000	29 NE 1000	64 NE 1000	143 NE 1000
SiO_2	46.86	45.11	45.34	46.88	45.11	53.06
TiO_2	3.18	3.19	3.33	3.10	2.69	1.23
Al_2O_3	16.61	17.57	16.41	16.74	13.17	17.52
Fe_2O_3	12.58	11.56	12.65	12.61	12.85	11.42
MnO	0.18	0.15	0.19	0.21	0.19	0.26
MgO	4.22	4.65	6.59	3.73	10.83	1.84
CaO	9.43	9.76	10.73	8.20	12.66	4.95
Na_2O	3.50	3.53	3.29	4.91	2.20	5.84
K_2O	1.60	1.34	1.21	1.74	0.80	2.57
P_2O_5	0.80	0.73	0.58	0.92	0.43	0.67
Loss on ignition (wt%)	1.29	2.05	0.78	0.23	—	0.65
Cr	68	43	187	13	566	—
Co	60	37	50	35	58	30
Ni	42	31	95	12	233	6
Cu	231	40	34	30	89	20
Zn	26	97	108	121	94	151
Rb	790	22	22	37	13	59
Sr	42	853	687	802	447	834
Y	296	28	31	32	25	37
Zr	85	267	280	352	194	593
Nb	435	75	73	99	46	141
Ba	—	362	345	452	222	578
U	—	1.70	—	—	—	—
Th	—	5.21	—	—	—	—
Ta	175	4.21	—	—	—	—
V	31	266	207	120	219	—
Sc	—	21	30	18	43	6
Hf	—	6.20	—	—	—	—
La	—	48.30	40.70	60.50	—	83.40
Ce	—	102	86.30	129.40	—	167.30
Nd	—	50	38	54.50	—	57.80
Sm	—	9.85	7.80	10.60	—	9.80
Eu	—	3.10	2.50	3.30	—	3
Gd	—	—	7	9.20	—	7.90
Yb	—	2.11	2	2.50	—	3
Lu	—	0.33	0.28	0.36	—	0.44
Tb	—	1.15	—	—	—	—
$^{87}Sr/^{86}Sr$	0.703090	0.702885	0.702855	0.702840	0.702951	0.702876
$^{143}Nd/^{144}Nd$	0.512842	0.512922	—	—	0.512896	—
$^{208}Pb/^{204}Pb$	40.037	39.901	—	—	40.020	—
$^{207}Pb/^{204}Pb$	15.770	15.760	—	—	15.769	—
$^{206}Pb/^{204}Pb$	20.781	20.545	—	—	20.735	—

modifying effects with a geochemical filter. The samples considered in subsequent discussion of source differences all show concordant trends for rubidium, potassium, barium and yttrium (for heavy REEs) when plotted against an immobile element such as niobium. A routine 30 min 6M HCl leach was conducted on all samples prior to strontium and neodymium isotopic analysis.

Temporal variations

The alkali basalts have approximately parallel chondrite-normalized REE distributions (Fig. 8) but variable degrees of light REE enrichment (65–170 × chondrite) relative to heavy REE enrichment (6–12 × chondrite). An increase in the degree of incompatible-trace-element enrichment occurs through time. For

TABLE 1 (*cont.*)

Rock number	168	203	215	190	237	238
Volcano	SW	SW	SW	SW	SW	SW
Height (m)	1150	1248	1380	1845	1983	1988
SiO_2	48.22	46.53	44.64	45.11	46.10	46.21
TiO_2	2.85	2.98	3.57	2.77	2.59	2.31
Al_2O_3	15.29	16.77	12.92	14.20	14.06	12.18
Fe_2O_3	12.99	12.37	14.68	12.24	12.58	12.56
MnO	0.19	0.21	0.21	0.18	0.17	0.18
MgO	4.42	3.42	10.32	8.75	8.14	11.92
CaO	8.05	7.85	11.49	11.43	10.53	10.36
Na_2O	4.07	3.71	2.52	2.43	2.61	2.11
K_2O	1.42	1.88	0.95	0.93	0.81	0.82
P_2O_5	0.60	0.79	0.46	0.46	0.40	0.38
Loss on ignition (wt%)	1.90	2.34	1.47	1.52	2.56	1.77
Cr	30	—	310	402	385	763
Co	47	39	59	50	60	60
Ni	21	8	189	190	175	271
Cu	31	21	58	75	65	54
Zn	85	107	95	91	103	93
Rb	24	26	15	19	16	11
Sr	591	809	534	524	418	343
Y	27	34	22	23	22	22
Zr	257	309	194	199	176	184
Nb	68	86	48	54	42	42
Ba	373	501	247	271	211	201
U	—	—	—	—	—	0.79
Th	—	—	—	—	—	2.89
Ta	—	—	—	—	—	2.39
V	144	60	268	210	305	142
Sc	23	16	41	36	35	25
Hf	—	—	—	—	—	4.41
La	50.20	—	33.8	37.5	25.60	26.60
Ce	107.70	—	72.2	79.6	54.60	57.40
Nd	45.80	—	33.9	36.3	26.50	27.90
Sm	8.90	—	7.31	7.5	5.80	6.15
Eu	2.80	—	2.38	2.44	1.90	2.04
Gd	7.80		7.01	7.22	5.60	—
Yb	2.40	—	1.91	1.95	1.80	1.83
Lu	0.35	—	0.28	0.28	0.26	0.24
Tb	—	—	—	—	—	0.87
$^{87}Sr/^{86}Sr$	0.702837	—	0.702826	0.702867	0.702831	0.702818
$^{143}Nd/^{144}Nd$	0.512909	0.512854	0.512824	0.512873	0.512963	0.512943
$^{208}Pb/^{204}Pb$	40.072	40.065	40.093	40.058	39.822	39.846
$^{207}Pb/^{204}Pb$	15.783	15.789	15.795	15.796	15.744	15.749
$^{206}Pb/^{204}Pb$	20.764	20.741	20.839	20.586	20.488	20.518

example, an increase in $(La/Yb)_n$ ratio occurs during the activity of each volcano (see Fig. 10*d* later). The significant increase in $(La/Yb)_n$ through time in primitive rocks can be explained either by variation in the degree of partial melting or by mantle source heterogeneity. If this variation is caused by differences in partial melting a relatively low degree of partial melting is necessary. In comparing St Helena with Bouvet and Ascension Weaver *et al.* (1987) note that zirconium is not behaving as an incompatible element in the St Helena volcanic rocks. Hence the degree of partial melting is sufficiently low to fractionate the Zr/Nb ratio. The use of ratios of highly incompatible elements is therefore necessary to confirm whether the $(La/Yb)_n$ variations represent source differences, because highly incompatible elements are only likely to be fractionated at very small degrees of partial melting (*eg* Kay 1984).

TABLE 1 (*cont.*)

Rock number	216	173	114	117	119	13
Volcano	SW	SW	SW	SW	SW	SW
Height (m)	2150	2180	2275	2370	2581	2603
SiO_2	45.60	45.07	50.86	47.11	46.96	43.42
TiO_2	3.23	1.97	1.86	3.14	2.99	3.09
Al_2O_3	15.59	9.44	17.32	16.55	16.92	13.11
Fe_2O_3	13.57	12.40	11.42	13.13	12.46	14.30
MnO	0.19	0.17	0.19	0.19	0.20	0.16
MgO	6.36	13.26	2.56	4.95	4.53	8.81
CaO	9.91	11.75	6.21	8.83	8.25	10.80
Na_2O	3.72	2.03	5.69	3.41	4.16	2.65
K_2O	0.52	0.68	2.20	1.30	1.63	0.93
P_2O_5	0.49	0.33	0.85	0.68	0.80	0.44
Loss on ignition (wt%)	0.60	1.24	0.98	1.19	1.47	1.44
Cr	136	952	—	78	14	699
Co	52	73	41	44	43	61
Ni	82	392	2	49	23	298
Cu	69	88	14	52	40	72
Zn	108	97	142	104	108	98
Rb	1	14	46	22	30	12
Sr	583	410	800	675	653	443
Y	26	20	36	34	32	22
Zr	228	155	449	282	318	190
Nb	54	42	125	76	88	47
Ba	268	430	552	377	430	225
U	—	0.78	—	—	—	0.95
Th	—	2.85	—	—	—	3.21
Ta	—	2.26	—	—	—	2.58
V	171	218	14	146	145	285
Sc	34	46	12	27	18	39
Hf	—	3.94	—	—	—	4.55
La	—	28	—	—	60.80	29.50
Ce	—	57.10	—	—	122.80	63.20
Nd	—	32.0	—	—	49.50	31
Sm	—	6.03	—	—	9.30	6.71
Eu	—	1.99	—	—	2.80	2.20
Gd	—	—	—	—	8.10	—
Yb	—	1.56	—	—	2.50	1.68
Lu	—	0.22	—	—	0.36	0.25
Tb	—	0.82	—	—	—	0.94
$^{87}Sr/^{86}Sr$	0.702846	0.702927	0.702846	0.702890	0.702867	0.702890
$^{143}Nd/^{144}Nd$	0.512885	0.512909	0.512895	—	0.512878	0.512895
$^{208}Pb/^{204}Pb$	39.881	—	39.940	40.085	40.104	40.057
$^{207}Pb/^{204}Pb$	15.717	—	15.757	15.786	15.780	15.768
$^{206}Pb/^{204}Pb$	20.654	—	20.620	20.831	20.824	20.764

Th/Ta or Ba/Nb ratios (Fig. 6), which are commonly used to demonstrate source differences in OIB suites, show a very limited range in St Helena volcanic rocks. The small range suggests that increases in $(La/Yb)_n$ may be caused by a decrease in the degree of partial melting during shield activity. It must be remembered, however, that if different source components contain similar ratios of highly incompatible elements interactions between them will be difficult to distinguish. There is a likelihood that this effect is important in St Helena magmatism as it is known from comparative studies of S Atlantic OIB suites (Weaver *et al.* 1987) that St Helena is depleted in highly incompatible large-ion lithophile elements relative to other S Atlantic OIBs. If mixing is occurring between an OIB component and a depleted component with similar ratios of highly incompatible elements (*eg* Ba/Nb is 5–6

TABLE 1 (*cont.*)

Rock number	14	106	140	139	74	75
Volcano	SW	SW	SW	SW	SW	SW
Height (m)	2930	3091	3100	3100	3175	3195
SiO_2	49.86	54.06	44.31	44.32	45.12	44.33
TiO_2	2.87	1.98	3.19	3.16	3.29	3.08
Al_2O_3	17.03	14.53	13.88	13.31	15.23	14.64
Fe_2O_3	14.38	12.14	15.21	14.67	13.04	13.11
MnO	0.23	0.21	0.16	0.17	0.19	0.20
MgO	4.10	3.11	7.08	8.97	6.50	7.80
CaO	6.89	6.61	10.03	11.09	9.81	10.24
Na_2O	5.04	4.00	1.65	2.73	2.98	2.54
K_2O	2.16	1.69	0.82	0.95	0.88	1.10
P_2O_5	1.05	1.02	0.39	0.46	0.58	0.58
Loss on ignition (wt%)	0.67	0.65	3.30	0.17	1.66	2.17
Cr	20	1	404	437	151	237
Co	43	54	61	69	46	46
Ni	28	10	190	241	84	103
Cu	33	34	33	83	67	63
Zn	126	134	145	110	112	101
Rb	44	42	12	18	21	18
Sr	1047	753	451	541	581	562
Y	38	37	178	25	30	28
Zr	462	409	204	193	264	245
Nb	122	112	52	51	66	68
Ba	611	492	427	276	327	307
U	—	—	1.15	1.13	1.75	—
Th	—	—	3.62	3.67	4.51	—
Ta	—	—	2.93	2.90	3.67	—
V	88	51	—	310	240	323
Sc	12	12	8	41	30	32
Hf	—	—	4.97	5.06	5.94	—
La	80.60	80.50	39	32.70	41.60	41.30
Ce	168.70	168.90	82.40	67.50	83.40	87.30
Nd	64.50	67.10	51.90	34.80	41.60	38.30
Sm	11.60	12.00	14.30	7.55	8.99	7.60
Eu	3.70	3.50	5.39	2.47	2.83	2.40
Gd	9.70	10.00	—	—	—	6.90
Yb	2.90	2.90	13.10	1.80	2.22	2.10
Lu	0.40	0.43	2.28	0.27	0.34	0.29
Tb	—	—	3.35	1.05	1.16	—
$^{87}Sr/^{86}Sr$	0.702934	0.702916	0.702870	0.702864	0.702901	0.702913
$^{143}Nd/^{144}Nd$	0.512870	0.512854	0.512921	0.512885	0.512871	0.512898
$^{208}Pb/^{204}Pb$	40.016	40.051	—	39.850	40.113	40.104
$^{207}Pb/^{204}Pb$	15.775	15.779	—	15.739	15.786	15.789
$^{206}Pb/^{204}Pb$	20.736	20.721	—	20.491	20.809	20.844

for St Helena and 4–5 in normal-type mid-ocean ridge basalt (MORB) then it cannot easily be discerned by use of these elements. Isotopic studies offer the best opportunity for resolving this problem because interaction between magmas displaying the extreme (end-member) isotopic characteristics of St Helena and isotopically depleted material should be readily apparent. As a result the greatest emphasis is placed on isotopic determinations for trying to establish the degree of source heterogeneity.

Isotope data

Previous isotopic data for St Helena have been reported for strontium (Grant *et al.* 1976),

TABLE 1 (*cont.*)

Rock number	63	175	131	85	86	93
Volcano	SW	SW	SW	SW	SW	SW
Height (m)	3350	3400	3400	3450	3450	3450
SiO_2	55.18	53.10	48.95	46.94	47.67	57.02
TiO_2	0.80	0.96	2.70	2.76	2.89	0.25
Al_2O_3	17.68	17.16	17.28	16.12	16.03	18.72
Fe_2O_3	10.41	8.40	12.64	12.74	12.99	7.77
MnO	0.26	0.17	0.17	0.22	0.18	0.25
MgO	1.41	2.02	3.08	4.14	5.73	0.41
CaO	4.11	3.38	6.81	8.35	9.29	2.66
Na_2O	5.87	6.68	4.22	3.72	4.03	7.63
K_2O	2.88	3.19	1.76	1.50	1.43	3.90
P_2O_5	0.36	0.60	0.90	0.82	0.69	0.15
Loss on ignition (wt%)	2.30	0.54	1.46	0.79	0.13	1.31
Cr	—	25	—	73	158	—
Co	24	10	44	46	56	28
Ni	4	8	10	49	75	5
Cu	—	122	18	52	24	12
Zn	140	82	109	128	116	153
Rb	63	773	30	31	27	112
Sr	731	34	748	681	643	490
Y	38	674	76	35	31	38
Zr	640	175	367	329	300	841
Nb	169	575	100	88	81	163
Ba	827	5.40	963	407	367	679
U	3	21	—	—	—	—
Th	13.10	10.50	—	—	—	—
Ta	8.42	13	—	—	—	—
V	—	4	85	216	174	—
Sc	7	14.10	16	22	29	2
Hf	12.50	108	—	—	—	—
La	95.80	190	—	—	50.20	108
Ce	173	64.20	—	—	104.30	192
Nd	65.20	11.10	—	—	43.70	54
Sm	11.30	3.06	—	—	8.50	8
Eu	3.40	—	—	—	2.60	2
Gd	—	3.31	—	—	7.50	6
Yb	3.40	0.47	—	—	2.30	3.70
Lu	0.53	1.26	—	—	0.33	0.56
Tb	1.39		—	—	—	—
$^{87}Sr/^{86}Sr$	0.703083	0.703008	0.702896	0.702835	0.702931	0.702953
$^{143}Nd/^{144}Nd$	0.512879	0.512872	—	0.512854	—	—
$^{208}Pb/^{204}Pb$	40.059	40.154	—	40.055	40.134	0.512901
$^{207}Pb/^{204}Pb$	15.796	15.804	—	15.767	15.806	15.790
$^{206}Pb/^{204}Pb$	20.818	20.801	—	20.846	20.823	40.116
						20.858

strontium and neodymium (Hofmann & White 1982; Zindler *et al.* 1982), lead (Gast 1969; Oversby & Gast 1970; Sun 1980) and strontium, neodymium and lead (Cohen & O'Nions 1982; White 1985; Hart *et al.* 1986; Newsom *et al.* 1986) but comprehensive combined strontium, neodymium and lead isotope data and trace-element data have been unavailable previously.

Strontium and neodymium isotopes

Figure 9 shows the new Sr–Nd isotope data from this study relative to fields for other oceanic basalts. $^{143}Nd/^{144}Nd$ is more variable compared with age-corrected $^{87}Sr/^{86}Sr$. The new results are in good agreement with previously reported values. However, isotopic compositions as extreme as those reported for trachyte 39512 by

TABLE 1 (*cont.*)

Rock number	103	129	149	150	151	152
Volcano	SW	SW	SW	SW	SW	SW
Height (m)	3600	3600	3600	3600	3600	3600
SiO_2	59.74	57.08	59.78	58.11	59.30	52.99
TiO_2	0.07	0.51	0.07	0.03	0.08	0.75
Al_2O_3	19.37	18.07	19.25	18.78	19.22	13.56
Fe_2O_3	4.33	9.12	4.35	3.63	4.39	7.73
MnO	0.17	0.23	0.17	0.18	0.22	0.22
MgO	0.02	0.51	—	—	0.08	0.91
CaO	0.90	3.14	0.88	0.48	0.78	2.99
Na_2O	9.09	7.50	9.08	8.81	7.26	4.85
K_2O	4.98	3.54	4.95	4.75	4.99	2.33
P_2O_5	0.04	0.23	0.05	0.06	0.07	0.28
Loss on ignition (wt%)	1.63	0.90	1.50	3.12	2.53	1.01
Cr	—	—	—	—	—	—
Co	22	29	9	10	10	36
Ni	5	2	0	4	5	6
Cu	3	13	0	5	6	10
Zn	140	129	178	201	190	137
Rb	94	87	170	172	159	60
Sr	150	527	36	14	77	598
Y	54	39	40	47	36	48
Zr	916	767	1197	1436	1127	622
Nb	222	180	157	167	153	159
Ba	967	740	311	61	408	752
U	—	—	8.84	—	—	—
Th	—	—	33	—	—	—
Ta	—	—	11.30	—	—	—
V	—	—	—	—	—	—
Sc	—	3	—	3	2	6
Hf	—	—	19.50	—	—	—
La	115.50	—	125	—	—	89.60
Ce	224.80	—	189	—	—	185.10
Nd	73.50	—	48.50	—	—	66.70
Sm	12.80	—	7.30	—	—	12
Eu	2.90	—	1.12	—	—	3.70
Gd	10.50		—	—	—	10
Yb	5.20	—	4.29	—	—	4.10
Lu	0.76	—	0.65	—	—	0.58
Tb	—	—	1.03	—	—	—
$^{87}Sr/^{86}Sr$	0.702891	0.702903	0.702906	0.704697	0.703201	0.702898
$^{143}Nd/^{144}Nd$	0.512907	0.512885	0.512874	0.512839	0.512839	0.512843
$^{208}Pb/^{204}Pb$	—	40.057	40.167	—	—	—
$^{207}Pb/^{204}Pb$	—	15.790	15.813	—	—	—
$^{206}Pb/^{204}Pb$	—	20.788	20.891	—	—	—

REE data for samples with analysed thorium, uranium and tantalum undertaken by INAA; remaining determinations by ICP. $^{87}Sr/^{86}Sr$ data are adjusted to 0.7080 for the Eimer & Amend standard and corrected to initial ratios at the interpolated age. $^{143}Nd/^{144}Nd$ data normalized to $^{146}Nd/^{144}Nd$ and adjusted to 0.51262 for BCR-1. Measured values were:

$^{87}Sr/^{86}Sr$ — Eimer & Amend 0.70805; SRM 987 0.71029

$^{143}Nd/^{144}Nd$ — BCR-1 0.51267; La Jolla 0.51189.

Within-run errors for Sr typically 0.002% and for Nd 0.004%. Fifteen analyses of SRM 981 indicate mass fractionation of 1.1 per mil per mass unit for Pb. Major element oxides in wt%, trace elements in ppm.

Cohen & O'Nions (1982) were not found ($^{206}Pb/^{204}Pb$, 19.68±0.02; $^{87}Sr/^{86}Sr$; 0.70277; $^{143}Nd/^{144}Nd$, 0.51296±0.00002). This discrepancy occurs despite the analysis of a range of compositions from throughout the lava pile. It is not inconceivable that sample 39512 was collected from neighbouring Ascension Island as the isotopic compositions are in good agreement

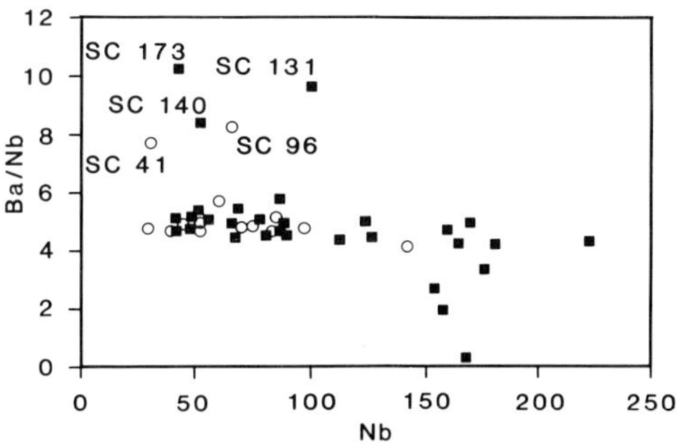

FIG. 6. Niobium versus Ba/Nb diagram for St Helena volcanic rocks. Named samples are those eliminated from subsequent discussion because of alteration.

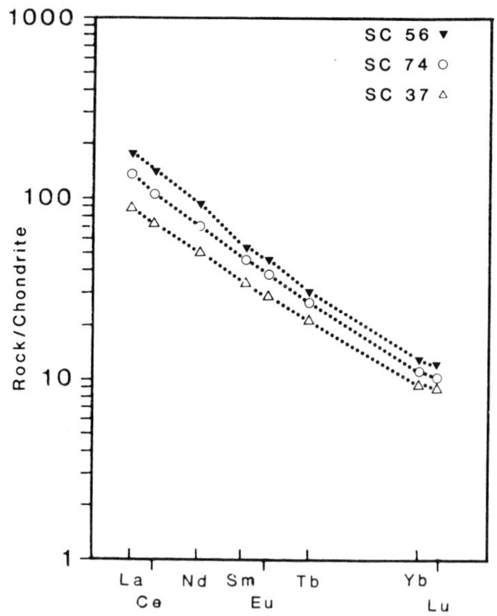

FIG. 8. Chondrite-normalized (Boynton 1984) REE abundances for a range of St Helena basalts defining the range in REE abundance for rocks containing more than 5% MgO (INAA data).

(Cohen & O'Nions 1982), and Melliss also visited Ascension Island. It is evident that evolved compositions (mugearites–phonolites) have higher initial ratios of $^{87}Sr/^{86}Sr$ at approximately constant levels of $^{143}Nd/^{144}Nd$. Similar features have been reported elsewhere (*eg* Palacz & Wolff, this volume). Such increases are considered to be high-level processes and were discussed by Grant *et al.* (1976) who thought that contamination of differentiated magmas in a high-level magma chamber by varying proportions of meteoric material and seawater was the most likely explanation. As these differences have no bearing on source region geochemistry, they are not discussed further in this paper and general descriptions of $^{87}Sr/^{86}Sr$ variations in the source region do not include these rocks.

The temporal variations in strontium and neodymium isotopes are shown in Figs 10a and 10b respectively. It is evident that during subaerial activity an overall decrease in $^{143}Nd/^{144}Nd$ and an overall increase in $^{87}Sr/^{86}Sr$ occurred. This overall pattern is repeated for both volcanoes; for example basalts of the NEBB and NE(BB) possess higher $^{143}Nd/^{144}Nd$ and lower $^{87}Sr/^{86}Sr$ ratios than the overlying shield. In detail, however, the situation is more complex. SC 8 at the top of the shield reverses this general trend. The situation in the SW volcano is even more complex. The SWLS appears to show a poor trend similar to that of the NE shield, but towards its top two samples—SC 237 and SC 238—have relatively high $^{143}Nd/^{144}Nd$ values and may represent a similar feature to SC 8. The SWMS exhibits the greatest degree of heterogeneity. No trends are discernible, but if the average values are compared with the values for the younger SWUS, SWEF and SWLI a decrease in $^{143}Nd/^{144}Nd$ and an increase in $^{87}Sr/^{86}Sr$ can be seen.

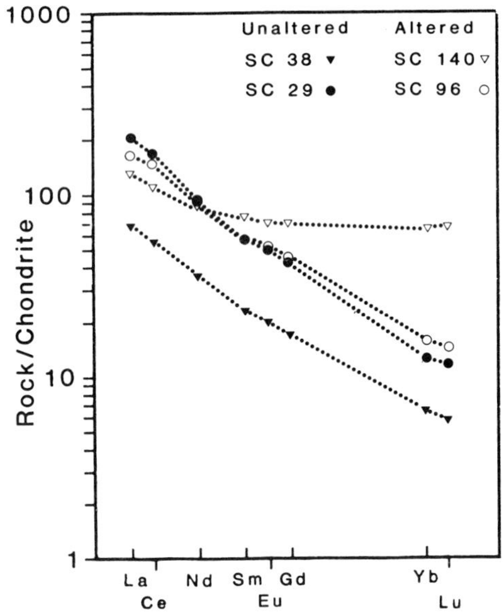

FIG. 7. Chondrite-normalized (Boynton 1984) REE abundances for selected altered samples compared with fresh samples: ankaramite SC 140 relative to fresh ankaramite SC 38; hawaiite SC 96 relative to fresh hawaiite SC 29.

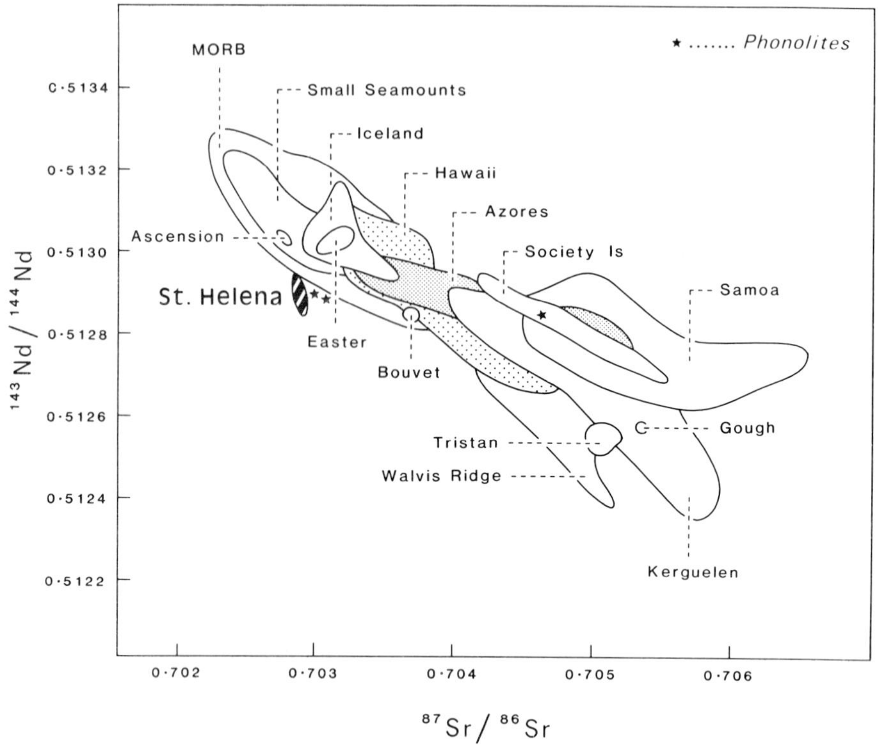

FIG. 9. $^{143}Nd/^{144}Nd$ versus $^{87}Sr/^{86}Sr$ diagram showing the field of new St Helena data relative to other OIBs and MORBs. Data sources are referenced in Zindler & Hart (1986).

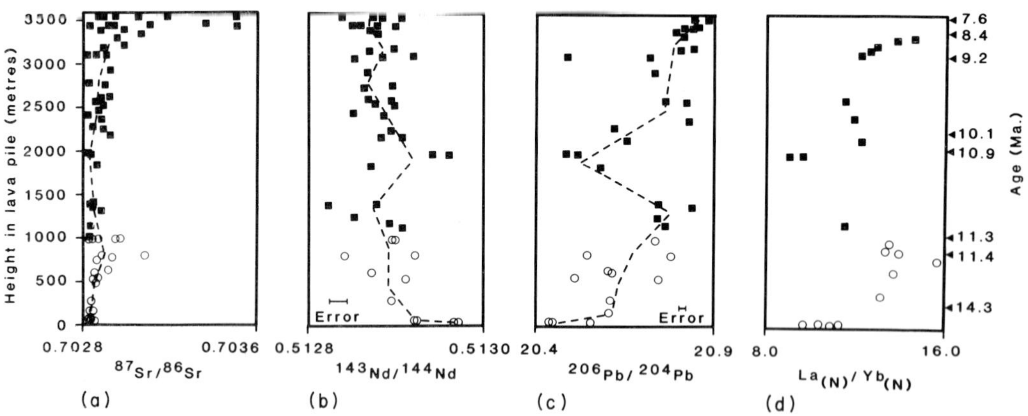

FIG. 10. Diagram summarizing temporal geochemical variations in the St Helena rock suite. (*a*) $^{87}Sr/^{86}Sr$ versus calculated height in the lava pile. (*b*) $^{143}Nd/^{144}Nd$ versus height. (*c*) $^{206}Pb/^{204}Pb$ versus height. (*d*) $(La/Yb)_n$ versus height (rocks containing more than 5% MgO). The broken line represents a moving average obtained by subdivision of the main shield units. 2σ error bars are represented by symbol size unless indicated.

Lead isotopes

Lead isotope ratios of St Helena volcanic rocks are compared with some other MORBs and OIBs in Fig. 11. A larger range is reported than by Sun (1980). A good correlation exists in Pb–Pb space (Fig. 12). This pattern is consistent with temporal strontium and neodymium isotopic variations (Fig. 10). It is seen that samples with unradiogenic lead compositions have the most radiogenic neodymium compositions and occur at the beginning or towards the end of shield activity. The isotopic variations are sympathetic with those registered by the trace elements. Thus Figs 10*a*–*c* closely parallel Fig. 10*d* in which $(La/Yb)_n$ is plotted against time. Similarly $^{143}Nd/^{144}Nd$ correlates positively with the Sm/Nd ratio.

Discussion

Development of HIMU characteristics

The extreme isotope characteristics of the St Helena magmas which exemplify the HIMU mantle component have been the subject of much discussion in recent years. The controversy revolves around the production of associated low $^{87}Sr/^{86}Sr$ ratios and radiogenic lead compositions. The enrichments in ^{208}Pb and ^{206}Pb indicate thorium and uranium increases relative to lead over 1.5 Ga ago (Tatsumoto 1978). The mantle segments that have experienced these fractionations have remained physically discrete from other mantle components that experienced different trace-element fractionation events. Early ideas to explain the high U/Pb ratios invoked extraction of lead from the mantle during core evolution (Vollmer 1977; Vidal & Dosso 1978). However, more recently it has been proposed that the HIMU characteristics may have evolved from subduction of altered oceanic crust (Chase 1981; Hofmann & White 1982). Proponents of this model argue that subduction zone processes may effect the removal of lead from the subducting oceanic crust (*eg* Zindler *et al.* 1982; Palacz & Saunders 1986) and could also produce a decrease in Rb/Sr to counteract the effect of seawater interaction. Others argue that because rubidium is not fractionated from uranium in the continental crust it is necessary that it be removed by other processes (Menzies & Murthy 1980: Hart *et al.* 1986). They argue that mobilization and removal of rubidium and lead from a mantle segment during metasomatic processes could also explain the source differences. Although this study improves constraints on the composition of the HIMU end-member, it does not contribute directly to the debate about its origin and evolution.

Interactions between mantle components beneath St Helena

A compositionally heterogeneous source region is indicated by the isotope and trace-element data. There is a continuum of isotopic compositions between volcanic rocks with high $^{143}Nd/^{144}Nd$ (greater than 0.51294) and relatively unradiogenic strontium ($^{87}Sr/^{86}Sr < 0.70280$) and lead ($^{206}Pb/^{204}Pb < 20.4$) and those with lower $^{143}Nd/^{144}Nd$ (less than 0.51285) and more radiogenic strontium ($^{87}Sr/^{86}Sr > 0.70295$) and lead ($^{206}Pb/^{204}Pb > 20.8$). These variations in geochemistry may be ascribed to differing contributions to the ascending melts from two distinct end-members. The first is that with the HIMU characteristics described above and the greatest degree of incompatible-trace-element enrichment. The second component is relatively depleted and has less radiogenic strontium and lead isotope compositions. In isotopic space this lies on a vector towards the depleted MORB mantle of Zindler & Hart (1986). Positive correlations between parent/daughter ratios and isotopes (*eg* Sm/Nd and $^{143}Nd/^{144}Nd$) are also consistent with two-component mixing but do not exclude the possibility that the trends could represent mantle isochrons. In the latter case the trends represent a previous separation of the components from a homogeneous parent. The correlation on a Pb–Pb isotope diagram (Fig. 12) could similarly be explained as a secondary isochron resulting from a previous mixing event as suggested by Chase (1981). Chemostratigraphic variations are of value in resolving this problem.

Overall coupled temporal variations in trace-element and isotopic characteristics are similar for both the NE and SW volcanoes. The lavas change from high to low $^{143}Nd/^{144}Nd$ ratios whilst strontium and lead compositions become more radiogenic and trace-element enrichment increases during the activity of each volcano. Because of these secular changes it is argued that all geochemical trends in primitive St Helena volcanic rocks can be accounted for by recent mantle-mixing events occurring immediately prior to eruption. If the trends represented mantle isochrons, random correlations of geochemical parameters with time would be expected unless chemical zoning of

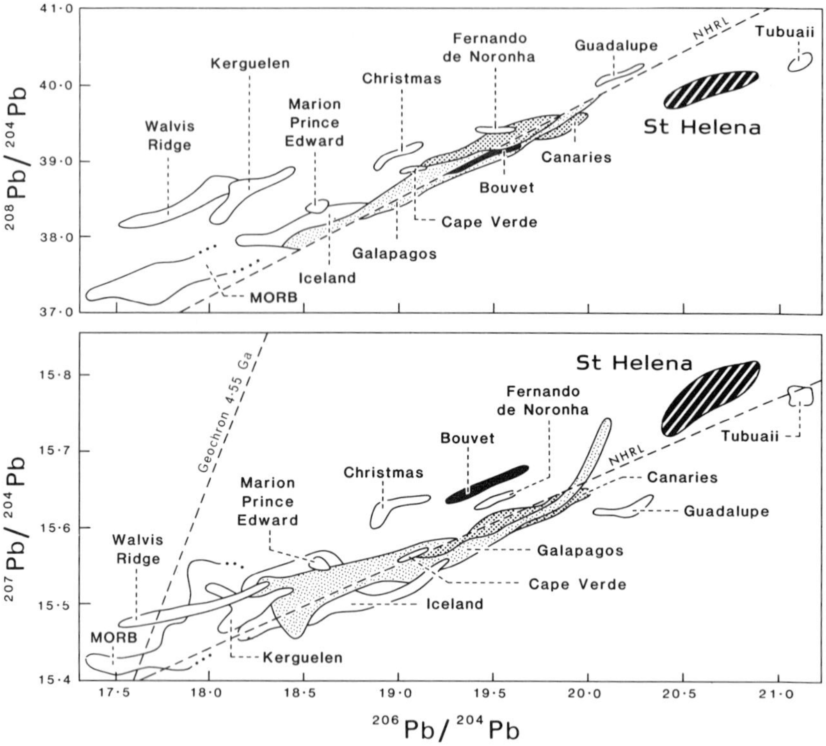

FIG. 11. $^{207}Pb/^{204}Pb$ versus $^{206}Pb/^{204}Pb$ diagram and $^{208}Pb/^{204}Pb$ versus $^{206}Pb/^{204}Pb$ diagram showing new St Helena data in relation to other OIBs and MORBs. Data sources are referenced in Zindler & Hart (1986).

the plume had occurred during its ascent (*cf* Duncan *et al.* 1986). It is then necessary to invoke preferential removal of the depleted component from the plume early in activity because it has become coincident with the zone of highest temperature. Fortuitous repetition of this zoning would also be necessary. It is difficult to reconcile the repeated regular geochemical variations with this alternative and the simpler model in which correlation reflects mixing of independent sources is favoured.

The remainder of this discussion will address the problem of the relative contribution of the two components in relation to the geometric framework of magma generation. Thus, their physical characteristics, how they become mixed and how these processes may vary over time will be considered.

Some constraints on the geometric relationship of the two components are provided by integrated studies of S Atlantic magmatism based on dredged rocks from along the adjacent Mid-Atlantic Ridge. Schilling *et al.* (1984) identified an enriched component indicated by high La/Sm and Nb/Zr ratios in MORB 700 km from and at a similar latitude to St Helena. More significantly Hanan *et al.* (1986) showed that a radial decline in radiogenic lead content of recent Mid-Atlantic Ridge basalts occurred with distance from St Helena (once the effect of mixing between the Mid-Atlantic Ridge basalts and other OIB sources such as Tristan de Cunha had been removed). This component can be identified over a distance of 3000 km. Studies by Fitton *et al.* (1983) have indicated the presence of a HIMU component in the basalts of the Cameroon line which occurs 1000 km to the NE of St Helena. Similarly a HIMU component has been reported in Cape Verdes volcanic rocks (Davies *et al.*, this volume). Whilst it is not known whether these HIMU domains are independent or spatially related to St Helena, these studies indicate the widespread nature of the St Helena component and suggest that a regional maximum occurs at St Helena. Thus a geochemically anomalous mantle source region existed beneath St Helena during volcanic activity. This anomaly was co-

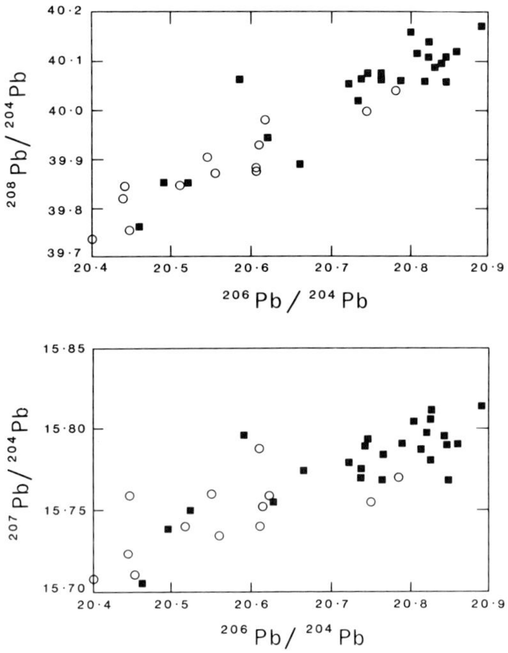

FIG. 12. $^{207}Pb/^{204}Pb$ versus $^{206}Pb/^{204}Pb$ and $^{208}Pb/^{204}Pb$ versus $^{206}Pb/^{204}Pb$ for the St Helena rock suite. The data indicate a source separation age of 2.4±0.2 Ga if interpreted as a mantle isochron.

incident with the thermal anomaly responsible for the volcanism. The chemically anomalous material was supplied throughout the eruptive history of St Helena, suggesting that the diapir was continually replenished rather than an isolated 'blob' that was passively heated. The studies of Hanan *et al.* (1986) and calculated hot-spot tracks (Morgan 1983) also support the longevity of a plume.

If the presence of a plume is accepted, then the potential zones in which mixing of the two components may occur can conveniently be divided into those that are primary and a direct consequence of the plume's presence and those that are secondary and occur subsequent to the departure of melts from the source region. In considering these interactions it is important to remember that the efficiency of mixing is likely to decline through time in response to a declining thermal flux. The decreasing volume of magmatic products during the transition from shield building to minor parasitic activity is the surface expression of this.

Interaction between the two end-members in the plume is likely to be influenced by the proportion of each component and their melting characteristics. The plume may possess either a high proportion of HIMU domains to surrounding matrix (small (metre) scale heterogeneity within the plume) or complete internal isolation of HIMU domains (large (kilometre) scale heterogeneity between the plume and its envelope). Because the HIMU signature associated with the plume increases with time it is not thought that variations in the volumetric proportions of the two components in the plume can account for the observed temporal variations. An increase in the proportion of HIMU to depleted mantle domains with time is counter-intuitive to the decline in thermal flux, which is likely to be coupled to a decreasing HIMU chemical flux.

We therefore appeal to preferential fusion of the HIMU component under the influence of a waning heat supply. Although knowledge of differences in solidi and melting behaviour of the HIMU and depleted mantle components is limited, this alternative satisfies the observations most readily. A primary interaction may also occur between the diapir and the surrounding envelope of relatively depleted material. At the inception of activity of a shield, a high heat flux would cause melting of the surrounding envelope. Clearly as the heat flux decreases the contribution to ascending melts will become largely restricted to the diapir. This effect will be accentuated if the diapir is stationary because only relatively refractory depleted material will remain adjacent to it.

The fact that the ascending diapir is not passive must also be considered; hence direct assimilation of depleted lithosphere may occur. Erosive thinning and subsolidus incorporation of the lithosphere have been suggested as efficient processes when the effect of secondary convection cells associated with an ascending plume is modelled (*eg* Yuen & Fleitout 1985). The importance of this process is difficult to establish, but it is evident that its efficacy is likely to decrease with time, which is again consistent with the results.

Secondary interactions can occur as melts formed by the combination of primary processes ascend. These interactions will occur at the base of, and within, the lithosphere. As melts ascend through the lithosphere incompatible elements will be scavenged from the surrounding depleted lithosphere as described by Navon & Stolper (1987). If the melt pathways possess a fixed geometry during the activity of a shield unit it is likely that mixing will decrease as the elements in the adjacent lithosphere become progressively depleted, or as the composition of the wall rocks approaches that of the ascending melt. Such a model is strongly dependent on the scale of the pathways and whether they are of a transient nature.

The general model that is envisioned draws upon all the alternatives outlined above, but we place particular emphasis on primary effects. We believe that the principal effects generating the mixing trends are interactions within the plume and with the surrounding envelope. This produces the clearest causal link between a declining thermal flux and the time-related geochemical variations. Following egress of melts from the source region secondary interactions between ascending melt and lithosphere may occur. These effects are superimposed on the mixing trends and obscure the main variations. This is believed to be a less important process as it does not satisfactorily explain the temporal geochemical variations unless lithospheric melt pathways are established throughout the activity of a shield and the geochemical contribution of the wall rock decreases throughout this time.

The possible processes by which the overall trends are generated can now be applied to samples that are discordant to the main trends. It is thought that the occurrence of samples reversing the trends at the top of shield units (*eg* SC 8 (NEMS) and SC 237, 238 (SWLS)) can most easily be explained as precursors to the next shield unit. It is necessary to invoke an overlap of high-level (crustal) magma conduits and mixing between the last melts of one phase of activity and the first of the next. Alternatively, if switching of melt pathways with renewed incorporation of adjacent depleted material is an important process, this could be invoked. Another alternative is that the eradication of heterogeneities could be effected by the establishment of long-lived high-level magma chambers. Thus, the difference between the relatively uniform SWUS and SWEF in which evolved lavas occur and the heterogeneous SWLS in which primitive types dominate could be explained if magma chambers in the SWLS are more transient and a more rapid throughput of material from depth to the surface occurs. This difference also corresponds to a difference in style of activity of the two shields: the SWUS has fewer discontinuities and pyroclastic horizons than the SWLS which represents less-stable shield activity.

The preliminary data from the new comprehensive sample set of St Helena volcanic rocks show a coherent pattern in isotopic and trace-element results. During the development of each shield an increase in incompatible-trace-element enrichment occurs which is coupled to a decrease in $^{143}Nd/^{144}Nd$ ratio and to strontium and lead isotope compositions that become progressively more radiogenic. This indicates a greater contribution from HIMU domains (with $^{87}Sr/^{86}Sr > 0.7030$, $^{143}Nd/^{144}Nd < 0.51282$ and $^{206}Pb/^{204}Pb > 20.8$) within the compositionally heterogeneous source region with time. The geochemical trends result from interactions between a dominant HIMU component and a depleted component which occurs in the asthenosphere and/or the lithosphere. An increase in both the elemental and isotopic signature of the HIMU component occurs during the development of each shield. This is likely to be achieved by a decreasing thermal flux acting on the base of the lithosphere. A larger HIMU signature results in the erupted products through the vertically integrated combination of mixing processes between source and surface.

ACKNOWLEDGMENTS: The cooperation of the Governor and people of St Helena is gratefully acknowledged. The authors thank J. P. Davison, G. R. Davies and D. C. Gerlach for helpful discussions. We thank P. Guise, R. Green and A. Grey for technical assistance and R. C. Boud and S. Brockbank for cartographic help. Special thanks are due to F. Buckley and N. Rogers for assistance with ICP and INAA preparation respectively, and the technical staff at Royal Holloway and Bedford New College for ICP analysis. DJC gratefully acknowledges receipt of a Natural Environment Research Council (NERC) research studentship. Isotope research facilities at Leeds University are supported by NERC (GR3/5932).

References

BAKER, I. 1968. Unpublished PhD thesis, University of London.

——, GALE, N. H. & SIMONS, J. 1967. Geochronology of the St. Helena volcanoes. *Nature* **215**, 1451–1454.

BOYNTON, W. V. 1984. Cosmochemistry of the rare earth elements; meteorite studies. *In*: HENDERSON, P. (ed.) *Rare Earth Element Geochemistry*. Elsevier, Amsterdam.

CHASE, C. G. 1981. Ocean island Pb: 2 stage histories and mantle evolution. *Earth and Planetary Science Letters* **52**, 277–284.

CHEN, C. Y. & FREY, F. A. 1985. Trace element and isotopic geochemistry of lavas from Haleakala volcano, East Maui, Hawaii. Implications for the origin of Hawaiian basalts. *Journal of Geophysical Research* **B10**, 8743–8768.

COHEN, R. S. & O'NIONS, R. K. 1982. Identification

of recycled continental material from Sr, Nd and Pb isotope investigations. *Earth and Planetary Science Letters* **61**, 73–84.

DALY, R. A. 1927. The geology of St. Helena Island. *Proceedings of the American Academy of Sciences* **62**, (2), 31–92.

DARWIN, C. 1876. *Geological Observations*, 2nd edn. Smith, Elder and Co., London.

DUNCAN, R. A., MCCULLOCH, M. T., BARSCZUS, H. G. & NELSON, D. R. 1986. Plume versus lithospheric sources for the melts at Ua Pou, Marquesas Is. *Nature* **322**, 534–538.

EMERY, K. O. & UCHUPI, E. 1984. *The Geology of the Atlantic Ocean*. Springer, New York.

FITTON, J. G., KILBURN, C. R. J., THIRLWALL, M. F. & HUGHES, D. J. 1983. The 1982 eruption of Mount Cameroon, West Africa. *Nature* **306**, 327–332.

GAST, P. W. 1969. The isotopic composition of lead from St. Helena and Ascension islands. *Earth and Planetary Science Letters* **5**, 353–355.

GRANT, N. K., POWELL, J. L., BURKHOLDER, F. R., WALTHER, J. V. & COLEMAN, M. L. 1976. The isotope composition of strontium and oxygen in lavas from St. Helena, South Atlantic. *Earth and Planetary Science Letters* **31**, 209–223.

HANAN, B. B., KINGSLEY, R. H. & SCHILLING, J.-G. 1986. Pb isotope evidence in the South Atlantic for migrating ridge–hotspot interactions. *Nature* **322**, 137–144.

HART, S. R. 1984. A large-scale isotope anomaly in the Southern Hemisphere mantle. *Nature* **309**, 753–757.

——, GERLACH, D. C. & WHITE, W. M. 1986. A possible new Sr–Nd–Pb array and consequences for mantle mixing. *Geochimica et Cosmochimica Acta* **50**, 1551–1557.

HOFMANN, A. W. & WHITE, W. M. 1982. Mantle plumes from ancient oceanic crust. *Earth and Planetary Science Letters* **57**, 421–436.

KAY, R. W. 1984. Elemental abundances relevant to the identification of magma sources. *Philosophical Transactions of the Royal Society of London, Series A* **310**, 535–547.

LABRECQUE, J. L., KENT, D. V. & CANDE, S. C. 1971. Revised magnetic polarity time scale for the Late Cretaceous to Cenozoic. *Geology* **5**, 330–335.

MACDONALD, K. A. & KATSURA, T. 1964. Chemical composition of Hawaiian lavas. *Journal of Petrology* **5**, 82–133.

MANHES, G., MINSTER, J. F. & ALLEGRE, C. J. 1978. Comparative uranium–thorium–lead and rubidium–strontium study of the Saint Severin Amphoterite: consequences for early solar system chronology. *Earth and Planetary Science Letters* **39**, 14–24.

MELLISS, J. C. 1875. *St Helena: A Physical, Historical and Topographical Description of the Island*. L. Reeve & Co., London.

MENZIES, M. & MURTHY, V. R. 1980. Nd and Sr isotope geochemistry of hydrous mantle nodules and their host alkali basalts: implications for local heterogeneities in metasomatically veined mantle. *Earth and Planetary Science Letters* **46**, 323–334.

MORGAN, W. J. 1971. Convection plumes in the lower mantle. *Nature* **230**, 42–43.

—— 1983. Hotspot tracks and the early rifting of the Atlantic. *Tectonophysics* **94**, 123–139.

NAVON, O. & STOLPER, E. 1987. Geochemical consequences of melt percolation: the upper mantle as a chromatographic column. *Journal of Geology* **95**, 285–309.

NEWSOM, H. E., WHITE, W. M., JOCHUM, K. P. & HOFMANN, A. W. 1986. Siderophile and chalcophile element abundances in oceanic basalts, Pb isotope evolution and growth of the Earth's core. *Earth and Planetary Science Letters* **80**, 299–313.

OVERSBY, V. R. & GAST, P. W. 1970. Lead from oceanic islands. *Journal of Geophysical Research* **75**, 2097–2114.

PALACZ, Z. A. & SAUNDERS, A. D. 1986. Coupled trace element and isotope enrichment in the Cook–Austral–Samoa islands, southwest Pacific. *Earth and Planetary Science Letters* **79**, 270–280.

POTTS, P. J., THORPE, O. W., ISAACS, M. C. & WRIGHT, D. W. 1985. High precision instrumental neutron activation analysis of geological samples employing simultaneous counting with both planar and coaxial detectors. *Chemical Geology* **48**, 145–155.

RICHARD, P., SHIMIZU, N. & ALLEGRE, C. J. 1976. $^{143}Nd/^{144}Nd$, a natural tracer, and application to oceanic basalts. *Earth and Planetary Science Letters* **31**, 296.

SCHILLING, J. G., THOMPSON, G., KINGSLEY, R. & HUMPHRIS, S. 1984. Hotspot-migrating ridge interaction in the S. Atlantic. *Nature* **313**, 187–191.

STILLE, P., UNRUH, D. M. & TATSUMOTO, M. 1986. Pb, Sr, Nd and Hf isotopic constraints on the origin of Hawaiian basalts and evidence for a unique mantle source. *Geochimica et Cosmochimica Acta* **50**, 2303–2319.

SUN, S.-S. 1980. Lead isotopic study of young volcanic rocks from mid-ocean ridges, ocean islands and volcanic arcs. *Philosophical Transactions of the Royal Society of London, Series A* **297**, 409–445.

TATSUMOTO, M. 1978. Isotopic composition of lead in oceanic basalts and its implication to mantle evolution. *Earth and Planetary Science Letters* **38**, 63–87.

THOMPSON, R. N. 1982. Magmatism of the British Tertiary volcanic province. *Scottish Journal of Geology* **18**, 50–107.

VIDAL, P. & DOSSO, L. 1978. Core formation: catastrophic or continuous? Sr and Pb isotope chemistry constraints. *Geology* **13**, 15–18.

VOLLMER, R. 1977. Terrestrial lead isotopic evolution and formation of the Earth's core. *Nature* **270**, 144–147.

WALSH, J. N. & HOWIE, R. A. 1980. An evaluation of the performance of an inductively coupled source spectrometer for the determination of the major and trace constituents of silicate rocks and minerals. *Mineralogical Magazine* **43**, 967–974.

——, BUCKLEY, F. & BARKER, J. 1981. The simultaneous determination of the rare-earth elements in rocks using inductively coupled plasma source spectrometry. *Chemical Geology* **33**, 141–153.

WEAVER, B. L., WOOD, D. A., TARNEY, J. & JORON, J.-L. 1987. Geochemistry of ocean island basalts from the South Atlantic: Ascension, Bouvet, St Helena, Gough and Tristan de Cunha. *In*: FITTON, F. G. & UPTON, B. G. J. (eds) *Alkaline Igneous Rocks*. Geological Society of London Special Publication **30**, pp. 233–267.

WHITE, W. M. 1985. Sources of oceanic basalts: radiogenic isotopic evidence. *Geology* **13**, 115–118.

YUEN, D. A. & FLEITOUT, L. 1985. Thinning of the lithosphere by small-scale convective destabilisation. *Nature* **313**, 125–128.

ZINDLER, A. & HART, S. R. 1986. Chemical geodynamics. *Annual Review of Earth and Planetary Sciences* **14**, 493–571.

——, JAGOUTZ, E. & GOLDSTEIN, S. 1982. Nd, Sr and Pb isotopic systematics in a three component mantle: a new perspective. *Nature* **58**, 519–523.

D. J. CHAFFEY, R. A. CLIFF & B. M. WILSON, Department of Earth Sciences, University of Leeds, Leeds LS2 9JT, UK.

Hygromagmaphile element distributions in oceanic basalts as fingerprints of partial melting and mantle heterogeneities: a specific approach and proposal of an identification and modelling method

J.-L. Joron & M. Treuil

SUMMARY: On the basis of the properties of hygromagmaphile (HYG) elements we propose a specific method of identifying and modelling the evolution of mantle geochemical properties using simple mass balance equations and taking into account a large volume of data obtained for basaltic series belonging to characteristic oceanic domains. This method uses both binary and ternary HYG element ratio diagrams to distinguish between the effects of partial melting and mantle heterogeneity. It is assumed that batch partial melting is the simpler and more suitable process of basaltic magma generation, which produces linear correlations on binary HYG element ratio diagrams. Melting of a homogeneous source is projected on ternary HYG element ratio diagrams as a single point approximating the composition of the mantle source.

We show that basalt HYG element ratios are closely dependent on four main factors, namely the nature of melting (batch partial melting), the degree of partial melting, the composition of the mantle and the mineralogy of the mantle.

Partial-melting trends and complex mantle heterogeneities are clearly indicated by binary diagrams, and a simple and regular suboceanic mantle HYG element array is obtained in ternary diagrams. Taking into account the close relation with isotopic trends, the HYG element array is considered to be the fingerprint of ancient mantle differentiation. It can be reproduced by a simple fractional crystallization process, and mostly excludes a large-scale and intensive chemical mixing process. These results are in good agreement with a marble cake mantle model, limiting convection effects to a mechanical stirring of the upper mantle, but they do not necessitate any recycling of crustal material into the suboceanic mantle.

The main results and arguments of a mainly personal and long-term study of mantle heterogeneities based on the geochemical properties and distribution of hygromagmaphile (HYG) elements in oceanic basalts are described. The report emphasizes the specific aspects of an identification and modelling method applied to a large number of results that we have obtained through Deep Sea Drilling Project (DSDP) legs and French oceanographic missions. Our approach uses the more regular and simple behaviour of HYG elements during magmatic processes determined by their structural properties and highly stable character in silicate melts (Treuil 1973*a,b*; Treuil *et al.* 1979). These properties are as effective during partial melting or other magmatic processes as during the fractional crystallization of a silicate melt.

Although trace-element distributions have been well established and broadly documented for volcanic rock series related by fractional crystallization processes, we meet with strong difficulties and have very few examples of convincing results concerning studies of trace-element distributions correlated to other magmatic processes, especially partial melting (Minster *et al.* 1977; Minster & Allègre 1978; Albarede 1984; Prinzhofer & Allègre 1985). In a first step, it was assumed that HYG element ratios in volcanic rocks are similar to the ratios of these elements in the solid source of the primary magmas (Gast 1968; Schilling 1973; Hofmann *et al.* 1986). We have emphasized that this estimate does not allow a precise identification and modelling of mantle heterogeneities without taking into account well-defined correlations between (HYG) elements analysed in over-sampled basaltic rock series. Such correlations may be used to decipher and model the respective contributions from magmatic processes and due to source properties (Joron & Treuil 1977; Joron *et al.* 1978; Treuil *et al.* 1982). Following this approach, and using simple and suitable diagrams pointing out these

From SAUNDERS, A. D. & NORRY, M. J. (eds), 1989, *Magmatism in the Ocean Basins*, Geological Society Special Publication No. 42, pp. 277–299

correlations for a large number of basaltic samples related to the main oceanic settings, we propose here to demonstrate and discuss HYG element fingerprints of mantle heterogeneities.

Assumptions and geochemical basis

Trace-element distribution in volcanic rocks: a simple model

Partial melting and fractional crystallization are the two main physical processes that must be taken into account during any assessment of magma generation and differentiation. Magma mixing, gaseous transfers and 'contamination' have also been pointed out as important contributors to isotopic, chemical and mineralogical characteristics in volcanic rocks. Nevertheless, with regard to trace-element distributions in whole-rock samples, specific mixing patterns cannot be clearly identified without first taking into account geochemical trends clearly related to source properties and melting and crystallization processes.

We initially assume that trace-element distributions in basaltic volcanic rocks integrate the effects of (i) their distribution in the mantle source; (ii) fractionation through partial melting and (iii) fractionation through fractional crystallization. The mechanism by which partial melting occurs, however, is poorly constrained. Physical considerations about melt rheology, percolation, hydrofracturing etc., as well as indirect geochemical results, lead us to prefer a stepped closed–system process (batch partial melting) (Minster *et al*, 1977; Joron *et al.* 1978; Treuil *et al.* 1982; Prinzhofer & Allegre 1985). We show below that our geochemical results agree with this assumption.

Another long-term topic of discussion concerns the concept and choice of partition coefficients. To avoid thermodynamical, kinetic, chemical and physical limitations, it is preferable to consider mass balance distribution coefficients (effective distribution coefficients). We have shown, by studying fractional crystallization in volcanic rock series, that 'bulk effective distribution coefficients' are more constrained than the different mineral distribution coefficients and can be considered as constant during each main step of the evolution course of the magmatic processes (Treuil 1973*a,b*; Treuil & Joron 1975; Treuil *et al.* 1979; Villemant *et al.* 1981).

Bearing in mind these remarks the concentration ratio C_1/C_2 of two elements E_1 and E_2 in a volcanic rock sample is given by the expression

$$\frac{C_1}{C_2} = \frac{C_1^{\circ}}{C_2^{\circ}} \times \frac{[F_f + D_2^f(1 - F_f)]}{[F_f + D_1^f(1 - F_f)]} F_c^{(D_1^c - D_2^c)} \qquad (1)$$

where C_1° and C_2° represent the concentrations of the two elements E_1 and E_2 in the solid source of the primary magma, D_1^f and D_2^f are the effective bulk solid–liquid distribution coefficients of E_1 and E_2 during melting, D_1^c and D_2^c are the effective bulk solid–liquid distribution coefficients of E_1 and E_2 during fractional crystallization and F_f and F_c are the relative weight fractions of melt in the melting and crystallization processes respectively. Even from this simplest formulation we can expect, for any element, complicated patterns resulting from the summed contributions of chemical source composition C°, mineralogical source composition D^f, degree of partial melting F_f, extent of fractional crystallization F_c and mineralogy and proportion of the crystallizing phases (D^c). It is difficult to unravel these different contributions, and to distinguish them from any possible 'mixing effects'.

Simplified distribution law for hygromagmaphile elements in basaltic rock series

As we quoted above, because they exist as highly stabilized species in the liquid phase, HYG elements are characterized by low effective bulk distribution coefficients during both partial melting and fractional crystallization: $D^f \ll 1$, $D^c \ll 1$. This property allows simplifications of the general expression (1).

Firstly, the simplification is more easily done for the fractional crystallization term $F_c^{(D_1^c - D_2^c)}$ than for the partial melting term. If F_c is not too small ($F_c > 0.5$) and as $D_1^c - D_2^c$ is the difference of two small numbers ($D_1^c - D_2^c \ll 1$) we satisfy $F_c^{(D_1^c - D_2^c)} \approx 1$. largely within the limits of the analytical errors. For example, for $F_c = 0.5$ and $D_1^c - D_2^c = 0.01$, we have $F_c^{(D_1^c - D_2^c)} = 0.993$. These conditions can easily be satisfied by using basaltic samples and checking for low extents of fractional crystallization by relatively high transition element contents in rock samples (*eg* scandium and cobalt concentrations higher than 30 ppm and up to 40 ppm respectively). In these conditions we may omit the fractional crystallization term from the expression (1) and consider

$$\frac{C_1}{C_2} = \frac{C_1^{\circ}}{C_2^{\circ}} \times \frac{[F_f + D_2^f(1 - F_f)]}{[F_f + D_1^f(1 - F_f)]} \qquad (2)$$

Secondly, let us consider that basaltic magmas are generated through small degrees of partial melting of the mantle sources ($F_f < 0.2$) or even very small degrees of melting ($F_f < 0.01$) (McKenzie 1985). So, only for highly HYG (H.HYG) elements ($D^f_2 < 0.01$ and $D^f_1 < 0.01$), we have

$$\frac{F_f + D^f_2(1 - F_f)}{F_f + D^f_1(1 - F_f)} \approx 1$$

It remains to check that the variations in this term are not significantly larger than either the analytical errors or variations in the element concentration ratio C^o_1/C^o_2 in the mantle source. Correlations between the H.HYG elements thorium, uranium, tantalum, niobium, lanthanum and cerium in oceanic basaltic samples will satisfy the oversimplified relation

$$\frac{C_1}{C_2} \approx \frac{C^o_1}{C^o_2} \quad (3)$$

These principles have been successfully applied to the N Atlantic ridge basalts sampled during various French, American, French–American and IPOD surveys (VEMA and FAMOUS surveys, Legs 37, 45–46, 49, 51–52). Distinct linear correlations passing through the origin are observed for samples taken in the neighbourhood of the Azores triple junction and for typical MORB tholeiites taken along the ridge at 22°N (Fig. 1). These correlations demonstrate that the above assumptions and equation (3) are justified in these cases at the scale of sampling and for the analytical errors considered, and that the two different basaltic populations (Azores triple junction and 22°N) are generated in two chemically distinct mantle regions. Moreover, the observed correlations contradict a simple melt-mixing model between two end-members for explaining the chemical characteristics of these basalts, as well as a mixing model between a depleted mantle source generating 22°N tholeiites and an enriched mantle source for explaining the observed chemical variations of the Azores triple junction (Joron & Treuil 1977; Joron *et al.* 1976, 1978; Bougault *et al.* 1979; Wood *et al.* 1979*c*, *d*; Joron *et al.* 1982).

According to the preceding remarks, equation (3) cannot be satisfied if we consider lower HYG (L.HYG) elements such as the heavy rare earth elements (REEs), zirconium or hafnium, nor if we attempt to distinguish small differences in mantle composition. Then we must consider at least relation (2). Relation (2) indicates that HYG element ratios in basalts depend on three factors:

(1) chemical heterogeneities of their mantle sources (C^o_1/C^o_2);
(2) mineralogical heterogeneities of their mantle sources (D^f_1, D^f_2);
(3) extent of partial melting (F_f).

Therefore, the problem is to decipher between these three main contributions by examining HYG element distributions in selected basaltic samples as previously indicated.

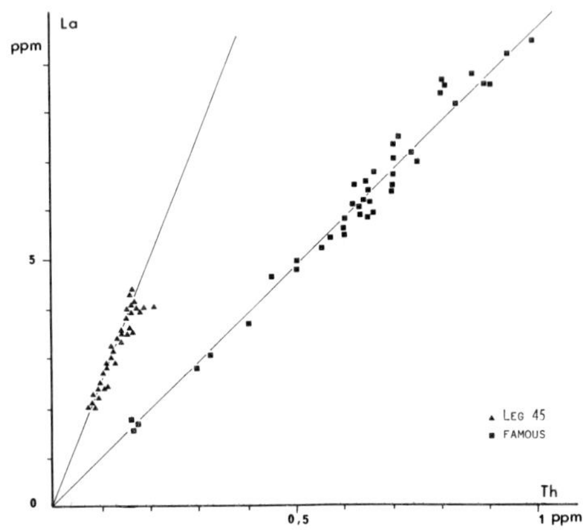

FIG. 1. Lanthanum versus thorium plots for North Atlantic Ridge FAMOUS zone basaltic samples and Leg 45 MORB tholeiites: ■, FAMOUS zone (*ie* 36°N, S flank of the Azores triple junction); ▲ Leg 45 (22°N, S Kane transform fault). Data sources are given in the text.

Discriminating between partial melting and mantle source heterogeneities by using $C_*/C_1 = C_*/C_2$ binary hygromagmaphile ratio diagrams

Basic equations and assumptions

Taking equation (2) into account and considering four HYG elements such as *, 1, 2 and 3, then C_*, C_1, C_2, C_3 represent their concentrations in relatively aphyric and weakly differentiated basaltic samples (see above), * denotes the highest HYG elements (H.HYG), generally thorium and in the limit uranium, 1 denotes slightly less high HYG elements, essentially tantalum, and in the limit niobium, lanthanum and cerium, praseodymium and neodymium, 2 and 3 denote lower HYG elements (L.HYG) such as intermediate and heavy REEs, hafnium and zirconium, C°_{*}, C°_{1}, C°_{2}, C°_{3} represent the concentrations of these elements in mantle sources, D^{F}_{*}, D^{F}_{1}, D^{F}_{2}, D^{F}_{3} denote their bulk solid–liquid effective distribution coefficients during partial melting and F_f denotes the degree of partial melting.

We obtain the following relations between these HYG element concentration ratios in basaltic samples:

$$\frac{C_*}{C_1} = \left[\frac{D^F_2 - D^F_1}{D^F_2 - D^F_*} \times \frac{C^\circ_*}{C^\circ_1}\right] + \left[\frac{D^F_1 - D^F_*}{D^F_2 - D^F_*} \times \frac{C^\circ_2}{C^\circ_1}\right] \times \frac{C_*}{C_2} \quad (4)$$

$$\frac{C_*}{C_1} = \left[\frac{D^F_3 - D^F_1}{D^F_3 - D^F_*} \times \frac{C^\circ_*}{C^\circ_1}\right] + \left[\frac{D^F_1 - D^F_*}{D^F_3 - D^F_*} \times \frac{C^\circ_3}{C^\circ_1}\right] \times \frac{C_*}{C_3} \quad (5)$$

Equations (4) and (5) relate element concentration ratios in basaltic samples generated from a homogeneous mantle source. They are viable only for HYG elements because of the above fractional crystallization statements. Moreover, as the D^F terms involve only differences and ratios between small and relatively constant effective distribution coefficients of elements belonging to a homogeneous geochemical family, we can consider that (4) and (5) are linear equations between concentration ratios of HYG elements in basaltic samples generated from the same homogeneous mantle source:

$$\frac{C_*}{C_1} = B_2 + A_2 \frac{C_*}{C_2} \quad (4')$$

$$\frac{C_*}{C_1} = B_3 + A_3 \frac{C_*}{C_3} \quad (5')$$

As stated above, for a set of HYG elements such as

$$D^F_* < D^F_1 \ll D^F_2 \text{ and } D^F_3 \ll 1$$

we can make the following remarks about these linear correlations.

(1) The terms $(D^F_2 - D^F_1)/(D^F_2 - D^F_*)$ and $(D^F_3 - D^F_1)/(D^F_3 - D^F_*)$ must be positive, lower than but not very different from unity, and so might satisfy $0 < B_2 \approx B_3$
(2) In an H.HYG-depleted mantle source, C°_*/C°_1 is small, and so B_2 and B_3 are approximately zero and the correlation lines expressed by (4) and (5) originate close to the origin (Fig. 2).
(3) In an H.HYG-enriched mantle source, C°_*/C°_1 is higher and the B_2 and B_3 coefficients satisfy $B_2 \approx B_3 \gg 0$; the correlation lines originate considerably above the origin.
(4) The slope of the correlation lines defined by A_2 and A_3 in (4) and (5) will also depend closely on the nature of the melting mantle source: A_2 and A_3 will be higher in an

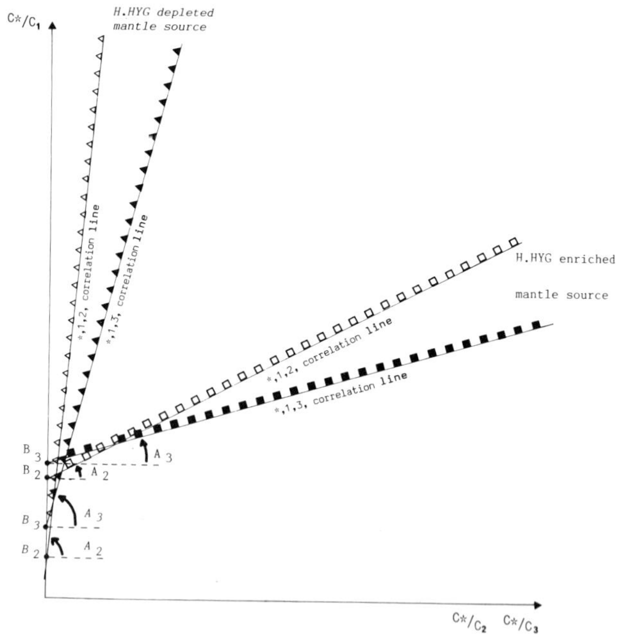

FIG. 2. Theoretical correlation lines corresponding to batch partial melting equations (4), (5), (4′) and (5′) in binary HYG element concentration ratio C_*/C_1, C_*/C_2 and C_*/C_3 diagrams.

H.HYG depleted environment (high C_2°/C_1° and C_3°/C_1°) than in an H.HYG-enriched environment (low C_2°/C_1° and C_3°/C_1°); in both cases A_2 and A_3 must be positive.

(5) In summary, the model will be validated by checking that analytical results effectively correlate and fit the theoretical curves in binary C_*/C_1, C_*/C_2 and C_*/C_1, C_*/C_3 diagrams (Fig. 2).

Analytical results and validation of the model

We have first considered examples of basaltic series from distinct but typical geotectonic settings and characterized by high concentrations of transition metal elements. Then we have compared several series belonging to the main oceanic fields, making comparisons with isotopic measurements for examples where isotopic and trace-element measurements have been made on the same samples.

Figures 3*a* and 3*b* illustrate the results obtained for two selected series (Joron *et al.* 1980*a*, *b*; Treuil *et al.* 1982). These two series include a typical H.HYG-depleted tholeiitic mid-ocean ridge basalt (MORB) series, East Pacific Rise and a transitional basaltic series from the Manda Hararo Rift (Ethiopia). For these two series we obtain good and clearly distinct linear correlation curves which can be related to partial-melting fingerprints applied to two distinct mantle sources. Furthermore, these distinct trends can be related to different isotopic properties. There is no doubt that comparisons between the isotopic properties and HYG element distributions in basalts is one of the most promising geochemical tools for investigations of magma genesis and mantle properties.

$^{143}Nd/^{144}Nd$ and $^{87}Sr/^{86}Sr$ ratios have been determined on Manda Hararo samples (L. Turpin, pers. comm.; Joron *et al.*, to be published). For Manda Hararo we observe that all the transitional basalt samples which are on the same correlation line have the same neodymium (0.512894 ± 9) and strontium (0.703 96) ratios. Three other samples which have a more tholeiitic and H.HYG-depleted character have a slightly higher neodymium ratio (0.512 909 ± 7) but not significantly lower strontium ratio (0.70393). Nevertheless there is a much greater difference for HYG element ratios than for neodymium ratios. This can be explained by a relatively recent separation or isolation of the two different mantle sources of these basalts. If we compare these Manda Hararo results

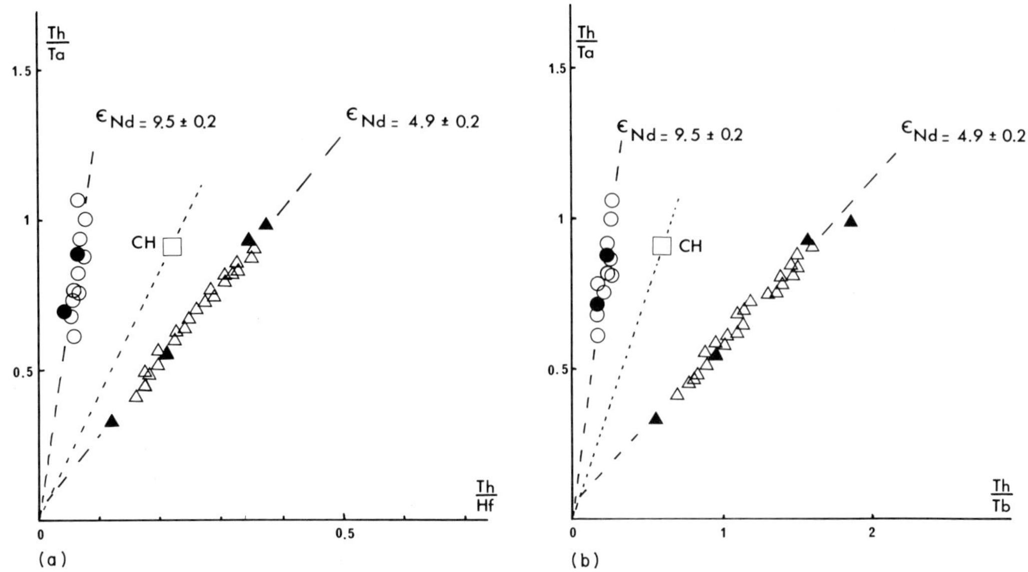

FIG. 3. (*a*)Th/Ta versus Th/Hf plots and (*b*) Th/Ta versus Th/Tb plots for the East Pacific Rise (near 12°50′N *Cyatherm* cruise) MORB tholeiites and Afar Manda Hararo Rift (Ethiopia) transitional basalts: ○, ●, East Pacific Rise samples; △, ▲, Manda Hararo samples; ●, ▲, samples for which we have isotopic data (●, Prinzhofer (1987), Prinzhofer *et al.* (in prep.); ▲, Turpin (pers. comm.)); CH, chondritic average (see the text).

with values given for East Pacific Rise samples ($^{143}Nd/^{144}Nd$ = 0.513110 ± 10 and $^{87}Sr/^{86}Sr$ = 0.702650 ± 20) (Prinzhofer *et al.*, in prep.) we note significant differences, *ie* this MORB tholeiite series shows a well-defined isotopic value.

Comparing basalts of the oceanic domain, we observe a good correlation between HYG element and isotope ratios within the spectrum of H.HYG-depleted MORB tholeiites and H.HYG-enriched alkali basalts. Figure 4 shows such a correlation using results for Zabargad Island alkali basalts and tholeiitic samples of the Red Sea floor in the neighbourhood of this island (Petrini *et al.* 1988).

Generally, for all the oceanic samples analysed, we can also check that good and distinct linear correlation curves can be distinguished for alkaline and tholeiitic series of Atlantic, Pacific or Indian oceanic basalts (Figs 5 and 6). If we consider the relative positions of the different correlation curves we may note the following:

(1) The correlation curves for tholeiitic and alkaline samples are located on either side of representative chondrite analyses. These positions agree with what can be expected from the theoretical equations (4) and (5) for primary magmas generated respectively from H.HYG-depleted and H.HYG-enriched mantle sources relative to a primitive chondritic mantle.
(2) The slopes and the ordinate axis intercepts of the different curves vary in agreement with equations (4) and (5) and our preceding remarks: tholeiitic series curves are characterized by a high positive slope and a low ordinate intercept; in contrast, alkaline series curves are characterized by a low positive slope and a generally higher ordinate intercept.
(3) The spectrum of these different curves cannot be definitively associated with a mixing model, in view of the restrictive number of end-members to explain mantle heterogeneities and possible crustal contamination.
(4) We shall show further that the evolution of the characteristics of these correlation lines is closely correlated with isotopic data.

In summary, we have confidence that each of these linear correlation lines is related to a homogeneous mantle source and that its characteristics are directly related to mineralogical and chemical properties of this source through melting processes. These results also suggest that batch partial melting is a suitable model, effectively describing the processes that govern the generation of mantle-derived magmas.

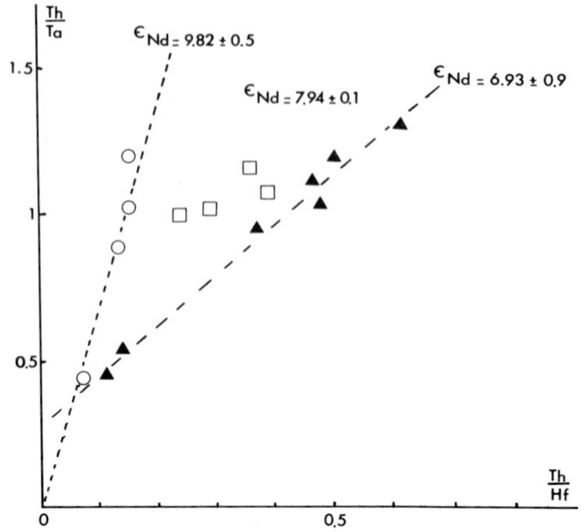

FIG. 4. Th/Ta versus Th/Hf plots for Red Sea Zabargad Island basaltic dykes (□ and ▲) and Red Sea axial ridge tholeiites (○) (Nereus deep) (ϵ_{Nd} data from Petrini *et al.* 1988).

The different partial-melting correlation lines in these binary HYG element diagrams demonstrate a complex heterogeneous mantle. We shall try now to establish the signifance and space–time relationships of these heterogeneities.

Fingerprints of mantle geochemical evolution using hygromagmaphile element ratio (C_*/C_1, C_*/C_2, C_*/C_3) ternary diagrams

Basic theoretical considerations

In previous sections we have demonstrated that the use of a bivariate representation of HYG element ratios measured in basalt samples allows us to decode the fingerprints of partial melting and mantle heterogeneity. This approach, however, remains a qualitative one and does not give direct estimates of mantle chemical composition. Considering equations (4) and (5), it can be stated that along each partial melting line the composition of the mantle source is located on the lower part of the line below the lowest C_*/C_1, C_*/C_2, C_*/C_3 ratios of basaltic samples. The insolvable nature of the coefficients of equations (4) and (5) precludes any more quantitative analysis.

In order to overcome this problem, we propose to identify the locus of the solid sources graphically by projecting each melting line in

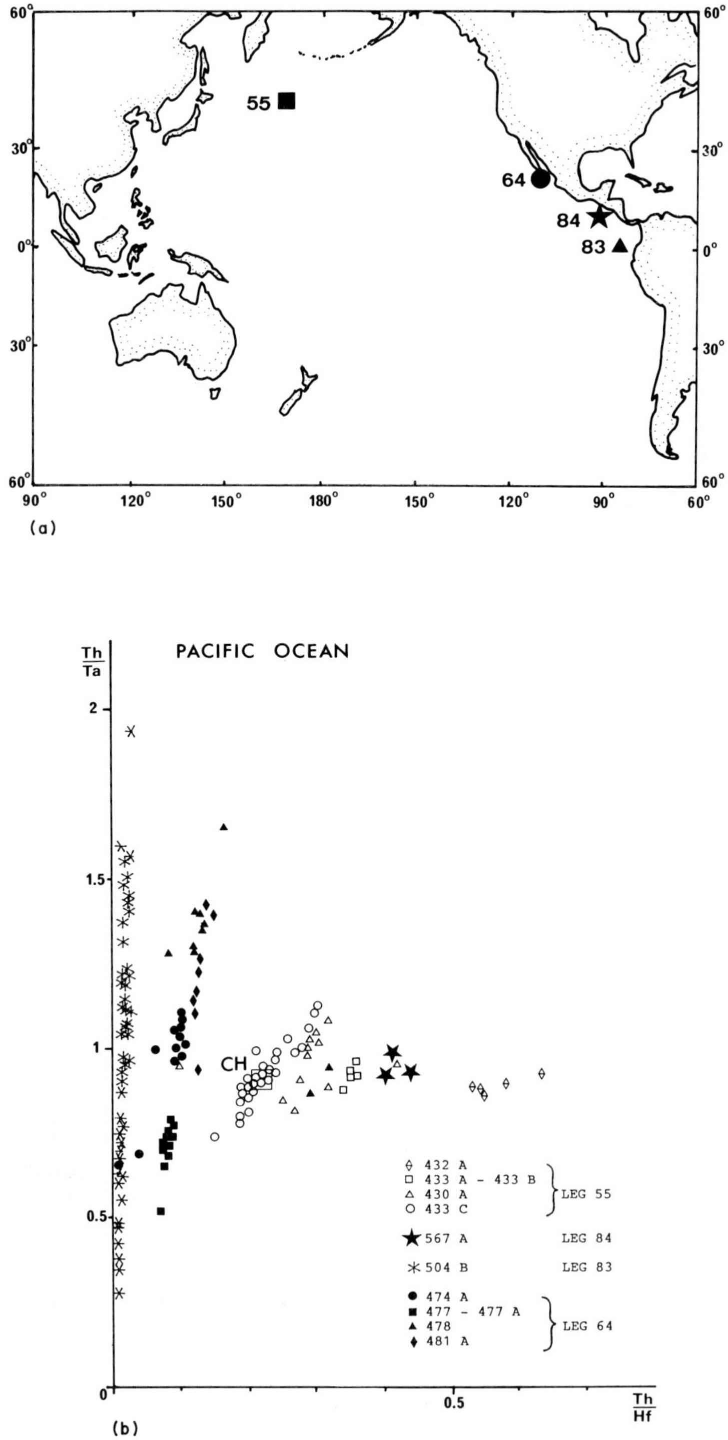

FIG. 5. (*a*) Location map of Pacific oceanic crust basaltic samples (numbers correspond to the different DSDP Legs); (*b*) Th/Ta versus Th/Hf plots and (*c*) Th/Ta versus Th/Tb plots for these different typical Pacific Ocean basaltic samples.

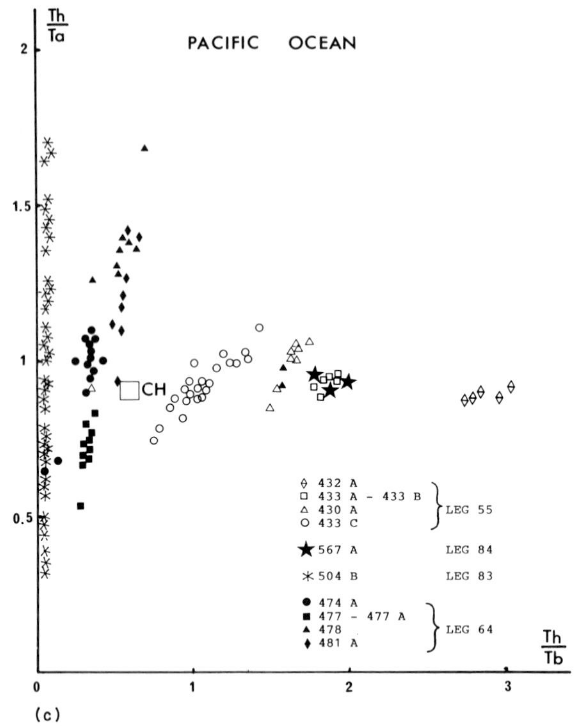

FIG. 5. (*cont.*)

the neighbourhood of a single point. This projection can be achieved to a good approximation by using ternary diagrams in which the variations of HYG element ratios C_*/C_1, C_*/C_2, C_*/C_3 are plotted. Following the same assumptions as in the previous sections, it can be demonstrated that in such a ternary diagram each pair of partial melting lines obtained in binary HYG ratio diagrams is projected on a linear correlation curve, the equation for which is

$$(B_3 - B_2)\left(\frac{C_*}{C_1}\right)^{\mathrm{T}} = A_2B_3\left(\frac{C_*}{C_2}\right)^{\mathrm{T}} - A_3B_2\left(\frac{C_*}{C_3}\right)^{\mathrm{T}} \quad (6)$$

with $$\left(\frac{C_*}{C_1}\right)^{\mathrm{T}} + \left(\frac{C_*}{C_2}\right)^{\mathrm{T}} + \left(\frac{C_*}{C_3}\right)^{\mathrm{T}} = 1 \quad (7)$$

where A_2, A_3, B_2 and B_3 are the same coefficients as in equations (4), (4′) and (5), (5′). Also

$$\left(\frac{C_*}{C_1}\right)^{\mathrm{T}} = \frac{C_*/C_1}{C_*/C_1 + C_*/C_2 + C_*/C_3}$$

$$\left(\frac{C_*}{C_2}\right)^{\mathrm{T}} = \frac{C_*/C_2}{C_*/C_1 + C_*/C_2 + C_*/C_3}$$

$$\left(\frac{C_*}{C_3}\right)^{\mathrm{T}} = \frac{C_*/C_3}{C_*/C_1 + C_*/C_2 + C_*/C_3}$$

It can easily proved that this correlation line has a very restricted extent in the neighbourhood of the theoretical point representing the solid source if the selected HYG elements satisfy the conditions defined above, namely

$$D_* < D_1 \ll 1 \text{ and } D_3 < 2D_2 < 1$$

The coordinates of the linear melting curve in the ternary diagram can be written

$$\left(\frac{C_*}{C_1}\right)^{\mathrm{T}} = \left[1 + \frac{1}{A_2}\left(1 - \frac{B_2}{C_*/C_1}\right) + \frac{1}{A_3}\left(1 - \frac{B_3}{C_*/C_1}\right)\right]^{-1} \quad (8)$$

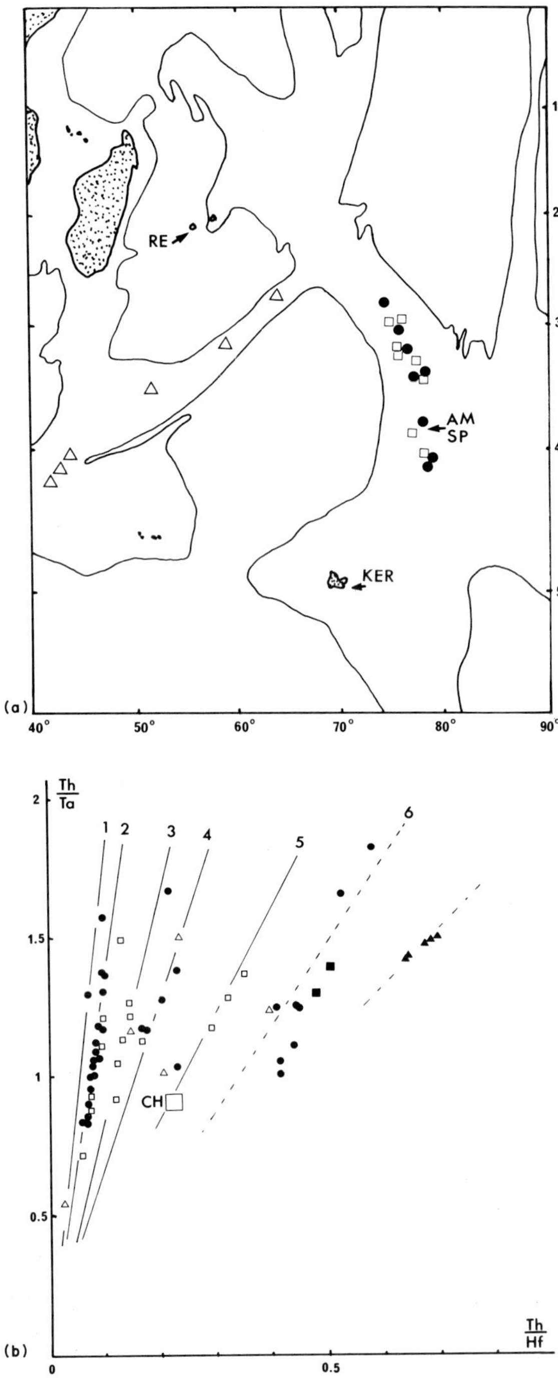

FIG. 6. (*a*) Location map of Indian oceanic crust basaltic samples dredged along the SE Indian Ridge and the SW Indian Ridge: △, samples taken during the MD 34 N/O Marion Dufresnes cruise; □, samples taken during the MD 37 N/O Marion Dufresnes cruise; ●, samples taken during the *Hydroamsterdam* N/O Jean Charcot cruise. Oceanic island samples: KER, Kerguelen; RE, Reunion; AM, Amsterdam; SP, St Paul. (*b*) Th/Ta versus Th/Hf plots and (*c*) Th/Tb versus Th/Tb plots for these different typical Indian Ocean basaltic samples.

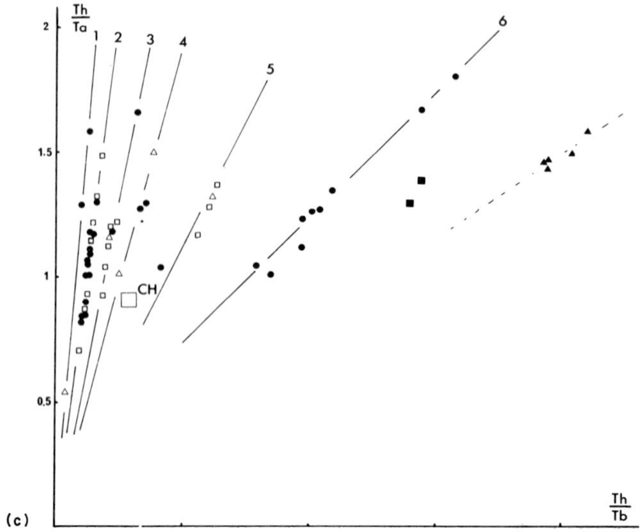

FIG. 6. (*cont.*)

$$\left(\frac{C_*}{C_2}\right)^{\mathrm{T}} = \frac{1}{A_2}\left(1 - \frac{B_2}{C_*/C_1}\right)\left[1 + \frac{1}{A_2}\left(1 - \frac{B_2}{C_*/C_1}\right) + \frac{1}{A_3}\left(1 - \frac{B_3}{C_*/C_1}\right)\right]^{-1} \quad (9)$$

$$\left(\frac{C_*}{C_3}\right)^{\mathrm{T}} = \frac{1}{A_3}\left(1 - \frac{B_3}{C_*/C_1}\right)\left[1 + \frac{1}{A_2}\left(1 - \frac{B_2}{C_*/C_1}\right) + \frac{1}{A_3}\left(1 - \frac{B_3}{C_*/C_1}\right)\right]^{-1} \quad (10)$$

In equations (8), (9), and (10) coefficients A and B must be expressed as a function of D^{F}, the effective bulk distribution coeficients, and $C°$, the HYG element content of the source. If we assume that $D^{\mathrm{F}}_* < D^{F}_1 \ll 1$ *and* $F_{\mathrm{f}} > D_1$ (F_{f} denotes the degree of partial melting) then we can use the approximation

$$\frac{C^{\circ}_*}{C_*}\,\frac{C_1}{C^{\circ}_1} \approx 1$$

and we get

$$\left(\frac{C_*}{C_1}\right)^{\mathrm{T}} \approx \frac{C^{\circ}_*/C^{\circ}_1}{C^{\circ}_*/C^{\circ}_1 + C^{\circ}_*/C^{\circ}_2 + C^{\circ}_*/C^{\circ}_3} = \left(\frac{C^{\circ}_*}{C_1}\right)^{\mathrm{T}}_{\circ}$$

$$\left(\frac{C_*}{C_2}\right)^{\mathrm{T}} \approx \frac{C^{\circ}_*/C^{\circ}_2}{C^{\circ}_*/C^{\circ}_1 + C^{\circ}_*/C^{\circ}_2 + C^{\circ}_*/C^{\circ}_3} = \left(\frac{C_*}{C_2}\right)^{\mathrm{T}}_{\circ}$$

$$\left(\frac{C_*}{C_3}\right)^{\mathrm{T}} \approx \frac{C^{\circ}_*/C^{\circ}_3}{C^{\circ}_*/C^{\circ}_1 + C^{\circ}_*/C^{\circ}_2 + C^{\circ}_*/C^{\circ}_3} = \left(\frac{C_*}{C_3}\right)^{\mathrm{T}}_{\circ}$$

This demonstrates that all the points of the linear melting curve have coordinates very close to those of the solid source; the lowest values of the C_*/C_1, C_*/C_2 and C_*/C_3 ratios in binary diagrams are also the closest to the solid source locus in the ternary diagram (Fig. 7).

When F_{f} is decreasing, the extent of the melting curve is restricted to a limit point with coordinates

$$\left(\frac{C_*}{C_1}\right)^{\mathrm{T}}_{l} = \left(\frac{C_*}{C_1}\right)^{\mathrm{T}}_{\circ}\left[1 + \frac{\dfrac{C^{\circ}_*}{C^{\circ}_2}\left(\dfrac{D^{\mathrm{F}}_2 + F}{D^{\mathrm{F}}_1 + F} - 1\right) + \dfrac{C^{\circ}_*}{C^{\circ}_3}\left(\dfrac{D^{\mathrm{F}}_3 + F}{D^{\mathrm{F}}_1 + F} - 1\right)}{\dfrac{C^{\circ}_*}{C^{\circ}_1} + \dfrac{C^{\circ}_*}{C^{\circ}_2} + \dfrac{C^{\circ}_*}{C^{\circ}_3}}\right]^{-1} = \left(\frac{C_*}{C_1}\right)^{\mathrm{T}}_{\circ}[1 + \alpha_1]^{-1}$$

$$\left(\frac{C_*}{C_2}\right)^{\mathrm{T}}_{l} = \left(\frac{C_*}{C_2}\right)^{\mathrm{T}}_{\circ} \times$$

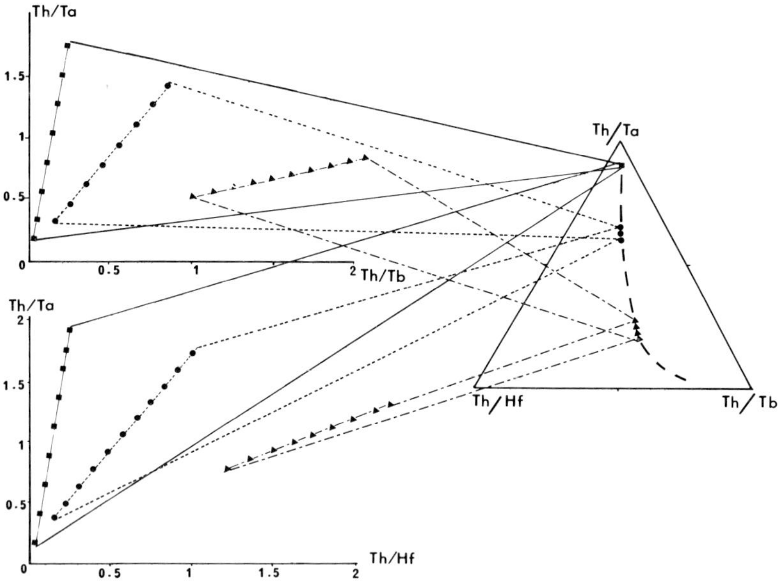

FIG. 7. Theoretical HYG element ratio ternary diagrams showing the representative correlation lines of batch partial-melting equations (4), (5), (4′) and (5′) in binary diagrams, and the projections of the linear batch partial-melting equations (6) − (10) on the ternary diagram. In such HYG element ratio diagrams, linear equations (6) and (7) must be represented by correlation lines restricted to very small areas. Note that the diagram is equivalent to 1/Ta, 1/Tb, 1/Hf plots. The ratios are normalized to chondritic average data. The following normalization values have been used (Bougault 1980; Bougault *et al.* 1985; Th = 0.028 ppm; Ta = 0.031 ppm; Hf = 0.128 ppm; Tb = 0.047 ppm.

$$\left[1+\frac{\frac{C_*^\circ}{C_1^\circ}\left(\frac{D_1^F+F}{D_2^F+F}-1\right)+\frac{C_*^\circ}{C_3^\circ}\left(\frac{D_3^F+F}{D_2^F+F}-1\right)}{\frac{C_*^\circ}{C_1^\circ}+\frac{C_*^\circ}{C_2^\circ}+\frac{C_*^\circ}{C_3^\circ}}\right]^{-1} = \left(\frac{C_*}{C_1}\right)_\circ^T [1+\alpha_2]^{-1}$$

$$\left(\frac{C_*}{C_3}\right)_1^T = \left(\frac{C_*}{C_3}\right)_\circ^T$$

$$\left[1+\frac{\frac{C_*^\circ}{C_1^\circ}\left(\frac{D_1^F+F}{D_3^F+F}-1\right)+\frac{C_*^\circ}{C_2^\circ}\left(\frac{D_2^F+F}{D_3^F+F}-1\right)}{\frac{C_*^\circ}{C_1^\circ}+\frac{C_*^\circ}{C_2^\circ}+\frac{C_*^\circ}{C_3^\circ}}\right]^{-1} = \left(\frac{C_*}{C_3}\right)_\circ^T [1+\alpha_3]^{-1}$$

It should be noted that α_1 can be above unity as F_f decreases, but, provided that $D^F{}_3 < 2D^F{}_2 + F < 1$ (which is always satisfied if $D^F{}_3 < 2D^F{}_2$) then $\alpha_2 < 1$ and $\alpha_3 < 1$. Also $\left[\frac{C_*}{C_3}\right]_1^T$ and $\left[\frac{C_*}{C_2}\right]_1^T$ are not very different from $\left[\frac{C_*}{C_3}\right]_\circ^T$ and $\left[\frac{C_*}{C_2}\right]_\circ^T$, and as $\left[\frac{C_*}{C_3}\right]_1^T + \left[\frac{C_*}{C_2}\right]_1^T + \left[\frac{C_*}{C_1}\right]_1^T = 1$ this limit always stays close to the solid source point.

We can check these principles on well-located basalt samples for which we have both trace-element and isotopic measurements: the transitional Manda Hararo Rift series and the East Pacific Rise tholeiite series (Figs 3*a* and 3*b*). Figure 8 clearly demonstrates that points plotting on the same partial melting line in HYG element ratio binary diagrams effectively project on the same spot in an HYG element ratio ternary diagram. For each melting line, samples have identical $^{208}Pb/^{204}Pb$ and $^{143}Nd/^{144}Nd$ ratios corresponding to the same homogeneous mantle source, and between the two different series, well identified by their different partial melting lines, we note substantial differences in $^{143}Nd/^{144}Nd$ ratios, according to what can be expected from H.HYG-element-depleted and H.HYG-element-enriched mantle

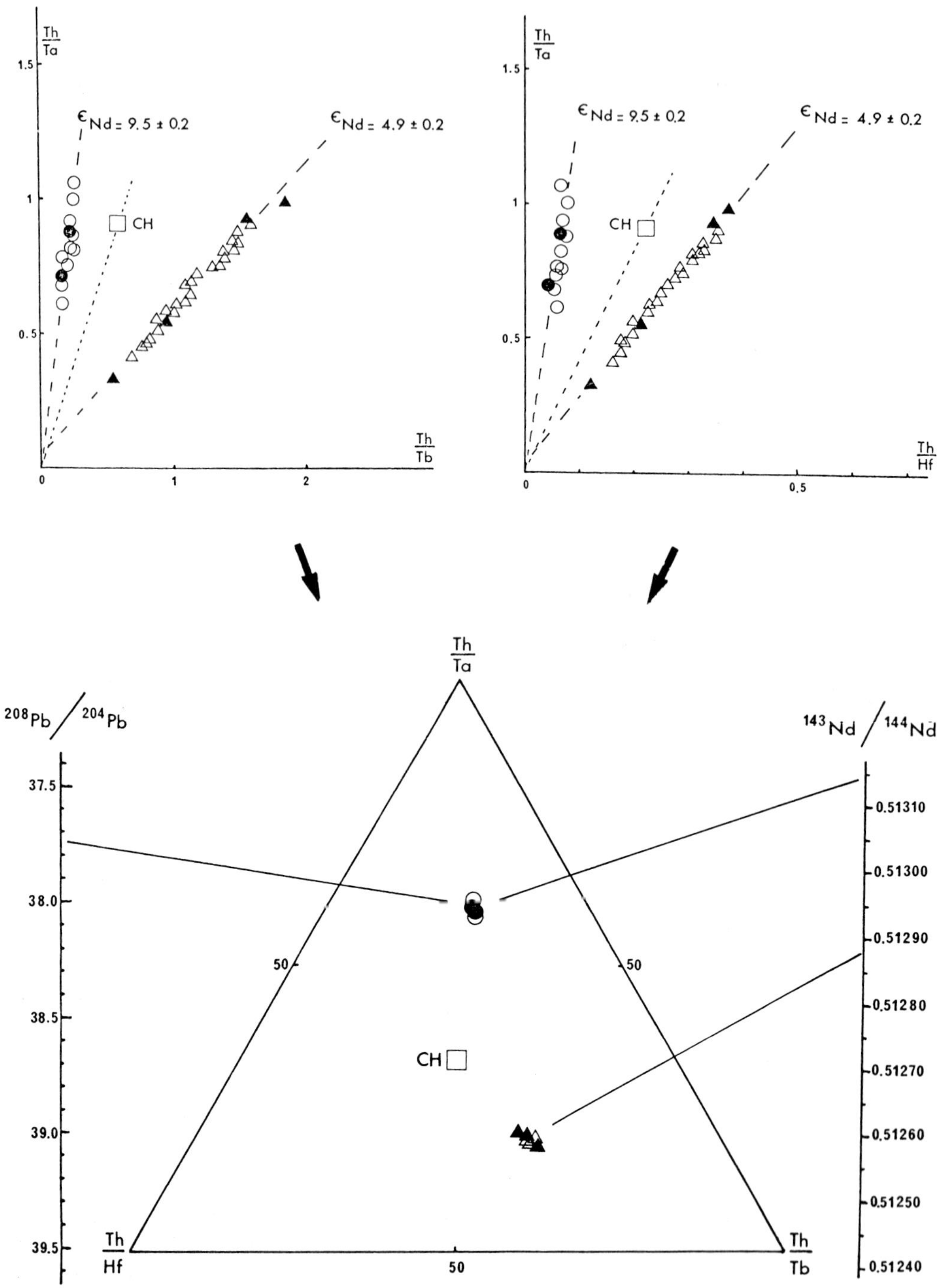

FIG. 8. Ternary diagram Th/Ta, Th/Hf, Th/Tb plots of the Manda Hararo Rift transitional basaltic series and the East Pacific Rise tholeiitic series. The two spots correspond to projected correlation lines obtained in binary diagrams for these series (Figs 3*a* and 3*b*). On each side of the ternary diagram $^{143}Nd/^{144}Nd$ and $^{208}Pb/^{204}Pb$ ratios illustrate the isotopic characteristics of the series (see values in the text).

sources. The simultaneous construction of binary and ternary diagrams is therefore necessary and in practice sufficient for checking that the foregoing principles and conditions are satisfied. These principles and results demonstrate that binary and ternary HYG element ratio diagrams allow partial melting and mantle source properties to be identified and give the basis of an efficient method for mantle geochemical investigations.

Practical representation and the evolution of the compositional properties of mantle sources

Practical construction

We propose the simultaneous use of binary HYG element ratio diagrams of the type C_*/C_1, C_*/C_2 and C_*/C_1, C_*/C_3 and ternary HYG element ratio diagrams of the type $(C_*/C_1)^T$, $(C_*/C_2)^T$ and $(C_*/C_3)^T$. The ternary diagrams are adjusted by normalization to chondrite average values (C^{ch}) (Bougault 1980; Bougault *et al.* 1985) in order to refer to the primordial Earth composition. In a ternary of diagram, of this type, *ie*

$$\left(\frac{C_*/C_1}{C^{ch}/C_1^{ch}}\right)^T_N, \left(\frac{C_*/C_2}{C^{ch}/C_2^{ch}}\right)^T_N, \left(\frac{C_*/C_3}{C^{ch}/C_3^{ch}}\right)^T_N$$

the chondrite average plots in the middle. Such a diagram is equivalent and can be constructed by plotting the values of

$$\left(\frac{1/C_1}{1/C_1^{ch}}\right)^T_N, \left(\frac{1/C_2}{1/C_2^{ch}}\right)^T_N, \left(\frac{1/C_3}{1/C_3^{ch}}\right)^T_N$$

Analytical results

The locations of the main sampling sites are shown in Figs 5*a*, 6*a*, and 11*a*. The data were obtained for oceanic basalts mainly taken during the IPOD surveys (Bougault *et al.* 1978, 1979, 1981, 1984; Joron *et al.* 1979, 1980, 1981; Wood *et al.* 1979 *a*, *b*, *c*, *d*, Cambon *et al.* 1980, 1983; Maury *et al.* 1982; Saunders *et al.* 1982; Etoubleau *et al.* 1983, Tual *et al.* 1985) and during French and foreign oceanic surveys conducted in the Indian Ocean and in the Afar and Issas region (Joron *et al.* 1980*a*, *b*, Maury *et al.* 1985). All the results are plotted in Figs 5, 6 and 8–11. This sampling allows a study and a comparison of the major oceanic regions (Pacific, N Atlantic, S Atlantic, Indian Ocean, and the Red Sea–Afar–Gulf of Aden). We shall restrict the illustration of the method by considering the results for the Pacific region for the elements thorium, tantalum, hafnium and terbium.

Figure 9 illustrates our results and corresponds to sampling of ridge, island and seamount volcanics. We note an excellent confirmation of the principles stated above, and make the following remarks.

(1) The points located on the same melting lines in the binary diagrams are grouped in a very small area in the ternary diagram.
(2) In the ternary diagram, all data are located on a single regular curve, which illustrates the evolution of the composition of the different solid sources of the mantle with respect to thorium, tantalum, hafnium and terbium.
(3) The regular curve, which is not time- or space-dependent, can therefore be considered as a fingerprint of a very simple trend affecting the suboceanic mantle sampled by magmatic processes.
(4) This curve is not linear and does not pass through the representative point of the chondritic average.
(5) The different types of basalts, and hence the mantle sources which gave rise to them, plot regularly from the Th/Ta apex in the order H.HYG-depleted MORB tholeiite sources, MORB transitional basalt sources close to a chondritic composition and H.HYG-enriched alkali basalts of the oceanic islands and seamounts.
(6) The other oceanic regions (Figs 10 and 11) present similar curves and evolutions, although small but significant differences are observed and must be carefully discussed. For example, the Pacific curve presents the greatest extension toward the upper apex (the strongly depleted part of the diagram).
(7) Along these curves, the isotope ratios are closely correlated, ranging from low $^{87}Sr/^{86}Sr$, $^{208}Pb/^{204}Pb$ and high $^{143}Nd/^{144}Nd$ ratios for H.HYG-depleted MORB tholeiites, to high $^{87}Sr/^{86}Sr$, $^{208}Pb/^{204}Pb$ and low $^{143}Nd/^{144}Nd$ ratios for H.HYG-enriched islands and seamount alkali basalts (Figs 10 and 11). This confirms the overlap between such trace–element representations and the isotope ratios, and by extension the ability of HYG elements to describe the chemical evolution of the mantle according to the principles and method presented here.
(8) The close correlation between HYG elements and isotope ratios implies that this curve does not reflect recent processes related to magmas genesis but is the effect

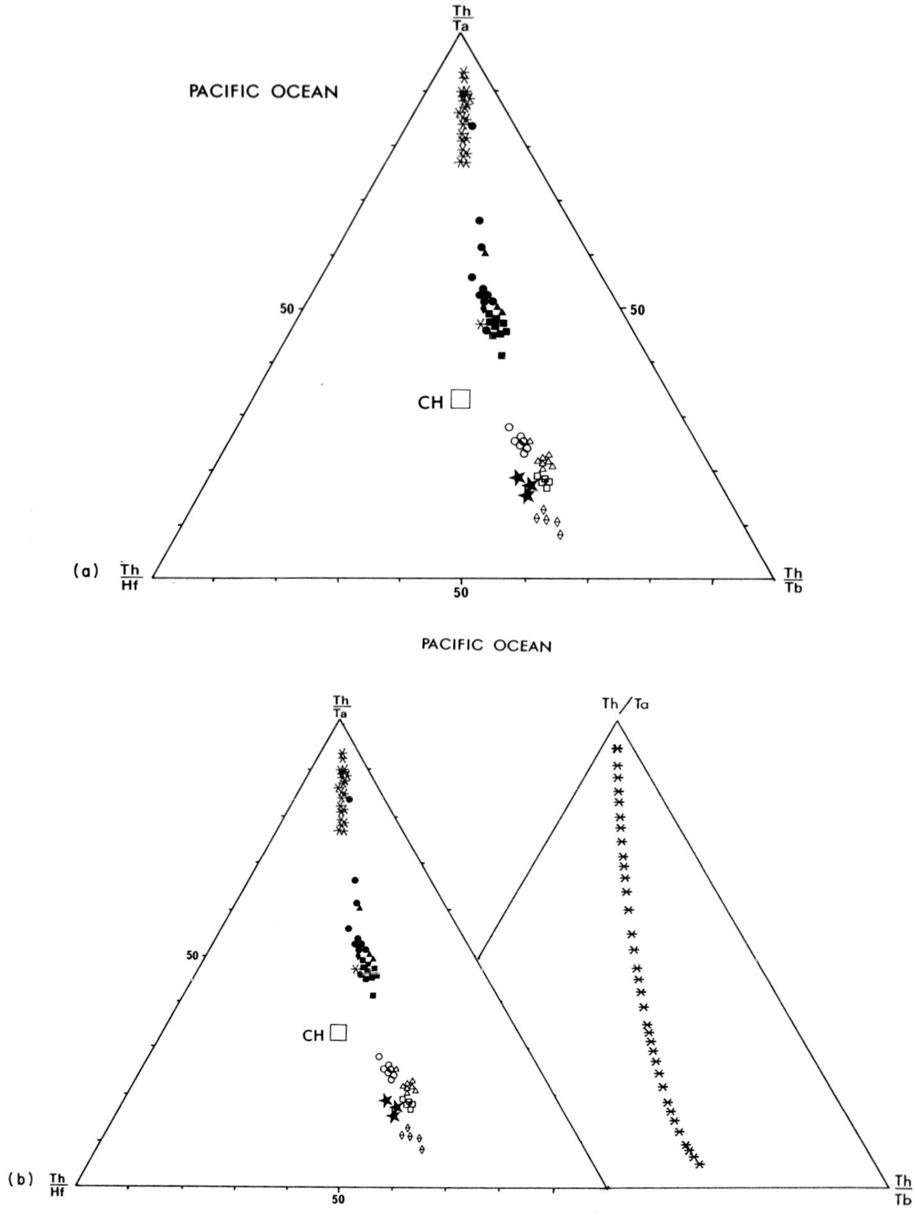

FIG. 9. (*a*) Th/Ta, Th/Hf, Th/Tb chondritic-average-normalized plots of Pacific oceanic crust basaltic samples. The symbols are the same as in Figs 5*b* and 5*c*. (*b*) Modelling of the Pacific suboceanic HYG element array. The best fit of a fractional crystallization model: $D_{Th} = 0.010$ (specified value, see text); $D_{Ta} = 0.025$ (calculated); $D_{Hf} = 0.70$ (calculated) $D_{Tb} = 0.80$ (calculated); F_d (limit) = 0.0004 (calculated).

of a more profound and ancient differentiation and zonation of the mantle.

(9) Conversely, important differences are obtained for basalts or andesites from converging plate boundaries. Figure 12 (Vanuatu Island arc samples) provides an illustration of this, and demonstrates that mantle metasomatism or crustal recycling into the mantle can be clearly identified and distinguished from suboceanic mantle evolution curves in these ternary HYG element ratio diagrams.

(10) Lunar basalts (Fig. 13) also plot on this curve, at an intermediate position near the

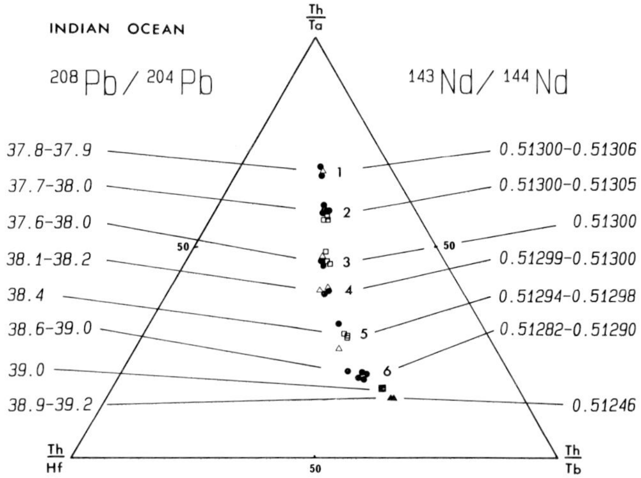

FIG. 10. Th/Ta, Th/Hf, Th/Tb chondritic-average-normalized plots of Indian Ocean basaltic samples. Symbols as in Figs 6*b* and 6*c*. The isotopic data are given for each spot (Dosso & Murthy 1980; Dupre & Allègre 1983; Hamelin *et al.* & Allegre 1985; Hamelin 1985; Dosso *et al.* 1988).

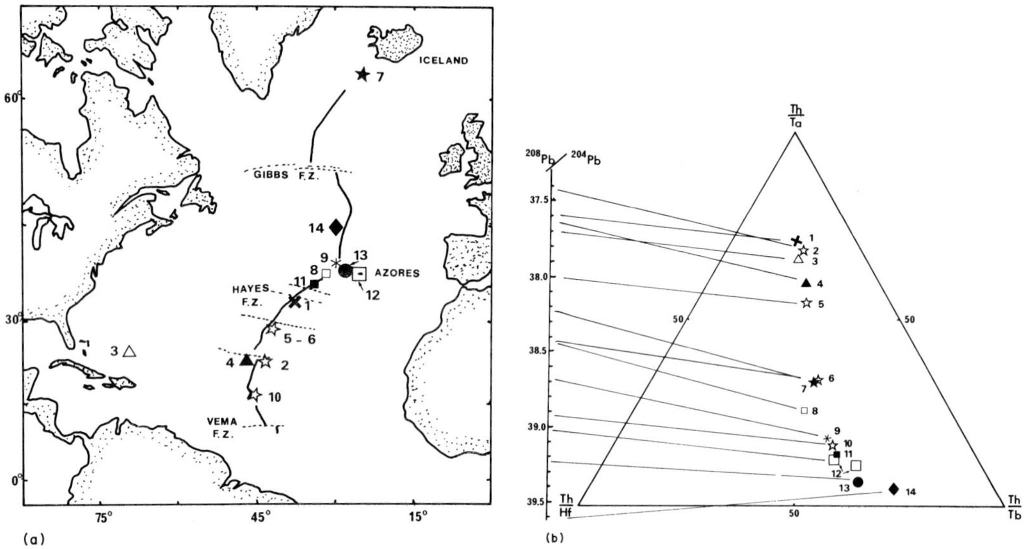

FIG. 11. (*a*) Location map of N Atlantic oceanic crust basaltic samples dredged or drilled along the Ridge, Azores triple junction and old crust: 9, Mapco I cruise; 1, Mapco II cruise; 2, 5, 6, 10, Vema cruise (2, Dr4; 5 and 6, Dr3; 10, Dr6); 4, Hole 395, Leg 45; 3, Hole 411, Leg 52; 7, Hole 409, Leg 49; 14, Hole 410, Leg 49; 8, Hole 412, Leg 49; 11, Hole 413, Leg 49; 12, Terceira Island; 13, Corvo Island. (*b*) Th/Ta, Th/Hf, Th/Tb chondritic-average-normalized plots of these typical N Atlantic oceanic crust basaltic samples, with isotopic data obtained on the same samples (Dupre & Allegre 1980; Dupre *et al.* 1981; Dupre *et al.* 1982; Hamelin *et al.* 1984).

chondrite value. Their location on the curve supports the hypothesis that the parent bodies of these samples had similar compositions and differentiation histories.

(11) Different chondritic and basaltic achondrite samples (Fig. 13) are relatively dispersed near the chondrite average plot. Thus, even if we take a chondrite average as a reference for the global Earth primitive composition, we must bear in mind

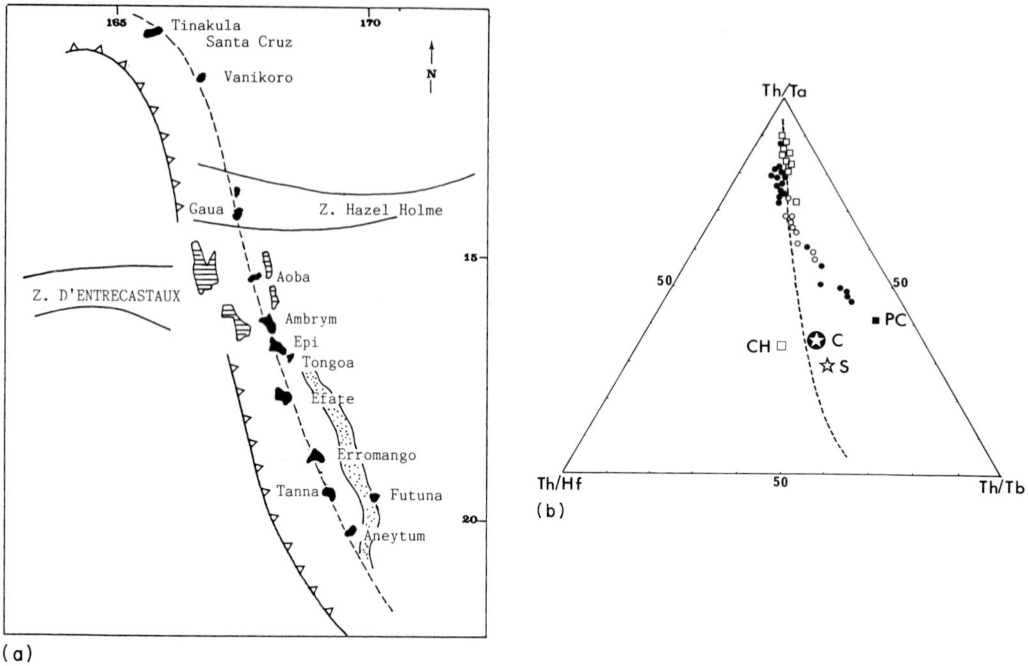

FIG. 12. (*a*) Location map of Vanuatu Island Arc basaltic samples. (*b*) Th/Ta, Th/Hf, Th/Tb chondritic-average-normalized plots showing a very different trend from that obtained for diverging plate boundary basaltic samples: □, Tanna; ○, Efate; ● Tongoa; CH, chondritic average; C, crustal average; S, shales average; PC, mantle contaminant end-member.

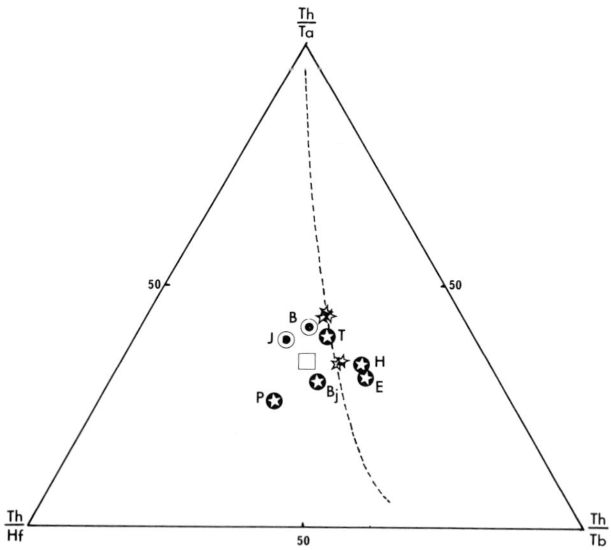

FIG. 13. Th/Ta, Th/Hf, Th/Tb chondritic-average-normalized plots for some lunar and meteoritic samples: ☆, lunar samples; ◉ achondrite samples (J, Juvinas; B, Bereba); ✪, chondrite samples (E, Estacado; P, Pultusk; Bj, Bjürbole; T, Tillaberi; H, Holbrook); □ chondrite average; ----, the Pacific suboceanic mantle trend.

the complex history of the parent bodies of these samples.

An attempt to model suboceanic mantle geochemical evolution by means of HYG element ratio ternary diagrams

Our purpose here is to show that the curves obtained in the ternary diagrams can be used for quantitative modelling of mantle geochemical evolution. We shall use, as an example, the Pacific array described in Fig. 9.

The model will be constrained by (i) the shape and regularity of the curve, (ii) the extent of the curve, (iii) the position of the curve relative to the chondrite average, which is considered as the bulk Earth (primordial) composition, (iv) the upper limit that can reasonably be considered for the effective global distribution coefficient of thorium, the most HYG element analysed which is taken as reference in the present method (with $D^\circ_* = D^\circ_{Th} \leq 0.010$) and (v) the correlation with isotope ratio variations, emphasizing that this evolution path is the fingerprint of an ancient and possibly long-term mantle differentiation. Assuming these constraints, let us consider how some simple and classical theoretical models (Fig. 14) could fit the evolution curve. These models can be mixing, partial melting or fractional crystallization.

The simplest model could be a simple binary mixing model between the two extreme end-members of a H.HYG-depleted MORB tholeiitic source and a H.HYG-enriched alkali basalt source. The curve which represents such a binary mixing relation (Fig. 14) does not fit the Pacific evolution path.

The same demonstration can be used to rule out a three end-member mixing model, taking into account the two end-members quoted above and as the third end-member either the average chondritic composition or the mean shales, or crustal, composition (Fig. 14).

These mixing models are also ruled out by the use of reverse ratio ternary diagrams: C_1/C_*, C_2/C_*, C_3/C_* (equivalent to C_1, C_2, C_3 diagrams, and also chondrite average normalized). In such a diagram (Fig. 15) a two-end member mixing model is represented by a straight line. Consequently, mixing models between H.HYG-depleted MORB tholeiite, H.HYG-enriched alkali basalt or chondritic mantle sources (Fig. 15) are unable to reproduce the Pacific curve (Fig. 16). Models of mixing between shales (or crustal) averages and H.HYG-depleted or even

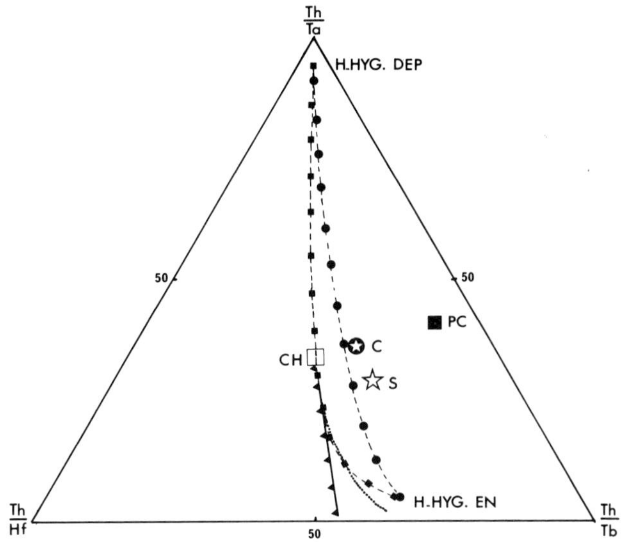

FIG. 14. Theoretical trends for the different mixing, partial melting and crystal fractionation models (see text for comments): curve a, ■, mixing between enriched and depleted poles, depleted and chondritic poles, and enriched and chondritic poles (the curves of the different mixing models cannot be distinguished); curve b, ▲, liquid compositions in the batch melting of a chondritic mantle, curve c, ···· liquid compositions in a crystal fractionation process; curve d, ●, bulk solid compositions in a crystal fractionation process. The solid evolution curve, corresponding to a fractional melting model, becomes nearly identical with the mixing model between chondrite average and H.HYG-depleted end-members. Other symbols as in Fig. 12*b*.

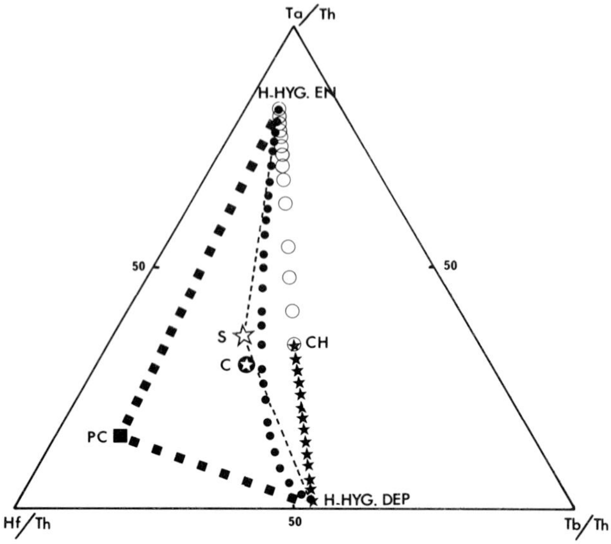

FIG. 15. Theoretical evolution trends for the different mixing, partial melting and fractional crystallization models, in inverse HYG element ratio ternary diagrams: ○, mixing between enriched and chondritic poles; ★, mixing between depleted and chondritic poles; ---, mixing between enriched and shales poles and depleted and shales poles; ■, mixing between enriched and mantle contamination end-members and between depleted and mantle contamination end-members; ●, bulk solid compositons in a crystal fractionation process. Other symbols as Fig. 12*b*.

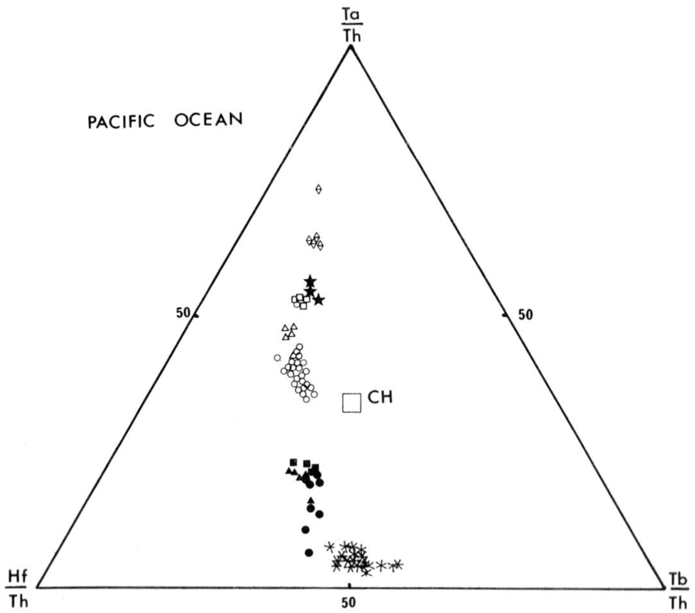

FIG. 16. Ta/Th, Hf/Th, Tb/Th chondritic-average-normalized plots for Pacific oceanic crust basaltic samples. Symbols as in Fig. 5.

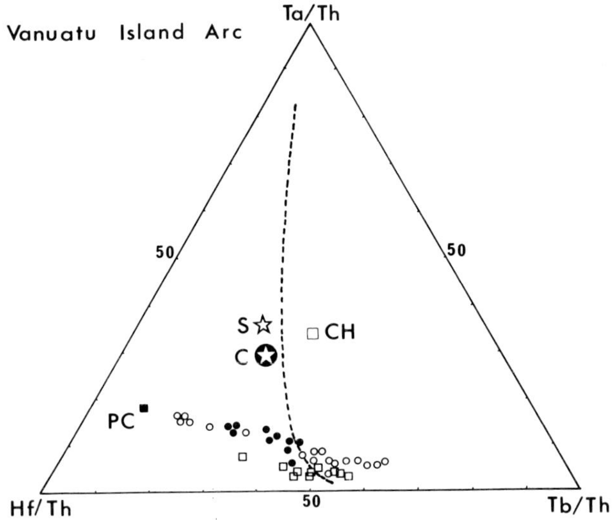

FIG. 17. Ta/Th, Hf/Th, Tb/Th chondritic-average-normalized plots for Vanuatu Island Arc samples. Symbols as in Fig. 12*b*.

H.HYG-enriched mantle sources also do not fit the Pacific evolution path. Even a complicated four end-member mixing relationship (shales (or crustal) average, chondrite average, H.HYG-depleted, H.HYG-enriched mantle sources) is unable to fit the complete HYG element mantle array perfectly. Additionally, Fig. 17 demonstrates that the main Vanuatu evolution path is consistent with a two end-member mixing model at least in terms of thorium, tantalum, hafnium and terbium space. In the cases of converging plate boundaries, HYG element mantle arrays can be explained by mixing processes involving a contaminant end-member (PC) the composition of which might be a hydrous carbonaceous silicate melt resulting from crustal material recycled into the mantle by lithospheric subduction.

The non-linear shape of the curve also rules out a closed-system partial melting model (batch partial melting). As shown above, such a model would produce a linear correlation in this diagram (Fig. 14).

The extension of the curve and the fact that it does not contain the chondrite average also rules out the model involving the fractional melting of a primitive mantle having a chondritic composition and involving evolution of the composition of the residual solids (Fig. 14).

On a theoretical basis, the fractional melting model of a primitive chondritic Earth and the subsequent compositional evolution of the liquids may produce the observed spread of data. Single-stage extraction and mixing of such liquids, however, are not consistent with the isotope data which indicate ancient differentiation of the mantle.

Finally, the most simple model able to fit this mantle HYG array perfectly is a fractional crystallization process involving the evolution of the successive solids from a chondritic liquid domain of the primitive mantle (Fig. 14). However, we cannot rule out more sophisticated or complex models, such as zone melting or combinations of simple physical processes. To go further, it will be necessary to extend the modelling to other HYG elements and to integrate more thoroughly with isotopic and geophysical investigations.

The main points of the preceding discussion are, first, that the regular mantle HYG array does not require complex physico-chemical process to have governed suboceanic mantle evolution and, second, that large-scale mixing models involving the main possible mantle or crustal end-members are not able alone, to account for the mantle geochemical variation. This conclusion is consistent with an upper-mantle marble cake model (Allègre & Turcotte 1986) showing that mechanical stirring operates without important chemical exchanges between the remaining 'slabs' of the ancient mantle zonation. With regard to this model, however, the preceding modelling does not suggest any important crustal recycling into suboceanic mantle.

Mantle chemical composition and zonation: a first quantitative estimate

This ancient differentiation can be considered as a progressive solidification of a more primitive mantle that has remained partly liquid. The solidification, accompanied by vigorous convection and heat transfer, can be marked by a settling of the solid crystallizing phases across a boundary layer, retention within a stagnant layer allowing their extraction from the liquid. This solid–liquid segregation is supported by results of convection experiments (Brandeis *et al.* 1984).

The composition of the liquids and solids during such a process obeys Rayleigh's law:

$$C_L^\circ = C^{ch} F_d^{\,D^\circ - 1}$$

for the liquid and

$$C^\circ = D^\circ C^{ch} F_d^{\,D^\circ - 1}$$

for the solids D° represents the effective bulk distribution coefficient during this primeval solidification of the mantle; it must not be confused with the effective bulk distribution coefficient D^F operating during the partial melting of the mantle. F_d represents the degree of solidification and hence the fraction of residual liquid. C^{ch} is the concentration in the original chondritic mantle, calculated here from the chondritic average and taking extraction of core material into account.

The variations in the ratios of the elements in the crystallizing solid obey the following laws:

$$\frac{C_*^\circ}{C_1^\circ} = \frac{D_*^\circ}{D_1^\circ}\frac{C_*^{ch}}{C_1^{ch}} F^{(D_*^\circ - D_1^\circ)}$$

$$\frac{C_*^\circ}{C_2^\circ} = \frac{D_*^\circ}{D_2^\circ}\frac{C_*^{ch}}{C_2^{ch}} F^{(D_*^\circ - D_2^\circ)}$$

$$\frac{C_*^\circ}{C_3^\circ} = \frac{D_*^\circ}{D_3^\circ}\frac{C_*^{ch}}{C_3^{ch}} F^{(D_*^\circ - D_3^\circ)}$$

The fit of the model, taking into account the characteristics of the experimental curve and the constraints listed above, allows the values of the parameters to be accurately determined (Fig. 9b):

$$D_*^\circ = D_{Th}^\circ = 0.010$$
$$D_1^\circ = D_{Ta}^\circ = 0.025 \pm 0.005$$
$$D_2^\circ = D_{Hf}^\circ = 0.70 \pm 0.05$$
$$D_3^\circ = D_{Tb}^\circ = 0.80 \pm 0.05$$
F_d varying from 1 to 0.0004

The variations in the element abundances in the upper mantle during this differentiation can also be calculated (Table 1). For these calculations the primeval mantle composition is deduced from chondritic average values and a pure mass balance effect of core segregation, taking into account that all the HYG elements remain in the mantle during this process. Furthermore, the value of $D_*^\circ = D_{Th}^\circ$ is taken to be equal to 0.010; higher values would be incompatible with the strong and ubiquitous HYG character of this element and with results of effective partition coefficient measurements (Treuil *et al.* 1979). Lower values would be incompatible with the extension of the experimental curve towards the H.HYG-depleted sources.

The calculated values of the effective bulk distribution coefficients are a good control of the coherence of the model, since they agree with published or estimated partition coefficients (Treuil *et al.* 1979; Villemant *et al.* 1981; Lemarchand *et al.* 1987). Relatively high values obtained for hafnium and terbium may be explained by a mineralogical control involving more hydrous- or clinopyroxene-type minerals than the present mantle mineralogy, as can be suspected in an incompletely degassed mantle.

The extreme value of the degree of solidification and hence of the ultimate fraction of residual liquid, is estimated at 0.0004 relative to the total mass of initial liquid. If we consider that this process has affected the whole upper mantle, the residual liquid would account for approximately 5% of the total mass of the present continents. This value is hence compatible with the mass of a primitive crust likely to be derived from extraction of this residual liquid.

TABLE 1. *Variations in element abundances in the upper mantle*

	Th	Ta	Hf	Tb
Mean of the chondrites	0.0280	0.0310	0.1280	0.0470
Primeval mantle (chondritic type)	0.0420	0.0465	0.1920	0.0705
Depleted type mantle	0.00042	0.00116	0.1344	0.0564
Intermediate type mantle	0.0401	0.1036	0.535	0.1417
Most strongly enriched mantle	0.97	2.39	1.40	0.27

Values in parts per million.

According to this model the crustal growth over geological time would have been 20 times greater than that of the protocrust. The last estimate is limited by the fact that flotation of light minerals may have contributed to crustal accretion. It is clear that such a low degree of solidification could only be achieved with the large amount of initial liquid involved in the model applied to the total upper mantle.

Conclusions

In the preceding discussion we have argued that basalt genesis in the mantle is governed by a batch partial-melting process. This assumption is confirmed by linear correlations obtained using binary HYG element ratio diagrams applied to different oceanic basaltic series and by the good agreement between these correlations and isotopic compositions. Moreover, these correlations result from relatively large variations of partial melting (F_f varying from 20% to 1%) and indicate complex chemical and mineralogical mantle heterogeneities.

In ternary HYG element ratio diagrams, the very limited fields of the different partial-melting trends are consistent with the theoretical principles of the method. These restricted fields also imply that, although the degree of melting might be small, it is not generally very much smaller than the lowest HYG effective global distribution coefficients, *ie* $F_f > 0.01$. Large variations in partial melting and low degrees of partial melting do not contradict the findings of McKenzie (1985). In this diagram each spot gives the precise mantle source locus of the corresponding melting trend.

These principles are also supported by the simple and regular suboceanic mantle HYG element arrays, closely correlated with isotopic variations, attesting the precise relationship with mantle chemical zonation and its ancient differentiation. The mantle evolution fingerprint does not necessitate complicated chemical mixing models and restricts the effects of mantle convection to limited mechanical stirring. This is in good agreement with the 'marble cake' mantle of Allègre & Turcotte (1986) but does not provide evidence of recycling of crustal material into the suboceanic mantle. The most realistic model, that of fractional crystallization and upper-mantle zonation by phenocryst cumulation, must be confirmed by further analysis and other geophysical and geochemical data. The calculated effective bulk distribution coefficients are in good agreement with known partition coefficients but involve a mineralogy different from that of the actual suboceanic mantle.

We have emphasized that the assumptions and principles of this method must be checked by the simultaneous use of both binary and ternary HYG element ratio diagrams applied to a large number of data. This allows an independent and fruitful comparison with isotopic data which is now one of the more promising challenges of future geochemical investigations.

ACKNOWLEDGMENTS: We are much indebted to Louis Raimbault, Laurent Turpin and Andrew Saunders for their critical review of the manuscript, their help in computing and English, and for all their very fruitful remarks and suggestions. Catherine Leroy and Anne Morel have kindly typed the manuscript. This work was supported by ATP 'Geodynamique' and 'IPOD' INSU–CNRS and IFREMER Institute and by CEA (DAMN) Grants with the constant and encouraging trust of Dr V. Ziegler and Dr J.Dardel and permanent friendship and scientific cooperation from H. Bougault.

References

ALBAREDE F. 1984. Inversion of batch melting equation and the trace element pattern of the mantle. *Journal of Geophysical Research* **88**, 10–573.

ALLÈGRE, C. J. & TURCOTTE, D. L. 1986. Implications of a two-component marble-cake mantle. *Nature* **323**, 123–127.

BOUGAULT, H. 1980. *Contribution des Eléments de Transition à la Comprehension de la Genèse des Basaltes Oceaniques*. Thèse Doctorat, Université Paris VII, 220 pp.

——, CAMBON, P., CORRE, O., JORON, J.-L. & TREUIL, M. 1979. Evidence for variability of magmatic processes and upper mantle heterogeneity in the axial region of the Mid-Atlantic Ridge near 22° and 36°. *Tectonophysics* **55**, 11.

——, JORON, J.-L., MAURY, R., BOHN, M., TARDY, M. & BIJU-DUVAL, B. 1984. Basalts from the Atlantic crust, West of the Barbados ridge (site 543, Leg 78): geochemistry and mineralogy. *In*: BIJU-DUVAL, B., MOORE, J.-C. *et al.* (eds) *Initial Reports of the Deep Sea Drilling Project*, **78**, 401–408.

——, ——, TREUIL, M. & MAURY, R. 1985. IPOD Leg 82. Local versus regional mantle heterogeneities. Evidence from hygromagmaphile elements. *In*: *Initial Reports of the Deep Sea Drilling Project* **82**.

——, MAURY, R.-C., EL-AZZOUZI, M., JORON, J.-L., COTTEN, J. & TREUIL, M. 1981. Tholeiites from sites 454–456. Basaltic andesites and andesites from sites 458–459 of Leg 60 of DSDP. Geochemistry, mineralogy and low partition coefficient elements. *In*: Hussong, D. M., Uyeda, S. *et al.* (eds) *Initial Reports of the Deep Sea Drilling Project* **60**, 567–677.

——, Treuil, M. & Joron, J.-L. 1978. Trace elements from 22°N and 36°N in the Atlantic Ocean: fractional crystallization, partial melting and heterogeneity of the upper mantle. *In*: Melson, W.-G., Rabinowitz, *et al.* (eds) *DSDP Leg 45. Initial Reports of the Deep Sea Drilling Project* **45**, 493–506.

Brandeis, G., Jaupart, C. & Allègre, C. J. 1984. De l'existence de couches stagnantes à la base des chambres magmatiques convectives. *Bulletin PIRPSEV* **84**, 54 pp.

Cambon, P., Joron, J.-L., Bougault, H. & Treuil, M. 1980. Leg 55 Emperor Seamounts: trace elements in transitional tholeiites, alkali basalts and hawaiites; mantle homogeneity or heterogeneity and magmatic processes. *In*: *Initial Reports of the Deep Sea Drilling Project* **55**.

——, ——, ——, & —— 1983. Basalts from the East Pacific Rise. An example of typical oceanic crust depleted in hygromagmaphile elements. *In*: Lewis, B.T.R., Robinson, P. *et al.* (eds) *Initial Reports of the Deep Sea Drilling Project* **65**, 623–633.

Dosso, L. & Murthy, V. R. 1980. A Nd isotopic study of the Kerguelen Islands: inferences on enriched oceanic mantle sources. *Earth and Planetary Science Letters* **48**, 268–276.

——, Bougault, H., Beuzart, P., Calvez, J. Y. & Joron, J.-L. (1988). The geochemical structure of the South East Indian Ridge. *Earth and Planetary Science Letters* **88**, 47–59.

Dupre, B. & Allègre, C. J. 1980. Pb–Sr–Nd isotopic correlation and the chemistry of the North Atlantic mantle. *Nature* **286**, 17–22.

——, & —— 1983. Pb–Sr isotope variation in Indian Ocean basalts and mixing phenomena. *Nature* **303**, 142–146.

——, Lambert, B. & Allègre, C. J. 1982. Isotopic variations within a single oceanic island: the Terceira case. *Nature* **299**, 620–622.

——, ——, Rousseau, D. & Allègre, C. J. 1981. Limitations on the scale of mantle heterogeneities under oceanic ridges. *Nature* **294**, 552–554.

Etoubleau, J., Corre, O., Joron, J.-L., Bougault, H. & Treuil, M. 1983. Costa Rica Rift. Variably depleted basalts in the same hole. *In*: Cann, J. R., Lanseth *et al.* (eds) *Initial Reports of the Deep Sea Drilling Project* **69**.

Gast, P. W. 1968. Trace element fractionation and the origin of tholeiitic and alkaline magma types. *Geochimica et Cosmochimica Acta* **32**, 1057.

Hamelin, B. & Allègre, C. J. 1985. Large scale regional units in the depleted upper mantle revealed by an isotope study of the South West Indian Ridge. *Nature* **315**, 196–199.

——, Dupre, B. & Allègre, C. J. 1984. Lead–strontium isotopic variations along the East Pacific Rise and the Mid Atlantic Ridge: a comparative study. *Earth and Planetary Science Letters* **67**, 340–350.

——, —— & —— 1985. Pb, Sr, Nd isotopic data of Indian Ocean Ridges: new evidence of large scale mapping of mantle heterogeneities. *Earth and Planetary Science Letters* **76**, 288–298.

Hofmann, A. W., Jochum, K. P., Seufert, M. & White, W. M. 1986. Nb and Pb in oceanic basalts: new constraints on mantle evolution. *Earth and Planetary Science Letters* **79**, 33–45.

Joron, J.-L. & Treuil, M. 1977. Utilisation des propriétés des éléments fortement hygromagmaphiles pour l'étude de la composition chimique et de l'hétérogénéité du manteau. *Bulletin de la Société Géologique de France, N° special Géochimie* **19**, (6), 1197.

——, Bollinger, C., Quisefit, J.-P., Bougault, H. & Treuil, M. 1979. Trace elements in basalts at 25°N, old crust, in the Atlantic Ocean: alteration, mantle and magmatic processes. *In*: Donhelly, T. W. *et al.* (eds) *Initial Reports of the Deep Sea Drilling Project* **51**, **52**, **53**, 1087–1098.

——, Bougault, H., Maury, R.-C. & Stephan, J.-F. 1981. Basalt from the Cocos plate, site 487, Leg 66: petrology and geochemistry. *In*: Watkins, J. S., Moore, J. C. *et al.* (eds) *Initial Reports of the Deep Sea Drilling Project* **66**, 731–734.

——, ——, Treuil, M. & Allègre, C.-J. 1976. Etude géochimique des roches magmatiques de la zone Famous et de l'Archipel des Açores. *Bulletin de la Société Géologique de France* **18** (7), 4, 811–818.

——, ——, Wood, D. A. & Treuil, M. 1978. Application de la géochimie des éléments en traces à l'étude des propriétés et des processus de genèse de la croûte océanique et du manteau supérieur. *Bulletin de la Société Géologique de France* **20** (4), 521.

——, Briqueu, L., Bougault, H & Treuil, M. 1980. East Pacific Rise. Galapagos spreading center and Siquieros fracture zone. Hygromagmaphile elements: a comparison with the North Atlantic. *In*: Rosendath, B., Hekinian, R. *et al.* (eds) *Initial Reports of the Deep Sea Drilling Project* **54**, 725–735.

——, Jaffrezic, H. & Treuil, M. 1982. Géochimie du manteau: distribution des éléments en trace dans les magmas basaltiques. 1. Corrélations entre éléments fortement hygromagmaphiles: exemple de l'Atlantique Nord. *Journal of Radioanalytical Chemistry* **71**, 1–2, 333–346.

——, Treuil, M., Jaffrezic, H. & Villemant, B. 1980*a*. Etude géochimique des éléments en traces dans les séries volcaniques du rift d'Asal. Identification et analyse des processus d' accrétion. *Bulletin de la Société Géologique de France* **22** (7), 6, 851–862.

——, ——, ——, ——, & Richard, O. 1980*b*. Géochimie des éléments en traces du magmatisme de l'Afar et de la mégastructure Mer Rouge–Afar–Golfe d'Aden. Implications pétrogenétiques et géodynamiques. *Bulletin de la Société Géologique de France* **22** (7), 6, 945–957.

Lemarchand, F., Villemant, B. & Calas, G. 1987.

Trace element distribution coefficients in alkaline series. *Geochimica et Cosmochimica Acta* **51**, 1071–1081.

Maury, R.-C., Bougault, H., Coutelle, A., Guennoc, P., Joron, J.-L. & Pautot, G. 1985. Présence de ferrobasite tholéitique dans la fosse Jean Charcot (26°15′N, 35°22′E): signification dans le contexte géodynamique de la Mer Rouge. *Comptes Rendus de l'Academie des Sciences, Paris* **300**, II, 16.

——, ——, Joron, J.-L., Girard, D., Treuil, M., Azema, J. & Aubouin, J. 1982. Volcanic rocks from Leg 67 sites: mineralogy and geochemistry. *In*: Aubouin, J., Von Huene, R. *et al.* (eds) *Initial Reports of the Deep Sea Drilling Project* **67**, 557–576.

McKenzie, D. 1985. The extraction of magma from the crust and mantle. *Earth and Planetary Science Letters* **74**, 81–91.

Minster, J. F. & Allègre, C. J. 1978. Systematic use of trace elements in igneous processes. Part III: Inverse problem of batch partial melting in volcanic suites. *Contributions to Mineralogy and Petrology* **68**, 37.

——, Minster, J. B., Treuil, M. & Allègre, C. J. 1977. Systematic use of trace elements in igneous processes. Part II: Inverse problem of the fractional crystallization process in volcanic suites. *Contributions to Mineralogy and Petrology* **61**, 49–77.

Petrini, R., Joron, J. L., Ottonello, G., Bonatti, E. & Seyler, M. (1988). Basaltic dykes from Zabargad Island, Red Sea: petrology and geochemistry. *Tectonophysics* **150**, 229–248.

Prinzhofer, A. & Allègre, J. C. 1985. Residual peridotites and the mechanisms of partial melting. *Earth and Planetary Science Letters* **74**, 251–565.

——, Lewin, E. & Allègre, C. J. (in prep.). Stochastic melting of the marble-cake mantle. Evidence from local study of the East Pacific Rise at 12°50′N. *Earth and Planetary Science Letters*, submitted.

Saunders, A. D., Fornari, D. J., Joron, J.-L., Tarney, J. & Treuil, M. 1982. Geochemistry of basic igneous rocks, Gulf of California, DSDP Leg 64. *Initial Reports of the Deep Sea Drilling Project* **64**, 2, 595–642.

Schilling, J. G. 1973. Iceland mantle plume: geochemical study of Reykjanes ridge. *Nature* **242**, 565–571.

Treuil, M. 1973*a*. *Critères pétrologiques, géochimiques et structuraux de la genèse et de la différenciation des magmas basaltiques. Exemple de l'Afar*. Thèse Doctorat d'Etat, Université d'Orleans.

——, 1973*b*. Critères volcanologiques, pétrologiques et géochimiques de la genèse et de la différenciation des magmas basaltiques. Exemple de l'Afar. *Bulletin de la Société Géologique de France* **15** (7), 401–644.

—— & Joron, J.-L. 1975. Utilisation des éléments hygromagmaphiles pour la simplification de la modélisation quantitative des processus magmatiques. *Rendiconti della Societa Italiana di Mineralogia e Petrologia* **31**, 125–174.

——, —— & Jaffrezic, H. 1982. Géochimie du manteau: distribution des éléments en trace dans les magmas basaltiques. 2. Proposition d'une méthode d'identification des effets de sources et de distinction de ceux de la fusion partielle et de la différenciation des magmas. Exemples des domaines d'expansion océanique. *Journal of Radioanalytical Chemistry* **71**, 1–2, 235–238.

——, ——, ——, Villemant, B. & Calas, G. 1979. Géochimie des éléments hygromagmaphiles, coefficients de partage minéraux/liquide et propriétés structurales de ces éléments dans les liquides magmatiques. *Bulletin de la Société Française de Minéralogie et de Cristallographie* **102**, 402–409.

Tual, E., Jahn, B.-M., Bougault, H. & Joron, J.-L. 1985. Geochemistry of basalts from hole 504 D (DSDP Leg 83), East Pacific Rise. *In*: Anderson, R. N., Honnorez, J.,Becker, K. *et al.* (eds) *Initial Reports of the Deep Sea Drilling Project*, **83**, 201–214.

Villemant, B., Jaffrezic, H., Joron, J.-L. & Treuil, M. 1981. Distribution coefficients of major and trace elements: fractional crystallisation in the alkali basalt series of Chaîne des Puys (Massif Central, France). *Geochimica et Cosmochimica Acta* **45**, 1997–2016.

Wood, D. A., Joron, J.-L. & Treuil, M. 1979*a*. A re-appraisal of the use of trace elements to classify and discriminate between magma series erupted in different tectonic settings. *Earth and Planetary Science Letters* **45**, 326.

——, ——, ——, Norry, M. & Tarney, J. 1979*b*. Elemental and Sr isotope variations in basic lavas from Iceland and the surrounding oceanic floor: the nature of mantle source heterogeneity. *Contributions to Mineralogy and Petrology* **70**, 319–339.

——, Tarney, J., Varet, J., Saunders, A. D., Bougault, H., Joron, J.-L., Treuil, M. & Cann, J. R. 1979*c*. Geochemistry of basalts drilled in the North Atlantic by IPOD, Leg 49. Implications for mantle heterogeneity. *Earth and Planetary Science Letters* **42**, 77–97.

——, Varet, J., Bougault, H., Corre, O., Joron, J.-L., Treuil M., Bizouard, H., Norry, M. J., Hawkesworth, C. J. & Roddick J. C. 1979*d*. The petrology, geochemistry and mineralogy of north Atlantic basalts: a discussion based on IPOD Leg 49. *In*: Luyendyk, B., Cann, J. K. *et al.* (eds) *Initial Reports of the Deep Sea Drilling Project* **49**, 597.

Jean-Louis Joron & Michel Treuil, Laboratoire de Géochimie comparée et systématique, U.A. 196 CNRS, Institut de Physique du Globe de Paris et UER des Sciences de la Terre, Université Pierre et Marie Curie, 4 place Jussieu, 75230 Paris Cédex 05, and Groupe des Sciences de la Terre, Laboratoire Pierre Süe, CNRS, Centre d'Etudes Nucléaires de Saclay, 91191 Gif sur Yvette Cédex, France.

Evaluation of rare gas data in relation to oceanic magmas

D. E. Fisher

SUMMARY: The isotopic compositions of the rare gases measured in oceanic rock samples present stimulating but incomplete information concerning the time and extent of magma degassing processes. Isotopic abundances of gases trapped from suboceanic mantle source regions are related to the depleted/undepleted mantle concepts, while atmospheric or crustal components form an important part of the data set. A reasonable story can be presented for the rare gas component from normal mid-ocean ridge basalt mantle source regions, but oceanic islands and island arc basalts result from more complex sources; the rare gas evidence is not yet sufficiently free of ambiguities and experimental uncertainties to provide a firm constraint to various models.

In this paper I attempt to present the basic concepts of rare gas elemental and isotopic geochemistry in relation to oceanic magmatism and mantle evolution. The idea is to outline what information about the suboceanic mantle and crust is potentially available from such studies, what has actually been learned so far, and what areas of disagreement and argument exist. I shall be concerned mainly with delineating mantle and subducted crustal sources for oceanic magmas, with particular emphasis on the concept of the depleted and undepleted mantle system, and the complications introduced into the interpretation of the rare gas data by the presence of an atmospheric-like component.

The origins of the field of oceanic rare gas studies are twofold. In 1968 clear indications of excess radiogenic argon were observed trapped in mid-ocean ridge basalt (MORB) glasses (Dalrymple & Moore 1968; Funkhouser *et al.* 1968; Noble & Naughton 1968). This was attributed to mantle production of radiogenic ^{40}Ar and its subsequent entrapment under the eruptive conditions of high hydrostatic pressure and low temperatures found on the ocean floor. Excess (primordial) ^{3}He was found in ocean waters the next year (Clarke *et al.* 1969), but it was not until 1975 that a successful search for this nuclide in basalts was established (Lupton & Craig 1975), and I will begin the story there.

Mid-ocean ridge basalts, oceanic island basalts and seamounts

Helium

Helium is a transient element on Earth. It comes in from above with cosmic rays and from below as it leaks upward from the mantle and crust, and it disappears out into space via gravitational loss. ^{3}He is a primordial nuclide (with miniscule radiogenic production) while ^{4}He is both primordial and radiogenic; the $^{3}He/^{4}He$ ratio observed in a given reservoir is therefore indicative of its origin.

The ratio is remarkably constant in a large number of MORB samples from various localities, and is generally referred to as $8R_a$ (± 0.5) where R_a is the atmospheric ratio (1.4×10^{-6}). Since the radiogenic $^{3}He/^{4}He$ ratio is about 10^{-8}, it is clear that the MORB values cannot be the result of radiogenic production plus atmospheric contamination. The MORB values are instead attributed to primordial ^{3}He residual in the mantle and trapped in the resulting eruptives.

The next major experimental result came when oceanic island basalts (OIBs) were found to have even higher values, about 20–50 R_a, indicating a higher proportion of primordial gas, which has been interpreted as evidence for a primitive undepleted mantle source—or at least one less depleted in volatile content than the MORB source (Craig & Lupton 1976; Kaneoka & Takaoka 1978; Kurz *et al.* 1982*a*, *b*). This conclusion seems incontrovertible. Details of the helium isotope distributions in Iceland argue for variable contributions from such a plume source rather than crustal contamination as the primary control of the helium signature (Kurz *et al.* 1985), but this cannot be used to argue against models of oceanic island formation largely as a mixture of subducted lithosphere plus a normal MORB source (Ringwood 1982; Zindler *et al.* 1984) since the very volatile helium may have been decoupled from the less volatile elements. It should be noted that many trace-element data can be interpreted as favouring the crustal mixing concept, and evidence involving recycling of the atmospheric/crustal components of the rare gases does

From Saunders, A. D. & Norry, M. J. (eds), 1989, *Magmatism in the Ocean Basins*, Geological Society Special Publication No. 42, pp. 301–311.

exist and is discussed further in the following sections.

In order to make a more complete story it would be of great use to correlate the helium isotope evidence with that from the other rare gases but, as discussed below, the evidence obtained from them is even more equivocal. In addition, Ozima & Zashu (1983) point out that in at least one geographical region, the East Pacific Rise (EPR) as sampled by CYAMEX glasses, there is no correlation between helium and the other rare gases, suggesting origins in different sources or at least in sources of differing dimension.

Neon

Neon isotope anomalies have been found in MORB, but the agreement between different laboratories is not good either on the data or on the interpretation. Poreda & Radicatti (1984) found enrichments of ^{20}Ne and ^{21}Ne relative to ^{22}Ne in Mid-Atlantic Ridge (MAR) samples, while Barraclough (1986) found similar results in samples mainly from the Pacific and Indian Oceans, but including some Atlantic MORBs. However, Kirsten & Richter (1981) found atmospheric isotopic values (with one possible ^{21}Ne excess) in MAR and EPR samples, and Ozima & Zashu (1983) found atmospheric values in their EPR and Deep Sea Drilling Project samples. It is difficult to judge whether these differences are real or experimental (but note that the Poreda & Radicatti (1984) data include measurements made in two different laboratories, a factor which argues for their acceptance).

Mass 21 enrichments would be expected to be nucleogenetic, produced by (α,n) reactions on ^{18}O and (n,α) reactions on ^{24}Mg, with the nuclear projectiles resulting from natural radioactivity in the mantle source, but the observed anomalies do not correlate well with uranium measured in the same MORB; although of course there is ample opportunity for U/Ne fractionation in the mantle–MORB generation processes. Still, it must be admitted that we do not yet know what these data mean.

The ^{20}Ne enrichments are probably the signature of a mantle component, the observed variations being due to varying amounts of an admixed atmospheric component. It has been suggested (see the section on Island arcs) that observed He/Ne ratios may reflect mixing of this atmospheric component (Craig *et al.* 1978; Sano *et al.* 1986), although some of the variations seem to be too large to be explained in this manner and may represent a fractionation process. Further information by Sarda and coworkers has been published in the form of abstracts only and will not be reviewed here. To summarize, at the present time the neon data are intriguing but largely uninformative.

Argon

The argon story is potentially more useful than that arising from the light rare gases but is also more difficult to assess. Both of these aspects arise from the facts that argon is a gas heavy enough to be retained in the Earth's atmosphere and it has a large radiogenic component. Since the cosmic ray contribution to the argon terrestrial abundance is negligible, and since the atmosphere provides a permanent sink for degassed argon (if we ignore for the moment possible recycling effects), internal Earth isotopic ratios might be compared with those in the atmosphere and calculations made relevant to the 'age' of the atmosphere, rates of terrestrial degassing, and 'residence' or 'transit' times of the argon isotopes as they travel through the mantle. Unfortunately all such calculations are complicated by parent–daughter fractionation parameters which are unknown and by the presence of variable amounts of atmospheric argon which have somehow become incorporated into the samples and which, at least so far, cannot be accounted for quantitatively.

Three sets of representative argon data in MORB and seamount glasses are summarized in Table 1. At first glance the data show a bewildering variety of isotopic ratios, from less than 1000 to greater than 20 000, but generally they are much greater than the atmospheric ratio of 295. The absolute abundances of ^{36}Ar are also variable, generally correlating inversely with the $^{40}Ar/^{36}Ar$ ratios. When different samples from one dredge haul are measured, this inverse correlation approaches a straight-line relationship, as shown in Fig. 1, and indicates that each of the samples in any one dredge haul has trapped a constant amount of mantle argon with a specific $^{40}Ar/^{36}Ar$ value and has incorporated in addition a variable amount of atmospheric argon. (See Fisher (1986) and references therein for a discussion of this point.) Marty *et al.* (1983), using a combination of crushing and heating experiments, found that the trapped mantle argon ($^{40}Ar/^{36}Ar = 10\,524$) in one MORB glass is concentrated in vesicles, while the argon dissolved in the glass phase ($^{40}Ar/^{36}Ar = 385$) displays evidence of atmospheric contamination.

All this suggests that the range in measured $^{40}Ar/^{36}Ar$ ratios may result from a unique value in the MORB mantle source region, analogous

TABLE 1. *Representative argon data in glasses*

Reference	No. of samples	$^{40}Ar/^{36}Ar$ range (average)	^{36}Ar
MORBs and seamount			
Sarda *et al.* 1985	31	980–24 400 (9000)	0.3–7.94
Fisher 1986	19	825–15 000 (5500)	0.7–11
Ozima & Zashu 1983	10	693–5564 (2600)	2.1–9.1
OIBs			
Hart *et al.* 1983	5	302–436 (346)	2.4–4.5
Allegre *et al.* 1983	3	313–1149 (618)	1.6–12[a]

Units for ^{36}Ar are 10^{-10} cm^3 g^{-1}
This table is *representative*, not all inclusive, intended to show the range in values of both absolute abundances and isotopic ratios. Oceanic samples from subduction zones and abyssal planes generally exhibit lower isotopic ratios and higher absolute abundances than MORBs (Ozima & Zashu 1983).
[a]Includes correction to published values (Staudacher, pers. comm.).

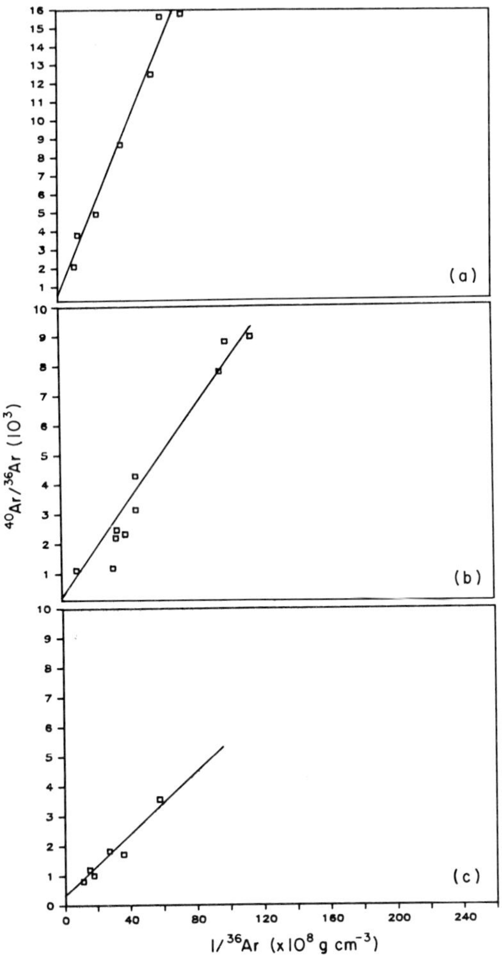

FIG. 1. $^{40}Ar/^{36}Ar$ versus $1/^{36}Ar$ for three dredge hauls of glass MORB from the EPR (from Fisher 1986). Sample i.d. nos: (a)^{32}P, (b)^{40}P, (c)^{44}P.

to the MORB $^3He/^4He$ value, modified in the case of argon by varying amounts of atmospheric 'contamination'. (The word does not imply post-eruptive contamination but simply the introduction of an atmospheric-like component into the samples by any means. Since the $^{36}Ar/^3He$ ratio for the atmosphere is approximately 45 000, it is understandable that the atmospheric component is more noticeable for argon than for helium.)

I have suggested that a MORB mantle source value of $^{40}Ar/^{36}Ar \approx 15\,000 \pm 3000$ combined with varying amounts of atmospheric argon to produce lower observed ratios would account for the observations (Fisher 1986). At the present time, nearly all published analyses are equal to or lower than the suggested upper boundary of 18 000; whether the one exception (published in both Allegre *et al.* 1983 and Sarda *et al.* 1985) is meaningful or not requires confirmation and duplication of the experimental datum. One vacuum-crushed sample also gives a higher value (Jambon *et al.* 1985) but if I read the error bars correctly they extend well into my quoted range. One or two other values reaching the range above 20 000 have been reported in abstracts but the details are not yet published. The difference between 1.8×10^4 and 2.3×10^4 is, in any case, not at all important for any of the arguments which follow.

Calculations comparing the trapped mantle and atmospheric values to determine the rate of degassing were pioneered by Ozima and coworkers (Ozima & Kudo 1972; Ozima 1973, 1975; Ozima & Alexander 1976). It has been my contention that slow degassing continuing throughout geological time can satisfactorily explain the argon data, albeit with large uncertainties (Fisher 1978). Allégre *et al.* (1983) contend that the xenon data imply sudden degassing early in the Earth's history, but I have argued that the xenon data actually allow up to about 65% later (slow) degassing (Fisher 1985*a*); the possibility also exists that the ^{129}Xe anomaly on which their calculations are based may be due in part to mass fractionation in addition to iodine decay, which would further dilute the early degassing conclusion (see the next two sections). It seems reasonable that both early and late degassing have been important, but precise calculations of the extent of each are not definitive at the present time.

Upper limits to mantle residence times of radiogenic ^{40}Ar can be estimated by comparing trapped ^{40}Ar with potassium contents measured in MORB glasses. (The upper limit follows from the fact that the Ar/K ratio in the glasses must be equal to or greater than the corresponding source ratio.) The result is an upper limit of 1 billion years, in good agreement with the results of calculations based on other isotopic systems and providing evidence that the degassing of the mantle source region continued long after any early Earth degassing 'catastrophe' (Fisher 1985*a*). An upper limit of 2 billion years has been obtained for glasses from the Mariana Trough region, leading to the same conclusion about continued late degassing (Sano *et al.* 1986). The particulars of my calculation have been attacked and defended; for details see Staudacher & Sarda (1987) and Fisher (1987).

The ^{40}Ar/^{36}Ar ratios in OIBs are distinctly different from those in MORBs (Table 1). One interpretation is that these data are analogous to the OIB helium numbers, and reflect a non-depleted mantle source with a higher content of primordial gas, in this case ^{36}Ar (Kaneoka & Takaoka 1980; Allégre *et al.* 1983; Hart *et al.* 1983). There is a difference, however: the characteristic high ^{40}Ar/^{36}Ar values of MORB glasses (more than 10 000) are all associated with ^{36}Ar values lower than those observed in OIBs (less than 10^{-10} cm^{3}g^{-1} versus more than 10^{-10} cm^{3}g^{-1}), while the ^{3}He values show the opposite effect, being lower in OIB than in MORB. The ^{3}He numbers undoubtedly reflect less efficient trapping of the mantle component in the OIB samples (which were dredged at comparatively shallow depths), leading to the suggestion that the higher ^{36}Ar (and concomitant atmospheric-like ^{40}Ar/^{36}Ar) in OIB is actually due to nothing more meaningful than the less efficient trapping of mantle gas together with an increased atmospheric component. This same trend (lower isotopic ratios and higher absolute abundances of ^{36}Ar) is seen in crystalline MORB (Fisher 1986) and in subduction zone and abyssal plain glasses (Ozima & Zashu 1983) and island arc volcanics (Sano *et al.* 1986), where it is most probably due to the atmospheric 'contamination' explanation offered here. If this argument is accepted, the same trend cannot be taken as evidence of a non-depleted source for the OIB. For details of the argument and a full list of references see Fisher (1985*b*; 1986).

Xenon

Xenon is potentially the most useful gas of all because it has nine stable isotopes and two radiogenic processes contribute to its isotopic abundance ratios: ^{129}Xe arises from the decay of the extinct nuclide ^{129}I, and ^{134,136}Xe (and to a lesser extent some lighter isotopes) arise from the spontaneous fission of ^{238}U (and possibly to the spontaneous fission of ^{244}Pu), although the presence in MORB of the latter component has not been demonstrated, despite an early assertion to this effect (Staudacher & Allégre 1982; Allégre *et al.* 1983; Fisher 1985*a*; Fisher 1987)). At this stage in our investigations the interpretation of xenon data has proved highly contentious because the abundance of xenon is so low that it is difficult to measure precisely, or at least (as detailed below) different laboratories get different results. In addition we must keep in mind, when analysing the data and preparing models, the tendency of xenon to adsorb on surfaces and thus to be incorporated into solids much more easily than any of the other rare gases. This leads to the possibility that atmospheric–oceanic–sediment–mantle recycling is likely to be more important for xenon than for argon, neon or helium. Recent work indicates a significant atmospheric/crustal component of helium in OIB (Condomines *et al.* 1983; see the section on rare gases and other isotopic systems) and of neon, argon and/or xenon in MORB (Marty *et al.* 1983; Ozima & Zashu 1983; Poreda & Radicatti 1984; Fisher 1986) and back-arc basaltic glasses (Sano *et al.* 1986; section on rare gases and other isotopic systems).

The first MORB isotopic excesses in xenon were described by Fisher (1979), who used them only to demonstrate that the radiogenic/fissiogenic ratio ^{4}He/^{136}Xe was higher than the production ratio, indicating an elemental fractionation predicted by Alexander (1976) and Ozima & Alexander (1976) in the rising magma. Staudacher & Allégre (1982) found the first ^{129}Xe excesses; others have been reported by Allégre *et al.* (1983), Fisher (1986) and Barraclough (1986). These are generally in the range 1–6% (normalized to ^{132}Xe or ^{130}Xe), although individual temperature fractions with excesses up to 12% have been reported (Allégre *et al.* 1983). However, other investigators have reported only atmospheric ratios in MORB, within error ranges of typically 1%–3% (Takaoka & Nagao 1978; Kirstein & Richter 1981; Ozima & Zashu 1983). Thus it must be admitted there is room to doubt that the claimed anomalies have been demonstrated unequivocably to be real, although it is my personal feeling that they are.

This feeling is based primarily on Fig. 7 of Fisher (1986), in which all measured MORB $^{129}Xe/^{130}Xe$ values correlate at least roughly with total $(^{130}Xe)^{-1}$. This is the behaviour, analogous to that discussed above for argon, that would be expected if the measured xenon consists of a mixture of atmospheric and mantle components, the latter with a $^{129}Xe/^{130}Xe$ value higher than atmospheric.

Staudacher & Allégre (1982) and Allégre *et al.* (1983) have calculated that the atmosphere must have formed within the first 10–25 Ma of Earth history and that the depleted mantle sampled by the MORBs has been separate from the undepleted mantle sampled by the OIBs ever since, on the basis of their measured ^{129}Xe excesses. Their model has been questioned by Ozima *et al.* (1985) and Fisher (1985*a*; 1986), while Kaneoka (1986) has pointed out that the significance of the calculations is obscure until the ubiquity of the MORB ^{129}Xe anomalies has been established. Since three groups (Takaoka and Nagao 1978; Kirsten & Richter 1981; Ozima & Zashu 1983) have found no xenon anomalies in their MORB samples, such ubiquity cannot be said to have been convincingly established. However, three other groups (Staudacher & Allegre 1982; Allegre *et al.* 1983; Barraclough 1986; Fisher 1986) have found anomalies in MORBs with a world-wide distribution. There are many MORB glasses with insufficient amounts for interlaboratory comparison studics, but until such studies are made on at least some specimens the question of the reality and extent of the MORB xenon anomalies must remain open and will provide continuing fuel for heat (if not light).

Krypton and mass fractionation processes

Krypton has been neglected in this paper so far, just as it has been largely neglected in oceanic studies. Owing to the experimental difficulty of cleanly separating krypton from both argon and xenon, krypton data do not automatically fall out of other rare gas experiments. Generally one sacrifices the krypton in order to measure the argon and xenon most precisely, since there is less information to be gained from krypton isotope analyses in terrestrial materials because of the lack of any radiogenic component. (There is actually a small fissiogenic component at ^{86}Kr, with a 1% fission yield, and an even smaller one (approximately 0.1%) at ^{84}Kr, but taken together with a trapped abundance ratio $Kr/Xe > 1$ the radiogenic contribution to krypton will be even harder to measure than that in xenon.) However, the krypton isotopes may eventually be of use in unravelling the possibility that mass fractionation may play some part in the observed isotopic abundances of MORB xenon, particularly with reference to the ^{129}Xe anomaly.

Measurement of the radiogenic He/Xe ratio in some MORBs has confirmed an earlier suggestion that elemental mass fractionation effects are important, enriching the light elements with respect to the heavy elements by up to an order of magnitude (section on xenon). This raises the possibility that such fractionation may affect the isotopic abundances as well, although of course to a much slighter extent.

At the present time there is just a hint of such light isotope enrichment in the published xenon data. Allégre *et al.* (1983) found an excess of ^{124}Xe in one sample and an excess of ^{126}Xe in one of two samples they measured. Staudacher & Allégre (1982) found excesses of ^{124}Xe in all six of their MORB samples, but errors in three of them overlap atmospheric values. Their ^{126}Xe data fluctuate above and below atmospheric. I found $^{124,126}Xe$ excesses in the one MORB in which I was able to measure these isotopes. All these data are less precise than the ^{129}Xe anomaly measurements because of the very low natural abundance of the light xenon isotopes, but the excesses mentioned are all beyond the quoted experimental errors.

If the light isotopes of xenon are enriched, the heavy ones must be depleted. No such depletion has been seen, but it would be masked by fissiogenic production of the heavy isotopes. In calculating the fissiogenic production, both Allégre and co-workers and I have assumed normal atmospheric ratios (Staudacher & Allégre 1982; Allégre *et al.* 1983; Fisher 1986), but this was purely an assumption.

Isotopic fractionation in other rare gases is problematical. It would not be recognized in helium because there are only two stable isotopes; the 'possibility exists' in neon (Poreda & Radicatti 1984); it has never been looked for in argon, probably because of the large tail from the normal MORB overabundance of ^{40}Ar extending into the ^{38}Ar peak (at least, this is the case in my laboratory). Krypton is the element where it might be seen most clearly, both because of the lack of radiogenic contributions and because of the existence of six stable isotopes spanning the mass range from 78 to 86. It is hoped that precise measurements of krypton in those MORB glasses in which ^{129}Xe excesses have already been found may soon be carried out. Previous studies are insufficiently precise to provide an answer (Fisher 1979; Ozima & Zashu 1983).

Island arcs

Measurements on lead isotopes in volcanic rocks from island arc systems have suggested the possibility that oceanic sediments subducted below some arcs may contribute to the rising magmas which form the arc system, an idea which is supported by recent ^{10}Be measurements (Tera *et al.* 1986, and references therein). Rare gas data support and amplify some details of this idea. The ^{3}He/^{4}He ratio for island arc volcanism is rather uniform and appears to be slightly less than the MORB value; a good average is probably $7R_a$ (Sano *et al.* 1986; Takaoka 1986). (Since helium is not as affected by atmospheric contamination and has lower machine backgrounds than argon, the sample isotopic ratios can be measured more precisely, but note that agreement between different laboratories is not always as good as the quoted errors. For example Sano *et al.* (1987) point out that their measurements on a suite of New Zealand samples give ^{3}He/^{4}He values approximately 15% lower than the results on a similar sample suite by Torgerson *et al.* (1982), although the estimated errors are of the order of 3%.) Takaoka (1986) has found that several Japanese volcanoes give ratios considerably lower, in the range 3–5. This could be due to atmospheric contamination in the collecting process, but the ^{4}He/^{20}Ne ratio is too high for this. The other possibilities would seem to be assimilation of crustal rocks as the mantle magma ascends to the surface or incorporation of subducted oceanic sediment into the ascending magma (Sano & Wakita 1985). The first possibility would give high (radiogenic) ^{40}Ar/^{36}Ar ratios, while the second would give low (atmospheric) ^{40}Ar/^{36}Ar ratios, since radiogenic products are lost from the sediments while atmospheric (oceanic) argon equilibrates.

In one of the volcanoes Takaoka determined a ^{40}Ar/^{36}Ar ratio of 303, leading to the conclusion that it is subducted sediments rather than MORB which contribute to the magma source of at least some island arc volcanics. However, note that the ^{4}He/^{20}Ne ratio, on the basis of which atmospheric contamination was ruled out, is not well known for oceanic sediments, and so this criterion was not considered.

The basins behind oceanic arcs are generally characterized by tholeiitic basalts which resemble but are not identical with MORB. Their generation is clearly linked to the subduction process but the details are not clear. The back-arc basin basalts (BABB) are similar to MORB in major-element chemistry and $\delta(^{18}O)$ but have greater abundances of large-ion lithophile elements and water and higher ^{87}Sr/^{86}Sr ratios (Saunders & Tarney 1984).

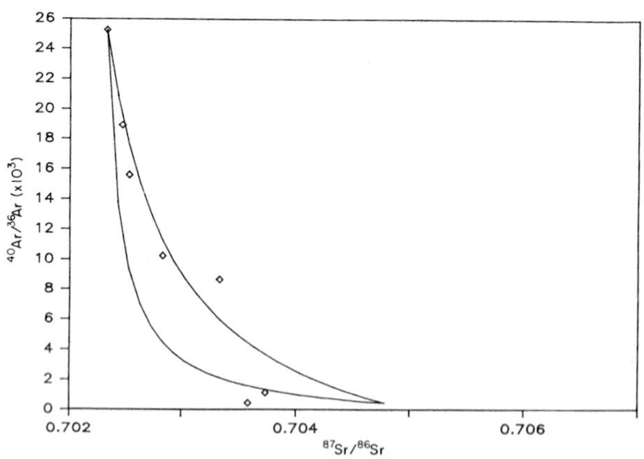

FIG. 2. ^{40}Ar/^{36}Ar versus ^{87}Sr/^{86}Sr for glass MORB samples. Curves are hyperbolic mixing curves of radii 0.05 and 0.2, with end-points given by sample AII 96 18–1 (upper left) and 'bulk Earth' (lower right). See Sarda *et al.* (1985) for details.

Several workers have measured the helium and argon isotopes in a few samples from back-arc basins. Poreda (1985) found that BABB from the Mariana Trough have helium isotope ratios essentially identical with those of MORB, while Sano *et al.* (1986) found values slightly lower than for MORB; the latter group also found argon isotope abundances somewhat lower than the MORB average but which could be well explained by mixing of atmospheric and MORB-type mantle argon. This indicates that the mantle underlying the basin is essentially identical with the MORB mantle source and that the differences seen in front-arc samples must be due to either atmospheric or sediment admixture into the mantle source region. Poreda (1985) also found that D/H ratios in the BABB indicate a water-rich component which presumably originates in a subducted slab, while higher $^3He/^4He$ ratios in the Lau Basin constitute evidence for an enriched mantle source (as discussed above), in this case perhaps one related to the high ^3He Samoan hot-spot which had been proposed previously.

Thus the two gases helium and argon give different types of information on the island arc processes. The helium isotopes indicate a MORB mantle source underlying the Mariana system and a hot-spot type of source under the Lau system, while the argon isotopes indicate that the MORB which are subducted have been flushed out (by hydrothermal circulation?) and do not contribute their MORB mantle gases to the island arc volcanic process; argon from the subducted sediments, however, is seen. A word of caution: these conclusions are reached on the basis of a very limited data set and should be regarded as preliminary. (This is particularly true of the argon story.)

Rare gases and other isotopic systems

The nature of the relationship of the rare gases to other isotopic systems such as strontium is open to question, I believe, and is more complicated than originally envisaged.

The Ar–Sr isotopic correlations have been discussed by Allégre *et al.* (1983, 1987) and Sarda *et al.* (1985), who find that the data lie amongst an envelope of hyperbolic mixing curves between end-points taken as the most radiogenic argon/least radiogenic strontium MORB sample and the bulk Earth. Figure 2 shows their data and the limits of the mixing curves they invoke. (Their papers show many more points per graph. I have included only those data which have been published in the literature. Inclusion of the other (unpublished) points does not change the results in any way.) They conclude that the coherency shown in this graph thoroughly supports the concept of two mantle reservoirs, with blobs rising upwards from the lower reservoir. The blobs contain less depleted material (the 'bulk Earth' component, with radiogenic strontium and atmospheric-like (non-radiogenic) argon), which mixes with material from the depleted mantle having highly radiogenic argon and less radiogenic strontium to form the MORB.

The Ar–Sr coherency can equally well be explained by another mechanism: mixing of depleted mantle with recycled (subducted) crustal material. In Fig. 3 I show the same data but with the 'bulk Earth' end-point replaced by components corresponding to oceanic sediments (lower curve) and MORB which have been altered by seawater interaction (upper curve). Clearly a combination of subducted sediment plus altered basalt provides an end-point and an envelope of mixing curves which fit the data as well as the bulk Earth end-point, and so these data cannot be used to argue for or against either model: data on a mixing curve cannot define the end-points. (Again, inclusion of the unpublished data plotted by Allégre and co-workers does not change the story.)

This muddling of models arises because the supposed bulk Earth or undepleted mantle source ('blob') has an argon isotope composition which is both unknown and supposed to be similar to that of the atmosphere. Thus it is impossible to distinguish between atmospheric and 'bulk Earth' components on the basis of argon isotope compositions alone. I apologize for stating such an obvious point so explicitly, but it seems to have been missed somehow. Helium, in contrast, shows demonstrated isotopic differences between the undepleted (primitive) mantle and the atmosphere, as well as between these and the depleted (normal MORB) mantle source. Helium is also not as subject to atmospheric recycling as is argon because of its more volatile nature which leads to less sorption by sediments and altered rock (Fisher 1986) and to the expectation of greater degassing from subducted materials, and also because of its lower atmospheric abundance. Even in the helium system, however, evidence of atmospheric or recycled crustal influences is seen.

The first study of helium and strontium isotopic systematics showed a good correlation in a limited region of the MAR (50°–53°N), but this

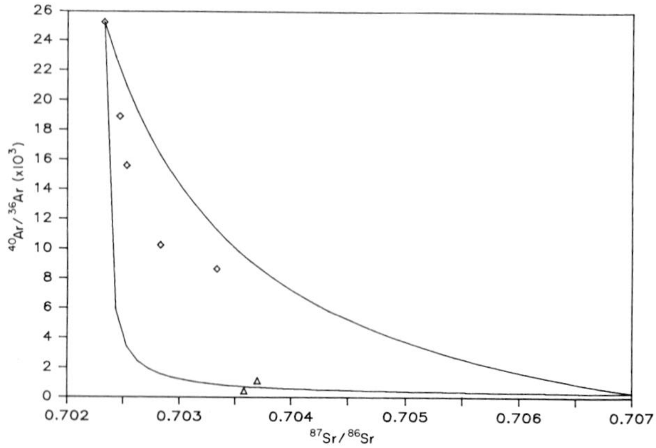

FIG. 3. The same data as Fig. 2 with the lower right end-point replaced by average altered MORB (upper curve) or average deep-sea sediment (lower curve).

relationship broke down in the region 33°–50°N (Kurz *et al.* 1982*b*). Kurz *et al.* (1982*a*) then pointed out that there is a basic difference in the parent–daughter separation processes between the helium and strontium (and Nd–Sm and U–Pb) systems: since ^{4}He is a radiogenic daughter of uranium and thorium, parent–daughter discrimination in the helium isotope system can arise by the degassing of helium from an He–U–Th source region. Combining helium measurements of oceanic islands with those of MORBs, these workers suggested that both an undepleted and a recycled crustal reservoir must be combined with a depleted source to produce the OIB.

Condomines *et al.* (1983) found no correlation between helium and either strontium or neodymium isotope ratios in basaltic glasses from Iceland, although the generally high $^{3}He/^{4}He$ ratios do indicate 'hot-spot' activity similar to other OIB. Comparison with oxygen isotope values suggested alteration processes involving new magmas and old basalts, and these workers cautioned that similar alteration processes may influence the helium isotope signatures seen elsewhere.

Poreda *et al.* (1986) summarized all previous He–Sr data for the MAR from 31°N to 71°N and showed that for such a large region there is no simple relation, owing presumably to the convoluted mixture of degassed subducted crust with both enriched and normal (depleted) MORB mantle sources. However, within specific localized regions of the MOR system, for example along the MAR from 50°N to the Gibbs Fracture Zone, a good correlation can sometimes be seen, and thus major mantle provinces and evolutionary histories can be delineated.

It remains to be seen whether Ar–Sr studies will be able to provide the same sort of useful results. It seems safe to say that correlations which reflect mixing of depleted and undepleted mantle sources in particular oceanic magmatic regimes are to be expected, but it is difficult to understand how the undepleted mantle component can be quantitatively evaluated in the presence of a (subducted?) atmospheric component. The demonstration of a subducted helium component in many samples, and the suspicion of it in others (Condomines *et al.* 1983), certainly implies the existence of an even larger atmospheric argon component, for reasons given above. These atmospheric or crustal components will certainly be added via different mechanisms for the volatile rare gases and the non-volatile strontium, neodymium and lead isotopes and will probably show differences even between helium and argon (and xenon). Fractionation effects will be vastly different depending on the different mechanisms, and so it will not be surprising to see a complex relationship developing as more data are gathered.

Conclusions

(1) Helium has so far proved to be the most useful of the rare gases for elucidating oceanic mantle structures and processes. The $^{3}He/^{4}He$ isotopic ratio in MORB is higher than atmospheric, indicating the presence of trapped primordial ^{3}He. The ratio is even higher in OIB, indicating the

presence of a more primitive, less depleted mantle source. It is lower in island arc basalts than in MORB, presumably because of the mixing of subducted, radiogenic (crustal) and atmospheric components with the uprising magma.

(2) Neon is the element in which nucleogenetic processes will most easily be seen, since it has no radiogenic component and has an abundance of suitable target nuclides (Mg, O etc.). Isotopic variations have been reported, and at least the mass 21 excess is probably nucleogenetic. The mass 20 excess may define a separate mantle component. Variations in the He/Ne ratios are difficult to understand quantitatively but may be related to both a mantle and an atmospheric component, with the possible added complication of mass fractionation. Further work is needed to confirm these measurements and to document their extent and interpretation. So far we have not added to our understanding of mantle processes with neon measurements, but there is great potential there once we have learned to understand the reasons for the anomalies.

(3) Argon and xenon have the greatest potential of all, because they are not lost from the atmosphere to space and because they have radiogenic components with quite different half-lives. But the observations are not yet completely understood. $^{40}Ar/^{36}Ar$ ratios are clearly much higher in MORB than in the atmosphere, reflecting radiogenic growth of ^{40}Ar in the mantle source. (The amount of radiogenic ^{40}Ar is much greater than can be accounted for by post-eruptional *in situ* production.) $^{40}Ar/^{36}Ar$ ratios in OIB are only slightly higher than the atmospheric ratio, a circumstance which has been ascribed to either a primitive mantle source or an atmospheric-like component. Probably both mantle and atmospheric components contribute to the observed values, as has been shown to be the case for helium in Icelandic basalts; at the present time it is not possible to separate these two components from each other. An atmospheric-like component is also seen in MORB glasses. Both helium and argon in island arc volcanic rocks indicate the presence of a subducted component.

(4) Excesses of xenon isotopes at masses 129, 134 and 136 have been reported. The heavy isotopes are fissiogenic, arising from uranium decay. The mass 129 anomaly has been ascribed to decay of the extinct nuclide ^{129}I and therefore provides information on the early degassing of the Earth. The quantitative interpretation at the present time remains equivocal, but the mass 129 anomaly, if confirmed and if proved to arise solely from ^{129}I decay, provides a source of information on the early history of mantle processes not available by any other method.

(5) Krypton has no measurable radiogenic component and not enough target nuclides for a nucelogenetic component to be observed. At the present time it has not been of great use in Earth studies but in future may prove to be a good indicator of isotopic mass fractionation effects.

(6) Helium data, when combined with other isotopic systems such as strontium, delineate major mantle source regions. In some oceanic regions (*eg* the MAR just N of 50°), helium and strontium correlate well. In other regions (Iceland) they do not.

(7) Argon and strontium isotope ratios fall within reasonable mixing envelopes, but the end-points are not well defined and thus the sources of the different argon components remain unknown.

ACKNOWLEDGMENT: Partial support was provided by the National Science Foundation.

References

ALEXANDER, E. C. JR 1976. Trapped helium and argon and formation of the earth's atmosphere. *Nature* **261**, 77.

ALLÉGRE, C. J., STAUDACHER, T. & SARDA, P. 1987. Rare gas systematics: formation of the atmosphere, evolution and structure of the Earth's mantle. *Earth and Planetary Science Letters* **81**, 127–150.

——, ——, —— & KURZ, M. 1983. Constraints on evolution of earth's mantle from rare gas systematics. *Nature* **303**, 762–766.

BARRACLOUGH, B. L. 1986. Noble gases in MORB glasses, Paper presented at Japan–US Seminar on Terrestial Rare Gases, Yellowstone, 1986.

CLARKE, W. B., BEG, M. A. & CRAIG, H. 1969. Excess ^3He in the sea. *Earth and Planetary Science Letters* **6**, 213–220.

CONDOMINES, M., GRONVOLD, K., HOOKER, P. J.,

MUEHLENBACHS, K., O'NIONS, R. K., OSKARSSON, N. & OXBURGH, E. R. 1983. Helium, oxygen, strontium and neodymium isotopic relationships in Icelandic volcanics. *Earth and Planetary Science Letters* **66**, 125–136.

CRAIG, H. & LUPTON, J. E. 1976. Primordial neon, helium, and hydrogen in oceanic basalts. *Earth and Planetary Science Letters* **31**, 369–385.

——, —— & HORIBE, Y. 1978. A mantle helium component in circum-Pacific volcanic gases. *In*: ALEXANDER, E. C. JR & OZIMA, M. (eds) *Terrestrial Rare Gases*, 3–16. Japan Scientific Societies Press.

DALRYMPLE, G. B. & MOORE, J. G. 1968. Argon-40: excess in submarine pillow basalts from Kilauea Volcano, Hawaii. *Science* **161**, 1132–1135.

FISHER, D. E. 1978. Terrestrial potassium and argon abundances as limits to models of atmospheric evolution. *In*: ALEXANDER, E. C. JR & OZIMA, M. (eds) *Terrestrial Rare Gases*, 173–184.

—— 1979. Helium and xenon in deepsea basalts as a measure of magmatic differentiation. *Nature* **282**, 825–827.

—— 1985*a*. Radiogenic rare gases and the evolutionary history of the depleted mantle. *Journal of Geophysical Research* **90**, 1801–1807.

—— 1985*b*. Noble gases from oceanic island basalts do not require an undepleted mantle source. *Nature* **316**, 716–718.

—— 1986. Rare gas abundances in MORB. *Geochemica Cosmochimica Acta* **50**, 2531–2541.

—— 1987. Reply to comments. *Journal of Geophysical Research* **92**, 2813–2817.

FUNKHOUSER, J. G., FISHER, D. E. & BONATTI, E. 1968. Excess argon in deep-sea rocks. *Earth and Planetary Science Letters* **5**, 95–100.

HART, R., DYMOND, J., HOGAN, L & SCHILLING, J. G. 1983. Mantle plume noble gas component in glassy basalts from Reykjanes Ridge. *Nature* **305**, 403–407.

JAMBON, A., WEBER, H. W. & BEGEMANN, F. 1985. Helium and argon from an Atlantic MORB glass. *Earth and Planetary Science Letters* **73**, 255–267.

KANEOKA, I. 1986. Noble gas signatures and geological environments. Preprint.

—— & TAKAOKA, N. 1978. Excess ^{129}Xe and high ^{3}He/^{4}He in olivine phenocrysts of Kapuho lava and xenolithic dunites from Hawaii. *Earth and Planetary Science Letters* **39**, 382–386.

—— & —— 1980. Rare gas isotopes in Hawaiian ultramafic nodules and volcanic rocks. *Science* **208**, 1366–1368.

KIRSTEN, T. & RICHTER, H. 1981. Abundance patterns of rare gases in submarine basalts and glasses. *Meteoritics* **16**, 341.

KURZ, M. D., JENKINS, W. J. & HART, S. R. 1982*a*. Helium isotopic systematics of oceanic islands and mantle heterogeneity. *Nature* **297**, 43–47.

——, ——, SCHILLING, J. G. & HART, S. R. 1982*b*. Helium isotopic variations in the mantle beneath the central North Atlantic Ocean. *Earth and Planetary Science Letters* **58**, 1–14.

——, MEYER, P. S. & SIGURDSSON, H. 1985. Helium isotopic systematics within the neovolcanic zones of Iceland. *Earth and Planetary Science Letters* **74**, 291–305.

LUPTON, J. E. & CRAIG, H. 1975. Excess ^{3}He in oceanic basalts. *Earth and Planetary Science Letters* **26**, 133–139.

MARTY, B., ZASHU, S. & OZIMA, M. 1983. Two noble gas components in a Mid-Atlantic Ridge basalt. *Nature* **302**, 239–240.

NOBLE, C. S. & NAUGHTON, J. J. 1968. Deep ocean basalts: inert gas content and uncertainties in age dating. *Science* **162**, 265–268.

OZIMA, M. 1973. Was the evolution of the atmosphere continuous or catastrophic? *Nature* **246**, 41.

—— 1975. Ar isotopes and earth–atmosphere evolution models. *Geochimica et Cosmochimica Acta* **39**, 1127–1134.

—— & ALEXANDER, E. C. 1976. Rare gas fractionation patterns in terrestrial samples and the earth–atmosphere evolution model. *Reviews of Geophysics and Space Physics* **14**, 385–390.

—— & KUDO, K. 1972. Excess argon in submarine basalts and an earth–atmosphere evolution model. *Nature, Physical Sciences* **239**, 23–24.

—— & ZASHU, S. 1983. Noble gases in submarine pillow volcanic glasses. *Earth and Planetary Science Letters* **62**, 24–40.

——, PODOSEK, F. A. & IGARASHI, G. 1985. Terrestrial xenon isotope constraints on the early history of the earth. *Nature* **315**, 471–474.

POREDA, R. 1985. Helium-3 and deuterium in back-arc basalts: Lau Basin and the Mariana Trough. *Earth and Planetary Science Letters* **73**, 244–254.

—— & RADICATTI, F. 1984. Neon isotope variations in Mid-Atlantic Ridge basalts. *Earth and Planetary Science Letters* **69**, 277–289.

——, SCHILLING, J. G. & CRAIG, H. 1986. Helium and hydrogen isotopes in ocean-ridge basalts north and south of Iceland. Preprint.

RINGWOOD, A. E. 1982. Phase transformations and differentiation in subducted lithosphere. *Journal of Geology* **90**, 611–644.

SANO, Y. & WAKITA, H. 1985. Geographical distribution of ^{3}He/^{4}He ratios in Japan. *Journal of Geophysical Research* **90**, 8729–8741.

——, NAKAMURA, Y., WAKITA, H. & ISHII, T. 1986. Light noble gases in basalt glasses from Mariana Trough. Preprint.

——, WAKITA, H. & GIGGENBACH, W. F. 1987. Island arc tectonics of New Zealand manifested in helium isotope ratios. *Geochimica et Cosmochimica Acta* **51**, 1855–1860.

SARDA, P., STAUDACHER, T. & ALLEGRE, C. J. 1985. ^{40}Ar/^{36}Ar in MORB glasses. *Earth and Planetary Science Letters* **72**, 357–375.

SAUNDERS, A. D. & TARNEY, J. 1984. Geochemical characteristics of basaltic volcanism within back-arc basins. *In*: KOKELAAR, B. P. & HOWELLS, M. F. (eds). *Marginal Basin Geology* Geological Society of London Special Publication **16**, 59–76.

STAUDACHER, T. & ALLÉGRE, C. J. 1982. Terrestrial xenology. *Earth and Planetary Science Letters* **60**, 389–406.

—— & SARDA, P. 1987. Comments on 'Radiogenic rare gases and the evolutionary history of the depleted mantle'. *Journal of Geophysical Research*, pp. 2810–2813.

TAKAOKA, N. 1986. Application of noble gas measurement to volcanology and implications to earth sciences. Paper presented at Japan–US Seminar on Terrestrial Rare Gases, Yellowstone, 1986.

—— & NAGAO, K. 1978. Rare gas studies of Cretaceous deep sea basalts. *Nature* **276**, 491–492.

TERA, F., BROWN, L., MORRIS, J., SACKS, I. S., KLEIN, J. & MIDDLETON, R. 1986. Sediment incorporation in island-arc magmas: inferences from ^{10}Be. *Geochimica et Cosmochimica Acta* **50**, 535–550.

TORGERSON, E., LUPTON, J. E., SHEPPARD, D. S. & GIGGENBACH, W. F. 1982. Helium isotope variations in thermal areas of New Zealand. *Journal of Volcanology and Geothermal Research* **12**, 283–298.

ZINDLER, A., STAUDIGEL, H. & BATIZA, R. 1984. Isotope and trace element geochemistry of young Pacific seamounts. *Earth and Planetary Science Letters* **70**, 175–195.

DAVID E. FISHER, Rosenstiel School of Marine and Atmospheric Sciences, University of Miami, Miami, FL 33149, USA.

Chemical and isotopic systematics of oceanic basalts: implications for mantle composition and processes

S.-s. Sun & W.F. McDonough

SUMMARY: Trace-element data for mid-ocean ridge basalts (MORBs) and ocean island basalts (OIB) are used to formulate chemical systematics for oceanic basalts. The data suggest that the order of trace-element incompatibility in oceanic basalts is Cs ≈ Rb ≈ (≈ Tl) ≈ Ba(≈ W) > Th > U ≈ Nb = Ta ≈ K > La > Ce ≈ Pb > Pr (≈ Mo) ≈ Sr > P ≈ Nd (> F) > Zr = Hf ≈ Sm > Eu ≈ Sn (≈ Sb) ≈ Ti > Dy ≈ (Li) > Ho = Y > Yb. This rule works in general and suggests that the overall fractionation processes operating during magma generation and evolution are relatively simple, involving no significant change in the environment of formation for MORBs and OIBs.

In detail, minor differences in element ratios correlate with the isotopic characteristics of different types of OIB components (HIMU, EM, MORB). These systematics are interpreted in terms of partial-melting conditions, variations in residual mineralogy, involvement of subducted sediment, recycling of oceanic lithosphere and processes within the low velocity zone. Niobium data indicate that the mantle sources of MORB and OIB are not exact complementary reservoirs to the continental crust. Subduction of oceanic crust or separation of refractory eclogite material from the former oceanic crust into the lower mantle appears to be required. The negative europium anomalies observed in some EM-type OIBs and the systematics of their key element ratios suggest the addition of a small amount (≤1% or less) of subducted sediment to their mantle sources. However, a general lack of a crustal signature in OIBs indicates that sediment recycling has not been an important process in the convecting mantle, at least not in more recent times (≤2 Ga). Upward migration of silica-undersaturated melts from the low velocity zone can generate an enriched reservoir in the continental and oceanic lithospheric mantle. We propose that the HIMU type (*eg* St Helena) OIB component can be generated in this way. This enriched mantle can be re-introduced into the convective mantle by thermal erosion of the continental lithosphere and by the recycling of the enriched oceanic lithosphere back into the mantle.

> That much good ensues and that the science is greatly advanced by the collision of various theories cannot be doubted. Each party is anxious to support opinions by facts. Thus, new countries are explored and old districts re-examined; facts come to light that do not suit either party; new theories spring up; and in the end, a greater insight into the real structure of the earth's surface is obtained.
>
> Henry Thomas De là Beche (1830)

It is now well established that chemical and isotopic heterogeneities of different scales, ranging from centimetres to thousands of kilometres, exist in the Earth's mantle. Such heterogeneities reflect variations in elemental abundance ratios generated by depletion and enrichment processes that have operated in the Earth's mantle throughout geological time.

In this paper we attempt to integrate the chemical and isotopic data from oceanic basalts in order to establish coherent relationships between isotopic compositions and the consistent (but variable) trace-element patterns observed in these basalts. These regularities are then interpreted in terms of the physical and chemical processes that have operated in the mantle.

Our discussion is preceded by an outline of some of the major issues in the chemical evolution and geodynamics of the Earth's mantle. This is followed by some ideas on forward modelling, developed by investigating the possible consequences of major physical and chemical processes in the mantle. Finally, we discuss factors which influence the chemical and isotopic compositions of oceanic basalts.

Major issues in the chemical evolution and geodynamics of the mantle

A first-order aim in the study of oceanic basalts is to improve our understanding of the chemical

From Saunders, A. D. & Norry, M. J. (eds), 1989, *Magmatism in the Ocean Basins*, Geological Society Special Publication No. 42, pp. 313–345.

and dynamic processes which have operated in the past and today within the Earth's mantle. To achieve this it is necessary to delineate the major issues, which requires the integration of data derived from different branches of Earth sciences. We need to do the following.

(1) Undertake complete chemical and isotopic characterization of the Earth's primitive mantle (the silicate sphere of the Earth) and understand the timing and processes involved in core growth, and related mantle differentiation.
(2) Characterize the behaviour of elements and the melting and extraction processes involved during magma genesis within different tectonic environments.
(3) Delineate the processes responsible for development of ocean island basalt (OIB) source characteristics.
(4) Document the secular variations of element abundances and their ratios in the mantle in order to understand mantle differentiation processes involved in crust formation.
(5) Firmly establish the extent of crustal recycling in the mantle and document its effects on the chemical and isotopic evolution of the mantle.
(6) Establish whether the chemical and isotopic characteristics shared between mid-ocean ridge basalts (MORBs) and OIBs reflect a complementary relationship with the continental crust.
(7) Clarify the geometric and dynamic relationship between the different mantle reservoirs, and evaluate dynamic processes involved in the genesis and evolution of these reservoirs.
(8) Determine the survival time of chemical and isotopic heterogeneities in the continental lithosphere and in the underlying convecting mantle.
(9) Determine the fate of subducted oceanic lithosphere: is it simply recycled and resorbed back into the upper mantle, or is part of it transported into the lower mantle during mantle differentiation processes?

Forward modelling of mantle differentiation and crustal formation

The combined processes of plate tectonics, mantle convection, magma generaton, crustal recycling and upper and lower mantle exchange contribute to the geochemical and isotopic evolution of mantle reservoirs. The nature of mantle convection processes through time (whole mantle or layered mantle) is critical to our understanding of the chemical and thermal evolution of the Earth.

The term 'reservoir' is used here in a general sense to refer to a part of the mantle which has a particular regional chemical and isotopic composition, whereas the term 'component' specifically refers to a reservoir (or many reservoirs) in the mantle with an isotopically distinctive composition (*eg* HIMU, EM, MORB). This use of the term 'component' is similar to that in Zindler & Hart (1986).

In essence each mantle reservoir carries an identifiable chemistry and isotopic fingerprint of the specific processes and environments which have acted upon it. These compositional fingerprints reflect the responses to such factors as partial melting under different $P-T-X$(CO_2, water rich, melts or fluids) conditions, sediment subduction, and recycling of oceanic crust and asthenosphere through the subduction zone environment.

Mantle differentiation processes through time

Our understanding of mantle differentiation associated with the Earth's accretion, core formation and the early history of mantle–crust fractionation relies upon chemical and isotopic studies of Archaean to modern volcanic rocks and other planetary bodies, petrological and chemical experiments carried out under high-temperature and high-pressure conditions, and numerical modelling of the thermal evolution of the Earth.

Even if some thermal models favour the presence of upper and lower mantle convection cells at present (*eg* Richter 1985), there is no obvious reason to argue against whole-mantle convection during the early history of the Earth. It is generally assumed that the early Earth's mantle temperature was higher (*eg* 2000 °C surface potential temperature) (Richter 1985), which would favour vigorous, and probably chaotic, whole-mantle convection with possible large-scale mantle melting. Consequently, it is very likely that the lower mantle would have been involved in the formation of the earliest enriched lithosphere, resulting in an incompatible-element-depleted character, *ie* a non-primitive fractionated lower mantle. At the same time, dense early-formed severely hydrothermally altered mafic to ultramafic crust and lithospheric mantle may well have been rapidly recycled back into the convective mantle by meteorite bombardment and lithosphere subduction. Vigorous convection in the early Earth's mantle would quickly stretch and mix this oceanic lithosphere with the convective mantle. Some of this lithosphere may have been subducted into the lower mantle. The magnitude of these activities decreased as the Earth's heat engine slowed down. Two-layer mantle convection may have eventually been initiated as a result of a density barrier imposed by temperature-dependent phase transformation at 670 km (*eg* Sawamoto 1987) or cumulation of refractory subducted lithosphere at the base of the upper mantle (Ringwood & Irifune 1988). However, it is possible that whole-mantle convection has continued throughout Earth's history (Gurnis & Davies 1986).

If the mantle differentiation scheme discussed above is realistic then the high $^3He/^4He$ ratios ($\geqslant$ 20 $\times$ atmospheric) observed in some OIBs do not indicate an origin from a primordial unfractionated lower mantle. These ratios may instead result from the cooling of the outer core. Similarly, the ^{129}Xe anomaly observed in some MORB samples (Allègre *et al.* 1987) is unlikely to result from the early (4.4 Ga) separation of the upper mantle from the lower mantle.

The stabilization of the Archaean continental crust was accompanied by the formation of thick refractory harzburgitic lithosphere. This subjacent mantle formed as a residual product of mantle melting. Occurrences of diamonds of Archaean age from kimberlite pipes from the Kaapvaal craton, South Africa (Kramers 1977; Richardson *et al.* 1984), suggest the early existence of Archaean continental lithosphere up to about 200 km thick. Such lithosphere could have been developed through underplating of residual mantle related to intraplate magmatic activity and/or depletion of the mantle wedge above a subduction zone.

Studies of modern island arc basalts suggest that subduction of the oceanic lithosphere, with or without associated sediment, can continuously modify the chemical, isotopic and mineralogical composition of the overlying mantle wedge. The release of fluids and hydrous silicate melts from the down-going altered oceanic crust can result in the enrichment of incompatible elements in the mantle wedge above the subduction zone. Stabilization of phlogopite in the mantle wedge provides a site for the introduced alkalis and barium (*eg* Wyllie & Sekine 1982) and it has been suggested that the formation of titanate minerals (such as sphene and rutile) in the subducted slab may have played an important role in the depletion of high field strength elements (*eg* titanium, niobium, tantalum) in arc basalts (*eg* Saunders *et al.* 1980). Recycling of this mantle wedge or the overlying lithosphere could introduce subduction-zone-related chemical and isotopic characteristics into the convecting asthenosphere. For example, the recycled oceanic crust might show a positive niobium anomaly relative to lanthanum (*ie* La/Nb $>$ 0.95 (the chondritic ratio), but La/Sm $<$ chondrite).

The upward migration of highly incompatible-element-enriched CO_2-rich melts from the low velocity zone (LVZ) beneath the continental and oceanic lithosphere (*eg* kimberlites, carbonatites, nephelinites) can be a continuous or episodic process responsible for local enrichments of incompatible trace elements (*eg* Green 1971). Although such enrichment processes should not be limited to the Archaean and early Proterozoic, the more vigorous melting possible in the Precambrian mantle would promote such enrichment processes. This type of enrichment can be superimposed upon earlier enrichment events related to subduction zone processes. Such enriched lithospheric mantle commonly has a refractory major-element composition, reflecting residual products of earlier melting event(s) (*eg* Frey & Green 1974).

Recycling of the continental lithosphere by delamination (McKenzie & O'Nions 1983) or thermal erosion, and the subduction of oceanic lithosphere (Hofmann & White 1982; Ringwood 1982), introduce compositionally distinctive sources into the convecting mantle. Such material could be preserved as coherent entities through billions of years of convection. Thermal reactivation of enriched ancient continental lithosphere by plume activity or rifting may also play an important role in the generation of some continental basalts. Additionally, it may also cause thermal erosion of the continental lithosphere and incorporate it into the convecting mantle.

There are different opinions regarding the fate of the subducted oceanic lithosphere. It could have been continuously stretched and eventually well mixed back into the convective upper mantle (*eg* Allègre & Turcotte 1986). Alternatively, because of the density barrier at 670 km caused by phase transformations, the subducted oceanic crust and refractory harzburgite lithosphere beneath it may buckle and accumulate at the base of the upper mantle and form a boundary layer (*eg* Ringwood 1982; Ringwood & Irifune 1988). However, subduction of the oceanic lithosphere into the lower mantle appears to be indicated by geophysical observations (Creager & Jordan 1984). In Ringwood's model, the megalith accumulates at the base of the upper mantle and warms up after some hundreds of millions of years. Partial melts of the former oceanic crust may fertilize the surrounding refractory harzburgite, while denser residues of former oceanic crust may sink into the lower mantle. Once thermally equilibrated, the harzburgite becomes intrinsically lighter than the surrounding and overlying mantle and ascends. These refractory peridotite diapirs are considered to give birth to mantle plumes or blobs that are enriched in incompatible elements due to equilibration with residual garnetite minerals, which are stable at depths of 400–600 km (Ringwood 1982).

Factors influencing chemical and isotopic characteristics of basaltic magmas

The distinctive chemical and isotopic characteristics of basalts from different tectonic environments can be attributed to the following factors:

(1) the source characteristics, which are a function of its previous history;
(2) the present tectonic environment of magma generation;
(3) magma generation conditions and processes;
(4) mixing of different mantle source regions.

As discussed earlier, core–mantle–crust differentiation and geodynamic processes have contributed to the formation of a chemically and isotopically heterogeneous mantle. Modern processes operating in the source regions produce further modifications to the chemical compositions of basalts.

In addition to the effects of source mineral composition, pressure, temperature and mineral/melt partition coefficients K_d, magma generation processes can also affect the composition of the resulting melt. Segregation of melts from a mantle diapir most probably involves mixing of migrating melts formed under different conditions (*eg* Oxburgh & Turcotte 1968; Hanson 1977; Langmuir *et al.*; McKenzie 1984). At present, numerical modelling of melt migration and separation is very dependent upon some basic assumptions. Detailed studies of ophiolites, alpine peridotites and ultramafic xenoliths can aid in evaluating the dynamic processes involved in melt generation and segregation (*eg* Nicolas 1986). If the mantle is veined as a result of melt migration and/or convective mixing of the recycled oceanic crust, low-temperature melting will selectively sample areas near these fertile veins. At higher temperatures the degree of partial melting increases, and the refractory non-veined peridotite becomes progressively more involved in melting (*eg* Sun & Hanson 1975*a, b*; Hanson 1977; Zindler *et al.* 1984). Finally, mixing of a diapiric plume with the asthenosphere and/or lithosphere through which it passes can be reflected in the erupted basalts. Hot and dense tholeiitic and picritic magmas may interact extensively with the lithosphere and crust to induce further melting. In contrast, owing to their volatile-rich character, alkali basalts rise rapidly to the Earth's surface, often carrying mantle xenoliths, and generally escape significant crustal contamination.

Chemical systematics of oceanic basalts

Since chemical and isotopic characteristics of basalts are governed primarily by their source character, their tectonic environment of magma generation and magma generation processes, it is logical to expect systematic compositional variations caused by these factors. Here we describe some chemical regularities observed in oceanic basalts (OIB and MORB) and relate the chemical variations in different groups of oceanic basalts to their isotope character.

Systematic relationships amongst incompatible trace elements

During partial melting of mantle peridotite the incompatible elements are concentrated into the magma. When two incompatible elements have identical mineral/melt K_d during partial melting their abundance ratio in the magma reflects that of their source, regardless of the degree of partial melting, *ie* their abundance ratios do not vary with increasing concentrations in the magma. In contrast, the ratios of incompatible elements having different K_d values will vary with the degree of partial melting. The more incompatible an element, the more it will be preferentially enriched in the melt. On the basis of this simple rule we can establish the degree of incompatibility of trace elements in oceanic basalts during magma generation. Obviously, in different tectonic environments with different physical conditions and mineral assemblages, the order of element incompatibility may change significantly. Thus, such an approach is very useful in evaluating the viability of possible mantle processes.

To present graphically the trace-element data of a sample for the purpose of pattern recognition it is convenient to normalize these data to the trace-element abundances of the primitive mantle composition. This type of plot, established by Sun *et al.* (1979), has been called a 'spidergram' by Thompson *et al.* (1983) and an 'extended Coryell–Masuda diagram' by Hofmann *et al.* (1986). We prefer that the diagram be called a mantle normalizing diagram.

This method of normalization is similar to the practice of chondrite normalization for the rare earth elements. However, as Newsom *et al.* (1986) and Hofmann *et al.* (1986) have pointed out, there are potential problems in such normalization plots if the relative incompatibility of elements changes with time. They found that the Ce/Pb and Nb/U ratios observed in OIB and MORB are nearly constant, 25±5 and 47±10 respectively, but quite different from the primitive silicate mantle ratios (9 and 30), indicating that processes operating in the early history of the Earth ($\geqslant$ 2 Ga) caused fractionation of cerium from lead and niobium from uranium. Rubidium and caesium show a similar behaviour. Oceanic basalts have nearly constant Rb/Cs ratios of about 80, which is distinct from the primitive mantle value of 20±10 (McDonough *et al.* 1987). Taking these observations into account we have calculated the normalizing mantle values for lead (used in Fig. 1 and other figures of this paper) on the basis of a Ce/Pb ratio of 25 and for caesium on the basis of an Rb/Cs ratio of 80. The primitive mantle values used here are given in Table 1, with the element concentrations in C1 carbonaceous chondrites and the three different types of oceanic basalts shown for comparison.

Figure 1 presents the primitive-mantle-normalized patterns (except for Cs and Pb) for typical N-type and E-type MORB and an oceanic alkali basalt (OIB) with $^{87}Sr/^{86}Sr \approx 0.7035$. The data suggest that the order of trace-element incompatibility in oceanic basalts is Cs ≈ Rb (≈ Tl) ≈ Ba (≈W) > Th > U ≈ Nb = Ta ≈ K > La > Ce ≈ Pb > Pr (≈ Mo) ≈ Sr > P ≈ Nd (> F) > Zr = Hf ≈ Sm > Eu ≈ Sn (≈ Sb) ≈ Ti> Dy ≈ (Li) > Ho = Y > Yb. This order is mainly based upon the depletion pattern of N-type MORB and the systematics of concentration ratios observed in E-type MORB and OIB (Sun & Hanson 1975*a*, 1976; Sun & Nesbitt 1977; Sun *et al.* 1979; Sun 1980; Hofmann *et al.* 1986; Newsom *et al.* 1986; Ryan & Langmuir 1987). The relative incompatibilities of these elements appear constant within oceanic basalts suggesting that the fractionation processes are simple and involve no drastic change in elemental behaviour. The systematic behaviour of some elements, *eg* Nb–Ta, Sr–Pr, Zr–Hf–Sm, Ti–Eu^{3+}, Ce–Pb, Nb–U, indicate that the main control of mineral/melt K_d for these elements during oceanic basalt genesis is the presence of residual clinopyroxene and garnet (Fujimaki *et al.* 1984; Jochum *et al.* 1986; Watson *et al.* 1987).

On the basis of these chemical systematics, when accurate analytical data are available for only a few critical trace elements (*eg* Y, Ti, Zr, Sr, P, Nb, K, Rb), the concentrations of other elements on this plot can be estimated quite accurately. In this respect it is somewhat surprising to observe that phosphorus and neodymium behave very similarly during MORB and OIB genesis (P/Nd = 74±13) (McDonough *et al.* 1985), since these elements substitute into different crystal sites (phosphorus substitutes for silicon whereas neodymiun substitutes for calcium). Additionally, it is suggested that large deviations from Nb/Ta = 17 and Zr/Hf = 36 in MORB and OIB are most probably due to analytical errors or contamination of niobium and tantalum from the tungsten carbide mill during sample preparation. Fractionation of Nb/Ta in island arc basalts due to residual titanates during magma generation is possible, but reliable data are needed to evaluate this possibility.

Neodymium and hafnium isotope studies of mantle-derived volcanic rocks indicate that there is a consistent behaviour between Sm–Nd and Lu–Hf throughout geological time (Patchett 1983). Since Sr ≈ Pr, Pb ≈ Ce and Hf ≈ Sm on the mantle-normalized plot (Fig. 1) and samarium, praseodymium and cerium are close to neodymium, the mixing of different mantle components for MORB and OIB (with different mantle-normalized patterns) is expected to generate linear trends in plots of isotope ratios, *eg* $^{143}Nd/^{144}Nd$ versus $^{87}Sr/^{86}Sr$, $^{143}Nd/^{144}Nd$ versus $^{176}Hf/^{177}Hf$ and $^{87}Sr/^{86}Sr$ versus $^{206}Pb/^{204}Pb$. The isotope data in the literature are consistent with this (*eg* Zindler *et al.* 1982; White 1985).

The chemical regularities observed in basalts can be used to obtain information on the composition of the source and the processes and mineralogy involved in magma generation. The relatively constant but low concentrations of potassium, coupled with high concentrations of niobium and lanthanum in some highly undersaturated alkali basalts, kimberlites and carbonatites, suggests that potassium in the melts could be buffered by residual amphibole, phlogopite and clinopyroxene (under very high pressure) during source enrichment processes and/or magma generation (*eg* Sun & Hanson 1975*a*; Clague & Frey 1982). Such residual potassium-bearing minerals could also cause depletions of rubidium, caesium, and barium in

TABLE 1. *Element concentrations (ppm) in C1 chondrite, primitive mantle, N-type MORB, E-type MORB and ocean island basalts (OIB)*

Element	C1 chondrite[a]	Primitive mantle[a]	N-type MORB	E-type MORB	OIB
Cs	0.188	0.032[b]	0.0070	0.063	0.387
Tl	0.140	0.005	0.0014	0.013	0.077
Rb	2.32	0.635	0.56	5.04	31.0
Ba	2.41	6.989	6.30	57	350
W	0.095	0.020	0.010	0.092	0.560
Th	0.029	0.085	0.120	0.60	4.00
U	0.008	0.021	0.047	0.18	1.02
Nb	0.246	0.713	2.33	8.30	48.0
Ta	0.014	0.041	0.132	0.47	2.70
K	545	250	600	2100	12000
La	0.237	0.687	2.50	6.30	37.0
Ce	0.612	1.775	7.50	15.0	80.0
Pb	2.47	0.185[b]	0.30	0.60	3.20
Pr	0.095	0.276	1.32	2.05	9.70
Mo	0.92	0.063	0.31	0.47	2.40
Sr	7.26	21.1	90	155	660
P	1220	95	510	620	2700
Nd	0.467	1.354	7.30	9.00	38.5
F	60.7	26	210	250	1150
Sm	0.153	0.444	2.63	2.60	10.0
Zr	3.87	11.2	74	73	280
Hf	0.1066	0.309	2.05	2.03	7.80
Eu	0.058	0.168	1.02	0.91	3.00
Sn	1.72	0.170	1.1	0.8	2.7
Sb	0.16	0.005	0.01	0.01	0.03
Ti	445	1300	7600	6000	17200
Gd	0.2055	0.596	3.680	2.970	7.620
Tb	0.0374	0.108	0.670	0.530	1.050
Dy	0.2540	0.737	4.550	3.550	5.600
Li	1.57	1.60	4.3	3.5	5.6
Y	1.57	4.55	28	22	29
Ho	0.0566	0.164	1.01	0.790	1.06
Er	0.1655	0.480	2.97	2.31	2.62
Tm	0.0255	0.074	0.456	0.356	0.350
Yb	0.170	0.493	3.05	2.37	2.16
Lu	0.0254	0.074	0.455	0.354	0.300

[a] The compositions of C1 chondrite and primitive mantle are from McDonough & Sun (in prep.)
Values for N-type and E-type MORB and OIB are based on a literature survey and internal consistency of elemental ratios.
[b] For mantle-normalized diagrams, the recommended normalizing values for lead and caesium are 0.071 and 0.0079, respectively.

these magmas relative to their neighbouring elements on a mantle normalizing diagram. Decreasing ratios of Ba/Nb, K/Nb, Rb/Nb, Cs/Nb and Rb/Sr with increasing lanthanum and niobium abundances are common in alkali basalts, nephelinites and kimberlites, whereas tholeiites commonly do not show such relative depletions. This suggests that phlogopite does not play a role during tholeiitic magma generation. Moreover, since phlogopite is not an early liquidus mineral in most silica-undersaturated basaltic magmas, it acts as a residual phase during the initial small degrees of mantle melting or during a metasomatic enrichment of the source region. If phlogopite is indeed the mineral holding back potassium, rubidium, and caesium in the source regions of silica-undersaturated nephelinites, non-micaceous kimberlites and melilitites (Sun & Hanson 1975*a, b*; Clague & Frey 1982), then the near constant Ba/Rb ($\approx$12) and Rb/Cs ($\approx$80) ratios observed in most OIBs requires that barium and caesium have similar phlogopite/melt K_d values relative to rubidium (Philpotts & Schnetzler 1970; Kuehner *et al.* 1981) or that

Hofmann, AW, 1988, Chemical Differentiation of the Earth: the relationship between mantle, continental crust, and oceanic crust: Earth and Planetary Science Letters, 90, pp 297-314

TABLE 1

Major and trace element abundances in the Earth's primitive mantle, the continental crust and MORB

Element	Primitive mantle	MORB			Estimated bulk continental crust		
		N-MORB avg. (n)	N-MORB normalized	percent std. dev.	TM1	TM1 normalized	TM2 normalized
SiO_2 (%)	45.96	50.45	1.098	1.8	57.3	1.247	1.262
Al_2O_3	4.06	15.255	3.757	8.0	15.9	3.916	4.433
FeO	7.54	10.426	1.383	14.8	9.1	1.207	0.995
MgO	37.78	7.576	0.201	12.2	5.3	0.140	0.093
CaO	3.21	11.303	3.521	7.2	7.4	2.305	2.336
Na_2O	0.332	2.679	8.070	13.3	3.1	9.337	10.542
Na_2O (MgO = 8.0%)	0.332	2.526	7.608	15.2			
TiO_2	0.181	1.615	8.925	34.0	0.901	4.978	4.420
La (ppm)	0.6139	3.895	6.345	41.4	16	26.06	30.95
Ce	1.6011	12.001	7.496	40.1	33	20.61	23.73
Pr	0.2419	2.074	8.574	39.0	3.9	16.12	17.78
Nd	1.1892	11.179	9.401	37.8	16	13.45	13.45
Sm	0.3865	3.752	9.708	37.0	3.5	9.06	9.57
Eu	0.1456	1.335	9.167	30.1	1.1	7.56	7.56
Gd	0.5128	5.077	9.901	36.5	3.3	6.44	7.02
Tb	0.0940	0.885	9.412	38.1	0.60	6.38	6.81
Dy	0.6378	6.304	9.884	34.5	3.7	5.81	5.80
Ho	0.1423	1.342	9.433	36.6	0.78	5.48	5.76
Er	0.4167	4.143	9.944	34.3	2.2	5.28	5.52
Tm	0.0643	0.621	9.663	34.7	0.32	4.98	4.98
Yb	0.4144	3.900	9.411	33.8	2.2	5.31	5.31
Lu	0.0637	0.589	9.246	33.8	0.30	4.71	4.71

Rb	0.5353	1.262	2.357	76.2	32	59.78	78.40
Cs	(0.0268)	0.01408	(0.53)	71.9	1.0	(37.3)	(63.43)
Sr	18.21	113.2	6.216	24.1	260	14.28	21.97
Ba	6.049	13.87	2.293	71.9	250	41.33	57.86
Hf	0.2676	2.974	11.116	42.7	3.0	11.21	11.21
Zr	9.714	104.24	10.705	40.0	100	10.29	10.29
Ta	0.0351	0.192	5.467	55.0	(1.0)	(28.49)	–
Nb	0.6175	3.507	5.679	55.1	11.0	17.82	17.82
U	0.0203	0.0711	3.496	52.3	0.91	44.76	61.49
Th	0.0813	0.1871	2.300	69.3	3.5	43.04	59.03
Pb	0.175	0.489	2.794	30.7	8.0	45.71	57.14
Y	3.940	35.82	9.091	31.2	20.0	5.08	5.58
Sc	14.88	41.37	2.784	9.8	30.0	2.02	2.02
Co	104	47.07	0.453	7.9	29.0	0.279	0.24
Ni	2080	149.5	0.072		105	0.050	0.014
Cu	28.0	74.4	2.657	29.1	75	2.679	2.14
Sn	0.150	1.382	9.213	32.4	2.5	16.67	–

The data for the primitive mantle and N-MORB average are taken from Jochum et al. [4]. The N-MORB average represents 26 fresh MORB glasses, defined as N-type MORB by their light-REE depletion [5]. The values for the primitive mantle are taken from Hart and Zindler [6] for the major elements, and their factor of 2.51 is used to obtain the mantle values of the refractory trace elements from the abundances in C1 charbonacous chondrites. The sources of the C1 chondrite data are: REE [7], Ba, Sr, Hf [8], Zr, Ta, Nb, Y, Sc [9], U [10]. Other elements were obtained as follows: Th = 4·U; Cu, Ni, Co [11], K = 12,700·U [12], Rb = Ba/12.3 [13], Cs = Rb/20 [14], Sn [15], and Pb from $\mu = 8.88$ [1]. Most of these values are believed to be correct within about 10%, except for Cs, for which different estimates differ by a factor of 4 [13,14,16].

The data sources for the two continental crust averages are: TM1 [17, table 3.5, p. 67]; TM2 [17, table 3.3, p. 62] (this is the so-called andesite model).

All normalized values are derived by division of the concentration through the primitive mantle value.

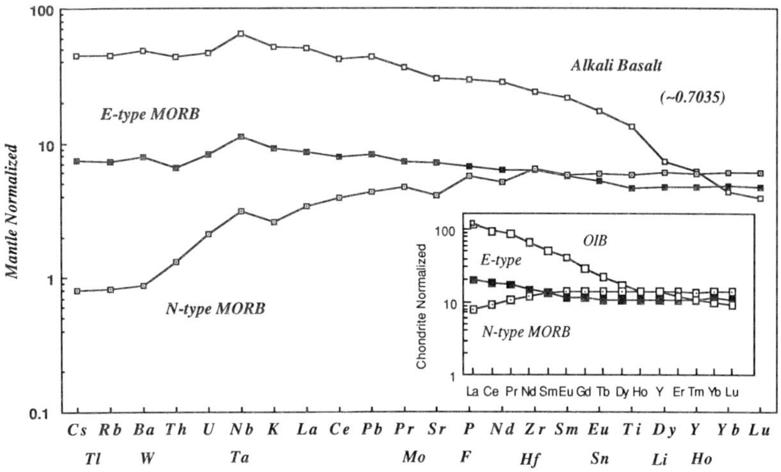

FIG. 1. Primitive mantle (modified) normalized abundance patterns for moderately to highly incompatible elements in average N-type and E-type MORB and oceanic alkali basalts with $^{87}Sr/^{86}Sr \approx 0.7035$. See Table 1 for normalizing values.

K_d are large for barium, rubidium and caesium when phlogopite is stable. Studies of uranium disequilibrium series isotopes in OIBs indicate that 20% uranium or more has been held back in the source, assuming that thorium is perfectly incompatible (*eg* Oversby & Gast 1968; Newman *et al.* 1984). Therefore the variable (by more than a factor of 2) K/U ratios observed in OIBs could be due to these combined effects in addition to original source variations.

Finally, constant elemental ratios suggest that zone refining melting or the chromatographic column model of Navon & Stolper (1987) and magma chamber processes such as mixing and extensive fractionation and replenishment (as discussed by O'Hara & Mathews 1981), which could fractionate incompatible elements efficiently, are not dominant processes operating during the generation of primitive oceanic basalts.

Modification of the 'mantle-normalized' rule and exceptions to the rule

Among OIB samples with La $\geqslant$ 50 ppm, decoupling and lowering of titanium, zirconium, strontium, phosphorus, potassium and thorium abundances from their corresponding rare earth elements (REEs) is common. A new set of relationships is developed for such rocks (Table 2). These features may be due to factors such as the saturation of minor minerals, changes in the residual mineral proportions (*eg* an increase in garnet abundance with depth), a change in mineral/melt K_d, mantle metasomatism and/or melting processes (*eg* if zone refining becomes important). For example, nephelinites from Oahu Island have Ti/Eu ratios less than half of that commonly observed in alkali basalts (3000 versus approximately 6000±700 respectively) and they are also depleted in niobium, tantalum,

TABLE 2. *Generalization of primitive-mantle-normalized abundance ratios (alkali basalt with $^{87}Sr/^{86}Sr \approx 0.7035$ to group I kimberlites)*

La (ppm)	Nb/La	K/La	Sr/Pr	Sr/Nd	P/Nd	Zr/Hf/Sm	Ti/Eu
30	1.2–1.5	0.7–1.0	≈1.0	≈1.0	1.0	1.0	≈0.8
50–100	(1.5)–2.0	≈0.4	(0.8)–1.0	≈1.0	1.0	(0.7)–1.0	≈0.8
>200[a]	1.2–2.0	⩽0.2	≈0.6	⩽0.6	≈0.5	⩾0.5	0.2–0.5

[a] In general we are not sure whether at low degrees of partial melting these ratios are reflecting source mineralogy or subsequent lower-pressure fractionation effects.

zirconium and hafnium. It is very likely that these depletions are due to saturation of a titanate mineral in the source region (Clague & Frey 1982).

Furthermore, an examination of the chemical and isotopic compositions of group I and II kimberlite samples (Kramers *et al.* 1981; Smith 1983; Muramatsu 1985; Muramatsu & Wedepohl 1985; Smith *et al.* 1985) reveals that in many respects group I kimberlites are comparable with Bouvet type OIB, whereas group II kimberlites are more akin to Dupal type OIB (Le Roex 1986). However, exceptions to these general comparisons are that both groups of kimberlites have higher La/Ce ratios, lower Sr/Nd ratios and commonly show zirconium and hafnium depletions relative to samarium. These differences are most probably due to melting under very different conditions (*eg* variable clinopyroxene/garnet ratios, $P-T$ conditions and/or K_d values). Detailed study of the fine features of the mantle-normalized diagram may provide useful information regarding the relative importance of garnet and clinopyroxene, in addition to the effects of other factors. It may also give insights into the possible process of melting of stretched veins of eclogite during OIB generation.

There are several reports of tholeiites and alkali basalts that are enriched in REEs and yttrium relative to other incompatible elements (*eg* western Victoria, Australia (Frey *et al.* 1978; McDonough *et al.* 1985), Norfolk Island (Green 1978) and the Hawaiian islands (Fodor *et al.* 1987)). Some samples have negative cerium anomalies and concave-downward REE patterns with relative enrichment of the middle REEs. Recently, detailed microprobe studies by Nagashima *et al.* (1986), Fodor *et al.* (1987) and Price (1987, pers. comm.) have detected very small amounts of REE-rich carbonates and phosphates of secondary origin. It is therefore likely that post-magmatic alteration is responsible for REE and yttrium enrichment in these unusual samples.

Trace element–isotope correlations in major types of mantle sources of oceanic basalts

Combined lead, strontium and neodymium isotope studies of oceanic basalts demonstrate that the observed isotope variations in these rocks cannot be generated by a single mantle process or mixing of two kinds of mantle components with well-defined isotopic characteristics (*eg* Sun & Hanson 1975*b*; Sun 1980; Cohen & O'Nions 1982; White & Hofmann 1982; Zindler *et al.* 1982; White 1985; Hart *et al.* 1986; Zindler & Hart 1986; Allégre *et al.* 1987). Different approaches have been used to estimate the minimum number of end-members required. Zindler *et al.* (1982) suggested that on a three-dimensional plot of lead, strontium and neodymium isotopes most oceanic basalts fall on or near a mantle plane defined by average N-type MORB ($^{206}Pb/^{204}Pb \approx 18.5$ $^{87}Sr/^{86}Sr \approx 0.7029$, $\varepsilon_{Nd} \approx +13$), St Helena ($\approx 20.8, 0.7029, +5$) and Kerguelen Island ($\approx 18.4, 0.7054, -1$). White (1985) subdivided oceanic basalts into five distinct groups on isotope plots: MORB group (including Iceland, Galapagos, Easter Islands), St Helena group (including Austral, Comores, Ascension), Kerguelen group (including Gough, Tristan da Cunha), Society group (including Marquesas, Samoa, Sao Miguel) and Hawaii island group. Most Society group samples clearly fall above the mantle plane of Zindler *et al.* (1982) with higher $^{87}Sr/^{86}Sr$ ratios at the same lead and neodymium isotope ratios. Zindler & Hart (1986) used different terminology to define the characteristics of mantle end-members: DMM (depleted MORB mantle), PREMA (prevalent mantle, including Iceland, Galapagos, the Easter Islands of White's MORB group and the Hawaiian islands), HIMU (long-term high U/Pb in the source, including St Helena, Tubuai, Mangaia), EM1 (enriched mantle type 1 for the Kerguelen type) and EM2 (enriched mantle type 2 for the Society Island type). Their EM1 samples from the Walvis Ridge clearly fall below the mantle plane of Zindler *et al.* (1982) with lower $^{87}Sr/^{86}Sr$ at the same lead and neodymium isotope ratios (Fig. 2).

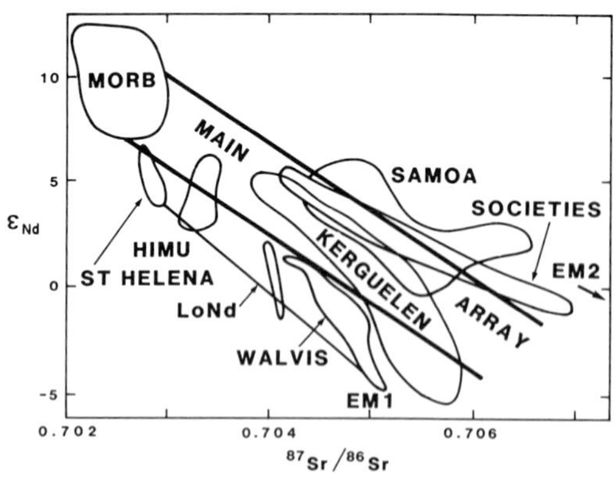

FIG. 2. Schematic diagram of the ε_{Nd} versus $^{87}Sr/^{86}Sr$ plot for OIB data to show the fields of different types of oceanic basalts and the end-member assigned by Zindler & Hart (1986). Modified from Hart *et al.* (1986).

Different mantle processes have been proposed for the generation of the different isotope groups and end-members. Since isotope variation is a function of parent/daughter abundance ratio and isolation time, the same mantle process operating at different times and with different degrees of elemental fractionation will result in variable but coherent lead, strontium and neodymium isotopic compositions. Consequently, the 'end-members' (or components) proposed by Zindler & Hart (1986) would not be expected to have fixed isotopic compositions.

It is not surprising to find that the different mantle processes responsible for generating a variety of *isotope* groups also create distinct *chemical* signatures. Therefore the variations in the mantle-normalized patterns for different mantle components may reflect the effects of the different processes and environments, as with isotopic data. A comparison of Nb/Pb, Ce/Pb, Nb/U, K/U, K/Nb and Rb/Cs ratios, ratios which encompass elements considered to behave similarly during generation of oceanic basalts, for different OIB end-members is presented in Table 3. The contrast between samples of the HIMU (or St Helena) and EM (or Gough) type is quite clear and there are consistent differences, especially for Nb/Pb and Ce/Pb ratios, in other OIB samples. Ratios involving potassium may be affected by the retention of potassium (rubidium, barium and caesium) in potassium-bearing residual minerals, as discussed earlier.

In Fig. 3 a representative HIMU-type sample (St Helena: 2882) and an EM-type sample (Gough island: G111) (Table 4) are used to illustrate the differences in mantle-normalized patterns for these two types of mantle source. These patterns are considered representative of their respective types on the basis of these plus other high quality data having good consistency with other samples from the same islands. This statement is also supported by data of critical element ratios presented in Table 3. The HIMU-type mantle with low $^{87}Sr/^{86}Sr$ (0.7027−0.7030) and very radiogenic $^{206}Pb/^{204}Pb$ (≥21) is characterized by a low Rb/Sr ratio (0.01−0.03) and a high $^{238}U/^{204}Pb$ ratio (≥30); such mantle materials tend to have low K/U and K/Nb ratios (related to potassium, rubidium and barium depletions). The mantle-normalized patterns of HIMU-type mantle (St Helena) display a continuous decrease in the abundances of the elements niobium to caesium, similar to typical MORB. In contrast, for EM-type OIB with $^{87}Sr/^{86}Sr$ ≥0.7040 and lower $^{206}Pb/^{204}Pb$ ratios than St Helena type mantle (<18.6 for EM1, 18.6−19.7 for EM2), the mantle-normalized pattern shows increased abundances of elements more incompatible than niobium (Cs, Rb, Ba, Th, U, K) and lead compared with the St Helena type. Because these two patterns

TABLE 3. *Contrasts of some element ratios observed in N-type MORB and HIMU-type and EM-type ocean island basalts*

	Nb/Pb	Ce/Pb	Nb/U	K/U	K/Nb	Rb/Cs	Nb/Th
N-type MORB 0.7025							
MORB	7.5	25	50	12700	260	80	20
HIMU (St Helena) type 0.7029							
St Helena	23	33	48	8000	180	100	13.3
Rurutu	18	≈30	—	7000	180	—	12.5
Mangaia	18	29	—	6000	160	—	10
Dunedin	20	≈30	~45	7000	170	—	17
Ua Pou (th)	19	38	43	8000	180	—	13
EM-type OIB 0.705−0.706							
Gough	10.4	20	43	18000	430	70	9.3
Kerguelen	8.5	14.5	43	(8000)[a]	270	—	11.5
Aitutaki	8.3	16	—	(5600)	(170)	—	6.3
Upolu	12.1	31	—	(9700)	(200)	—	—
Tutuila	12.2	≈25	—	(8300)	(250)	—	9.1
Tristan	15.9	24	41	11500	270	85	9.1
Sao Miguel	—	—	40	13700	340	90	9.2

[a] Low K/La suggesting potassium has been held back? Data sources: Sun & Hanson (1975*a*); White *et al.* (1979); Hofmann & White (1983); Hofmann *et al.* (1986); Newsom *et al.* (1986); Palacz & Saunders (1986); Weaver *et al.* (1986); this paper.

(Fig. 3) have similar REE abundances, the two distinctive patterns from caesium to niobium cannot be due solely to different degrees of partial melting. Moreover, such OIBs often have barium spikes and high La/Nb ratios (sometimes higher than the chondritic ratio). This latter point is illustrated in Fig. 4 which shows a linear correlation between Ba/Nb and La/Nb among OIBs (Weaver *et al.* 1986). These two ratios also correlate with $^{87}Sr/^{86}Sr$ in general. 'Exceptions' are known for other EM-type samples (*eg* group I kimberlites and Tutuila), which have Ba/Nb $\leqslant$ 8 and La/Nb $\leqslant$ 0.8, and such differences most probably reflect the multiple origins of EM-type source regions. However, we note that a general trend of low $^{87}Sr/^{86}Sr$, La/Nb and Ba/Nb to high $^{87}Sr/^{86}Sr$, La/Nb and Ba/Nb exists.

Lamproites from western Australia (Jaques *et al.* 1988) may also represent extremely enriched EM2 type. They have a highly fractionated mantle-normalized pattern (Fig. 5), especially from niobium to rubidium. These samples have low strontium and phosphorus abundances, a negative europium anomaly of up to 20% and high lead and zirconium abundances (similar to some western US minettes; Alibert *et al.* 1986), possibly reflecting a sedimentary component added to their source (*eg* Thompson *et al.* 1984). Often, but not always, they have depletions of niobium relative to lanthanum.

The generation of different mantle types and the niobium issue

OIBs with high K_2O ($\geqslant$2%) do not always have low U/Pb (μ) values ($\leqslant$16) but it is generally true that OIBs with low K_2O ($\approx$1%) always have high μ values ($\geqslant$20). This correlation must be related to the source character and partial-melting conditions. Several mechanisms have been proposed to generate the isotopic and chemical characteristics of different types of OIB mantle sources. The HIMU and low $^{87}Sr/^{86}Sr$ character St Helena-type OIBs have been attributed to the recycling of altered oceanic crust into the OIB source region (*eg* Zindler *et al.* 1982; Palacz & Saunders 1986), mantle enrichment through melt migration from the LVZ into oceanic and continental lithosphere (*eg* Sun 1980; Hart *et al.* 1986) or derivation of OIBs from the lower mantle which has lost lead by continuous core formation (*eg* Vidal & Dosso 1978). However, Newsom *et al.* (1986) showed that St Helena basalts have the same molybdenum (highly chalcophile) to praseodymium (highly lithophile) abundance ratios as MORBs and other OIBs, effectively arguing against continuous core formation. The EM-type character is generally considered to be related to processes associated with convergent margins, such as the recycling of sediments with oceanic crust (*eg* Hawkesworth *et al.* 1979; Cohen & O'Nions 1982; White 1985; Weaver *et al.* 1986). The formation of these OIB sources from remobilization or delamination of metasomatized enriched continental lithosphere has also been proposed (*eg* Richardson *et al.* 1982; McKenzie & O'Nions 1983; Hawkesworth *et al.* 1986).

Multiple origins for EM-type mantle is likely, including superposition of different processes. Menzies (1983) suggested that EM1 could be related to enrichment by the introduction of CO_2-rich silicate melt (*eg* kimberlite, nephelinite) whereas the high Rb/Sr, $^{87}Sr/^{86}Sr$ character of EM2 is similar to some modern island arc basalts (*eg* White 1985). Hart (1984) and Zindler

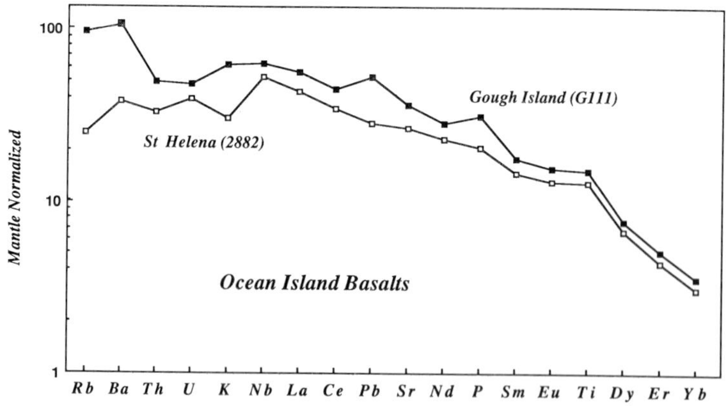

FIG. 3. A comparison of the mantle-normalized patterns of basalts from St Helena (2882), with a low $^{87}Sr/^{86}Sr$, and Gough Island (G 111), with a high $^{87}Sr/^{86}Sr$. See Table 1 for normalizing values. Data from Table 3.

TABLE 4. *Chemical and isotopic composition of selected basalts*

	St Helena 2882	Gough G 111	Chain 43°–45°N MAR	
			104–16	104–18
Major elements (wt%)				
SiO_2	45.50	47.73	50.53	48.85
TiO_2	2.80	3.30	1.35	1.12
Al_2O_3	13.54	15.53	14.88	16.10
FeO	11.10	10.67	9.39	7.84
MnO	0.11	0.14	0.18	0.16
MgO	10.40	8.37	8.22	9.22
CaO	10.69	8.71	10.99	12.31
Na_2O	2.52	2.89	2.60	2.42
K_2O	0.84	1.70	0.34	0.48
P_2O_5	0.43	0.66	0.13	0.15
Total	97.73	99.70	98.61	98.65
Trace elements (ppm)				
Rb	16.1	62.0	4.1	7.8
Sr	581	792	104	185
Ba	275	760	41	84
Nb	38.5	46.0	5.10	14.1
Zr	—	255	77	75
Y	—	26	33	24
La	31.7	41.2	5.92	8.58
Ce	64.8	84.5	14.8	19.7
Nd	32.3	40.2	11.0	11.50
Sm	6.68	8.08	3.59	3.06
Eu	2.25	2.66	1.30	1.10
Gd	6.20	7.20	4.83	3.66
Dy	4.75	5.44	5.84	4.13
Er	2.09	2.42	3.72	2.54
Yb	1.55	1.81	3.67	2.46
Lu	0.211	0.247	0.568	0.385
Pb	2.08	3.85	0.75	1.20
Th	3.01	4.52	0.51	0.85
U	0.86	1.05	0.19	0.27
$^{87}Sr/^{86}Sr$	0.70292	—	0.70317	0.70319
$^{143}Nd/^{144}Nd$	0.512857	—	0.513101	0.513086
$^{206}Pb/^{204}Pb$	20.896	18.552	19.332	19.351
$^{207}Pb/^{204}Pb$	15.791	15.606	15.563	15.570
$^{208}Pb/^{204}Pb$	40.131	39.045	38.826	38.878

Neodymium and strontium isotope data for Chain 43(104–16 and 104–18) and St Helena samples are from White & Hofmann (1982) (see their CH43 106–16 and 106–18 sample numbers); however, for 104–16 we report the strontium isotope ratio for an acid-leached sample (which was normalized relative to 0.70800 for the E&A standard).

& Hart (1986) emphasized that EM2 is almost exclusively restricted to the southern hemisphere. It could have originated from the delamination of subduction-zone-related metasomatized continental lithosphere of Gondawana or the recycling of the mantle wedge above the subduction zones. However, they do not preclude the earlier suggestion by Hart (1984) that this regional EM-type anomaly in the sourthern hemisphere existed since the Archaean and might be controlled by global convection patterns. Hawkesworth *et al.* (1986), on the other hand, favour the idea that the EM-type mantle in the southern Atlantic (Gough, Walvis Ridge, Tristan da Cunha) has a shallow origin due to the thermal remobilization and detachment of the continental lithosphere of southern Africa from South America during the breakup of Gondawana. More recently, Hart *et al.* (1986) pointed out that HIMU and EM1 mantle types are often spatially related and share the characteristic of lying below the mantle

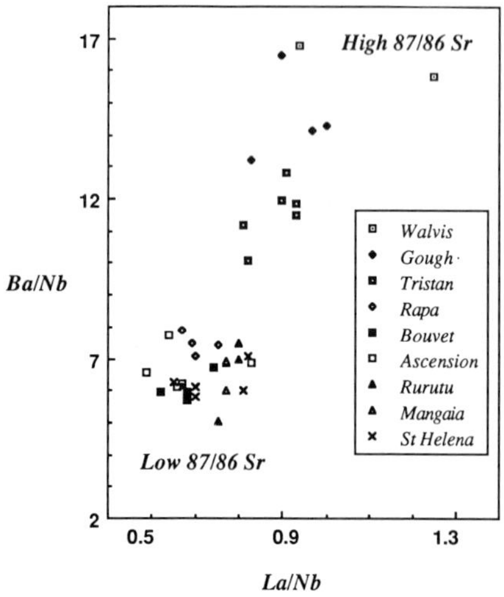

FIG. 4. Positive correlation between Ba/Nb and La/Nb ratios of HIMU-type (low $^{87}Sr/^{86}Sr$) and EM-type (high $^{87}Sr/^{86}Sr$) OIB. Data sources: Richardson *et al.* (1982); Palacz & Saunders (1986); Weaver *et al.* (1986); this paper.

array on the ε_{Nd} (vertical axis) versus $^{87}Sr/^{86}Sr$ plot (Fig. 2). They suggested a subcontinental lithospheric mantle origin for both HIMU and EM1 mantle types through mantle metasomatism by silicate melts.

The niobium issue

The La/Nb variation in some oceanic basalts can be seen in Fig. 6. N-type MORBs have high La/Nb>1.0 (La/Ta>17), E-type MORBs have lower La/Nb (<1.0), and HIMU-type OIBs have much lower La/Nb (0.6–0.7). Given that the oceanic crust once processed through the subduction zone could have a lower La/Nb (<1.0), it has been proposed that the low La/Nb ratios (relative to the chondritic ratio of 0.95) observed in some OIBs, especially HIMU-type (Figs. 4 and 6), reflect the mantle source character of the recycled oceanic crust which has experienced dehydration/partial melting in the subduction zone region (Weaver *et al.* 1986). However, this interpretation is not unique, because low La/Nb ratios in the mantle source could also be the result of mantle metasomatism and/or small degrees of partial melting, since niobium is more incompatible than lanthanum (Fig. 2). On the other hand, high La/Nb ratios (≥0.9) found in many (most?) EM1 and EM2 types of OIB could be due to the introduction of subducted sediment (with high La/Nb ratios, Fig. 6) into the OIB source, or it could be due to the presence of a residual niobium-bearing titanate mineral during magma generation. The effects of partial melting on La/Nb ratios can be assessed by several observations. Two MORB samples from 45°N in the Mid-Atlantic Ridge (Chain 43, 104–16 and 104–18) have identical lead, strontium and neodymium isotopic compositions (Fig. 7 and Table 4). The sample that is light REE enriched also has a much lower La/Nb ratio, consistent with its representing a lower percentage of partial melt derived from the same source as the light-REE-depleted sample. Figure 8 shows similar partial-melting effects on the La/Nb ratio of basalts from Loihi Seamount. The Loihi Seamount tholeiites and basanites have similar strontium and neodymium isotopic compositions. In contrast, the tholeiites have La/Nb ratios which are close to the chondritic value, whereas the basanites have La/Nb ratios as low as the HIMU samples of Fig. 6. These data argue strongly against the idea that the HIMU source must have a low La/Nb ratio inherited from the subducted oceanic crust. Either the recycled oceanic crust model is incorrect, or the La/Nb ratio of the recycled oceanic crust has been increased before final diapiric rise. The basalts

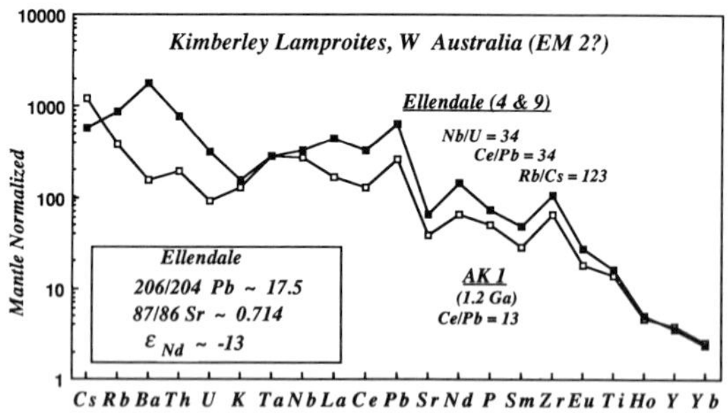

FIG. 5. Mantle-normalized diagram of EM2-type Miocene lamproites from W Kimberley and 1.2 Ga lamproites from E Kimberley, western Australia. Data source: Jaques *et al.* (in press). See Table 1 for normalizing values.

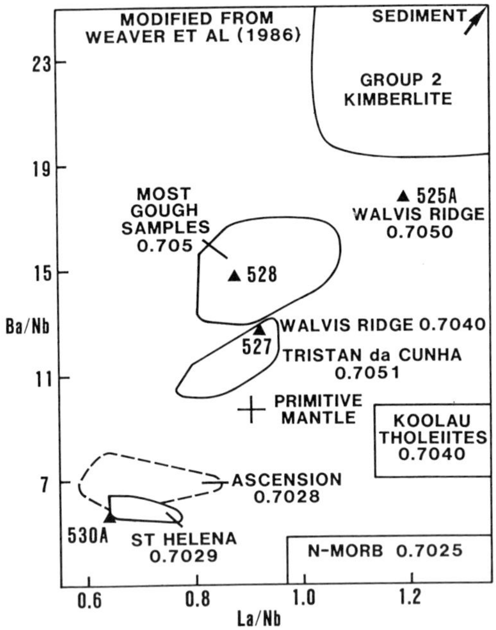

FIG. 6. Positive correlation between Ba/Nb and La/Nb ratios of basalts from St Helena, Ascension, Tristan da Cunha and Walvis Ridge (modified from Weaver *et al.* 1986). Data for Koolau tholeiites, N-type MORB, group II kimberlites and sediments are added for comparison.

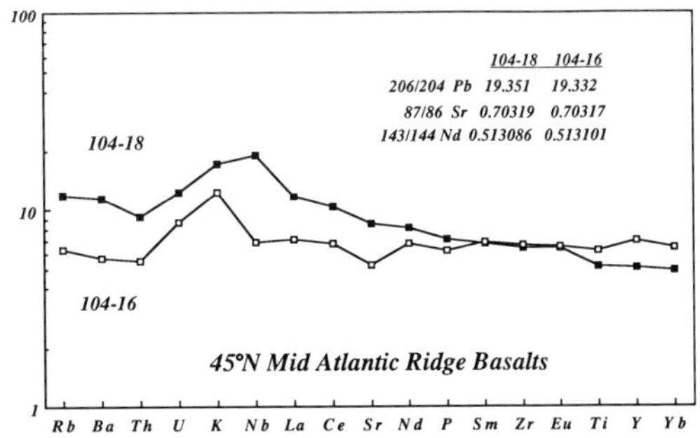

FIG. 7. Comparison of mantle-normalized patterns for two MORB samples from 45°N from the Mid-Atlantic Ridge. These basalts have similar lead, strontium and neodymium isotope compositions but different La/Nb ratios which can only be attributed to melting. See Table 1 for normalizing values.

from Haleakala, Maui, also show a consistent variation in La/Nb ratios with degree of partial melting: the basanites representing the lower degrees of partial melting have lower La/Nb ratios than the tholeiites which are derived by larger degrees of partial melting. However, in contrast to the Loihi basalts, the tholeiitic and alkalic basalts from Haleakala are derived from isotopically different mantle sources (Chen & Frey 1985).

In addition to reflecting the source character, high La/Nb ratios in OIBs can be produced if a residual titanate mineral is present during open-system processes in the source (*eg* metasomatic enrichment or partial melting) or during magma differentiation. Nephelinites and basanites of the Honolulu series from Oahu are good examples (Fig. 8). They show relative depletions in titanium, niobium, tantalum, zirconium and hafnium and have Ti/Eu ratios ($\approx$3000) less than half those normally observed in OIBs (about 6000, close to the chondritic ratio approximately 7800). Clague & Frey (1982) suggested that a residual titanate mineral is responsible for the niobium depletion relative to lanthanum. Additionally, it is interesting that high La/Nb ratios (1.6), similar to those found in the Honolulu series, are also found in some Koolau tholeiites (Fig. 8), although these two Oahu basalt series are derived from isotopically different sources (Roden *et al.* 1984). In contrast with the Oahu case, EM samples from Walvis Ridge and Gough (Fig. 4) with high La/Nb ratios have Ti/Eu ratios close to the chondritic ratio, suggesting that a titanate mineral does not play a role in the generation of these basalts. We therefore agree with Weaver *et al.* (1986) that the high La/Nb ratios of these EM-type basalts reflect a mantle source characteristic, possibly resulting from the involvement of subducted sediments. This conclusion is further supported by the high Ba/Nb, $^{87}Sr/^{86}Sr$ (Fig. 6) and $^{208}Pb/^{206}Pb$ commonly observed in EM-type OIBs and typical of sedimentary material. However, we would emphasize that sediment subduction, along with the oceanic crust, is not a unique solution for the generation of EM-type mantle. The addition of a fluid or melt derived from an altered subducted oceanic crust, with or without sediment, into the overlying mantle wedge or adjacent continental lithospheric mantle may also generate high Ba/Nb and La/Nb ratios.

Mid-ocean ridge basalt source connection of the HIMU-type mantle

HIMU-type OIBs are characterized by very low $^{87}Sr/^{86}Sr$ ratios (as low as 0.7027) similar to some N-type MORBs, high $^{206}Pb/^{204}Pb$ ratios ($\approx$21.0) but low inferred Th/U ratio in the source (approximately 3.3, see below). It is now generally accepted that the HIMU-type mantle

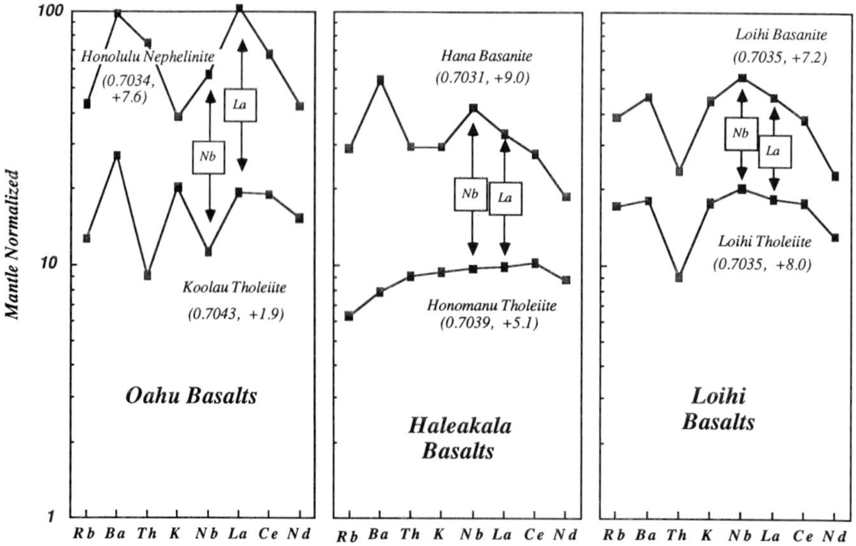

FIG. 8. Comparison of mantle-normalized patterns of Rb, Ba, Th, K, Nb, La, Ce and Nd for selected Hawaiian basalts. The samples used include Koolau tholeiite (WW 9980), Honolulu nephelinite (68 Pb 2), Honomanu tholeiite (C122), Hana basanite (65−11), Loihi tholeiite (24−7) and Loihi basanite (17−7). Data sources: Clague & Frey (1982); Frey & Clague (1983); Roden *et al.* (1984); Staudigel *et al.* (1984); Chen & Frey (1985). See Table 1 for normalizing values.

was originally derived from a MORB-type component more than a billion years ago; however, there is disagreement regarding the processes responsible.

On lead isotope ratio plots (Fig. 9) each HIMU island exhibits a small range of isotopic variation. To estimate time of isolation for the HIMU-type mantle, we use the lead isotopic composition evolution of the depleted upper mantle, as suggested by Zartman & Doe (1981), to calculate a two-stage lead isotopic evolution. Their version II lead isotopic growth curves for the depleted mantle are consistent with the available lead isotope data for Archaean and Proterozoic komatiites and basalts derived from depleted mantle, as well as Phanerozoic ophiolites. Feasible solutions to our calculations are evaluated by comparing the calculated μ_2 values

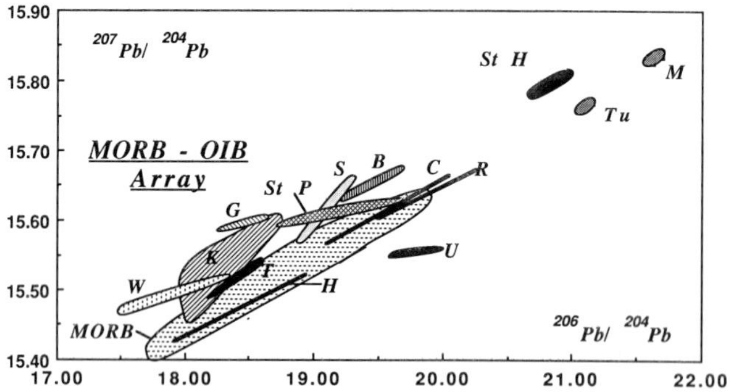

FIG. 9. Simplified $^{207}Pb/^{204}Pb$ versus $^{206}Pb/^{204}Pb$ plot for oceanic basalts from mid-ocean ridge (MORB), Walvis Ridge (W), Hawaiian Islands (H), Kerguelen (K), Gough (G), Tristan da Cunha (T), Society Islands (S), St Paul's Rocks (St P), Ua Pou (U), Ross Island (R), Canary Islands (C), Bouvet (B), St Helena (St H), Tubuai (Tu) and Mangaia (M). Data sources: Sun (1980); Cohen & O'Nions (1982); Richardson *et al.* (1982); Roden *et al.* (1984); Vidal *et al.* (1984); Palacz & Saunders (1986); Stille *et al.* (1986); Dupuy *et al.* (1987); Wright & White (1987).

($^{238}U/^{204}Pb$) needed to generate the observed HIMU $^{206}Pb/^{204}Pb$ and the resulting $^{207}Pb/^{204}Pb$ values with the measured μ and $^{207}Pb/^{204}Pb$ values. These results (Fig. 10) indicate that the isolation time must be greater than 1 billion years, for otherwise the μ value in the source would have to be higher than that measured in the sample, which is contrary to the inference that uranium is more incompatible than lead (Fig. 1). However, the calculated $^{207}Pb/^{204}Pb$ value with less than 1 billion years of isolation is lower than the measured value; therefore a longer isolation time is required. Using a reasonable μ_2 value and considering $^{207}Pb/^{204}Pb$ heterogeneity (±0.05) in the Precambrian depleted mantle, we obtain an isolation time of 2.0±0.2 Ga (Fig. 10). Since the HIMU source is related to the MORB source through mantle fractionation processes, it is not surprising that similar age information can be derived from the slope connecting HIMU data with the MORB field on Fig. 9, assuming that the slope represents an isochron. A mantle source isolation age calculated in this way for the other two oceanic islands—Tubuai and Mangaia—gives 1.7 Ga (Palacz & Saunders 1986), similar to the St Helena case. An isolation time of about 1.0±0.2 Ga is required to derive the most radiogenic $^{206}Pb/^{204}Pb$ ratio (19.99) in the Ua Pou (Marquesas Archipelago) tholeiites from a depleted MORB-type mantle, assuming a reasonable μ_2 value of 20±2 in the mantle source. This is the youngest isolation time calculated for HIMU-type OIBs.

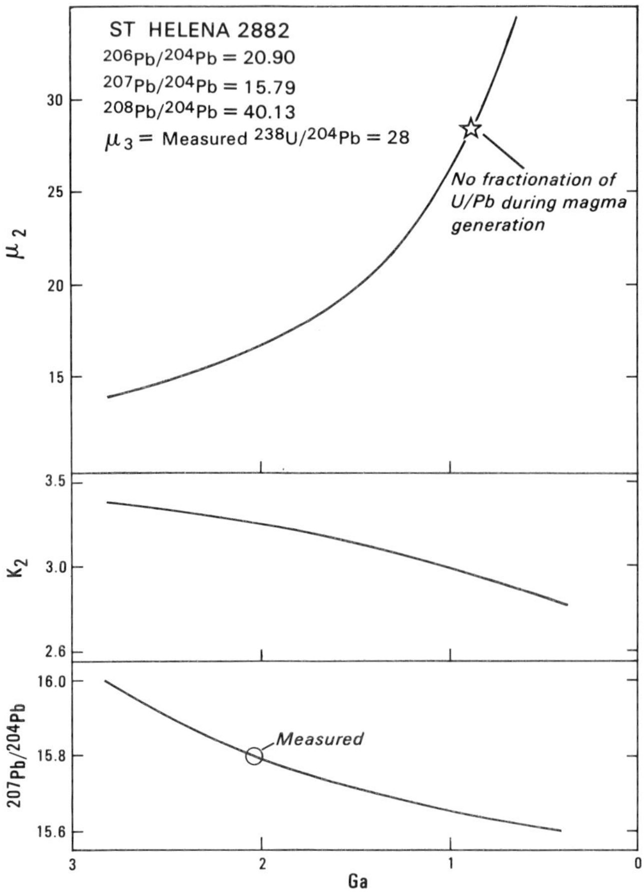

FIG. 10. Model calculations for the lead isotope evolution of a St Helena sample (2882) assuming that it was derived from a depleted upper mantle MORB-type source some time ago. When an isolation time is assigned, specific values of μ_2 and K_2 are required to match the measured $^{206}Pb/^{204}Pb$ and $^{208}Pb/^{204}Pb$ ratios. At 2.0 Ga, model calculations give the best solution to match the measured $^{207}Pb/^{204}Pb$ and μ values.

It is important to point out that if St Helena, Tubuai, Mangaia and Ua Pou samples are not derived from pure HIMU end-members but either (i) contain an EM-type component with low $^{206}Pb/^{204}Pb$ but high $^{207}Pb/^{204}Pb$ (*eg* Duncan *et al.* 1986), or (ii) are derived from a recently enriched high-μ reservoir (*eg* St Paul's Rock; Roden *et al.* 1984) followed by mixing with depleted asthenosphere (Fig. 9), then the age of the HIMU isolation could be considerably younger. Such an interpretation may apply to much of the non-HIMU OIB lead isotope data (see later).

Model calculations for the $^{208}Pb/^{204}Pb$ evolution of St Helena samples (Fig. 10) yield a Th/U ratio of 3.30±0.05 in the mantle source starting at 2.0 Ga ago. This value is lower than the measured Th/U ratios in St Helena samples (3.5–3.8) (Weaver *et al.* 1986; this paper), consistent with the fractionation of thorium from uranium during OIB magma generation. Similar conclusions are reached through $^{230}Th/^{232}Th$ studies on young OIB (*eg* Oversby & Gast 1968; Newman *et al.* 1984).

Recently, Dupuy *et al.* (1987) presented a strong case for a connection between the recycled oceanic crust and HIMU-type OIB sources on the basis of their study of Ua Pou Island, Marquesas Archipelago. They emphasized that Ua Pou tholeiites are highly relatively depleted in most incompatible elements (barium, rubidium, thorium, uranium, potassium, niobium and light REEs) and have mantle-normalized patterns (Fig. 11) showing mirror images of those for island arc basalts, *ie* these tholeiites have high Nb/La, Ce/Pb, U/Pb and Hf/Lu ratios. Dupuy *et al.* suggested that the mantle source of the Ua Pou tholeiites contains residual oceanic crust formed by dehydration and partial melting during subduction.

These unusual Ua Pou tholeiites with concave-downward REE patterns (Fig. 11) offer a rare opportunity to evaluate the recycled oceanic crust model. They have high nickel (155–550 ppm), TiO_2 (3.7%–4.1%) and cerium (61–78 ppm), comparable with those of other HIMU OIBs considered in this study. Consequently, based on the arguments presented earlier, it would be reasonable to expect more than 20% fractionation of the La/Nb ratio during magma generation. The measured La/Nb ratio (0.85) of these tholeiites is higher than those of other HIMU OIBs (St Helena, 0.69; Ascension, 0.65; Mangaia, 0.77) but similar to that of the EM-type OIB of Tristan da Cunha (0.86). After a 20% correction for the partial-melting effect (based on studies of Hawaiian nephelinites discussed earlier), the mantle source for Ua Pou tholeiites has La/Nb = 1.06, which is greater than the chondritic ratio (0.96), *ie* it actually shows niobium depletion relative to lanthanum. The lack of a positive niobium anomaly relative to lanthanum in their mantle source weakens the argument of a recycled oceanic crust model for the Ua Pou tholeiites, or at least makes it unnecessary.

The high Hf/Lu ratios (22–28) observed in the Ua Pou tholeiites have been used by Dupuy *et al.* to support the recycled oceanic crust model. However, since hafnium is much more

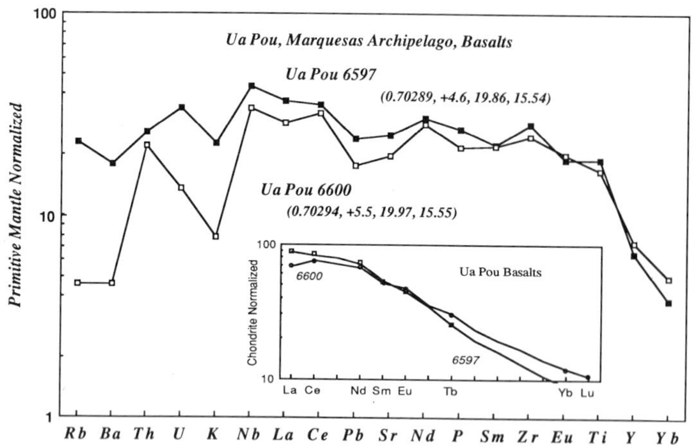

FIG. 11. Mantle-normalized diagram of two HIMU-type OIB samples from Ua Pou. They have concave-downward patterns in the light REE region. Data from Liotard *et al.* (1986) and Dupuy *et al.* (1987). See Table 1 for normalizing values.

incompatible than lutetium during OIB generation, this argument may not be valid. A comparison of Sm/Hf is more appropriate because these elements are similarly incompatible (Fig. 1). In fact, the Sm/Hf ratio of Ua Pou tholeiites (1.43) is equal to that of the primitive mantle (Table 1) suggesting that hafnium has not been fractionated from samarium by processes involved in the generation of HIMU-type mantle. Although we agree with Dupuy *et al.* that the very high Ce/Pb ratios of Ua Pou tholeiites (38–78) are very unusual for OIBs (25±5), these ratios are also much higher than those of other HIMU-type OIBs shown in Table 2. We do not have a good explanation for this anomaly.

The neodymium isotope data presented by Duncan *et al.* (1986) and Dupuy *et al.* (1987) require that the mantle source of Ua Pou tholeiites has an Sm/Nd ratio lower than chondrite, after isolation from the MORB-type source about 1 billion years ago. Combining this information with niobium and REE data, we suggest that this mantle has a concave-downward REE pattern with niobium depletion similar to the pattern of many Hawaiian tholeiites (*eg* Fig. 8).

In order to make a correct choice between the two processes proposed for generation of the HIMU-type mantle, *ie* through recycling of altered oceanic crust or migration of silicate melt from the depleted mantle into the lithosphere, critical evaluation of the two processes is required. Some MORB and OIB tholeiites (*ie* Iceland) have light-REE-depleted or flat patterns with positive niobium kicks on the mantle-normalized diagram (La/Nb less than the chondritic value). These data support the idea of recycling oceanic crust in the MORB and OIB sources and decreasing La/Nb through the subduction zone dehydration/partial-melting process. However, if the oceanic crust recycling model is correct for the HIMU character, then alteration and subduction of this crust (dehydration and partial melting) must decrease the La/Nb ratio by 20% or more and the U/Pb ratio by a factor of 2 without significantly (>20%) affecting the Rb/Cs, Rb/Sr, Th/U, Ce/Pb and Nb/U ratios. This does not seem likely. Taking Th/U as an example, a model calculation of the lead isotope evolution of the HIMU-type St Helena sample (2882, Fig. 10) requires that the Th/U actually increases slightly relative to the MORB source at about 2.0 Ga ago. Furthermore, dehydration and partial melting is an effective 'depletion' mechanism. In order to generate a St Helena-type OIB source character an enrichment process after subduction, such as megalith melting (Ringwood 1982), is required. At present, there are no experimental data to show how these trace elements (and the La/Nb ratio discussed earlier) will be fractionated during partial melting in the garnetite stability field. However, upward migration of CO_2-rich silicate melts such as kimberlite, carbonatite and nephelinite from the LVZ at the top of the depleted asthenosphere could be an effective mechanism to achieve the HIMU characteristics (including Th/U) and real geological observations are available to support this idea (*eg* Green 1971; Kay 1979; Menzies & Murthy 1980; Sun 1980; Hart *et al.* 1986). These CO_2-rich silica-undersaturated melts have low Rb/Sr ratios but high U/Pb ratios (see earlier) and match the characteristics of the HIMU type. Thus, the HIMU character can be fully satisfied by this melting process, since thorium is more incompatible than uranium, which is more incompatible than lead, and residual phlogopite can hold back more rubidium from the melt during small degrees of partial melting. The choice between oceanic and continental lithosphere for the HIMU component will be evaluated later.

Subduction zone processes and EM-type mantle

EM2-type mantle with high Rb/Sr and $^{87}Sr/^{86}Sr$ ratios is commonly considered to be a product of the migration of water-rich fluids or melts enriched in potassium, rubidium and barium associated with subduction zones (*eg* Menzies 1983; White 1985; Zindler & Hart 1986), whereas the chemical and isotopic character of EM1-type mantle, with high Ba/Nb and La/Nb ratios, has been interpreted by Weaver *et al.* (1986) as the product of sediment subduction. In this section we point out some weaknesses of the subduction-zone-related models, emphasize the need for alternative mechanisms and propose more definite tests for sediment involvement in OIB sources.

As discussed earlier, high Ba/Nb and La/Nb ratios observed in many EM-type OIBs are not necessarily diagnostic of sediment involvement in their source. Except for Gough, Walvis Ridge and Aitutaki (Table 3), there is a general lack of strong trace-element indication of sediment recycling. Mixing as little as 1% of pelagic sediment (with Pb=25 ppm, Pb/Ce=0.6) with subducted oceanic crust (or 0.2% for MORB-type peridotite mantle) is expected to change the Nb/U, Ce/Pb and Rb/Cs ratios considerably from the nearly constant values (47±10, 25±5, 80±20) for MORB and OIB (Hofmann & White 1983; Hofmann *et al.* 1986; Newsom *et al.* 1986). For example, the Ce/Pb ratio could be reduced to about 14. Furthermore, a mixture of about 0.5% average shale (Taylor & McLennan 1985, Table 2.9) and a model OIB source (based on the OIB data in Table 1) would produce a source capable of generating an EM-type OIB with consistent key element ratios as found in Table 3 (*eg* Nb/U=42, Ce/Pb=20, Nb/Pb=11, K/U=13 000, K/Nb=275, Nb/Th=11). Additionally, the introduction of such a small sediment component 1.0–2.0 Ga ago would also be consistent with the variation in strontium and neodymium isotopic compositions observed in EM-type OIB (*eg* Bouvet versus Gough islands).

However, if sediment subduction is invoked to explain the isotope evolution and the development of negative europium anomalies in some EM-type mantle (see below), then some modification of this simple mixing model is required. In this connection, it is interesting to note that the ratios Nb/U ≈ 34, Ce/Pb ≈ 13, Rb/Cs ≈ 123 (Fig. 5) of Miocene lamproites from western Australia (Jaques *et al.* 1988) are not very different from values for MORB and OIB mentioned above. Isotope and trace-element characteristics (including a 20% negative europium anomaly) of these lamproites (extreme EM2 types) strongly suggest the involvement of sediments in their mantle source (Nelson *et al.* 1986; Jaques *et al.* 1988). Since the EM-type OIBs in Table 3 require much less sediment involvement than these lamproites, the lack of a large amount of variation in Nb/U, Ce/Pb and Rb/Cs ratios may not be a strong argument against involvement of a minor amount of sediment (2% or less) in some EM-type mantle. However, mixing of small amounts of pelagic sediment with oceanic crust is not capable of generating the $^{208}Pb/^{206}Pb$ characteristics of the EM1-type mantle in the S Atlantic. Using lead isotope evolution curves for the depleted mantle and upper crust (sediment source) given in Zartman & Doe (1981), a model calculation for sediment–oceanic crust mixing at 1.5–2.0 Ga with 50% of the lead contributed by the 1% subducted sediment (as suggested by Weaver *et al.* 1986) requires a Th/U ratio of 5.0 to generate the observed $^{208}Pb/^{206}Pb$ ratio. This value is much higher than the value (approximately 3.7) estimated for the Tristan da Cunha source. In general, most oceanic basalts (EM, MORB, HIMU) come from source regions with Th/U ratios of 3.8 or less, as has been demonstrated by $^{230}Th/^{232}Th$ studies, or by a reduction of 15%±5% of the measured Th/U of fresh basalts to correct for the Th/U fractionation during melting (*eg* Oversby & Gast 1968). In our opinion, EM1-type OIB in the S Atlantic does not represent a mixture of sediment and oceanic crust. It is more likely to be derived from an enriched mantle wedge or lithosphere above the subduction zone, with some input from the subducted sediments.

Secondly, a subduction zone environment is not required for at least some EM1-type mantle. Enrichment can be achieved by silicate melt migration from mantle plumes with less depleted character than a MORB source (*eg* Bouvet, Hawaii or OIBs lying isotopically closer to the primitive mantle). EM1-type mantle generated in this way will have higher Th/Nb and Th/Ta ratios than the HIMU-type mantle, whereas Ba/Nb, Rb/Nb and La/Nb ratios could be similar to the HIMU type owing to a low alkali character controlled by residual potassium-bearing minerals and the fact that niobium is more incompatible than lanthanum. These characters are shared by some group I kimberlites with $^{87}Sr/^{86}Sr$ = 0.704–0.705 and some EM2 OIBs such as Sao Miguel (White *et al.* 1979; Le Roex 1986). Le Roex (1986) emphasized that the trace-element and isotope data for group I kimberlites cannot be explained by the sediment subduction model. Instead, they are probably evolved from less depleted or close to primitive mantle material.

Although the discussion presented above sounds negative towards the connection between sediment subduction and the formation of EM-type mantle, we do believe that sediment subduction has left an important fingerprint on the geochemistry of at least some EM-type OIBs (*eg* Walvis Ridge, Gough, Society Island, Samoan Islands). Further detailed geochemical studies of both EM1 and EM2 end-members are required to establish this case firmly. We suggest that accurate estimation of the europium anomaly in OIBs can be of special importance for the evaluation of the effect of the subduction of sediment and altered oceanic crust on the OIB source as well as on fO_2 during magma generation. Since plagioclase, which has a high K_d for Eu^{2+}, is neither a residual mineral during alkali basalt genesis nor an early liquidus mineral during magma differentiation, europium anomalies measured in primitive alkali basalts reflect the combined effects of source character and fO_2 during magma generation. Sun & Hanson (1975*a*) suggested that the 2%–5% positive europium anomaly commonly observed in OIBs is due to the existence of Eu^{2+} (≥10%) in addition to Eu^{3+} under low fO_2 mantle conditions and the fact that Eu^{2+} (≈Sr) is more incompatible than Eu^{3+} and its neighbouring Sm^{3+} and Gd^{3+}. Sediments derived from various terranes have different amounts of negative europium anomaly (*eg* MacLennan & Taylor 1981), with up to 35% europium depletion for modern crustal rocks. The available accurate REE data indicate that HIMU OIBs have a 4%−6% positive europium anomaly whereas EM2 OIBs from Samoan and Society islands show up to a 4%−6% negative europium anomaly (*eg* Sun & Hanson 1975*a*; Newsom *et al.* 1986; Palacz & Saunders 1986; McDonough & Sun, unpublished data), consistent with the sediment subduction model. In contrast, EM-type OIBs from Gough and Tristan da Cunha have a 2%−5% positive europium anomaly despite our belief that sediment has been involved in their mantle source. Either the sediment which was subducted and ultimately incorporated into their source has a small negative europium anomaly (*ie* from Archaean terranes) and/or more probably the sediment effect has been compensated by an increase of the Eu^{2+}/Eu^{3+} ratio during source modification (*eg* dehydration and partial melting of the subducted sediments) and magma generation under more reducing environments than for other OIBs. To test further the significance of the negative europium anomaly in some primitive OIBs, we propose that samples with a strong sediment lead isotope signature are most suitable, because modern sediments have a distinct and strong negative europium anomaly.

Continental lithosphere connection of the Dupal anomaly and the low neodymium array

Hart (1984) pointed out that EM-type OIBs are concentrated in the southern hemisphere, are globe encircling in extent, and are centred at latitude 30°S. He named this isotopically anomalous region the Dupal anomaly region and suggested that it may have been in existence in the southern hemisphere since Archaean times. In terms of the model, the occurrences of EM-type OIBs in the northern hemisphere, *eg* the Azores, Oahu Island (Koolau tholeiites) and the Japan Sea, and numerous occurrences of EM-type continental alkali basalts, tholeiites and kimberlites have to be considered as exceptions to the general rule or due to crustal contamination. Later, Zindler & Hart (1986) suggested that the enhanced rate of Pangean subduction into the Dupal mantle (Anderson 1982) may account for localization of EM2-type OIBs in the southern hemisphere. If this interpretation is correct, then EM2 anomalies in the southern hemisphere are a relatively young subduction-related phenomenon surrounding the Pangean continents, but are not circum-global. This is consistent with the absence of the Dupal anomaly in the Pacific mid-ocean ridge of the southern hemisphere (Macdougall & Lugmair 1986; White *et al.* 1987). Furthermore, Archaean and Proterozoic Dupal-like anomalies observed in komatiites and galena and feldspar in granites of the southern continents (Africa, Australia, India), where there is abundant early Archaean crust, are likely to be a result of crustal contamination. For example, some late Archaean komatiites (and associated NiS mineralization) of the Yilgarn Block, western Australia, have been contaminated by early Archaean felsic crust which resulted in Dupal-type lead anomalies (Sun *et al.*, in press). In contrast, Hawkesworth *et al.* (1986) suggested that the Dupal anomaly in the southern Atlantic (*ie* Walvis Ridge, Gough, Tristan da Cunha and the Discovery Table Mounts) may be of shallow origin and due to the thermal reactivation and detachment of the metasomatized lithosphere of southern Africa and South America before the breakup of Gondwana. The continental lithospheric delamination model of McKenzie & O'Nions (1983) is attractive because there is the opportunity to generate a variety of mantle types in the continental lithosphere within close proximity through subduction-zone-related and unrelated processes. The superposition of the products of different processes is possible. It also allows long-term isolation of the enriched mantle within the lithosphere before delamination and recycling back into the convecting mantle.

More recently, Hart *et al.* (1986) also seem to favour an ultimate origin for the Dupal anomaly from the continental lithosphere. They pointed out that EM1- and HIMU-type OIBs often show close association in occurrence and share the same character of long-term low Rb/Sr ratio compared with OIBs within the mantle array on an $^{87}Sr/^{86}Sr$ versus ε_{Nd} plot (Fig. 2). HIMU-and EM1-type OIBs have lower ε_{Nd} values than the mantle array. There appears to be a lower limit on this plot for OIB data defined by connecting the HIMU and EM1 end members. Hart *et al.* (1986) named this limiting line a 'LoNd' (low ε_{Nd}) array. (Since HIMU and EM1 are characterized by having lower $^{87}Sr/^{86}Sr$ values due to lower time-integrated Rb/Sr ratios than samples within the mantle array, it might be more appropriate to call the array a 'LoSr' (low $^{87}Sr/^{86}Sr$) array.) They suggested that this LoNd array was developed within the subcontinental lithosphere through the introduction of silicate melts of different chemical and isotopic compositions (p. 1556): '. . . EM1 end member being slightly modified primitive mantle, and the

HIMU end member being metasomatically produced, possibly during subduction episodes preceding lithosphere accretion.' It is interesting to note that many of the Mesozoic and Cenozoic carbonatites of worldwide occurrence, not limited to the southern hemisphere, also fall on or close to this LoNd array (*eg* Bell & Blenkinsop 1987; Dudas *et al.* 1987; Nelson *et al.* 1987). This is also true for some group 1 kimberlites (Kramers *et al.* 1981; Smith 1983).

The association of HIMU- and EM-type OIBs in some ocean island or island groups of the southern Pacific (Vidal *et al.* 1984; Palacz & Saunders 1986; Dupuy *et al.* 1987) and Walvis Ridge (Fig. 6) does not offer a firm argument for identifying their parentage (*eg* continental lithosphere). Alternative mechanisms can be suggested, *eg* mixing of different types of mantle components within the convective upper mantle. In addition, we would like to emphasize that the regional but isolated occurrences of HIMU (*eg* St Helena, Ascension, Cape Verde, New England Seamounts, Canary Islands, Ahaggar, Cameroon Line, Guadalupe, SW Pacific islands, eastern Australia, South Island, New Zealand, and Ross Island and Mary Byrd Land, Antarctica) or EM (Gough, Tristan da Cunha, Kerguelen, Samoa) without an accompanying LoNd counterpart are quiet common. The regional distribution tendency of the Dupal anomaly (S Atlantic, Indian and SW Pacific Oceans) and the HIMU anomaly (some of which are probably of shallow asthenospheric origin) also requires serious consideration in any genetic model.

The long isolation time (up to 2 Ga) required by the isotope data of OIBs has been a major consideration of many people in favouring the model of continental lithosphere delamination. In this respect, it is surprising to see that the OIB members of the LoNd array have well-defined isotopic correlations on strontium versus neodymium and lead versus strontium plots (Hart *et al.* 1986). Since the extent of low ε_{Nd} values below the main array is a function of Rb/Sr fractionation and isolation time, in addition to other variables such as Sm/Nd ratio, a well-regulated mantle LoNd array is not expected. Hart *et al.* (1986) pointed out that there are many EM1-type continental peralkaline rocks (*eg* Leucite Hills, Smoky Butte) which fall well below the LoNd array. The lack of an obvious influence of the low neodymium array by these extremely low ε_{Nd} components from the continental lithosphere formed in the Archaean (*eg* Dudas *et al.* 1987) may imply that (i) delamination of continental lithosphere with Archaean isotopic signatures is not feasible or very minor in modern times, (ii) metasomatized continental lithosphere formed at 2 Ga or less is abundant on the worldwide scale and dominates the scene, or (iii) delamination of continental lithosphere is sufficient but not necessary for OIB generation, and survival times of 2 Ga or less in the convecting mantle are possible (Gurnis & Davies 1986). The last option will be further discussed in later sections.

A major drawback to the continental lithosphere delamination model is that there is a general lack of OIB samples showing strong subduction-zone-related chemical and isotopic characteristics, which are commonly found in many continental mafic and ultramafic volcanic rocks. As shown in Figs 5 and 12, relative depletion of titanium, phosphorus, niobium, tantalum and strontium, but extreme enrichment of lead and the presence of up to 20% negative europium anomalies, reflecting sediment subduction effects (McLennan & Taylor 1981), have been observed in many post-orogenic and intraplate mafic to ultramafic peralkaline volcanics (group II kimberlites of southern Africa and lamproites from western Australia) which are considered to be derived from the continental lithosphere modified by subduction zone processes (*eg* Duncan *et al.* 1984; Thompson *et al.* 1984; Nelson *et al.* 1986; Jaques *et al.* 1988). Among these samples there is often a positive correlation between the size of the negative europium anomaly and the $^{87}Sr/^{86}Sr$ ratio (and Sr/Nd ratio) which is thought to reflect the effect of subducted sediment, with high $^{87}Sr/^{86}Sr$, large negative europium anomalies (about 30% but variable depending on the source rocks) and low Sr/Nd ratios (*eg* Alibert *et al.* 1986). In this connection, negative europium anomalies observed in Samoan and Society alkalic basalts can be considered as a good indication of sediment involvement in their mantle sources.

Detailed chemical and isotopic studies of Mesozoic Karoo basalts of southern Africa, Parana basalts of South America, African kimberlites and modern Dupal OIBs in the S Atlantic (*eg* Smith 1983; Smith *et al.* 1986; Duncan *et al.* 1984; Hawkesworth *et al.* 1984, 1986; Le Roex 1986) have offered an opportunity to evaluate different models suggesting their source relationships and mechanisms of magma generation. Based on hot-spot trace reconstructions and geochemical data, Le Roex (1986) concluded that group I kimberlites are not Dupal type (represented by Walvis Ridge and Gough Island) but are related to Bouvet and Marion type hot-spots, whereas group II kimberlites are related to the Dupal islands.

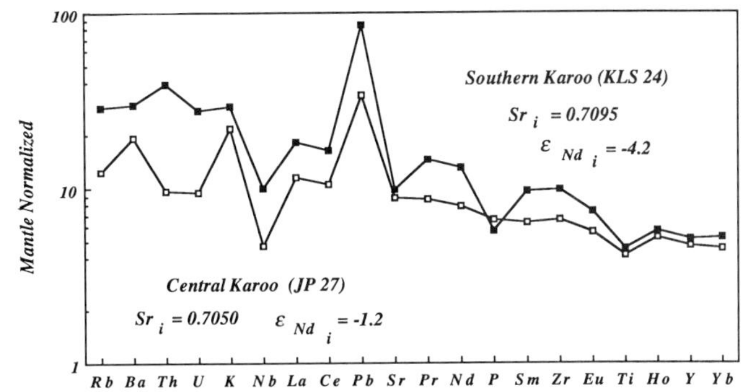

FIG. 12. Mantle-normalized diagram of two selected least-fractionated basalts from southern Karoo and central Karoo. They show strong signatures of sediment and subduction-zone-related processes. Data from Duncan *et al.* (1984). See Table 1 for normalizing values.

Some clinopyroxene megacrysts from the group I kimberlites (*eg* the Kimberley pool) actually have HIMU-type isotopic compositions (*eg* Kramers 1977). The group II kimberlites have higher $^{87}Sr/^{86}Sr$ ratios (0.707–0.710), lower ε_{Nd} values (−5 to −12), and less radiogenic lead isotopic compositions ($^{206}Pb/^{204}Pb \approx 17.2$–17.9) compared with the high P–Ti (HPT) basalts of Parana and northern Karoo as well as the Dupal islands of the southern Atlantic (Hawkesworth *et al.* 1986). As shown in Fig. 13, northern Karoo basalts have mantle-normalized diagrams (especially for Ce/Pb and La/Nb) similar to but still not quite the same as Dupal OIBs. If we accept the asthenospheric origin of Dupal OIBs and at least a partial asthenospheric origin for the group II kimberlites, it seems reasonable to suggest that the geochemical difference between them and the northern Karoo basalts is due to the involvement of the continental lithosphere (including crustal contamination) of southern Africa, which has a strong signature of sediment subduction combined with subduction zone processes (*ie* depletion in niobium, strontium, phosphorus and titanium, a negative europium anomaly and an enrichment in lead). Alternatively, this crustal signature may be derived from sediment subduction beneath Gondwana.

A magnification of this subduction-zone-related character is shown in the least fractionated Mesozoic southern Karoo basalts (Fig. 12) with about a 15% negative europium anomaly. These basalts have more radiogenic lead and strontium isotopic compositions but lower ε_{Nd} values than group II kimberlites and the high phosphorus, titanium northern Karoo and Parana basalts (Smith 1983; Mantovani *et al.* 1985; Hawkesworth *et al.* 1986). Additionally, the isotopic and chemical characteristics of southern Karoo basalts are shared by the relatively contemporaneous Ferrar dolerites of the Trans-Antarctic mountains and the Tasmanian dolerites (Hergt 1987). Hergt (1987) has shown

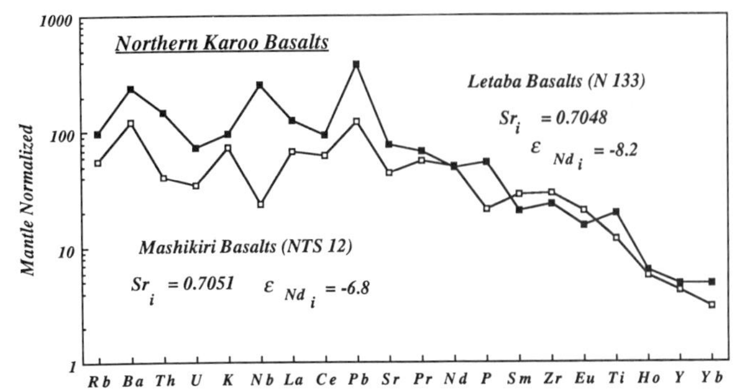

FIG. 13. Mantle-normalized diagram of two selected least-fractionated basalts from northern Karoo. They have similar $^{87}Sr/^{86}Sr$ ratios but lower ε_{Nd} values than Dupal-type OIB from the S Atlantic Ocean. Data from Duncan *et al.* (1984). See Table 1 for normalizing values.

that these dolerites have mantle-normalized patterns which mimic terrigenous sediments and have strontium, neodymium and lead isotopic compositions (0.710, 0.5123, 18.9, 15.64, 38.8 respectively) which strongly suggest a sediment component influence. Cox (1978) suggested that the Mesozoic Gondwana flood basalts are a product of back-arc volcanism related to subduction along the Pacific margin (*ie* close to the southern Karoo occurrence, Trans-Antarctic Mountain range, and Tasmania versus northern Karoo). If his model is favoured and correct, involvement of subduction zone processes as well as melting of the Gondwana lithosphere could be responsible for the crustal characteristics (*ie* low strontium, phosphorus, titanium, niobium and high lead) of these types of basalts.

Mantle plume activity induced by either thermal blanketing of the Gondwana supercontinent (Anderson 1982) surrounded by subduction of cold lithosphere or by lithosphere subduction in the back-arc environment could be the long-lasting hot-spot source for the Dupal anomaly in the southern Atlantic and Indian Oceans. Chemical and isotopic characteristics of the plume material can be modified by contributions from the subduction-zone-related processes (*eg* sediment subduction for low phosphorus, titanium Tasmanian and Ferrar dolerites and basalts of southern Karoo and Parana) or interactions with the continental lithosphere which could have been modified by this same subduction zone process (*eg* group II kimberlites). In this connection, we agree with Zindler & Hart (1986) that Pangean sediment subduction may also have left its influence on the EM2-type OIBs in the S Pacific region (*eg* Samoan and Society islands OIBs with negative europium anomalies and low Ce/Pb). Some residual effects from the recycled mantle wedge on the source character of the mid-ocean ridges, back-arc basins and volcanic arcs is also likely.

The alternative interpretation is that the Dupal anomaly in the S Atlantic is a result of thermal reactivation, remobilization and detachment of the southern African and South American lithosphere (*eg* Hawkesworth *et al.* 1986). This is inferred from the occurrences of HPT basalts in the northern Karoo and Parana which are chemically and isotopically similar to the Dupal OIBs. However, on this model there is difficulty in explaining the lack of a strong signature of subduction zone processes (observed in southern Karoo and some Parana basalts) in the Dupal OIBs. Furthermore, the long-lasting (more than 100 Ma) distinctive chemical and isotopic character of hot-spot traces of the Cameroon line – St Helena, Ninety East Ridge – Kerguelen and Walvis Ridge – Tristan da Cunha also favours a deep stable mantle origin. In this connection, it is worthwhile to point out that the HPT volcanism at about 130 Ma in Parana and Etendaka and the Jacupiringa carbonatite, on the E coast of Parana (Roden *et al.* 1985), could be related to the Walvis Ridge – Tristan da Cunha hot-spot trace.

On a worldwide scale, the lack of a strong subduction zone signature in OIBs and MORBs suggests that the effects of sediment subduction have been short-circuited in the volcanic arc, mantle wedge and continental lithosphere above it. The sediment does not subduct to as great a mantle depth (*eg* 500 km or more) as the oceanic crust (*eg* Sun 1980; Nakamura *et al.* 1985). Although earlier subduction zone effects have been detected in some young oceanic basins (*eg* Woodlark Basin, SW Pacific; Johnson *et al.*, in press) and back-arc basins (*eg* Saunders & Tarney 1979; Cohen & O'Nions 1982), the recycled wedge material in most cases is probably very refractory and depleted in incompatible elements. Thus it will have only a limited effect or no effect on MORB and OIB chemistry. Delamination and/or thermal erosion of continental lithosphere once situated above subduction zones should introduce a strong subduction zone signature into the convecting mantle. To make this model successful it is required that the continental lithosphere is generally formed by non-subduction-zone-related processes such as plume underplating (*eg* Ringwood 1982; McDonough & McCulloch 1987) or alternatively, after subduction ceased, it has been strong modified on a worldwide scale by upward migration of CO_2-rich silicate melts from the asthenosphere and/or intraplate magmatism (as observed in many mantle xenoliths).

Mixing of oceanic island basalt mantle components and interpretations of lead isotope data

On a combined lead, strontium and neodymium isotope ratio plot, the fields for five OIB groups defined by White (1985) all extend towards, and overlap with, the MORB group, suggesting mixing with a MORB-type component. Little mixing appears to have occurred between the four other groups. Detailed geochemical studies of seamount basalts from the N Pacific by Zindler *et al.* (1984) document small-scale heterogeneities in the mantle sources for the basalts.

However, systematic studies of the circum-global MORB system have shown large-scale regional heterogeneities and large-scale mixing of different mantle types (*eg* Schilling *et al.* 1985).

Since the isotopic characteristics of distinct mantle types is an integrated product of fractionation of parent–daughter elements and isolation time, no fixed end-member should be expected. Consequently, mantle mixing with four to five component fields in strontium, neodymium and lead isotope space will generate a volume of mixing products. The mantle plane suggested by Zindler *et al.* (1982) contains most of the OIB data, but it is only an approximation.

Often, more than one mantle type can be sampled in a single island or island group. For example, Li & Hart (in press) suggest that four mantle components (PREMA, recycled metasomatized continental lithosphere, ancient recycled oceanic crust and MORB source) are required to explain the lead, strontium and neodymium isotopic variations observed in Hawaiian basalts. Combined isotope and trace-element studies can offer important constraints on the geometrical relationships of different types of magma sources and magma generation processes. It is now well established that Hawaiian post-erosional alkali basalts have lower $^{87}Sr/^{86}Sr$ ratios but higher ε_{Nd} than the underlying shield-building tholeiites and associated pre-erosional alkali basalts (*eg* Chen & Frey 1985; Li & Hart, in press). In terms of the relative isotopic characteristics of alkali basalts versus tholeiites, the Hawaiian situation is the opposite of that observed for seamount basalts by Zindler *et al.* (1984), who suggested that the isotopic differences between the alkali basalts and tholeiites are due to different degrees of melting of a veined mantle. The isotope data suggest that interaction of the mantle plume (source for tholeiites) with the oceanic lithosphere and/or its immediately underlying asthenosphere is important, especially for the post-erosional alkali basalts (*eg* Tatsumoto 1978; Sun 1980; Chen & Frey 1985; McDonough *et al.* 1985). The inverse correlations between La/Ce, Nb/La and Ba/La with $^{87}Sr/^{86}Sr$ observed in many Hawaiian basalts have been interpreted as the result of a combination of mixing of different mantle sources and the effects of variable degrees of partial melting. In Figs 14*a* and *b*, $^{87}Sr/^{86}Sr$ versus La/Nb and Ba/La respectively are shown for the Hawaiian basalts. Post-erosional alkali basalts with low $^{87}Sr/^{86}Sr$ are mainly derived from a MORB-type source at the base of the Hawaiian oceanic lithosphere and/or in the asthenosphere immediately below it. Strong fractionation of Ba/La and La/Nb ratios from those of the MORB source is consistent with the idea that they have quite different degrees of incompatibility on the mantle-normalized diagram.

In view of the high frequency of inferred mantle mixing phenomena, it is desirable to reconsider the age significance of the lead isotopic data for OIBs. Earlier we suggested that lead isotope data for HIMU-type OIB indicated source isolation ages ranging from 1.0 Ga (Ua Pou) to 2.0 Ga (St Helena). Except for those OIB samples from individual islands having negative slopes on $^{207}Pb/^{204}Pb$ versus $^{206}Pb/^{204}Pb$

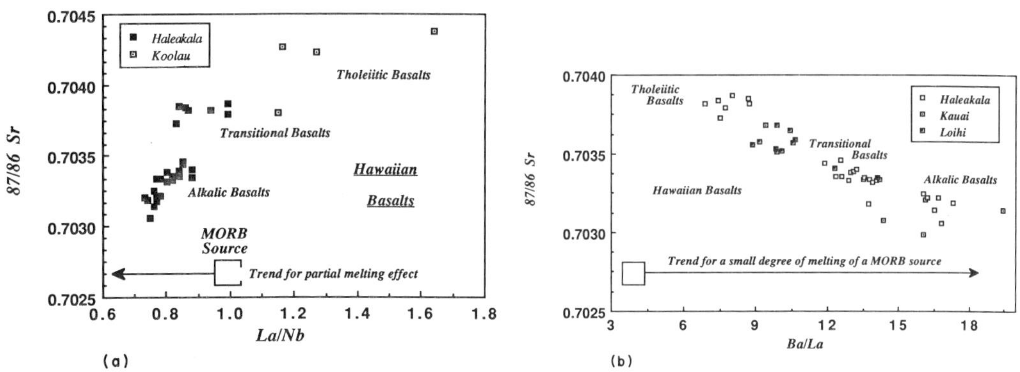

FIG. 14. Plots of (*a*) $^{87}Sr/^{86}Sr$ versus La/Nb and (*b*) $^{87}Sr/^{86}Sr$ versus Ba/La for selected Hawaiian basalts illustrating the combined effects of mixing of two types of mantle sources and different degrees of partial melting. The trend of changing La/Nb and Ba/La ratios caused by different degrees of partial melting is indicated by the arrow direction.

diagrams, indicating mixing of different types of mantle, the interpretation of other OIB data is not so straightforward. Some EM2-type OIBs from the Society Islands and Sao Miguel have high $^{207}Pb/^{204}Pb$ values and each island group exhibits a steep slope on a plot of $^{207}Pb/^{204}Pb$ versus $^{206}Pb/^{204}Pb$. At the high $^{206}Pb/^{204}Pb$ end this trend is constrained by continental sediments. It is likely that this high $^{207}Pb/^{204}Pb$ character is related to sediment subduction (see earlier), as suggested by White (1985). Even though White (pers. comm., 1987) no longer favours this interpretation, we consider that it may also apply to Kerguelen samples.

Good regression lines are also found for basalts from the Hawaiian Islands (0.94 Ga), Canary Island (1.7 Ga), Iceland (1.5 Ga), Ross Island (1.3 Ga) and the northeastern Pacific seamounts (1.8 Ga). Gough Island and the Discovery Tablemount basalts have a small range of $^{206}Pb/^{204}Pb$ and a shallow slope on the $^{207}Pb/^{204}Pb$ versus $^{206}Pb/^{204}Pb$ diagram and yield an age close to zero, whereas Walvis Ridge data give an age of about 0.8 Ga. Whether these regression lines have any age significance is not clear. For example, Stille *et al.* (1986) emphasized the mixing character of the Hawaiian basalts and suggested that approximately 0.9 Ga regression slope has no real age significance, although they felt that the general OIB–MORB trend reflects a 1.5–2.0 Ga isolation age from the MORB source for Iceland, Galapagos and Easter Island.

A good example to illustrate the non-unique solution of the 1.8 Ga Pb–Pb isochron of OIB data is given by the study of the St Paul's Rocks in the equatorial Atlantic (Roden *et al.* 1984). These peridotites have young Sm–Nd and Pb–Pb isochron ages ($\leqslant$ 200 Ma) corresponding to the time of metasomatism in the mantle. The high $^{206}Pb/^{204}Pb$ ratios (up to 19.8, see Fig. 9) and $^{238}U/^{204}Pb$ ratios in the rocks allow $^{206}Pb/^{204}Pb$ to evolve to high values quickly with time. Mixing of this type of mantle with a depleted MORB source can generate Pb–Pb slopes similar to those of the Canary and Ross Islands. However, it is important to point out that the St Paul's sample SE-6 with the least radiogenic lead ($^{206}Pb/^{204}Pb$ = 18.63) and with $^{87}Sr/^{86}Sr$ = 0.70347 has an unusually high $^{207}Pb/^{204}Pb$ ratio (15.61) for OIBs of this type with low $^{206}Pb/^{204}Pb$ and $^{87}Sr/^{86}Sr$.

On the basis of this analysis, we conclude that the available lead isotopic data for OIBs allow for mantle isolation ages ranging from a few hundred million years up to 2.0 Ga. Further research is required to assess the significance of the apparent ages better.

Mass exchange between the crust and mantle

The idea of continuous recycling of crustal material into the mantle has received attention ever since plate tectonic theory became well accepted. Armstrong (1968, 1981) suggested that the continental crust is mainly formed during the Archaean and has reached a steady state growth since then. He suggested that recycling of continental crustal material is required to keep lead and strontium isotope growth in the depleted mantle. In contrast, Hofmann *et al.* (1986) emphasized that the observation of constant Ce/Pb ratios (25±5) and Nb/U ratios (47±10) in MORBs and OIBs argues against significant amounts of sediment recycling into the convecting mantle since about 2 Ga ago, as sediment involvement will greatly affect these ratios (Fig. 6). We have also shown in this paper that, except for some EM-type OIBs, there is a general lack of sediment signature in the mantle-normalized diagram of OIBs and MORBs. We suggest that subduction of sediment at the convergent margins has been short-circuited within the mantle wedge above the subduction zone. This idea is consistent with a detailed geochemical study of alkali basalts in the Japan Sea region by Nakamura *et al.* (1985). Consequently, in order to keep the depleted MORB source isotopic growth as observed, input from OIB source and subcrustal lithosphere in the upper mantle and/or from the lower mantle is required.

Mass exchange between upper and lower mantle

Gast (1968) first suggested that the geochemical characteristics of MORB sources are mainly a product of extraction of small amounts of alkali basalts. In contrast, some lithosphere recycling models require resorption of the subducted oceanic crust into neighbouring fertile upper-mantle peridotite in order to regenerate the MORB source (Allègre & Turcotte 1986). It is obvious that a combination of these two processes is likely to be operative in the mantle. In terms of the model of lithosphere recycling within the upper mantle, the MORB source in the upper mantle is most likely to be the complement to the continental crust. This hypothesis can be tested by a mass balance calculation.

There are various model estimates of elemental abundances in the bulk crust, although these estimates vary by a factor of 2 (*eg* see the summary by Taylor & McLennan 1985). On a mantle-normalized diagram the continental crust has an enriched pattern which is in part a mirror image of the depleted MORB pattern. This complementary relationship between the crust and upper-mantle MORB source is commonly accepted as evidence for a mass balanced system. In Table 5 we use the primitive mantle composition presented in Table 1 and estimates of crustal abundances by Taylor & McLennan (1985) to perform some simple calculations. To satisfy crustal abundances of highly incompatible elements (*eg*, barium, rubidium, caesium, thorium and uranium), 25%–30% of the volume of the upper mantle must be primitive mantle. Using higher estimates of element abundances in the crust, up to 50 % primitive mantle is required. Furthermore, if the continental lithosphere (150–200 km thick) is globally or locally enriched in incompatible elements, then depletion of an even higher portion of the convecting mantle is required. Evidence for enriched continental lithosphere comes from studies of continental basalts and ultrapotassic rocks (*eg* Carlson 1984; Fraser *et al.* 1986; Nelson *et al.* 1986). These rocks always show strong crustal geochemical characteristics (relative depletions of titanium, niobium, phosphorus and strontium on a mantle–normalized diagram, high La/Nb ($\gg$ 1) and large negative europium anomalies) and isotopic systematics suggestive of crustal recycling in the early Precambrian.

According to the calculations in Table 5, the residual MORB source in the upper mantle should have a fertile major- and trace-element composition, similar to those of the fertile ultramafic xenoliths studied by Jagoutz *et al.* (1979). Despite all uncertainties involved in this calculation, it is clear that a major discrepancy between the calculated and observed La/Nb ratios requires special attention. We have argued that the lack of an obvious positive niobium anomaly in the average MORB and OIB sources and continental lithosphere requires a mechanism to fractionate niobium from lanthanum or preferentially withdraw former oceanic crust with low La/Nb ratios from the upper mantle. This can be achieved by subduction of oceanic lithosphere (Creager & Jordan 1984) or subtraction of refractory eclogitic material from the megalith into the lower mantle and/or at the base of the upper mantle (Ringwood & Irifune 1988). Preferential withdrawal of niobium from the upper mantle can also reduce the discrepancy between the calculated and measured Nb/U ratio of the depleted mantle shown in Table 4. A plot of La/Ta versus La/Sm ratios for MORBs from the literature, along with our estimates of the primitive mantle, is shown in Fig. 15. It is quite clear that transitional-type MORBs and some N-type MORBs with light-REE-depleted to slightly enriched patterns (La/Sm$\approx$0.8–1.8) have tantalum (and niobium) anomalies (La/Ta <17). These results are consistent with the idea of recycling and resorption of the subducted former oceanic crust into the asthenospheric mantle source for these MORBs.

Formation of heterogeneous mid-ocean ridge basalt sources

It appears that recycled former oceanic crust is not simply resorbed and mixed into neighbour-

TABLE 5. *Mass balance calculation assuming complementary relationship between the bulk continental crust and MORB source (upper mantle)*

	Rb	U	Nb	La	Ce	Pb	Nd	Sr	Sm	Al%	Ca%
Bulk crust[a] (0.6%)	32	0.91	8[b]	16	33	8	16	260	3.5	8.4	5.3
Primitive mantle	0.635	0.021	0.71	0.69	1.78	0.185	1.35	21.1	0.44	2.45	2.68
% Bulk mantle	30	25	6.5	13.5	10.8	24	6.9	7.2	4.6	2.2	1.3
% Upper mantle[c]	100	83	21.6	45.1	36.9	80	22.9	23.9	15.2	7.6	4.2
Depleted upper mantle after crust extraction	0.0	0.0039	0.59	0.40	1.20	0.041	1.10	16.9	0.40	2.16	2.44

	La/Nb	La/Sm	Sm/Nd	Ce/Pb	Nb/U
Calculated for UM after crust extraction	0.68	1.0	0.36	30	152
Observed in MORB	0.97	0.96	0.36	25±5	47±10

[a] Bulk crust = 0.4% bulk Earth = 0.6% bulk silicate Earth.
[b] A value of La/Nb = 2.0 is used for crustal abundance ratio.
[c] Upper mantle (UM) = 30% of the total mantle.

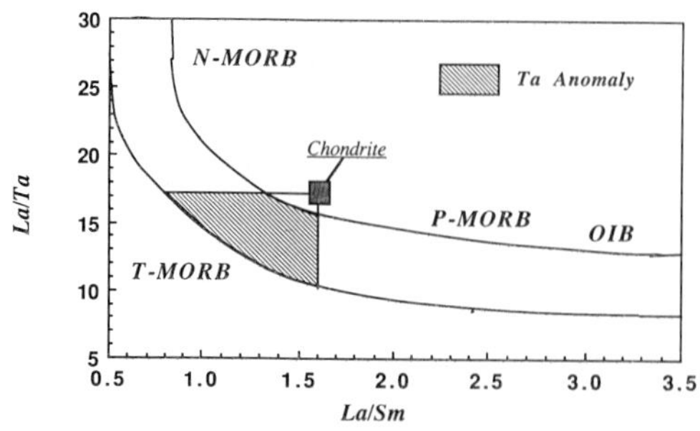

FIG. 15. Schematic diagram showing La/Sm versus La/Ta (Nb) ratios for N-type MORB, T-type MORB and plume-enriched MORB and OIB. The estimate for the earth's primitive mantle is from Table 1. The region which encompasses N-type MORB and T-type MORB with Ta (Nb) 'anomalies' (La/Ta < 17.5) is shaded.

ing fertile peridotite and refractory former oceanic lithosphere to regenerate a homogeneous MORB source. The heterogeneities in the trace-element and major-element abundances observed in MORB and inferred mantle sources call for a more sophisticated model. It is important to note that light-REE-enriched MORB from plume-ridge segments (*eg* the Azores platform) commonly show refractory residual characters in major elements (very low in FeO, Na_2O) compared with samples from normal ridge segments (*eg* Langmuir & Hanson 1980). The chemical and isotopic data for this type of MORB implies that the plume source is refractory in major elements (former lithospheric harzburgite?) and was fertilized before plume diapirism. Such enrichment could be achieved through upward migration of silica-undersaturated melts from the LVZ into oceanic lithosphere before subduction (Kay 1979). It could also be achieved through melting of the subducted oceanic crust (*eg* Ringwood 1982). In contrast with the case of the Azores platform, low-potassium tholeiites from Iceland often show high abundances of FeO and they have positive niobium kicks on the mantle-normalized plot (*eg* Wood *et al.* 1979). It is likely that the Iceland plume has received a larger amount of recycled former oceanic crust.

A uniformly depleted MORB reservoir which is homogenized through convective mixing may also have added to it a component from the recycled mantle wedge above the subduction zone, convective thermal erosion of the continental lithosphere and plumes or blob components derived from the lower mantle or base of the upper mantle where subducted former lithosphere may accumulate. This could result in the variation of chemical and isotopic characteristics in different ocean basins (*eg* the existence of the Dupal anomaly in the Indian Ocean). A veined and plum-pudding type of upper mantle with various scales of old and young heterogeneity can be generated this way.

Towards an integrated mantle-evolution model

The aim of an integrated trace-element and isotope approach is to improve our understanding of the chemical and dynamic processes operating in the Earth's mantle at present and throughout the geological record. Such an approach is very useful for evaluating the various mantle processes involved in the generation of OIB sources. In order to appreciate the chemical and isotope data fully, it is important to integrate this approach with geophysical and petrological studies of the Earth's mantle. In this regard, we are concerned with (i) the style and time-scale of mantle convection, (ii) the physical and chemical consequences of crust formation, recycling of altered oceanic crust (± sediment) and its resorption into the convective mantle and (iii) the connection between present-day regional chemical and isotopic heterogeneities (*eg* Dupal, HIMU) and ancient and more recent global tectonic history and mantle convective regimes.

Some controlling factors which can generate distinct features on the OIB mantle-normalized diagram are shown in Fig. 16. The effect of sediment subduction can superimpose a crustal signature (*eg* positive lead and negative europium, niobium and tantalum anomalies) onto the general pattern. Residual phlogopite may hold back not only potassium and rubidium but also barium and caesium. Partial melting under very high-pressure conditions (*eg* kimberlite generation) with changing mineral assemblage, mineral/melt K_d and melting process might greatly modify the mantle-normalized diagram rules.

Ringwood (1982) suggested that enrichment of refractory harzburgite in a subducted megalith by partial melting of the former oceanic crust near the 650 km phase transition can be a potential source for OIBs. It is expected that, during high-pressure melting, fractionation of some elements between minerals with either a garnetite or a perovskite structure and the coexisting melt can be quite different from melting under low pressure (Kato *et al.* 1987). Consequently, it is worthwhile pursuing detailed

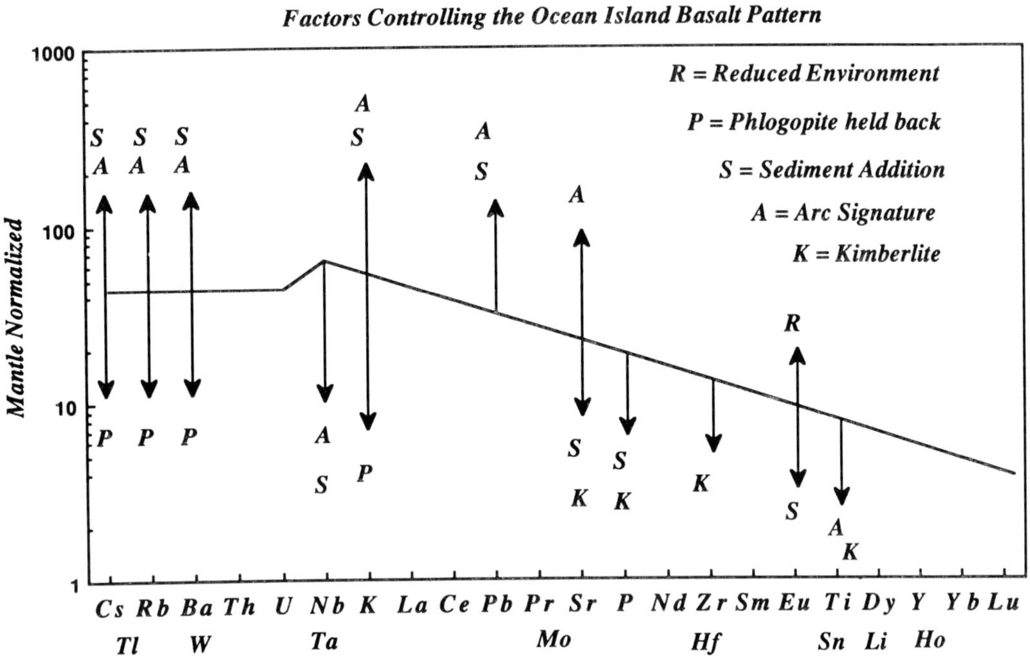

FIG. 16. Schematic description of the effects of various factors which can modify the mantle-normalized pattern of OIBs. For some elements enrichments or depletions may be due to a combination of factors; however, by using several different elements it may be possible to distinguish the relative roles of multiple effects. For example sediment addition (S) and arc signature (A) will both cause increases in caesium, rubidium, barium, potassium and lead; however, strontium is generally enriched in the arc signature, whereas it is generally depleted in the sediment addition.

studies of element ratio systematics (*eg* Ba–Rb–Cs, Nd–Sr, Nb–La, P–Nd, Ti–Eu, Sm–Zr) and the effect of changing mineralogy on K_d under different pressure and temperature conditions. Such differences in mineral assemblage and melting conditions may also be responsible for two types of light REE patterns observed in OIBs: (i) the predominant pattern showing a continuous increase from heavy REEs to light REEs with La/Ce = 0.46 (±10%) and (ii) a less common type showing concave-downward light REE patterns as in Ua Pou tholeiites (Fig. 11) and most Hawaiian tholeiites.

If whole-mantle convection was important in the early Earth's history, then we suggest that the lower mantle composition would also have been involved in crust formation and mantle differentiation (MORB–OIB). The total absence of OIBs with certain primitive mantle ratios (*eg* Rb/Cs ≈ 20, Nb/U ≈ 30, Ce/Pb ≈ 9) suggests that the different OIB-type sources have been previously well mixed with the depleted MORB reservoir (Hofmann *et al.* 1986). If some OIB sources are indeed derived from the core – mantle boundary, as a number of geophysicists seem to prefer, then there are three possible ways in which to explain these constant ratios: (i) the recycled oceanic lithosphere which sank to the core–mantle boundary about 2 Ga ago becomes the plume source, (ii) the upper and lower mantle are well mixed and (iii) mantle material that arises from the core – mantle boundary only initiates the plume activity, whereas the main chemical and isotopic characteristics of OIBs are derived from shallower mantle (1000 km or less).

It has been shown by some geophysicists (*eg* see a review by Hager & Gurnis 1987) that the global gravity pattern is consistent with whole-mantle convection, with viscosity increasing with depth across the phase transformation (10–100 times), and that oceanic lithosphere subduction can continue into the lower mantle (Creager & Jordan 1984). It has also been shown that the geoid pattern reflects plate tectonic patterns of the last 100–200 Ma (*eg* Hager & Gurnis 1987). This whole-mantle convection model has the property of faster convective mixing in the upper mantle than the lower mantle with a corresponding isolation time for chemical heterogeneities ranging from a few hundred million years (the upper mantle) to billions of years (the lower mantle). Although this situation could be crudely similar to the two-layer stratified mantle convection model, the exchange mechanism involving subduction of recycled oceanic crust is different from simple mass exchange between the upper and lower mantle.

The chemical and isotopic consequences of processing altered oceanic crust (with high U/Pb) through the subduction zone, of subjecting it to dehydration and partial melting (La/Nb decreased) and of its later resorption into the convecting mantle requires further evaluation. Could this mechanism fully explain the lead isotopic composition (with $^{206}Pb/^{204}Pb$ on the right-hand side of the geochron) and low Th/U (about 2.0) of the MORB source? Additionally, could it also generate the HIMU-type OIB source? It is true that niobium and tantalum kicks do exist in some MORB and OIB sources (see earlier discussion). This observation is at least partly consistent with the model of oceanic crust recycling. However, most light-REE-depleted MORBs also have niobium- (and tantalum-)depleted patterns (Fig. 14). If HIMU-type mantle sources have distinctly lower La/Nb (*eg* 0.7) than average N-type MORB (approximately 0.95) as a result of oceanic crust recycling, then one would also expect to find very different Ce/Pb, Nb/U and Rb/Cs ratios, which is not consistent with the available data (Table 3). Furthermore, an increase in the Th/U ratio in the HIMU source region, relative to MORB, has not been and could not be explained by this model.

Since the bulk continental crust has an La/Nb ratio higher by a factor of 2 than the OIB and MORB sources, and these sources have La/Nb ratios approximately equal to the primitive mantle value, then a niobium-rich component is missing from the upper mantle–crust system unless the data in Table 5 are biased or the continental lithosphere is a major reservoir for niobium and has a low La/Nb. We suggest that this niobium-rich component does not reside in the continental lithosphere but has been introduced into the lower mantle through lithosphere subduction and/or has been accumulated at the base of the upper mantle (Ringwood & Irifune 1988). The sequestering of this niobium-rich component is an irreversible process which continuously modifies the geochemical and possibly the geophysical character of the lower mantle. The subduction of oceanic lithosphere, possibly with a metasomatized enriched base, may introduce an enriched reservoir into the lower mantle.

The possibility that the continental lithosphere is a major reservoir for incompatible elements has been proposed previously (*eg* Brooks *et al.* 1976; Davies 1981; Carlson 1984). Although we also favour the idea that some parts of the lithosphere can be metasomatized, enriched by subduction zone processes and by the upward migration of silica-undersaturated melts from the asthenosphere, we have difficulty in quantitatively evaluating the volumetric importance of this mantle reservoir, especially when the effects of thermal erosion of the lithosphere are taken into account. However, if the continental lithosphere is indeed a major mantle reservoir for incompatible elements, then larger volumes of depleted mantle reservoirs (the whole mantle?) similar to the MORB source are required.

Relatively constant Ce/Pb (25 ± 5) and Nb/U (47 ± 10) ratios in OIBs and MORBs indicate either that they are produced through subduction zone processes (but how?) or that the subduction zone processes only introduce minor noise to the ratios (preferred). The effect of sediment subduction seems to have been short-circuited at the island arc and back-arc volcanic environments. Even in OIBs with a clear sediment signature, the trace-element and isotope data would limit the added sediment component to less than 1%.

Opinion in favour of a subduction zone connection of the Dupal anomaly, whether generated in the metasomatized mantle wedge above the subduction zone or the continental lithosphere, is gaining support. The Dupal anomaly in the S Atlantic and Indian Oceans seems to be related to the breakup of Gondwana. This type of isotopic and chemical anomaly can be expected in other regions with similar tectonic histories but not limited to the southern hemisphere as originally suggested by Hart (1984).

Lead and neodymium isotope data for HIMU-type OIBs require a source isolation age of 1.0–2.0 Ga with a long-term slightly light-REE-enriched character. The 2.0 Ga age may correspond to a major thermal event in the Earth's crust. Large amounts of HIMU- and possibly EM1-type OIB reservoirs might have been formed from the depleted asthenosphere through migration of silica-undersaturated melts from the LVZ into the oceanic and continental lithosphere. Pure HIMU character with low $^{87}Sr/^{86}Sr$ ratios (0.7026–0.7028) is observed in several ocean islands, suggesting that these HIMU sources originated from depleted MORB-type mantle. In view of the limited amount of continental lithosphere available for recycling and a general lack of crustal and subduction zone signatures in OIBs, we favour the idea that the HIMU source character (especially those with $^{87}Sr/^{86}Sr=0.7026-0.7028$) is generated by recycling of oceanic lithosphere.

Due to the problem of mantle mixing, no unique isotope isolation age can be established for EM1-type mantle with confidence, whereas trace-element and isotope data of the EM2-type mantle with $^{206}Pb/^{204}Pb \geq 19.0$ seems to suggest the involvement of recently subducted sediments. A range of source isolation ages from a few hundred million years to 2 billion years therefore seems likely for the EM-type reservoirs.

The three-dimensional geometrical relationship between the different OIB-type sources and the depleted upper mantle asthenosphere (MORB source) cannot be clearly established as yet. However, it is evident that post-erosional alkali basalts of the Hawaiian islands and Samoan islands were derived from shallower depths, whereas many OIB sources are derived from greater depths in the form of mantle plumes or isolated blobs.

In conclusion, in order to improve our interpretation of the chemical and isotope data, it is important to integrate a geochemical approach with other approaches based on information derived from high-pressure experimental petrology, the tectonic history of the Earth's crust, geophysical observations and numerical modelling. With a better understanding of the petrological and geophysical processes involved in the dynamic, chemical and isotopic properties of the continental lithosphere we shall have a more realistic flux input for the 'multi-box' models currently used in numerical calculations for the chemical and isotopic evolution of the Earth.

ACKNOWLEDGEMENTS: Constructive suggestions and comments from F. A. Frey, C. J. Hawkesworth, J. Hergt, B. M. Jahn, A. L. Jaques, M. T. McCulloch, Z. A. Palacz, R. Price, R. L. Rudnick, A. D. Saunders and W. M. White are gratefully acknowledged. S.-S. Sun publishes with the permission of the Director, Bureau of Mineral Resources, Geology and Geophysics, Australia.

References

ALIBERT, C., MICHARD, A. & ALBARÈDE, F. 1986. Isotope and trace element geochemistry of Colorado plateau volcanics. *Geochimica et Cosmochimica Acta* **50**, 2735–2750.

ALLÈGRE, C. J. & TURCOTTE, D. L. 1986. Implications of a two-component marble-cake mantle. *Nature* **323**, 123–127.

——, HAMELIN, B., PROVOST, A. & DUPRÈ, B. 1987. Topology in isotopic multispace and origin of mantle chemical heterogeneities. *Earth and Planetary Science Letters* **81**, 319–337.

ANDERSON, D. L. 1982. Hotspots, polar wander, Mesozoic convection and the Geoid. *Nature* **297**, 391–393.

ARMSTRONG, R. L. 1968. A model for the evolution of strontium and lead isotopes in a dynamic Earth. *Reviews of Geophysics* **6**, 175–199.

—— 1981. Radiogenic isotopes: the case for crustal recycling on a near-steady-state no-continental growth earth. *Philosophical Transactions of the Royal Society of London* **301**, 443–472.

BELL, K. & BLENKINSOP, J. 1987. Nd and Sr isotopic compositions of East African carbonatites: implications for mantle heterogeneity. *Geology* **15**, 99–102.

BROOKS, C., JAMES, D. E. & HART, S. R. 1976. Ancient lithosphere: its role in young continental volcanism. *Science* **193**, 1086–1094.

CARLSON, R. W. 1984. Isotopic constraints on Columbia River flood basalt genesis and the nature of the subcontinental mantle. *Geochimica et Cosmochimica Acta* **48**, 2357–2372.

CHEN, C-Y & FREY, F. A. 1985. Trace element and isotopic geochemistry of lavas from Haleakala Volcano, East Maui, Hawaii: implications for the origin of Hawaiian basalts. *Journal of Geophysical Research* **90**, 8743–8768.

CLAGUE, D. A. & FREY, F. A. 1982. Petrology and trace element geochemistry of the Honolulu Volcanics, Oahu: implications for the oceanic mantle below Hawaii. *Journal of Petrology* **23**, 447–504.

COHEN, R. S. O'NIONS, R. K. 1982. Identification of recycled continental meterial in the mantle from Sr, Nd and Pb isotope investigations. *Earth and Planetary Science Letters* **61**, 73–84.

COX, K. G. 1978. Flood basalts, subduction and the break-up of Gondwanaland. *Nature* **274**, 47–49.

CREAGER, K. C. & JORDAN, T. H. 1984. Slab penetration in the lower mantle. *Journal of Geophysical Research* **89**, 3031–3049.

DAVIES, G. F. 1981. Earth's neodymium budget and structure and evolution of the mantle. *Nature* **290**, 208–213.

DUDÁS, F. Ö. CARLSON, R. W. & EGGLER, D. H. 1987. Regional middle Proterozoic enrichment of the subcontinental mantle source of igneous rocks from central Montana. *Geology* **15**, 22–25.

DUNCAN, A. R., ERLANK, A. J. & MARCH, J. S. 1984. Regional geochemisty of the Karoo igneous province. *Special Publication of the Geological Society of South Africa* **13**, 355–388.

DUNCAN, R. A., MCCULLOCH, M. T., BARCZUS, H. G. & NELSON, D. R. 1986. Plume versus lithospheric sources for the melts at Ua Pou, Marquesas Island. *Nature* **303**, 142–146.

DUPUY, C., VIDAL, P., BARSCZUS, H. G. & CHAUVEL, C. 1987. Origin of basalts from the Marquesas Archipelago (south central Pacific Ocean): isotope and trace element constraints. *Earth and Planetary Science Letters* **82**, 145–152.

FODOR, R. V., BAUER, G. R., JACOBS, R. S. & BORNHORST, T. J. 1987. Kahoolawe Island, Hawaii: tholeiitic, alkalic and unusual hydrothermal(?) "enrichment" characteristics. *Journal of Volcanology and Geothermal Research* **31**, 171–176.

FRASER, K. J., HAWKESWORTH, C. J., ERLANK, A. J., MITCHELL, R. H. & SCOTT-SMITH, B. H. 1986. Sr, Nd and Pb isotope and minor element geochemistry of lamproites and kimberlites. *Earth and Planetary Science Letters* **76**, 57–70.

FREY, F. A. & CLAGUE, D. A. 1983. Geochemistry of diverse basalts types from Loihi Seamount, Hawaii: petrogenetic implications. *Earth and Planetary Science Letters* **66**, 337–355.

—— & GREEN D. H. 1974. The mineralogy, geochemistry and origin of lherzolite inclusions in Victorian basanites. *Geochimica et Cosmochimica Acta* **38**, 1023–1059.

——, —— & ROY, S. D. 1978. Integrated models of basalt petrogenesis: a study of quartz tholeiites to olivine melilitites from South Eastern Australia utilizing geochemical and experimental petrological data. *Journal of Petrology* **19**, 463–513.

FUJIMAKI, H., TATSUMOTO, M. & AOKI, K.-I. 1984. Partition coefficients of Hf, Zr, and REE between phenocrysts and groundmasses. *Journal of Geophysical Research* **89**, B662–B672.

GAST, P. W. 1968. Trace element fractionation and the origin of tholeiitic and alkaline magma types. *Geochimica et Cosmochimica Acta* **32**, 1057–1086.

GREEN, D. H. 1971. Composition of basaltic magmas as indicators of conditions of origin: application to oceanic volcanism. *Philosophical Transactions of the Royal Society of London* **268**, 707–725.

GREEN, T. H. 1978. Rare earth geochemistry of basalts from Norfolk Island, and implications for mantle inhomogeneity in the rare earth elements. *Geochemical Journal* **12**, 165–172.

GURNIS, M. & DAVIES, G. F. 1986. The effect of depth-dependent viscosity on convective mixing in the mantle and the possible survival of primitive mantle. *Geophysical Research Letters* **13**, 541–544.

HAGER, B. H. & GURNIS, M. 1987. Mantle convection and the state of the earth's interior. *Reviews of Geophysics* **25**, 1277–1285.

HANSON, G. N. 1977. Geochemical evolution of the suboceanic mantle. *Journal of the Geological Society of London* **134**, 235–253.

HART, S. R. 1984. A large-scale isotope anomaly in the Southern Hemisphere mantle. *Nature* **309**, 753–757.

——, GERLACH, D. C. & WHITE, W. M. 1986. A possible new Sr–Nd–Pb mantle array and consequences for mantle mixing. *Geochimica et Cosmochimica Acta* **50**, 1551–1557.

HAWKESWORTH, C. J., MANTOVANI, M. S. M., TAYLOR, P. N. & PALACZ, A. 1986. Evidence from the Parana of south Brazil for a continental contribution to Dupal basalts. *Nature* **322**, 356–359.

——, NORRY, M. J., RODDICK, J. C. & VOLLMER, R. 1979. $^{143}Nd/^{144}Nd$ and $^{87}Sr/^{86}Sr$ ratios from the Azores and their significance in LIL-element enriched mantle. *Nature* **280**, 28–31.

——, ROGERS, N. W., VAN CALSTEREN, P. & MENZIES, M. A 1984. Mantle enrichment processes. *Nature* **311**, 331–335.

HERGT, J. M. 1987. *The Origin and Evolution of the Tasmania Dolerites*. Unpublished PhD thesis, Australian National University.

HOFMANN, A. W. & WHITE, W. M. 1982. Mantle plumes from ancient oceanic crust. *Earth and Planetary Science Letters* **57**, 421–436.

——, & ——, 1983. Ba, Rb and Cs in the Earth's mantle. *Zeitschrift für Naturforschung* **38a**, 256–266.

——, JOCHUM, K. P., SEUFERT, M. & WHITE, W. M. 1986. Nb and Pb in oceanic basalts: new constraints on mantle evolution. *Earth and Planetary Science Letters* **79**, 33–45.

JAGOUTZ, E., PALME, H., BADDENHAUSEN, H., BLUM, K., CENDALES, M., DREIBUS, G., SPETTEL, B., LORENZ, V. & WÄNKE, H. 1979. The abundances of major, minor and trace elements in the Earth's mantle as derived from primitive ultramafic nodules. *Proceedings of the 10th Lunar and Planetary Science Conference*, (*Geochimica et Cosmochim Acta*, Suppl. 11) Vol. 2, 2031–2050.

JAQUES, A. L., SUN, S.-S. & CHAPPELL, B. W. 1988. Geochemistry of the Argyle (AK1) lamproite pipe, Western Australia. *Proceedings of the 4th International Kimberlite Conference*. Geological Society of Australia Special Publication, in press.

JOCHUM, K. P., SEUFERT, H. M., SPETTEL, B. & PALME, H. 1986. The solar-system abundances of Nb, Ta, and Y, and the relative abundances of refractory lithophile elements in differentiated planetary bodies. *Geochimica et Cosmochimica Acta* **50**, 1173–1183.

JOHNSON, R. W., JAQUES, A. L., LANGMUIR, C. H., PERFIT, M. R., STAUDIGEL, H., DUNKLEY, P. N., CHAPPELL, B. W., TAYLOR, S. R. & BAEKISAPA, M. 1987. Ridge subduction and forearc volcanism: petrology and geochemistry of rocks dredged from the western Solomon arc and Woodlark basin. *In*: TAYLOR, B. & EXON, N. F. (eds) *Marine Geology, Geophysics, and Geochemistry of the Woodlark Basin — Solomon Islands Region*, Circum-Pacific Council of Energy and Mineral Resources, Earth Science Series, in press.

KATO, T., IRIFUNE, T. & RINGWOOD, A. E. 1987. Majorite partitioning behavior and petrogenesis of the earth's upper mantle. *Geophysical Research Letters* **14**, 546–549.

KAY, R. W. 1979. Zone refining at the base of lithospheric plates: a model for steady-state asthenosphere. *Tectonophysics* **55**, 1–9.

KRAMERS, J. D. 1977. Lead and strontium isotopes in Cretaceous kimberlites and mantle-derived xenoliths from southern Africa. *Earth and Planetary Science Letters* **34**, 419–431.

——, SMITH, C. B., LOCK, N. P., HARMON, R. S. & BOYD, F. R. 1981. Can kimberlites be generated from an ordinary mantle? *Nature* **291**, 53–56.

KUEHNER, S. M., EDGAR, A. D. & ARIMA, M. 1981. Petrogenesis of the ultrapotassic rocks from the Leucite Hills, Wyoming. *American Mineralogist* **66**, 663–667.

LANGMUIR, C. H. & HANSON, G. N. 1980. An evaluation of major element heterogeneity in the mantle sources of basalts. *Philosophical Transactions of the Royal Society of London, Series A* **297**, 383–407.

——, BENDER, J. F., BENCE, A. E., HANSON, G. N. & TAYLOR, S. R. 1977. Petrogenesis of basalts from the FAMOUS area: Mid-Atlantic Ridge. *Earth and Planetary Science Letters* **36**, 133–156.

LE ROEX, A. P. 1986. Geochemical correlation between southern African kimberlites and South Atlantic hotspots. *Nature* **324**, 243–245.

LI, S. & HART, S. R. 1986. Sources of evolution of Hawaiian volcanism: isotopic constraints. *Earth and Planetary Science Letters*, in press.

LIOTARD, J. M., BARSCZUS, H. G., DUPUY, C. & DOSTAL, J. 1986. Geochemistry and origin of basaltic lavas from Marquesas Archipelago, French Polynesia. *Contributions to Mineralogy and Petrology* **92**, 260–268.

MACDOUGALL, J. D. & LUGMAIR, G. W. 1986. Sr and Nd isotopes in basalts from the East Pacific Rise: significance for mantle heterogeneity. *Earth and Planetary Science Letters* **77**, 273–284.

MANTOVANI, M. S. M., MARQUES, L. S., DE SOUSA, M. A., CIVETTI, L., ATALLA, L. & INNOCENTI, F. 1985. Trace element and strontium isotope constraints on the origin and evolution of Paraná continental flood basalts of Santa Catarina State (Southern Brazil). *Journal of Petrology* **26**, 187–209.

MCDONOUGH, W. F. & MCCULLOCH, M. T. 1987. The southeast Australian lithospheric mantle: implications for its growth and evolution. *Earth and Planetary Science Letters* **86**, 327–340.

——, —— & SUN, S.-S. 1985. Isotopic and geochemical systematics in Tertiary–Recent basalts from south eastern Australia and implications for the evolution of the sub-continental lithosphere. *Geochimica et Cosmochimica Acta* **49**, 2051–2067.

——, SUN, S.-S., RINGWOOD, A. E. & JAGOUTZ, E. 1987. Rb and Cs in the Earth and Moon. *Lunar and Planetary Science Conference Abstracts*, Vol. 18, 610–611. Lunar and Planetary Institute, Houston.

MCKENZIE, D. 1984. The generation and compaction of partially molten rock. *Journal of Petrology* **25**, 713–765.

—— & O'NIONS, R. K. 1983. Mantle reservoirs and ocean island basalts. *Nature* **301**, 229–231.

MCLENNAN, S. M. & TAYLOR, S. R. 1981. Role of subducted sediments in island-arc magmatism: constraints from REE patterns. *Earth and Planetary Science Letters* **54**, 423–430.

MENZIES, M. 1983. Mantle ultramafic xenoliths in alkaline magmas: evidence for mantle heterogeneity modified by magmatic activity. *In*: (eds), HAWKESWORTH C. J. & NORRY, M. J. *Continental Basalts and Mantle Xenoliths*, 92–110, Shiva, Cheshire.

—— & MURTHY, V. R. 1980. Nd and Sr isotope geochemistry of hydrous mantle nodules and their host alkali basalts: implications for local heterogeneities in metasomatically veined mantle. *Earth and Planetary Science Letters* **46**, 323–334.

MURMATSU, Y. 1983. Geochemical investigation of kimberlites from the Kimberley area, South Africa. *Geochemical Journal* **17**, 71–86.

—— & WEDEPOHL, K. H. 1985. REE and selected trace elements in kimberlites from the Kimberley area (South Africa). *Chemical Geology* **51**, 289–301.

NAGASHIMA, K., MIYAWAKI, R., TAKASE, J., NAKAI, I., SAKURAI, K.-I., MATSUBARA, S., KATO, A. & IWANO, S. 1986. Kimuraite, $CaY_2(CO_3)_4.6H_2O$, a new mineral from fissures in an alkali olivine basalt from Saga Prefecture, Japan, and new data on lokkaite. *American Mineralogist* **71**, 1028–1033.

NAKAMURA, E., CAMPBELL, I. H. — SUN, S.-S. 1985. The influence of subduction processes on the geochemistry of Japanese alkaline basalts. *Nature* **316**, 55–58.

NAVON, O. & STOLPER, E. 1987. Geochemical consequences of melt percolation: the upper mantle as a chromatographic column. *Journal of Geology* **95**, 285–307.

NELSON, D. R., CHAPPELL, B. W., CHIVAS, A. R. & MCCULLOCH, M. T. 1987. Geochemical and isotopic evidence for a subducted oceanic lithosphere origin for carbonatites. *Geochimica et Cosmochimica Acta* **52**, 1–18.

——, MCCULLOCH, M. T. & SUN, S.-S. 1986. The origins of ultrapotassic rocks as inferred from Sr, Nd and Pb isotopes. *Geochimica et Cosmochimica Acta* **50**, 231–245.

NEWMAN, S., FINKEL, R. C. & MACDOUGALL, J. D. 1984. Comparison of $^{230}Th-^{238}U$ disequilibrium systematics in basalts from three hot spot regions: Hawaii, Prince Edward and Samoa. *Geochimica et Cosmochimica Acta* **48**, 315–324.

NEWSOM, H. E., WHITE, W. M., JOCHUM, K. P. & HOFMANN, A. W. 1986. Siderophile and chalcophile element abundances in oceanic basalts, Pb isotope evolution and growth of the Earth's core. *Earth and Planetary Science Letters* **80**, 299–313.

NICOLAS, A. 1986. A melt extraction model based on structural studies in mantle peridotites. *Journal of Petrology* **27**, 999–1022.

O'HARA, M. J. & MATHEWS, R. E. 1981. Geochemical evolution in an advancing, periodically replenished, periodically tapped, continuously fractionated magma chamber. *Journal of the Geological Society, London* **138**, 237–277.

OVERSBY, V. M. & GAST, P. W. 1968. Lead isotope compositions and uranium decay series disequilibrium in recent volcanic rocks. *Earth and Planetary Science Letters* **5**, 199–206.

OXBURGH, E. R. & TURCOTTE, D. L. 1968. Mid-ocean ridges and geotherm distribution during mantle convection. *Journal of Geophysical Research* **73**, 2643–2661.

PALACZ, Z. A. & SAUNDERS, A. D. 1986. Coupled trace element and isotope enrichment in the Cook–Austral–Samoa islands, southwest Pacific. *Earth and Planetary Science Letters* **79**, 270–280.

PATCHETT, P. J. 1983. Importance of Lu–Hf isotopic system in studies of planetary chronology and chemical evolution. *Geochimica et Cosmochimica Acta* **47**, 81–91.

PHILPOTTS, J. A. & SCHNETZLER, C. C. 1970. Phenocryst-matrix partition coefficients for K, Rb, Sr and Ba, with application to anorthosite and basalt genesis. *Geochimica et Cosmochimica Acta* **34**, 307–322.

RICHARDSON, S. H., ERLANK, A. J., DUNCAN, A. R. & REID, D. L. 1982. Correlated Nd, Sr and Pb isotope variation in Walvis Ridge basalts and implications for the evolution of their mantle source. *Earth and Planetary Science Letters* **59**, 327–342.

——, GURNEY, J. J., ERLANK, A. J. & HARRIS, J. W. 1984. Origin of diamonds in old enriched mantle. *Nature* **310**, 198–202.

RICHTER, F. M. 1985. Models of the Archaean thermal regime. *Earth and Planetary Science Letters* **73**, 350–360.

RINGWOOD, A. E. 1982. Phase transformations and differentiation in subducted lithosphere: implications for mantle dynamics, basalt petrogenesis, and crustal evolution. *Journal of Geology* **90**, 611–643.

—— & IRIFUNE, T. 1988. Nature of the 650-km seismic discontinuity: implications for mantle dynamics and differentiation. *Nature* **331**, 131–136.

RODEN, M. F., FREY, F. A. & CLAGUE, D. A. 1984. Geochemistry of tholeiitic and alkalic lavas from the Koolau Range, Oahu, Hawaii: implications for Hawaiian volcanism. *Earth and Planetary Science Letters* **69**, 141–158.

——, HART, S. R., FREY, F. A. & MELSON, W. G. 1984. Sr, Nd and Pb isotopic and REE geochemistry of St. Paul's Rocks: the metamorphic and metasomatic development of an alkali basalt mantle source. *Contributions to Mineralogy and Petrology* **85**, 376–390.

——, RAMA MURTHY, V. & GASPAR, J. C. 1985. Sr and Nd isotopic composition of the Jacupiranga carbonatite. *Journal of Geology*, **93**, 212–220.

RYAN, J. G. & LANGMUIR, C. H. 1987. The systematics of lithium abundances in young volcanic rocks. *Geochimica et Cosmochimica Acta* **51**, 1727–1741.

SAUNDERS, A. D. & TARNEY, J. 1979. The geochemistry of basalts from a back-arc spreading centre in the East Scotia Sea. *Geochimica et Cosmochimica Acta* **43**, 555–572.

——, —— & WEAVER, S. D. 1980. Transverse chemical variations across the Antarctic Peninsula: implications for the genesis of calc-alkaline magmas. *Earth and Planetary Science Letters* **46**, 344–360.

SAWAMOTO, H. 1987. Phase diagram of $MgSiO_3$ at pressures up to 24 GPa and temperatures up to 2,200 °C — phase stability and properties of tetragonal garnet. In press.

SCHILLING, J. G., THOMPSON, G., KINGSLEY, R. & HUMPHRIS, S. 1985. Hotspot–migrating ridge interaction in the south Atlantic. *Nature* **313**, 187–191.

SMITH, C.B. 1983. Pb, Sr and Nd isotopic evidence for sources of southern African Cretaceous kimberlites. *Nature* **304**, 51–54.

——, GURNEY, J. J., SKINNER, E. M. W., CLEMENT, C. R. & EBRAHIM, N. 1985. Geochemical character of southern African kimberlites: a new approach based on isotopic constraints. *Transactions of Geological Society of South Africa* **88**, 267–280.

STAUDIGEL, H., ZINDLER, A., HART, S. R., LESLIE, T. M., CHEN, C.-Y. & CLAGUE, D. 1984. The isotope systematics of a juvenile intraplate volcano: Pb, Nd, and Sr isotope ratios of basalts from Loihi Seamount, Hawaii. *Earth and Planetary Science Letters* **69**, 13–29.

STILLE, P., UUNUH, D. M. & TATSUMOTO, M. 1986. Pb, Sr, Nd and Hf isotopic constraints on the origin of Hawaiian basalts and evidence for a unique mantle source. *Geochimica et Cosmochimica Acta* **50**, 2303–2319.

SUN, S.-S. 1980. Lead isotopic study of young volcanic rocks from mid-ocean ridges, ocean islands and island arcs. *Philosophical Transactions of the Royal Society of London* **297**, 409–445.

—— & HANSON, G. N. 1975*a*. Origin of Ross Island basanitoids and limitations upon the heterogeneity of mantle sources for alkali basalts and nephelinites. *Contributions to Mineralogy and Petrology* **52**, 77–106.

—— & —— 1975*b*. Evolution of the mantle: geochemical evidence from alkali basalt. *Geology* **3**, 297–302.

—— & —— 1976*a*.-Evolution of the mantle: geochemical evidence from alkali basalt: Reply. *Geology* **4**, 626–631.

—— & —— 1976*b*. Rare earth element evidence for differentiation of McMurdo volcanics, Ross Island, Antarctica. *Contributions to Mineralogy and Petrology* **54**, 139–155.

—— & NESBITT, R. W. 1977. Chemical heterogeneity of the Archaean mantle, composition of the Earth and mantle evolution. *Earth and Planetary Science Letters* **35**, 429–448.

——, —— & MCCULLOCH, M. T. 1988. Geochemistry and petrogenesis of siliceous high magnesian basalts of the Archaean and early Proterozoic. *In*: CRAWFORD, A. J. (ed.) *Boninites and Related Rocks*. George Allen and Unwin, London, in press.

——, —— & SHARASKIN, A. Y. 1979. Geochemical characteristics of mid-ocean ridge basalts. *Earth and Planetary Science Letters* **44**, 119–138.

TATSUMOTO, M. 1978. Isotopic composition of lead in

oceanic basalts and its implication to mantle evolution. *Earth and Planetary Science Letters* **38**, 119–138.

TAYLOR, S. R. & MCLENNAN, S. M. 1985. *The Continental Crust: Its Composition and Evolution*. Blackwells Scientific, Oxford, 312 pp.

THOMPSON, R. N., MORRISON, M. A., DICKIN, A. P. & HENDRY, G. L. 1983. Continental flood basalts... Arachnids rule OK? *In*: Eds. HAWKESWORTH, C. J. & NORRY, M. J. (eds) *Coninental Basalts and Mantle Xenoliths*, 158–185. Shiva, Cheshire.

——, ——, HENDRY, G. L. & PARRY, S. J. 1984. An assessment of the relative roles of crust and mantle in magma genesis: and elemental approach. *Philosophical Transactions of the Royal Society of London* **310**, 549–590.

VIDAL, P. & DOSSO, L. 1978. Core formation: catastrophic or continuous? Sr and Pb isotope geochemistry constraints. *Geophysical Research Letters* **5**, 169–172.

——, CHAUVEL, C. & BROUSSE, R. 1984. Large mantle heterogeneity beneath French Polynesia. *Nature* **307**, 536–538.

WATSON, E. B., BEN OTHMAN, D., LUCK, J. M. & HOFMANN, A. W. 1987. Partitioning of U, Pb, Hf, Yb, Cs, Re, Os between chromian diopsidic pyroxene and haplobasaltic liquid. *Chemical Geology* **62**, 191–208.

WEAVER, B. L., WOOD, D. A., TARNEY, J. & JORON, J. L. 1986. Role of subducted sediment in the genesis of ocean-island basalts: geochemical evidence from South Atlantic Ocean islands. *Geology* **14**, 275–278.

WHITE, W. M. 1985. Sources of oceanic basalts: radiogenic isotopic evidence. *Geology* **13**, 115–118.

—— & HOFMANN, A. W. 1982. Sr and Nd isotope geochemistry of oceanic basalts and mantle evolution. *Nature* **296**, 821–825.

——, —— & PUCHELT, H. 1987. Isotope geochemistry of Pacific mid-ocean ridge basalts. *Journal of Geophysical Research* **92**, 4881–4893.

——, TAPIA, M. D. M. & SCHILLING, J.-G. 1979. The petrology and geochemistry of the Azores Islands. *Contributions to Mineralogy and Petrology* **69**, 201–213.

WOOD, D. A., JORON, J.-L., TREUIL, M., NORRY, M. & TARNEY, J. 1979. Elemental and Sr isotope variations in basic lavas from Iceland and the surrounding ocean floor. *Contributions to Mineralogy and Petrology* **70**, 319–339.

WRIGHT, E. & WHITE, W. M. 1987. The origin of Samoa: new evidence from Sr, Nd, and Pb isotopes. *Earth and Planetary Science Letters* **81**, 151–162.

WYLLIE, P. J. & SEKINE, T. 1982. The formation of mantle phlogopite in subduction zone hybridization. *Contributions to Mineralogy and Petrology* **79**, 375–380.

ZARTMAN, R. E. & DOE, B. R. 1981. Plumbotectonics — the model. *Tectonophysics* **75**, 135–162.

ZINDLER, A. & HART, S. 1986. Chemical geodynamics. *Annual Reviews of Earth and Planetary Science* **14**, 493–571.

——, JAGOUTZ, E. & GOLDSTEIN, S. 1982. Nd, Sr and Pb isotopic systematics in a three-component mantle: a new perspective. *Nature* **298**, 519–523.

——, STAUDIGEL, H. & BATIZA, R. 1984. Isotope and trace element geochemistry of young Pacific seamounts: implications for the scale of upper mantle heterogeneity. *Earth and Planetary Science Letters* **70**, 175–195.

SHEN-SU SUN, Division of Petrology and Geochemistry, Bureau of Mineral Resources, Geology and Geophysics, G.P.O. Box 378, Canberra, ACT 2601, and Research School of Earth Sciences, The Australian National University, G.P.O. Box 4, Canberra, ACT 2601, Australia.

WILLIAM F. MCDONOUGH, Research School of Earth Sciences, The Australian National University, G.P.O. Box 4, Canberra, ACT 2601, Australia. *Present address*: Max-Planck-Institut für Chemie, Saarstraβe 23, D-6500 Mainz, FRG.

Tectonic controls on boninite genesis

B. J. Murton

SUMMARY: Boninites are rare high-MgO and compatible- element-rich basalts that are depleted in incompatible trace elements. Whilst their high magnesium number indicates equilibrium with upper-mantle peridotite, their depleted trace-element geochemistry indicates a highly refractory and depleted source. Many models account for boninite magmatism in destructive margin environments but none has so far explained their erratic spatial and temporal occurrence. Recent studies along the southern Troodos margin have revealed a history of boninite magmatism within a complex supra-subduction zone transform fault. The transform fault boninites form a suite that cross-cuts a less trace-element-depleted magmatic sequence forming the Troodos oceanic crust that was generated at a constructive plate margin.

Melting models demonstrate that the transform fault boninites were probably produced by a 10% melt of the upper mantle which had already been depleted by the extraction of the Troodos crust and mid-ocean ridge basalts (MORB). In contrast with the Troodos crustal sequence, however, the boninites are also enriched in two separate components, selected high field strength elements and large-ion lithophile elements. Such incompatible-trace-element enrichments are unlikely to have survived the earlier source depletion events. Therefore the enrichment of the transform boninites must have been added to their source immediately prior to or during their genesis within the supra-subduction transform environment.

The identification of rapid dilation and transtensional transform displacement along the southern Troodos margin suggests important tectonic factors in the formation and emplacement of boninite magmas. The requirement of an extensional regime and rifting of trace-element-depleted oceanic crust formed above a subduction zone (*eg* extension and rifting of an extended fore-arc region) explains the erratic and rare nature of boninite magmatism.

The Troodos massif, Cyprus, is one of several eastern Mediterranean Neo-Tethyan ophiolites. Ophiolites are now generally accepted as fragments of oceanic crust and lithosphere originally formed at a constructive plate margin. The Troodos massif is one of the best preserved structurally intact ophiolite complexes which, as a result of Tertiary uplift centred on Mount Olympus, reveals an annular outcrop pattern with the deepest sub-Moho levels exposed at the centre (Wilson 1959) and overlain by pelagic sediments that give a minimum palaeontological Campanian age (Mantis 1971).

Recent geochemical studies of the Troodos ophiolite extrusive suite have demonstrated a close affinity with destructive-margin basalts (Miyashiro 1973; Pearce 1975, 1980*a*; Smewing *et al.* 1975; Cameron *et al.* 1980; McCulloch & Cameron & Cameron 1983; Schminke *et al.* 1983; Moores *et al.* 1984; Rautenschlein *et al.* 1985). The geochemical evidence for a subduction-related setting is in contrast with the field and structural evidence of formation at a constructive plate margin and has led workers to the interpretation of a supra-subduction zone or fore-arc spreading model for the Troodos massif, possibly in response to oceanward migration of the Wadi-Bennioff zone (Moores *et al.* 1984; Pearce *et al.* 1984).

In contrast with the coherent oceanic spreading structure of the Troodos massif, the Limassol Forest Complex forms an anomalous ophiolite terrain to the S (Fig. 1). Separating the two areas is an E–W-trending graben, the Arakapas Fault Belt, that was initially interpreted by Moores & Vine (1971) as a fossil transform fault zone. Later studies by Simonian & Gass (1978) demonstrated that the Arakapas Fault Belt was in fact the northern wall of a wider transform domain that also included a large part of the Limassol Forest Complex. Structural and field studies (Murton 1986; Murton & Gass 1986) of the western Limassol Forest Complex revealed a complex history of transform fault activity along the southern Troodos transform fault that included transtension and the synmagmatic serpentinization and uplift of an upper-mantle block during geochemically distinct magmatism within the active transform domain.

Compared with the spreading sequence magmatism of the Troodos massif, here termed

From SAUNDERS, A. D. & NORRY, M. J. (eds), 1989, *Magmatism in the Ocean Basins*, Geological Society Special Publication No. 42, pp. 347–377.

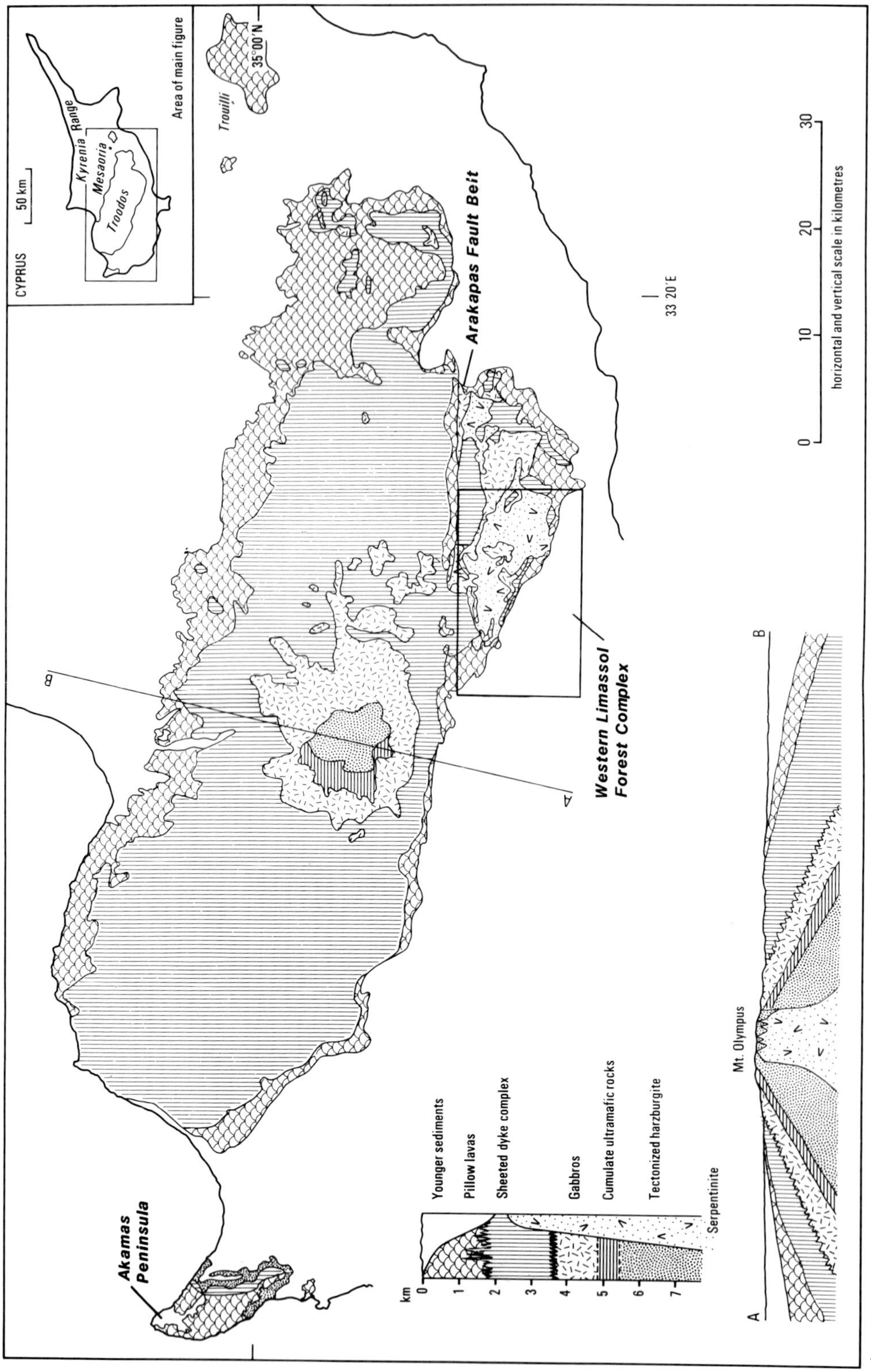
CYPRUS
50 km
Kyrenia Range
Mesaoria
Troodos
Area of main figure
Trouilli
35°00'N
Arakapas Fault Belt
33°20'E
0
10
20
30
horizontal and vertical scale in kilometres
Western Limassol Forest Complex
A
B
Akamas Peninsula
km
0
1
2
3
4
5
6
7
Younger sediments
Pillow lavas
Sheeted dyke complex
Gabbros
Cumulate ultramafic rocks
Tectonized harzburgite
Serpentinite
Mt. Olympus

(a)

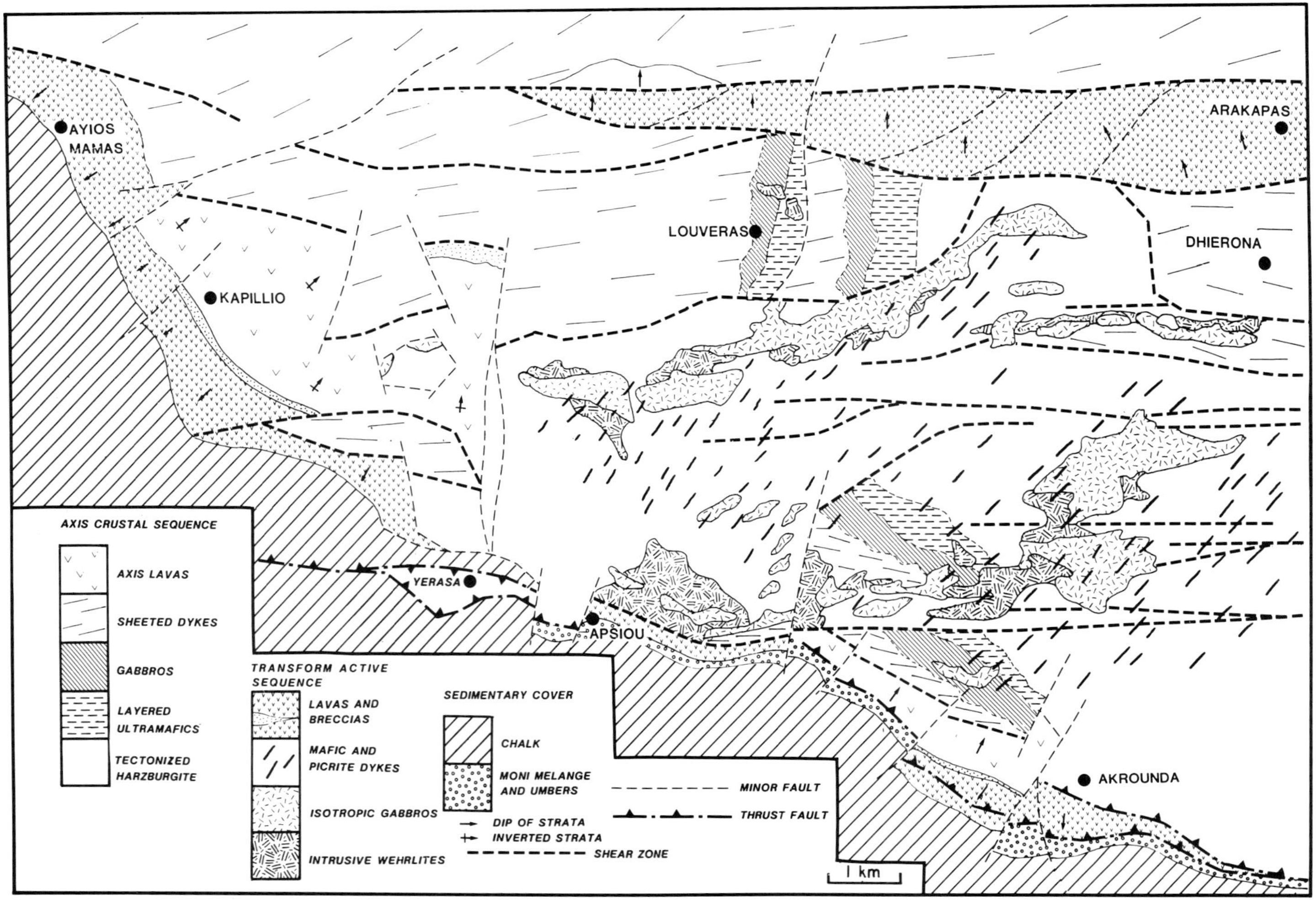

(b)

FIG. 1. (*a*) Location map of the Troodos massif and the southern Troodos transform fault showing the position of the western Limassol Forest Complex (after Wilson 1959). (*b*) Generalized geology of the western Limassol Forest Complex. The plutonic core complex is dominated both by a tectonized harzburgite of upper-mantle origin and by multiple mafic and ultramafic intrusions. Major serpentinite shear zones impose a strong E–W tectonic fabric and are of primary transform origin (after Murton 1986).

the Axis Sequence, the magmatic activity within the western Limassol Forest Complex transform fault zone is strongly incompatible trace element depleted with characteristic low-TiO_2 high-MgO basalts that resemble boninites. The Transform Sequence magmatism is further complicated by the addition of an enrichment phase that carries light rare earth elements (light REEs) and selected high field strength (HFS) elements and which has low ε_{Nd}. Both the Transform Sequence and the Axis Sequence have been ubiquitously enriched in large-ion lithophile (LIL) elements.

This contribution sets out to characterize the geochemistry of the Transform Sequence basalts and to compare them with accepted boninite compositions. The petrogenesis of the Transform Sequence basalts is then described and interpreted within the constraints of field and structural evidence of complex transform tectonics in a supra-subduction zone setting.

Geological and tectonic setting of the southern Troodos transform fault

The geological and tectonic history of the southern Troodos transform fault has already been described in detail (Murton 1986; Murton & Gass 1986) and so it is only briefly discussed here. Essentially the southern Troodos transform fault consists of two magmatic suites formed during separate tectonic episodes of ocean crustal generation. The Axis Sequence is oceanic crust and lithosphere formed at the Troodos plate constructive margin which was dismembered and sheared adjacent to and within the active transform domain. The Transform Sequence is oceanic crust and lithosphere that was formed within the active transform domain and which post-dates the ridge-generated Axis Sequence. The Transform Sequence consists of multiple ultramafic and mafic plutons and dykes and their associated extrusives. The intrusives were emplaced into a progressively serpentinized and synmagmatically uplifted upper-mantle block, within the active transform domain, that was also suffering transtensional transform deformation.

The disrupted Axis Sequence is consistently cross-cut by intrusive members of the Transform Sequence, clearly revealing the younger age of the transform magmatism, a point that is critical to the understanding of the petrogenesis of the complex. Synmagmatic deformation of the active transform domain is concentrated along high angle serpentinite shear zones that are interpreted as strike-slip faults, active during transform displacement. Structural analyses of mylonite microfabrics combined with the E–W orientation of principal serpentinite shears and the NE–SW orientation of synthetic serpentinite shears indicate a sinistral sense of transform shear. This sense of shear is further supported by the dominant NE–SW trend of multiple mafic and ultramafic dykes of the Transform Sequence, which are emplaced into the uplifted mantle lithologies and are at 40°–45° to the dominant N–S spreading fabric of the Troodos massif. The emplacement of the multiple Transform Sequence dyke swarms and sigmoidal-shaped Transform Sequence gabbro plutons (resembling megascale lithospheric tension gashes) into the active transform fault domain indicate up to 22% crustal dilation and hence a transtensional transform environment.

Geochemistry

Introduction

Compared with the extrusives of the Troodos massif Axis Sequence, those of the Transform Sequence are geochemically distinct, being more primitive, trace element depleted and variably light REE enriched. Both the Troodos extrusives and those of the southern Troodos transform fault are ubiquitously but variably enriched in LIL elements. When plotted on major and trace-element covariation diagrams, the Axis Sequence basalts follow a different paragenetic trend from the Transform Sequence basalts, indicating evolution from a different parental melt. The Transform Sequence intrusives, and in particular the mafic dykes that cut the uplifted serpentinized mantle block, are coeval with the Transform Sequence extrusives and were feeders to the transform fault volcanics.

Major element geochemistry

Mafic dykes

The mafic dykes have been divided into two groups, the Axis Sequence group, members of which form the Troodos sheeted dyke complex that has been disrupted within the active transform domain, and the Transform Sequence group. The latter has been further subdivided into four petrological subgroups on the following covariation diagrams: subgroup A, olivine-phyric picrite dykes; subgroup B, olivine + orthopyroxene + clinopyroxene-phyric with plagioclase in the groundmass; subgroup C, olivine + clinopyroxene + plagioclase-phyric; subgroup D, clinopyroxene + plagioclase-phyric (see Appendix 1).

Because of the altered state of the mafic dykes, only five major-element oxides (CaO, MgO, Na_2O, TiO_2 and Al_2O_3) show any magmatic trends (Fig. 2). These show paragenetic schemes that are consistent with the petrography of the mafic dykes. At values of MgO above 14 wt%, picrite and olivine + pyroxene-phyric dykes form a trend towards a point that lies approximately halfway between the mineral compositions of the olivine and pyroxene phenocrysts. This indicates olivine and pyroxene accumulation, in agreement with the petrography of the picrite and olivine + orthopyroxene + clinopyroxene-phyric dykes. Below 10 wt% MgO, the dykes form a trend that indicates plagioclase + clinopyroxene crystal fractionation.

The Transform Sequence and Axis Sequence dykes are only distinguished in two major-element oxide variation diagrams. In both cases the Axis Sequence forms a separate group with higher TiO_2, relative to either MgO and Al_2O_3, compared with the Transform Sequence. Although on Fig. 3 the Axis Sequence trend may be explained by plagioclase fractionation, the higher TiO_2 relative to MgO for the Axis Sequence dykes cannot be related to the Transform Sequence dykes by crystal fractionation alone and hence indicates evolution from a separate parental liquid.

Transform sequence lavas

The basaltic Transform Sequence lavas, described in the following section, are exclusively from within the active transform domain and have been sampled from the area of Kapillo and Akrounda (see location map, Fig. 1). They have been stratigraphically subdivided into upper (U), middle (M) and lower (L) subgroups and petrographically into olivine-phyric and aphyric subgroups (see Appendix 1).

Major-element variation within the Transform Sequence lavas is small (Fig. 4) compared with that in the mafic dykes. The lavas are primitive, having 14 wt% MgO, Mg(Mg + Fe) = 0.75 (mol%) and FeO*/MgO = 0.3, indicating equilibrium with mantle olivine of Fo_{90-91} (Roeder & Emslie 1970) and thus representing primary liquids. They show predominantly quench textures, are generally glassy and sparsely phyric, and hence resemble liquid compositions. They have low TiO_2 relative to MgO, indicating an affinity with the Transform Sequence mafic dykes compared with the Axis Sequence dykes. Although the complete volcanic sequence was sampled and is represented in Appendix 2, no systematic stratigraphic variation in lava composition was recorded.

Trace-element geochemistry

Due to the alteration of the mafic dykes, only those trace elements considered immobile are used in this section. The trace elements found most useful are titanium, chromium, nickel, zirconium and yttrium. Although titanium has been used as a major-element oxide in the preceding section, it is used here as a trace element. Examination and analysis of Transform Sequence lavas was restricted to those samples observed to be relatively unaltered (*eg* containing abundant fresh glass). Consequently, the abundances of LIL elements will be included in the sections below.

Mafic dykes

Although on most trace-element covariation diagrams the mafic dykes form a single continuous evolution suite, trace-element variation with titanium discriminates between the Axis Sequence and Transform Sequence dyke groups. On a titanium versus chromium covariation diagram (Fig. 5), the mafic dykes form two distinct fields, elongated subparallel to the chromium axis, in which the Axis Sequence has higher titanium relative to chromium compared with the Transform Sequence. In accordance with most mantle partial-melting models for the genesis of oceanic magmas involving a spinel or plagioclase lherzolite, the differences in Ti/Cr cannot be related by crystal fractionation and therefore indicate evolution from different primitive parental liquid compositions. This suggests different sources or source conditions during the extraction of the two dyke suites. The parallel evolution trends for the mafic dykes suggest either that the Axis Sequence was derived from a lower degree of partial mantle melting compared with the Transform Sequence, or that the Transform Sequence was derived from a mantle source that was already depleted in titanium relative to the source for the Axis Sequence. It is not possible, however, to distinguish between the two models on the basis of titanium versus chromium alone. The Transform Sequence dykes also have lower yttrium (*ie* also incompatible in basaltic phases)

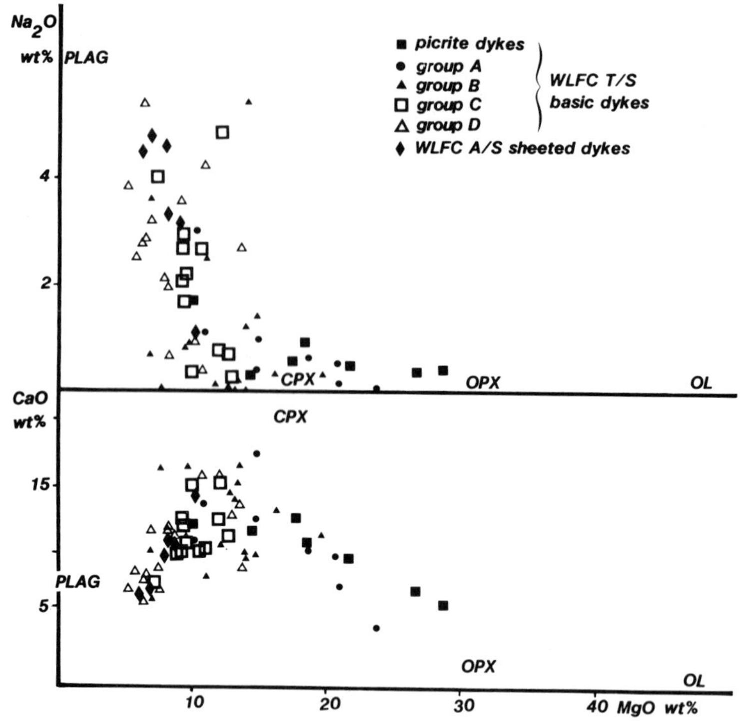
Na2O wt%
PLAG
picrite dykes
group A
group B
group C
group D
WLFC A/S sheeted dykes
WLFC T/S basic dykes
4
2
CPX
OPX
OL
CaO wt%
CPX
15
PLAG
5
OPX
OL
10
20
30
40 MgO wt%

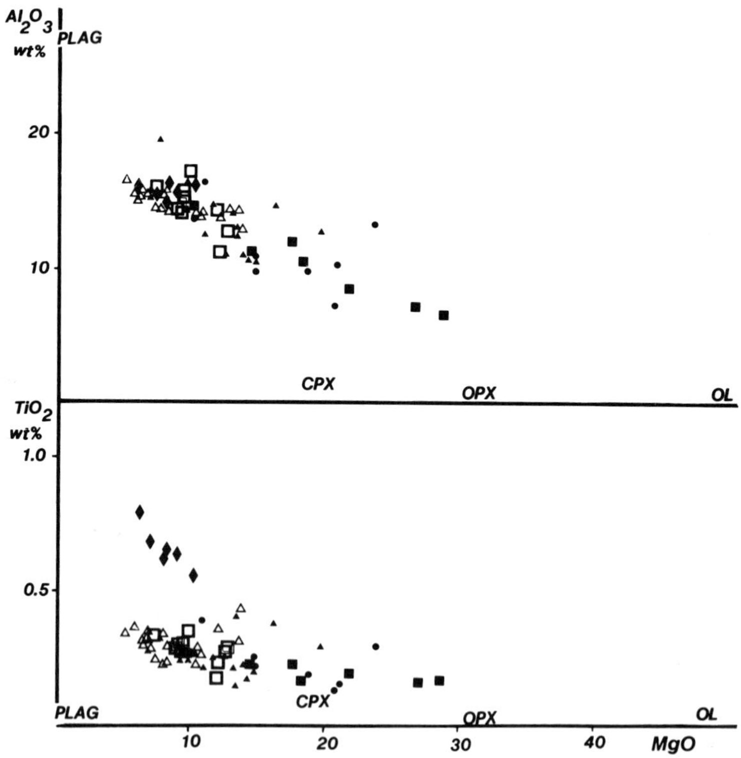
Al2O3 wt%
PLAG
20
10
CPX
OPX
OL
TiO2 wt%
1.0
0.5
CPX
PLAG
OPX
OL
10
20
30
40
MgO

relative to chromium compared with the Axis Sequence dykes, supporting the interpretation of a more depleted source or derivation via lower degrees of mantle partial melting for the Transform Sequence compared with the Axis Sequence.

Transform sequence lavas

It has already been shown that the lavas have a narrow range of major-element values. Their primitive composition is also reflected in their range of trace-element compositions, with most nickel values of 200–400 ppm and chromium values of about 1000 ppm. When compared with the mafic dykes on a titanium versus chromium covariation diagram (Fig. 6) the lavas form a tight group that is coincident with the most primitive non-cumulate Transform Sequence dykes, indicating cogenesis.

Table 1 shows the mean compositions for the Troodos Axis Sequence, the Arakapas Fault Belt and the Transform Sequence lavas. The low titanium abundance in the Transform Sequence lavas compared with the Troodos Axis Sequence, or mid-ocean ridge basalt (MORB) is reflected in low abundances of HFS elements (zirconium, yttrium and niobium). The Transform Sequence lavas have a very low abundance of both zirconium and yttrium (Fig. 7) compared with primitive N-type MORB which has 90 ppm zirconium and 30 ppm yttrium (Pearce 1980). The lavas also have Zr/Y ratios that vary from 1 to 3 and form trends that range from 10 to 26 ppm zirconium and 7 to 12 ppm yttrium. Although both zirconium and yttrium are HFS elements that are incompatible in lherzolite mineralogy, zirconium is slightly more incompatible than yttrium (Pearce & Norry 1979; Pearce 1980*b*). As a result Zr/Y ratios vary inversely with an increasing degree of mantle partial melting and crystal fractionation. Only the presence of residual garnet, and to a lesser extent clinopyroxene, in the source region will preferentially retain yttrium and hence reduce Zr/Y ratios in subsequent melts. The large variation in the Zr/Y ratio for the Transform Sequence lavas may be explained by residual garnet or amphibole in the source region or by large variations in very small degrees of mantle partial melting.

Such an interpretation, however, is not supported by the REE geochemistries of the lavas, which are all very similar (*ie* residual garnet will preferentially retain heavy REEs and large variations in the degree of mantle partial melting will greatly alter both the REE abundance and the La/Yb ratio). This suggests that the variation in Zr/Y ratio may be due to enrichment by zirconium relative to yttrium and the other trace elements.

Although the HFS elements and titanium are depleted in the lavas, the LIL elements (rubidium and strontium in particular) are not. Variation in both rubidium and strontium shows an order of magnitude difference in their abundances (Fig. 8) and is not consistent with the evidence that the lavas are primitive and have a narrow compositional range. When plotted against either incompatible or compatible trace elements, rubidium and strontium do not show any coherent trends. Although rubidium and strontium are very mobile during alteration, the Transform Sequence lavas analysed contain

FIG. 2. Na_2O versus MgO: this diagram shows a dispersed inverse relationship between MgO and Na_2O in which the overall evolution trend reflects a general change from olivine and pyroxene to clinopyroxene and plagioclase dominated fractionation for the Transform Sequence and Axis Sequence respectively; the picrite dykes lie on a line trending towards increasing olivine accumulation. CaO versus MgO: above 14 wt% MgO, the Transform Sequence dykes, including the picritic varieties, are dominated by olivine and pyroxene accumulation; between 14 and 12 wt% MgO the Transform Sequence dykes form a trend that is dominated by olivine fractionation; below 12 wt% MgO, clinopyroxene replaces olivine as the dominant fractionating phase and the evolution trend is dominated by clinopyroxene and plagioclase fractionation. Al_2O_3 versus MgO: between 14 and 10 wt% MgO, the Transform Sequence dykes are dominated by olivine and clinopyroxene fractionation. Between 10 and 5 wt% MgO, both the Transform Sequence and the Axis Sequence dykes follow an evolution trend dominated by clinopyroxene and plagioclase fractionation; the picritic dykes fall on a line trending towards olivine accumulation. Since both MgO and Al_2O_3 are considered immobile, the scatter in the correlation may be due to phenocryst phases forming non-liquid assemblages; this is supported by the petrographic evidence of phenocrystal and disequilibrium textured olivine and pyroxene in a large number of dykes. TiO_2 versus MgO: the Transform Sequence dykes form a dispersed trend, probably owing to both the very low values of TiO_2 and the presence of some cumulate phenocrysts assemblages, discrimination between the Transform Sequence and Axis Sequence mafic dykes is shown with the former having lower TiO_2, relative to MgO, compared with the latter for similar MgO. This difference between the Axis Sequence and Transform Sequence basalt groups reflects differences in parental liquid compositions.

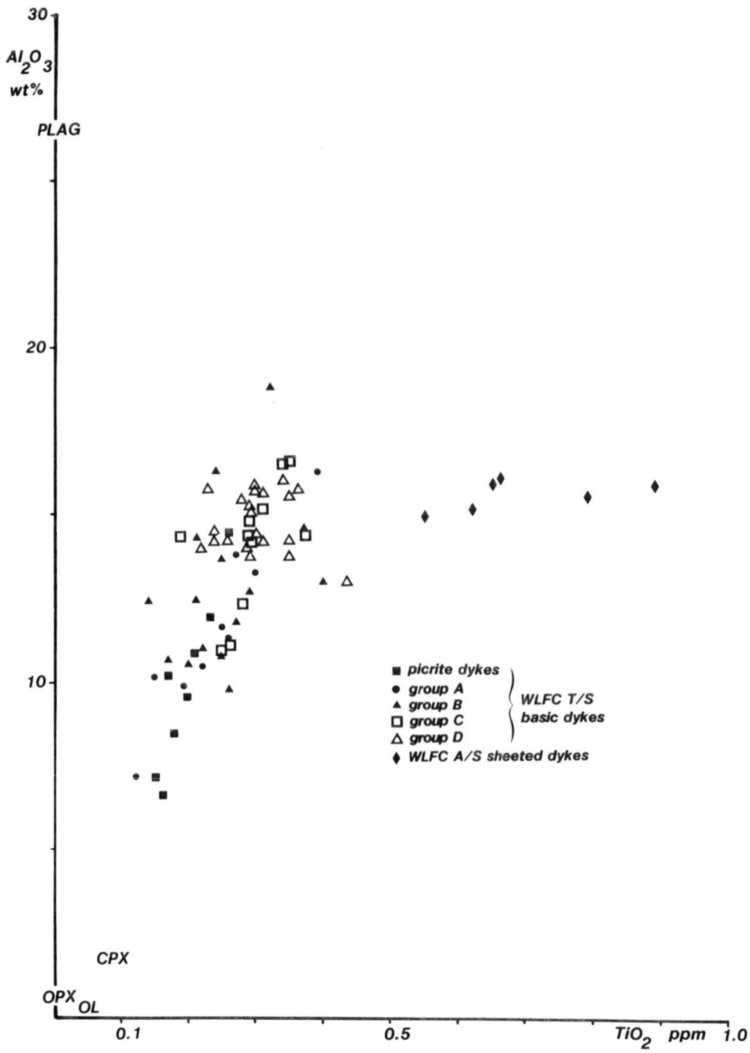

FIG. 3. Al_2O_3 versus TiO_2: this diagram further differentiates between the Axis Sequence and Transform Sequence mafic dykes. The Transform Sequence dykes have an evolution trend that is dominated by olivine and pyroxene fractionation. Only those samples with TiO_2 greater than 0.35 wt% show any evidence of plagioclase extraction. In contrast, the Axis Sequence dykes form a flat evolution trend with Al_2O_3 effectively buffered with respect to TiO_2, indicating approximately equal proportions of clinopyroxene and plagioclase fractionation.

FIG. 4. Major-element covariation diagrams, for the basaltic lavas from the Kapillio area, show a narrow and primitive range in composition. MgO is between 8.5 and 14 wt% for most of the lavas. The Al_2O_3 suggests pyroxene and olivine fractionation which is supported by the petrographic evidence.

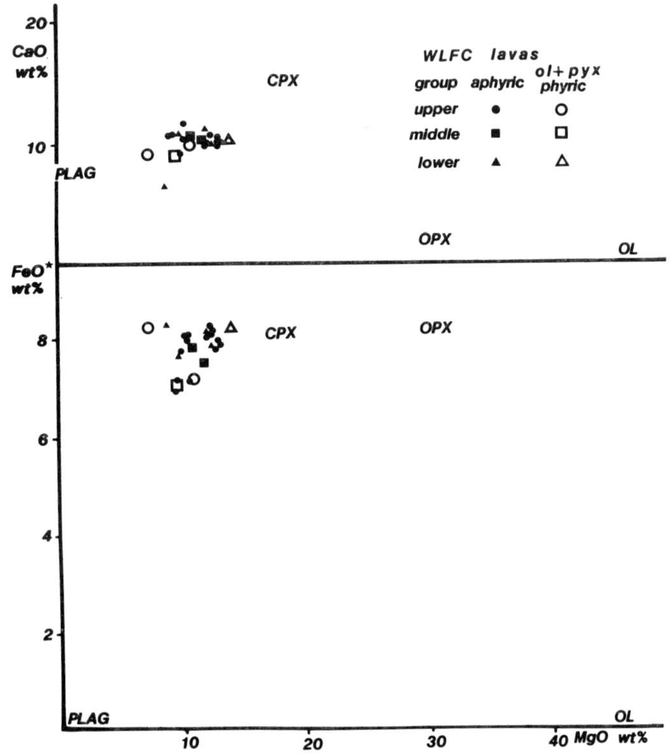
20
CaO
wt%
10
PLAG
CPX
WLFC lavas
group aphyric ol+ pyx phyric
upper
middle
lower
OPX
OL
FeO*
wt%
8
6
4
2
CPX
OPX
PLAG
OL
10
20
30
40 MgO wt%

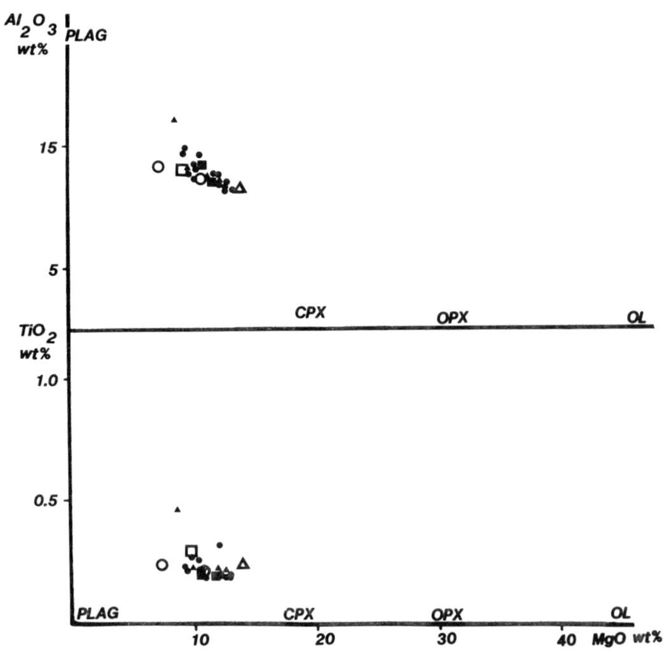
Al2O3
wt%
PLAG
15
5
CPX
OPX
OL
TiO2
wt%
1.0
0.5
PLAG
CPX
OPX
OL
10
20
30
40 MgO wt%

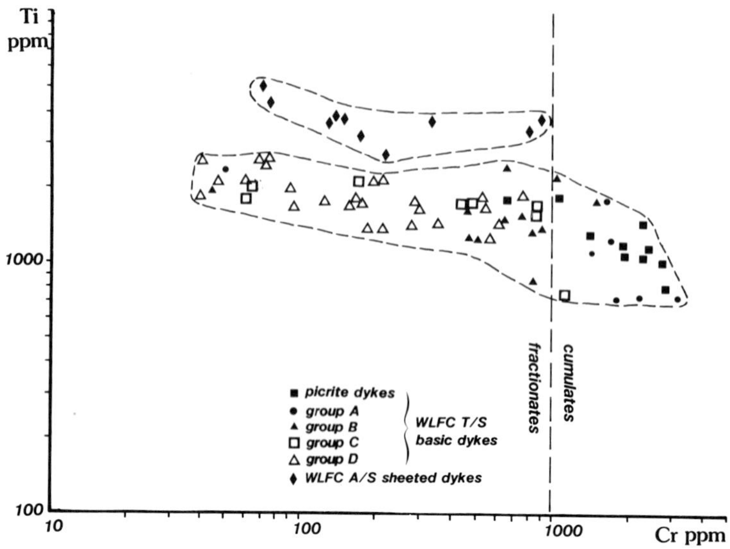

FIG. 5. Titanium versus chromium. There are two distinct suites, the Axis Sequence from the Transform Sequence. Both suites follow near-parallel evolution trends but start at parental liquids with higher and lower Ti/Cr respectively, reflecting different source melting conditions and parental liquid compositions. In principle, the more residual the mantle source, the lower the Ti/Cr of a subsequent melt. Higher degrees of partial melting of a mantle source have a similar effect. Fractionation in Ti–Cr space is both subhorizontal and, for different basaltic suites, approximately parallel.

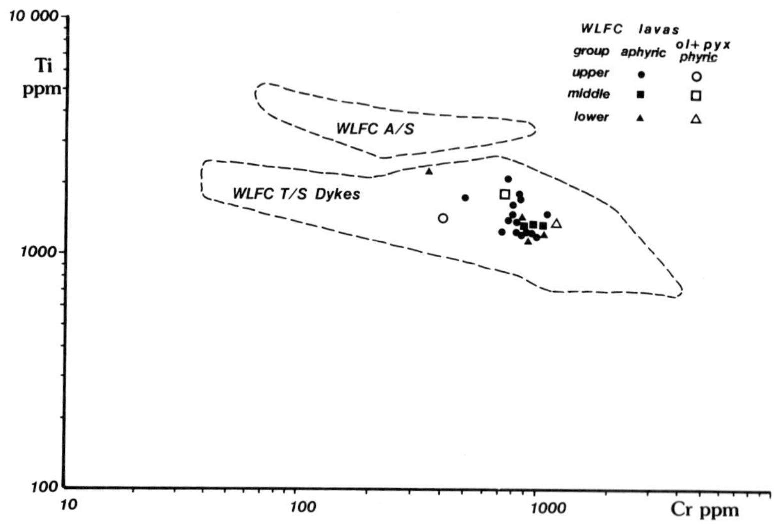

FIG. 6. Titanium versus chromium. The Transform Sequence lavas plot within the field defined by the Transform Sequence dykes and are coincident with the most primitive aphyric varieties. This indicates that the Transform Sequence dykes and lavas have a common parental liquid and hence are cogenetic.

abundant (more than 25%) fresh glass and are considered to reflect primary magmatic compositional variations. When plotted against one another (and potassium) both trace elements show strong positive correlations. The extreme variation (greater than one order of magnitude) of rubidium and strontium within the Transform Sequence lavas suggests that they are decoupled from major and trace-element systematics. The anomalously large variation in stron-

TABLE 1. *Mean compositions for the Arakapas Fault Belt basalts, primitive Arakapas Fault Belt basalts, Troodos Axis Sequence, N-type MORB and the Transform Sequence basalts used in the normalized element diagrams referred to in the text.*

	AFB basalt	AFB′	Troodos AS basalts	N-type MORB	TS basalts
Sr	155	138	88	120	148
K_2O	0.46	0.21	0.64	0.15	0.40
Rb	4	2.5	9	2	10
Th	0.19	0.22	0.22	0.20	0.28
Ta	0.16	0.17	0.13	0.18	0.08
Nb	1.8	2.5	—	3.5	2.1
La	—	—	—	3	1.4
Ce	1.17	1.12	4.9	9.0	2.4
Nd	0.77	0.42	5.4	7.7	1.2
Zr	11	13	30	90	17
Hf	0.35	0.35	0.62	2.4	0.41
Sm	0.54	0.41	1.89	3.30	0.40
TiO_2	0.28	0.22	0.43	1.5	0.23
Y	6.4	4	14.7	30	8
Yb	1.24	1.15	—	3.4	1.1
Ni	234	296	275	135	319
Cr	580	946	697	250	941
Gd	1.00	0.77	2.87	—	—
Tb	0.20	0.17	0.55	—	0.20
Tm	0.22	0.21	0.38	—	0.25
Lu	0.14	0.21	0.43	—	0.20

AFB, Arakapas Fault Belt; AFB′, primitive, Arakapas, Fault Belt basalts; AS, Axis Sequence; TS, Transform Sequence.
Major (wt%) and trace elements (ppm) were determined by energy-dispersive X-ray fluorescence (XRF) whole-rock analysis with the exception of chromium which was determined on a wavelength XRF spectrometer for greater accuracy. REEs were determined by instrumental neutron activation analysis. AFB basalt, mean of six glassy lavas from Simonian (1975); AFB′, average of the two most primitive AFB lavas), from Simonian (1975); Troodos AS, mean of nine lavas and dykes, from Smewing (1975); N-type MORB from Pearce (1980*a*).

tium and rubidium, their strong correlation with each other but not with their major and trace elements and their unusually high abundance compared with the other incompatible trace elements suggests that they were variably added to the lavas as part of a LIL-element enrichment phase, probably at source.

Rare-earth-element geochemistry

Chondrite-normalized REE diagrams are plotted for both Transform Sequence dykes and fresh glassy lavas (Figs 9 and 10). Because of the potential mobility of light REEs during alteration, the Axis Sequence sheeted dykes (which are at green schist facies) have not been analysed for REEs.

Transform sequence dykes

In general the Transform Sequence mafic dykes have REE abundances that range from 3 × to 10 × chondrite. The REE profiles (Fig. 9) have characteristic U-shaped REE patterns with increasing depletion of successively lighter REEs from lutetium to samarium (*ie* $(Yb/Sm)_n < 1$; the subscript n refers to chondrite-normalized values). From samarium to lanthanum the light REEs are successively more abundant with high Ce/Sm. Overall the REE profiles are substantially lower in REE abundance than MORB which forms a slightly light-REE-depleted pattern at *c.* 15 × chondrite. Between lutetium and samarium the successive depletion in light REEs is consistent with the trace-element evidence indicating that the Transform Sequence dykes were derived from either a high degree of mantle partial melting or an already depleted mantle source. Such light-REE-enriched profiles, however, cannot be generated from a chondrite REE pattern by partial melting of a spinel or plagioclase lherzolite source and suggest the addition of a strongly light-REE-enriched component to a heavy-REE-rich source (Hickey

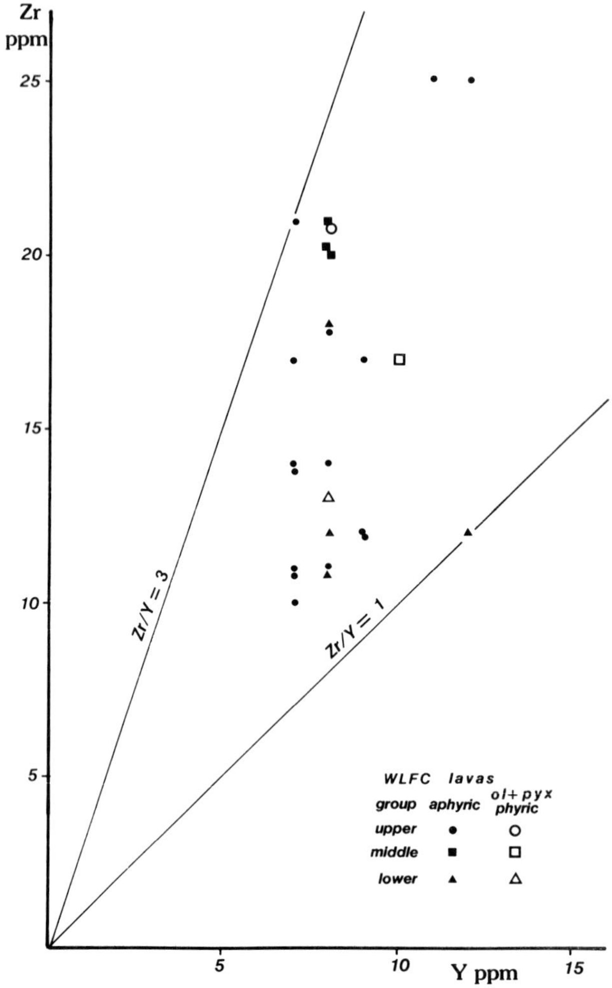

FIG. 7. Zirconium versus yttrium. Except for very small degrees of source partial melting, Zr/Y ratios do not vary greatly within a cogenetic suite. However, for the Transform Sequence boninites there is a variation of 2.5 times in zirconium but yttrium remains fixed. This suggests a decoupling between zirconium and yttrium systematics in the form of zirconium enrichment.

& Frey 1982; Cameron *et al.*, 1983; Cameron 1985).

Transform sequence lavas

REE profiles for the lavas (Fig. 10) closely resemble those for the Transform Sequence dykes. Absolute abundances are between 3 and 10 × chondrite. The lavas have a similar pattern of successive REE depletion from lutetium to samarium, with Ce/Sm<1, and similar light REE enrichment from samarium to lanthanum, with La/Sm > 1. Differences in the absolute abundances between various lava flows may reflect variable amounts of crystal fractionation and/or degrees of mantle partial melting. The similarity between the REE patterns for both the Transform Sequence lavas and the dykes is further evidence of their cogenetic nature.

Comparison between the Transform Sequence, Arakapas Fault Belt and Troodos Axis Sequence basalts

Within the southern Troodos transform fault there are two types of transform-related basalts: those of the Transform Sequence (*ie* from the Limassol Forest Complex) which have been

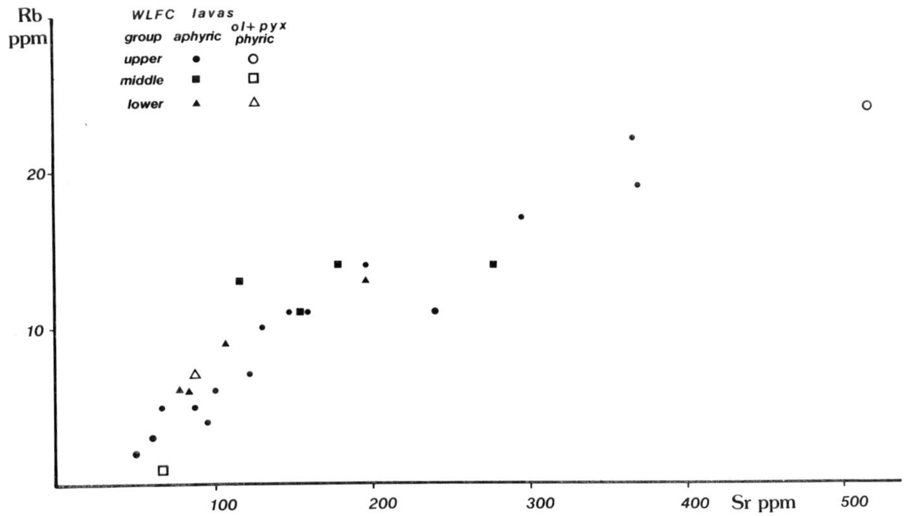

FIG. 8. Rubidium versus strontium. When plotted against either incompatible or compatible trace elements, rubidium and strontium do not show any coherent trends. However, when plotted against one another (and potassium) the LIL elements show strong positive correlations. Although rubidium and strontium are very mobile during alteration, the Transform Sequence basalts analysed contain abundant fresh glass and are considered to preserve primary magmatic compositions. Since rubidium and strontium are decoupled from other major and trace elements, their variation within the Transform Sequence lavas probably reflects independent LIL-element enrichment.

discussed in detail already and those of the Arakapas Fault Belt (*ie* from the northern wall of the southern Troodos transform fault) which have been characterized by Simonian & Gass (1978). Like the Transform Sequence basalts, those from the Arakapas Fault Belt have similar major-element compositions, are depleted in incompatible trace elements relative to the Troodos Axis Sequence and are probably cogenetic (Table 2). There are differences, however, between the two types of Transform Sequence basalt that are examined in this section and which are shown later to be important in the discussion of their petrogenesis.

Examination of the MORB-normalized multi-element diagram (Fig. 11) confirms that both the Transform Sequence and the Arakapas Fault Belt lavas are similarly depleted in trace

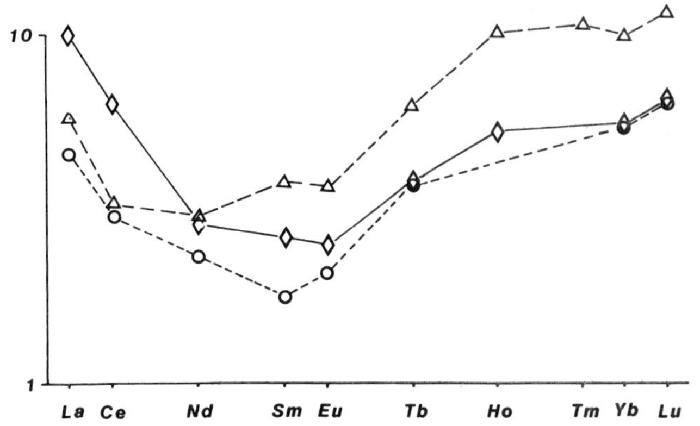

FIG. 9. REE (dykes): selected REE profiles for the mafic dykes, chondrite normalized. The REE profiles are depleted with respect to MORB and have a characteristic U shape. The REE were analysed by instrumental neutron activation at the Open University by Dr Nick Rogers.

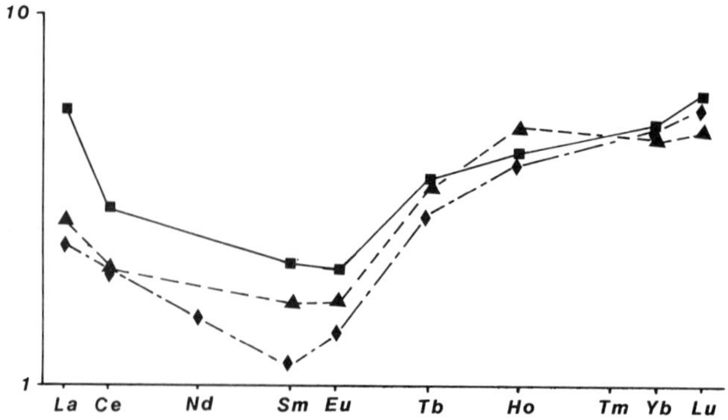

FIG. 10. REEs (lavas): selected REE profiles for the basaltic lavas, chondrite normalized. The REE profiles are similar in shape to those for the mafic dykes but have slightly lower absolute abundances. They are depleted with respect to MORB and also have the characteristic U shape.

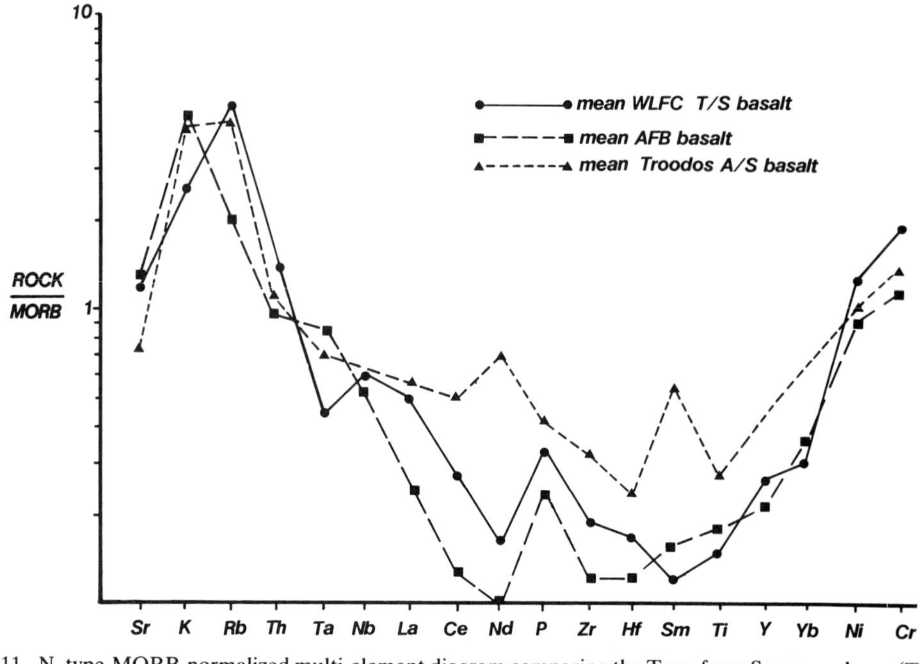

FIG. 11. N- type-MORB-normalized multi-element diagram comparing the Transform Sequence lavas (T/S), Arakapas Fault Belt lavas (AFB) (from Simonian 1975) and the Troodos Axis Sequence basalts (A/S) (from Smewing *et al*. 1975).

elements and are more depleted with respect to the Troodos Axis Sequence, which in turn is trace element depleted with respect to MORB. Further, all three types of Cyprus basalts are LIL element enriched. However, compared with the Arakapas Fault Belt basalts, those of the Transform Sequence have a higher La/Yb ratio (*ie* are light REE enriched) and are relatively HFS element enriched, although tantalum shows a depletion. Note, however, that tantalum is unreliable in the normalizing rock AFB′ (from Simonian 1975) because of contamination by sample preparation in tungsten carbide mills.

Detailed comparison between the REE profiles (Fig. 12) for the Troodos Axis Sequence and Arakapas Fault Belt basalts has already been made by Simonian (1975) and Simonian &

TABLE 2. *Mean Transform Sequence lavas compared to with representative boninite compositions.*

	Bonin Islands, Japan	Marianas Trench	Cape Vogel, Papua New Guinea	Marianas fore-arc region	Transform Sequence lavas
Major elements (wt%)					
SiO_2	58.46	57.33	56.80	58.00	52.50
TiO_2	0.10	0.14	0.33	0.23	0.22
Al_2O_3	13.37	9.97	11.9	14.45	15.8
FeO*	8.27	9.06	7.10	8.03	7.60
MnO	0.12	0.13	0.16	0.12	0.15
MgO	9.36	15.19	12.60	6.71	12.12
CaO	8.11	5.86	7.89	10.47	10.82
Na_2O	1.59	1.59	1.05	1.87	0.44
K_2O	0.70	0.93	0.41	0.35	0.52
P_2O_5	—	—	—	—	0.03
Trace elements (ppm)					
Sr	97	107	339	89	163
Rb	12	15	8	8	10
Th	—	—	—	—	0.28
Ta	—	—	—	—	0.4
La	1.27	1.27	1.47	0.75	1.35
Ce	2.57	2.93	3.51	2.10	2.85
Nd	1.65	1.85	2.09	1.72	2.85
Zr	25	36	29	28	15
Hf	0.69	0.79	0.58	0.62	0.41
Sm	0.43	0.51	0.60	0.57	0.40
Y	5	5	7	7	8
Yb	0.59	0.60	0.81	0.74	1.09
Ni	140	258	154	—	266
Cr	538	1386	715	197	848
Tb	0.10	0.12	0.14	0.16	0.19
Lu	0.10	0.10	0.13	0.12	0.20

Boninite data are from Hickey & Frey (1982). Elemental analyses for the Transform Sequence lavas are by energy-dispersive XRF except for chromium which was determined on a wavelength-dispersive XRF spectrometer. FeO*: total iron as FeO.

Gass (1978). Compared with N-type MORB, both the Troodos Axis Sequence and Arakapas Fault Belt basalts have lower REE abundances and successively increasing degrees of light REE depletion between samarium and lanthanum. The successive depletion in light REEs from N-type MORB to the Troodos Axis Sequence and to the Arakapas Fault Belt basalts is in agreement with their trace-element variation and supports the evidence of successively increasing degrees of mantle partial melting or derivation from an increasingly depleted source. Menzies & Allen (1974) suggested that the depletion in the Troodos Axis Sequence compared with N-type MORB is due to derivation from a more depleted source than MORB mantle and not larger degrees of partial melting. Simonian & Gass (1978), however, also suggested higher degrees of melting for the origin of the Arakapas Fault Belt basalts compared with the Troodos Axis Sequence.

Compared with the Arakapas Fault Belt basalts, the Transform Sequence basalts have both a similar Yb/Sm ratio and abundances of heavy REEs, supporting the trace-element evidence that both suites were derived from a similar source. However compared with the Transform Sequence basalts, none of the Arakapas Fault Belt basalts shows evidence of light REE enrichment. Therefore the Arakapas Fault Belt basalts may represent a 'non-light-REE and HFS-element-enriched' end-member of a depleted basalt group that includes the Transform Sequence lavas and dykes.

Basalt petrogenesis

Introduction

In the preceding sections it has been suggested that the Axis Sequence and Transform Sequence

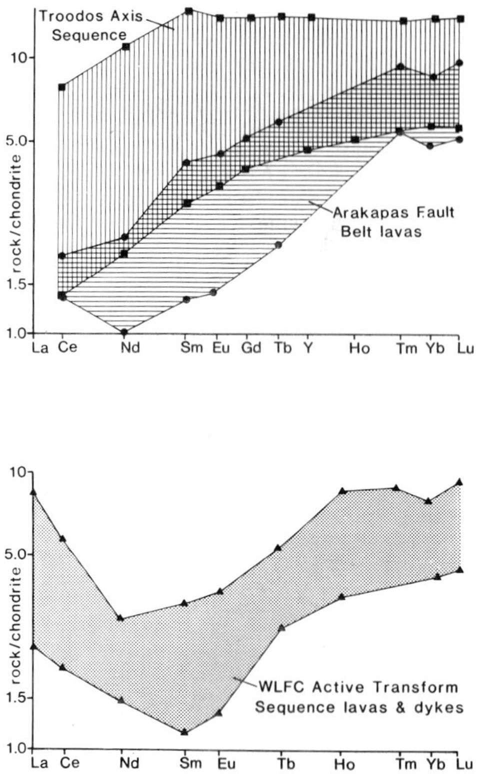

FIG. 12. Chondrite-normalized REE profiles for comparison between the Troodos Axis Sequence (from Smewing *et al.* 1975), Arakapas Fault Belt basalts (from Simonian 1975) and Transform Sequence basalts. The Arakapas Fault Belt basalts are depleted relative to the Troodos Axis Sequence; the Transform Sequence and Arakapas Fault Belt basalts are similarly depleted but only the Transform Sequence basalts are light REE enriched.

basalts form two separate groups derived from different sources, the latter being relatively depleted in incompatible trace elements. It has also been suggested that the Transform Sequence is enriched in light REEs, zirconium, yttrium and hafnium relative to the Axis Sequence and that both groups are ubiquitously enriched in LIL elements. In the following sections the Transform Sequence basalts are compared with boninites and a model is suggested for their relative depletion in incompatible elements that involves multiple stages of melt extraction from a MORB source. A model is also suggested for the selective enrichment of the Transform Sequence basalts in light REEs and HFS elements in which the enrichment phase is likened to either an ocean island basalt (OIB) or a mature pelagic sediment.

A comparison with boninites

When plotted on basalt discrimination diagrams (Pearce & Cann 1973; Pearce 1975, 1980*b*; Pearce & Norry 1979) the Axis Sequence basalts coincide with the Troodos Axis Sequence and fall within the field of island arc tholeiites. Many studies in the past have identified the Troodos Axis Sequence basalts as being subduction zone related (Miyashiro 1973; Pearce 1975, 1980*a*; Smewing *et al.* 1975; Cameron *et al.* 1980; McCulloch & Cameron 1983; Robinson *et al.* 1983; Schminke *et al.* 1983; Rautenschlein *et al.* 1985). In contrast, the Transform Sequence basalts lie outside the majority of basalt discriminant fields above.

Instead, the extremely low abundances of incompatible elements (zirconium, yttrium, niobium and TiO_2) and the moderately high MgO content of the Transform Sequence lavas are similar to those of boninites from Cape Vogel, Papua New Guinea (Dallwitz *et al.* 1966; Dallwitz 1968; Jenner 1981), the Mariana inner trench wall (Dietrich *et al.* 1978; Hawkins *et al.* 1979; Sharaskin & Dobretsov 1980) and the Mariana fore-arc region, DSDP site 458 (Meijer 1980; Meijer *et al.* 1981). Table 2 gives an average composition of the Transform Sequence lavas, and it is compared with a number of 'boninitic' compositions published by Hickey & Frey (1982).

The strongly incompatible-element-depleted nature of boninites is interpreted as the result of hydrous partial melting of residual mantle peridotite at less than 10 kbar (Cameron *et al.* 1979; Duncan & Green 1980; Hickey & Frey 1982; Cameron *et al.* 1983; Walker & Cameron 1983; Cameron 1985). The depleted geochemistry of the Transform Sequence basalts agrees with the above interpretation for boninite petrogenesis. Such a model, involving shallow mantle source melting, is supported by studies of ophiolitic mantle sequences. In both the Bay of Islands Ophiolite Complex, Newfoundland (Malpas 1978), and the Oman Ophiolite Complex (Browning 1982) the residual mantle harzburgites grade downward into increasingly more fertile lherzolite. In the former example, the lower limit to the residual mantle harzburgite is located at about 10 km below the petrological Moho (Malpas 1978). This implies that the depth of mantle partial melting for the genesis of boninites was as shallow as 14–15 km (*ie* between 5 and 7 kbar).

In their study of boninite petrogenesis Hickey & Frey (1982), Cameron *et al.* (1983) and Cameron (1985) recognized both a ubiquitous LIL-element enrichment and a variable light

REE and zirconium enrichment. In agreement with Pearce (1982) and Pearce *et al.* (1984), both Hickey & Frey (1982) and Cameron (1983) interpret LIL-element enrichment as volatile release during oceanic slab subduction and subsequent metasomatism of the mantle wedge.

Although both Hickey & Frey (1982) and Cameron (1983) recognize a separate silicate-melt-related enrichment of light REEs and zirconium in some boninites, the source for such a melt phase remains unknown.

Incremental melting

In the following sections it is suggested that the Axis Sequence and Transform Sequence basalts represent two magmatic groups that were derived from a 12% melt, followed by a subsequent 10% melt, of an already depleted MORB source. Although there are a number of possible melting processes (Shaw 1970) incremental batch melting is the most simple and has been used in the following calculations. The calculated degrees of partial melting are considered only relative to an assumed initial source P_0 that compares favourably with actual upper-mantle compositions. Mass balance calculations suggest that the final mantle residue following N-type MORB, Axis Sequence and Transform Sequence extraction resembles an extremely depleted tectonized harzburgite.

It has already been suggested that the Transform Sequence basalts were derived from a more depleted source or from a higher degree of partial melting compared with either the Axis Sequence or N-type MORB. In the past, variations in oceanic mantle partial melting have been modelled in terms of chromium versus yttrium covariation with melting curves for lherzolitic upper mantle calculated assuming that phases enter the melt in the ratio 3:1:4:4 for olivine, orthopyroxene, clinopyroxene and plagioclase respectively (Pearce 1980*a*; 1983) and using distribution coefficients listed in Pearce & Norry (1979) and Pearce (1980*b*, 1983). The positions of N-type MORB (Pearce 1980*a*; 1983; Bryan & Thompson 1981; Fisk *et al.* 1982) the Axis Sequence dykes and the Transform Sequence dykes and lavas are plotted on a chromium versus yttrium covariation diagram in Fig. 13. The position of the depleted MORB source P_0 assumes MORB melt extraction at an average of 20%. The position of C_3 chondrite is plotted for reference.

In Fig. 13, the Axis Sequence basalts intersect the incremental melting curve from P_0 at between 45% and 50%, implying exceptionally hot thermal conditions in the source region. In contrast, the requirement for such an anomalous thermal condition has been used by Menzies & Allen (1974), Pearce (1980*a*), Duncan & Green (1980) and Pearce *et al.* (1984) to argue in favour of a multiple-melting model for the genesis of similarly depleted magmas. Rather than using P_0 as a starting composition, the position of the source for the Axis Sequence is taken as the residue following MORB extraction. Assuming similar phase melting ratios and conditions as above (Pearce 1980) the Axis Sequence intersects the melt curve from the depleted source P_1 at *c.* 12%. This value is in general agreement with the models of Menzies & Allen (1974) and Pearce (1980) and implies a cooler less anomalous thermal regime than for the preceding case (Duncan & Green 1980; Pearce *et al.* 1984).

Since the Transform Sequence basalts do not intersect with the melt curve from P_0 (the assumed MORB source) they cannot be generated from it. Instead, they intersect with the melt curve from P_1 (the first-stage-depleted source) at between 50% and 55%. By applying a similar thermal argument that favours batch melting, P_2, which is the residue from P_1 following the extraction of the Axis Sequence, is taken as the source for the Transform Sequence basalts. In this situation, the melt curve from P_2 intersects with the Transform Sequence basalts at *c.* 10% partial melting, the lower degree of partial melting from P_2 compared with P_1 implying a cooler thermal regime and one that is more compatible with a transform environment.

Major-element considerations

It has often been argued that successive basalt extractions will rapidly deplete a lherzolite source in basaltic components (especially CaO and Al_2O_3), thus making further basalt genesis impossible. In the following calculations (and in agreement with Menzies & Allen 1974; Allen 1975), the source for the Axis Sequence basalts is taken as an average of three tectonized plagioclase lherzolites that outcrop on the Troodos massif. The composition of P_1 is given in Table 3. P_{2a} and P_{2b} are the various calculated residues from P_1, following Axis Sequence and Transform Sequence extraction at 12% and 50%–55% melting respectively, and are calculated by simple mass balance. Their compositions are also given in Table 3 together with the compositions of the primitive liquid end-members of the Axis Sequence and Transform Sequence basalts, termed here L_1 and L_2 respectively.

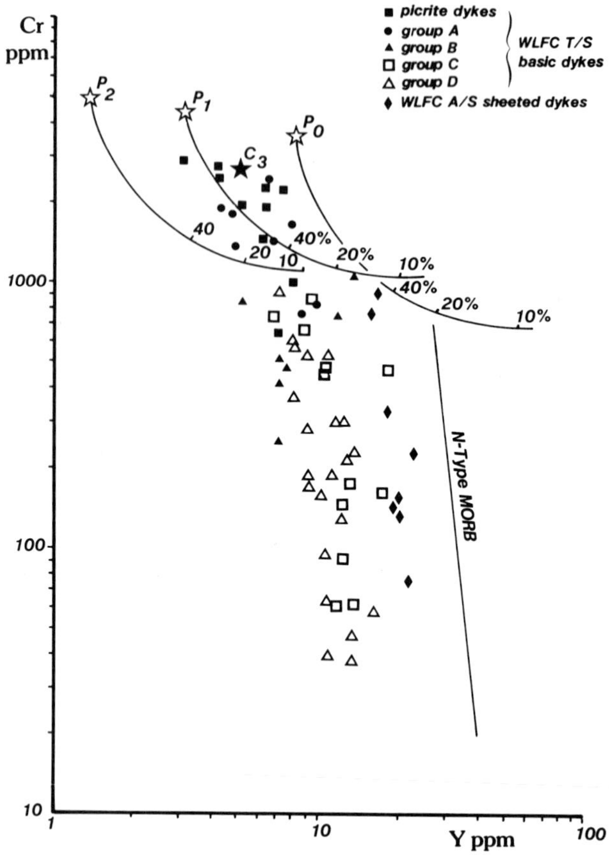

FIG. 13. Chromium versus yttrium covariation diagram in which incremental batch melt proportions are modelled. The positions of N-type MORB and C^3 chondrite are given for reference. Melt curves are based on those of Pearce (1980) and assume the same phase ratios entering the melt in each case. The source P_0 refers to a calculated MORB source (see text); P_1 is the residue following MORB extraction (at *c*.20% melting) and is taken as the source for the Troodos Axis Sequence; P_2 is the residue from P_1 following Troodos Axis Sequence extraction (at *c*.12% melting) and is taken as the source for the Transform Sequence.

The residue composition from P_1 is given by

$$C_{P_2} = (C_{P_1} - C_{L_1}\, F)\,\frac{1}{(1-F)}$$

where: C_{P_1} is the concentration in the source, C_{P_2} is the concentration in the residue, C_{L_1} is the concentration in the liquid extracted (*ie* primitive Axis Sequence or Transform Sequence basalts) and F is the melt fraction.

The excess of major elements in the calculated composition of P_{2a} demonstrates that it is feasible to extract the Axis Sequence from P_1 at 12% melting. The calculated composition of P_{2b}, however, has a deficit in CaO and Al_2O_3, indicating that the Transform Sequence basalts cannot be generated from P_1 at the estimated 50%–55% melting. This supports the argument made above that an anomalously hot thermal regime for the single-stage generation of the Transform Sequence basalts is geologically unlikely. By taking P_{2a} as the source for the Transform Sequence basalts at a modelled melt proportion of 10%, the composition of the calculated residue P_3 (given in Table 3) has an excess of major elements, supporting the suggestion that the depleted residue P_{2a} is a possible source for the Transform Sequence basalts.

The incremental batch melting model is further supported when P_3, the final residue after batch melting, is compared with depleted tectonized harzburgite from the mantle sequence (Table 4) (Menzies & Allen 1974; Allen 1975; Smewing *et al.* 1975). Further, the calculated composition of P_0 (Table 4) also compares

TABLE 3. *Calculated compositions of sources and residues after extraction of Axis Sequence and Transform Sequence melts.*

	L_1	L_2	P_1	P_{2a}	P_{2b}	P_3
SiO_2	47.47	51.00	43.18	42.59	37.97	41.42
TiO_2	0.60	0.30	0.02	−0.06	−0.16	−0.10
Al_{O3}	17.17	13.50	3.67	1.83	−2.88	0.52
Cr_2O_3	0.14	0.14	0.65	0.72	0.99	0.78
FeO*	8.14	8.74	8.39	8.42	8.16	8.34
MnO	0.03	0.04	0.25	0.28	0.39	0.31
MgO	11.07	14.00	40.74	44.78	58.60	48.01
CaO	15.37	12.30	3.49	1.87	−2.38	0.71

L_1 and L_2 are primitive Axis Sequence and Transform Sequence basalts respectively and are averages from the most primitive sparsely phyric samples analysed. P_1 is an average of three Troodos plagioclase lherzolites that are taken as the source for the Troodos Axis Sequence (Menzies & Allen 1975). P_{2a} and P_{2b} are the calculated residues following the bulk extraction of the Axis Sequence at 12% melting and the Transform Sequence at 50%−55% melting. P_3 is the calculated residue following the bulk extraction of the Transform Sequence from P_{2a}. Although the above melting model is consistent for both the major- and trace-element abundances, TiO_2 is deficient in both P_{2a} and P_3. This apparent anomaly can be accounted for by the presence of 0.04% ilmenite or 0.08% titaniferous magnetite in the initial source P_1. Small amounts of titaniferous spinel are likely to result in TiO_2 being very heterogeneous and make TiO_2 difficult to determine on the basis of only the three plagioclase lherzolite samples available.

favourably with suggested upper-mantle compositions and hence is not an unreasonable starting point for the batch melting model. The calculated major-element composition of P_0 has been derived by adding 20% of primitive MORB (Bender *et al.* 1984) to P_1, calculated using the simple mass balance equation:

$$C_{P_0} = C_{P_1}(1 - F) + C_{L_1}F$$

where C_{P_0} is the concentration in the proposed MORB source, C_{P_1} is the concentration in the MORB source residue (*ie* the Troodos plagioclase lherzolites), C_{L_1} is the concentration in primitive liquid (*ie* N-type MORB) and F is the melt fraction.

TABLE 4. *Suggested upper-mantle compositions compared with the compositions of the final calculated residue following incremental batch melting and the calculated N-type MORB source P_0.*

	1	2	3	4	5	6	P_0	P_1	P_3	H
SiO_2	42.86	45.16	44.62	44.36	44.25	37.46	44.28	42.89	41.42	43.72
TiO_2	0.33	0.71	0.14	0.13	0.07	0.25	0.26	0.02	<0.01	0.01
Al_2O_3	6.99	3.54	3.66	3.59	3.25	4.48	5.80	3.64	0.52	0.47
Cr_2O_3	0.18	0.43	0.33	0.43	0.43	0.51	0.53	0.65	0.78	0.39
FeO*	9.29	8.45	8.17	9.00	7.88	8.83	8.62	8.31	8.34	8.19
MnO	0.14	0.14	0.13	0.18	0.11	0.15	0.18	0.18	—	0.15
NiO	0.20	0.20	0.27	0.27	0.26	0.37	0.21	0.25	0.31	0.27
MgO	35.07	37.45	38.98	38.51	39.50	43.50	35.62	40.48	48.01	45.99
CaO	4.37	3.08	3.31	3.31	3.25	3.95	4.46	3.47	0.71	0.77
Na_2O	0.45	0.57	0.26	0.11	0.23	0.46	0.21	0.06	<0.01	<0.01
K_2O	<0.01	0.13	<0.01	0.13	0.01	0.04	0.02	0.01	—	<0.01
Total	99.91	99.94	99.87	100.02	99.35	100.00	100.00	100.00	100.00	100.00

1, Kilbourne Hole (Carter 1970); 2, pyrolite (Ringwood 1966); 3, Rhonda (Dickey 1970); 4, Orthris, Greece (Menzies 1974); 5, Tarreyres, France (Hutchinson *et al.* 1975); 6, average of mantle nodules (Jagoutz *et al.* 1979). P_0, calculated N-type MORB source; P_1, average of three Troodos plagioclase lherzolites (Allen 1975); P_3, calculated residue after extraction of Transform Sequence; H, average of eight tectonized harzburgites from Troodos (Allen 1975).

Light rare earth element, hafnium and zirconium enrichment

The Transform Sequence basalts are heavy REE depleted compared with the Troodos Axis Sequence. Their REE profiles are U-shaped with $(La/Sm)_n > 1$, $(Yb/Sm)_n < 1$, and they are enriched in zirconium, yttrium and hafnium. In the following sections the zirconium and hafnium enrichment is quantified and its relationship with the light REE enrichment is examined.

Enrichment factors

Since the only apparent difference between the Arakapas Fault Belt and Transform Sequence basalts is the addition of a light REE, zirconium and hafnium enriched phase to the latter, the Arakapas Fault Belt basalts are taken as a non-enriched end-member of the Transform Sequence suite. The amount of this enrichment phase that has been added to the Transform Sequence is qualified by subtracting the composition of the average primitive Arakapas Fault Belt basalt (termed here AFB′) from an average of the Transform Sequence lavas. From this method the enrichment phase has $(La/Sm)_n = 4.2$ and $(Ce/Sm)_n = 1.3$.

Figure 14 is a multi-element diagram on which the means of six primitive Transform Sequence lavas are normalized to AFB′. Positive deviations from the normalizing values are interpreted as enrichment while negative deviations represent depletion relative to AFB′. The light REE enrichment that characterizes the Transform Sequence basalts is also associated with variable enrichment in yttrium, titanium, zirconium, hafnium, rubidium and strontium and a slight depletion in niobium.

The normalized multi-element diagram profile for the Transform Sequence basalts, above, suggests that hafnium, zirconium and yttrium are associated with the light REE enrichment. On the basis of the procedure outlined by Cameron (1985), the proportion of light REE enrichment is estimated by calculating the abundance of cerium that has been added to AFB′ to generate the degree of cerium enrichment observed. Figure 15 illustrates zirconium, hafnium and Hf/Yb covariation with cerium enrichment. All three covariation diagrams show a strong positive correlation, demonstrating a close relationship between zirconium,

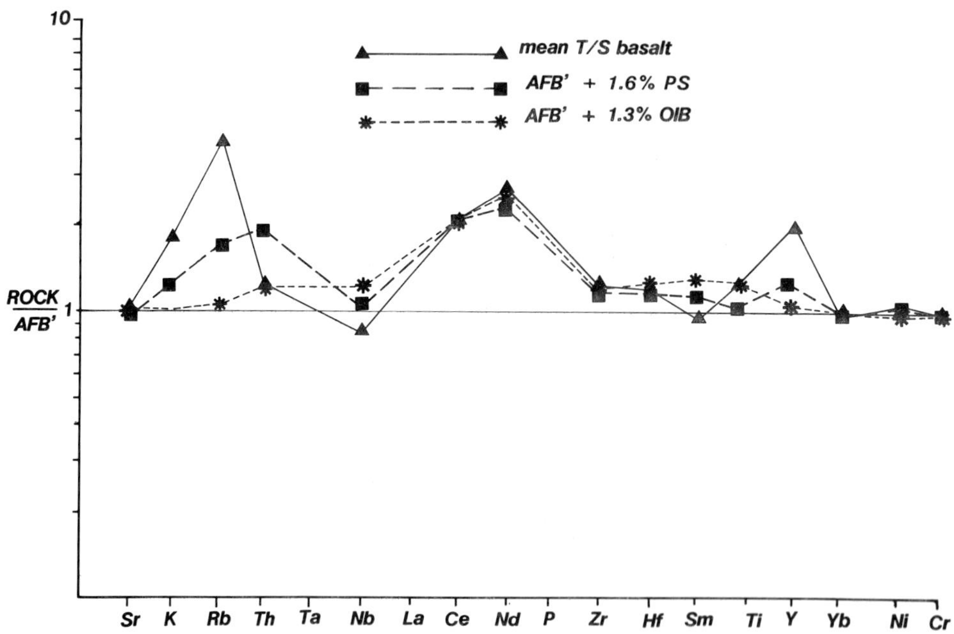

FIG. 14. Multi-element diagram normalized to AFB′. The Transform Sequence basalts are enriched in yttrium, hafnium, zirconium and light REEs as well as LIL elements compared with AFB′, which is taken as the non-enriched end-member. The Transform Sequence basalt profile is closely modelled by the bulk addition to AFB′ of 1.6% pelagic red clay or 1.3% OIB.

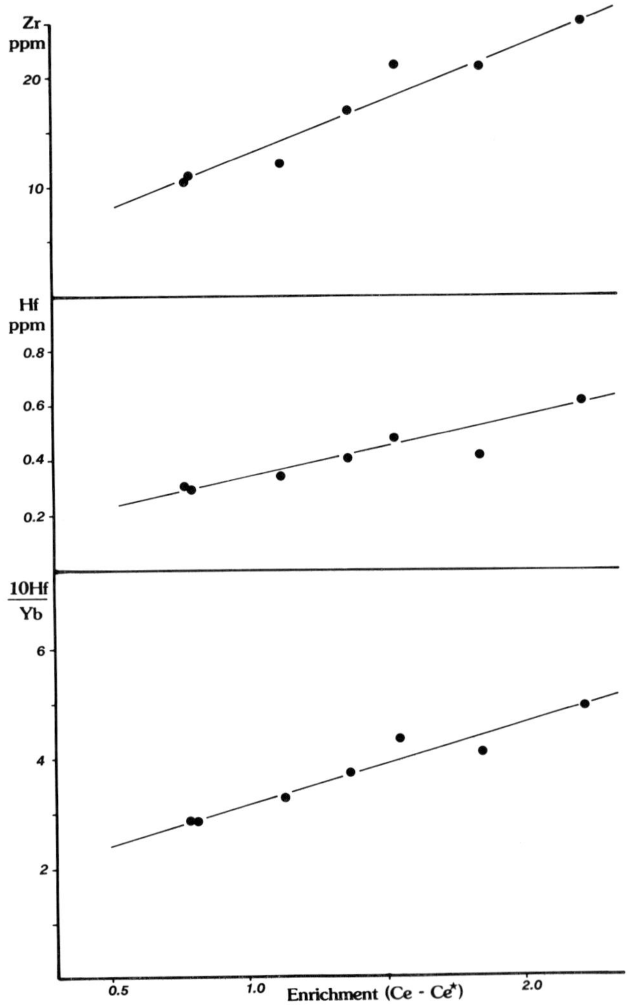

FIG. 15. Plot of zirconium, hafnium and Hf/Yb against enrichment. The enrichment is relative to cerium and is based on the difference between the actual cerium present compared with the calculated pre-enrichment amount. The correlations indicate that the enrichment in cerium (*ie* light REEs) is genetically related to the enrichment in HFS elements and suggest a silicate melt origin for the enrichment phase.

hafnium and light REE enrichment processes.

Although there is no correlation between LIL-element abundance and zirconium, hafnium and light REE enrichment, Fig. 14 suggests an excess of LIL elements compared with the ubiquitous LIL-element enrichment that effects the Arakapas Fault Belt, Troodos Axis Sequence and Transform Sequence basalts alike. This may be due to an extremely heterogeneous single-stage LIL-element enrichment or to a second stage of LIL-element enrichment that is related to the addition of the light REE and zirconium, hafnium component.

Nature of the enrichment phase in light rare earth elements, zirconium and hafnium

Although LIL elements and, to a lesser extent, light REEs are mobile in fluid phases, HFS elements and heavy REEs are most mobile in silicate melts (Pearce 1975; 1982; Schneider & Eggler 1986). Since the enrichment phase that affects the Transform Sequence basalts carries light REEs and selected HFS elements it is most probably a silicate melt.

In the multi stage melting model described above, any pre-existing ancient enrichment of incompatible elements in the initial mantle source would all have been removed by the second melting stage. The Transform Sequence basalts are the third and last stage of melt extraction from an already depleted source, yet they show the largest effect of enrichment, demonstrating that the enrichment in HFS elements and light REEs was introduced after the extraction of the Axis Sequence melt and prior to or during the generation of the Transform Sequence melt and is thus most likely to be related to the supra-subduction transform environment.

The enriched phase that was added to the primitive incompatible-trace-element and light-REE-depleted end-member (*ie* the AFB′) has already been calculated to have $(La/Sm)_n = 4.2$ (*ie* to be strongly light REE enriched). Figure 16 shows ε_{Nd} versus Nd^{-1} with values for the Troodos Axis Sequence taken from Rautenschlein *et al.* (1985) and for the Transform Sequence from Rogers *et al.* (in prep.). The data show a general trend of decreasing ε_{Nd} with decreasing abundances of neodymium. This means that the most depleted trace-element samples, being those which show the greatest effect of HFS-element and light REE enrichment, are pulled towards low values of ε_{Nd}. This suggests that the siliceous enrichment component has a very low or negative ε_{Nd}, indicating derivation from an ancient reservoir, in agreement with Cameron *et al.* (1983). The tectonic evidence and batch melting model, however, demonstrate that an ancient enriched component residing in the source would have been removed by the first or second stage of melting, suggesting that the enrichment was introduced into the source during the final

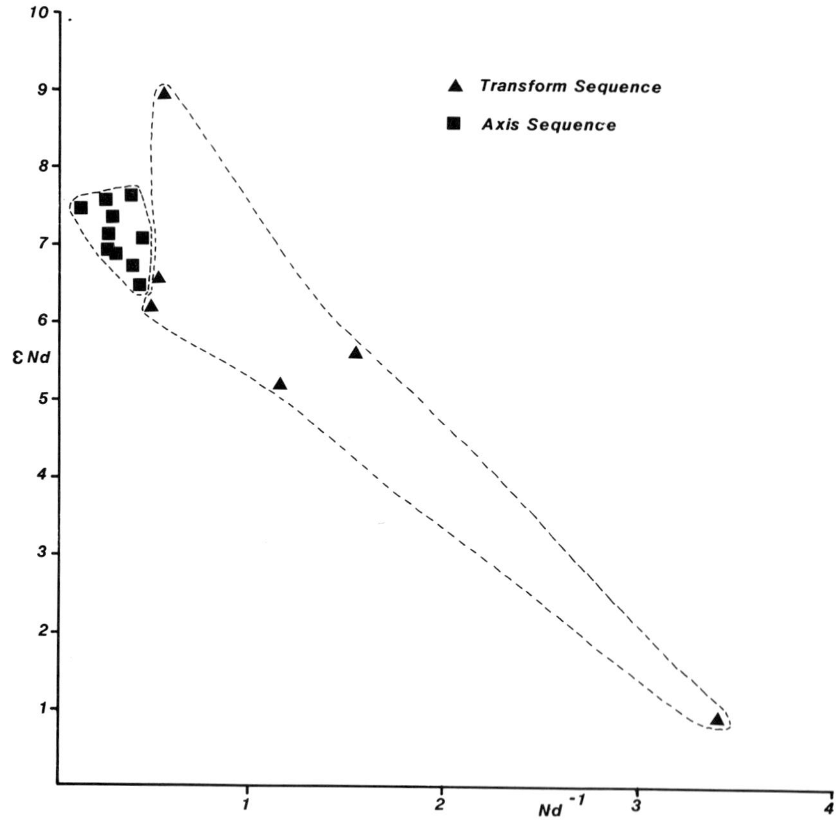

FIG. 16. Covariation between ε_{Nd} and Nd^{-1} shows values for the Troodos Axis Sequence (from Rautenschlein *et al.* 1985) and for the Transform Sequence (from Rogers *et al.*, in prep.). These data show a general trend of decreasing ε_{Nd} with decreasing abundances of neodymium. This means that the most trace-element-depleted samples, being those which show the greatest effect of HFS-element and light REE enrichment, are pulled towards values of low ε_{Nd}, suggesting that the siliceous enrichment component has a very low or negative ε_{Nd}, indicating derivation from an ancient reservoir.

melting stage generating the Transform Sequence boninites. There are only two sources for such a low ε_{Nd} enrichment component that can be considered appropriate in the inter-oceanic supra-subduction zone setting of the Transform Sequence boninites: (i) an OIB component that coincides with the subduction zone and (ii) pelagic sediment. Both may have low values of ε_{Nd}.

A primitive OIB (from La Palmas, Canary Islands) (Palacz, pers. comm.) has $(La/Sm)_n = 3.5$ and $(La/Yb)_n = 18$. Similarly, an average pelagic sediment (Atlantic red clay deposited under conditions of low productivity and below the CCD) (Taylor & McLennan 1985) has $(La/Sm)_n = 3.3$ and $(La/Yb)_n = 7.4$. Using the absolute values of cerium in the pelagic sediment and the OIB the amount needed of each, added in bulk to AFB′ to produce the average observed cerium enrichment, has been calculated. This gives an absolute value of light REE enrichment as a proportion of either OIB or pelagic sediment:

$$E = \frac{\mathrm{Ce} - \mathrm{Ce}^*}{\mathrm{Ce_E}}$$

where E is the proportion of enrichment phase present, Ce − Ce* is the mean abundance of cerium added to AFB′ and $\mathrm{Ce_E}$ is the actual cerium abundance in the enrichment phase. From this equation the average light REE enrichment in the Transform Sequence basalts represents either 1.3% OIB or 1.6% pelagic red clay, added in bulk to AFB′. To test which candidate for the enrichment phase most closely resembles the actual trace-element enrichment observed, they have both been added in bulk to AFB′ by simple mass balance:

$$C_1 = C_0\,(1 - E) + C_E E$$

where C_0 is the concentration in AFB′, C_E is the concentration in the enrichment phase, C_1 is the concentration in enriched basalt and E is the proportion of enrichment phase added.

In Fig. 14, 1.6% pelagic red clay and 1.3% OIB have been added to AFB′ in an attempt reproduce the characteristic normalized multi-element diagram profile of the Transform Sequence basalt. The compositions of AFB′, mean Transform Sequence basalt, OIB, pelagic sediment and the calculated enriched basalts are given with the results in Table 5.

At first glance, the modeled enriched Transform Sequence basalt compositions do not fit the actual values observed. However, when the multi-element profile is subdivided into its constituent components many of the differences can be explained. The LIL-element component is an especially poor match, but it has already been shown that these elements are mostly added as a ubiquitous enrichment which is both heterogeneous and added separately from the light REE and HFS-element low-ε_{Nd} component. The light REE and HFS-element component is closely matched by both models, with the exception of the positive yttrium peak which is not provided for by the OIB model. There is a misfit, however, with niobium which forms a slight negative anomaly in the 'enriched' Transform Sequence boninites. Although the pelagic sediment component goes some way towards matching this negative niodium anomaly by dilution effects, the OIB component produces a slight positive niobium peak and is clearly inadequate as a model for the enriching phase. That niobium becomes depleted with the addition of the enriching component suggests that AFB′, taken as the non-enriched end-member, has itself been heterogeneously enriched to some extent and that the negative niobium anomaly merely reflects the depletion of the source. Although tantalum follows the systematics of niobium, the AFB′ analyses were made by Simonian (1975) following sample preparation in tungsten carbide mills that caused tantalum contamination.

From trace-element considerations alone, the best model for the siliceous enriching phase appears to involve the bulk addition of pelagic red clay sediment to the depleted source for the Transform Sequence boninites. However, such bulk mixing models are simplistic in the extreme. In reality the added component is more likely to be heterogeneous, being derived from variable degrees of partial melting from either the added pelagic sediment or the OIB source. Further, it is not known whether the enrichment is added to the depleted Transform Sequence source and then remobilized in part during the subsequent generation or the Transform Sequence source and then remobilized in part during the subsequent generation of the Transform Sequence or whether it is added directly to the Transform Sequence melt. Given these additional undeterminable variables, the only possibility is to attempt to model the enrichment component in the somewhat simplistic way above. Despite these obvious limitations, however, the bulk addition of pelagic sediment to the Transform Sequence boninites gives a result that models most of the light REE, HFS-element and low-ε_{Nd} enrichment observed in these rocks. Further, the subduction of pelagic sediment is greatly favoured by the tectonics of subduction roll-back as invoked to generate

TABLE 5. *Compositions of oceanic island basalt, pelagic sediment, primitive Transform Sequence basalt and oceanic-island-basalt-enriched and pelagic-sediment-enriched primitive Arakapas Fault Belt basalt.*

	OIB	PS	TS′	E_1	E_2
Sr	818	18	148	147	136
K_2O	0.86	—	0.39	0.22	—
Rb	21	110	10	3	4.22
Th	4.40	13.2	0·28	0.27	0.43
Nb	57	14	2.1	3.2	2.68
Ta	4.58	1.00	0.08	0.23	0.18
La	47	42	1.4	—	—
Ce	102	80	2.4	2.4	2.4
Nd	56	41	1.2	1.14	1.1
Zr	276	150	17	16	15.2
Hf	6.82	4.1	0.41	0.43	0.40
Sm	10.4	8.0	0.40	0.54	0.53
TiO_2	4.39	0.77	0.28	0.28	0.23
Y	31	40	8	4.3	5
Yb	2.22	3.80	1.1	1.16	1.13
Ni	52	230	319	293	295
Cr	49	90	941	934	932
Tb	135	1.30	0.20	0.32	0.19
Lu	0.34	0.55	0.20	0.23	0.21

PS, pelagic sediment; TS′, Transform Sequence basalt; E_1, OIB-enriched AFB′; E_2, pelagic-sediment-enriched AFB′.
Major elements are in weight per cent; trace elements are in parts per million.
REEs analysed by neutron activation, trace elements by energy-dispersive XRF except for chromium, by wavelength-dispersive XRF.
OIB is from La Palmas (Z. Palacz, *pers. comm.*). The pelagic sediment is an average pelagic Atlantic red clay (McLennon & Taylor 1985). E_1 is the calculated product following bulk addition of 1.3% OIB to AFB′. E_2 is the calculated product following bulk addition of 1.6% pelagic red clay to AFB′. The composition of AFB′ is given in Table 2. The TS′ is an average of six of the most primitive, sparsely phyric and glassy samples.

supra-subduction oceanic crust (*ie* a condition of minimum subduction erosion compared with the opposite in which a large accretionary prism is generated). The conclusion that pelagic sediment is responsible for the light REE, HFS-element low-ε_{Nd} component is in agreement with Rautenschlein *et al.* (1985) for a similar but less evident enrichment in the Troodos Axis Sequence, but is in contrast with the ancient OIB source of enrichment, preexistent in the source region, as envisaged by Cameron *et al.* (1983).

Discussion

Comparison with oceanic transform fault zone magmatism

In contrast with the transform fault effect of observed in modern oceanic basins, when compared with the Troodos Axis Sequence the Transform Sequence basalts show a reverse effect. Instead of being more alkalic than their spreading ridge counterparts, the Transform Sequence basalts are less alkalic and more depleted in incompatible trace elements. Whereas basalts erupted within oceanic transform fault zones are generally more evolved (Hekinian & Thompson 1976; Schilling *et al.* 1983; Bender *et al.* 1984; Langmuir & Bender 1984) the Transform Sequence basalts have a very narrow and primitive compositional range and resemble primary parental liquids. Although the transform fault effect has been interpreted as a result of decreasing degrees of partial melting, both basalts erupted within oceanic transform fault zones and their adjacent spreading centres are considered to have been derived from the same source (Bender *et al.* 1984; Langmuir & Bender 1984). However, this is in contrast with the Transform Sequence basalts which are derived from a mantle source that has already had the adjacent Troodos Axis Sequence extracted from it.

Structural and other evidence indicates that

the Transform Sequence magmatism occurred within an oceanic transtensional and leaky transform fault zone, developed above an inter-oceanic subduction zone (Murton 1986; Murton & Gass 1986). The depleted geochemistry of the Transform Sequence basalts indicates shallow melting of an already depleted source that has had the Axis Sequence extracted from it. This shallow level of mantle melting and primitive magmatic activity suggests rapid extension within the transform fault zone. Such a rapid extension may be the result of the development of an extensional relay zone that laterally offsets a transform fault and which dilates at the spreading rate of the adjacent ridges (Holcombe *et al*. 1973).

The southern Troodos transform fault differs from oceanic transform fault zones in two fundamental ways. Firstly, the Troodos spreading system, and hence its related transform offsets, were formed above a subduction system. Secondly, unlike most oceanic transform fault zones which experience transcurrent transform displacement, the southern Troodos transform fault suffered a major component of transtension that resulted in the formation of anomalous oceanic crust within the active transform domain. The interaction between this supra-subduction setting and the transtensional transform tectonics may provide an explanation for the boninite geochemistry that characterizes the Transform Sequence magmatism in the southern Troodos transform fault.

The Troodos Axis Sequence basalts are significantly more trace element depleted than their mid-ocean basalt counterparts. Various models for supra-subduction zone spreading explain this relative trace-element depletion in supra-subduction zone basalts by identifying their source as oceanic upper mantle that has already been depleted by the exraction of MORB oceanic crust (Cameron *et al*. 1980; Pearce 1980*a*; McCulloch & Cameron 1983; Robinson *et al*. 1983; Schminke *et al*. 1983). As a consequence, type upper mantle both beneath supra-subduction zone oceanic crust and immediately adjacent to a supra-subduction zone transform fault is further depleted by the extraction of the supra-subduction zone oceanic crust. This composition of upper mantle will lie close to the modelled source for the excessively trace-element-depleted Transform Sequence boninite magmas found within the southern Troodos transform fault.

The physical conditions required to melt this highly refractory upper-mantle source are a high-temperature thermal regime and a lowering of the peridotite solidus either by a decrease in hydrostatic pressure or by the introduction of a hydrous phase (Mysen & Boetcher 1975). The highest temperatures encountered within a transform fault zone are likely to be at a ridge–transform intersection. However, the structural evidence for the western Limassol Forest Complex suggests that the Transform Sequence magmatism was syntectonic with active transform displacement and hence occurred within the active transform domain between the two constructive-margin axes (Murton 1986). Therefore the conditions for melting of the depleted upper mantle beneath the southern Troodos transform fault were more probably met by a lowering of the peridotite solidus. The structural and geochemical evidence suggests that this was achieved by a combination of processes. The initial result of transtension across the active transform domain of the southern Troodos transform fault was the disruption of the adjacent axis crustal sequence and uplift of an upper-mantle block by about 5 km (Murton 1986). Although this happened relatively quickly, it is unlikely that such an amount of uplift alone would be sufficient to produce adiabatic melting of the upper mantle. The addition of a hydrous fluid phase, however, that would further lower the peridotite solidus is evidenced both by the addition to the Transform Sequence magma source of a LIL-element-enriched component and by deep serpentinization of the upper mantle during Transform Sequence magmatism. The source for similar LIL-element-bearing hydrous phases is generally considered to be the dehydration of the subducted oceanic slab. A further factor involved in the melting of the Transform Sequence magma source beneath the southern Troodos transform fault would be the component of heat transmitted to the upper mantle by the addition of a silicate melt phase carrying the in light REEs, H, hafnium, zirconium and yttrium. Since this silicate phase enrichment component lies close to the composition of pelagic red clay sediment, the source is again likely to be related to the subducted oceanic slab.

Why selective enrichment of the Transform Sequence magmas?

The transition from transcurrent transform displacement, as recorded by the Arakapas Fault Belt, to transtensional transform displacement, as recorded by the western Limassol Forest Complex, and the subsequent formation of the anomalous Transform Sequence crust within the active transform domain was probably coincident with the onset of rotation that affected the Troodos microplate. The earliest measurable rotation of the Troodos microplate has been determined as late Cretaceous, approximately coincident with the formation age of the Troodos oceanic crust (Clube *et al.* 1985; Clube & Robertson, 1986). Arcuate strike-slip faults and amphibolite mylonites that were active during the rotation of the Troodos microplate (Swarbrick 1980; Roddick & Spray 1981) contain disrupted fragments of the southern Troodos transform fault indicating reactivation of the transform fault as a microplate margin (Murton, in prep.). As a result of the changing stress fields during the onset of rotation, the southern Troodos transform fault was effectively rifted transtensionally. This transtensional environment, involving dilational asthenospheric shearing, may have enhanced and focused fluid pathways from the subducted oceanic slab upwards into the upper mantle underlying the active transform domain of the southern Troodos transform fault. This focusing effect would have selectively enriched the source of the transtensional Transform Sequence magmas in components derived from the subducted slab. Because of their strongly trace-element-depleted nature, even the addition of small amounts of the silicate phase enrichment component would have a marked effect on the REE profiles of the Transform Sequence basalts. The tectonic control on the fluid pathways followed by the enrichment phase is likely to have resulted in an extremely heterogeneous addition of the component to the Transform Sequence magmas, with some areas receiving large amounts of component and other areas receiving very little or none at all (*eg* the Arakapas Fault Belt basalts). The effect of the enrichment component, however, would be significantly diluted and hence much less apparent if mixed with less trace-element-depleted basalts such as those of the Troodos Axis Sequence.

Conditions for boninite magmatism

Based on the evidence from the southern Troodos transform fault, the essential conditions for boninite genesis are (i) a mantle source that is extremely depleted in the incompatible trace elements, (ii) an extensional tectonic regime in which the source can be melted and (iii) a process whereby the peridotite solidus can be sufficiently lowered to allow source melting. The depleted source of the boninites must be a highly refractory residue that has had at least MORB and one other significant proportion of basaltic melt extracted from it. Although no present-day supra-subduction zone spreading centres have been identified, the oceanward migration of a subduction zone (*ie* the Marianas Trench) and the lengthening of the fore-arc region must involve oceanic crustal generation above the subducted plate. Such a situation will leave a residual upper-mantle composition that could, if melted, generate the characteristic trace-element-depleted geochemistry of boninite magmatism. The next condition required is rifting of an extended fore-arc region either by a sudden increase in the rate of oceanward migration of the trench or by changes in the convergence vectors of the colliding plates. This tectonic requirement is especially significant since it will have the effect of localizing boninite magmatism both spatially and temporally. This may explain the sporadic discontinuous outcrop of boninites along the Marianas inner trench wall and within the Marianas fore-arc region. The final condition required to generate boninite magmatism is the lowering of the peridotite solidus by the introduction of a hydrous fluid phase such as may be released by dehydration of the subducted oceanic slab. Only when all these conditions are met simultaneously will boninite magmatism be produced. Therefore, boninites will remain rare and volumetrically insignificant compared with other destructive-margin magmas.

ACKNOWLEDGMENTS: The ideas in this manuscript benefited greatly from discussions with K. J. Fraser, F. McDermott, S. Berhe, R. Taylor, C. J. MacLeod, N. Rogers and R. Price. The studies on the southern Troodos transform fault were initiated by Ian Gass. REE and isotope analyses were undertaken by N. Rogers and major and trace-element analyses were made by J. Watson at the Open University. Field work logistics in Cyprus were aided by the Cyprus Geological Survey and the project was funded by the Natural Environment Research Council.

References

ALLEN, C. G. (1975). *The Petrology of a Portion of the Troodos Plutonic Complex, Cyprus*. PhD thesis, Cambridge University, 315 pp.

BENDER, J. F., LANGMUIR, C. H. & HANSON, G. N. (1984). Petrogenesis of basalt glasses from the Gamago Region, East Pacific Rise. *Journal of Petrology* **25**, 213–254.

BROWNING, P. (1982). *The Petrology, Geochemistry and Structure of the Plutonic Rocks of the Oman Ophiolite*. PhD thesis, Open University (unpublished).

BRYAN, W. B., THOMPSON, G. & LUDDON J. N. (1981). Mid-Atlantic ridge and Kane fracture zone. *Journal of Geophysical Research* **86**, 11815–11836.

CAMERON, W. E. (1985). Petrology and origin of primitive lavas from the Troodos ophiolite, Cyprus. *Contributions to Mineralogy and Petrology* **89**, 1–17.

——, MCCULLOCH, M. T. & WALKER, D. A. (1983). Boninite petrogenesis chemical and Nd–Sr isotopic constraints. *Earth and Planetary Science Letters* **65**, 75–89.

——, NISBET, E. G. & DIETRICH, U. J. (1979). Boninites, komatiites and ophiolitic basalts. *Nature* **280**, 550–553.

——, —— & —— (1980). Petrographic dissimilarities between ophiolitic and ocean-floor basalts. *In*: PANAYIOTOU, A. (ed.) *Ophiolites, Proceedings of the International Ophiolites Symposium, Cyprus, 1979*, 182–192.

CARTER J. L. (1970). Mineralogy and chemistry of the Earth's upper-mantle based on the partial fusion–partial crystallization model. *Geological Society of America Bulletin* **81**, 2021–2034.

CLUBE, T. M. M. & ROBERTSON, A. H. F. (1986). The palaeorotation of the Troodos microplate, Cyprus, in the Late Mesozoic in Early Cenozoic plate tectonic framework of the Eastern Mediterranean. *Surveys in Geophysics* **8**, 375–437.

——, CREER, K. M. & ROBERTSON, A. H. F. (1985). Palaeorotation of the Troodos microplate, Cyprus. *Nature* **317**, 522–525.

DALWITZ, W. B., GREEN, D. H. & THOMPSON, J. E. (1966). Clinoenstatite in a volcanic rock from Cape Vogel, Papua Vogel, Papua: a discussion. *Proceedings of the 23rd International Geology Congress, Prague*, Vol. 2, 229–242.

—— (1968). Chemical composition of clinoenstatite-bearing volcanic rocks from Cape Vogel area, Papua: a discussion. *Proceedings of the 23rd International Geology Congress, Prague*, Vol. 2, 229–245.

DICKEY, J. S. JR (1970). Partial fusion products in Alpine type peridotites, Serrania de la Rhonda and other examples. *Mineralogical Society of America Special Paper* **3**, 33–49.

DIETRICH, V., EMMERMANN, R. & PUCHELT, H. (1978). Geochemistry of basaltic and gabbroic rocks from the west Mariana Basin and the Mariana Trench. *Earth and Planetary Science Letters* **39**, 127–144.

DUNCAN, R. A. & GREEN, D. H. (1980). Role of multistage melting in the formation of oceanic crust. *Geology* **8**, 22–26.

FISK, M. R., BENCE, A. E. & SCHILLING, J.-G. (1982). Major element chemistry of Galapagos rift zone magmas and their phenocrysts. *Earth and Planetary Science Letters* **61**, 171–189.

HAWKINS, J. W. JR, BLOOMER, S., EVANS, C. & MELCHOIR, J. (1979). Marianas arc–trench system: petrology of inner trench wall. *Eos (Transactions of the American Geophysical Union)* **60**, 968 pp.

HEKINIAN, R. & THOMPSON, G. (1976). Comparative geochemistry of volcanics from rift valleys, transforms and aseismic ridges. *Contributions to Mineralogy and Petrology* **57**, 145–162.

HICKEY, R. L. & FREY, F. A. (1982). Geochemical characteristics of boninite series volcanics: implications for their source. *Geochimica et Cosmochimica Acta* **46**, 2099–2115.

HOLLCOMBE, T. L., VOGT, P. R., MATTEWS, J. E. & MURCHISON, R. R. (1973). Evidence for sea-floor spreading in the Caymen Trough. *Earth and Planetary Science Letters* **20**, 357–371.

JAGOUTZ, E., PALME, H., BADDENHAUSEN, H. BLUM, K., CENDALES, M., GERLIND, D., SPETTEL, B., LORENZ, V. & WANKE, H. (1979). The abundances of major, minor and trace-elements in the Earth's mantle as derived from primitive ultramafic nodules. *Proceedings of the 10th Lunar and Planetary Science Conference*, 2031–2030.

JENNER, G. A. (1981). Geochemistry of high Mg andesites from Cape Vogel, Papua New Guinea. *Chemical Geology* **33**, 307–332.

LANGMUIR, C. H. & BENDER, F. J. (1984). The geochemistry of oceanic basalts in the vicinity of transform faults: observations and implications. *Earth and Planetary Science Letters* **69**, 107–127.

MALPAS, J. M. (1978). Magma generation in the upper-mantle, field evidence from ophiolite suites and application to the generation of oceanic lithosphere. *Philosophical Transactions of the Royal Society of London. Series A* **288** 527–546.

MANTIS, M. (1971). Paleontological evidence defining the age of the Troodos Pillow Lava Series in Cyprus. *Kypriakos Logos III* **15–16**, 202–208.

MCCULLOCH, M. T. & CAMERON, W. E. (1983). Nd–Sr isotopic study of primitive lavas from the Troodos ophiolite, Cyprus: evidence for a subduction-related setting. *Geology* **11**, 727–731.

MEIJER, A. (1980). Primitive are volcanism and a boninite series: examples from the western Pacific island arcs. *In*: HAYNES D. E. (ed.) *The Tectonic and Geologic Evolution of Southeast Asian Seas and Islands*. American Geophysical Union Memoir **23**, 269–282.

——, ANTHONY, E. & REAGAN, M. (1981). Petrology of volcanic rocks from the fore-arc sites. *Initial Reports of the Deep Sea Drilling Project* **60**, 709–730.

MENZIES, M. A. (1974). *Petrogenesis of the Makrirraki Ultramafic Complex*. PhD thesis, Open University (unpublished).

—— & ALLEN, C. R. (1974). Plagioclase lherzolite-residual mantle relationships within two eastern Mediterranean ophiolites. *Contributions to Mineralogy and Petrology* **45**, 195–213.

MIYASHIRO, A. (1973). The Troodos ophiolitic complex was probably formed.in an island arc. *Earth and Planetary Science Letters* **19**, 218–224.

MOORES, E. M. & VINE, F. J. (1971). The Troodos Massif, Cyprus, and other ophiolites as oceanic crust: evaluation and implications. *Philosophical Transactions of the Royal Society of London* **268**, 443–466.

—— &, ROBINSON, P. T., MALPAS, J. & XENOPHANTOS, C. (1984). The tectonic setting for the evaluation of the Troodos Complex, Cyprus. *Geology* **12**, 500–503.

MURTON, B. J. (1986). Anomalous oceanic crust formed in an oceanic transform fault: evidence from the Western Limassol Forest Complex. *Journal of the Geological Society, London* **143**, 845–854.

—— & GASS, I. G. (1986). Western Limassol Forest Complex, Cyprus. Part of an Upper Cretaceous leaky transform fault. *Geology* **14**, 255–258.

MYSEN, B. & BOETTCHER, A. C. (1975). Melting of hydrous mantle: Phase relations of natural peridotites at high pressures and temperatures with controlled activities of water, carbon dioxide and hydrogen. *Journal of Petrology* **16**, 520–548.

PEARCE, J. A. (1975). Basalt geochemistry used to investigate post tectonic environments on Cyprus. *Tectonophysics* **25**, 41–67.

—— (1980*a*). Genesis and eruptive environment of lavas from Tethyan ophiolites. *In*: PANAYIOTOU, A. (ed.) *Ophiolites, Proceedings of the International Ophiolites Symposium, Cyprus, 1979*.

—— (1980*b*). REE values for various OIB etc. from lead isotope study of young volcanic rocks from mid-ocean ridges, ocean islands and island arcs. *Philosophical Transactions of the Royal Society of London, Series A* **297**, 409–445.

—— (1982). Trace-element characteristics of lavas from destructive plate boundaries. *In*: THORPE, E. S. (ed.) *Andesites*. Wiley, New York, 525–548.

—— (1983). Role of the sub-continental lithosphere in magma genesis of active continental margins. *In*: HAWKESWORTH, C. J. & NORRY, M. J. (eds) *Continental Basalts and Mantle Xenoliths*. Shiva, 267 pp.

—— & CANN, J. R. (1973). Tectonic setting of basic volcanic rocks determined using trace-element analyses. *Earth and Planetary Science Letters* **19**, 290–300.

—— & NORRY, M. J. (1979). Petrogenetic implications of Ti, Zr, and Nb variations in volcanic rocks. *Contributions to Mineralogy and Petrology* **69**, 33–47.

——, LIPPARD, S. J. & ROBERTS, S. (1984). Characteristics and tectonic significance of supra-subduction zone (s.s.z) ophiolites. *In*: KOKELLAR B. P. & HOWELLS M. F. (eds) *Marginal Basin Geology*. Geological Society of London Special Publication **16**, 77–94.

RAUTENSCHLEIN, M., JENNER, G. A., HERTOGEN, J., HOFMANN, A. W., KERRICH, R., SCHMINCKE, H. U. & WHITE, W. M. (1985). Isotopic and trace-element composition of volcanic glasses from the Akaki Canyon, Cyprus: implications for the origin of the Troodos ophiolite. *Earth and Planetary Science Letters* **75**, 369–383.

RINGWOOD, A. E. (1966). Chemical evolution of the terrestrial planets. *Geochimica et Cosmochimica Acta* **30**, 326–339.

ROBINSON, P. T., MELSON, W. G., O'HEARN, T. & SCHMINKE, H. U. (1983). Volcanic glass compositions of the Troodos ophiolite, Cyprus. *Geology* **1**, 400–404.

RODDICK, J. G. & SPRAY J. F. (1981). Evidence for Upper Cretaceous transform fault metamorphism in West Cyprus. *Earth and Planetary Science Letters* **55**, 273–291.

ROEDER, P. L. & EMSLIE, R. F. (1970). Olivine–liquid equilibrium. *Contributions to Mineralogy and Petrology* **29**, 275–289.

ROGERS, N., MACLEOD, C. J. & MURTON, B. J. Petrogenesis of boninitic lavas from the Limassol forest complex, Cyprus, in press.

SCHILLING, J. G., ZAJAC, M., EVANS, R., JOHNSTON, T., WHITE, W., DEVINE, J. D. & KINGSLEY, R. (1983). Petrologic and geochemical variations along mid-Atlantic ridges from 29°N to 73°N. *American Journal of Science* **283**, 510–586.

SCHMINKE, H.-U., RAUTENSCHEIN, M., ROBINSON, P. T. & MEHEGEN, J. M. (1983). Troodos extrusive series of Cyprus: a comparison with oceanic crust. *Geology* **11**, 405–409.

SCHNEIDER, M. E. & EGGLER, D. H. (1986). Fluids in equilibrium with peridotite minerals: implications for mantle metasomatism. *Geochimica et Cosmochimica Acta* **50**, 711–724.

SHARASKIN, A. Y. & DOBRETSOV, N. L. (1980). Marianites: the clinoenstatite bearing pillow lavas associated with ophiolites of the Mariana Island Arc. *In*: PANAYIOTOU, A. (ed.) *Ophiolites, Proceedings of the International Ophiolites Symposium, Cyprus, 1979*, 166–175.

SHAW, D. M. (1970). Trace-element fractionation during anatexis. *Geochimica et Cosmochimica Acta* **34**, 237–243.

SIMONIAN, K. O. (1975). *The Geology of the Arakapas Fault Belt Area, Troodos Massif, Cyprus*. PhD thesis, Open University.

—— & GASS, I. G. (1978). Arakapas fault belt, Cyprus: a fossil transform belt. *Bulletin of the Geological Society of America* **89**, 1220–1230.

SMEWING, J. D., SIMONIAN, K. O. & GASS, I. G. (1975). Metabasalts from the Troodos massif, Cyprus: genetic implications deduced from petrography and trace-element geochemistry. *Contributions to Mineralogy and Petrology* **51**, 49–64.

SWARBRICK, R. E. (1980). The Mamonia complex of S. W. Cyprus: a Mesozoic continental margin and its relationship to the Troodos complex. *In*: PANAYIOTOU, A. (ed.) *Ophiolites, Proceedings of the International Ophiolites Symposium, Cyprus, 1979*, 86–92.

TAYLOR, S. R. & MCLENNAN, S. M. (eds) (1985). *The Continental Crust*. Blackwell Scientific, Oxford, 312 pp.

WALKER, D. A. & CAMERON, W. E. (1983). Boninite primary magmas: evidence from the Cape Vogel Peninsula, P. N. G. *Contributions to Mineralogy and Petrology* **83**, 150–158.

WILSON, R. A. M. (1959). The geology of the Xeros–Troodos area, Cyprus. *Geological Survey Memoir* **1**, 135.

B. J. MURTON, Department of Earth Sciences, Open University, Milton Keynes, Bucks MK7 6AA, UK.

Appendix 1

Petrographic tables for various analysed mafic dykes and lavas.

Sample	Ol	Opx	Cpx	Plag	Amph	Glass	Spinel	AI	Group
Mafic Transform Sequence dykes cutting the mantle sequence									
43/CY83	—	P5–G3	P2 G25	G65	—		M2	6	C
52/CY83	G10	—	P<1 G30	P2 G55	—		M<1	7	D
70A/CY83	P5	P25 G5	P15 G10	P25 G15	—		—	7	C
85A/CY83	P1 G8	—	P8 G25	P3 G55	—		M4s	8	D
85B/CY83	—	—	P10 G30	G70	—		M<1	10	B
85C/CY83	—	—	P2 G20	P3 G75	5s		—	5	D
100I/CY83	—	—	P<1 G34	P2 G63	—		—	3	D
100II/CY83	G5	G25	G65	G5	—		C<1	7	B
104A/CY83	—	G5	G35	G60	—		—	3	D
113/CY83	P5 G3	P6 G6	P10 G70	—	—		—	1	A
134B/CY83	P3 G1	P<1	P3 G40	P3 G52	5s		M<1	6	D
150A/CY83	—	P2 G10	P6 G70	P2 G10	10s		—	9	C
150B/CY83	P10	—	P8 G45	G18	—		M4s	8	A
150C/CY83	G10?	P5 G10	P5 G70	—	—		G15s	10	B
150A1/CY83	P1	P5 G13	P5 G30	G30	—		M15s	8	C
152B/CY83	P22 G5	P5 G10	P20 G30	G2	2s		C<1	6	A
156B/CY83	—	—	P10 G50	P5 G35	—		—	7	D
299A/CY83	—	—	P10 G20	P16 G50	3s		M<1	8	D
347A/CY83	G5	P5	G30	G60	—		—	8	B

Sample	Ol	Opx	Cpx	Plag	Amph	Glass	Spinel	AI	Group
Transform Sequence lavas									
113/CY84	—	13G	2P 80G	—		10	–	0–1	U
143/CY84	8P	10G	3P 72G	—		—	C<1	4	U
153B/CY84	8P	17G	35G	—		—	C<1	7	M
158/CY84	—	5G	2P 67G	—		60	—	2	L
159/CY84	3P	3G	4P 32G	—		25	—	0–1	L
170/CY84	4P	—	36G	—		60	—	0–1	L
171/CY84	2P	—	6P 66G	—		62	—	0–1	L
172/CY84	3P	5G	3P 75G	—		20	—	0–1	M
174/CY84	2P	5G	1P 6G	—		15	—	4	M
187B/CY84	3P	—	2	—		90	—	0	U
187C/CY84	3	—	1	—		60	—	1	U
189A/CY84	3	—	2	—		—	—	9	U
189B/CY84	4	—	—	—		30	—	3	U
189C/CY84	—	—	2	—		5	—	4	U
189D/CY84	—	—	3	—		60	—	1	U
191/CY84	12	—	5	—		25	C2	6	U
192A/CY84	20	12	4	—		10	C2	1	U
192B/CY84	12	8	3	—		20	C1	0–1	U
193/CY84	3	—	2	2		—	—	9	U
197/CY84	8	—	3	—		10	—	2	U
199A/CY84	15	1		—		12	C2	0–1	U

Mineral phases: Ol, olivine; Opx, orthopyroxene; Cpx, clinopyroxene; Plag, plagioclase; C, chromite; M, magnetite; Amph, amphibole (usually actinolite). Textural terms: P, phenocryst phase; X, xenocryst phase; G, groundmass phase. Numbers after abbreviated terms refer to estimated abundance in volume per cent. Petrographical groups: A, B, C and D referred to in the text and on geochemical diagrams. Stratigraphical groups: U, upper lavas group; M, middle lavas group; L, lower lavas group. Other terms used: A I, alteration index, which is proportional to the secondary mineralization caused by low-grade hydrothermal alteration or weathering; 0, very low; 10 complete; s, secondary mineralization; –, phenocryst phase not present.

Appendix 2
Range in compositions of various mafic dykes and lavas. Major and trace elements were determined by energy-dispersive XRF whole-rock analyses with the exception of chromium which was determined on a wavelength XRF spectrometer for greater accuracy.

	Mafic dykes					Lavas			
	Picritic	A	B	C	D	Sheeted	Lower	Middle	Upper
Major elements (wt%)									
SiO_2	42–52	38–40	39–51	46–52	45–58	44–53	49–52	51–52	50–54
TiO_2	0.13–0.26	0.14–0.39	0.14–0.39	0.18–0.56	0.23–0.57	0.43–0.85	0.21–0.47	0.20–0.22	0.19–0.33
Al_2O_3	5.6–14.5	7–16	11–17	7–15	14–17	15–16	12–18	12–14	11–15
FeO*	8–9	7–12	7–10	7–9	7–9	7–9	7–8	7–8	7–8
MnO	0.12–0.38	0.14–0.17	0.15–0.20	0.10–0.17	0.13–0.18	0.05–0.11	0.12–0.38	0.15–0.17	0.13–0.17
MgO	10–31	10–21	7–16	7–13	5–13	7–12	9–15	11–12	10–13
CaO	5–13	7–9	11–16	10–13	9–16	7–14	5–10	10–11	10–11
Na_2O	0.4–1.7	0.2–0.3	0.0–1.4	0.4–4.1	0.1–4.7	1.0–5.0	0.4–0.9	0.6–1.0	0.6–3.4
K_2O	0.0–0.2	0.0–0.0	0.0–2.5	0.0–1.7	0.0–2.3	0.0–0.2	0.3–0.5	0.4–0.6	0.1–1.2
P_2O_5	0.02–0.9	0.00–0.00	0.00–0.08	0.00–0.04	0.00–0.08	0.00–0.01	0.00–0.03	0.03–0.04	0.03–0.04
Trace elements (ppm)									
Rb	1–6	2–3	2–13	0–8	2–8	1–3	6–14	13–14	2–24
Sr	2–232	2–132	3–120	56–450	10–390	44–145	79–188	117–276	51–515
Y	3–7	5–10	3–13	6–17	8–13	14–21	8–12	8	7–11
Zr	8–18	13–20	12–18	17–37	18–34	22–42	11–18	20–21	11–25
Nb	1–5	2–5	2–4	3–7	3–6	2–3	1–3	1–2	3–1
Pb	1–3	0–6	4–15	5–20	5–16	1–4	12–7	5–6	3–10
Th	0–2	0.5–0.9	0.5–1.2	0.5–0.9	1.0–2.1	12–3	0	0	0–3
Cr	630–2800	450–2308	45–1060	61–860	47–530	77–906	350–1060	943–1033	510–1137
Cu	14–88	12–194	73–106	8–106	3–190	6–370	111–140	113–124	40–120
Zn	32–60	44–66	50–80	21–52	26–63	33–230	69–72	69–82	9–72
Ga	6–12	8–12	8–12	10–14	12–14	1–50	8–12	8–12	8–12

* FeO total iron as FeO.

Index

References to tables appear in *italics*, those to figures in **bold**